It Came From Outer
an RAF Blazer!

Martin Mobberley

It Came From Outer Space Wearing an RAF Blazer!

A Fan's Biography of Sir Patrick Moore

Springer

Martin Mobberley
Denmara
Cockfield
Suffolk, UK

ISBN 978-3-319-00608-6 ISBN 978-3-319-00609-3 (eBook)
DOI 10.1007/978-3-319-00609-3
Springer Cham Heidelberg New York Dordrecht London

Library of Congress Control Number: 2013939729

Cover illustration: Patrick Moore at his home 'Farthings' at 39 West Street, Selsey, posing with his 12.5-inch Newtonian reflector on 15 June 1970. Photograph by Jon Lyons/Rex Features.

Printed on acid-free paper

Springer is part of Springer Science+Business Media (www.springer.com)

Dedicated to the memory of Barbara Mobberley (1929–2006). For 48 years of my life a remarkably devoted mother, who always encouraged me in my astronomical endeavours and was always excited when she answered any telephone call from Patrick to me. She passed away a few hours after the crystal clear moonrise lunar eclipse of September 2006.

Preface

In January 1968, only a few days after my tenth birthday, I picked up a pocket-sized book entitled *The Observer's Book of Astronomy* by Patrick Moore. Looking back, it is no exaggeration to say that this little book changed my whole life. I instantly became obsessed with astronomy, with watching the book's author on TV and with buying his other books. Just 11 months later, due to NASA's Apollo 8 mission, Patrick started becoming one of the most famous people on UK Television. Seven months after that he was on BBC screens every night when Neil Armstrong and Buzz Aldrin landed on the Moon. If I had not already been hooked by his books, I was certainly hooked on the man by then. The late 1960s seemed to be an era when every young British boy's favourite TV hero was named Patrick! Patrick Troughton was *Dr Who*, Patrick Macnee was John Steed of *The Avengers*, Patrick McGoohan was John Drake in *Danger Man*, and Patrick Moore was Mr Astronomy and Space on the BBC.

What was the astronomer Patrick's real attraction? Well, he was one of the few adults that a young boy like me could identify with: he had a child's unbounded enthusiasm for space, planets, aliens and rockets all wrapped up in a man's (very large) body. To my young mind this seemed like the way to go. Patrick appeared to be a giant-sized schoolboy with none of the millstones that normal adults have. He was his own boss, with no wife and screaming kids, and so was free to totally immerse his life in a fascinating hobby: that was the sort of life I wanted! As a 10 year old boy I looked around at other men of Patrick's age, married men, and they all seemed to be buried under a veritable infinity of commitments and hassles. In many cases they seemed to always have a sort of simmering irritation with their nagging other half, or with the dreaded mother-in-law, the butt of most 1960s jokes! All other adult males seemed to have lost the excitement and sense of wonder they had enjoyed as children. But there was Patrick, with big telescopes in his garden and permanent enthusiasm about what he was going to point them towards that night. Somehow, he had not gone astray and he had kept his childhood dreams alive. He carried no emotional baggage. What a great life!

Of course, as a ten year old boy in the male dominated 1960s, blissfully removed from adulthood, you don't always interpret the world of grown-ups correctly, but after the Apollo 11 Moon landing I became hooked on astronomy and space. I noticed that Patrick, like me, was an only child. I also found out, some years later, that Patrick had joined the British Astronomical Association (BAA), aged 11; so had I. At the age of 25 I was invited to become a post holder in the BAA Lunar Section (the photographic co-ordinator); so was Patrick, 35 years earlier, when he was made the BAA Lunar Section secretary, also at the age of 25! Patrick was a BAA President and a BAA Goodacre medallist: so, many years later, was I. As the years went by there seemed to be more and more strange parallels which I couldn't seem to ignore. For example, our mothers had the same birthday, June 27th, and our fathers had a distinguished military background. Patrick had a bad left knee from a wartime accident, sustained when he was 21. I permanently damaged my left knee in a cycling accident at the same age. He had one very good eye and one weak eye: Snap! Patrick claimed that he first came to prominence in the BAA after making some observations of the lunar 'sea' known as the Mare Crisium, in the late 1930s. Again, ditto, except it was 1981 in my case, when my photograph of that feature won the BAA Lunar Section's photographic competition. We were even the same height, although definitely NOT the same weight!

Now, I am not a religious or superstitious person, but these similarities kept my interest in Patrick going, and, throughout my adult years, I became more and more interested in everything he did, as well as buying all (well, almost all!) his astronomy books. I started to wonder if the same part of life's 'quantum matrix' (my term – and no, I can't explain it) had been used to program his and my DNA! This ridiculous belief was only strengthened when I realised that, back in 1968, my parents and I had chosen a house in Suffolk in the same week that Patrick and his mother had chosen their thatched cottage, *Farthings*, in Selsey, just a stroll from a friend of Patrick's who lived in a big property called *The Old Mill House*. Less than 100 yards from our Suffolk home, where I am typing these words, there is another cottage called *Farthings* and, a few yards away, another property called *The Old Mill House*. Bizarre coincidences no doubt, but just a little bit spooky too!

Anyway, needless to say, from my childhood years onward I had started compiling a scrap book about Patrick, which just got bigger and bigger. Every book or article that Patrick wrote, I tried to get hold of… etcetera, etcetera. In many ways this book is the end result of that scrapbook.

I first met Patrick 'in the flesh' on September 26, 1970, at a BAA Lunar Section meeting in London. I was 12 and he was 47. Apart from a cheery "Hello there" from Patrick (which put me on cloud nine) we did not converse; but I was in awe, even though he wasn't an official speaker on that day. His Churchillian physical presence was one thing, but his crystal clear voice was like a mobile public address system! Patrick never used a microphone and just laughed if one was offered; he was clearer without one than everyone else was using one. He was a powerhouse of activity at all BAA meetings in that era. Everyone else seemed half asleep compared to his enthusiasm and energy levels. Other speakers mumbled and got confused and droned on and on for ages. He always excited and inspired the audience. He delivered the goods; he was wired up; he was plugged in; he was smoking!

If there was any sign of audience scepticism or negativity during those 1970s era meetings, or any hitch with the slide projector, Patrick would adopt his bulldog expression, pull an extraordinary grimace and immediately thump both his massive fists so firmly onto his hips (actually, rather higher up than his hips) that it must surely have hurt. This extreme hands-on-hips 'stance of defiance' seemed to stay with Patrick his whole life and seemed to make him look even bigger than he already was. Another of Patrick's characteristics, signifying total enthusiasm and 100 % commitment to the BAA, was the very manner he would leave his seat on the front row of the audience to take his place on the stage, prior to delivering a talk. He might seem bored when sitting waiting for his turn, but when called by the President (or Chairman) he would literally explode from his seat and propel, at great speed, his enormous bulk forward, heading, as a crow flies, directly for the stage. His stance would be one of a charging rhinoceros, with his head down, and his facial expression grim and fixed rigid, as if he was about to embark in a fight to the death! Then, as soon as he arrived on the stage, the audience, perhaps shocked by his deadly serious expression, would soon be laughing as he cracked an opening joke, such as "Mr President, Ladies and Gentlemen, after the Lord Mayor's show comes the dustcart!" He would then promptly deliver a self-deprecating talk about a field of astronomy in which he would modestly claim his contribution was insignificant. He always spoke very fluently and very clearly, and he never, ever, used written notes. The performance was always loud, perfect and enthusiastic, without any hesitation whatsoever; it was also, invariably, hilarious! A short while ago someone asked me if I could think of a modern character whose manner most resembled the Patrick Moore, in his prime, that I remembered from the astronomy meetings during my teenage years. Well, no-one quite fits the bill, but I'd say the modern character who comes closest in general demeanour and humour to the 1970s Patrick Moore is the London Mayor Boris Johnson; except Patrick, in his prime, was louder, larger and spoke a lot faster.

Patrick would always wear a blue blazer, with a shirt and tie; sometimes an RAF tie and sometimes a BAA tie. Admittedly he owned more than one blazer by his later years and even a sports jacket in his younger years. One vivid turquoise blazer that Patrick owned he claimed he had purchased when abroad in a gloomy shop, thinking it was "Oxford Blue"! You would never, *ever*, see Patrick wearing a coat, or a jumper, or a hat. It was always just a blue blazer, a shirt and a tie, at every UK venue. On one very rainy day, at a BAA London meeting, I remember someone asking Patrick if he had brought an umbrella along? 'Heaven forfend!' he cried, eyebrow raised, 'I'd lose it in minutes!' Even in the depths of winter and outdoors he would be dressed in a blazer, shirt and tie. In hot countries, and in his later years indoors, he took to wearing huge colourful short-sleeved shirts and tent-like smocks, but I do not think in his adult years he ever owned a coat or wore a pullover, even though his mother once insisted he took a woolly jumper to Siberia. It was extremely rare to see Patrick carrying any form of case either. He never referred to written notes and he had no interest in bureaucratic paperwork; so a case was superfluous.

His huge blazer was his filing cabinet and contained plenty of spiral bound notepads, pencils, rubbers, illuminated pens, his pipe, and blank card templates for rough sketching Mars, Jupiter or Saturn onto, whenever he was near a telescope. The only time you ever saw him with a case was if he had a huge number of photographic slides to project. Then he would bring a blue and battered suitcase along, so bulky he could not possibly lose it. Inside the case it would be 99 % empty, except for a few boxes of slides. I saw him with that same case, on rare occasions, throughout the 1970s, 1980s and 1990s and I have been told that he owned it way back into the 1950s too!

I wrote to Patrick shortly after I first met him at that September 1970 meeting, trying to get him to visit the school astronomy club which I had founded. He never actually made it (he said his housekeeper, Mrs Denny, had just died) but I was amazed that someone that famous would even write back to a 12 year old child – instantly! In later years I learned that Patrick always answered every letter he received, within days, and during the 1970s and 1980s they piled in at more than 100 every week! To cope with this volume he used blank postcards inserted into his 1908 Woodstock typewriter and he answered most queries with just a few lines, typed in seconds. All postage costs (for thousands of replies each year) were funded by him and not by the BBC. He was simply not interested in bureaucratic hassles and claim forms, just in providing the perfect answer, quickly and concisely. Amazing!

I used to bump into Patrick at BAA meetings throughout the 1970s and he always had a kind word for me. Then, in the 1980s, I started serious astronomical observing and got to know him much better. We served on the BAA Lunar Section Committee and the BAA Council together and, from 1982 onwards, he was keen to show my photographs on *The Sky at Night*. From the early 1990s he started asking me to supply photographs for his books and programmes on a regular basis and even to write occasional technical chapters for him. Then he asked me to write an entire book in his Springer 'Practical Astronomy Series' followed by a second book after that! When I followed in his footsteps and became the BAA President, he invited me onto *The Sky at Night*; I could scarcely believe it! I was the sole guest on the programme three times during 1998–1999 and an occasional guest on numerous episodes after that. It seemed like destiny and I was proud that, like him, I had kept my childhood dreams alive.

As my scrapbooks on Patrick became bigger, I started collecting more and more little-known information about him, as well as compiling a complete list of his books. In 2002, I started writing it all up into a draft biography. Then, in 2003, Patrick suddenly published his own autobiography entitled *80 Not Out*. Rarely have I ever been so curious about any publication, even though I felt I knew almost as much about Patrick as he did himself and even though his book might make mine obsolete. As it turned out, I needn't have worried. Yes, it was an entertaining read, but to me huge chunks were mysteriously missing or plain untrue! The book did not really give the reader any idea of how much amateur astronomy had governed Patrick's life. The BAA, which totally dominated and shaped Patrick's world from the age of 11, barely got a mention and neither did things that had gone wrong in his life. The book was mainly about his TV career, his political views, trips abroad, bureaucracy, and the famous people he had met; but there was little about his obsession with observing the Moon, his flawed belief that its craters had a volcanic origin, 'Transient Lunar Phenomena', or his BAA politics. Neither was there anything about the numerous foes he had branded as 'serpents'! I decided to put this right and, as planned, cover these aspects of his life in a much more accurate book. Even in 2003, I judged my own manuscript on Patrick to be far superior to his autobiography. There are obviously overlaps with his *80 Not Out*, but I would venture to suggest that my book is far more comprehensive on the events that will be of most interest to the amateur astronomer reading this book. My alternative biography is not about *The Sky at Night*, it is about Patrick, the British amateur astronomer, warts and all. I have said relatively little about Patrick's cricket, 'amateur dramatic' and musical accomplishments as it is really just the astronomical/back-garden observer side of his life that interests me. Also, there simply is not space to write any more about him – he just did too much in his life!

Patrick had his good and bad points, leading some to class him as a confusing enigma. One lifelong acquaintance said that he seemed to verge on the inscrutable at all times! He was a man of his word, totally loyal to his real friends, a truly tireless observer at the telescope and a tireless charity worker too. He must have raised millions for UK charities associated with Cancer Research, Cystic Fibrosis and a host of other worthy causes, often by simply giving public talks and donating the thousands of pounds worth of ticket money. The money never entered his bank account; he just told them to write a cheque to the charity in question. Whenever he gave such talks, for charities, or for local astronomical societies, up and down the country, he waived any fee and did not even claim travel expenses or overnight accommodation costs. Staggering! Perhaps even more amazing was that when he was driving hundreds of miles around the country in his beaten up wrecks of cars, to give a free lecture, and the best lecture anyone would ever hear, he would happily pick up bedraggled strangers, hitching a lift when their car had broken down. Their jaws would drop to the floor when they saw who was offering them a free taxi service.

Barely a month went by without Patrick being involved in a local charity event with either *The Lord's Taverners* or Sussex Country Cricket club in the 1960s, 1970s and 1980s. I should add that the BAA itself is a registered charity, devoted to providing assistance to amateur astronomers for no financial gain. Undoubtedly the time Patrick donated to the BAA over the years, as a section director, meetings recorder, President and speaker, was equivalent to many man-years of any normal person's effort. Yet it was all given totally free, and from a man who was a household name for more than half a century.

Beyond this, his personal generosity to children, sick or healthy, was astounding. If he was at an astronomy meeting with a bookstall and he saw a young child looking at one of his books he would buy it for them, with no hesitation! If a parent wrote to Patrick saying their child was seriously unwell but was interested in astronomy, Patrick would not only write to the child, he would send them a load of books or, on several occasions, buy them a telescope and deliver it in person to the hospital, without a moment's delay! On one occasion, when a child's parents brought their excited astronomy-mad youngster to Patrick's door, but found that he'd had to dash out to a TV commitment, Patrick placed adverts in national newspapers to trace the parents and re-arrange the visit! As well as his generosity to those who were seriously ill, and to children, he was incredibly generous to his friends. Many future astronomers and scientists, such as Peter Cattermole, Iain Nicolson, John Mason, and many others, asked for a look through Patrick's telescopes one day, as young men, and were astounded to find they were welcome at any time, on any day or night (by Patrick and his mother). Like Dr Who's assistants they suddenly found they had entered a magical world in a different continuum, where their own personal Time Lord could show them the wonders of the Universe, and they could spend a day with a TV star, for free! Patrick's homes, in East Grinstead, Armagh and Selsey, were like a Victorian oasis of fun and sanity, well away from the angry and competitive outside world. Also, the number of rounds of drinks Patrick purchased in his lifetime must have run into countless thousands. It was impossible to be in Patrick's house for more than 30 seconds before he offered you a drink, or a meal, or your own big bedroom for the night, if it was getting late. Everyone who left his house seemed to leave with a book or something of value. Complete strangers, with their Space-mad kids, would nervously knock on his door while passing through Selsey, and be welcomed in, fed and watered and shown the telescopes! They would be treated to a memorable day that they and their children would never forget. Loyalty to what he saw as his true friends, fans and admirers was at the very top of Patrick's list.

At complete odds to this, if you were a friend, but criticised Patrick, even slightly (especially behind his back), you made an enemy for life. For his loyalty and immense generosity he expected unswerving loyalty in return, like a mediaeval baron. True friends did not ever criticise or envy one another; that was a code he lived by. Also, if a true friend was ever asked to review one of his books, the review would have to be glowing: what other kind of book review could any decent person possibly give to a dear friend? Patrick also had a frustrating habit of exaggerating countless stories about his early life. His RAF career, his alleged lunar discoveries and the people he had met; all these tales got a bit more colourful as the years ticked by. The most popular stories got more and more out of control until, in some cases, he simply blatantly lied because, being so famous and so well-loved, he could get away with it and infuriate his opponents. There was certainly a dark side to the Moore, if not a dark side of the Moon! Perhaps I should qualify this by explaining that the lunar dark side is constantly changing as the Moon orbits the Sun; there is no permanent dark side. Apologies to Pink Floyd!

Patrick was, quite simply, a phenomenal public speaker and an unbelievably prolific author: on one occasion he wrote a major book entitled The Unfolding Universe in 10 days! Surely he must have been one of the most prolific popularisers of science of all time. Realistically, only Isaac Asimov can be compared to him in this context. For more than half a century Patrick churned out a new book every few months or even weeks! Also, unlike almost every other so-called TV astronomy expert, Patrick was a real amateur scientist and observer and kept his feet firmly on the ground. If he was at home and the sky was clear, his eye would be at the telescope eyepiece and he would be making a sketch or an observation to send to the BAA. He was one of the most prolific British visual observers of all time. In terms of the number of observations of different categories of object, he may well have been unique. One only has to leaf through his personal observing logbooks to verify his awesome output and his sheer enthusiasm for looking through his telescopes and drawing what he saw. This lasted from childhood into his late seventies and only waned due to immobility and poor health. Thus he was an astronomer's astronomer, not just a 'TV personality' with a pushy agent, craving fame for fame's sake. On the negative side, he could be a spoilt, sulking, overgrown schoolboy, a woman hater, a modern teacher hater (and don't ask about women teachers), totally opinionated and an out-and-out racist on some occasions. With a distinctly Victorian style of upbringing, including parents who had lived in the British Colony and Protectorate of Kenya prior to his birth, many of Patrick's negative traits could never be reversed. But one thing is for sure: he was a one-off and refreshingly different. While his name was not unique, there will never be another Patrick Moore quite like him. Boring he was not!

Without a doubt there are people who saw Patrick more frequently than I: his close friends in Selsey, BBC producers, his many godsons and, in later life, his carers. However, I very much doubt whether any of them would have the mental stamina to write a biography of this size about Patrick. I also doubt whether any of them understood him better than I did. As a spoilt, single, only-child and lifelong bachelor myself, who also joined the BAA aged 11, and have been obsessed by amateur astronomy ever since, I feel I have a better basis than most to write about the great man. I hope this book gives a more complete view of 'Patrick the Amateur Astronomer' and fills in some of the huge voids he deliberately created in his autobiography.

To create this work, I have spoken to scores of people, trawled through hundreds of BAA journals and meeting reports, hundreds of BAA Lunar Section circulars, over 200 of Patrick's own books and hundreds of hours of archival videotape footage (some acquired at great personal expense). In addition I have trawled through 40 years worth of personal letters from Patrick to myself, as well as half a century of newspaper cuttings, and 70 years worth of his observing logbooks, stored in his home, which he was always happy for me to read and photograph. I have also distilled my own personal memories of many conversations with the great man, on the telephone, at astronomy meetings, and at his Selsey home. I hope it has all been worth it! For me, every visit to Patrick's beloved 'Farthings' was like a pilgrimage to a holy shrine. His inspiration shaped my life and now I am writing about the man himself. He was, surely, the last of the great English visual 'gentlemen astronomers'; observing like a wealthy Victorian amateur, with Victorian standards, but living in the twentieth and twenty-first centuries. No cooking, cleaning, washing, gardening or DIY chores for him. His Mum, housekeepers, friends and, in the final years, various carers and neighbours, did those tedious, enthusiasm-sapping things!

In addition, as Patrick was, surely, the greatest astronomy populariser of all time, who would begrudge the man a few negative eccentricities? Not me. I loved his good points and I loved his bad points just as much! Patrick survived for over half a century in the fickle TV business because he had no interest in money or material possessions and because people simply liked him and respected his principles, even if they didn't always agree with him. Although he was undoubtedly sexist by today's standards, women liked him too. They realised that, deep down, he was a harmless pussycat who had boundless mental energy, but would never hurt a fly. Patrick rarely took himself too seriously either, which was another endearing feature. The bottom line is, he was refreshingly different, not just a photocopied, politically correct, brainless clone, like so many vacuous celebrities these days. He was, I repeat, unique, and surely that is a good enough reason to write an honest biography about him. Outer Space is mind-bogglingly huge and we were incredibly lucky in having such a huge personality to explain it all to us, in plain English, and with a sense of humour.

With Patrick now departed, I can only say that our Universe, to me at least, seems a much poorer place. I hope this work allows his many fans to relive the excitement they felt whenever he appeared on TV, once a month, after the opening bars of Sibelius' *At the Castle Gate* had faded out…

To his many fans this was Patrick's Universe and we just lived in it. Now we have to somehow re-adjust. It won't be easy.

Cockfield, UK Martin Mobberley

Acknowledgements

Over the past 30 years, I have spoken to many amateur astronomers about Patrick's life. Only a few knew their comments would be logged and meticulously recorded! Nevertheless I am grateful to all those who have helped me gather data on Patrick for this book, whether they were aware of the fact or not, and whether they divulged one small snippet of data, or a lifetime of recollections. Certainly, amongst those who knew him, swapping stories about the great man was a favourite pastime in observatories, at astronomy meetings or in bars! I always had a notebook and pen with me on such occasions. Sadly, many of the older astronomers I consulted are no longer alive, but I was very privileged to have spoken to them, or received letters and e-mails from them. In alphabetical order I would especially like to thank:

Paul Abel, George Alcock, Geoff Amery, Laurence Anslow, Ron Arbour, Rossie Atwell, Rodney Austin, Richard Baum, Tom Boles, Mike Brown, Denis Buczynski, Charles Capen, Peter Cattermole, Tom Cave, Horace Dall, Tom Dobbins, Paul Doherty, Jane Fletcher, John Fletcher, Peter Foley, Dudley Fuller, Neville Goodman, Ken Goward, David A. Hardy, Cdr Henry Hatfield, Alan Heath, Andy Hollis, Guy Hurst, Jim Hysom, Nick James, Ken Kennedy, Pete Lawrence, Bill Leatherbarrow, Chris Lintott, Peter Louwman, Ron Maddison, Mike Maunder, John Mason, John C. McConnell, Richard McKim, Pieter Morpurgo, Terry Moseley, Gerald North, Damian Peach, John Pedler, Callum Potter, Douglas Richardson, Harold Ridley, Steve Ringwood, J. Hedley Robinson, Leif J. Robinson, Colin Ronan, Ian Russell, Robin Scagell, Ian Sharp, Bill Sheehan, Reg. Spry, Gordon Taylor, David Tucker and, last but not least, Dave Tyler.

Thanks also, with respect to photographic archive help, to the British Astronomical Association, the Royal Astronomical Society and their late librarian Peter Hingley, Clare Hindson (Press Association), James Hoyle (British Pathé), Simon Rowland (BBC Photo Library), Robin James (BBC Motion Gallery), Paul Carson (Belfast Telegraph), Prof. Nye Evans (Keele University), Sarah McMahon (Random House), Mrs K. Goodway and Brad Barnes (Peterborough Telegraph).

In addition, I am indebted to my father, retired Squadron Leader Denys Mobberley AFC, who, like Patrick, joined the RAF during World War II and was therefore of invaluable help in researching Patrick's wartime RAF record.

I must also thank the staff at Springer New York, especially Harry Blom, Maury Solomon and Jennifer Satten, for their support and belief in this project. I am very grateful to Karen Holland too, who enthusiastically read through the first and final drafts of the manuscript and provided numerous helpful comments and suggestions.

At the production stage I am indebted to Lesley Poliner of Springer NY and also to Karthikeyan Gurunathan and Rekha Udaiyar of Springer and SPI Technologies, India for their hard work on such a large manuscript.

Last, but not least, thanks to Patrick himself for entertaining me on TV, in books and in person for the last 45 years, as well as, on 100 or more occasions, revealing a few more pieces of the jigsaw of his amazing life.

Cockfield, UK Martin Mobberley

Some of Patrick's Favourite Expressions

Best spoken rapidly, in the Queen's English, with the right eye half closed, the other eyebrow raised, mouth slightly pouted and both hands on hips!

"Frankly, we just don't know."
"Our Milky Way Galaxy looks like two fried eggs clapped back to back."
"Astrology proves one thing and one thing only: there's one born every minute."
"Personally, I wouldn't recommend anything less than a three inch refractor or a six inch reflector. Binoculars are far better than a small telescope."
"A pillar and claw mounting is as steady as a blancmange."
"Astronomy attracts cranks like moths to a lamp."
"Every nut thinks every other nut is a nut."
"Ban all women teachers."
"Introducing Astronomy to the National Curriculum will kill it stone dead."
"Every child psychiatrist is as mad as a field of March hares and as mad as a room full of hatters."
"You can't beat a Henry Wildey mirror."
About any long term friend who held a post in the British Astronomical Association:
"This man is a Stalwart, a Backbone and a Mainstay of the BAA."
"Someone once said that I appear to have been rather hastily constructed. Someone else once said that I resemble an unmade bed. I am not stock size: the Selsey boatyard makes my size 13 shoes."
"According to Allan Chapman my suits are cut with a circular saw. I am not prepared to deny this."
"When it comes to practical things I have two left hands. I cannot even knock a nail into a piece of wood without bending the nail, splitting the wood and then watching the head of the hammer fly off. Ask the paper boy: he had to do it for me."
"I used to have a really nice singing voice as a boy, but then my voice didn't break: it shattered!"
"I do sing, very loudly, and I sound like a Corncrake with Croup!"
While helping himself to a mass of second helpings piled a foot high on his dinner plate:
"Despite my size and despite my bulk I have the appetite of an anorexic sparrow. I only have one main meal like this a day…"
"I've made wine: rose-petal, banana and coffee. I used to make rice wine too, but the vat exploded and turned the airing cupboard into radioactive waste."
"Well, it looks like I'm the final speaker: after the Lord Mayor's Show comes the dustcart!"
"Listen to this." Patrick bangs his foot like a monster sledgehammer twice on the stage.

"In that time the Sun has lost 4 million tons in mass. But don't worry, there's plenty left!"
"This feature on the Moon is called the Straight Wall. It's called that because it's NOT straight and it's NOT a wall."
"Policemen should catch criminals, not chase motorists."
"I always wanted a family life but Herr Hitler killed my girl."
"I was destined for Cambridge, but a chap called Hitler came along."
"I pottered about in the war, flying things."
"Harold Wilson? Worst thing that ever happened to this country!"
"Without Margaret Thatcher we'd be right in the cart."
"How did I sleep last night? Horizontally. I've tried vertically; doesn't work!"
"I used to have a pear tree in the garden. One night it blocked my view of Jupiter. Suddenly it became a small stumpy thing."
"By the time I got home Jupiter was tree'd out!"
"Observed the Moon tonight; sadly in jellyfish seeing"
"Spode has set up his HQ on Saturn tonight!"
[To astronomers of Patrick's era, the term Sod's law was too rude a term to use when things went wrong in an observing session, so Spode's law and Spode were used instead.]
"Frankly, some of my early lunar occultation timings may be inaccurate. A Cuckoo Clock is not the best timepiece and that bird's always been a DAMN LIAR!"
"Musically I only have two attributes: perfect pitch and perfect timing: no credit to me whatsoever."
"As a general rule I don't listen to any music if it was written after 1895."
"What would be the best plan if we heard a comet was going to strike the Earth? I would advise people to recite, very slowly: Our Father, who art in heaven………"
"I have actually seen a UFO. It turned out to be pollen drifting in front of the telescope."
"What's my idea of Hell? Bowling to a left-hander on a dead wicket, with a Pakistani umpire."
"Actually, yes, I have had an 'Out of this World experience': I once found myself in Bradford."
"I might be accused of being a dinosaur, but remember this: dinosaurs ruled the Earth for a very long time!"
About anyone famous he had ever met, however, briefly:
"Ohhh Yersss, I knew him, knew him very well."
"I was once driving up Duncton hill in the Ark and was overtaken by a dog."
(The Ark was Patrick's clapped out Ford Prefect)
About Creationists, who believe the world was created by God 6,000 years ago and who dispute Darwinian evolution:
"If ignorance is bliss, they must be very happy."
On once visiting Utah a local dignitary greeted Patrick with the words "Welcome to the Mormon state. We are quite different from the rest of America. You will find no swearing or drinking or wild women here".
To which Patrick replied instantly:
"It's hardly worth coming is it?!"
"When studying a phenomenon for which a perfectly simple and straightforward explanation is available always do your best to dig up an alternative solution which is infinitely less plausible!"
"The Halley's Comet Society, which I helped found, has one purpose and one purpose only: to meet on licensed premises. It is the only totally useless society in the world apart from, of course, the European Commission and the United Nations."
"There is no difficult situation on the Earth that the UN can't make worse."
"I take that particular theory, not with a grain of salt, but with an entire salt mine."
"Voyager 2 got to Uranus less than two minutes late after a journey of billions of miles. British Rail please note."
"My politics are slightly to the right of Atilla the Hun."

"The disc on Voyager 2 contains a recording of a pop group, so that intelligent life will know they should give Earth a very wide berth indeed."
On the rare occasion in his younger days that he was laid low with health problems:
"It's most frustrating but I am currently Hors de combat!"
Regarding the planet Venus:
"You can read a newspaper by it and it casts shadows."
"I have never done a days work in my life and I have never taken a holiday."
"I would love to have gone to the Moon, but it would take a very massive rocket to launch me!"
"I must be the only TV Presenter to have swallowed a fly, live, on air, on an early *Sky at Night*. As my mother said 'Awful for you dear, but so much worse for the fly.' It buzzed all the damn way down. ALL THE DAMN WAY!"
"I met Orville Wright and I met Neil Armstrong. Their lives overlapped, but they never met each other."
"Alien Life? Well, there's no sign of intelligent life anywhere in our Solar System, except, possibly, on the Earth, and I'm not too sure about that."
"If an alien did land in my garden I would say 'Welcome to Earth, tea or coffee? Please follow me to the nearest TV studio!'"
About various politicians and science ministers: "That man is as mad as a coot and concrete from the neck up."
"There's only one man we really need in parliament today: Guy Fawkes!"
"Inside a black hole, all the ordinary laws of science break down, and all the ordinary laws of common sense break down. Just like in the House of Commons!"
Concerning his belief that advanced life must be commonplace, given the sheer size of the Universe:
"Somewhere in the Universe there could be a complete carbon copy of Anthony Wedgwood Benn – although I sincerely hope not."
About any astronomical trouble maker, who is liable to make a fool of himself:
"Like Churchill, I recommend a policy of Masterly Inactivity."
"The only good German is a dead German."
Patrick: "Do you know what to do if you see a German drowning?"
Guest: "No, I don't Patrick."
Patrick: "Good!"
"If I saw the entire German nation sinking into the sea I'd help push it down."
"A Kraut, is a Kraut, is a Kraut!"
About any publication that criticised Patrick's views or appearance:
"I'm sure that particular magazine is enjoyed by both its readers."
About the Archbishop of Canterbury (for refusing to condemn blood sports):
"That bewhiskered old coot has neither the brains, nor the ability, to run a whelk stall!"
About anything that seems to be a scam:
"This has all the fine smells of rotting fish!"
About any suggestion that he should make money from doing TV commercials:
"Frankly, I would infinitely prefer to be found dead in a ditch!"
"The only people who get National Lottery funding these days are one-legged, single parent, Nigerian lesbians. I know, I tried getting some for South Downs Planetarium; ermmm money that is, not one-legged, single parent, Nigerian lesbians."
To anyone entering his house:
"Help yourself to a drink: Colonel Iremonger's rules apply in this house!"
"The English, the English, the English are the best; it's as simple as that."
"At my age I do what Mark Twain did. I get my daily paper, look at the obituaries page and if I'm not there I carry on as usual."

Contents

Chapter 1
The Pre-War Years

Patrick Alfred Moore was born, at a house named Innamincka, in Cannon Lane, Pinner, Middlesex, on the fourth of March 1923. As the reader of this book you are more than welcome to think of him as Patrick Caldwell-Moore, if you prefer, but Caldwell was not used by any descendant after his grandfather and was absent from Patrick's birth certificate. Patrick was the only child of Captain Charles Trachsel (pronounced Troxel) Moore, MC and Mrs Gertrude Lilian White. The house name, Innamincka, came from a tiny outback settlement in Australia, roughly 1,000 kilometres northeast of Adelaide. Why the Australian connection? Well, at the time of Patrick's birth his parents were living in the Pinner house of his "Mad Uncle George", the naval engineer and inventor George Thomas Macfarlane (an expert in winch, windlass and glider design) who had worked in Australia, visited Innamincka, and knew Patrick's father. Patrick once claimed (on Radio 4 in 1979) that he was born 'in a thunderstorm', although whether this was true (from Patrick, or his mother) will never be known! However, he also added that, unlike for Glendower's birth (Shakespeare's Henry IV part I), there were no celestial indications of Patrick's own entry into the world and the ground did not tremble.

On Patrick's mother's side, the non-military side, and the only side that really interested him, his family tree had a leaning towards music and the stage and his great-grandfather was also a noted historian. Patrick's maternal grandparents were Julius A. and Josephine C.C. White, born in 1854 and 1856 respectively. For the first 28 years of the twentieth century they lived in a very large house at 17 Alleyn Park in Camberwell, Central London (near to today's Sydenham Hill Underground station) and they also had a holiday home in Bognor Regis. Julius White was a very wealthy solicitor. The Whites had ten people living in their house at the turn of the twentieth century, namely, five children, plus a Governess (Amy Scotts), two servants (Maria Woodford and Matilda Lager) and two family friends. The friends were a 13 year old girl (Eleanor A. Wyatt) and a 25 year old man (Harold G.B. Gulley). The latter would ultimately marry one of Patrick's aunts. Patrick's mother Gertrude was the fourth of the five children at 17 Alleyn Park. In order the children were Reginald, Josephine (after her mother), Gwendoline, Gertrude and, 11 years

M. Mobberley, *It Came From Outer Space Wearing an RAF Blazer!: A Fan's Biography of Sir Patrick Moore*, DOI 10.1007/978-3-319-00609-3_1,

later, another son, Leslie. Patrick's grandparents' servant Maria Woodford (born in 1870) would serve three generations of the family and was, according to Patrick, the most devoted and dependable housekeeper anyone could have. She served Patrick's grandparents from the 1890s, Patrick's parents, and then, for a short time after the second World War, Patrick and his mother.

Patrick's mother's brother, Reginald, 7 years her senior, was a Gilbert & Sullivan actor who had given up employment in his father's legal profession for being a performer, in the early 1900s. Patrick often recalled that he had once accompanied Reginald on the stage in the late 1940s and even named two of his black cats 'Ptolemy' simply because Uncle Reg. had played Ptolemy in an Egyptian themed theatre production called *Amasis*.

Patrick's mother Gertrude was born on June 27th 1886 in Islington, London and was always proud to claim she was a cockney. However, unlike most cockneys, she was educated at Ravensfield College at Hendon and trained, in Italy, as an opera singer under Vincenzo Sabatini. With such a love of music, performing, and opera in Patrick's family it is hardly surprising that he could project his own voice with such volume at astronomy meetings in later life and had such musical talent. Patrick claimed his mother once met Florence Nightingale, although he never elaborated on where and when this encounter might have occurred. Gertrude was even offered the lead part in an Italian grand opera, during 1914 (aged 28), but with the outbreak of World War I she was forced to return to England where, in 1917, she married Charles her fiancée.

Patrick's father Charles was a military man through and through. He was born on September 14th 1885, the son of a Scottish Father (William Roger Caldwell-Moore, a chartered accountant, born in 1851) and a miserable, humourless German mother (Celina Emily Trachsel, born, in Germany, in 1854). Patrick usually claimed his miserable grandmother Celina was French-Swiss and, indeed, she and Patrick's father were fluent in French. However, she was actually born in Germany and Trachsel *is* a German surname, although one which many Swiss people have too. However, Patrick's Germanic ancestory was rarely a subject for debate: he hated Germans and Patrick and his father fought them in consecutive World Wars! This was not a subject that Patrick was usually prepared to discuss. He hated any implication that there was a German side to his family. However, in February 2008 he did bring this subject up, briefly, in his study and admitted he had looked into his 'German ancestory' and was not happy about it. Patrick told me and another amateur who was there (Jamie Cooper) that if you looked back beyond his grandmother Celina, you arrived at a rather disturbing "Count Von Trachsel" who was 100 % German.

At the time of Patrick's father's birth (1885) the Caldwell-Moore's lived at 45 Miranda Road, Upper Holloway, in north London. Patrick's paternal grandparents, William and Celina, had four children in total. Charles was the youngest, but the first to leave home (for the Navy). Charles had three elder sisters: Celina (after the mother), Sarita and Adele Caldwell-Moore. In the 1890s, with four young children, the Caldwell-Moore's left 45 Miranda Road and moved a mile north to a bigger house: 2 Park Villas, Bloomfield Road, Hornsey. [Interestingly, though purely coincidental, the next occupants of 45 Miranda Road would be a family with the surname Alcock: a legendary surname in British amateur astronomy as George Alcock would discover five comets and five novae from Peterborough]. At 2 Park Villas the Caldwell-Moore's employed a full-time Cook, Mrs Fanny Brompton.

Patrick's father, Charles Trachsel Moore (he does not seem to have used the Caldwell part of his father's surname, despite adopting his German mother's maiden name of Trachsel) was originally commissioned, as a teenager, in the Royal Navy, but shortly after the outbreak of World War I he transferred to the Royal Army Service Corps and was awarded the Military Cross for his bravery in the Battle of Arras. After transferring to the British Expeditionary Force Tank Regiment, Charles was twice mentioned in dispatches. In 1917 a German gas attack on the French battlefield finally finished his military career and his health was impaired for the rest of his life, but he was at least able to leave the battle front, return to England, and marry Patrick's mother. Patrick's parents, the son of an accountant and the daughter of a solicitor, Charles and Gertrude, were married at St Stephen's Church, South Dulwich on September 5th 1917. They were both 31 years old and single; this was quite unusual at a time when most couples were married by their early twenties. The witnesses to the marriage were Gertrude's parents and her brother Reginald, plus Charles' father William. The main wedding present from Gertrude's parents was the giant grandfather clock that still dominated the hallway of Patrick's house at Selsey in the early years of the twenty-first century and could occasionally be heard chiming in *The Sky at Night* broadcasts! After the war, Patrick's parents emigrated for 3 years to Mombasa, Kenya where, from his military connections, Charles had secured a job as an accountant, but in late 1922, with Patrick on the way, Charles and Gertrude left Kenya, returned to England and Charles became a chartered accountant (like his father William), first living in Middlesex and then in Sussex. Before his World War I injury, Patrick's father was a very keen sportsman. He won numerous swimming trophies, was a keen amateur boxer, and a county hockey player for Middlesex.

At this point I would like to clarify and re-iterate this 'Caldwell' business which first emerged when Patrick produced his list of 109 deep sky objects in later life. Patrick rarely used the Caldwell part of his surname and neither did his father, but his paternal grandfather William appears to have preferred the Caldwell-Moore surname to just plain 'Moore'. Indeed, although, in his later years, Patrick liked to say he was a Caldwell-Moore, his birth certificate states he is definitely a plain Patrick Alfred Moore. In addition, there is no mention of his extra and elusive fictional third initial which he very occasionally mentioned after a drink or two, namely a 'T' for Tremayne, making him Patrick Alfred Tremayne Caldwell-Moore! For most people he will always be, simply, Patrick Moore. Extra adornments, real or imagined are unnecessary! I told Patrick on one occasion that I thought he had made the 'T' part of his initials up, so they were P.A.T. and he did not deny it: he just laughed!

Despite being born in Pinner (now part of Greater London) Patrick's first 6 years were spent at his maternal grandmother Josephine's holiday home in Glencathara

Fig. 1.1 A young Patrick with his mother and (background) grandmother at his grandmother's holiday home in Bognor (Probably taken in 1926 or 1927 by Patrick's father. Re-photographed at Farthings with Patrick's permission in 2002)

Road, Bognor Regis in Sussex and then, from 1929, he lived at East Grinstead (also in Sussex). Patrick's parents only lived in Pinner for the first 6 months of his life. On one occasion, while living in Bognor, aged 4, Patrick declared that he wanted to be a writer when he grew up: what an accurate prediction, if true! The Bognor house where Patrick spent his first 6 years still exists; it is situated on the junction between Glencathara road and Nyewood lane. For 36 years, from 1929 to 1965 (apart from much of the later War Years), Patrick's home address would be 'Glencathara' (named after the same Bognor Road), Worsted Lane, East Grinstead, Sussex (telephone no. East Grinstead 322). This large house belonged to Patrick's wealthy maternal grandmother who lived in an even larger adjoining house to the south-east, linked by a passageway. Glencathara had been built onto Grandmother Josephine White's East Grinstead house (she had moved, with housekeeper Maria, out of Camberwell when Grandfather Julius had died). Patrick's grandmother's house was the first house on the left (west) side of Worsted Lane and the connected house where Patrick and his parents lived was the second property on that same side. The adjoining house tenancy plan enabled Patrick's mother to look after her own mother, as well as her husband Charles, and Patrick himself. In fact, Josephine White had three adjacent houses purpose built in Worsted Lane; one of the houses was her property, one was for her daughter Gertrude (Patrick's mother) and one for Patrick's aunt Gwen, under a tenancy agreement.

The White family kept the Bognor Regis holiday home until the start of World War II, when Patrick's grandmother died, which pleased the young Patrick as he loved life by the seaside and he loved riding on the donkeys at Bognor. Fortunately for the donkeys he was much smaller in those days! He resolved that one day he would return to live on the south coast. Like almost everything else that Patrick resolved he would keep that promise. In fact the only reason that the Moore's moved from Bognor was that the widowed grandmother White's asthma appeared to be worsened by the sea air, hence the move inland. All three houses in Worsted Lane were tied together with a complex will designed to ensure that Patrick's mother and aunt would never be homeless as long as they survived, but ultimately the properties were left to all five of Josephine's children. Glencathara was a huge house for a single child family, featuring five large bedrooms, a drawing room, a dining room and a room solely reserved for the young Patrick. Worsted Lane is situated on the A22 road, a mile to the southeast of East Grinstead town centre and was quite a rural environment in the 1930s and 1940s. After the Second World War, Patrick's room at Glencathara, originally Patrick's nursery, became his study and the place in which he planned his days as a schoolteacher and then his career as a prolific author and TV presenter, until 1965. In that year he would move to Northern Ireland. By an extraordinary twist of fate Glencathara was only a few hundred yards from a keen amateur astronomer's observatory, which we will come to very shortly.

Patrick: The Child

Patrick usually said that his earliest recollection was of the General Strike of 1926, when he was three and the family lived in Bognor. He distinctly remembered his parents driving him (they had a car in 1926?!) into Bognor town to buy him a new pair of shoes during the strike, since Bognor, as a non-union town, was not involved. It had not then become Bognor Regis.

Patrick's mother, when aged 92, described the young Patrick in a 1978 newspaper interview thus: "He was the untidiest, oddest little devil as a boy and he hasn't changed. Life with him is a bit strange but not bad when you stop wondering about what could happen next." For Patrick's sixth birthday, in March 1929, he wanted a cuckoo clock, not a telescope, and his mother drove him all the way from Bognor to London's Oxford Street to choose one. It would never keep particularly good time (Patrick would often say "that damn bird's a liar") but he would always claim, for the rest of his life, that he could never write a book without hearing it's tick-tock in the background.

As an aside to all this, at Patrick's 83rd birthday party I met Patrick's cousin (his mother's sister's daughter Eileen Tanner), who was 3 years older than Patrick. I asked her what she remembered of the boy Patrick and she replied: "He was always coming out with long words and scientific stuff. We never understood what he was talking about as a child and still don't!"

On one particularly rainy day in 1929, the year the family had moved to live next to his grandmother at East Grinstead, Patrick's mother introduced her 6 year old son to a small book entitled *The Story of the Solar System* by G.F. Chambers (published in 1898). This book fired Patrick's imagination and inspired him to "devour its companion volume straight away" and the rest, as they say, is history! The companion volume was entitled *The Story of the Stars*. Shortly after arriving at East Grinstead, in the loft in Patrick's grandmother's house, his late grandfather Julius' 1892 Remington typewriter was unearthed. This find would shape Patrick's destiny.

It was obvious from an early age that Patrick was keen on writing and typing, as well as playing the piano. Aged seven he learned to type (using the middle finger of each hand – a technique that he used throughout his life) with that 1892 Remington typewriter. In 1931, his father bought him a second hand (1908 vintage) Woodstock 5 typewriter, which he still used well into the twenty-first century! His father bought him a solid writing desk too, for his eighth birthday present. As with the typewriter, that desk would still be the one that Patrick was using well into his eighties.

Patrick often said that his 'First World War hero' father was everything he was not: "athletic, practically minded and very strong, and he would have liked a son who was the same… but he got me!" Unlike Patrick's Mother, his Father had: "No interest in Astronomy whatsoever…..We got along……sort of." Interestingly, if you ever visited Patrick's house and had a good snoop around, you would find loads of photographs of his mother, but not a single one of his father on display. Indeed, there was no evidence of his father ever existing, anywhere in the house, on first glance. If you mentioned Patrick's father, or that side of his family, he would usually change the subject, immediately. However, on one occasion, when Patrick himself mentioned his father, unprovoked, I expressed interest in this obviously brave military man. At this point he directed me towards a cupboard in his study, inside which was an old painting of his young father in military uniform. Facially, Captain Charles Trachsel Moore looked nothing like even a young Patrick. He had a lean, thoughtful face, with a long slender nose and close set eyes and eyebrows that almost joined in the middle, along with a very small moustache. His hair was brown, with a central parting, and the instant impression was of a distinctly aristocratic military man who could easily have passed for a Frenchman. In fact, Patrick's father's portrait reminded me vividly of a French character played by Nicholas Lyndhurst in the BBC time travelling comedy 'Goodnight Sweetheart!', when Lyndhurst played a French Colonel, Henri Dupont, who coincidentally resembled his TV character Gary Sparrow. Anyway, after a while Patrick asked me to put his father's portrait back in the cupboard, and that was the end of that investigation.

At the age of seven Patrick read his first science fiction novel in an old 1908 copy of the magazine for boys called *Young England*. There was a story in the magazine by the journalist and author Fenton Ash (real name Frank Atkins) entitled *A Son of the Stars* in which two young boys travelled to Mars. Patrick was really hooked on Space Travel from that moment on. He bought a star map, soon acquired a small pair of binoculars (on loan) and started devouring any space fiction he could find. Patrick actually tried writing his first astronomy book aged eight. He told his relatives that he was: "Going to write an astronomy book in simple language for the young, and so that mother can understand it too!" However, at only eight, even Patrick never managed to complete the work.

The 7 year-old Patrick, for 1 year, had attended the prestigious Dulwich preparatory school in South London (a school well known to his mother's side of the family and only a stone's throw from his grandparents' home). Unfortunately, health problems (a "silly crooked heart – the family curse", according to Patrick) and his dislike of school life resulted in Patrick's parents having him privately tutored at home. In later years, after Patrick's mother had died, he constantly talked about how he was "a crock" as a child: too unwell to go to school. However, this author remembers one press interview with his mother when she stated that it was amazing to see him so full of energy on TV, because "as a child he was thoroughly lazy…he only put up with going to school for one year." Anyone who knew Patrick would tell you that he never, EVER, did anything that he did not want to: he must surely have been like that as a child too! Indeed, a picture in the hallway of his Selsey home 'Farthings' showed a very stubborn looking young master Moore. Although Patrick would later be able to sketch the Moon and planets well through his telescopes, one of his worst childhood subjects was art (despite his mother's sketching and painting skills), while his best natural talent was for music. Coincidentally he had an art teacher called Mr Moore, who one day asked his young pupils to "sketch a towel draped over a chair". Patrick mis-heard him and thought he had asked them to "sketch a cow draped over a chair". The resulting sketch was one that his mother treasured for many years. After some abortive attempts to teach Patrick to draw, and to paint, the art teacher eventually told Mrs Moore that "your son tries hard but I think he is more of a musician than an artist". Patrick was allowed to miss art lessons and practice his piano work instead. Once a week there was an elementary science lesson and Patrick vividly recalled being told of the discovery of the ninth planet Pluto by the science teacher. Half a century later, Patrick would collaborate with the discoverer, Clyde Tombaugh, on the story behind that piece of celestial detective work.

From the age of eight, Patrick was deemed too unwell to attend a preparatory school and so a tutor was found. That tutor, who along with Patrick's mother taught him for the next 7 years, was the Reverend John Missen (born in 1889) of Coleman's Hatch, Hartfield, in East Sussex. He was the vicar of Coleman's Hatch for half a century. Patrick often said that if not for the teachings of John Missen he would never have achieved anything in life. In 1974, the Reverend Missen described the young Patrick as: "Never depressed and always faithful to his friends". He also said: "If Patrick didn't know any particular subject well he'd play the fool and he did it so well you realised he was certainly not a fool whatever else he was!"

Just before Patrick's ninth birthday he hand typed a copy of the astronomer W.H. Pickering's 1907 book about the Moon (*The Place of Origin of the Moon*), almost 60,000 words of it, from a library book acquired for him by Major Arthur Everard Levin of the British Astronomical Association (BAA). Major Levin was the BAA President from 1930–1932 and, remarkably, lived at Patrick's future village destination of Selsey, in the same street where Patrick would eventually live! By another fortunate coincidence, he was also a military acquaintance and a friend of Patrick's father. Patrick performed this marathon typing exercise because there was no way he could afford to buy the book and because the loan, from the Royal Astronomical Society Library, expired in 1 month! It seemed destined for Patrick to become a Moon fanatic, Moon book author and manic typist from that moment on; and so it proved. Patrick's 1908 Woodstock and, later, his 1911 Underwood (for use during

Woodstock breakdowns) would serve him well for the next 80 years! Patrick did not own a telescope aged nine but he did have his good pair of binoculars which were regularly pointed at the night sky. Encouraged by a copy of Annie and Walter Maunder's book 'The Heavens and their story', Patrick was soon becoming very familiar with the constellations. Astronomy and music seemed to have an equal fascination for Patrick in those early years and he taught himself to read and write music fluently aged 9, and was composing whole pieces only 2 years later!

Patrick claims to have observed the celebrated 1933 White Spot on Saturn (discovered by the stage and screen comedian Will Hay), although he did not acquire a telescope until 1934, so he must have seen it through the local Hanbury refractor (which I mention a bit further on). Strangely though, in *Guide to the Planets*, Patrick claims he did see the White Spot with his own telescope. In addition, I have studied the local Hanbury Observatory sketches kept by W.S. Franks (much more on him later) who was an avid observer and, while there are sketches of Saturn in his logbook, there is just one of Saturn in 1933, made on August 9th at 10 p.m.; that drawing does not show the famous White Spot. This is not the first time some of Patrick's claims have failed to stack up, chronologically or otherwise!

When Patrick was 10 his parents bought him a bicycle (not a telescope), with no brakes, for his birthday present! No brakes?! Did they want to get rid of him perhaps…..? Surely not?! In later years (even 40 years later) Patrick would claim that the same bicycle had to be constantly hidden from the dustman by him or his mother. If the bike was anywhere near the dustbin it would have been hurled into the bin lorry "without a qualm".

Some time later, on a trip to London to see his maternal grandparents in 1933, Patrick spotted an antique astronomical orrery for the bargain price of 30 shillings, on sale in London's Caledonian market. With money from his parents he purchased the orrery and would keep it in his study for the rest of his life. An orrery is a clockwork model of the solar system in which the planets rotate around the Sun at the correct relative speeds. But 1933 was also a year of great sadness for Patrick because the family cat, Ptolemy, contracted cat flu and died. The young Patrick was totally distraught as of their two cats, Ginger and Ptolemy, it was always the pitch black Ptolemy that followed the budding astronomer everywhere he went, whether it was in the daytime, or at night to learn some more constellations. Ginger survived, but it took Patrick many months to overcome Ptolemy's demise. Seventy years later another jet black cat, literally identical in appearance, took up residence in Patrick's home at the other end of his life. The octogenarian Patrick would name the new moggie Ptolemy too, as it was, in every respect, spookily identical to his long lost childhood friend. That second Ptolemy would lie, faithfully, on Patrick's bed, when he died, on December 9th 2012.

A couple of years after the death of the original Ptolemy a feral cat gave birth to four kittens under the Moore's East Grinstead garden shed. Three of the kittens were given good homes but the fourth, named Rufus, made it clear that it was staying with Patrick and his parents. Rufus would outlive Ginger and survive for 20 years, right through the War years and almost up to the point where Patrick would become a book author.

Patrick applied to join the prestigious British Astronomical Association in 1934 when (like this author, some 35 years later) he was only 11 years old. At the time of his application the BAA had a mere 830 members and only 570 of them were based in England. The official procedure was that two people of good character, or established BAA members, proposed the new member and, if the council was happy, the member was duly elected. Patrick managed to enlist the help of the aforementioned family friend, the well-known amateur astronomer and British Army officer Major A.E. Levin, as his proposer and was duly elected to the association on November 28th 1934, aged 11 years and 8 months. Patrick's seconder was J.T. Foxell, another BAA member. Foxell was a friend of Levin and an expert on predicting lunar occultations of bright stars, such as Regulus and Antares, as well as calculating cometary orbits. Both Levin and Foxell were prominent figures in the BAA computing section where abilities in mathematics were, obviously, crucial. Ironically, mathematics was always going to be Patrick's weakest subject, with music his strongest and it is true to say that he was in awe of the BAA's computing experts.

As I mentioned a few paragraphs ago, by an extraordinary coincidence, while Patrick was a child in the East Grinstead of 1934, Major Levin lived in the same street (West Street in Selsey) where Patrick would take up residence from 1968; Levin owned a house called Elleray, located a stone's throw from Patrick's house 'Farthings' some 34 years in the future! Patrick told me that Levin's West Street house number was 87, but confusingly that would now place it in the sea. When the average age of a BAA member was 50-something, obtaining membership aged 11 was quite an achievement. In the decades to come though Patrick would, single-handedly, reduce the average age of the BAA's membership considerably, simply by attracting his many fans to join up. In 1934, even asking to join the BAA at the age of 11 was highly unusual. However, Major Levin had been the President of the Association 2 years earlier, as well as being the Computing Section Director, and so with him and Foxell as proposer and seconder Patrick's election was never in any doubt. Two weeks after Patrick was officially elected to the BAA he made a note in his 'Brockhurst Observatory' observing log book (whose name I will explain shortly) stating that on December 13th 1934 he had been officially told he was now an elected BAA member.

Major Levin owned a 6 inch refractor at his home in Selsey and knew the Moore's neighbours, the Hanburys, owners of another quality 6 inch refractor; we will learn more about them soon. In those days, and throughout the twentieth century, it was the custom for the BAA President to offer to shake the hand of any newly elected member if they were present at a London meeting. Patrick claimed that, shortly after being elected, he was taken to a BAA meeting and shook the hand of the President (1934–1936) Sir Harold Spencer Jones, who was also the Astronomer Royal at the time. As the BAA was mainly an association of elderly bearded military men, doctors, or members of the clergy, it must have been an unusual sight to see an 11 year old boy shaking the President's hand.

The 11 year old Patrick also decided it was time to invest in a telescope of his own, but as a child a 6 inch refractor was clearly out of the question. On a trip to London in 1934, accompanied by his mother, he acquired a good quality 3 inch refractor for the sum of £7–10 shillings (£7.50 in today's money) from Broadhurst Clarkson. It was of 1910 vintage (so, roughly 24 years old) but in good condition. The telescope suppliers were recommended by the BAA's Dr W.H. Steavenson and survive today as Broadhurst, Clarkson & Fuller (BC&F); they were based at 63 Farringdon Road in London at the end of the twentieth century and in Lingfield, Surrey, today. This was to be the only telescope Patrick owned until 1950. Steavenson had been the BAA President from 1926 to 1928, the period including the UK total solar eclipse of 1927. He was the youngest ever BAA President and became a President of the Royal Astronomical Society too. When the stage and screen comedian Will Hay had discovered the White Spot on Saturn from his south London address of 45, The Chase, Norbury, a year earlier, Steavenson, a stone's throw away at 70 Idmiston Road, West Norwood, had been the first person Hay phoned. Remarkably, when the 11 year old Patrick's parents asked Steavenson's advice on a telescope for Patrick, Steavenson actually travelled the short distance from West Norwood to East Grinstead to meet Patrick, and he recommended the 3 inch refractor from Broadhurst Clarkson. Although £7.50 might not seem much these days, it was, in fact, a substantial sum for an 11 year old in 1934, equivalent to 2 weeks wages for a working man. Almost 40 years later BC&F would supply Patrick with the mounting and the rotating top for his largest telescope. In passing, it is worth noting that Patrick's observing logs always refer to that 3 inch refractor as a 3⅛ inch refractor.

The new telescope, with its three foot focal length (roughly f/12) was used heavily by the young Master Moore. However, as with so many telescope purchases there was, initially, a fly in the ointment. The instrument was supplied with a shakey table-top 'pillar and claw' mounting which, according to Patrick, was 'as stable as a blancmange'. Patrick's parents purchased a solid wooden extendable tripod, on which to re-mount the instrument, for 30 shillings (£1.50). Who knows, if not for that extra refinement the young boy, whose name became synonymous with astronomy, might not have acquired the patience to continue with his hobby and the whole future of astronomy might have been changed for ever? Although Steavenson, who was blind in one eye, was never a household name in the UK, in 1934 he had contact with two household names: the 45 year old Will Hay, who was just approaching his peak years of fame, and the 11 year old Patrick, whose fame would start 23 years in the future. In later years Patrick would say that Dr W.H. Steavenson was the man he had respected most in his early years in the BAA.

A number of people have told me that, when answering questions from audiences, Patrick has occasionally claimed he observed with Will Hay, as a young BAA member, at Hay's Norbury address, where Hay lived from 1927 to 1934. This could only have been made when Patrick was 11, as Hay had separated from his wife Gladys by 1935 and then moved to Hendon. However, when I have tried to clarify Patrick's encounters with Hay and his claim to have observed with him he just used to say "Yes, I knew Hay, as an astronomer" and trying to get any further would lead to him stonewalling. As we shall see, at various points in this book, investigating the stories about people who Patrick claimed to have known in his early life often leads to some very murky waters indeed.

Hanbury Observatory

By a lucky coincidence Patrick's Glencathara home, in East Grinstead's Worsted Lane, was directly opposite a large country estate, called the Brockhurst Estate, which featured a well equipped, privately owned, astronomical observatory. The Brockhurst Estate was owned by Frederick J. Hanbury FLS who was a millionaire associated with the pharmaceutical firm of Allen & Hanbury and also a Fellow of the Linnean Society (which specialised in promoting the biological sciences). F.J. Hanbury had joined his father Cornelius' huge company in 1872 and by 1916 was the chairman. As Allen & Hanbury was the prime manufacturer of Cod Liver Oil and throat pastilles in Great Britain it meant that F.J. Hanbury was a very wealthy man. However, his main interest in later life was in growing orchids (he was a world authority on them) and he was also a keen amateur astronomer and naturalist. Hanbury was so wealthy that he employed a local obsessive astronomer, called W.S. Franks, to run the small observatory full-time. To say that William Sadler Franks appeared unusual would be a gross understatement! He was decidedly eccentric, but he was also, without a doubt, the main astronomical influence on the 11 year old Patrick.

Fig. 1.2 Patrick's first astronomical mentor, William Sadler Franks (1851–1935), who ran the Brockhurst Observatory near to Patrick's East Grinstead home (Photographed around 1932. Photographer unknown. RAS archive picture by kind permission of the late Peter Hingley)

The middle class Moore's knew the extremely wealthy Hanburys and, one evening in early 1934, when the family was invited for a meal at the Brockhurst Estate, the young Patrick was introduced to William Sadler Franks. He knew that Patrick was applying for membership of the BAA and invited Patrick to join him in the Hanbury observatory, whenever he wished. Franks could never have dreamed that taking the young Patrick 'under his wing' and teaching him all he knew about astronomy would not only inspire Patrick, but ultimately generations of astronomers, over the next 78 years and well into the twenty-first century!

The octogenarian who inspired Patrick in 1934 and 1935 was actually born in Newark, Nottinghamshire on April 24th 1851, 106 years *to the day* before Patrick broadcast the first ever *Sky at Night* programme live, on BBC1, on April 24th 1957. Actually, Patrick's dates for both events, alternated between April 26th and 24th; but he was adamant that Franks was born on the same day and month as the first *Sky at Night*, which *was* Wednesday April 24th 1957. This was just 1 day after another favourite event in Patrick's calendar, namely St George's Day. As a proud Englishman Patrick always liked to celebrate that patriotic day and with the date occurring so close to the first ever *Sky at Night* broadcast it all seemed like destiny. W.S. Franks' first major astronomical paper was written in 1878, when he was 27, and entitled 'A Catalogue of the Colours of 3,890 Stars', which he communicated to the Royal Astronomical Society in 1878, via the legendary observer, the Rev. T.W. Webb. Franks became Director of the Star Colours Section of the Liverpool Astronomical Society, the most prestigious amateur astronomy club in the country at that time. In 1891 he joined the British Astronomical Association (the year after it was formed) and became its 'Star Colours' section director. He held the post for 3 years until he was succeeded by G.F. Chambers, the man whose 1898 book would be purchased by Patrick's mother and ultimately inspire the 6 year old Patrick: it's a small world indeed! Franks moved to Crowborough in Sussex in 1892, where he worked with the legendary astrophotographer Isaac Roberts until 1904. Roberts had a massive telescope (for its day) with a 20 inch (50 centimeters) mirror. After Roberts died, in 1904, Franks continued for another 2 years at Crowborough, then went to live in Uxbridge and, finally, in 1910 (aged 59) moved to F.J. Hanbury's Brockhurst observatory at East Grinstead as the Observatory Director. This was still 24 years before he would meet the young Patrick at the same observatory. Francks purchased a house half a mile away, on the Lewes Road and named it 'Starfield' after Isaac Roberts Observatory.

Fig. 1.3 A picture of Brockhurst Observatory, taken by Patrick on March 4th 1939, his 16th birthday. Patrick was the Observatory Director while still a teenager! (Re-photographed in 2002 with Patrick's permission)

So, in 1934, the 83 year old Franks took Patrick under his wing and the course of British amateur astronomy was changed forever. Throughout his life, Patrick would take many young astronomers under his wing too: many became leading amateur and professional observers. The Brockhurst Observatory was modest by modern standards, but very well equipped by the standards of the day. Just prior to Patrick's arrival it briefly featured a massive 24 inch (60 centimeters) reflector, but its main instrument was a fine Cooke 6⅛ inch refractor (155 millimeters) of 82 inches (208 cm) focal length. Franks was a meticulous observer and a stickler for facts and calculated the observatory's position as latitude 51° 7′27″ N, longitude 2.27 s E, altitude 435 feet. As Patrick later mentioned, on numerous occasions, Franks was a familiar figure in East Grinstead and, almost every day, he could be seen cycling from his house to the observatory at Brockhurst. In Patrick's oft used words "It is no disrespect to say that he looked remarkably like a garden gnome". He was under 5 feet tall in Patrick's estimate (Patrick recalled being a similar height to Franks when he was only aged 11) always wore a skull cap and had a white beard! Even up to the age of 83 he would regularly attend BAA meetings in London. Tragically, in early 1935, when he had known Patrick for barely a year, he suffered a bad cycling accident. Although the accident did not kill him, he was badly shaken and was never to recover. He died some months later on June 19th 1935. The 12 year old Patrick was devastated. Not only had he enjoyed his evenings under the stars at Brockhurst, he had been a regular visitor to Franks' home.

In 1936 the young Patrick won a staggering £87 on the Football Pools, equivalent to a win of maybe 10,000 pounds these days! The family spent the winnings on a holiday in Belgium with £7 and 10 shillings being spent on a xylophone for Patrick and, surprisingly (to me at least) NOT on more astronomical equipment. Back in East Grinstead and taking a break from music, Patrick decided to put his own 3 inch refractor to good use and spent many evenings studying the Moon. In 1937, at the advanced age of 14, he proposed his first scientific paper, to be delivered to a BAA audience, entitled 'Small Craterlets in the Mare Crisium'. The Mare Crisium is a distinctive smooth circular lunar 'sea' easily visible in a small telescope. The BAA secretary (F.J. Sellers) wrote back to Patrick thanking him for his proposed paper and saying that he noted Patrick was 14 'although I suppose that doesn't matter'. In later years Patrick would say: "They did a double-take when a fourteen year old turned up to speak. They thought I was probably an old man". Strangely there is no account of Patrick actually delivering his talk in the BAA meeting reports of that era, although not everything was always recorded, especially if the meetings recorder missed a contributor's name or if the Journal editor decided the contribution was relatively minor. There was certainly much interest in small Mare Crisium craterlets in that era though. At the June 24th 1936 BAA meeting Mr Robert Barker had mentioned 'Some small craters in the Mare Crisium, easily seen with a 3-inch telescope' and 'it seemed curious that they had not been previously recorded'. Maybe Patrick's proposed paper was inspired by those comments and the related BAA Lunar *Memoir* published in that year?

Despite the death of Franks all hope of using the Hanbury Observatory's splendid Cooke refractor was not lost. Brockhurst's owner, F.J. himself, was still alive (although also in his eighties) and still liked to show his visiting friends the stars, the planets and the Moon through his telescope, so he kept the observatory going. However, at his age he was increasingly in need of a new 'Observatory Director'. In 1937, Patrick was asked by the ageing Hanbury to take over this role. Patrick was not intended to carry out the relentless observing schedule of Franks (who was obsessed with estimating star colours, sketching planets and measuring double star separations), just to show Hanbury's guests the best sights through the telescope, when the clouds parted. How many 14 year olds can boast of being an Observatory Director? Patrick put his access to Hanbury's 6-inch refractor to good use and used it to observe the Moon, the planets and the bright comet discovered by Finsler in the same year. The telescope must have seemed enormous to the teenage Patrick. Unlike today's compact German Equatorial Mountings, typically used for modern short-focus refractors, the 82-inch focal length Brockhurst refractor was slung between the two tapering mahogany pillars of a huge English Equatorial mounting; one end of the mount was anchored into the observatory floor and the other was positioned well above head height. The late Franks himself had admitted (in the Monthly Notices of the RAS) that the Brockhurst refractor's clock drive was 'somewhat antiquated' and 'irregular in action' but the telescope was blissfully equipped with enormous 20-inch (50 centimeters) diameter setting circles in R.A. and Dec., enabling an object's position to be dialled up to an accuracy of 2 seconds in time and 10 arcseconds in declination. This setting circle accuracy proved to be a joy for the young Patrick, when he wanted to find new objects to observe. Patrick has mentioned in recent years that being an Observatory Director meant that his Woodstock typewriter was suddenly a lot busier, as the Hanbury Observatory received quite a lot of postal correspondence, both from amateur and professional astronomers. So his two fingered teenage typing ability became even more frantic.

Fig. 1.4 Patrick aged about 13, in the mid 1930s, around the time he became Brockhurst Observatory Director! (BAA archival image)

In December 1937 the BAA Journal dropped through the Moore's *Glencathara* letterbox and, on page 80, it contained an article which must surely have fascinated the young Patrick. In an article by H.P. Wilkins the discovery of a possible new lunar sea (or mare) was announced right on the very edge of the Moon. Due to the orbital characteristics of the Moon, which we shall learn more about later, Wilkins had been able to glimpse a feature normally beyond the Eastern (Classical orientation) limb, so not usually visible from the Earth. Wilkins named the feature Mare X (although it had actually been seen well before his observation, way back before 1906 in fact, by foreign observers). Patrick would glimpse this feature again in 1940 and, when collaborating with Wilkins in earnest, after the war, in 1946. Wilkins would ultimately become the next mentor for Patrick. In later years Patrick would claim that he discovered both the Mare Orientale and the crater Einstein! In fact, Orientale may have been independently discovered by Wilkins in 1937, and Wilkins may well have been the first British observer of the Mare, but Patrick's claim to be Orientale's discoverer was just a story that became exaggerated over many years. As we shall see, Patrick would become a fanatically keen *observer* of the feature, with Wilkins, in the late 1940s and early 1950s, but nothing more; he certainly could not claim to have discovered it.

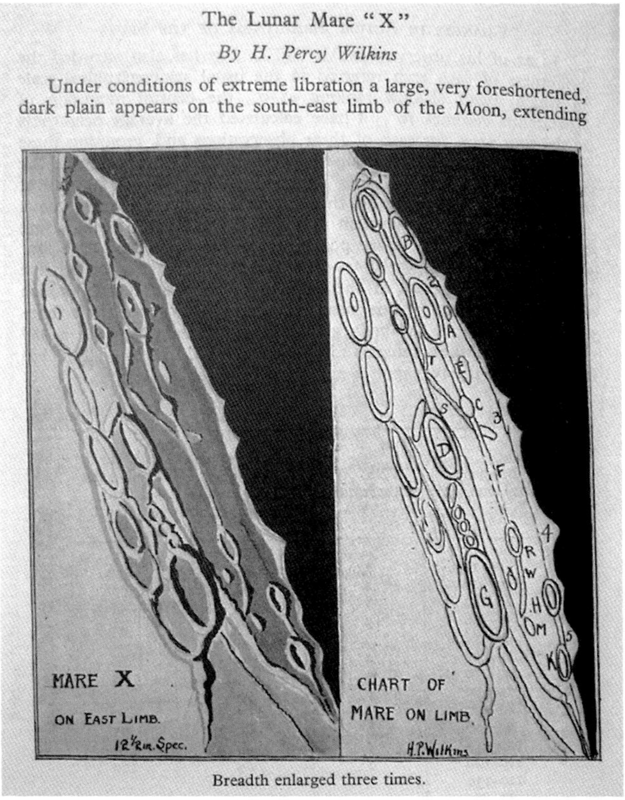
The Lunar Mare "X"

By H. Percy Wilkins

Under conditions of extreme libration a large, very foreshortened, dark plain appears on the south-east limb of the Moon, extending

Fig. 1.5 H.P. Wilkins' paper on the 'Mare X' (Mare Orientale) appeared in the December 1937 BAA Journal. Initially it appeared that the Mare might have been a new discovery; however, it had been seen and recognized as a Mare (lunar 'sea') as early as 1906. At best, it was an independent British discovery by Wilkins in which Patrick never played a part, despite claiming it was his own discovery in his later years! Note that although the east limb is indicated, due to IAU politics, the east limb became the west limb, so the eastern sea ended up on the Moon's western limb! (BAA Journal image)

Wilkins was not a publicity shy man. Indeed, he sought publicity whenever he could and one could imagine that in the late 1930s, before the War, from a newspaper reader's viewpoint, there may have appeared to be two famous amateur astronomers in the UK: the comedian Will Hay and H.P. Wilkins. The August 27th 1938 *Saturday News Chronicle* devoted an entire page to Wilkins, with a banner headline proclaiming: **AMATEUR's GLIMPSES of MOON's OTHER SIDE – Finds more Craters, Peaks and Dark Plain 100 miles wide**. It went on to describe how Wilkins, using a home made 12½ inch reflector, had claimed to have spied new peaks and craters. We will have a lot more to read about Wilkins and his friendship with Patrick, after the War years.

Saturday NEWS CHRONICLE August 27, 1938

AMATEUR'S GLIMPSES OF MOON'S OTHER SIDE

Finds More Craters, Peaks, and Dark Plain 100 Miles Wide

Mr. H. P. Wilkins with his home-made telescope

IF you are an amateur astronomer like, say, Will Hay, one of life's ambitions is to discover something new in the moon and preserve it for posterity by naming it after a friend.

The International Astronomical Union, which meets periodically in the world's capitals, controls the naming of newly-discovered mountains and craters in the moon, just as strictly as the L.C.C. maintains watch on London street names.

Now a Welshman living at Barnehurst, Kent, Mr. H. Percy Wilkins, late of Llanelly, claims to have spied new peaks and craters after looking through his home-made telescope for 25 years.

Two of the craters he has named after his friends, and the great honour has fallen to him of having the names accepted by the International Astronomical Union.

"Scientists have accepted the theory that the moon sways a little on its axis from time to time," Mr. Wilkins explained to the News Chronicle.

Fig. 1.6 Patrick's second mentor, Hugh Percy Wilkins, in the August 27th 1938 Saturday News Chronicle, fully milking the limelight of his Mare X discovery claim. The newspaper report reads as follows: 'AMATEUR'S GLIMPSES OF MOON'S OTHER SIDE. Finds More Craters, Peaks, and Dark Plain 100 Miles Wide. If you are an amateur astronomer like, say, Will Hay, one of life's ambitions is to discover something new in the moon and preserve it for posterity by naming it after a friend. The International Astronomical Union, which meets periodically in the world's capitals, controls the naming of newly-discovered mountains and craters in the moon, just as strictly as the L.C.C. maintains watch on London street names. Now a Welshman living at Barnehurst, Kent, Mr H. Percy Wilkins, late of Llanelly, claims to have spied new peaks and craters after looking through his home-made telescope for 25 years. Two of the craters he has named after his friends, and the great honour has fallen to him of having the names accepted by the International Astronomical Union. "Scientists have accepted the theory that the moon sways a little on its axis from time to time" Mr Wilkins explained to the News Chronicle. Appearances of the moon which are normally hidden from view become visible at that period. **SPOTS THAT MOVE** "I have seen, as it were, part of the other side of the moon. I have detected a dark half-obscured plain, hitherto unknown to the ordinary observer. It is about 260 miles long and 100 miles wide. There are several regions already known to be partly obscured, it may be by some form of water, vapour or fog. There are dark spots over the moon's face that definitely move about. They can hardly be shadows, because they do not behave like shadows, and the late Professor Pickering attributed them to vast swarms of insect life. But I am inclined to attribute the existence of these dark patches to some form of fungus created by vapour or fog". **ST. PAUL'S AS A PIN-POINT** Mr Wilkins is a telescope instrument maker and his big telescope, which is 12½ inch in diameter and 12½ feet long, is made of curved plywood for lightness so that he can prop it up easily in his back garden. It brings the moon within a distance of about 800 miles of the earth, but he doubts whether the Queen Mary or St Paul's Cathedral if placed in the moon would look bigger than a pin-point. Mr Wilkins has sliced the moon into 25 different maps and carefully marked his discoveries. Each section shows the geography of the moon in wonderful detail.' [N.B. Quite why the reporter thought Wilkins' telescope was 12½ feet long is a total mystery! –M.M.]

It might be thought, from the newspaper coverage of Wilkins' attempts to see around the lunar limb, that he was the only BAA observer involved in this type of work, but he was not. Indeed, in the 1930s there was an informal group of observers within the BAA and the RAS who referred to themselves as 'The Circle' or 'Barker's Circle' who were all keen on peering around to the far side of the Moon, just beyond the normal limb, when libration conditions were favourable. These included men such as Robert Barker, Bill Fox and, the master of lunar artistry at that time, Leslie Ball (who would later illustrate many of Patrick's books). Richard McKim has researched the history of this group of friends and listed the full membership in his 2013 paper in the BAA Journal. He found that other members included R.E. Diggles, E.F. Emley, H. Simmons, Chas. F.O. Smith, H.E. Wooldridge and Ben Burrell and, later, Dr S.M. Green. Richard McKim has also established that Wilkins was not even allowed into this circle of the best lunar observers until after World War II. However, these men were modest and not obsessively publicity-hungry, unlike Wilkins, the future lunar mentor of the young Patrick Moore.

The aforementioned newspaper article hinting at Wilkins' sighting of a dark plain, namely the Mare X/Mare Orientale, was published despite the fact that the BAA Lunar Section Director Thomas Logie MacDonald (1901–1973) had, in response to Wilkins' 1937 BAA Journal paper, carried out some research of his own. MacDonald announced at the 1938 April BAA meeting that the feature had been observed previously, by observers abroad; this was an announcement that must surely have frustrated Wilkins intensely. Specifically, from the BAA meeting report, MacDonald's statement is recorded thus:

> Even on the best maps there were large regions which were incomplete. Mr H. Percy Wilkins had published a paper in the Journal early in the session describing an almost new Mare, which was mainly on the invisible hemisphere. This had only been described once before, by Franz, who named it the Mare Orientalis.

So, as early as 1938 the BAA and Wilkins knew, categorically, that Wilkins' Mare X had been seen before his observation. Indeed, Wilkins' preferred name for it, the Mare Orientale, is, quite obviously, almost identical to Franz' Mare Orientalis! MacDonald, a Scottish politician and chairman of the BAA's 'West of Scotland Branch' can surely not have been Wilkins' favourite person at that time.

Around this time Patrick had been destined for Eton and, ultimately, for Clare College at Cambridge University. This latter fact has been verified, and his Cambridge place to study geology was still available for his entire life! However, with his sickly health not improving his home tutoring had become permanent, and running the Brockhurst Observatory right next to his home would be far easier if he stayed with his parents, rather than headed for Eton. Reading between the lines, it is also obvious that Patrick's affection for his home life, his mother, and Brockhurst, weighed against a life at boarding school, could only result in one outcome anyway! Just how much Patrick's 'sickly' state of health actually influenced matters he never discussed, but how many mad keen amateur astronomers, with unique access to a fine 6 inch refractor, would prefer to go off to a boarding school, especially if they were a spoilt only child and were likely to be bullied by the notorious prefects of that era?

It was in 1938 that Patrick first realised how strongly he felt about animals. Invited with some friends to walk into a field on the Brockhurst estate, armed with some air rifles, they encouraged Patrick to have a go at shooting a rabbit. Fortunately Patrick missed the animal and he was immediately 'sick to the stomach' at what he had tried to do. After all, the animal was hardly any different to the Moore's beloved family cats Ginger and Rufus. He would be a strong opponent of animal cruelty from that moment on.

Of course, with the elderly Mr Hanbury being the sole owner of the Brockhurst estate, it was obvious that Patrick's status as 'Observatory Director' might be somewhat limited; and this, unfortunately, proved to be the case. On March 1st 1938 Mr F.J. Hanbury died, aged 86, and Brockhurst Observatory was eventually put up for sale. Patrick's last observation with the big refractor was 1 year later and was a sighting of the comet Jurlov-Achmarov-Hassell, which had a head with a subtle greenish appearance. Patrick, by then 16, racked his brains to see if there was any way he could acquire the fine 6⅛ inch Cooke refractor, but the asking price of £40 was beyond his means and his access to a decent telescope disappeared. If only he still had that £87 football pools money! However, he was allowed to keep the observatory observing steps and many of W.S. Franks observing notebooks; he would treasure both for the rest of his life.

In passing, the reader may be interested to know that a quarter of a century later, in the 1960s, 1970s and up to his death in 1985, the man who decimated the British railway system in the 1960s, Doctor Beeching (Baron Richard Beeching) lived on the Brockhurst estate. The reader may also be interested to know precisely where the Hanbury Observatory that the teenage Patrick used was located. Well, from Franks' notes to the RAS, and Patrick's recollections, I place it roughly a quarter of a mile northwest of Patrick's home on Worsted Lane. Relative to the modern farmland I reckon it would have been very close to the northeastern tip of Farm Close, and easily visible in the distance from the Moore's back garden, at least in the 1930s. This was a long walk from the main house on the Brockhurst estate, which was across the Lewes Road, and Patrick once told me that the Observatory had its own telephone installed, so that guests at the house could be informed if viewing conditions were good. Patrick visited the area out of nostalgia in 2001 and told me that the site of the old observatory was now covered by a clump of trees, near to a path, which fits in with my calculations.

It was later, in 1939, that Patrick detected a slight lack of focus in his right eye, so he went to the optician for a check-up. The optician confirmed that he needed a weak lens for that eye and recommended a pair of spectacles. This idea did not appeal to Patrick at all, and he suggested that he would prefer to wear a monocle. The optician thought that a monocle would look very odd on a 16 year old, but Patrick was adamant, and from that day on he used a monocle, if only to hide a slight squint in that eye when he was tired. Another somewhat unusual accessory for a 16 year old to own was a pipe! One day in 1939, shortly after his 82 year old grandmother Josephine had died, he decided to rummage through some of his grandfather's belongings in the attic of her adjoining home at East Grinstead. It was there that Patrick found, in bits, his grandfather's old Meerschaum pipe. At 16 he had tried smoking cigarettes, four in total, but invariably bit them in half! So, he had the Meerschaum repaired and cut an interesting figure as a 16 year old, pipe smoking, monocle wearing, amateur astronomer!

Chapter 2
The War Years

With the Second World War starting in September 1939 and access to a decent telescope denied him, Patrick decided to join the East Grinstead LDV (Local Defence Volunteers). The LDV initials acquired a few unfortunate interpretations such as 'Leap, Duck and Vanish' and, a year later, it was renamed 'The Home Guard'. Patrick admitted, on more than one occasion, that he was probably the 'Pike' (the BBC *Dad's Army's* 'stupid boy' character, played by Ian Lavender) of his day! One reason why Patrick became a Home Guard private was that his father, retired Army Captain Charles Moore M.C., now employed as an accountant, had been elected Platoon Commander of the East Grinstead Home Guard. So, Patrick's dad was the Captain Mainwaring of his day! Unlike private Pike though, Patrick was active in both the Home Guard and the ARP (the Air Raid Precautions volunteer organization). One of Patrick's enduring friends, Pat Clarke, served with Patrick during his Home Guard days. Once, on a TV programme in 1974, he recalled how, despite their orders to search the skies for German paratroopers, our Patrick had them stargazing too. On one occasion a Home Guard member rushed into the platoon H.Q. and shouted that he had spotted a bright German flare and the invasion was starting. Everyone started getting guns and ammunition out of cupboards in a mad panic. Patrick rushed out and was back in seconds. "You bloody idiot, that's not the Germans, it's Venus" bellowed a frustrated teenage Patrick! He was amazed anyone, especially in his informal astronomy class, could be that stupid! Patrick occasionally mentioned another Home Guard military commander too, a retired Major J. H. Marr, MC, DSO. Apparently the Major stated: "Good Grief! I'd go home to lunch permanently if young Moore was given a rifle and some ammo, doncha know! What?! What?!"

Despite the war, BAA London meetings were still held whenever possible and the BAA meeting reports from that era show that the young Patrick was keen to attend when he could, although with more and more young men being drawn into the armed services the attendances were low and Patrick would have been a young face amongst mainly middle-aged or elderly men, some of distinctly Victorian origin. The first wartime BAA meeting in which Patrick's presence is officially

M. Mobberley, *It Came From Outer Space Wearing an RAF Blazer!: A Fan's Biography of Sir Patrick Moore*, DOI 10.1007/978-3-319-00609-3_2,

recorded is the meeting of May 29th 1940. He would have been 17 years old. The BAA meetings, until August 1942, were held in an upper room of Sion College on the Embankment. The BAA Library was housed in an adjacent room at Sion College. When the War began the meeting start time was moved forward from 17:00 to 15:00, but still on the last Wednesday of the month, except during the summer recess. However, the BAA Council was soon forced to modify this impossibly early timing so that tea was served at 16:00 with the meeting commencing at 16:30 from February 1940. The aerial bombardment of Britain obviously affected BAA observers. Yes, there was a nighttime blackout, which meant light pollution was reduced, but the skies were increasingly cluttered by barrage balloons, exhaust trails, shrapnel and blazing aircraft and many residents of the capital were sleeping in the London Underground.

That earliest recorded contribution from Patrick, described in the BAA meeting minutes for May 29th 1940, at Sion College on the London Embankment, may be of interest to the reader. It came during a discussion on variable stars and the question is attributed to Mr P.A. Moore, so it definitely was Patrick. There were no other P.A. Moore's in the BAA in that wartime year of 1940. The 17 year old Patrick asked the following question: "Does the Variable Star Section take any account of the brightest irregular and long-period variables such as alpha Orionis, alpha Herculis and alpha and gamma Cassiopeia?" Mr F.M. Holborn, deputising for the Variable Star Section director W.M. Lindley, who was unable to attend due to military duties, replied: "No, the Variable Star Section does not observe the brightest variables. The most brilliant object studied is omicron Ceti, but I do Gamma Cassiopeia unofficially." Patrick replied: "Gamma Cassiopeia is really the star I am thinking about most, and I should like to know if naked-eye observation is considered adequate for these stars?" Mr Holborn affirmed that the teenage Patrick must use the naked eye for estimating the magnitude of such a bright star.

The next BAA meeting, namely the one scheduled for the last Wednesday in June 1940, had to be cancelled, and even the October 1940 AGM was postponed *sine die* due to the sustained bombing of London by the Luftwaffe during the period known as 'The Blitz'. Due to the aerial bombardment hazard, the devastation in many parts of London, and the impracticality of winter travel during a blackout, the BAA cancelled all further Ordinary meetings until January 1941, when the aerial warfare over London had abated. Also, due solely to the war, the BAA's membership had already dropped by 46 to only 860 members. Just when Patrick had decided to regularly participate in BAA London meetings, Hitler had intervened.

Patrick's Home Guard and ARP duties did not wipe out his observing though, as even a cursory scan through his vast collection of observing notebooks will show. Any object in the night sky was a target for his 3⅛ inch refractor. For example, his notes record a good observation of Comet Rigollet on March 8th 1940 and variable stars were always being recorded. Nevertheless, the Moon was Patrick's passion. On October 15th 1940 Patrick, observing with his 3⅛ inch refractor from his parents' garden, was presented with a very favourable libration (lunar tilt, towards the Earth) of the features on the edge of the disc. Patrick claimed he noted a large crater (it would later be named Caramuel and eventually officially designated as Einstein) and on the very edge of the disc, beyond some mountains, Patrick saw the suggestion of a dark region (a lunar 'sea') and a very prominent crater with high walls and a prominent central peak. It would be 7 years later that Patrick would have a clear night, a favourable libration and the experience to pursue this observation further. He had, of course, been trying to see (Patrick would later phrase it as 'discovering'!) Wilkins' 'Mare X', described in that Wilkins' 1937 BAA Journal article (and his 1938 Newspaper article) and mentioned previously as Mare Orientalis as far back as 1906. While Patrick's later claim that he had 'discovered' the new sea was certainly a gross exaggeration, persistence and determination had meant that he was one of the few observers to study the feature well in the 1940s and 1950s.

In fact, I have studied Patrick's lunar notebook for October 15th 1940 and although there are indeed some nice sketches of these limb regions during a favourable libration there is nothing that corresponds to Caramuel/Einstein. In addition, the dark feature he has sketched close to the Mare Orientale location is more likely the Lacus Autumnae which is also a very foreshortened feature close to the Orientale region, but not the same thing. In addition, Patrick has not drawn attention to these features in his notebook in any way. Patrick actually committed his October 15th 1940 claims to print in his book *The Wandering Astronomer* (pages 158–160) and included the comment that he was on leave from Bomber Command in the RAF, which is certainly not true (we will come to that in a few more lines). However, as a minimum, his sketches, aged 17, of these limb regions, made with a small aperture refractor, are impressive. Also, undoubtedly, he was studying those crucial regions because he had boundless enthusiasm for glimpsing the Moon's far side. So let's not judge him too harshly.

On a humorous 50th anniversary *Sky at Night*, on April 1st 2007, Patrick would cynically say to his younger self (impressionist Jon Culshaw) that the Americans would 'eventually re-discover the Mare Orientale'. Nevertheless, a year later, after a definitive paper by Richard Baum and Ewen Whitaker appeared in the BAA Journal, he made an apology for wrongly claiming to have discovered the Eastern Sea, citing it as simply not knowing about foreign observations of it....Hmmmm.... Whatever you say Patrick!!

Of course, the War was well underway when Patrick claims to have made his 1940 Orientale observation and he often stated that he certainly wanted to go to Cambridge University in the coming year to study geology, but "Hitler changed my plans." This brings us nicely onto his RAF war service.

The RAF Records

On a number of occasions Patrick has claimed that he joined the RAF in 1939, aged 16, faking his age and his medical history into the bargain. However, MOD records do not bear this out. These records state that Patrick Alfred Moore (born March 4th 1923), Service Number 1800747, enlisted in the RAF Volunteer Reserves on December 31st 1941, in other words, just over 2 months before his *nineteenth* birthday and well after he could have taken up his geology degree at Cambridge. On more than one occasion he claimed that he swapped with another young (healthy) recruit at the RAF medical. The healthy recruit took two medicals and the bright 16 year old Patrick took two aptitude tests, or so Patrick claimed! In fact, the RAF records and Patrick's own jottings in his lunar notebooks agree precisely with respect to him enlisting 2 months before his 19th birthday. His third lunar notebook covering 1941–1946 records, under the heading 'January 1st – 2nd 1942' (the days immediately after his official enlistment) the following note:

> Went to Euston House, where I was accepted as an RAF pilot and formerly attested – so it seems unlikely that I shall be able to do much lunar work after this spring.

The next entry in his lunar notebook records the following:

> Jan – Feb 1942: A period of snow and cloud. On duty with the Home Guard & ARP. Attended Feb BAA meeting where the Rev. Dr. Davidson informed me that actions are to be taken against the inefficiency and slackness of MacDonald, the Lunar Section Director.

This author cannot help wondering if MacDonald's announcement to the BAA that Wilkins had not been the first discoverer of the Mare Orientale/Mare X, at that April 1938 meeting, 4 years earlier, had turned Wilkins, and his future apprentice Patrick, against MacDonald. The BAA meeting minutes for February 25th 1942 record that Patrick asked the Reverend Dr Martin Davidson (BAA Comet Section Director) a question, following Davidson's lecture about recent comets. Patrick asked: "Is it possible that there may be another large planet beyond Neptune?" The Reverend Dr Davidson thought not, or its influence would have been noted on the orbits of comets.

In fact, in early 1942, despite the War, Patrick managed to attend two consecutive BAA London Wednesday meetings, at Sion College on the Embankment: the meeting of February 25th (mentioned above and in Patrick's notebook) and March 25th. He would have turned 19 by the time of the second meeting. These would probably have been Patrick's last BAA meetings at the Sion College venue. On the evening of March 2nd/3rd 1942 a Total Lunar Eclipse had taken place and had been at a favourable altitude from the longitude and latitude of the UK. At that March 25th BAA meeting Patrick told the members that he had seen the eclipse from East Grinstead and had concentrated on the crater Linné, as well as Dionysius and Censorinus, to see if 'the wave of intense cold sweeping over the Moon' affected them. He also observed two stars being occulted by the Moon during totality, all using his 3-inch refractor. Unfortunately the weather had deteriorated after mid-eclipse, as seen from his East Grinstead home, but at least he had observed it and relayed his views to the BAA members present at the meeting. From the cloudy skies of Britain most time-critical astronomical events end up being thwarted.

According to MoD records Patrick was finally called up for RAF service a few months later, specifically, on July 13th 1942 as an AC2 (Aircraftsman Second Class). Again, this information tallies with the notes in his lunar notebook:

> I joined the RAF VR (Volunteer Reserve) to train as a pilot on 1942 July 13th and from that date of course, all lunar work must be suspended for the duration of the War.

Patrick would have been 19 years and 4 months old, so certainly *not* 16! Around early 1942, before his official call-up to active service, Patrick spent many nights as a volunteer ambulance driver, during and after air raids. From the late 1970s the middle-aged Patrick would claim that he became very friendly, during this wartime period, with a local East Grinstead girl who was a nurse and a fellow ambulance driver. For the first time ever, in his 2003 auto-biography he gave her a name: Lorna. More on this tale shortly! Patrick said that he became a regular pipe smoker on his night-time ambulance shifts as the nicotine kept him awake. His favourite pipe tobacco, then, and for the rest of his life, was 'Three Nuns', although he rarely smoked beyond his sixties.

From July 13th 1942 Patrick only managed to get home to East Grinstead every 2 or 3 months, namely, while on RAF leave. His RAF training took him first to St John's Wood in London and then to RAF stations at Ludlow in Shropshire, Cosford and Stretton in Staffordshire, Paignton in Devon, Sywell in Northamptonshire and Manchester. After completing his basic training with the Initial Training Wing (ITW), Patrick was promoted to the rank of LAC (Leading Aircraftsmen) in late 1942. But despite being incredibly busy with his RAF studies he did not neglect astronomy and made many observations while in the military, mainly of bright variable stars which were easy naked eye or binocular targets (stars such as alpha and gamma Cassiopeia and epsilon Aurigae were his favourites). One of his observing logs records that on December 9th 1942, while at RAF Paignton, he gave a talk to the Air Training Cadets on star recognition; he was little more than a cadet himself and was already instructing his colleagues in navigation! Patrick had four visits to RAF Paignton in total during his training period. He was based at the Tenbani hotel there but visited the Palace Hotel too at various times. Almost all of the hotels in the town were being used by the RAF at that time. By analyzing all the observing notebooks Patrick compiled while in the RAF, his early locations, official addresses, and the types of object he observed in the night sky, can be summarized as shown in the first table below (Table 2.1).

The Fianceé: Fact or Fiction?

In early March 1943 Patrick was back at East Grinstead in a brief leave period between spells at RAF Paignton in Devon. He was able to celebrate his twentieth birthday, with family, friends and (so he claimed) the enigmatic Lorna, who was now, allegedly, his fianceé. His 20th birthday, on March 4th, occurred just 1 day after the Bethnal Green tube disaster, when 173 people were crushed to death on the stairs leading to the London Underground station. However (again, according solely to Patrick, on the few occasions he spoke of the War years), disaster was to strike him personally 4 months later. But just before then, on June 30th 1943, Patrick managed to attend another BAA London meeting. From September 1942 the BAA meeting venue had changed from Sion College on the Thames Embankment to the Royal Astronomical Society premises at Burlington House, Piccadilly. The meeting

Table 2.1 RAF stations, observing locations, objects observed, and official addresses for Patrick from 1942 July to 1943 November

Date	RAF station/location	Objects observed
1942 July	St John's Wood, London	Variable stars
1942 August	Ludlow, Shropshire	Variable stars
1942 August	Cosford, Staffordshire	Variable stars & partial solar eclipse
1942 September	Stretton, Staffordshire	Variable stars
1942 September	Cosford, Staffordshire	Variable stars
1942 October	East Grinstead, Sussex	Mainly variable stars/planetary
1942 October	Ludlow, Shropshire	Variable stars
1942 October	Paignton, Devonshire	Variable stars, aurorae, meteors
1942 November	Cosford, Staffordshire	Variable stars, meteors
1942 November	Paignton, Devonshire	Variable stars
1943 January	East Grinstead, Sussex	Mainly variable stars/planetary
1943 February	Paignton, Devonshire	Variable stars, comets
1943 March	East Grinstead, Sussex	Mainly variable stars/planetary
1943 March	Paignton, Devonshire	Variable stars, comets
1943 April	Sywell, Northamptonshire	Variable stars
1943 April	East Grinstead, Sussex	Mainly variable stars/planetary
1943 May	Manchester, Lancashire	Variable stars
1943 June	East Grinstead, Sussex	Mainly variable stars/planetary
1943 July	Cranwell, Lincolnshire	Variable stars
1943 November	East Grinstead, Sussex	Mainly variable stars/planetary
Mail:- 1800747 LAC Moore. G Flight, 1 Squadron, RAF Heaton Park, Manchester		
1943 November	Manchester, Lancashire	Mainly variable stars

report for June 30th 1943, at the new Burlington House venue, recorded him simply as Mr P.A. Moore, that is, without an RAF rank. At that wartime meeting the BAA President F.J. Hargreaves was in the chair, flanked by the secretaries Holborn and Macintyre. During the afternoon the 20 year old Patrick made sure his presence was noted. As was quite common in that wartime era The Reverend Dr Martin Davidson read a paper on behalf of a member who could not be present. In this case the member was Colonel Edgeworth (1880–1972) and the paper was a truly ground-breaking one which then (and now) has largely been forgotten, except when the solar system's Kuiper belt is correctly referred to as the Edgeworth-Kuiper belt.

So, it was truly a moment in astronomical history when Dr Davidson read out Colonel Edgeworth's paper entitled 'The Evolution of our Planetary System' to that wartime BAA audience. The 20 year old Patrick was soon asking questions once the paper was delivered. "Does Colonel Edgeworth's Theory provide in any way for the retrograde satellites and comets?" Patrick asked. The Rev. Dr Davidson replied that there was no attempt to explain this in the paper. Patrick continued: "It seems to me that were the asteroids formed in the manner described, Saturn and the other large planets would also be responsible for rings of asteroids, but so far as I know none have been discovered." Shortly after, on the subject of the origin of the Moon, Patrick added another point, namely: "If the Moon was formed in this way, is it not curious that Venus has no satellites? I should have thought the conditions would have been very similar." Dr Davidson agreed that the lack of a Venusian moon was puzzling. Anyway, that was the end of Patrick's contribution to that particular BAA meeting and as he left Piccadilly on that summer evening, heading back to Sussex, he could have had no idea what lay in store for his part of the country, courtesy of the Luftwaffe.

Nine days later, on July 9th 1943, death dealing blows were struck at the heart of Patrick's home town of East Grinstead, shortly after 5 p.m. on Friday afternoon, when one of about ten enemy raiders swept in from the coast to cause havoc in the shopping centre, resulting in a large number of casualties amongst men, women and children. The majority of these were in the Whitehall cinema, where a bomb had scored a direct hit. Six years earlier, the 14 year old Patrick had played a xylophone solo in that same establishment. It was in that Whitehall Cinema that the death toll was heaviest; 184 people had been watching a Hopalong Cassidy cowboy film when the air raid sirens went off. It was quite common for children to fill the cinema after school ended on Friday and July 9th was no exception. A warning appeared on the screen, about the air raid, but few people took any notice. The later speculation was that the one Luftwaffe pilot who became separated from the other planes decided that he would find another target before he returned home. Supposedly aiming for a train entering the railway station, one bomb hit the cinema and others landed on several shops in the High Street and in London Road. As a result of the raid 108 people were killed and 235 were seriously injured. It was the largest loss of life in any air raid in Sussex and, for no obvious reason, Patrick's quiet town of East Grinstead had been the target. Surely, his lifelong hatred of Germans was cast in stone from that point onwards.

Within a few minutes of this ruthless attack on an open town, civil defence workers, including police, troops and members of the Home Guard, many well-known to Patrick, had arrived on the scene. Members of the public also helped in various heroic tasks. The combined services accomplished many feats of skill and daring, and worked feverishly throughout the late afternoon and night. There were many harrowing scenes as children and women were recovered from the debris. A newspaper office was used for a mortuary, and later the bodies were taken to a garage where they were left for identification purposes. Less than half of the victims had been identified by Sunday.

Patrick claimed that his fiancée Lorna was killed in early July 1943 during an air raid, and on some occasions he claimed that she was killed while inside her ambulance. Frankly, like many of Patrick's wartime stories, his account was slightly different every time! Sometimes a V-bomb was responsible and sometimes he would revert to the air raid story. He also repeatedly claimed that it was the day his life ended and he never, ever, gave a lot more detail than that. In one newspaper interview he claimed her death was a week before their wedding. In a few interviews he was reported as saying she was killed in the East End, not East Grinstead. When quizzed he would always say, very rapidly, and in a highly agitated manner, that the war was, "a long time ago and it's best to forget it and she was the only girl for me". Indeed, dozens of interviewers over the years, on radio, TV and in the tabloid press, tried to extract more information from Patrick, but none was ever forthcoming.

He claimed he couldn't even recall her first name in one radio interview. The conversation almost always ended with those same words, "it's a long time ago now, a long, long time ago" or, "best to forget it".

The July 9th 1943 bombing date fits perfectly with the few chronological facts that Patrick ever let slip. However, it should be stressed that there really is considerable doubt over the very existence of Lorna, whose surname has never been revealed and whose first name (real or not) was mentioned, for the first time ever, in Patrick's biography, published in 2003. Indeed, prior to the late 1970s a fiancée killed in the war was never, ever, mentioned by Patrick to the press. He always stated that he was a lifelong bachelor and would remain one. Crucially, Lorna was not mentioned, or even hinted at, in Patrick's 'This is Your Life' programme in 1974 and Patrick, under questioning, has stated that there were never any photographs of her, taken by anyone! [I paid £300 for a studio copy of that TV programme and I once found the 'This is Your Life' big red book in Patrick's house. All of the red book's pages had been removed and simply replaced with half a dozen photographs of Patrick alongside the more famous guests.......] It was only after his mother's death, in January 1981, that this story of a wartime fiancée received wide publicity and Patrick did himself no favours by stubbornly refusing to give any details of the girl, even in his own biography. On one radio interview in 1999 he almost walked out of the studio when the subject was persistently raised. It seemed to really stress him when it was pointed out that if he was only 20 when the tragedy occurred, and if he really had wanted a family, there was plenty of time left to find someone else. "Second best's no good for me" he'd repeatedly bark.

Patrick and his parents knew many people who were killed in the East Grinstead July 1943 raid. Indeed, many of his former East Grinstead Home Guard and ARP colleagues were involved in the aftermath, injured, or even killed. Some have wondered if Lorna was loosely based on Private Joan Barber of the Auxiliary Territorial Service (ATS) who was, indeed, an ambulance driver, 3 years older than Patrick, and was known to Patrick and his family. Her service number was W/75120. She died in the July 9th East Grinstead raid and was buried, by Patrick's childhood tutor and rector, the Rev. John Missen, at the Holy Trinity Church, Coleman's Hatch. However, Patrick has always said that Lorna was the same age as him, namely 20 at the time of the raid, whereas Private Barber was 23.

A lifelong correspondent with Patrick, and a man born in the same year, the late American telescope maker Tom Cave (1923–2003), visited Patrick in 1944 just before the Normandy landings and exchanged correspondence with him for 60 years. While waiting for D-Day Cave was able to attend a few BAA meetings in London too. However, Cave was adamant that Patrick (and his parents) never mentioned the wartime loss of his fiancée, even when he visited East Grinstead a matter of months after Lorna's alleged demise. In addition, Colin Ronan (1920–1995) was a lifelong friend of Patrick's and, during the BAA Centenary year of 1990 (when he was the BAA President and I was the incoming Papers Secretary), I mentioned the death of Patrick's fiancée to him. His memories were the same as Tom Cave's. Colin's comments were along the lines of: "You know, I joined the BAA in 1938 and sometimes saw Patrick during the War Years at BAA meetings and we've been the closest of friends ever since. He's always declared himself a life long bachelor.

He never said a thing about this wartime sweetheart. I read the ghastly story in a newspaper about 40 years later and couldn't believe it. I rang Patrick up but he just changed the subject straight away – what a frightfully strange business!" I should add that I have scoured all of Patrick's observing logbooks from the 1940s, searching for any mention of Lorna. He often used these logbooks as a diary, noting if others were with him when he observed, along with other facts. No female, of any name, is mentioned at any point, apart from his mother and housekeeper, and there is no hint of a family tragedy either.

With no real evidence of Lorna's existence maybe we should just leave it at that? Maybe, 30 years later, Patrick just got sick and tired of the 1970s press (and a few vocal psychiatrists) asking why he was not married, thereby insinuating he might be homosexual, or simply an overgrown mother's boy. In the mid 1970s, like Arthur C. Clarke more than 20 years later, Patrick was, briefly, accused (in mischievous rumours) of maybe having ulterior motives with respect to the teenagers and scout groups who he was mentoring in astronomy. These wild and malicious theories were totally and utterly unfounded, but for a few weeks in his local area some people were giving Patrick very strange looks. The world seems to be full of small-minded and envious trouble makers who simply cannot accept that a life of total celibacy can be a very happy one. His unmarried status did him no favours at this time and it was shortly afterwards that the story of a wartime fiancée was first revealed to the press. This, to me, seems the most likely reason for the Lorna myth. One final thought that I leave the reader with is this. In interviews prior to the late 1970s Patrick always used to say that he had no time for marriage because "I'm married to the Moon, I'm married to the lunar surface….." Well, Freudian slip or not, 'Lunar' and 'Lorna' sound very similar and Patrick was always a big fan, consciously and sub-consciously, of word association, riddles and pseudonyms. So, my view is that 'Lorna' means 'Lunar' and the girl of that first name never even existed………. However, there was actually another girl who Patrick became totally infatuated with, as we shall see much later in this book, during the 1950s.

[Reluctantly, given the cynical times we now live in, and some twenty-first century revelations about unmarried celebrity TV perverts (and even priests), I feel I should add a few more words here, before returning to the War Years. Patrick spent a lifetime helping and encouraging children and teenagers, especially those who were, like he had been, unwell in childhood. In effect, he was giving these children the same encouragement that he had received from W.S. Franks. As these children grew older they always retained a great affection for Patrick and were always welcome in his home, years later, as adults. Decades after they first met him, they would return to see him, along with their wives and children, and many asked him to be their own childrens' godparent. Some of the children Patrick taught during the 1950s were still visiting him when they themselves were pensioners and some even moved to West Sussex because he lived there! Basically, Patrick's teaching work with young people was 100 % genuine and they always held him in the highest esteem.]

There are no indications that Patrick attended any BAA meetings in the year following his attendance at the June 30th 1943 meeting. As a young man Patrick almost always made his presence known by asking questions, which were duly recorded in the minutes. He invariably signed his name in the meetings register too. In fact, Patrick's absence at BAA meetings was entirely predictable, as from July to November 1943 he was on an intensive course at RAF Cranwell and, after a short trip home in November 1943, he joined G Flight 1 Squadron of RAF Heaton Park at Manchester. He then set off with many other recruits for his final training as an RAF navigator in Canada, under the Empire Air Training Scheme, primarily at RAF Moncton in New Brunswick. The second table of his RAF astronomy observations is shown below. This details his period in Canada along with the observations made in his observing notebooks, cross-correlated to his locations and official RAF mail addresses, from December 1943 to the end of the war. As we can see, from that month of December 1943, until June 1944, Patrick was initially in Canada. Then, on his return, he was mainly in northern England, well away from the BAA meetings in London (Table 2.2).

Table 2.2 RAF stations, observing locations, objects observed (nearly all being naked eye or binocular variable stars), and official addresses for Patrick from 1943 December to the end of the war

Date	RAF station/location	Objects observed
Mail:- 1800747 LAC Moore. MPO 304. RCAF, Ottawa, Canada		
1943 December	Halifax, Nova Scotia	Mainly variable stars
1943 December	Moncton, New Brunswick	Mainly variable stars
1943 December	Montreal, Quebec	Mainly variable stars
1943 December	Moncton, New Brunswick	Mainly variable stars
Mail:- 1800747 LAC Moore. MPO 211. RCAF A Block, Hamilton, Ontario		
1944 January	Hamilton, Ontario	Mainly variable stars
1944 January	Toronto, Ontario	Mainly variable stars
1944 January	Hamilton, Ontario	Mainly variable stars
1944 February	Oakville, Ontario	Mainly variable stars
1944 March	Hamilton, Ontario	Mainly variable stars
Mail:- Pilot Officer Moore. MPO 304. RCAF, Ottawa, Canada		
1944 June	Moncton, New Brunswick	Mainly variable stars
1944 June	Harrogate, Yorkshire	Mainly variable stars
1944 July 3rd	East Grinstead	First lunar work for 10 months!
Mail:- P/O Moore. A Wing 10 Course, 1 Officers Mess, ACOS, Credon Hill, Hereford		
1944 August	Hereford, England	Mainly variable stars
1944 August	Wellington, Shropshire	Mainly variable stars
1944 August	Hereford, England	Mainly variable stars
1944 August	East Grinstead, Sussex	Numerous objects
Mail:- P/O Moore, Room 157, Queen Hotel, RAF Harrogate, Yorkshire		
1944 September	Harrogate, Yorkshire	Mainly variable stars
Mail:- P/O Moore, Officers Mess, RAF Millom, Cumberland		
Sept'44 – Sept'45	Millom, Cumberland	Mainly variable stars
" " " "	Barrow, Lancashire	" " "
" " " "	Lossiemouth & Elgin, Moray	" " "
" " " "	Aberdeen & Inverness, Scotland	" " "
" " " "	St Davids, Pembroke	" " "
" " " "	Lindholme/Doncaster, Yorkshire	" " "
" " " "	Sheffield, Yorkshire	" " "
" " " "	Oxford	" " "
October 1945	Home to East Grinstead, Sussex	" " "

Although Patrick's familiarity with the night sky and the points of the compass singled him out as being destined to be a navigator he did get his pilot's wings. However, he was the first to admit that he was not destined to fly planes. In an oft repeated story he would recount an occasion when his Commanding Officer watched him landing a Tiger Moth solo, after 9 or 10 hours of instruction. In the evening in the 'Officers Mess' the CO approached him. In Patrick's words: "He gave me an old-fashioned look and said, 'Kid, if you were one of our fighter pilots the Germans would have awarded you an Iron Cross: please stick to being a Navigator!' I bought him a drink in a marked manner". Not surprisingly it was a navigator that Patrick would become. Patrick told me that he often used to entertain his colleagues by playing on the piano in the Officer's Mess and that he spent more time practising his bowling in the cricket nets than on the parade ground. In the Foreword to Patrick's 1961 book entitled *Conquest of the air: The Story of the Wright Brothers* he describes his first RAF trip in the Tiger Moth trainer, with a young officer instructor. 'For a hectic half-hour he hurled me all over the sky, evidently doing his best to tie knots in the exhaust smoke. When it was over I felt somewhat dizzy, but at least I knew what flying was like.'

While in Canada, Patrick claimed to have met no less a person than Orville Wright, the very first man to fly in a 'heavier-than-air machine'. Like many of Patrick's wartime stories the account varied slightly each time, as did the year and country of the encounter.

Over the years Patrick's account became more detailed, with many more recollections of numerous questions he had supposedly asked Orville Wright. Yet, in Patrick's short book about the Wright Brothers, he simply said 'I will always be glad that I once met him.'

Sometimes Patrick said that he had met Orville while he was in Canada, training with the RAF, and sometimes he was in the USA, having hitched a ride on an RAF flight. In the latter version of events Patrick claimed he was attending a fortunately timed meeting of an American society, of which he (Patrick) was a member, on the day that Orville was a guest of honour. I have searched for possible dates when this might have occurred and have found just one, whereas Patrick could never provide any specific dates or details whatsoever, claiming that he could not remember, every time I raised the subject! The one date I came up with was Friday December 17th 1943, the 40th anniversary of the famous Kitty Hawk flight. Patrick had just started his RAF training period in Canada that month and a gala evening in honour of Orville Wright was being held at the Hotel Statler, in Washington D.C., on that day. Numerous dignitaries from the world of aviation had been invited and, bearing in mind Orville's well known sadness at the use of aircraft in war, the evening was themed 'Aviation for Peace'. It was broadcast on the NBC Radio Network. Orville did not enjoy publicity and had only agreed to attend because he had been promised that President Roosevelt would be there. The President was scheduled to announce that the Kitty Hawk Flyer would be returned from the Science Museum in London, to the Smithsonian Institute in Washington, after the War. As it turned out Roosevelt did not turn up and Orville was asked to present an award to a military man, General Hap Arnold, which he was not at all happy with.

Anyway, I have digressed a bit, and my point here is that during Patrick's 9 months in Canada, the December 17th 1943 Washington gala is the most likely event where the publicity shy Orville Wright might have been at the same venue as Patrick, if he had hitched a ride with senior RAF Officers attending the gala event. Orville's brother Wilbur had died in 1912 and Orville Wright himself did very little flying after 1920. He allegedly told Patrick, in person, that he was saddened to see aircraft used in warfare and it had dented his flying enthusiasm. Patrick described him as quiet and unassuming but said that he liked him immensely. In his eighties Patrick added a further twist to the story, claiming that in his final question to Orville he had asked him if he thought that men would land on the Moon? Despite Patrick's inability to recall precisely when and where he met Orville Wright he claimed that he remembered his reply, regarding men landing on the Moon, word for word, namely: "I reckon so, but it'll sure mean some hard work."

As Canada is a good place from which to view aurorae the young RAF trainee Moore was always keen to be up in the air at night, armed with a camera, just in case there was a dramatic display. Frequently he was a night time passenger on an Avro Ansen trainer plane, even in his days off, as it got him above the clouds on moonless nights, when the aurora is best seen. On Patrick's second wartime trip from Canada to the USA he managed to wangle another invite to an unspecified science meeting, and to a small reception afterwards, where he claims to have met albert Einstein. Patrick found him to be "Exactly what I expected; unworldly, communicative and blissfully unaware of his unique status". Patrick often recounted asking Einstein if he could explain the infinity of the Universe in non-mathematical terms, and Albert had apparently replied that he had never been able to do so satisfactorily. Einstein was an expert violinist and, at the reception, was encouraged to demonstrate his musical skill, but he wanted a piano accompaniment and, according to Patrick, his young self jumped at the chance. Patrick accompanied Einstein on the piano as he played Saint-Saëns' Swan. Surely, a claim to fame that few people could ever hope to equal? Well, maybe; however, like almost all of Patrick's claims to have met an astonishing number of great men of science, there is absolutely no independent evidence of the Einstein musical pairing...... In the twenty-first century a friend of Patrick's created an amusing photographic montage of Einstein playing a violin, next to Patrick playing the piano. The picture usually sat on top of the *Marshall & Rose* baby grand piano in Patrick's music room. Even Patrick was amazed that people thought this picture was genuine, as clearly it was a middle-aged Patrick in the image, looking barely younger than Einstein, who was 44 years his senior!

If I sound sceptical about these meetings with famous people it is simply because, over the years, Patrick's stories tend to grow and grow, from actually seeing someone at a distance to knowing them very well indeed. His tales of meetings with the great Mars observer E.M. Antoniadi, when he was pressed for more details, literally evaporated into thin air. Patrick was a great entertainer and raconteur and hated to disappoint anyone with a boring tale. The dates of the meeting with Einstein would vary between 1939 and 1944, depending on who Patrick was telling the story too. Patrick often mentioned Einstein in his early books, but it was only well after his mother's death, and, specifically, after 1986, that the 'accompanying Einstein' story was related to the media, or divulged in his books. Now, 1986 was the year of the death of a former BAA President (1954–1956), Reginald (Reggie) Waterfield (1900–1986). Waterfield had genuinely met many famous people in his life, including Will Hay, and was a stickler for painstaking scientific research. He was, however, not a great fan of Patrick's and once described him, in later life, as "a beached whale"! However, the two men knew each other very well. Waterfield often used to relate, at BAA meetings, that one of *his* proudest moments was when he met Einstein, in the 1940s, at a small reception in New York, and was encouraged to play the piano as an accompaniment to Einstein on the violin. Sounds familiar? Waterfield's account of that meeting bore staggering similarities to Patrick's own account, which Patrick only announced to the media shortly after Waterfield's death, on June 10th 1986. Waterfield's friend Harold Ridley once told me that "some of Patrick's stories sound like Xerox copies of Reggie's life". I will say no more!

In June 1944 Patrick and numerous other RAF trainees returned to England from Canada on the good ship 'Empress of Japan', swiftly renamed 'Empress of Scotland' after the Japanese joined the war on the enemy side! Patrick claimed that even a basic knowledge of astro-navigation told him while on deck that the ship was sailing unusually far north, near to Greenland. When he mentioned this to a naval officer he was greeted with a stony silence. Apparently this course (to avoid U boats) was highly classified data, and despite Patrick being a trained navigator the naval officer seemed most unhappy that he had deduced this Top Secret information simply by staring at the sky!

For most of the war, according to Patrick, he was a full time navigator with RAF Bomber command, based at various RAF bases, mostly in northern England and mostly with Wellington bombers. From the information available it appears he spent most of the last *year* (not years) of the war at RAF Millom in Cumberland, with time also spent at RAF Barrow in Lancashire and RAF Harrogate in Yorkshire. Trying to observe the Moon setting, when he was in the bomber at night, became something of an obsession for Patrick, although he could not explain why when I asked him about this. He never actually witnessed an aerial 'moonset'. However, he did add that he vividly remembered seeing a rare 'lunar rainbow' one night while on a mission.

Fig. 2.1 A stern-faced Patrick (*centre*, *back row*) and the other five members of his six man Wellington Crew in 1945. On the *right*, in the *front row*, is his colleague the radio operator Guest Harding Dempster (later Flight Lieutenant Dempster) (Official RAF photograph. Photographer unknown)

In Patrick's observing notes for November 27th 1944, made from RAF Millom, he notes that the BAA's Auroral and Zodiacal Light Section Director, Mr W.B. Housman (the Director since 1928), lived relatively nearby at Seaton Observatory, Workington. Patrick's notebook comments read: 'Work on auroral effects upon magnetism is to be commenced shortly. Compass swinging (air over Walney Island) carried out on Nov 21st at 9 hours.' Patrick was obviously endlessly enthusiastic about all aspects of his hobby even while on active duty. The war had one big advantage in that the whole country was blacked out and so skies were very dark. This enabled Patrick to carry out many naked eye observations, including some of the Zodiacal Light and Gegenschien, which are almost impossible to see when skies are even slightly bright. Also, when in his Wellington bomber, navigating, he would again look out for aurorae and usually had a film camera with him!

From the limited information available it appears that Patrick's main role in the final year of the war was as a navigator in a six man Wellington crew, patrolling the north Atlantic air space and available for bombing raids on U Boat bases and even mainland Germany. In the 1960s and 1970s he let slip to a few journalists that his Wellington crew were trained to be part of a Pathfinder squadron, flying in low-level, to mark targets with parachute flares. The Atlantic clasp on his Air Crew Europe Star medal confirms his involvement in RAF air crew activities over the Atlantic. When asked in later years if he ever took part in bombing raids, Patrick never, EVER, answered the question, except with: "I spent the war pottering around, flying things". Would he want to bomb Germans, after what had happened in East Grinstead on July 9th 1943? Would he relish or be revulsed by the prospect? We shall never know for sure as he refused to talk about it. When really pressed he would say "I don't talk about the War ever because of........" dramatic pause and finger tapping of the nose...... "Military Secrets"!!! I even heard him say that 65 years after the war ended! One story Patrick often told to numerous people, including myself, usually late at night, after a few drinks, was that he was once dropped behind enemy lines and hit by shrapnel, while hanging from a parachute. He claimed he still managed to escape under fire, and the 'Secret Army' French Resistance got him back to the UK. Many of his colleagues were, apparently, killed and they had been known as the 'The Ten', because there were ten of them! Again, a lot of finger tapping on the nose and a wink: "Military Secrets....can't say another word". According to the Royal Air Force Personnel Management Agency at RAF Innsworth, there is no record of him serving in any *specific* RAF squadron as either an airman or an officer. This could mean that he never became accepted for regular activities (the RAF had plenty of staff by the end of the war and Patrick was very young and not very healthy) or it could mean he was assigned to dogsbody duties.

Patrick said that he looked forward to his periods of RAF leave when he could return to his beloved East Grinstead home and play a game of tennis. On such occasions Patrick (unlike most astronomers) loved the British 'double summer time' as it enabled him to "play tennis till midnight". As with many who served in the RAF, Patrick considered himself lucky to survive the war, as he knew so many people who did not make it through. On one flight in a Wellington he sustained a serious injury to his left leg and knee caused, allegedly, by shrapnel from enemy fire. He also claimed that he sustained a serious spinal injury and was told by RAF medics that his back might only hold up into his thirties. Regarding another occasion, he said he only just managed to land from a mission in an RAF Wellington. It was his responsibility to get four men and the injured pilot back to safety. Sometimes this story changed in the tabloid press, with the plane even becoming a Lancaster (which he told me he'd never flown!) and he had to clamber, Hollywood style, over the bodies of the dead crew to wrestle with the controls. Patrick said he was injured and lost many of his teeth in the crash landing. From that moment on he would, apparently, have to wear a full denture. [N.B. Horrendous though it sounds it was not uncommon for young adults, with a few dental problems, to have all their teeth removed and replaced with dentures in the pre-NHS 1940s. Dental work was relatively crude and highly expensive for the working classes of that era.] Excluding Patrick's 'Biggles' style war hero yarns there is one independently verified wartime story in which Patrick really did almost lose his life; but we will come to that later.

One positive thing that did emerge from Patrick's war-time experience was that he had an excellent fluency in French by late 1945. For much of his time with Bomber command he flew with a Belgian pilot, so Patrick learned to speak French with an Anglo-Flemish accent! This would serve him in good stead as a splendid French teacher during the 1950s. Even before the war, some French fluency was already present, inherited from his father and 'French-Swiss' grandmother.

Patrick's lunar observations suffered badly in 1942, 1943 and 1944. Indeed, between 1942 July 13th and 1944 July 3rd there are no entries whatsoever in his lunar notebook. On the latter date (just after the end of the 'Project Neptune' assault phase of the allied invasion of northwestern Europe) there is simply a note saying:

Looked at the Moon for the first time in Ten Months!

The RAF records show that Patrick had finally graduated at New Brunswick, and was commissioned as a Pilot Officer, on June 2nd 1944, serving in 'General Duties (Aircrew Branch)'. In other words, he was only officially commissioned in the final year of the war, just 4 days prior to D-Day. In that era, prior to 1947, an additional six digit commissioned service number was issued to the officer and this number is sometimes seen in parentheses after the commission date. Thus Patrick's full service identity now became 1800747 Patrick Alfred MOORE (165462). When asked where he was on D-Day Patrick would, on many occasions, snap "In Denmark" and quickly change the subject. Well, as he was returning from Canada to Yorkshire at that time and Denmark was under Nazi occupation, this seems somewhat unlikely!! Frankly, whoever asked that question got a different answer. Sometimes it was Denmark, sometimes Norway and sometimes he was disguised as a fisherman, on a Dutch fishing vessel. Unfortunately, the RAF records and his own observing notes make a total mockery of these entertaining yarns!

There are many tragic events in warfare which make one wonder what would have happened if someone who had been killed at a young age had, instead, survived. On D-Day itself another BAA member of Patrick's generation and one equally as skilled in observing and sketching the Moon (more so in truth) was killed on his first day of active military service. His name was Samuel Morris Green

(1921–1944) and such were his achievements, even 2 months before he would have turned 23, that a full obituary appeared in the Monthly Notices of the Royal Astronomical Society. The lunar observing world and even Patrick's future may have been very different if Green had survived: we will simply never know.

Patrick's RAF commission was confirmed by his presence at the BAA London meeting of Wednesday August 30th 1944, at Burlington House in Piccadilly, where he was recorded in the minutes, for the first time, as 'Pilot Officer Moore'. In all earlier entries he was simply 'Mr Moore'. For such a young man, in an era when the BAA was dominated by distinctly Victorian looking gentlemen, the youthful Patrick certainly had a lot to say at that August 1944 meeting and it is worth mentioning what the minutes recorded of his thoughts on that wartime Wednesday afternoon in London. If nothing else, they reveal that Patrick, even aged 21, had already decided that the lunar craters were volcanic, not meteoritic, in origin; a view he would unswervingly hold, against all the evidence, even into his late seventies! We will revisit his obsessive, unswerving view of the lunar cratering process many times in this book. The President, in that year of 1944, was the distinguished planetary observer F.J. Hargreaves and he invited the tall, bearded and bespectacled BAA stalwart F.J. Sellers, who bore a distinct resemblance to the modern day entertainer Rolf Harris, to present a paper on behalf of Patrick's future mentor (who was not present) H.P. Wilkins. It was a brief talk, in which a detailed sketch of the lunar crater Plato, made by Wilkins, was shown, along with its main features, some elusive and others not. Members were invited to comment on the features indicated on Wilkins' drawing. The 21 year old Pilot Officer Patrick Moore was not shy in coming forward and had a great many comments to make. Patrick began as follows: "I would like to say that I looked at the Moon four nights ago for the first time for a year. I noticed an increased dark appearance of Plato's floor, but I was unfortunately unable to make a drawing. Visibility was mediocre and I could see no detail, but the depth of hue was unmistakeable. Could the mist have been caused by the emission of CO_2, which is known to be a late stage in the activity of a volcano?" After some musings from the President and Mr Sellers about Patrick's suggestion, Patrick continued: "I expect a mist composed of CO_2 would be low-lying. There are many recorded instances of lunar fogs other than those in Plato – in Schickard and Theophilus, for instance." Later, Patrick added: "I think the difficulty lies in finding a satisfactory alternative to CO_2". In answer to theories about ice on the Moon Patrick suggested: "Surely if the surface were composed of snow or ice the albedo of the Moon would be far higher than it is?" Later at that same meeting Patrick makes the additional comment: "Surely it is reasonable to suppose that there is also a layer of volcanic ash, since the lunar surface exhibits so many signs of great volcanic activity in the past?" However, most of the members present, even in those pre-spaceprobe years, when the origin of the lunar craters was unknown, seemed unimpressed by Patrick's arguments that the lunar craters were volcanic. This did not put the young Patrick off though as he pressed home his view adding: "Is it not curious that so many [craters] are completely circular?" Also, in answer to a 'tidal pocket' explanation by Mr Sellers, Patrick stressed "I should not have expected such circular craters" and "Meteors striking the surface of the Moon would have to strike the surface normally". The President disagreed with Patrick here, explaining that it was the explosion on impact that caused a circular crater, not the impact itself. Patrick countered "That would be reasonable if meteoritic bombardment occurred when the Moon had lost its atmosphere." Clearly, in a room of much older, and much more experienced observers Patrick was very confident, and even precocious, in his views. He certainly liked his voice to be heard.

Interestingly, immediately after this discussion the next speaker at that August 1944 meeting was a man who would become a life long friend of Patrick's, and arguably his closest ever friend in the BAA: the future historian, author and BAA Journal editor Colin A. Ronan (1920–1995), then Lieutenant Ronan. By 1946 he would have risen to the rank of Major Ronan. The subject of Colin Ronan's talk was particularly interesting, bearing in mind what would eventually happen to Patrick. Colin addressed the audience on the subject of making astronomy more popular to the public by making educational astronomical films! One almost wonders if Colin could see Patrick's future role laid out in a crystal ball. Well, outside the BAA and back in his RAF role Pilot Officer Moore was soon promoted to Flying Officer, on December 2nd 1944 in fact, but only 1 month later, Patrick, and his entire Wellington bomber crew, almost lost their lives.

In January 1945, Patrick, and the six-man Wellington Crew he had recently been assigned to (a picture of which always hung in his study) were flying at an altitude of 19,000 feet. Suddenly, the plane stopped responding to the pilot. Everything had iced up and the plane went into a near vertical dive. As the pilot grappled with the controls, and the plane plunged through the 4,500 foot level, the decision was made to bail out. The radio operator, Flight Sergeant Guest Harding Dempster (later Flight Lieutenant Dempster, RAF service numbers 1575608/177185) handed the navigator Patrick a parachute; but, with horror, they realised there was something wrong with it. The ripcord appeared to have already been pulled. It was very unlikely that the parachute would work. There was a frantic decision as to what to do. The rest of the six man crew were preparing to bail out. There were only two options: they stay on the plane and try to make a landing, or Patrick and one other would have to share a parachute. According to Dempster the crew decided that 'because of the spirit Patrick had engendered amongst the crew', the problem with his parachute was their problem, not his alone. They would all stay on the plane and attempt a crash landing. Remarkably, the pilot managed to achieve a successful landing at St David's RAF base in Pembrokeshire. Patrick had experienced the closest shave of his life. This is the only flying incident in Patrick's war years which can definitely be verified as true. It was certainly dramatic, but I suspect it triggered a lot of other 'boy's own' adventure stories worthy of Biggles which Patrick would enjoy telling to close and trusted friends in later years. A few weeks after that incident Patrick and his Wellington colleagues were moved from St David's to RAF Lindholme near Doncaster where they were trained in dropping 'window' (silver foil) to confuse the German radar ahead of RAF bombing raids.

Catch Patrick in the right mood (after a few drinks) and he would sometimes reveal a bit more about his RAF period, even if one had to allow for significant embellishment. He once told me that during the war he was flying in an aircraft at

30,000 feet and did not notice that his oxygen line had become disconnected. "If another member of that Wellington crew hadn't noticed there would have been no books and no *Sky at Night*" he would comment. Another thing Patrick mentioned was "FIDO......straight from the fires of hell." I wanted to know more about this: "A rabid dog?" I queried. "No, Fog, Intensive Dispersal Of.....I once landed a Wellington using FIDO, and frankly it is not something I would want to do again, to put it mildly." Patrick explained that, during the War the RAF developed a system for burning off mist and fog on runways. The amounts of fuel involved (100,000 gallons per hour!) were designed to change the local weather by the amount of heat generated and make a fogbound runway more visible. Unfortunately it was like landing a plane "inside an active volcano" according to Patrick, but with improving radar techniques FIDO did not last very long. At other times Patrick would hint at special self defence lessons he was given for secret RAF military operations in case he was behind enemy lines. "You never forget how to look after yourself once you've had that kind of British military training" he would say, in a manner incredibly reminiscent of Foggy Dewhirst in *Last of the Summer Wine* (along with *Dad's Army*, *Yes Minister*, and *'Allo, 'Allo*, one of the few TV shows he actually liked). "Came in useful in my 70s, in the BBC car park, when two black youths tried to mug me. I broke the one Sambo's arm like a twig". Now and again he would also throw in the comment "Take my word for it, those German concentration camps were unspeakable; you don't forget scenes like that". Patrick occasionally claimed that he had been to the camp at Dachau, 10 miles northwest of Munich, in 1945, but refused to go into any more detail.

After a few more drinks a few more wartime stories would emerge. "Frankly, my entry into the RAF disrupted the allied war effort considerably. The little incident of us accidentally dropping practice bombs onto an ornamental rock garden in Barrow-in-Furness has been blamed on me, but I have to say it wasn't entirely my fault!" He also claimed that one of his RAF colleagues was so useless at astro-navigation that when they were flying over the Irish Sea, the colleague swore that his sextant measurements showed they were over Cape Horn! A similar favourite yarn and one that he repeated on the Michael Parkinson show (alongside his mother) in 1974, was that he was the only RAF trainee navigator who had been convinced they were over Norwich when the plane was actually flying over Bristol! In a short article inside the *Radio Times* during 1969 (the May 8th edition) Patrick claimed that during the war years he was involved in the so-called 'Department of Bright Ideas', which dreamed up crazy plans like building a raft the size of England in the North Sea, to confuse the enemy radar. Yet this was a tale he never repeated in his later years! Instead, he would claim he was a personal friend of the wartime weapons designer Col. Robert Stuart Macrae, the inventor of the sticky bomb and a helmet which fired bullets and "damn near blew my own head off". One thing Patrick was never short of was ripping yarns!

Despite serving in the RAF during the later war years, Patrick was becoming well known at BAA meetings as an enthusiastic lunar observer, and his presence was increasingly recorded in the minutes. At the April 25th 1945 meeting, 3 months after his near brush with death in the Wellington over Pembrokeshire, a question was asked by another BAA member, referred to as 'Mr Cox' in Patrick's lunar notebook, but as Mr R.E. Diggles in the official meeting report! The question, as Patrick remembered it, was:

> By observations of the rays (on the Moon) is it possible to plot approximately positions of craters on the invisible side of the Moon?

The question was referred to Patrick who replied:

> No – all the rays observed to pass over the limb are members of known systems – Tycho, Copernicus, Anaxagoras.

In fact, although Patrick's own record of this exchange is similarly recorded in the BAA minutes, the question was actually re-directed to Mr P.M. Ryves, by the President P.J. Melotte, but before Mr Ryves could respond Pilot Officer P.A. Moore had interjected! Sometimes Patrick's recollections of events could be very different to the actual situation.

After that meeting, Patrick agreed to some co-operative lunar work with another BAA member, R.H. Whittome at Peterborough, who intended to examine the Mare Crisium region. Patrick's notebook also records that he was keen to examine what he described as "my new craters" in the Mare Humorum and the ruined ring south of Heraclides.

With Patrick having suffered back, knee and dental injuries (by whatever mishaps) during his brief active period with the RAF, and with his existing heart condition, he was put on somewhat lighter duties for much of 1945. Instead of being a navigator he was made an 'adjutant', a term which meant that he assisted senior officers by communicating their orders and dealing with their correspondence, something that Patrick would have been very well suited to.

After World War II hostilities ended, in the late summer of 1945, Patrick spent a brief period as an RAF Area Meteorological Officer. On numerous occasions Patrick admitted that his knowledge of meteorology was highly limited and his main source of information was a sodden clump of seaweed nailed to the door of his office! With the end of the war, Patrick's father (now in very poor health) was able to step down from his role as East Grinstead Home Guard Platoon Commander and he managed to get an accountancy job close to home, as the accountant at the Felbridge Place Hotel in the town.

Patrick's lunar notebook records the following on October 3rd 1945:

> Demobilized from the RAF with the rank of Flying Officer. This means that serious and regular observations of the moon can be resumed at once. A proper program of work must be made out.

According to MoD records Patrick officially left the RAF on December 5th 1945, although this was not the final farewell. The discrepancy of 2 months between 'demob' and leaving was not unusual. Many airmen had a lot of leave allocation due to them when they left the forces at the end of the war. Sometimes Patrick would describe himself as having been 'invalided' out of the RAF. After a 6 month break when he considered writing for a living, and even accepting a grant to take his planned geology degree at Cambridge, he eventually scrapped both plans and managed to be re-commissioned as an Acting Pilot Officer (Training), on June 13th 1946! This was purely a teaching role with the ATC though. On December 1st 1946 Patrick was confirmed as a Pilot Officer (no longer acting) and he was made 'Flight Commander' at the East Grinstead ATC. The 'Commanding Officer' plaque he so often had in the background on later *Sky at Night* programmes (next to a favourite snoozing place for his cat Jeannie) hailed from his ATC commander days. He held this training post until September 5th 1947. After a few drinks in later life Patrick could easily promote himself, in a moment of fantasy, to having been an RAF Flight Lieutenant, or even a Squadron Leader, to account for the plaque. In reality he only ever had ATC cadets under his command and in the real RAF he only rose to the rank of Flying Officer Moore! Tut-tut, naughty Patrick!

Fig. 2.2 Patrick with his ATC colleagues at East Grinstead in December 1947, shortly after his father's death. Patrick is in the *front row*, to the *left* of the baton holder (Local 1940s newspaper photograph. Photographer unknown)

Incidentally, in the same month that Patrick was demobilized from the RAF, his British Interplanetary Society fellow member Arthur C. Clarke's famous paper, *The Future of World Communications* (predicting the Geostationary Satellite), was published in the October 1945 Wireless World. Arthur had genuinely been demobilized from the RAF with the rank of Flight Lieutenant; he had not made it up! Arthur's address at that time had returned to his original Somerset one, namely: Ballifants, Bishops Lydeard, near Taunton. As we shall see in the coming pages Arthur was, undoubtedly, a huge influence on Patrick. Both men had many shared interests, but were very different. Five years Patrick's senior, Arthur was a technical man, good at maths and physics. Unlike Patrick he had resumed his studies after the war, gaining a first class degree from Kings College London, writing a technical book on Interplanetary Flight in 1950, and becoming the assistant editor at *Physics*

Abstracts, all while building his science fiction writing career. As the war ended Arthur had played a pivotal role in restructuring a new British Interplanetary Society, along with Eric Burgess, Kenneth Gatland, Phil Cleator and Ralph Smith; but while the idea of future space travel obviously excited Patrick, observing the Moon through a telescope, as part of the BAA, would always be his first love.

While serving as Flight Commander for the East Grinstead Air Training Cadets Patrick was already working his way back into regular observing. Just one week after he took that post on there was a total lunar eclipse, on December 8th 1946. Frustratingly, the skies were cloudy, and only the briefest gap appeared at 18:37 GMT, some 18 minutes after totality had ended. Nevertheless, despite the Moon's altitude of just 25 degrees in the eastern sky, Patrick observed the lunar eclipse until clouds returned at 18:48 GMT. In Patrick's 3-inch refractor the eclipsed portion seemed a dark purple, the umbral border was definitely sharp and, judging by this partial phase, it seemed like a dark eclipse. Patrick's observation was described by H.P. Wilkins a few weeks later, at the 1947 New Year's Day BAA meeting. At the 1947 February 26 BAA meeting, following a talk by Wilkins about F.H. (Harry) Thornton's observation of a possible meteor strike in the lunar crater Plato, Patrick gave a short presentation about his 10 years of observing the naked eye variable star Gamma Cassiopeia. During Patrick's talk he claimed to have independently detected the star's significant brightening, as a 13 year old, back in December 1936! Patrick explained that over the past 10 years he had averaged 70 magnitude estimates of the star each year.

Two years after Patrick's War demob his father became seriously ill (towards the end of 1947) and Captain Charles Moore had to relinquish his hotel accountancy post. Patrick was needed at home to help his mother cope with his bedridden father; hence he left the ATC. On December 15th 1947 Patrick's father died at home. He was only 62 years old. The family Doctor, N.B. Shaw, diagnosed the cause of death as "left lobar pneumonia with Delirium Tremens." Delirium Tremens is an alcohol withdrawal symptom. As Charles Moore's health had deteriorated, he had become a heavy drinker. Captain Charles Moore was cremated, at Streatham Crematorium, on December 18th 1947; exactly 1 week before Christmas Day. After his father's death, Patrick still maintained close links with the ATC (and the local Scout group) and even attended the Christmas 1947 party. A picture permanently in his study showed him at that function and was marked 'Junior Service Club ATC Army Cadets'.

At that point, Patrick decided to consider his options outside the military (presumably his father's pride in Patrick's military duties were no longer an issue) and it was time to think seriously about becoming more heavily involved in astronomy. It was time to acquire some better telescopic equipment too, as a priority. Patrick could have resumed his plan to get a degree at this point, but for whatever reason he chose not to. Servicemen whose degrees were interrupted by the War were entitled to a grant, but Patrick said he intended to "pay his own way, but it did not work." I would venture to suggest that Patrick liked his East Grinstead home and his mother's cooking too much. He was still only 24 years old and with his father no longer around the close bond between mother and son (and Rufus the cat) must surely have been even stronger in that bereavement period. Patrick had always been closest to his mother, so he was unlikely to leave her now. They would live together for another 33 years, until she died, aged 94, at the very start of 1981.

Chapter 3
The Post War Years

As we have already seen, one of Patrick's top priorities after the war was to become an active amateur astronomer again and, in particular, play more of a role in the British Astronomical Association. The Association was thriving in the post-War years: membership had climbed from 850 in 1939 to 1600 by 1947. Strangely, Patrick told me that he also made a conscious decision to permanently drop the Caldwell part of his Caldwell-Moore surname at this time: "I was lazy and signing Caldwell-Moore was tedious; I hardly ever used the Caldwell part after I left school". Why he would say that is a mystery, because there is no evidence he, or his father, used his grandfather's double barrelled surname at all and if he used it at school, well, he only attended during one school year! In the BAA, he was recorded as Mr P. A. Moore, or just Patrick Moore, which was how he liked to be known. This, of course, was some time before he would become famous and known affectionately by everyone as just 'Patrick'.

According to Patrick, roughly 2 years after the war ended (and so around the time of his father's terminal decline, in 1947) he acquired an alt-azimuth mounted 12½ inch aperture reflecting telescope with a focal length of 72 inches and a mirror made by Henry Wildey (1913–2003), a master mirror-maker who would become the BAA's Curator of Instruments 4 years later. [I will stick to imperial measurements here as there were no metric units in use in Britain at that time.] The 12½ inch reflector would live in Patrick's East Grinstead back garden and was a huge instrument for that era. Patrick could not afford to have it equatorially mounted and so he purchased it as an alt-azimuth instrument, with a view to 'upgrading it in due course'. He never got around to the upgrade: it would stay as an alt-az for the whole of Patrick's life, still in it's housing well into the twenty-first century!

Patrick had a double-ended run-off shed made for the new telescope, fabricated from asbestos panels on a wooden frame. The shed warped badly and so Patrick had a new, wooden, double-ended shed made for the telescope in the late 1950s. Patrick decided his new telescope deserved a name and he duly named it 'Oscar'! In later years he claimed he could not "for the life of me" remember why he gave it that name. However, as with so many of Patrick's dates, there are huge discrepancies

M. Mobberley, *It Came From Outer Space Wearing an RAF Blazer!: A Fan's Biography of Sir Patrick Moore*, DOI 10.1007/978-3-319-00609-3_3,

regarding when he acquired and started using the 12½ inch. In the later years of his life the 12½ inch was listed as being used 'from 1947 or 1948' and then, eventually, right back from 1945. In fact, the 12½ inch originally belonged to Percy Wilkins, who we will meet frequently in the coming pages. Patrick had seen this telescope many times in Wilkins' back garden, initially on a very crude mounting and then on an English mounting with an oversized tube (to cope with a planned aperture increase). When Wilkins decided to upgrade to a 15¼ inch reflector, around late 1949, Patrick acquired the wreck which was the original 12½ inch. So, despite Patrick's varied recollections, the acquisition of the 12½ inch Newtonian was not 1945, 1946, 1947 or 1948, but late 1949! Wilkins named the date of 'first light' on his 15¼ inch Newtonian as being in February 1950, with the big new mirror being ground and polished by the late Mr W. MacIntyre; he mentioned all this at the May 1950 BAA meeting.

In truth, after the War, Patrick acquired a number of telescopes. As well as using his trusty 3⅛ inch Broadhurst-Clarkson refractor, he owned a 3 inch 'Cary' refractor and an 8½ inch alt-azimuth Horne & Thornthwaite Newtonian (not to be confused with the 8½ inch With-Browning he acquired many years later). He also used the 6 inch refractor of Henley Fort Observatory at Guildford in Surrey. From time to time a 6½ inch reflector is also mentioned in his mid-1950s logbooks, especially when observing Venus or Mercury. In fact, Patrick's 'first light' through the 8½ inch Horne & Thornthwaite reflector at East Grinstead would not occur until April 20th 1950 and 'first light' through the restored, re-tubed and rehoused 12½ inch, complete with newly aluminised Henry Wildey mirror, did not occur until November 11th 1951; a far cry from the 1945–1948 dates quoted by Patrick. It appears that it took Patrick some considerable time to get both telescopes into working order and to finalise the run-off shed for the 12½ inch reflector. According to Patrick there was some dispute about the sturdier mount for the 12½ inch too. Patrick had liked the simple alt-az design featured in a BAA Journal article 10 years earlier (Jan. 1940) written by the Rev. W. Rees Wright. So he had asked the telescope craftsman Ron Irving (1915–2005) to make that style of alt-az mounting, but was not expecting an additional hefty carriage charge to be added at the last moment! So, surprisingly, Patrick's largest usable telescope was a 3 inch refractor until 1950, when he was aged 27. For superior light grasp he had to travel well away from East Grinstead at night, but such was his enthusiasm that this was not a big problem.

Apart from telescopes, another essential post-War acquisition was transportation, and a Triumph motorbike was acquired for the princely sum of 9 pounds. It was already 18 years old and produced clouds of sparks and choking black smoke. Not surprisingly, Patrick decided to name that piece of equipment 'Vesuvius' and he still owned it well into the 1970s when it was stolen, much to Patrick's dismay. By all accounts Patrick on his motorbike, with his RAF flying jacket and goggles, was the living embodiment of the fictional character Biggles! But with a motorbike he could travel to see Percy Wilkins and also to use the 6 inch refractor at Henley Fort Observatory. Incidentally, in 2008 I was told by the planetary geologist Peter Cattermole that Patrick had another smaller "pop pop runabout" motorbike that he used to let Peter ride in the 1960s. This motorbike also produced much smoke and Patrick named that vehicle 'Etna!'

The Moon

Patrick's first love in the night sky was, and would always remain, the Moon. Although maps of the Moon had been drawn by a number of observers over the years, the first lunar space probes would not fly past our nearest neighbour until 1959. Admittedly, lunar photography had been carried out by a number of observatories ever since the 1890s. The main professional observatories in France (Paris-Loewy and Paris-Puiseux), Belgium and the USA (Lick) produced good photographic atlases of the Moon, and in 1904 the astronomer W.H. Pickering produced an atlas in which he showed each region under five different conditions of illumination. However, the young Patrick wanted to help map the Moon in more detail, especially the more neglected regions, and the best way to get into this work was through the BAA's Lunar Section.

It should be emphasised that mapping the Moon from the Earth was not simply a case of putting a camera onto a telescope and taking a photo! The films of the 1940s were nowhere near as light sensitive as those available in the latter twentieth century. In addition, the Moon's features only really show themselves under specific low angles of illumination. At Full Moon, with the Sun shining directly onto the lunar surface, the Moon is a bleached and boring world, with no shadows and just different shades of grey. However, when the Moon is a crescent, half, or even gibbous (three-quarter's full) the shadows near the terminator reveal the true, rugged nature of the terrain. In addition, 1940s film was grainy and you needed very long focal lengths to resolve the finest features. Very long focal lengths meant that you only covered a small region of the Moon with each photograph and the image would be faint. Also, and perhaps most importantly, the Earth's atmosphere is a turbulent seething mess and distorts all attempts to see fine detail, except on the most perfect and calm (often misty) nights. The human eye can detect fine details in brief glimpses of 'good seeing', but the slow film of the 1940s required exposures of many seconds at long focal lengths; thus the visual observer, even with a back-garden telescope, had an overwhelming advantage. As a rough guide, a skilled lunar observer/artist, at a decent telescope, can detect features smaller than half a mile in size, whereas a photographic exposure of several seconds will rarely record features smaller than 1 or 2 miles across on the lunar surface, a quarter of a million miles away. Thus, despite the existence of photographic atlases, the visual lunar mapper still had a serious role to play in the 1940s and early 1950s, patiently waiting for clear skies and tantalisingly brief glimpses of fine detail when the Earth's atmosphere co-operated. All this would soon change though, as the NASA Moon mission plans prompted big observatories in the USA to use better emulsions, techniques, and new photo-visual methods during the late 1950s.

Another factor here that was mentioned earlier (Patrick's Oct. 15th 1940 observation) is that although the Moon's rotation on its axis is gravitationally locked to the Earth (that is, we always see the same side) the fact that the Moon's orbit is not a perfect circle (and other factors) means we can sometimes see a bit more of the Moon than normal. In fact, we can see a total of 59 % of the lunar surface from the Earth by using these 'librations' that tilt the Moon's limb regions towards us.

Obviously these areas are very foreshortened, but with no way of seeing the Moon's far side in the pre space-probe era, peering at the Moon's edges at favourable times was the only way of seeing that other side. What an exciting period to live in! The amateur, in his or her back garden, with simple equipment, and just one pair of eyes, reigned supreme.

During the 1930s, when Patrick was a very young BAA member, the driving force behind all of the BAA's lunar studies was Walter Goodacre (1856–1938), who had produced his own map (in 1930) of the lunar surface, along with an account of each major feature. Goodacre directed the section for 42 years from 1896 to 1938, but it fell into something of a decline during the war years, under the new Director, T.L. MacDonald. This was not entirely MacDonald's fault, but he was never in the same enthusiastic league as Goodacre and comments in Patrick's observing notebooks make it clear that even the teenage Patrick hoped MacDonald would soon be replaced. After the war, in 1946, a much more vigorous Director was appointed: yes, it was that publicity seeking Welshman H. Percy Wilkins, who was obsessed with drawing and mapping the Moon and getting as much kudos from it as possible. By 1924 he had published a lunar map 60 inches in diameter and, 8 years later, he started an even larger map to a scale of 300 inches to the Moon's diameter. This latter map, with Patrick's help, was completed in 1951, with major revisions continuing until 1954. This all sounds incredibly impressive until you see the maps produced. They were horrendously detailed and cluttered line drawings (the term 'drawn by a demented spider covered in ink' has been used) and much of the fine detail was painful to interpret and even misleading. Wilkins was a civil servant at the Ministry of Supply and had a crude observatory, first at 127 Eversley Avenue, Barnehurst, in Kent (housing his 12½ inch reflector) and later, from the late 1940s, at 35 Fairlawn Avenue, Bexleyheath, also in Kent. At this latter site he eventually installed that much larger reflecting telescope with its 15¼ (387 millimetres) mirror. Ultimately, he would embark on making a 22 inch mirror, but died before it was completed. In the post-war years Patrick occasionally visited Wilkins' Civil Service office at the Ministry of Supply in London, before heading off to the last Wednesday in the month BAA meetings. In later years Patrick came to despise all penpushers, bureaucrats and 'jobsworth' Civil Servants and, even where Wilkins was concerned, I heard him say: "His workday office was full of maps and charts and photographs of the Moon……but then, as a Civil Servant, I suppose he must have had plenty of spare time on his hands!" Upon Wilkins' appointment as Lunar Section Director, in 1946, he had installed the 23 year old Patrick as his main helper; a prestigious post for one so young. This was just prior to Patrick's father's death and well before Patrick had moved Wilkins' own 12½ inch reflector's components to East Grinstead. Of course, the BAA post was entirely voluntary, as the BAA was an organisation of advanced amateur astronomers, not professionals, but Patrick was delighted by the appointment. In fact, only 2 years later, on December 19th 1948, Patrick would be officially installed as the BAA's Lunar Section secretary. Patrick got on well with Wilkins, his wife Hilda (who spoke in the strongest imaginable Welsh accent), Wilkins' attractive teenage daughter Eileen, and even with Wilkin's cat 'Buttons'. It was as if Patrick had a second family, but one with a like-minded astronomical father figure, a sister, and a huge telescope to look through.

Being based at Bexleyheath (now part of Greater London, near Dartford) Wilkins was only 25 miles to the north of Patrick's home at East Grinstead. Both men were within easy reach of BAA main meetings and council meetings. Despite not having a car at that time, Patrick had acquired his 'Vesuvius' motorbike just after the war ended, was within relatively easy travelling distance of Wilkins' observatory, and often joined him in his lunar studies on clear nights. There were plenty of phone calls, if the weather looked favourable and the lunar phase was right, between East Grinstead 322 and Bexleyheath 6506.

Fig. 3.1 Patrick with H. Percy Wilkins and Wilkins' 15¼ inch reflector in his garden at 35 Fairlawn Avenue, Bexleyheath. Wilkins' cat was called Buttons (Photograph probably taken around 1948 by the late James (Doc) Quincy Gant (1906–1989))

Wilkins' main goal was to draw a map of the Moon which was of a higher resolution than any previously issued. Since Goodacre's 1930 book, the International Astronomical Union had issued a map (in 1935) drawn by W.H. Wesley and M.A. Blagg, but it did not show the smallest features visible through a good telescope at high magnification and it was defective and incomplete at the lunar limb. Wilkins, with help from Patrick, set out to produce his aforementioned map to a scale of 300 inches to the Moon's diameter, with particular emphasis on the neglected limb regions. Wilkins would not complete his final tinkering with the limb regions of the map until 1959, some 13 years after taking over as Lunar Section Director.

In that year the Moon's far side would be imaged by the Russians and Wilkins would then promptly die, on January 23rd 1960, just after an appearance on Patrick's *Sky at Night* programme! Also in 1959, Gerard Kuiper at the Yerkes Observatory in America would bring out a superb photographic lunar atlas, which in virtually all respects would make the Wilkins and Moore map totally redundant! In later years Patrick would throw into conversation the fact that his Moon maps were used by NASA in mapping the Moon, prior to the Moon landings. In fact some of his drawings of the limb regions were requested by Soviet researchers, and the Wilkins map was acquired by many organisations; however, neither were of any real use by the time of the Moon landings. The Wilkins map would prove to be a lifetime's obsession, to no avail. It was superseded as soon as it was published, and was too cluttered to be of any value whatsoever.

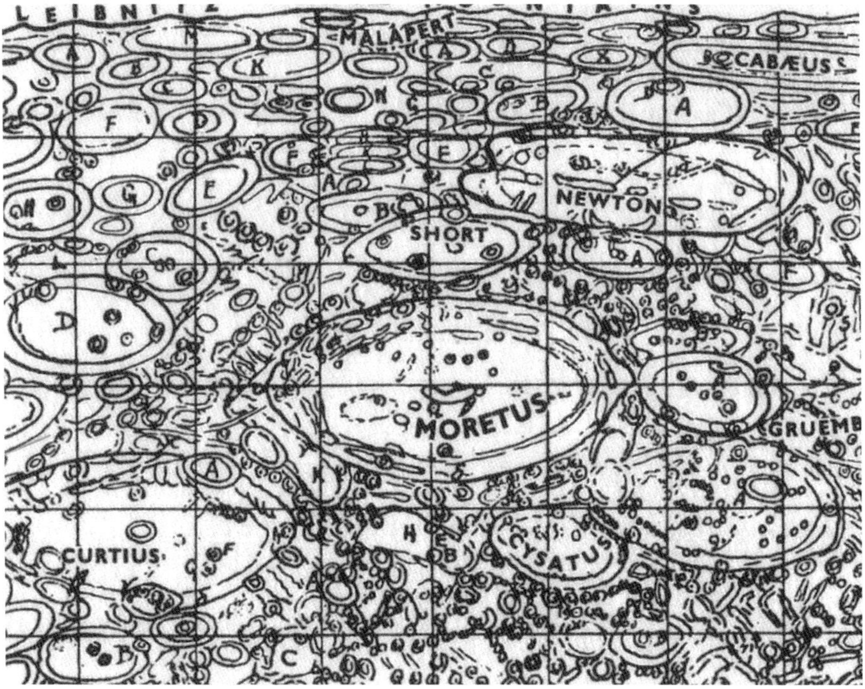

Fig. 3.2 The horrendously over-complicated detail on Wilkins' Moon Map, in the region of the crater Moretus

However, back to 1946, and the young Patrick was excited by his new role in the Lunar Section and the prospect of making new discoveries with the enthusiastic Wilkins. Ever since the death of his childhood mentor W.S. Franks he had needed another senior astronomer to feed his enthusiasm. Patrick had few friends of his own age in East Grinstead, with the exception of two chums from his teenage years, Ian Corrie and Pat Clarke (Ian's grandson Adam would become one of Patrick's favourite godsons in later life). Indeed, due to his heart condition, absence from schooling, and his private tutoring, Patrick had no proper friends until he was aged 15! He would always be a single-minded individual who had no time for fools, or those with opinions different from his own. He wanted friends with the same interests as himself and he found these in the BAA and the British Interplanetary Society. In the Lunar Section he would work closely with established artists like Leslie Ball and Harry Thornton and with Robert Barker and Arthur Lenham. Closer to his own generation he swiftly established friendships with experienced observers like Dai Arthur, Keith Abineri and Richard Baum, as well as with arguably his closest BAA friend, the historian Colin Ronan. Undoubtedly his post-war friendship with Arthur C. Clarke was a critical factor too. Although Clarke was not in Patrick's league as an observer at the eyepiece, his knowledge of electronics, radio, and the likely future direction of space rocketry, was a major influence, even without Clarke's superb science fiction writing skills. Clarke was also living proof that you could earn money as a young author.

Patrick's enthusiasm for observing the lunar limb regions was fired when, shortly after his new collaboration had started, in late 1946 in fact, he observed a crater right on the limb, which he called 'Crater X'. As I touched on earlier, this was the same crater that was subsequently named as 'Einstein'. On another occasion in late 1946, although in recent years Patrick much preferred the date of 16th December 1948, Wilkins and Moore observed the Moon, under good conditions, when the ESE lunar limb was tilted some 7 degrees towards the Earth. This was the same illumination and phase as Patrick had enjoyed in October 1940, when using his 3⅛ inch refractor. But now he also had access to Wilkins' 12½ inch Newtonian. A range of prominent lunar peaks, the Cordillera mountains, were an impressive sight that night, but there seemed to be a much darker 'lunar sea' (a smooth, dark area) just glimpsed beyond the mountains. Wilkins and his young apprentice dared to wonder if the Cordillera peaks were, in fact, a huge ring of mountains surrounding that truly gigantic 'lunar sea' on the Moon's far side. Actually, as we saw earlier, this was not a new observation as Wilkins had already published his observation of this 'Mare X' in the December 1937 BAA Journal, a decade earlier. However, due to the rarity of a favourable illumination, favourable libration, clear skies, and steady atmospheric seeing, all sightings of this feature were exciting, and in the intervening decade few others had made good observations of this 'Mare X'. This time, Wilkins and Patrick communicated the information to professional astronomers, calling it the Mare Orientale or 'Eastern Sea'. The name was, ultimately, adopted by the International Astronomical Union (IAU) although, ironically, at the start of the space probe era the IAU reversed east and west on the Moon to match terrestrial maps: so the Eastern Sea is now on the west of the Moon!!

However, according to Patrick's lifelong friend, the author and astro-historian Richard Baum, the Mare Orientale was first found *and named as such* by Julius Franz sometime before 1906, as his Book *Der Mond* names it. It is also featured in a published lunar atlas dated 1923 and Patrick has long admitted to owning Franz' book! Not only that, but (as we saw earlier) in the BAA Journal (Vol 48,7 page 271) it is clearly reported that the new Lunar Section Director in 1938, Mr T.L.

MacDonald, announced at the April 1938 BAA meeting, only a year after Wilkins' 'discovery' of his Mare X, that Franz had seen it and named it the Mare Orientalis. Patrick definitely would have known this. Still, Patrick and Wilkins studied it, sketched it and drew the IAU's attention to it and it may have been that their efforts, especially Wilkins', were a factor in its official naming. Actually, references to the same feature, (but not named Orientale) go back as far as 1871. The publication called *Investigation of Lunar Phenomena undertaken at Harvard College Observatory in 1871* by Nathaniel Shaler accurately describes a huge crater beyond the Rook Mountains, which is where Orientale is. So, essentially, Patrick did NOT discover Mare Orientale and neither did Wilkins or Julius Franz, but Patrick certainly drew more attention to it than anyone else!

By an amazing twist of fate, Patrick would find himself broadcasting, live, on BBC1, when the first images from the Moon's far side arrived from a Russian spaceprobe in 1959. As the decades rolled by Patrick would seem to care less and less about the historical facts regarding the Mare Orientale and the crater Einstein. In reality he discovered neither, but, by his eighties, he was claiming he discovered both, before Richard Baum and Ewen Whitaker, his former colleagues, wrote a paper on its real history in the BAA Journal. In reality he and Wilkins probably raised more awareness of the Mare Orientale than any other people in the pre-spaceprobe era, but that was all. Having been through *all* of Patrick's lunar observation notebooks I can confirm that the only decent sketch of the Orientale region I spotted was the one made on December 16th 1948. There are no other good drawings of it! That sketch was made 11 years after Wilkins' paper about the feature in the BAA Journal.

With Patrick's obsessive enthusiasm and Wilkins' publicity-craving approach and experience driving the BAA lunar section, the membership increased dramatically: from ten members in 1945 to 130 in 1952! It should be stressed that Wilkins was not the only person in the BAA who knew the surface of the Moon somewhat better than the back of his own hand! Perhaps the best lunar observer of those years was Harry Thornton, who made an observation on April 15th 1948 which especially interested the young Patrick. He sent the observation to the BAA Lunar Section and it may well be a genuine sighting of a small meteor impacting the Moon; this was the second such event reported by Thornton, who was a compulsive lunar observer. The observation reads as follows:

> When I was examining Plato, I saw at its eastern rim, just inside the wall, a minute but brilliant flash of light. The nearest approach to a description of this is to say that it resembled the flash of an AA shell exploding in the air at a distance of about ten miles. In colour it was on the orange side of yellow.... My first thought was that it was due to a large fall of rock, but I changed my opinion when I realized that, close as it seemed to the mountain wall, it was possibly over half a mile away.

Unexplained lunar observations like this would fascinate Patrick for the rest of his life.

Patrick was an enthusiastic attendee of all the BAA's London meetings in the post war years and this was to remain the case up to the late 1990s, when immobility finally curtailed his monthly trips to the capital. From September 1942, and up to 1970, the Association's meetings were held in the Upper Library of the Royal

Astronomical Society's building in Burlington House, Piccadilly, on the last Wednesday of each month, from October to June. The BAA council meetings were (and still are) held in the downstairs council chamber of the RAS. As a keen attendee of all the London meetings, and a friend of many on the council, Patrick was soon becoming very well known to the most prominent people in the BAA.

Richard Baum, a friend of Patrick's since 1946, remembered those days in some e-mails to this author:

> The first time [I met Patrick] in person was at a BAA meeting in Burlington House in 1946. He sailed across a room full of greybeards hand outstretched. One was made to feel welcome straightaway. F. J. Sellers, Robert Barker, H. P. Wilkins, D.W.G. Arthur they were all there.
>
> I remember my second attendance at a BAA meeting. It was in 1948. I recall two events, a talk by Peter Lancaster Brown on comets, and the walk back to Euston along Piccadilly in the company of H. P. Wilkins, Patrick Moore, Robert Barker and D. W. G. Arthur. It was a dark, damp and cool evening. Dai Arthur, I remember, was carrying rolls of lunar photographs – at the time he had access to the great Paris Atlas and was making black and white prints from its large sheets and selling them to members who wanted them at around five shillings each! Soon after he joined the lunar mapping team at Tucson. He is now retired and lives in a trailer I believe, outside Tucson, his interest being model railways. There was a magic to those days which is strangely lacking today or is it me getting old?! Patrick at this time (1948) and for sometime thereafter was characterized by a navy blue suit that fitted where it touched! Needless to say it caused much amusement. Still I wish you had known him then. He was a different person. Terrifically enthusiastic, energetic and kindly disposed. Each time I see him now I see him as he was – not as time and fame have made him. I shall always hold a special affection for him in my heart. I still have his very first letter to me dated late 1946.

Some snippets from the BAA meeting records of the late 1940s may interest the reader at this point. For example, Patrick attended the January 28th 1948 BAA meeting at Burlington House, when the experienced lunar observer Robert Barker delivered a talk about the dark radial bands on the inner eastern wall (classical directions) of the crater Aristarchus. Barker insisted that these bands were changing with time, as there were now at least five bands visible, not seen by historical lunar observers. Various prominent BAA members, including the president F.M. Holborn, telescope expert W.H. Steavenson and H.P. Wilkins joined in the discussion. The young Patrick was sure there had been a change in the crater's appearance since historical times, and he was recorded as saying: "I should like to add that I have seen the bands with a 3-inch telescope, and it is quite impossible that Elger or Neison could have overlooked them if they were as conspicuous then as now." It is fascinating to note that also present at that 1948 January 28th meeting in Piccadilly was Patrick's friend, the future science fiction writing legend Arthur C. Clarke, aged just 30. In answer to a paper delivered by Mr H.W. Cox, Arthur described how, in the war, he had used a photomultiplier tube with a 3-inch telescope to monitor light levels on the Moon. Arthur added that anyone who knew their way around London's Lisle Street could pick up the necessary photomultiplier equipment for 5 pounds. He also wondered if photomultipliers could be used to measure rotation periods of asteroids and planetary moons. Clearly, and as you would expect, Arthur C. Clarke was years ahead of his time!

If we now delve into the meeting minutes one month later we find Patrick, H.P. Wilkins and Arthur C. Clarke were also recorded as contributing to the February

25th 1948 BAA meeting at Burlington House. Another influential figure present at that meeting was Dr J.G. Porter, who I will have more to say about shortly. At this particular meeting Dr Porter read out a paper by Bernard Lovell, a 34 year old researcher at Manchester University, whose name would forever become linked with radio astronomy. Lovell's paper discussed the detection of meteors by radar and Arthur C. Clarke queried Dr Porter on whether narrow beam high precision radar had been tried to pin down a meteor radiant. H.P. Wilkins also presented a paper on the outstanding work of the young lunar and planetary observer Dr S.M. Green, whose life had been cut tragically short by the War. Patrick asked what magnification Green had used for observing Mars, and Wilkins replied that he thought the power was no greater than 300×. It is certainly fascinating delving back into these records to see the details of a meeting where Patrick, aged almost 25, and Arthur C. Clarke, aged 30, were listening to a speaker read a paper written by the 34 year old Bernard Lovell. All three of these men would live at least 60 more years and become legends in the fields of amateur astronomy, science fiction and radio astronomy. Of course, Bernard Lovell and Jodrell Bank (and Patrick) would become famous in 1957 after the 250 foot Jodrell Bank 'Mk I' telescope was constructed; but even in the late 1940s Bernard Lovell was encouraging amateurs, including Patrick, to spend a night at Jodrell Bank, lying in a field counting meteors crossing the sky, while the primitive Yagi array 'Searchlight' radio telescope detected their ionisation trails. Patrick made a number of trips to the Jodrell Bank field around this time and there were various society trips there too, such as the joint Royal Astronomical Society/Manchester Astronomical Society visit of July 1st 1949.

In 1948, the British Astronomical Association's Computing Section Director, the aforementioned Dr J.G. (Guy) Porter, was elected as their President. He was a mathematics and chemistry teacher at Medway Technical College in Kent, although during his term as President he became a Principal Scientific Officer at the Nautical Almanac Office in Bath and then at the Royal Greenwich Observatory, Herstmonceux. This change to being an astronomer was largely achieved by his Ph.D. which was written on the subject of collating and reducing the prolific George Alcock's meteor observations; there will be loads more on Alcock later. Porter was also the author of a number of science and astronomy papers and books, including his most popular book, 'The Night Sky', published by Winchester publications, also in 1948. In many ways Dr Porter was Patrick's predecessor as the BBC's resident astronomer. Upon becoming President of the BAA the BBC invited Porter to present a monthly radio programme on Astronomy. The series was appropriately called 'The Night Sky' and ran, every month, from 1948 to 1961. It started off as a 15 minutes broadcast on the BBC Home Service (equivalent to the modern Radio 4) but had shrunk to 10 minutes duration by its later years. The typical broadcast time was 10.15 p.m., but as this was way beyond the average 1950s childrens' bedtime, complaints from parents occasionally prodded the BBC to give it an earlier timeslot. Patrick was an avid listener to Guy Porter's radio broadcasts, but was fully aware that he was more than capable of presenting such programmes himself. He already had a reputation as a superb public speaker; conveying enthusiasm and explaining difficult concepts to beginners had always been Patrick's main strengths.

Porter was not the only famous name on the BAA council: the 1930s and 1940s stage and screen comedian Will Hay (1888–1949) was, without doubt, the most famous BAA council member (he was elected to the council for two consecutive years, so served from October 1945 to October 1947). Patrick's post war BAA period overlapped Hay's briefly, prior to Hay's death from a third stroke at the age of 60. Hay was a BAA member from 1932 to 1949. As we have seen, Hay discovered Saturn's White Spot in 1933 and was an active planet and comet observer. With Patrick's family connections to the stage, seeing Will Hay at occasional BAA meetings in London, when Hay's failing health allowed, must have been an exciting experience. However, the comedian Hay and the astronomer Hay were two completely different people. Hay was a highly accomplished man with an engineering background and a fluency in many languages. He was a pilot and a sailor, with a boat in the Norwegian fjords and a Norwegian mistress. In private he was a quiet, serious and studious man who never discussed his comedy roles in astronomical circles, where he wanted to be known for his observing accomplishments. Hay did not suffer fools but mingled with BAA members of great wisdom; men like Dr W.H. Steavenson (1894–1975), Dr R.M. Fry (1896–1980), F.M. Holborn (1884–1962), Dr J.G. Porter (1900–1981) and Steavenson's close friend Dr R.L. Waterfield (1900–1986).

Hay died of a stroke in his new Chelsea flat on Easter Monday (April 18th) 1949, aged just 60. Only 3 days earlier he had given a speech at a meeting of the show business charity organisation called the Water Rats and he had arranged a date with some friends at the Savoy on the day he died. He never made the appointment. His death was announced to the BAA by the President and BBC radio's astronomy voice, Dr J.G. Porter, at the 1949 April 27th Ordinary Meeting of the BAA. This was held just after the brief Annual Conversation Meeting and Exhibition in the upper library at Burlington House. Patrick told me he was at that meeting, although I have no independent proof of this. The President's comments were recorded as:

> Members will have heard with deep regret of the passing of Will Hay. William Thomson Hay became a member of our Association in 1932 June and his ability as an astronomer will be known to you all. Much prominence has been given to his discovery of the white spot on Saturn in 1933 and his descriptions of micrometers and a chronograph and other devices for observational work will be found in our Journal. A painstaking observer he was always ready to help those less experienced than himself; and, indeed, those of us who were privileged to have known him will remember him best in that way – as a great friend of all who were interested in astronomy.

The BAA's most famous member was dead and those astronomers who were thrilled by his rare appearances at BAA and RAS (Royal Astronomical Society) meetings must have been greatly saddened. However, they would only have to wait 8 years before a new BAA star was born and Patrick became a TV astronomer!

In late 1947, 2 years before Hay's death, with Patrick having left the RAF for good, he decided that he would try becoming a teacher. Hopefully NOT the sort of teacher Will Hay had emulated in his famous "Fourth Form at St Michael's" routine or in his films! Patrick was, for a time, simply living with his mother (and dying father) and initial attempts at earning money from writing had failed. He was accepted for a post at St Andrew's school in Horsell, Woking and taught there for a short while. While at

St Andrew's Patrick became a member of the local Horsell Amateur Dramatics Society (HADS) and took part in five productions. Rumour has it that Patrick only needed to read through his lines once to know them off by heart. Amateur Dramatics was another one of Patrick's hobbies and no doubt stood him in good stead for speaking live to camera in the years ahead. While he played various parts, such as eccentric vicars, he invariably ended up in a pantomime style 'Demon King' role in his later years. In the late 1940s he also made himself available for tutoring and lecturing and was one of the first lecturers at the new Wansfell College situated in Piercing Hill at Theydon Bois in Essex. Remarkably, Patrick would present annual lectures at that college, often with his good friend Colin Ronan or other friends, for 50 years, from 1949 to 1999, until problems with his knee slowed him down.

In 1948 Patrick heard of a new school being set up at Holmewood House, Langton Green, near Tunbridge Wells. He liked the sound of being in on a new venture from the start, plus Tunbridge was only 12 miles away (Woking was 25), clinched a job there and was a schoolteacher at Holmewood until 1953. He was initially employed as a History and French teacher, but also taught a bit of basic Maths and Science too (schoolboy Maths was all Patrick could cope with). However, it was his extra-curricula activities that most excited the boys at Holmewood. Patrick wrote the school song, set up a school band, conducted nighttime observing sessions with his refractor and formed a school Chess Club! As always, Patrick was possessed with a manic level of mental energy and enthusiasm. A former Holmewood teaching colleague who was the 'Beech Housemaster' and would become the Deputy Headmaster, Sandy Helm, said about him: "The aggravating thing about Patrick was that he had a photographic memory and would only need to glance at a piece of writing! He would mark written work, set exam papers, have them marked, reports written etc., while the rest of us were struggling into the early hours!"

Fig. 3.3 A young Patrick (central teacher) in the late 1940s at Holmewood House school, Tunbridge Wells. The lady sitting next to him was a co-founder of the school, Mary Trotter, who appeared on his *This is Your Life* programme a quarter of a century later. According to Patrick, the young boy behind him and to *his left* (directly behind Mary Trotter) was David Whiting, the stepson of Air Chief Marshall Sir Hugh Dowding, who believed in Flying Saucers, the supernatural and fairies! Holmewood House photograph

Some amusing tales from his school teaching days at Holmewood were detailed in Patrick's 2003 autobiography '80 not out'. Some time ago (on June 7th 2004) I came across a pupil of Patrick's from those earliest teaching days at Woking; it was Douglas Richardson, who Patrick taught French in those post-War years. I asked Douglas if it was obvious that Patrick was destined for stardom? His reply was: "No, not at all. However, he was undoubtedly an incredibly enthusiastic person at everything he did, not just at sharing his passion for astronomy, but at teaching French too. Patrick was a good teacher and, because of him, I continued my interest in languages in the subsequent years." Another pupil, at Holmewood, and one for whom Patrick was his housemaster, was David Whiting, the stepson of the RAF Battle of Britain chief Lord Dowding, who lived at Tunbridge Wells. I will have more to say about the remarkable Lord Dowding later.......

One of the more amusing tales from Patrick's Holmewood days involved his occasional nightly excursions onto a balcony, where he took his trusty 3 inch refractor. The only route onto the balcony was via a window from one of the school's bathrooms. One night, Patrick was out on the balcony with his refractor and decided, as it had clouded over, that it was time to come indoors. Unfortunately (or, maybe, fortunately!!) the school's amply proportioned matron had decided to take a bath while Patrick was on the balcony. His return, via the bathroom window, was, according to a grinning Patrick, "straight out of a Carry On film"! According to fellow teachers at Holmewood, Patrick seemed to work 20 hours a day, and even found time to dabble in writing some science fiction stories which was where he felt his true future might lie.

In 1949, while at Holmewood, Patrick had been indulging in occasional bouts of writing when he had the time. He had always wanted to be an author and, since 1946, his friend Arthur C. Clarke had been selling science fiction short stories for a modest fee. In 1949 Patrick was asked to write a 15 minutes script for a short forward-looking film being produced at the Taplow Film Studios in Buckinghamshire. Interestingly, this was where Gerry Anderson, creator of the Stingray, Fireball XL5, and Thunderbirds puppets, would start his film producing career some 6 years later. Patrick's script described a possible flight to the Moon around 1980, launched from Woomera, Australia. Launches from Woomera would feature in a number of Patrick's Sci-Fi books in the coming years. The story was filmed by a Mr A. Smith-Morris and produced by a chap called Peter Collins for his 'Pete Collins Space Show'; it was compered by a certain Raymond Witch. The 'Pete Collins Space Show' exhibited the short film at venues all around the UK for the next few years, but attempts by Patrick to track down a copy in the 1960s and 1970s failed miserably. In fact, the film 'Destination Moon', made a few years later, was probably the best full length Moon mission film of that era. Patrick gave the drafts of his earliest Sci-Fi efforts to the boys at Holmewood School, and apparently they eagerly read them in the dormitories after lights out, using torches under the blankets.

On June 6th 1949, while using the 6 inch refractor at Henley Fort Observatory (near Guildford), at a magnification of 325×, Patrick's attention was drawn to some bands in the lunar crater Hippalus A. These bands, to his knowledge, had never been reported by other observers. Six months later Patrick communicated his observations at the December 28th 1949 BAA meeting at Burlington House in

Piccadilly. In Patrick's write-up of his talk, in the February 1950 Journal, he made the following comment:

> The cause of the bands is naturally of the highest importance I favour the theory that they are due to radial cracks in the lunar surface which, as the Sun rises over them release sufficient gas to support a very low type of vegetation; but much close study will be necessary before we can come to any definite conclusions as to their origin.

Even in 1949/1950 this was an extraordinary and eccentric suggestion by Patrick, because the Moon, even then, was regarded as a barren rocky world, with any thoughts of primitive extraterrestrial vegetation being largely confined to speculations about Mars! Nevertheless, in the first years of the 1950s Patrick was becoming more and more highly regarded as a keen and reliable observer of not only the Moon, but the planets too. His, and Wilkins', enthusiasm for potentially newspaper headline grabbing events (such as grass growing on the Moon) rather than the more formal methods of BAA observers, had earned him many admirers and quite a few critics. Patrick was not the only BAA observer looking for evidence of change on the Moon in 1949 either. As we saw earlier, in the previous year the highly respected (though controversial) BAA observer Robert Barker had declared that the bands on the wall of the bright crater Aristarchus had become more prominent in the twentieth century than in the nineteenth century. Barker stated in the Journal that: 'the first real lunar physical change has been discovered by members of this Association, and this change is proceeding in evolutionary form today, and can be seen through small telescopes'. Well, an alternative explanation was simply that more observers, of a romantic disposition, were now observing with bigger and better telescopes, but that theory was far less exciting.

In respect of observable lunar changes it is likely that Robert Barker's controversial views were as much an influence on the young Patrick as the views of Wilkins, and probably even more so. Eleven years earlier, in the July 1938 Journal, Barker had listed the lunar regions of Aristarchus, Schroter's valley, Copernicus, Cassini, Goldschmidt, Mare Crisium and Plato as suspected of showing long term changes. Yet, in the modern era, we know that these changes were as much illusions as the Martian canals; but then Robert Barker had observed them too! Like many BAA members of the 1940s and 1950s Robert Barker was a fascinating character and distinctly Victorian, being born in 1873; so he was a full 50 years older than Patrick. Barker had been a music critic of the *Morning Post* and the *Manchester Guardian*, as well as being a first class sprinter and cyclist. Remarkably, even in his seventies he would often cycle the 20 miles from his home in Crossbrook road, Cheshunt, to BAA meetings at Burlington House, and then cycle back again! Sadly, just like the young Patrick's mentor W.S. Francks, a cycling accident, when Barker was 83, would curtail his physical activities in 1956. However, he lived a further 10 years, and Patrick was always keen to acknowledge Barker's encouragement of his lunar work, even during his teenage years. Nevertheless, I cannot help thinking that Patrick's future obsession with unexplained, illusory, and unproven lunar phenomena may well have been seeded, in those teenage years, by Barker's belief in such things.

Whenever the Moon was the subject of a lecture at a BAA meeting Patrick was likely to make some contribution to the discussion. This was certainly the case at the

BAA meeting held in Burlington House, Piccadilly, on Wednesday February 22nd 1950, the day before a general election. Dr J.G. Porter, the BAA President and the BBC's radio voice of astronomy, mentioned at that February meeting that the BBC were concerned that the current large sunspot (thought to be responsible for the recent geomagnetic disturbances and aurorae) might affect their ability to broadcast the general election results. He lamented that "It seems a remarkable thing that you cannot keep politics out of astronomy!" [Incidentally, the election result was that Clement Attlee's Labour Party achieved a narrow victory over Winston Churchill's Conservative Party.] After this political humour Dr Porter then invited Mr D.W.G. Arthur to comment about crater chains on the Moon, a subject that strongly divided the supporters of the opposing meteoritic bombardment and volcanic cratering theories. H.P. Wilkins had much to say on this subject, and his young apprentice Patrick, 10 days prior to his 27th birthday, addressed the following comment to the BAA members present: "It is very possible that crater-chains occur in considerable numbers close to the Moon's limb. For instance, there is an excellent example south of Demonax. Others could probably be detected with small instruments, and this seems a fruitful field for investigation."

It is worth saying a few words about D.W.G. (Dai) Arthur here, as he was of the same generation as Patrick and worked closely with him, for many years, in the post-war BAA Lunar Section. Four months after that February 1950 BAA meeting Wilkins would launch the Lunar Section publication 'The Moon' and install Dai Arthur as its Editor. At that time Dai Arthur's address was in England, at 35 Vastern Road, Reading, Berkshire. However, a few years later, like another Lunar Section man, Ewen Whitaker, Dai Arthur would emigrate and join the University of Arizona's Lunar & Planetary Laboratory lunar mapping team. Dai would end his career at Flagstaff, studying Martian Viking lander images, but in retirement he would take up model railways as a hobby and not astronomy. In the BAA of the late 1940s and 1950s young Dai Arthur had a reputation for copying and re-distributing rare lunar data and lunar maps, for sale to BAA members. He self-published five issues of his own magazine, called *Contributions to Selenography*, mostly containing lunar crater measurements he had made from glass negatives, obtained from Mount Wilson and Lick Observatories. Patrick was certainly not the only Moon-obsessed young man in the BAA of that era.

As we shall see throughout this book, Patrick had a romantic 1940s/1950s science fiction side to him that wanted desperately to believe in changes on the Moon, an atmosphere on the Moon, and unexplained events bordering on pseudo-science. During 1950 Patrick had even translated a controversial paper by his French counterpart Audouin Dollfus (1924–2010), entitled 'Research into an atmosphere in the neighbourhood of the Moon', so that BAA members could study it. However, in general, Patrick was careful enough to keep on the right side of the pseudo-science line, *especially* after he became famous. Historically, one of Patrick's benchmarks for proof of change on the lunar surface was the apparent change in the Mare Serenitatis crater Linné, described by the great observer Schmidt in 1866 and, allegedly, confirmed by Secchi, Neison and even Sir John Herschel. I will describe all this in more detail in Chap. 8. In a twenty-first century e-mail to this author, Patrick's contemporary and long-term friend Richard Baum commented about him in relation to this subject:

> Back in the early 1950s I purchased 18 volumes of the Astronomical Register for seven guineas (from Heffers in Cambridge). On arrival I discovered they contained a paper by Maedler (Madler) presented at the 1868 meeting of the British Association for the Advancement of Science in Norwich, in which Maedler plainly stated that Linné was then no different to how he saw it earlier in the [19th] century. Patrick was flabbergasted when I told him – I have his letter dated 1954 still. I had unwittingly destroyed one of his selenographical props!

But despite the young Patrick's leanings towards non-scientific lunar changes, he was well respected in the BAA. In 1951, with the BAA Jupiter Section Director (Millar) becoming too ill to continue, a new Director (Alexander) took over and appointed Patrick and Bill Fox as Assistant Directors. Patrick had started submitting regular Jupiter observations to the BAA 3 years earlier, in 1948, and would continue sending in sketches even into the 1990s. By this time Patrick's 3⅛ inch refractor and 3 inch Cary refractor had been supplemented by his much more powerful 8½ inch Horne & Thornthwaite reflector; specifically, first light was on April 20th 1950. Patrick's lunar notebook entry for that date reads as follows:

> The first observations with the 8½ inch reflector but conditions were mediocre and observations hurried as the Moon was descending below trees. A thin crescent terminator. Macrobius and eastern Mare Crisium examined.

Meanwhile, despite now having a large aperture at his disposal, Patrick was still struggling with finalising an observatory building for the even larger 12½ inch reflector. Throughout the late 1940s and 1950s Patrick became an increasingly regular visitor to H. Percy Wilkins home at Bexleyheath. He now had a long period of collaboration with Wilkins and they were great friends, despite the difference in their ages. Wilkins was born on December 4th 1896, and so was more than 26 years Patrick's senior.

On the night of November 11th 1951 Patrick's double ended run-off shed for the 12½ inch reflector was finally working and Patrick enjoyed 'first light' through the new instrument. In the final pages of Patrick's fifth lunar notebook (1948–1952) the following comments were recorded for that night:

> The 12½ inch reflector in action at last and my first lunar view with it! Although conditions were far from good, with continual interruptions from cloud, the superior light grasp compared with the 8½ inch is most noticeable and the telescope will give me greatly increased scope. Phase was almost full and for the first time in years I saw Drygalski right on the terminator.

A comment just below that entry reads as follows:

> Nov 15th 1951. Letter from Prof. Haas (USA) received. My Kies A sketch is to be published in the next Strolling Astronomer and he is asking for more observations.

Apart from the extra light grasp of the larger Newtonian, with two decent working telescopes at East Grinstead Patrick could position them to minimise trees (and the *Glencathara* house) blocking out specific parts of the sky: it gave him much greater sky coverage.

Fig. 3.4 Patrick and his 12½ inch Newtonian 'Oscar' at East Grinstead (Probably photographed around 1952. Copied with Patrick's permission in 2003)

The BAA meeting held on Wednesday February 27th 1952 started in a very sombre mood, due to the death of King George VI; all the members present stood for a few moments in silent tribute at his passing. While he had been the Duke of York he had, for 11 years, been the BAA's patron. However, there was a celebration too, because Miss Lydia Brown, the Association's secretary, had been serving the BAA for 25 years; 21 in her secretarial role and, initially, 4 years as the librarian. In those 25 years she had only missed one BAA meeting and would finally retire in 1970. On more than one occasion Patrick would declare that "Lydia Brown *is* the BAA". The members present applauded Miss Brown's remarkable period of service. Many notable BAA names were at that February 1952 meeting, including Patrick and Arthur C. Clarke, the telescope expert W.H. Steavenson (once a great friend of Will Hay) and various lunar observers. Those Moon watchers present included the lunar section director H.P. Wilkins, section member D.W.G. Arthur, and the keen observer Robert Barker. The President, Gerald Merton, invited Steavenson to address the meeting on the subject of 'Lunar Shadows'. Steavenson had been using a modified form of a 'Dawes solar eyepiece' on his telescope to exclude bright lunar regions and enable the eye to peer into the shadows cast by lunar craters, in the hope of glimpsing faint details. The audience was obviously fascinated and there was much discussion about the technique. Arthur C. Clarke thought the best way to proceed was to use photography to study the regions in the shadows, especially near lunar canyons, where specular reflection should increase any likelihood of glimpsing detail.

There was also some disagreement between the young Patrick and the experienced veteran Dr Steavenson when Patrick stated that, 18 months earlier, he had been looking at the crater Philolaus, near the North Pole, and saw quite a distinct border to the shadow, with a purplish cast. Dr Steavenson was adamant that Patrick must have seen an effect caused by the telescope or by the Earth's atmosphere, but Patrick argued that he had checked other shadows in the area, which were normal, had received confirmation of the observation from America (!) and he already knew that most of the colour on the Moon was illusory. Dr Steavenson was still unhappy with Patrick's observation though and was recorded as saying: "You saw it near the fringe of the shadow, which is the last place in which you could expect to see the true illumination. The effects of contrast would be too great." Patrick could surely not have been happy with such a degree of public scepticism from such a respected authority on optics who, in the 1950s, only had one equal in the BAA, namely Horace Dall. However, the conversation soon slid much further into wild speculation when Robert Barker (keen, as we have seen, on lunar changes) mentioned a strange observation he had made in the shadows cast by the mountainous walls on the rim of the crater Plato. Barker claimed that Mr W.E. Fox at Newark had confirmed the brownish-orange appearance he had seen and could only think that some atmospheric condition on the Moon was responsible, or 'some lowly, fast-growing form of lichenous growth.' Fascinating! Once again, I think it is not hard to see where Patrick's own belief in lunar change was fuelled during the 1950s! Later that same afternoon Mr P.M. Ryves, who often gave an entertaining account of the night sky at the BAA meetings, asked if anyone had seen the bright meteor, reported by the press a few days earlier? Patrick's rather surprising reply was: "Yes, I did, but I made no note of it. It appeared to be almost as bright as Venus." Quite why such a meticulous observer as Patrick would not make a note of such an unusually bright meteor is a mystery to me.

Between 1951 and 1956, while still a teacher at Holmewood in the earlier years, Patrick and H.P. Wilkins made numerous weekend trips to Paris to use the massive 33 inch Meudon refractor (32.7 inch, to be precise) which had far more light grasp (though, in hindsight, no more resolving power) than either of their telescopes. Some of Patrick's log book entries for these periods at Meudon are highly entertaining, such as the following entry:

> April 24th 1953. Observed with Mme [Mademoiselle] Hermann. Odd things went on but we eventually persuaded the dome to move the right way, but periodically motors whirred, mysterious lights flashed on and off, the platform creaked alarmingly, and on one occasion there was a small fire in the telescope pillar due to a short circuit. These things happen! Naturally the driving clock went on strike, but we managed.

Perhaps their most successful trip was in the Easter holidays in late March and early April 1952, when they made a number of detailed sketches of the lunar surface, with the Meudon refractor, under good conditions. During their week long stay they enjoyed three clear nights. They were assisted by a Monsieur Bertaud who was on hand to control the dome and drive of this most user-unfriendly monster! The telescope had a length of some 52 feet! Patrick and Wilkins made sketches of the craters Aristarchus, Birt (and the 'Straight Wall'), Cassini A, Conon, Guericke, Kraft, Marius, Plato, Riccioli, Sirsalis and Wargentin during the clear nights. On the 3rd of April 1952, at Meudon, Patrick and Percy Wilkins were observing the lunar crater Plato under excellent observing conditions, but there was no sign of the tiny craterlets on the crater floor. These were a small set of craterlets, the largest of which are rougly 2 kilometres in diameter. Four hours later the experienced observer Thomas Cragg was observing the same crater Plato, also under excellent conditions, but saw no sign of the craterlets either. Patrick was well aware of similar anomalies on the Moon, reported over many years, but had never seen such a good example. In later years he would become devoted to the dubious study of what he called 'Transient Lunar Phenomena', a term which he invented and became shortened to TLP. These days, most observers would attribute the disappearance of Plato's craterlets to simply be a result of the Earth's turbulent air or 'bad seeing'. But Patrick was convinced there was a more mysterious and exciting explanation.

On the same night (April 3rd 1952) as the 'disappearance' of the Plato craterlets Patrick and Percy Wilkins also observed some features within the crater known as Cassini A. Both men had a liking for somewhat humorous and even romantic descriptions of some of the features they observed and they likened the crater Cassini A to a washbowl. To quote from the 1961 second edition of their book 'The Moon': [we discovered] 'a white, very shallow crater within which is a minute central pit, like a plughole, the whole strongly resembling a washbowl'. In the 1960s, NASA Orbiter pictures showed the feature in its true light, namely a double crater, which our intrepid duo had not appreciated in 1952. Wilkins and Patrick described their 1952 Meudon adventure to a fascinated audience at the BAA London meeting of April 28th, a few weeks after they had returned.

One of the main reasons for Patrick and Percy Wilkins' trips to Meudon was to disprove a certain Ralph Baldwin's theories that the craters were caused by asteroid impacts; we shall return to this theme later. The central peaks seen in some of the craters would be expected to have tiny summit pits or craterlets if they were volcanic. The chance of them having tiny summit craterlets due to impacts was very low indeed, unless small asteroids kept hitting central peaks. An oft used term that Patrick would bellow during lectures, to describe this unlikely event in later years, was: "The bomb doesn't fall into the bucket every time!" This would always get a laugh from a sympathetic audience. Wilkins, assisted by the young Patrick, claimed that they had discovered over 50 summit pits on central peaks while observing from Meudon, and he used this to bolster the case for a volcanic Moon; in other words a Moon whose craters were formed by volcanic activity, not meteoritic or asteroidal impact activity. Later observations showed that many of these claimed summit pit objects, less than 1 mile in diameter, were merely tricks of the eye. Wilkins was keen on having many of his and Patrick's observations (sometimes referred to as discoveries) officially named by the IAU (International Astronomical Union).

Wilkins would propose a crater name, but in no instance was that name agreed to! Indeed, Wilkins, in the second (1961) edition of his and Patrick's book 'The Moon' named one crater as 'Moore' after Patrick! To quote from Wilkins, his citation for crater 'Moore' read as follows:

> The crater "formerly" known as Agatharchides A, lies to the north west of Hippalus on the most westerly of three great valley clefts, on the western side of the Mare Humorum, looking like a small Aristarchus, complete with two dusky bands. "Discovered" by Patrick in 1949 with a 6 inch refractor.

East and West are now the opposite way around on the Moon, but the crater Wilkins was referring to is, in modern co-ordinates, located at 28.3 degrees West, 23.3 degrees South; but, as it is 20 kilometres in diameter it certainly cannot have been 'discovered' by Patrick or anyone in the twentieth century. Twenty kilometres is a dead easy target! Agatharchides A is, of course, the same crater that Patrick had called Hippalus A when he suggested, in 1949, that vegetation was causing the bands he had spotted inside the crater.

Incidentally, some time ago Patrick was asked to clarify whether there was a crater 'Moore' on the Moon. He replied: "You either have to be dead or have been within 12 miles of the lunar surface to have a crater named after you – apart from nearly being dead, I don't apply!"

There seemed to have been a fame-seeking, romantic and science-fictional side to Wilkins, a disturbing desire to find extra-terrestrial activity on the Moon and (like Barker) to see glows and unexplained phenomena. This, undoubtedly, excited the young Patrick, who had various disturbing pseudo-scientific beliefs at the time and was keen to believe that unexplained events might be happening on the Moon and even Mars. Patrick's interest in science fiction and fantasy and the Flying Saucer reports of the late forties and early fifties would be something that he would desperately try to bury in later years, becoming incredibly edgy if anyone ever mentioned them; but he would always retain some unorthodox views about the Moon and the origin of its craters, right up to his late seventies. As I touched on earlier, another obsession of Percy Wilkins' was that of trying to get the IAU (International Astronomical Union) to accept *all* his suggestions for crater names on the Moon, not just getting one crater named 'Moore'. As far as Wilkins was concerned if he named a feature on his 300 inch hand-drawn map the IAU should ratify it, without question, and it should appear on *all* lunar maps. In 1948 he proposed 22 new crater names and these requests were turned down flat by the IAU who stated: 'The formations to which he wishes to give names are, in general, small or observable near the limb, in mediocre conditions. Most of them are already designated by letters, which were adopted by the IAU as definitive'. In 1952, and in 1955, Wilkins would try again. This time the IAU adopted a formal resolution to block Wilkins and others: 'The commission recommends to the Union that at the present time – and particularly pending the completion of the proposed photographic map of the Moon – no official recognition shall be given to additional lunar nomenclature'.

Let it not be thought though that Patrick only observed the Moon. It was his top priority in the 1950s but he observed everything in the night sky. On April 28th 1952, while observing the planet Mars with his 12½ inch reflector 'Oscar' at East

Grinstead, Patrick was able to glimpse its tiny satellite Phobos. This was an impressive observation, so close to the bright planet. Indeed, when Patrick announced this at the April BAA meeting, 2 days later, his elderly friend the veteran observer Robert Barker said that he had failed to see Phobos in 40 years of observing. This spurred Patrick and Percy Wilkins on to see the other Martian satellite, Deimos; a feat they both claimed to have achieved, with Wilkins 15¼ inch reflector, on May 17th. As always, Patrick was a sheer powerhouse of energy and enthusiasm for astronomical observation in the early 1950s, years before he became a TV presenter and while still a full-time schoolteacher.

Patrick never shied away from the donkey work involved in preparing material for BAA Observing Sections either. For example, in the October 1952 BAA Journal, on page 282, a Jupiter Section report compiled by the director A.F.O'D. Alexander states: 'It is a pleasure to present this excellent short report, the completion of which so soon after the apparition is entirely due to the zeal, industry and experience of W.E. Fox, who, moreover, shares with P.A. Moore and E.J. Reese the credit of providing the bulk of the observational data….' An additional section to the report, compiled by Patrick, covered the observation of the Galilean satellites by Jupiter Section members, although only F.M. Bateson, M.B.B. Heath, F. Perry, E.J. Reese (in the USA) and Patrick himself observed them. Again, Dr Alexander praised the 29 year old Patrick on page 284 of that October 1952 BAA Journal: 'P.A. Moore, with special interest and experience in this branch of the work, has done a real service to the Section in preparing such an interesting report (which I am pleased to present) on the brightness, colour, surface markings and shadows of the Galilean satellites…. this report should encourage more attention being paid to Jupiter's moons.'

As well as observing, teaching, and giving talks at BAA meetings, Patrick was becoming in demand as an astronomy speaker at local functions by the early 1950s; and as the proud owner of a motorbike the travel involved was not a problem. Patrick had a small collection of photographic glass slides (known as lantern slides in that era) which he invariably took with him. Thus, in November 1952, he gave a well publicised talk entitled 'The Stars' at the Fordcombe Village Hall, near Tunbridge Wells, just 10 miles from East Grinstead. The audience on that occasion was the local Women's Institute. While the talk went well the female audience would, perhaps, have been less than impressed by Patrick's attempts at survival some weeks earlier, when his mother had taken a short break. Left on his own, with notes on how to cook the food left for him, he had put a tin can of tomato soup directly above the gas flame, while nipping out to the observatory. The resulting explosion sounded like "A Soviet A-Bomb test" and when Patrick inspected the mess to the kitchen he was horrified. A frantic phone call to his teaching colleague Sandy Helm enabled a re-decoration of the *Glencathara* kitchen to be completed, just before his mother returned. Phew!

Shortly after returning from one of his 1952 trips to Paris, Patrick was travelling back from Wilkins' observatory one night when his motorcycle wheel caught in a rut, catapulting him off the motorbike and into the road. He was badly bruised, semi-concussed and suffered a serious gash above his left eyebrow (the non-monocle eye). Patrick managed to remount his motorbike and continued his journey home to mother, antiseptic, bandages, and sympathy. The bad gash above his eyebrow needed numerous stitches and the resulting scar would be a notable feature for the rest of his life. On a number of occasions Patrick would frequently attribute the wound to a "wartime bombing injury, surrounded by a sea of flak....only just limped home". Then he would admit the truth! However, on some occasions, after a few whiskys, the story was subtly modified: "Crashed the damn motorbike trying to avoid a camel that ran into the moonlit road. Ugly brute had escaped from the local circus!" Patrick could never be accused of being a boring storyteller! Twenty years later he would reprimand the impressionist Mike Yarwood for not having quite got that eyebrow and scar right when doing his Patrick Moore impersonation.

Chapter 4
The First Books and O'Neill's Bridge

Although Patrick was reasonably happy at Holmewood school he still desired to earn a living from what he loved, namely by writing about astronomy or science fiction. Then, in 1952, he had a lucky break. Patrick was already known as a keen speaker at BAA meetings. He was also (since 1948) a fully paid up member of the British Interplanetary Society (BIS) too; that society was interested in promoting space travel which was, quite obviously, a distinctly science fiction concept at the time. Arthur C. Clarke was a big name in the BIS and had become the Chairman after the end of the War. Clarke was 5 years and 3 months older than Patrick and was as well known in BIS and science fiction circles (writing short stories) as Patrick had become in the BAA. Clarke had actually joined the BAA as well, on June 27th 1945 (Patrick's mothers 59th birthday), and so had witnessed Patrick speaking in his enthusiastic manner about the Moon at many BAA London meetings. Patrick was always mainly an astronomer, whereas Clarke was a space travel and science fiction enthusiast. Nevertheless, Patrick had recently applied to join the BIS council and would serve on it from 1952 to 1958. Arthur invited Patrick to give one of his BAA style lunar talks at a BIS meeting in 1951. Patrick agreed, and his talk, entitled 'Guide to the Moon', was also heard by an agent for a New York newspaper, sitting in that BIS audience. The agent phoned Patrick's mother (while Patrick was teaching at Holmewood) asking her if he could quote Patrick in the New York Press. Fortunately Patrick's mother said yes and Patrick's talk received much publicity. Crucially, Eric Swenson, the head of the New York publishers W.W. Norton, saw the newspaper article. Swenson had been looking for an author to write a book about the Moon and, after checking Patrick's credentials, and finding he was the secretary of the BAA Lunar section, he started some wheels in motion. As a result, some months later, in 1952, Patrick was approached by Robin Warren Fisher of the publishers Eyre and Spottiswoode (the UK associates of those New York publishers W.W. Norton) to write a major book entitled *Guide to the Moon*. Fisher thought a friendlier version of Goodacre's 1930 book could probably attract decent sales.

Patrick was surprised and delighted and, after a friendly chat in Robin Warren Fisher's London office, he clinched the commission. The book was completed in

M. Mobberley, *It Came From Outer Space Wearing an RAF Blazer!: A Fan's Biography of Sir Patrick Moore*, DOI 10.1007/978-3-319-00609-3_4,

good time and eventually went into print the next year. At last, he was earning serious money from writing about astronomy. In fact, in later years, when Patrick was asked which book was his first, he would not cite *Guide to the Moon*, or even his earliest foray into science fiction; he said that his first book was actually Gérard De Vaucouleurs *Physique de la Planète Mars*, published by Albin Michel in 1951 and translated from French to English by Patrick. He claimed that he performed the English translation 'around 1949', making it his first 'book'. But regardless of all this, Arthur C. Clarke's invitation to speak at a BIS meeting had eventually led to Patrick clinching his first astronomical book deal, with Patrick as the author, not just as a translator. So, you might ask, when did Patrick A. Moore first meet Arthur C. Clarke? Well, probably about 15 years earlier, but like many of Patrick's recollections from the 1930s to 1940s you got a different answer every time, and 'meet' often means that Patrick was simply, probably, in the same room as the person. As I mentioned earlier, Arthur was more than 5 years Patrick's senior (d.o.b. December 16th 1917 compared to March 4th 1923) and Arthur had moved to London in the summer of 1936 from which point he started attending BIS meetings. Arthur had joined the BIS in mid-1934 (aged 16), the same year that Patrick joined the BAA (aged 11). Patrick used to claim he first encountered Clarke when he was 13 or 14, so Clarke would have been about 19. If that is so it would have been just after Clarke moved from Somerset to London, and would have taken place at a London BAA or BIS meeting.

I should add that, in keeping with Patrick's tendency to exaggerate dates and how very young he was when he did something, as soon as Arthur died, in 2008, Patrick started saying he was only 12 when he first met Clarke. However, his original claim that he was 13 or 14 is far more likely, as Clarke was then living in London. It also ties in more closely with Clarke's own recollections that he was probably at least 19 or 20 when he first attended a meeting where Patrick was also in attendance. One of Patrick's favourite stories to emerge whenever the subject of Arthur was raised was that on the day of one early 1950s London meeting (BAA or BIS Patrick couldn't recall) he went with Arthur to the Science Museum in London. Patrick claimed that Arthur had received some publicity about his feeling that maybe British and Soviet rocket designers should collaborate. So, as Soviet nuclear tests were taking place in that era, and Arthur was recognised as a BIS member at the Science Museum entrance, a member of the staff insisted on checking Arthur's briefcase in case it contained a nuclear bomb!

Already a good writer of papers and an enthusiastic speaker in amateur astronomy circles Patrick decided to turn his hand to serious writing, big time, with a view to giving up teaching. Patrick's major output in the 1950s, undoubtedly influenced by the BIS/Arthur C. Clarke connection, would be science fiction, but he would increasingly turn back to astronomy after he became the BBC *Sky at Night* presenter in 1957. His first published book, which actually just beat *Guide to the Moon* off the printing press, was a jolly sci-fi jaunt entitled *Master of the Moon* which was published in late 1952. It was no sci-fi classic, but the very fact that Patrick could summon up the willpower to write readable astronomy and science-fiction books must have been a great confidence booster. *Master of the Moon* was Patrick's second submission to a publisher called *Museum Press*, following an initial short story submission which, while it was not rejected, was never published either. Death rays, a megalomaniac, a hollow Moon and gigantic spiders all featured in the somewhat schoolboy level work of fiction which would become Patrick's first book.

Patrick's success as an author was of great interest to his colleagues at Holmewood House and, during 1952, the Holmewoodian school magazine reported: 'Mr P. A. Moore is to be congratulated on his election to the Council of the Inter-Planetary Society, and on the fact that his serious work on the Moon is to be published shortly in London and New York simultaneously. His exciting novel of adventure has been appreciated by many boys between 9 and 90, and another novel has been accepted for publication in the spring. Judging from his increasing 'fan mail' his work is receiving international recognition.'

Arthur C. Clarke had his first book length novel published at the end of 1951 as well; it was entitled *The Sands of Mars*. On January 10th 1952 the BBC's voice of astronomy and prominent BAA member Dr J.G. Porter had invited Arthur to take part in a programme on the BBC Radio Light Programme entitled 'Space Travel'. After discussing recent science fiction films Dr Porter and four science fiction enthusiasts praised Arthur's new book very highly; a fact which, surely, cannot have escaped Patrick's attention. Arthur had also been a prolific writer of S.F. short stories prior to 1952. Undoubtedly, Patrick would be inspired by the literary successes of his friend Arthur in the coming years. Patrick had far less story-telling ability (he had no patience developing the characters or the dialogue) but far more science popularising ability. Arthur C. Clarke was not the only science-fiction writing pal of Patrick's in the 1950s either. His BAA friend Henry Brinton (who he would move to within a stone's throw of in the village of Selsey, in 1968) churned out quite a few works of science fiction himself, including: *Death to Windward* (1954), *One down and two to slay* (1954), *Drug on the market* (1957) and *Purple 6* (1962) as well as a couple of amateur astronomy works and several collaborative efforts with Patrick. One of these future collaborations would be on the subject of Guy Fawkes, not exactly a man associated with astronomy! However, once Patrick became famous Henry knew that even a 50 % share in a Patrick Moore book would net more income than a dozen solo Brinton titles! Born in Wolverhampton in 1901, Brinton had started writing in 1951 and his friendship with Patrick meant he would eventually write a television play and appear on quite a few children's TV science programmes too. Patrick always managed to push a lot of good work in the direction of his most loyal friends. Incidentally, in later life one of Patrick's favourite expressions was: "Guy Fawkes, the one man we need in Parliament today!"

Just before Patrick's first non-fiction book *Guide to the Moon* was started, the former BAA President (1936–1938), the Reverend Dr Martin Davidson, had suggested that the BAA Section Directors should create a popular astronomy book with the proposed title *Astronomy for Everyone* and this must also have encouraged Patrick to think about the prospect of writing books full-time himself. Suddenly there were lots of author opportunities opening up. In addition, since October 1951, Patrick had been elected to the BAA Council, so suddenly he was serving on the councils of the BAA and the BIS, aged only 28. In an era when the BAA was dominated by elderly greybeards from the late Victorian era this was quite an achievement, although not unique. Perhaps surprisingly, given his science fiction and rocketry leanings, Arthur C. Clarke had preceded Patrick onto the BAA council by 2 years and served from October 1949 to October 1952 (so Arthur missed serving on the same council as Will Hay by just a few years). The BAA Council for the October 1951 to October 1952 session must have been a fascinating one, with the newly elected Patrick sitting around the table with many future BAA colleagues, such as Colin Ronan, Horace Dall, Dai Arthur, Neville Goodman and Arthur C. Clarke. If only a film of one of those council meetings existed!

On the subject of astronomical book opportunities, in 1952 the former BAA President and BBC 'voice of astronomy', Dr Porter, was just completing a major book on comets and meteor streams; this must surely have been an additional inspiration to Patrick. With all these writing opportunities opening up, and a small income arriving from authoring, Patrick would eventually resign from his Holmewood teaching post at Christmas in 1953. He would never look back. In truth, it was only Patrick's loyalty to the headmaster John Collings (who was not a well man due to his wartime injuries) and the rest of the close-knit Holmewood staff that had retained him in teaching for so long. At last, with the school now running smoothly, he would not be letting his teaching colleagues down by leaving.

As soon as *Guide to the Moon* was completed Patrick set about several other books. I would, at this juncture, like to remind the reader that PC's and *Microsoft Word* did not exist in the 1950s! Neither did Tippex correction fluid or electric typewriters. Every sentence typed in those days had to be correct as regards spelling and grammar. An error meant the page was torn up. Also, as anyone who has used Patrick's typewriters will confirm, those keys needed pounding, hard! I am not surprised that his fingers were the size of bananas and riddled with arthritis in later life. In addition, many modern authors, if stuck for data, simply '*Google*' a query and use *Wikipedia* to cut-and-paste data into the page. In the 1950s writing a book was clearly an order of magnitude more painful than in the twenty-first century, although it was rather more lucrative too. As soon as something becomes easy everyone jumps on the bandwagon.

Jumping ahead a couple of years, in 1954 Patrick would manage to churn out another couple of jolly sci-fi novels entitled *The Frozen Planet* and *The Island of Fear*. Childish though they appear today, young sci-fi addicts of the 1950s did buy them and it was useful income. Patrick's next serious astronomy book was his *Guide to the Planets*. The book which would help make him famous (by sheer luck) was 2 years in the future and would be entitled *Suns, Myths and Men*. This was a small book about the history, myths and legends of astronomy, but it would play a crucial role in Patrick's future in a few years time. The book was inspired by Patrick's amateur astronomer friend J.C. Reynolds and illustrated by a talented young artist called David Hardy who would help with many of Patrick's books in the coming years, as well as providing illustrations for the *Sky at Night* programme. David was introduced to Patrick by a mutual friend, Mike Guest. David just managed to complete the eight vital illustrations for Patrick in the 5 days before starting his compulsory National Service. If not for that fortunate timing the whole course of history might have been changed, because the first *Sky at Night* producer was suitably impressed with Patrick's book *Suns Myths and Men* and those Hardy illustrations enhanced it greatly. David would ultimately appear as a guest on the *Sky at Night* in February 2003, 49 years after illustrating the book!

Bizarrely, after the 1970s, Patrick shunned David Hardy completely, believing he had pinched a couple of book contracts from under his nose. At Athens airport, in April 2006, David told me how he had confronted Patrick with his own version of events in the 1990s. It was, in fact, Patrick who had encouraged David to write those books, but Patrick had forgotten. On realising this was true, in an instant Patrick resumed his friendship with David as if nothing had happened! They then brought out a joint book, *Futures: 50 years in Space*, to celebrate their 50 years partnership! Patrick's life was littered with him distrusting other authors for having 'stolen his ideas'.

However, also in 1954, Patrick would take a huge gamble with another book, written under a very dodgy pseudonym. He would deny having written that book for the rest of his life. But more about that in a few pages time, because so much happened in Patrick's life around 1953, the year he decided to become a full time author, it is perhaps best to resume the story at that point, or slightly earlier.

In 1952, with Patrick at the dawn of his writing career, his friend Eric Turner had written an article in the June BAA Journal entitled *Astronomy for Britain's youth.* Turner was concerned that young people were simply not being encouraged and trained to take up serious observing within the BAA. The average BAA member was 50 years old. Turner, with help from Ernest Noon (an expert on the BAA's Lantern Slide collection), Patrick, and Richard Baum, launched the Junior Astronomical Society on March 21st 1953. The first, somewhat informal meeting, was held on the steps of the Royal Astronomical Society! The JAS, under the name SPA (Society for Popular Astronomy) survives to this day and Patrick was in at the start, as he frequently was with so many good astronomical ideas.

A few weeks after the JAS was launched, specifically, between April 17th and April 25th 1953, H.P. Wilkins and Patrick spent a further week at the Meudon Observatory in Paris and enjoyed a remarkable seven successive clear nights. Wilkins sketched 63 lunar formations and Patrick sketched 59; an impressive body of work for 1 week with the 33-inch refractor. The lunar regions that Wilkins concentrated most upon were Torricelli, Atlas, Posidonius, Archimedes, Pitatus, Galle, Reinhold, Sinus Aesteum and Gassendi. Patrick's main efforts were directed at Gartner, Tobias Mayer, Aliacensis, Grove, the Straight Wall, Linné, and the planet Saturn. Patrick's contemporary Audouin Dollfus (1924–2010) observed with them on a few occasions. In many ways Dollfus would become the French equivalent of Patrick Moore, as a Gallic astronomy populariser, in the decades ahead. Patrick was able to record a total of 92 objects (craterlets and pits) in the region of the Straight Wall during that April 1953 trip, which surpassed any previous number observed in that area. Patrick always enjoyed telling a story about a mistake he once made, while converting a rough sketch to a final ink drawing, in that era. It was late at night and he was being sustained by a cup of strong black coffee. Unfortunately, he mistook the Indian Ink bottle for his coffee! "Tasted very similar" he quipped "but a lot harder to remove the stains from my dentures!"

At the May 27th 1953 BAA meeting Wilkins apologised for Patrick being unable to attend but he described their week's work in Paris to the members present. He added that all the features they had observed had already been added to his enormous (and, frankly, ludicrously cluttered) 300-inch diameter lunar map, and they would also be included in a forthcoming book. Wilkins presentation obviously impressed the BAA audience; Patrick and Percy had achieved much in terms of lunar sketches in that busy April week. However, that happy reception by the BAA members was the last major occasion that Wilkins would be highly praised at a meeting of the Association. Things would soon slide rapidly downhill in his relations with the BAA, as we shall see very shortly.

Now, returning to the subject of that meticulous observer and astronomical wordsmith Richard Baum, a co-founder of the JAS; well, the man himself told me some years ago of an interesting episode around this time. The episode in question involved Patrick's scepticism that anyone seeing 'banded' markings on Venusian cloud belts was really seeing genuine features in the planet's atmosphere. In early 1953 Patrick had been on a Venus observing high, having claimed to have seen the elusive and highly controversial 'Ashen light' (the glowing dark side) on no fewer than five separate occasions during the March evening apparition of that year, between the 16th and the 31st of that month. As few others had seen it, it was quite a claim and would establish him as a leading Venus observer, if the observations were believed. I should add that these days the feature is considered to be an illusion, possibly caused by a badly baffled telescope, scattering stray light from the brilliant crescent.

According to Richard, the young pre-BBC Patrick was booked to give a Venus talk to the Astronomical Section of the Chester Society of Natural Science, Literature and Art at the Grosvenor Museum, Chester, on August 24th 1953. After Patrick's rousing lecture a quite heated debate ensued about whether the banded atmospheric markings could be seen visually by observers. Patrick had never seen any such features, so as far as he was concerned they did NOT exist, especially as he had seen the 'Ashen light', which was regarded as a truly elusive sight. However, his dogmatic stance was not, in 1954, backed up by any long-established reputation. He was not famous in 1954, so people were apt to challenge Patrick's views. They were not challenging a British institution back then, just a tall young man in a badly fitting blazer, with a loud voice and unshakeable views.

Patrick banged on about Percival Lowell and the spurious streak-like markings reported by Lowell and others at the meeting. According to Richard Baum, Patrick 'damned them wholesale', simply because he, Patrick, had not seen these features. From the audience Colin Reid, a very quiet man but 'possessed of a razor sharp intellect' (Baum's comments), pulled Patrick up by mentioning the F. E. Ross ultraviolet (UV) photographs taken from Mount Wilson, in 1927. This development had Patrick rattled, as Colin obviously knew his facts. Those UV photographs showed banded and 'Y' shaped markings very similar to those sketched by some visual observers, but invisible to Patrick. Colin was a spectroscopist at the UK Atomic Energy plant at Capenhurst, Cheshire, and he challenged Patrick, as well as Richard Baum and John Hutchings, to have their UV sensitivity tested by himself, in a scientific experiment, in a darkened room the next day. Working at the UKAE plant, he knew where he could borrow a spectrograph and how it could be employed to really test these observers' eyesight. Baum was already sure he could clearly see banded features in the Venusian (sometimes called Cytherean) cloud belts through his telescope. The experiment was carried out to see if any atmospheric banding could be seen by observers with good UV sensitivity. Reid inserted some genuine Venus UV pictures from 1927, transmitted at UV wavelengths, to see who could see the markings. Numerous carefully thought out trials were carried out during the 45 minutes test, just before midday and before Patrick had to return to Sussex. The comments of the observers were carefully recorded. Reid decided he would announce his findings when Patrick was next in Chester, in April 1954 some 8 months later, by which time he would have properly calibrated and analysed the results.

The Bridge

One month before Patrick's talk at Chester, in late July 1953, the science editor of the *New York Herald Tribune*, John J. O'Neill, made a remarkable claim regarding a feature on the Moon. The claim would result in his surname going down in lunar observing history, for all the wrong reasons! On July 28th 1953 O'Neill observed a fan of light which appeared to be shining under a bridge of rock on the edge of the Mare Crisium (remember, the Mare Crisium had, supposedly, been the subject of the teenage Patrick's first scientific talk in 1937). If this bridge were real it would have to be some 12 miles in length! As far as the media were concerned H.P. Wilkins was the man who knew about the Moon and, as we have seen, he never wasted an opportunity to publicise his efforts, in Britain, or even in America.

A few months earlier, on March 2nd 1953, just 2 days before Patrick turned 30, the British Pathé News reel, shown in all the cinemas of that era, actually featured Percy Wilkins, assisted by his daughter Eileen, using his 15¼ inch Newtonian to map the Moon. The commentator, a 30 year old Eamonn Andrews, mistakenly called it an 18 inch telescope. In fact, Percy Wilkins had mentioned his imminent Pathé News reel appearance at the February 25th 1953 BAA meeting, at which Patrick had spent much time describing his own observations of the January 29th/30th total lunar eclipse, as had one of the BAA's youngest members, a young master B.R. Tanner, who had observed with his father. Patrick had, and I quote, kept "a close watch on certain formations such as Linné, which might be expected to show variations due to the wave of intense cold sweeping across the Moon's surface." Patrick also showed some lunar eclipse slides taken at Uppingham school at that meeting. Anyway, I've digressed a bit here, but it was nearer the end of that February meeting that Wilkins mentioned his Pathé News reel filming. What Wilkins actually said was this: "I have made my début – I believe that is the right word – as a film star. [Laughter] It is a 'short', being an item in Pathé Pictorial No. 442. What proportion of what was taken is going to be shown I do not know. In the film

I look like a gentleman of leisure. The film people brought an awful lot of apparatus, but it was all very grand. My daughter comes out with a chart and walks to the telescope through which I am observing. It was a cold night, and in place of the chart a cup of tea would have been more acceptable!" In answer to Robert Barker asking about make-up Wilkins replied "No they had to 'shoot' me as I was; but I am fully aware that a little make-up would make me look more handsome. [Laughter]" He also added, when the President E.H. Collinson asked if the BAA were mentioned, "I wrote it down for them. These film people have notoriously bad memories."

Now, at that time, and for the last 7 years, Patrick had been Wilkins' young apprentice, observing alongside him on many clear nights, and accompanying him on their trips to Paris. Yet Wilkins had not mentioned Patrick at all in the filming, or invited him along (I checked this with Patrick in 2007). Maybe he knew that the young and good looking Patrick would have stolen the show. In appearance Wilkins was so gaunt he looked ill, and almost skeletal. Now, film or TV publicity can be a very dangerous thing and it can do strange things to people's egos. For a brief period those unused to such attention can think they are famous and invincible and they can crave more and more attention. It's a dangerous drug. Anyway, largely by H.P. Wilkins deliberately attracting a lot of media publicity his face and telescope were now appearing on the silver screen in that year of 1953, not just in Britain, but in the USA as well.

So, it was hardly surprising that O' Neill wrote to H.P. Wilkins about the extraordinary 'bridge' feature; his letter arrived from New York on August 2nd or 3rd 1953, according to Wilkins. Observing from Bexleyheath in Kent, Wilkins waited until the next suitable illumination of the feature, on August 26th, and was able to confirm it with his 15¼ inch reflector. To quote Wilkins' comments from the Christmas BAA meeting of 1953/1954 in January 1954: "Well, there it was! At least, there was the appearance of a bridge with the sunlight streaming under it and with the shadow of the arch cast on the surface of the plain". Other observers, in the BAA and the USA, confirmed that the feature *looked* like a giant archway, but explained that this was merely a trick of the light at a very specific angle of illumination. In a bizarre and totally surreal twist, O'Neill promptly died while Wilkins 'confirming' observation was being sent to him in the post!

Unfortunately, Patrick's publicity seeking mentor was then interviewed by the BBC's Bernard Forbes on the radio, about a range of dubious issues, including the subject of Flying Saucers. In the interview, which Wilkins claimed was subjected to very bad editing, he expressed the view that O'Neill's bridge was a real feature and might even have been made by aliens! Not surprisingly this aspect of the interview was latched onto by the press and it appeared in many American newspapers that were keen to stress the discovery had been made by a recently deceased New Yorker. In time-honoured tradition (the BAA is no different to any other pack of human beings) Wilkins' BAA colleagues lost no time in venting their spleens on him and kicking the man when he was down. Wilkins had an incurable romantic side to him which, coupled with his desire to be noticed, had briefly made him look very, very silly indeed. Rather than rallying to a friend who had made a serious error of judgement, most of his BAA friends started to desert him, although Patrick stayed loyal.

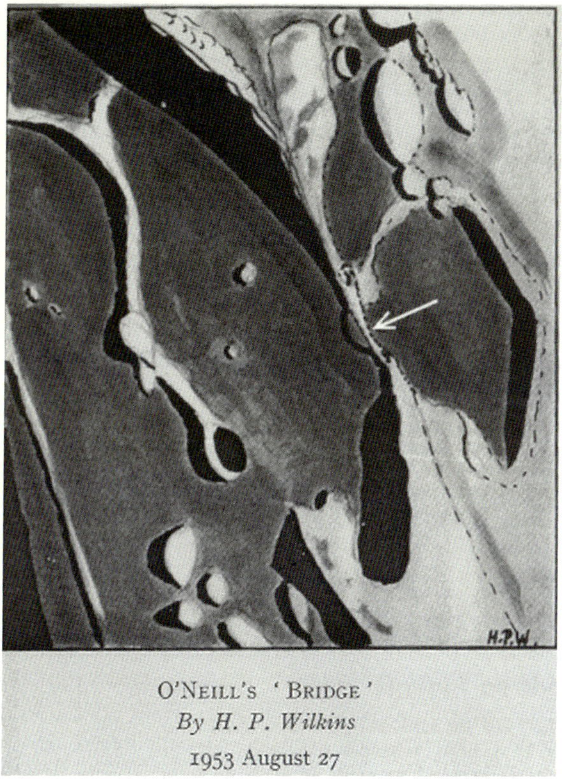

Fig. 4.1 O'Neill's Bridge sketched by H.P. Wilkins on 1953 August 27th with his 15¼ inch reflector, as it appeared in the 1954 February BAA Journal (page 119). Note the clear indication of a long bridge with a shadow. Wilkins showed this sketch at the BAA's 1953/54 BAA Xmas meeting, amid much skepticism. South is up in all these sketches

This was a difficult position for Patrick to be placed in. His friend and mentor was looking ridiculous, but he was still his friend and mentor and surely you stick by your best mates when times are hard, rather than squealing with glee that a friend has made an error that you can pounce on? Undoubtedly Wilkins craved publicity and just loved the media attention, but he had been an enthusiastic Lunar Section Director too, giving many talks at BAA meetings and he had written up many papers for the BAA Journal. He was also a skilled lunar observer, even if his drawings were ludicrously cluttered. However, the film and radio publicity generated about this alleged alien bridge on the Moon had, for most in the BAA, meant that Wilkins had crossed the line. He was looking like a publicity mad laughing stock, and only Patrick was staying loyal.

Patrick was also getting significant publicity himself by late 1953, because people liked his book *Guide to the Moon*, priced at 16 shillings, and it was even being mentioned in national newspapers read by millions every day. For example, in the *Daily Express* for Saturday October 10th 1953 the columnist Bernard Wicksteed devoted a quarter of a page to Patrick's book. The section was headlined 'PLEASE BE careful when you smoke on the Moon', which was a reference to Patrick's coverage of possible mists that appear in craters. Wicksteed was implying that if the mists were methane vapour a cigarette might cause an explosion! The newspaper article started with: 'Well, what times we live in! Here's a chap who has brought out a guide book, for travellers to the Moon. He certainly meant to get in first'. In conclusion Wicksteed says: 'By the time people really go to the Moon, Mr Moore's guide may be a collector's piece, but it is fascinating reading now'. It was this glowing review, in a newspaper read by millions, that finally convinced Patrick that he could safely resign from his Holmewood House teaching post a couple of months later.

Strangely, the school's Holmewoodian magazine for the last quarter of 1953 seemed to think that Patrick's reason for leaving was due to his own poor health, whereas Patrick's story was that the school was running smoothly and he could survive as a writer. Maybe it was six of one and half a dozen of the other? Anyway, the magazine reported: 'Parents and boys alike will be sorry to hear that Mr Moore will have to leave Holmewood owing to ill-health. He will, however, keep closely in touch with the school and all will hope that he will soon be well again. Mr Moore will be able to devote more time to writing, and it is pleasing to record that his book 'Guide to the Moon' is a great success, both in Britain and America, and has already run into three editions. We wish him all success in his future work.' Patrick honoured his promise to keep in contact with the school, including when he became famous. In July 1971 he filmed a *Sky at Night* there and in the summer of 1982 he took part in a celebrity Holmewood Academic Challenge, along with Jan Leeming, Sandy Gall, and Simon May. Three years after that he would present the prizes on Speech Day, and in 1990 he opened the new Science and Pottery labs at Holmewood.

More Wilkins, More Bridge, More Venus

According to many of Patrick's contacts from the 1950s, especially those living in the USA, Patrick's loyalty to Wilkins, in 1953 at least, went deeper than just their BAA lunar partnership, which had existed since the end of the War, and their regular correspondence, which had started around 1940. Patrick had become a close friend of the family and was infatuated with Wilkins' attractive 19 year old daughter Eileen, even though, at 30, Patrick was almost 11 years her senior (March 4th 1923 vs. January 25th 1934). According to the late Charles (Chick) Capen (1926–1986) of the USA's Lowell Observatory and to the late Tom Cave (1923–2003), a famous US telescope maker and lifelong friend of Patrick's, Patrick was 'madly in love' with Eileen Wilkins. Even in the 1980s, Chick recalled how the very mention of Eileen Wilkins caused Patrick to instantly burst into tears! Whether the wartime girlfriend Lorna ever existed or not, it was Eileen Wilkins in the 1950s who Patrick really had deep emotions for, and Percy Wilkins was her father. [Patrick once told me that he and his friends were astonished around this time when their 'confirmed bachelor' friend Arthur C. Clarke wrote to them all from a business and scuba-diving trip in the USA, in June 1953, saying that he had suddenly met, and was going to marry, a beautiful young Florida woman who had a 2 years old son. Whether this had any bearing on Patrick's emotions at this time I simply don't know, but the timing is interesting.]

At that BAA 1953/1954 Christmas meeting, in early January 1954, Wilkins attempted to explain his controversial O' Neill's Bridge views to a highly amused audience and cited Patrick's observations thus:

> About a month afterwards [after August 26th], I mentioned the bridge to Mr Moore, who said that he also had seen it with his 12 inch telescope. I now show on the screen the drawings made by myself, confirmed by Mr Moore, and that made at Hampstead.

Wilkins concluded his talk at the meeting with the line:

> Somebody rang me up and said "I know what the bridge is. It is a Flying Saucer that has crashed and lodged across the two mountains. So now we know everything."

In fact, I have studied Patrick's *Seventh Lunar Notebook* of that era in detail and there are a number of observations of the region, all somewhat sceptical and certainly not supporting Wilkins' view. Here is Patrick's observational record, from page 26, for the early hours of Christmas Eve 1953, just before the Christmas meeting where Wilkins stated that Patrick had seen O'Neill's Bridge:

> December 24th 0h – 0h 30. Conditions atrocious; mist, drifting cloud. However, Oscar performed extremely well. [Oscar being Patrick's 12.5 inch Newtonian – M.M.]
>
> O'Neill's Bridge. An interesting rock-formation, though not particularly significant or important. The ridge rises steadily, S.W. of Proclus, and there is a gap of, perhaps, 1½ miles where it is unsupported, forming a natural arch; but doubtless it is only one of many such formations, previously undetected owing to their small size. So far as my own view goes, it might not exist at all; but as HPW is certain about it, I suppose it does!

An accompanying sketch by Patrick, timed at 00h15m and with powers of 350× and 400×, is included in the log book. In passing it is worth mentioning that Patrick's big Newtonian 'Oscar' had been completely overhauled in 1953, culminating in the installation of a new improved mirror by Henry Wildey that month (December). In the Monthly Notices of the RAS Patrick mentioned that he planned erecting a dome for the 12.5 inch reflector in 1954, but it never happened.

Meanwhile, 7 weeks later, the 1954 February 13th edition of the UK magazine *Illustrated* made much of Wilkins' observations of the mythical bridge, with their artist providing a dramatic representation of the structure which would look at home in a 1950s science-fiction film. Their artwork sat alongside the crazy caption 'IN THE TELESCOPE, THE TWENTY- MILE MOON BRIDGE IS JUST A SPECK'.

The week after, on February 21st 1954, at 01h10m GMT, again using his 12½ inch reflector 'Oscar' at 350× magnification, Patrick made the following entries on page 36 of his seventh Lunar notebook:

> Studied for as long as possible using 350 on Oscar. The area was clear enough but I was unable to give confirmation of the form and size of the Bridge. The speck of light seen by O'Neill was there, but so far as I am concerned, it might just as well have been an ordinary peak. There seems to me to be a break in the ridge N. of Lavinium, but this may have been due to tricks of the light.

Again, a sketch was completed by Patrick. After a slight digression, a few lines later he also comments:

> The whole area is highly complex, and I wanted to follow the progress of the terminator right up to the Bridge, mapping the surrounds meanwhile. Unfortunately, altostratus cloud developed rapidly covering the Moon at 1.10 and by 1.30 it was quite clear that there was no hope of any further observing. So I still cannot say from personal observation that the Bridge exists. I am beginning to doubt whether even Oscar is powerful enough for me to make useful observations of it. We must hope for good weather at Cambridge in March.

This was a reference by Patrick about his plan to use the 25 inch Newall Refractor at Cambridge to solve the 'Bridge' enigma. Wilkins and Patrick had occasionally used the telescopes at Cambridge before (both the 25 inch Newall refractor and the 30 inch 'Steavenson' reflector). The Newall refractor was moved to Penteli Observatory near Athens, Greece, in 1957.

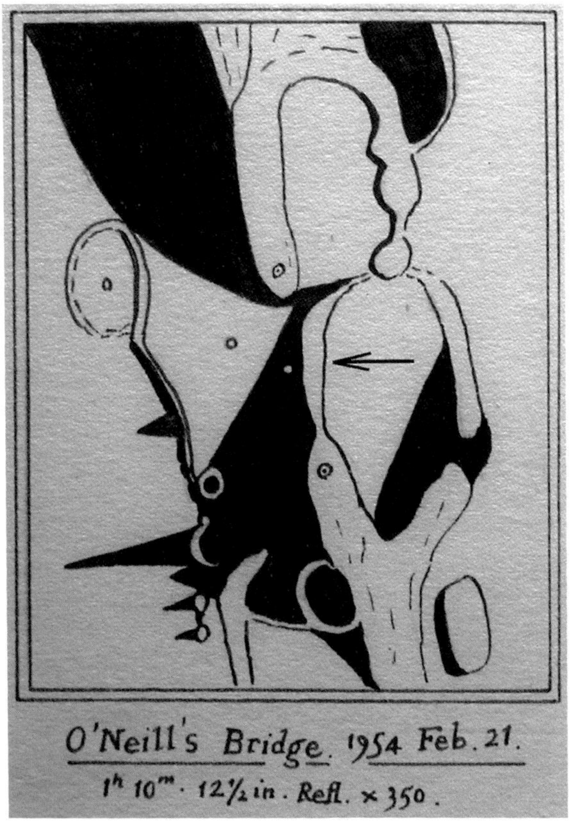

Fig. 4.2 Patrick's 1954 Feb 21st sketch of the O'Neill's Bridge region. *Arrow* added by MPM. (From BAA Lunar Section records)

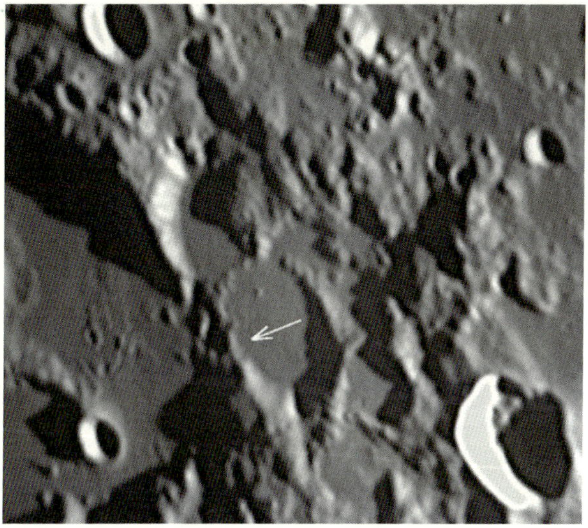

Fig. 4.3 A digital image of the controversial O'Neill's bridge region taken by the author, using a 250 millimetres f/6.3 Newtonian on 2005 September 21st at 02:29 UT. Lumenera LU075M camera plus Powermate, projected to f/50. This shows the disputed region without any visual observer bias! (Image: Martin Mobberley)

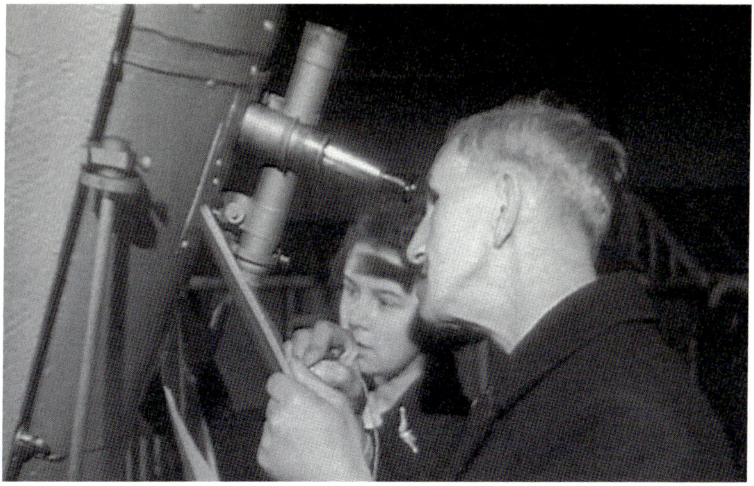

Fig. 4.4 H. Percy Wilkins and his daughter Eileen, observing the Moon with Wilkins' 15¼ inch Newtonian, as featured in a British Pathé cinema newsreel shown on March 2nd 1953 (Reproduced with permission from British Pathé)

It would appear that the Newall refractor was not able to be used, because the next O'Neill's Bridge entry in Patrick's log book is dated 20 hours GMT, 1954 April 6th, back with the 12½ inch 'Oscar' at 200×. Another sketch was made along with the following observations:

> At 20 hours I managed to use Oscar to make a sketch of the O'Neill's Bridge area. It looks most unbridgelike! I can see the main crater in Olivinium, which casts an arc of shadow. The crater in Lavinium is much less prominent, and under this lighting it looks very much like a peak or a mound. The shadow cast by Olivium is very distinctive. To the W. is the famous winding ridge and the low rounded mound-hill. No time to check properly as the trees intervened and the drawing had to be left in a somewhat unfinished condition. [Sent to David Grundy].

In the April 1954 BAA Journal, Wilkins committed his somewhat revised views to paper. He now stated that his own observations 'clearly show that, while there is an arch or 'bridge' at this point, it has not got a span of 12 miles but, as I have maintained since the discovery was announced, the actual gap, arch or bridge has an aperture of about 2 miles only'. The matter was still grumbling on by June 1954, when Patrick clinched his own first BBC Radio broadcast (in French!), but by then many members' thoughts were turning to something more exciting, involving Sun and Moon: a solar eclipse. Patrick only secured one more observation of the 'Bridge' region before then. Made with only his 3 inch refractor (called 'Egbert') at 120×, he declared the sketch 'virtually worthless due to the poor conditions'. As an aside, I once asked Patrick about his 1954 radio broadcast in French. His reply was typical of Patrick and probably true: "Technically, yes, I broadcast in French, but I understand that 99 % of the French speaking population on the other side of the Channel thought I was speaking German." In fact, his role in that broadcast was a very minor one, as the main BBC Foreign Service guest that day was the Astronomer Royal, Sir Harold Spencer Jones, who was talking about the Venice Observatory. The Astronomer Royal knew Patrick, because he was a former BAA President.

Meanwhile, what happened to those Venus UV sensitivity results conducted at Chester, by Colin Reid, shortly after the O'Neill saga originally started: namely, the scientific test of August 25th 1953? Well, 8 months later, on the evening of Wednesday, April 21st 1954, Patrick found himself back at Chester for a 3 day astronomy meeting. According to Richard Baum, on that evening at Colin's house, Patrick, Richard and Colin inspected Venus themselves through Reid's 9 inch refractor. Although observing conditions were poor, Reid and Baum were able to see 'a dusky Y shape', but Patrick was not. Reid then announced that the scientific experiment he had carried out in August had shown the same results. Basically, Patrick could not see UV markings in the scientifically controlled test, but the other three observers (Baum, Hutchings and Reid) could. Patrick had been conclusively proven wrong but he was still very loathe to accept this. As Richard Baum has pointed out to me in explaining all of this, Patrick did mention this experiment briefly in a footnote in his 1955 *Guide to the Planets* but stated:

> Possibly, therefore, the broader linear features result from misinterpretation of much vaguer streaks high in the Cytherean atmosphere – in which case only those observers whose eyes are very sensitive to short wavelengths can hope to glimpse them in the ordinary way. However, it will be necessary to make more accurate experiments before coming to any definite conclusions.

But all reference to that experiment was totally erased from any subsequent books! Patrick simply did NOT like being proved so comprehensively wrong, especially when he had been so adamant. It did appear that Patrick had rather poor violet sensitivity, something he did not seem keen to advertise. Nevertheless, such was Patrick's enthusiasm about Venus that he would be appointed as the BAA Mercury & Venus Section Director in one year's time. In years to come, Patrick would admit to me that he had "virtually no UV sensitivity, but great sensitivity at the red end… as confirmed by Henry Hatfield's tests…"

The First Totality

The total solar eclipse of 1954 June 30th was something that Patrick had been looking forward to for many years. In 1927, when Patrick was four and living in Bognor, the Moon's shadow had swept across a small strip of the UK from Criccieth (in North Wales) to Hartlepool on June 29th. The duration of totality was a mere 24 seconds! Even Patrick never claimed he travelled to that one!

Twenty-seven years later another total solar eclipse was passing conveniently close to the UK. The *next* total solar eclipse over the UK would not take place until August 11th 1999; but Patrick wanted to see many more total solar eclipses before he reached the advanced age of 76! Unfortunately, although total solar eclipses occur about every 15 months, on average, they rarely occur near the UK and the tracks often cross vast expanses of sea, desert, jungle, or simply inhospitable (and often cloudy) regions. However, the track of the 1954 June 30th eclipse was very favourable. The Moon's shadow was predicted to cross over the whole of the Faroe Islands and the northernmost Shetland Island of Unst. It would be almost 3 minutes long and high up in the sky too. However, the weather prospects were expected to be better on the west coast of Sweden, especially at the sites of Lysekil and Stromstad, and so these sites were chosen for the joint BAA/RAS expedition. Patrick was one of the first to enrol for the trip! He dearly hoped that Percy Wilkins and his wife and, especially, his pretty daughter Eileen (now 20), would be going on the trip, but it was not to be. Wilkin's O'Neill's bridge controversy had netted him a lucrative lecture tour in the USA, at the start of the eclipse track, and the Wilkins family were headed for America.

A party of 140 BAA and RAS members, friends and relatives set sail from Tilbury on the SS Patricia on June 24th and arrived in Gothenburg, Sweden on June 26th. The party stayed at a hotel in Lysekil but most observed from the reserved site at Stromstad, which was guarded by the local police in full colour uniform. Despite some thin cirrus cloud everyone enjoyed a good view of the eclipsed Sun, corona and prominences. The site at Stromstad was a patch of wasteland near to a building site, but it had good views of the harbour. It was from here that Patrick had a good, if not a perfect view, of his first total solar eclipse. He was ecstatic. Not only had the sea trip been an extremely civilised affair (as opposed to the cattle-truck airlines of today), he had seen the eclipse and been surrounded by his many BAA friends. Patrick took numerous pieces of equipment to the eclipse site. As well as binoculars, a thermometer, and a camera, he had also brought along an enormous roll of white paper, to lay along the ground and so make the ghostly 'shadow bands' more visible. Upon seeing Patrick's personal eclipse kit one eminent RAS astronomer started using his film to photograph the equipment. When Patrick queried this the astronomer commented, in a somewhat sarcastic and condescending tone, that: "I may possibly live to witness another total solar eclipse, but never will I see so peculiar an arrangement as this Mr Moore!" Patrick would see many more total solar eclipses over the next 44 years; the Caribbean eclipse of Feb 26th 1998 would be his last, but he would also sit under the Moon's shadow for the cloudy August 11th 1999 Cornwall eclipse.

Fig. 4.5 Patrick (background, *centre*) on 1954 June 30th at the Swedish total solar eclipse BAA expedition site. He is standing behind some *white* sheets to be used in shadow band experiments (Image from the BAA archives, taken by the late E.H. Collinson (1903–1990))

By the end of the Swedish eclipse trip the heated debate about O'Neill's bridge had rumbled on for over a year in the BAA. It might have died down but, as we have seen, H.P. Wilkins travelled to the USA in late June 1954, with his wife and daughter, both to see the solar eclipse from the American end of the track and to re-emphasise his views on the lunar bridge in a lecture tour. On the way to Atlanta, on June 11th, Wilkins added to his notoriety by reporting a UFO sighting to the US press. He claimed that he had seen two egg-shaped objects, like polished metal dinner plates, reflecting the sunlight. They were later joined by a third object and all were about 50 feet wide! There was no doubt that Patrick's mentor was cashing in on UFOs in a big way. Wilkins boarded a Northwest Airlines 'stratocruiser' and was able to see the eclipse from 20,000 feet over Minneapolis. The eclipse was of 76 seconds duration from the USA, and he had an excellent view. From there he was ultimately bound for the Astronomical League's national meeting in Madison, Wisconsin. On his way there he used the 60 inch reflector at Mount Wilson in California to study the Mare Crisium again and decided to re-emphasise that O'Neill's bridge did exist, but was only 1.5 miles long, certainly not the 12 miles that had originally been claimed! This may have been an attempt at a halfway solution to salvage his reputation. As the furore continued the BAA challenged Wilkins to present his evidence and, ultimately, Wilkins put his case in a paper delivered on the last Wednesday of November 1954, the traditional post-AGM date of the BAA meetings in the RAS meeting rooms at Burlington House, Piccadilly. This put Patrick in a tricky position as he had certainly seen the O'Neill's bridge feature and would have liked it to be real and even artificial, but he knew it was much more likely to be a trick of the light.

At the November BAA meeting Wilkins put his evidence to the assembled audience. His views were strongly challenged by a number of highly regarded observers including Frank Maloney, Keith Abineri, Harry Thornton and Dr W.H. Steavenson. As before, Wilkins stated that he had first seen the bridge on August 26th 1953, using his 15 inch reflector at Bexleyheath with magnifications of up to 300×. He also stated that, about a month later, he had mentioned the bridge to Patrick who confirmed he had seen the feature with his 12½ inch reflector. There was much discussion about two craters in the same region, whose appearance, at the limit of observation, might have a bearing on what was being seen. Patrick never publicly backed Wilkins view at the meeting, but he showed a sketch of his own, made that month (November 1954), with his 12½ inch reflector. Patrick's only recorded comments were: "I am inclined to disagree with everybody! My own drawing was made in November with my 12½ inch reflector and shows the two craters as complete rings, whereas Mr Lenham and others make them incomplete." [As an aside, I might add that Patrick made an amusing quip about his 12½ inch reflector at the following BAA meeting, on 1954 December 29, namely: "I have been using an altazimuth stand for some years, which accounts for my greying hair and haggard appearance. Some observers have done splendid work with altazimuth instruments, but an equatorial makes things much easier and should always be incorporated if possible."]

Other reputable observers of the time, namely Elliott, Goss and Sharpe at Hampstead Observatory, confirmed the existence of the bridge even *after* that stormy November 1954 BAA meeting! In later years Patrick would say that he had seen the feature, but had made no comment as to its reality as a bridge, artificial or otherwise. Nevertheless, he stressed at the time that Wilkins was one of the most skilled lunar observers and should not be ridiculed; he was entitled to his views. Unfortunately, despite the Hampstead 'confirmations' the prominent BAA observer and past-president Steavenson ridiculed Wilkins and the baying pack of critics could smell the publicity hungry Wilkins' blood. Much to Patrick's dismay, Wilkins, the last of the Moon mappers and the BAA Lunar Section Director, would never be taken seriously again. This is hardly surprising when you take his Flying Saucer claims into account as well. All this was doubly embarrassing as Patrick had collaborated with Wilkins on a book (published by Faber and Faber) entitled: *The Moon – a complete description of the surface of the Moon, by H. Percy Wilkins and Patrick Moore*. It was published in 1955. After a further year of ridicule and sarcasm Wilkins resigned from the Lunar Section and the BAA in 1956, although he remained a close friend of Patrick's. Indeed, they even collaborated on another book together, published by Eyre & Spottiswoode in 1956 and entitled: *Making and using a telescope; the home assembly and applications of astronomical equipment.* Fortunately there was no mention of Flying Saucers in that book (but wait until we get to Chap. 5 and Patrick's alter ego...). Possibly the last straw for Wilkins was Dr David Dewhirst's review of Wilkins' book *Mysteries of Space and Time* in volume 66 of the 1956 *BAA Journal* (pages 179–182). There had been scathing reviews by the BAA's Henry King already (more on him later) but Dewhirst (the Solar Section Director) openly lambasted Wilkins saying he had misrepresented the nature of astronomy, ignored hard-won knowledge and done a disservice to his colleagues by misleading the public. From such a respected and normally mild mannered astronomer this was surely the killer blow.

Wilkins went on to found the 'International Lunar Society' and became its first President. Patrick, still loyal to his mentor, joined the new organisation too, but much to his utter dismay Wilkins' daughter Eileen became engaged shortly after the new society was founded, but not to Patrick. Unknown to her, Patrick was totally devastated. She was married in 1957, to a diplomat, and became Mrs Eileen Coombes. Eileen's decision to marry another man was a mentally scarring emotional blow from which the painfully shy (where women were concerned) Patrick never really recovered. It occurred at a time when Patrick was about to become famous and he seems to have made the mental decision to have nothing whatsoever to do with women from that moment on.

By the time of her marriage Patrick had known Eileen Wilkins for more than a decade and had seen her grow into a very attractive young woman. However, she had presumably never seen Patrick as more than a sort of quirky uncle figure and had chosen another, with, on the face of it, far better prospects. Patrick was, after all, a struggling author who lived with (and was devoted to) his Mum. In a fascinating photograph in the author's possession, taken at the Manor House Centre at Crayford in Kent on August 20th 1957, Patrick and Percy Wilkins and the Mayor of Crayford are shown at a meeting of the International Lunar Society, where they were welcoming the American Lunar expert 'Doc' James Q. Gant. Wilkins' wife, daughter and brand new husband are seated only a few yards to Patrick's left and Patrick's head is turned sharply away from them and the camera, as if to snub them! Everyone else is looking directly at the camera but Patrick is looking directly away from Eileen Wilkins. [N.B. also in the picture is a 25 year old John Wall, the future inventor of the Crayford focuser!] By this time Patrick was, of course, on TV. However, I have seen other photographs of Patrick from that era where his head was turned to the right (*his* right), hiding his weaker eye from the camera. In later years he would frequently wear a monocle in his right eye and he seems to have been self conscious about that weaker eye's appearance on a number of occasions. So, maybe he was not snubbing Wilkins' daughter, her new husband and her mother after all?

Interestingly, Patrick's friend Arthur C. Clarke got married and then promptly separated in 1953 and, according to Patrick, Arthur had declared that writing and marriage were totally incompatible. So maybe that was just an extra factor in Patrick's decision to stay away from women from then on? It is doubtful that Eileen Wilkins had the slightest idea of how deep Patrick's emotions towards her went and how much her marriage distressed him. Having said that, I don't know any friend of Patrick's who could ever see him forming a relationship with anyone, except with his devoted mother.

Chapter 5
Desmond Leslie, Cedric Allingham and Science Fiction

Although Patrick was pleased to have several books in print, one on astronomy, three sci-fi novels and one more astronomy book in the pipeline (*Guide to the Planets*), it was blindingly obvious to him that writing science fiction and astronomy books was not a way to earn a lucrative living. However, since 1947, writing books about Flying Saucers had become a highly lucrative source of income for those with a vivid imagination! Wilkins had also netted a very nice pay packet from his US lecture tour about O'Neill's Bridge. The American amateur astronomer George Adamski had made huge amounts of money from cashing in on the Flying Saucer sightings of the late 1940s, claiming he had been onboard a UFO and met with aliens. Patrick, through his fascinations with some very bizarre groups of UFO hunters had met the eccentric, yet wealthy and talented, Desmond Leslie in the 1950s. Leslie knew Adamski and thought his stories were real. Patrick saw an opportunity to jump on the Adamski gravy train. In addition, Immanuel Velikovsky's books entitled *Worlds in Collision* and *Ages in Chaos* were generating huge publicity and huge sales, despite being utter pseudo-scientific garbage. Yet another author, Truman Bethurum, claimed he had an affair with a beautiful female alien from the planet Clarion (Aura Reins). Truman's wife filed for divorce because of the affair and won!!! The first legally verified interplanetary sex scandal in history! Patrick's friend Desmond Leslie was an extraordinary character himself, but Patrick, at least had his feet on the ground where con men and Flying Saucer hoaxers were concerned. Leslie, who remained on good terms with Patrick until his death in 2002, had an extraordinary life and background. It is worth digressing a bit just to analyse Desmond Leslie's life as he played a critical part in Patrick clinching the *Sky at Night* role which would set him up for life.

A year older than Patrick, Desmond Leslie was born into a wealthy Anglo-Irish family in County Monaghan and educated at Ampleforth College; he spent a year at Trinity College Dublin. His father was Sir Shane Leslie, a first cousin of Winston Churchill. Like Patrick, Leslie was in RAF Bomber Command, but from 1942 to 1945. During that period he wrote his first novel, *Careless Lovers*, which was a wartime

M. Mobberley, *It Came From Outer Space Wearing an RAF Blazer!: A Fan's Biography of Sir Patrick Moore*, DOI 10.1007/978-3-319-00609-3_5,

bestseller. He met his first wife, a German-born (!) actress and singer, Agnes Bernelle, at the end of the war and he married her in August 1945.

In that year of 1945 Desmond Leslie became involved in films and made a production in Ireland entitled *Stranger at My Door*. In 1949, he co-directed the film *Another Shore* in which his wife starred. The Leslies lived in London until 1963 during which time Desmond Leslie developed an interest in electronic music. A number of his composures were used in BBC productions (including Dr Who) and he continued his friendship with Patrick during those years. In 1963, he famously punched the BBC's drama critic, Bernard Levin, in front of a live audience of millions, on the television programme *That Was The Week That Was*. Levin had given a bad review to Leslie's wife's one-woman show *From Brecht to Brecht, Cabarets of Savagery and Delight*, one week earlier on the programme. In fact, as Leslie must have known, it was not solely his wife Agnes that had caused the review to be so bad, but Leslie himself! He had his own design of custom loudspeakers used for his wife's show and mounted them beneath the stage. Unfortunately this made the sound of his wife's heavily accented voice virtually inaudible unless you were in the front row. Leslie had failed to check out the acoustics. This was a huge error of judgement for a man who considered himself a pioneer of electronic music. On that classic *That Was The Week That Was* programme the entertainer Lance Percival announced that ten studio guests were about to discuss some points with Bernard Levin. As the camera cut to a seated Levin the tall figure of Desmond Leslie was seen standing over him. In a very upper class voice Leslie stated "One minute Mr Levin, before you begin, it won't take a minute, would you stand up a second?" A startled Bernard Levin stood up and it was clear to see that he was a good 8–10 inches shorter than Desmond Leslie who positively towered over him! Leslie continued "Mr Levin, your review of *Savagery and Delight* was not a review; it was a viscious attack on an un......." As Levin protested it was obvious that TV studio officials were quickly homing in on Desmond Leslie, so Leslie cut to the chase and with the words "There's just one tiny thing that......." he aimed a swipe at Levin who fell backwards out of shot as several studio officials walked Leslie away, to a single set of hand claps from an audience member. Ten million BBC viewers saw the assault live, on air! Levin quickly returned to his stool with the words "That wasn't part of our Saturday evening show. Can we concentrate on non-violence you and I?" As Patrick was a non-violent man for all of his life, and was appalled by violence, he must have been shocked by his friend's behaviour, although when I once mentioned this incident Patrick claimed not to have known anything about it.......!

The Leslies moved back to the family mansion, 'Castle Leslie' in County Monaghan, Ireland in 1963 and during their 24-year marriage, which ended in 1969, they had three children: two sons, Mark and Sean, and a daughter Wendyl. Desmond Leslie also had a relationship with a woman called Jennifer Fibbs and they had a daughter, Antonia. In 1972 Desmond Leslie and Patrick would co-author the spoof book, *How Britain won* the *Space Race*! Like Patrick's friend Wilkins, Leslie was a Flying Saucer believer.

There is another interesting connection here too. Desmond Leslie also knew the wartime Commander-in-Chief of RAF Fighter Command, Air Chief Marshal Sir Hugh Dowding who, believe it or not, had a belief in Flying Saucers *and* the paranormal! Dowding, or to give him his full title, Air Chief Marshal Sir Hugh Caswall Tremenheere Dowding, first Baron Dowding GCB, GCVO, CMG, was a member of *The Ghost Club* and *The Fairy Investigation Society* and once stated that Fairies were "essential to the growth of plants and the welfare of the vegetable kingdom". Extraordinary! Flying Saucers and the paranormal became Dowding's main interests in the 1950s. The reader may recall me mentioning that in Patrick's Holmewood House teaching years Dowding's stepson, David Whiting, was a pupil at the school and Patrick was David's housemaster. It's a small world! Patrick's future friend the comedian Michael Bentine had an interest in both subjects too, and was convinced that paranormal and psychic phenomena were genuine. In later years, and especially after Patrick's mother's death, he would, in his bereaved condition, develop spiritualist tendencies too.

Although Patrick was a serious amateur astronomer he was fascinated by UFO stories, and as a struggling author he could clearly see there was money to be made from Flying Saucers, either by writing about them or by making science fiction films; and who better to make a film with than his friend, the eccentric film producer Desmond Leslie! So, in 1954, Desmond, with a small amount of help from Patrick, made a definitely B-grade short film at 'Castle Leslie' in County Monaghan entitled 'Them and the Thing'. No doubt the film was inspired by the recently released Hollywood movie 'It came from Outer Space'. Rather than using a hurled lamp-shade, a car hubcap, or a genuine crockery Flying Saucer, Desmond Leslie put a Spanish Renaissance shield from the family castle on the end of a fishing line and instructed one of his sons to shine reflected sunlight from a mirror onto it, to make it glow like an alien space ship. Various friends from the neighbouring village of Glaslough helped with the film, which was almost one hour in length. Remarkably, 56 years later, in 2010, the reel of film containing 'Them and the Thing' was re-discovered and the Irish Film Institute said that they would show it as part of their Archive Home Movie Heritage Festival. Patrick, in 2010, claimed he had no memories of making the film "whatsoever"! Leslie had decided that, as Patrick was a UFO sceptic, he would appear in the film as an innocent bystander, totally unaware of the invasion of planet Earth taking place around him. Needless to say, the film never made any money.

Anyway, back to the plot and let us return to the not-yet-famous struggling book author (not film producer) that was Patrick Moore of 1954. Patrick had always had a prankster side to him and was genuinely fascinated by the thought of intelligent life elsewhere in the Universe. However, unlike Wilkins and Leslie, Patrick NEVER claimed he believed in UFO's, but he certainly appreciated their pseudo-science money-making potential and was even prepared to write a spoof novel about them. In 1954, man had not sent space probes to Mars or Venus and some amateur astronomers thought that aliens might actually exist in the solar system. Patrick was not one of these, but he did see an opportunity to make much more money from his writing talents and he certainly did not want to have to get a day job.

Like many modern day TV celebrities Patrick was, to be honest, totally unemployable in any job where he would have to obey a boss and do tedious work. In his later life Patrick would often say that apart from his period in the RAF, 'pottering about flying things', he had not done a day's work in his life and did not intend to. He was never going to do anything he found boring for a living; he would rather be dead! In the early 1950s Patrick really had been a struggling author. He was not then famous and his writings did not bring in much income. Another little-known fact was that in 1954 Patrick had been an active member of the Liberal Party and had big plans to be elected as the Liberal M.P. for East Grinstead at the 1954 General Election. Unfortunately, this venture did not succeed, leaving an income based solely on astronomical writing.

Another factor here is that Wilkins had also decided to cash in further on his O'Neill's Bridge/'Flying Saucer on the Moon' theme. Wilkins brought out two books in 1954, entitled *Our Moon* and *Flying Saucers on the Moon*. Note well, the first book was published with the author's name H.P. Wilkins, the second with the most subtle pseudonym imaginable, H.T. Wilkins. Swapping the P for a T did not fool anyone, although Patrick swore blind they were different people!

Digressing slightly here, one of Patrick's favourite films of all time was the 1953 release 'Invaders from Mars' directed by William Cameron Menzies and starring Helena Carter and Arthur Franz. It chillingly depicts a very young amateur astronomer's horror when a Martian spacecraft lands and starts to turn his parents and neighbours into automatons; maybe this film too was an inspiration for his next, rather dubious writing project.

So, despite all the controversy over O'Neill's bridge, Patrick decided he would invent an account of his meeting with aliens from another world and jump onto the bandwagon too. There was BIG money to be made in the Adamski era. This would also be an ideal opportunity to prove to those who claimed to have met aliens, that anyone could invent this kind of rubbish! But first he had to change his name; unlike Percy Wilkins, Patrick did NOT want to be officially associated with Flying Saucers and become a laughing stock within the BAA. So, he decided to become Cedric Allingham! The book was published, by Frederick Muller of London, at the end of 1954 and was priced at ten shillings and sixpence (52½ pence in post-1971 decimal currency). The title of the book was *Flying Saucer from Mars: The Facts – presented by Cedric Allingham*.

The front and back inside covers of the book give a good summary of its contents:

> Since the existence of Flying Saucers was first announced in 1947, much has been written about them. In a book published in 1953, George Adamski of Palomar, California, described his meeting with a being from Venus. Cedric Allingham, author of this new book, is no mystic, but a trained scientific observer with a considerable astronomical knowledge. He claims that in February 1954 he saw a Flying Saucer land in north Scotland and not only photographed it from close range but made direct contact with its occupant who indicated that he came from the planet Mars. Mr Allingham's description is fully confirmed by an independent witness, James Duncan. This book has been written not as a sensational story, but as a sober, scientific document. It contains a selection of the author's remarkable and hitherto unpublished photographs.

Fig. 5.1 The cover of Patrick's '*Flying Saucer from Mars*' book, written under the pseudonym of Cedric Allingham and marketed by the now extinct publishers Frederick Muller (A photo of the author's personal copy)

On the back cover of the book is the following information about Patrick's alter-ego, the author:

> Cedric Allingham, born 1922 in Bombay, only son of a wealthy British textile manufacturer, was educated privately until he was 10, when his father retired and bought a house near Durban. Allingham's education was continued at schools in South Africa and England until he was 17, when he entered a sanatorium with a serious illness which left him on his back for nearly 2 years. In 1941 he joined the Army (RAOC) and was posted to the Middle East where he spent most of the war. It was during that time that he became interested in astronomy through "identifying the stars for want of something better to do in the desert when it was too hot to sleep at night". Since the war and the tragic death of his parents – whose ship was torpedoed only a week before the German surrender, Allingham has lived a nomadic life in a caravan in which he likes to travel the country and in which he also makes long trips to the continent. Most of his time is spent in writing thrillers (which he publishes under a pseudonym) and in bird watching. He also owns a cottage in Yorkshire where he occasionally retires to do his serious writing and observe the Moon and planets through his 10-inch reflecting telescope. Since his unique experience, related in this book, he has started to make a collection of Flying Saucer sightings and is planning to carry out further extensive research during 1955 in California, where a considerable number of Saucers have been reportedly seen in recent months.

The inside cover of the book contained a picture of the author, standing by his '10 inch reflector'. Anyone who knows Patrick will be in no doubt whatsoever that this is actually a picture of his own 12½ inch alt-azimuth reflector, with Henry Wildey optics, mounted in the garden of the family home, *Glencathara*, at East Grinstead. Every unique detail pins that telescope down as Patrick's, without a shadow of a doubt. There can also be little doubt that the man posing next to the telescope looks rather odd. Some have suggested that he is heavily disguised, wearing a false nose/spectacle/moustache combination (of the sort that joke shops sell) and, possibly, a false ear too! Admittedly, for the '1950s Patrick in disguise' theorists the jaw and hairline look a bit like Patrick's, except for the hair at the front being somewhat 'flattened'. An investigation by Christopher Allan and Stewart Campbell of Edinburgh, for the UFO journal *Magonia*, some 30 years later, led them, via the publisher Frederick Muller, to the door of a Surrey journalist called Peter Davies. They claimed Davies admitted rewriting the book to hide the original author's style. Davies also claimed that he was the heavily disguised man standing by Patrick's telescope at East Grinstead and he admitted to being an old friend of Patrick's; but he never actually said that Cedric Allingham (the author) was Patrick.

Fig. 5.2 The only good picture of the mysterious figure called Cedric Allingham, from the inside cover of 'Flying Saucer from Mars'. Compare the *telescope* with Patrick's 12½ inch Newtonian shown in the next figure. A number of people have been suggested as the real Cedric Allingham, even Patrick in disguise! However, this is certainly not Patrick. My money is on the chap being Patrick's mate George Hole, slightly disguised, who Patrick consulted on all equipment matters during the 1950s. Hole sported a moustache in the early 1950s and was a slight man. No good pictures of Hole appear in any of Patrick's books, but those in his 1940s and 1950s scrapbooks are very similar to the Allingham character shown here

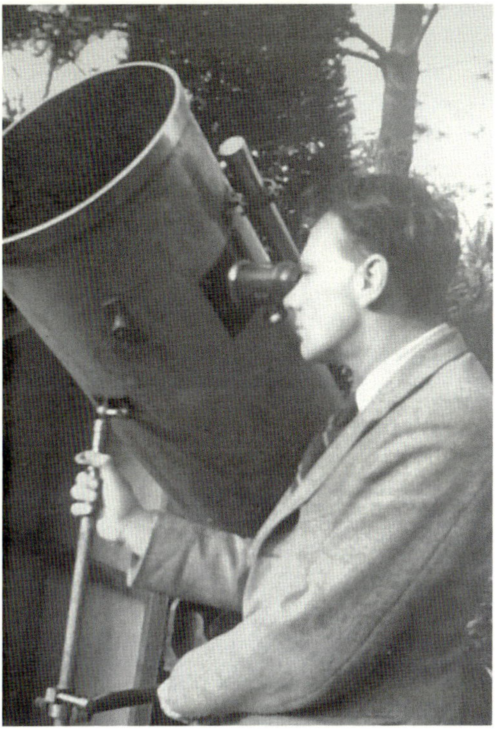

Fig. 5.3 This picture of Patrick with his 12½-inch Newtonian 'Oscar' was one he sent to many publishers and societies and friends during the mid-1950s. Compare this *telescope* with the one seen next to Cedric Allingham in the previous figure and there can be zero doubt that it is the same reflector! (Photographer unknown)

The publishers claimed that Allingham had died in 1956 and was buried in Switzerland, but there is even an account of Allingham giving a public lecture. In fact, Patrick claimed to have witnessed the lecture himself, as he had been invited to a UFO society meeting by the World War II RAF eccentric and UFO fanatic Lord Dowding, to hear Cedric Allingham speak! Boy, would I have liked to have been in that audience!!! The problem here for me is this: if Davies was paid to disguise Patrick's writing style, he did an appalling job. The style is 1950s Patrick Moore through and through. Plus, it was Patrick's publisher too. Also, if you want to hide the author's identity you don't put a picture of his telescope in the book, or advertise Wilkins' and Moore's books in the same tome! Various theories abound as to who the man appearing on the front and back cover really was; however, my favourite theory is that it is a young George Hole (1913–1986) who was quite a celebrity in the Brighton and Hove area. He was a Brighton fireman during the Second World War and had an observatory near there from the late 1930s. At first he owned a 14 inch (35.5 centimetres) reflector at nearby Patcham, but after the war he owned a monstrous 24 inch (61 centimetres) reflector of 12 feet in length (3.7 metres) which, as we shall soon see, appeared on an infamous *Sky at Night* programme in 1961.

George Hole's telescopes were featured on cinema screen Pathé newsreels up and down the country soon after they were completed in April 1942 (the 14 inch) and March 1949 (the tube and mount of the eventual 24 inch, at that time equipped with a smaller 18½ inch mirror). Hole's 14 inch reflector was mounted on a massive fork, constructed from a unique material amongst amateur equatorial mountings: it was made from concrete. Even to this day Hole is remembered in his local Brighton: a local number 607 Scania Omnidekka bus was named after him in July 2003. George Hole was a close friend of Patrick's throughout the 1950s, 1960s and 1970s and Patrick was actually a director of Hole's telescope making company based in Middle Road, Brighton. Hole even made a scale model of the Isaac Newton telescope which was presented to the Queen. He became the BAA's Instruments & Observing Methods Section Director in 1961, a post he held until 1973. In the late 1950s, and for the rest of his life, George Hole did not sport a moustache. However, during much of the Second World War, and in the early 1950s, he did. If you see a picture of George Hole from the 1940s (and there were a few tucked away in albums at Patrick's house) there is little doubt that the man with the moustache and glasses, posing as Cedric Allingham in Patrick's spoof book, is actually that same good friend George Hole, albeit slightly disguised. I have always thought that Hole is an excellent surname for a man with a huge aperture telescope!

So, what about the name Cedric Allingham itself, where did that come from? I have had a number of discussions with the BAA Mars Section Director Richard McKim on this subject. Richard has suggested that, as Patrick was especially keen on amusing pseudonyms that sound like statements (e.g. R.T. Fishal), C. Allingham actually means 'Calling 'Em', in other words 'Calling the Aliens'. This sounds like a very plausible theory to me, and right up Patrick's street. Certainly, no anagrams of the name seem to mean anything else that can be construed as extra-terrestrial.

Of course, Patrick, at the time of *Flying Saucer from Mars*, was no more than a very enthusiastic amateur astronomer with a few published books to his credit. He was certainly not famous. There must have seemed no risk in this little bit of 'fun'; it would bring in some money and it would show the likes of George Adamski and other Flying Saucer authors that anyone could write this sort of garbage and the gullible public would fall for it. The book contained the usual out of focus Flying Saucer/lampshade shots, typical of that era, as well as a letter from a witness, clearly in Patrick's handwriting. In one picture the space alien seems to be wearing braces to hold up his space trousers! In many ways, Patrick's *Flying Saucer from Mars* was simply a spoof of Adamski's *Flying Saucers have landed.*

Patrick did not simply introduce the book out of nowhere either. Well before the book's publication he (as Allingham's contact) spread sufficient publicity to get rumours going that a claimed alien encounter had occurred at Lossiemouth in Scotland. He even invited the eccentric UFO enthusiast and author Leonard G. Cramp to meet him and Desmond Leslie, at the Athenaeum club in Pall Mall, to talk about Allingham's claims (I will mention Cramp a few more times in the coming pages). On page 61 of Mr Cramp's very obscure book entitled 'The A.T. Factor: Piece for a Jigsaw Part III' he makes the following comments:

> In March 1954 I received a letter from Patrick Moore inviting me to join him and Desmond Leslie at their club. I was seated beside Patrick and, over dinner, despite his claimed scepticism about UFOs, we had a very pleasant evening. However, I was surprised to learn that only a week before he had met Cedric Allingham, author of *Flying Saucer from Mars*, and he was keen to discuss this book with me.

Well, meeting Allingham would not have been tricky for Patrick; he only had to look at himself in the mirror! A few lines later Cramp goes on to say:

> At that time the extraordinary thing was that, other than Patrick Moore, nobody else had met Allingham and all subsequent attempts to locate him failed. It was said he had been very ill, but where did that information originate? It could hardly explain his total disappearance. It was generally suspected that *Allingham* was a pseudonym...........

Hmmm! Was Mr Cramp so thick he couldn't work out that Patrick was Allingham? Or maybe he was just being polite?

Such was Patrick's sense of the outrageous that he went one step further and *reviewed his own book* (!) in the April 1955 Journal of the BAA. Yes, you did read that correctly: he actually reviewed his own spoof book. The procedure then, as now, was that books were sent by the publishers or authors to relevant journal editors and those editors then passed the books to knowledgeable authorities to review. Patrick had been outspoken in his criticism of Flying Saucers and had made sure that a copy of his own book was sent to the BAA editor, Neville Goodman, to review. Being a prominent member of the BAA Patrick knew the editor well (he was only 6 years younger than Neville) and he made sure that Neville knew he was keen to debunk any Flying Saucer books that were received. Of course, Patrick knew that Neville had just received his (Allingham's) own book, but he would not know that two other Flying Saucer books had also been received by the BAA. Neville sent Patrick the three books to review and Patrick reviewed them all, with considerable bias towards his own (!) in that April 1955 issue of the BAA Journal. The two other books received were entitled *Flying Saucers from Outer Space*, by Donald E. Keyhoe, and *Space, Gravity and the Flying Saucer*, by the aforementioned Leonard G. Cramp, who Patrick had met at the Athenaeum Club 13 months earlier.

Patrick's review of his own book, plus the two others, was published as follows:

> After the mass of nonsense that has been produced during the past few years one might have imagined that the subject of Flying Saucers had been exhausted. That is not the case. Saucer books continue to appear; and though a scientific Journal is not the proper place to review them, a few words should be said in order to dispel any lingering doubts as to their authenticity. The three volumes listed above are equally absurd in their various ways. Mr Cedric Allingham, for instance, says that he met a Martian on the Scottish coast and obtained information showing clearly that members of BAA observational sections are wasting their time. He has also produced photographs of the saucer that put Mr Adamski's completely in the shade. As a matter of fact, it must be said, in all fairness, that Mr Allingham's book is very cleverly written, and he rejects telepathy, Atlantis, astrology and all the rest of the outdated clap-trap usually to be found. In fact, Mr Allingham is not a crank. Unless we are prepared to accept his Martian, the inference is obvious. Had the book been advertised as science fiction it would have been well worth reading! If we wish to plumb the depths of credulity we can turn to Mr Keyhoe. And lastly, Mr Leonard Cramp tells us just how the saucers work – by creating their own gravity; there is even a chapter on human levitation

and yogism. Have we at last rejected the end of Flying Saucery? Apart, possibly, from a conducted tour of Venus, there can surely be very little to add now. Fifty years hence these three books will probably have a certain interest from a psychological viewpoint; but otherwise – silence is best – P.M.

Yes indeed Patrick! Sixty years hence these books, one in particular, do have a certain interest from a psychological viewpoint: the psychology of Patrick Moore!

Although the alter-ego of Cedric Allingham shows Patrick the prankster at his best, the whole episode would haunt him for the rest of his life. From the late 1950s and, increasingly throughout the 1960s and 1970s, Patrick would become a household name, with barely a blot on his character. In the years to come he would attain a reputation as a British amateur scientist who you could trust, albeit a somewhat eccentric one, and certainly not a Flying Saucer con-artist. Those who knew Patrick well were keen to forgive him for the entertaining Cedric Allingham episode, but Patrick became deeply embarrassed by the whole thing, threatening to sue anyone who claimed he was Allingham. On one occasion, on a Radio Sussex phone-in, during 1986, Patrick threatened to walk out of the studio if the Allingham allegation was mentioned by the interviewer again, and yet in the Radio Times for the 7th to the 13th of May 1977 he brazenly drew attention to the existence of Allingham and the man's claims of having met a Martian on the Scottish Coast! In a Discovery Channel programme about UFO sightings from Britain, filmed in 2000 (Britain's Secret UFO Hunters), Patrick is filmed implying that people will believe anything they want to believe, just after he is accused of being Cedric Allingham. However, the crucial question "But that IS, surely, your telescope that Allingham is standing next to?" was never asked!

Patrick, however, did admit to two other books which he published under a pseudonym: entitled *Bureaucrats and how to annoy them* and *The Twitmarsh File* by R.T. Fishall. Yes, it is meant to sound like 'Artificial'. Another amusing pseudonym he sometimes used was R. Hugh Hall-Wright, along with L.E. Fantine! Patrick would use pseudonyms on various occasions throughout his entire life when writing to pseudo-scientific Journals or, when in Armagh, writing to highly religious newspapers. His view was that the readership and editorship of such publications was so deluded that they would believe in anything. The Aetherius Society, formed in London by George King in 1955, believed it had received messages and had experienced contact with extra-terrestrials. A number of bizarre letters from alleged sympathetic scientists appeared in the pages of its journal *Cosmic Voice* and Patrick and his BAA friends were strongly suspected as being the originators of most of them. The letters' authors were, for example, Dr's Houla and Huizenass or, sometimes, Dr L. Puller. Other obviously concocted names writing to *Cosmic Voice* from time to time were Dr Dominic Fidler, Professor Huttle-Glank, Mr E. Ratic, Mr Egon Spunrass and, sometimes, Dr N. Ormuss (think about it!) and his trusty assistant Waathervan. Anyone who knows Patrick is fully aware that this sort of hilarious letter writing prank was the hallmark of his humour throughout the 1950s, 1960s and 1970s and he could churn such stuff out at an incredible rate, as could a few of his friends.

So, with *Flying Saucer from Mars: The Facts* published, and generating a healthy income, Patrick found time to churn out another three sci-fi novels in 1954/1955 namely: *Destination Luna, Quest of the Spaceways* and *Mission to Mars*. A fourth Sci-Fi yarn entitled *Out into Space* was co-written by Patrick and A.L. Helm in 1954 and published by Museum Press. I will digress slightly here to explain that A.L. (Alex) Helm was an occasional author and translator of astronomy books; for example, he translated Gunter D. Roth's book 'The Amateur Astronomer & his Telescope' in 1963. He was also the exact same person as Sandy Helm, Patrick's teaching colleague at Holmewood House School. Occasionally Helm observed with Patrick, such as during the January 29th/30th 1953 total lunar eclipse. Alex (Sandy) Helm was also the future husband of the artist Patricia Cullen who would eventually illustrate many of Patrick's books and also provide many diagrams for his *Sky at Night* programmes in the years to come. Patrick would be the best man at Sandy and Patricia's wedding in 1956. Anyway, returning to the Sci-Fi book theme, I should add that Patrick's good friend Richard Baum supplied a Saturn sketch for that joint Moore/Helm *Out into Space* book. The novel *Mission to Mars* was, unlike the Allingham Mars escapade, written under Patrick's real name and was also the first in a Martian quintet of novels by Patrick to be written between the years 1955 and 1959. The other four novels in the Mars quintet would be entitled: *The Domes of Mars* (1956); *The Voices of Mars* (1957); *Peril on Mars* (1958); and, finally, *Raiders on Mars* (1959). The hero of Patrick's Martian quintet was a certain Maurice Gray, a 16 year old sent to Woomera, Australia after the death of his parents in an air disaster (note the similarity with Allingham's torpedoed parents). Maurice discovers that his missing uncle is stranded on Mars after travelling there on the spaceship Hermes. Needless to say, Maurice follows him there, in another spaceship called Ares. As well as the *Domes of Mars* sequel to *Mission to Mars*, Patrick would churn out two more novels in 1956. These were *Wheel in Space* and *World of Mists*. Patrick was not the only prolific Sci-Fi author about in the 1950s. Isaac Asimov and, as we saw earlier, Patrick's friend Arthur C. Clarke were also hard at it, as were about a dozen other writers.

In the mid to late 1950s Patrick certainly seemed more interested in writing fiction than astronomy. Maybe part of the reason for this was that another BAA member, J.B. Sidgwick, had produced two comprehensive mammoth tomes in the mid fifties which seemed to have covered everything! These were entitled *Amateur Astronomers Handbook* and *Observational Astronomy for Amateurs*. While these were, without doubt, incredibly exhaustive reference works (the former was nearly 600 pages long) they were arguably in the category 'dry as dust', to quote a phrase Patrick often used. Another British author called Henry King (a future BAA President from 1958 to 1960) wrote the classic 'The History of the Telescope' in 1955. King lived at Slough, in a house he named 'Stellafane', at 18 Sutton Avenue. Without a doubt King and Patrick despised each other with a passion. While Patrick would not write anything as ambitious for many years he could certainly write books that oozed enthusiasm. It would be some years before he appreciated just how popular his astronomy books could be.

Despite Patrick's fascination with science fiction and Flying Saucer stories, he was keen to ridicule them, wherever possible, in public. He was also fascinated by the gullibility of the public in believing such drivel and amused by the failings of the press in reporting scientific events. He kept extensive scrapbooks of such newspaper articles. At the January 26th 1955 BAA meeting he gave an amusing talk entitled 'Astromania or Science Friction' which was recorded by him in the March 1955 BAA Journal, and which is reproduced, in full, below. At this meeting, the President, Dr Reggie Waterfield (the man who really did accompany Einstein in a duet) invited 'Mr Moore' to tell the audience about his collection of astronomical cuttings from newspapers. The 31 years old Patrick responded as follows:

> I think a suitable title for my talk would be 'Astromania, or Science Friction'. On the whole, newspaper reporters are a much-maligned race. Even in astronomy, a subject totally unfamiliar to most of them, they seldom make really bad mistakes. Every now and then, however, one comes across a real gem; and as I have collected newspaper cuttings for the past twenty years or so, I have accumulated a number of decidedly amusing specimens. All these particular examples of 'Astromania' come from my own collection of cuttings – apart from two which I have seen – and I am not going to deal with the usual Crazy Gang, which at once disposes of all astrologers, Flat Earthers, people who see the Moon nodding gently up and down, and Russian scientists [Velikovsky] who picture the planet Venus bouncing about the Solar System rather in the manner of a celestial table-tennis ball. However, it will be worthwhile to mention the good lady who was interviewed by the Sunday Pictorial in August 1951. When asked about the Moon, she revealed that 'it was a radiating station for all vibrations', and added as an afterthought, that she had been there.
>
> Genuine reporting slips can be most enlightening. For instance, a Russian report in 1938 stated that three Soviet astronomers were going up 30,000 feet in a balloon to get a 'close-up' view of a solar eclipse; and an Australian columnist stated in a Sydney journal that polecats had been observed on Mars. Needless to say he really meant pole caps, and in this case, I am rather inclined to blame the printer.
>
> Now for a case of real precision. According to the Daily Graphic of May 14, 1935, Professor Velyaminov, of Moscow Observatory, determined the temperature of Nova Herculis 1934 as 64,032 degrees Fahrenheit. Doubtless the spare 32 degrees was most significant. On 1938 August 30, the Daily Mirror reported that Professor Rossiter, of Bloemfontein, had discovered 287 new stars. Presumably he meant 287 double stars; I think that a glut of novae such as this would tax the observational skill even of Dr Steavenson!
>
> The giant rarefied star Epsilon Aurigae was in the news some 16 years ago, and on 1938 January 14 the Daily Mail stated that the whole star 'contained only enough material to fill the Albert Hall'. Unfortunately nothing was said about density.
>
> So much for the stars. We now come to the Solar System, where the journalistic field is even wider. In 1938 a Roumanian professor was said to have discovered 'a new planet with a small tail' in the constellation Lynx – I am still uncertain just what this was; presumably a comet – and even editorial comment does not always improve matters, as the following example will show. The Evening News of 1951 April 13 contained a letter from a Londoner stating that 'during a cycle ride last night, I saw a bright, amber-coloured star rise and set within about two hours', and the attached editorial comment added that 'the supposed new planet, observed recently by the United States Navy, has now been identified as the asteroid Thalia, which comes within 200,000,000 miles of the Earth about every five years'. As a matter of fact, Thalia was in opposition about that time; but I am not clear just how the United States Navy comes into it.
>
> The following extract is from the Daily Mirror of 1934 September 28th: 'An electric blue ball of eight feet long shot across the sky of North London yesterday. This phenomenon was reported to the Royal Observatory at Greenwich, where it is believed to have been a comet'.

This was also referred to by the Morning Post, a few days later, which printed a letter from a correspondent with the rather appropriate name of Belam: 'The comet was like a white-hot wire, with the anterior end thickened like a knob'.

I suppose that it is excusable to confuse comets with meteors, but I cannot account for the Daily Mail article of 1947 December 23. A bright comet, 1947n, had been discovered at the Cape, and the Mail reporter went up to 18,000 feet in a Lancaster bomber to have a look at it. His article was headed: 'Comet 1947n blazes for Five Minutes', and read: 'The comet shone a bright orange, sitting like a new Farthing balanced on the edge of a fleecy white cloud-bank through which the aircraft had soared'. Other people supported him. A police constable, described by the Daily Mirror as 'an amateur astronomer', reported that the comet 'was in the west, and had a halo. It looked like a bright fuzzy ball', while the local paper in my own home town, East Grinstead, printed a letter from a man who had seen it looking 'like a large golden star, surrounded by a golden aura'.

Unfortunately, however, Comet 1947n was never visible to the unaided eye in England, and the above reports show clearly just what lively imagination can do.

It would take an encyclopaedia to deal with Flying Saucers, and I do not propose to try, but now and then one reads a really remarkable article on the subject of space-flight. One appeared in Picture Post of August 1951, written by a Mr Derek Wragge Morley. Some of the statements were distinctly unusual, and the illustrations were to match; I particularly noted a banana-yellow Saturn surrounded by rings of technicolour red, and provided with a wealth of satellites, one of which even had a small tail. One typical example will sum up the text. The writer stated that Phoebe was 'the brightest of all Saturn's moons'. What he obviously had done was to look up the satellite magnitudes, and, finding that Phoebe had the highest numerical value, assumed it to be the brightest object. Returning now to pure astronomy, I must quote the Daily Mail report of 1935 January 30, just after a total lunar eclipse. It read: 'While the Moon was in eclipse last night, the shyest star in the sky had its big night out. Practically every astronomical telescope in Europe was trying to follow the movement of NZC 1327, a star so small that it can never be seen by the naked eye. It can be watched by a powerful telescope only during a total eclipse of the Moon…'

May I end with two personal reminiscences? The first concerns the Cambridge review of 1954 March 6, which carried a comment on a book of mine [Patrick's 'Guide to the Moon'], written by a Dr Felix Pirani. After complaining that I described the Moon in some detail 'even while suffering the disability of being unable to get within a quarter of a million miles of it', Dr Pirani went on: 'His distance from his subject does not deter the author from giving a chapter each to means of travel to the Moon, to the possible nature of the inhabitants….' And so on. What he had done, of course, was to read the index, see a chapter headed 'Life on the Moon', and draw his own conclusions! It is only fair to add that the editor subsequently added an apology. And lastly, in November 1952, I was ill-advised enough to give a lantern slide lecture in the village hall of Fordcombe, in Kent. Sure enough, I was given a paragraph in the local paper. 'At the annual meeting of the Fordcombe Women's Institute, at the Village Hall, a talk on "The Stars" was given by **MRS** P. Moore.

Even 2 years before he would start broadcasting for the BBC, Patrick was an entertaining speaker! The May to June issue of the *Journal of the British Interplanetary Society* (Volume 14 no 3) contained more proof of Patrick never taking himself very seriously. In a small piece entitled 'Know your council' Patrick described himself in a highly amusing tone, adding that he had, so far, written 13 books, but was threatening more!

The reader of this book may already be amazed at Patrick's mental stamina for anything intellectual and his willingness to volunteer for roles in any society that reflected his interests, whether national or local. In May 1955 Patrick spotted a letter from a Mr Eichner in the *East Grinstead Observer* newspaper, proposing the founding of a local Chess Club. Patrick, and two other East Grinstead Chess players, replied to Mr Eichner's letter and he attended an inaugural planning meeting on the 15th of June, at Eichner's home (called 'The Rocks') in nearby Ashurst Wood, just a mile down the A22 road from Patrick's mother's house. The East Grinstead Chess Club was duly formed, with Patrick as its secretary, and the Chess champion of the local East Grinstead Grammar school (P. Merrett) as Patrick's assistant. Patrick would eventually become the President! How on Earth did he find the time? On the subject of East Grinstead Grammar School, two more of its students would be encouraged in astronomy by Patrick during the mid 1950s. The first of these was no less a person than the future planetary geologist Peter Cattermole, then living at 117a London Road in the town. Patrick proposed Peter for membership of the BAA in 1955 and he was elected on November 30th of that year. Six years later Patrick would successfully propose Peter as a Fellow of the Royal Astronomical Society. The other East Grinstead Grammar School student who was encouraged by Patrick in 1955 was the 16 year old Brian Warner, who would go on to become the Astronomy Department head at the University of Cape Town, from 1972 to 2004, and the Emeritus professor after that.

A Book Review from Hell!

When the March 1955 BAA Journal arrived at Patrick's East Grinstead home, and he turned to page 171, his blood, literally, ran cold. Even decades later Patrick told me that within seconds he felt physically sick. Patrick was an up and coming 32 year old author and observer and his reputation in the BAA was one of the most important things in his life. In addition, book writing was his sole source of income. He was, therefore, horrified at the scathing review his book *Suns, Myths and Men* had received at the hands of the respected author Dr Henry C. King who would, in 3 years' time, become the BAA President and the Director of The London Planetarium; but, of course, by then Patrick would be presenting his own TV programme, in part at least due to that very same book! Henry King (1915–2005) was, at 40, some 8 years older than Patrick and had joined the BAA in 1933, as an 18 year old. Even the fact that King was therefore a BAA member of longer duration than Patrick, regardless of the 8 year age difference, would niggle him until King's death in 2005. Patrick had served on the BAA Council for three consecutive years, from October 1951 to October 1954, but for the 1954/1955 session he had been forced (under the BAA's bye-laws, promoting a healthy influx of new councillors) to take a year off. During that session Henry King had materialised on the Council and had been given Patrick's book to review. In previous years Patrick (from the inside) may have seen King's onslaught coming in advance, but it had come like a bolt from the blue.

When King did not like a book, his reviews were nothing short of thermonuclear, but in those days such reviews were published, regardless of the dripping sarcasm. However, one thing that should be borne in mind when assessing Dr King's ruthless reviews is that his classic and definitive book, *The History of the Telescope*, had also been published in 1955. It was the result of more than a decade of painstaking research and perfectionism and had been peer reviewed by telescope experts and academics until he was happy that no stone had been left unturned, either in the text, or in the illustrations. Patrick's high speed writing technique was completely different, namely: type manically for a few months, in a humorous style, and hope your memory for the facts is roughly correct. If there are errors, well, life's too short to worry about them, and it's better to start the next book project instead. King's review of Patrick's *Suns, Myths and Men* occupied two pages in the Journal and contained, amongst many other cringe-worthy lines, the following statements:

> The general reader may, of course, enjoy the sparkle and froth, unaware that the author, with his forced visual images and vague generalities, is flirting with science. Here lies the danger, for the uninitiated are deluded into thinking that what is grist to Mr Moore's mill is the essence of scientific thought.
>
> In playing at science one can, like playing with fire, get burnt; and Mr Moore, however slightly, keeps burning himself.
>
> In the reviewer's opinion, popular scientific exposition demands the same care and attention as does the preparation of a scientific treatise; simplification can so easily lead to falsification. Expression must be concrete, clear and unmistakeable – the writer must be honest with himself and his public. One cannot help feeling that once an otherwise serious and talented astronomer discards these and similar precepts, he sows the seeds of irrationality.

Well, clearly Patrick was appalled by the lambasting of his latest book and he was allowed to respond. His reply was published in the July 1955 BAA Journal on page 310, under the title 'A Reply to a Reviewer'. Patrick had this to say to Dr Henry King:

> I have received a number of letters strongly criticizing Dr H.C. King's remarkable attack on me and my book, Suns, Myths and Men, and I feel therefore bound to comment, although this sort of thing does not really warrant any reply. Though the 'review' does read like a violent personal attack, I am sure that the unpleasant tone is due only to loose wording, and no more need be said. Neither do I mind his objections to my literary style; doubtless it differs from his own! However, it is only too easy to distort the meanings of isolated sentences and phrases by quoting them out of context, and this is what Dr King has done throughout, with amusing results. When he is more specific, he burns his fingers rather badly, and two instances of this will suffice. The remark 'but perhaps education was not Mr Moore's intention' is not perhaps as courteous as might be wished, and Dr King will forgive my saying that it does not make sense. He then takes me to task for saying that in a Christian country Anaxagoras would probably have been burned at the stake. Surely Dr King has heard of Giordano Bruno? His offence was fundamentally similar, though of course different in detail, and his fate is a fact, not merely opinion or misplaced emphasis. One should always be honest about facts, and to attempt to conceal them in this way is unscientific, though of interest psychologically. There is no point examining the review further, since these two examples are typical of the rest; and I do not want to offend Dr King in any way. Like King Gama, he is right to be outspoken, and from my point of view at least there is certainly no harm done. The book has now been favourably reviewed by the Astronomer Royal, who describes it as being 'suitable for the general reader' and has no adverse criticism. It seems possible that the opinion of Sir Harold Spencer Jones carries a little more weight than that of Dr King! – PATRICK MOORE

Following on immediately below Patrick's response, Henry King was also allowed to comment. His reply started with:

> I was sorry to hear that my review of Suns, Myths and Men was interpreted as an attack on Mr Moore himself. Such was not my intention; in any case I should not express my personal grievances (if they existed) in a review, but should communicate them personally. Indeed, I have the strongest admiration for Mr Moore's energy, enthusiasm and abilities as an observer.

However, Henry King ended his own response by stating:

> I cannot endorse vague, unqualified, and sweeping generalizations in history and science. Mr Moore's implication in the book seemed clear enough, and in my opinion his emphases were misplaced. – H.C. KING

Remarkably, in the June 1955 Journal (page 253), King also ripped apart Patrick's mentor H.P. Wilkins' book 'Clouds, Rings and Crocodiles: By Spaceship round the Planets' in a similar fashion, using sentences such as:

> The reviewer considers that the book is a work of fiction, that it contains so much nonsense as to have little educational value, and that most of the drawings are largely the work of a vivid imagination. If this book represents the new educational approach, the reviewer is profoundly grateful that he learnt the foundations of astronomical knowledge over twenty-five years ago. – H.C. KING.

Needless to say, H.P. Wilkins was also granted a right to reply in the Journal (October 1955 page 361) where he mentioned that King's review 'is of the type for which he has unfortunately become rather notorious'; an obvious reference to the mauling given to his friend Patrick's book. However, after the O'Neill's bridge fiasco, the older Wilkins was not going to be offered any mercy by Henry King who stated, in reply:

> I am sorry that Dr Wilkins thinks that I have purposely distorted his text and been guilty of dishonesty and 'common rudeness'. Whilst he is of this impression nothing is gained by writing further – except to add that I would not retract a single word of the review. – H.C. KING.

After such a bad review in the BAA Journal it is, perhaps, surprising, that Patrick carried on writing books. For many young authors Dr King's words would have been the death knell. However, Patrick persevered, although it is true to say that in the following years he tended to concentrate on planetary observing guides (a subject he was familiar with) and churning out his unique brand of 'Boy's Own' science fiction, while giving historical matters a rather wide berth. Nevertheless, I am convinced that Henry King's sarcasm was a spur to Patrick. He knew that he could really irritate King by churning out more and more enthusiasm-packed books, in the same style; and that is exactly what he did.

Mainly Mercury and Venus

Although Patrick's real enthusiasm was for observing the Moon, he was a keen observer of anything in the night sky, especially the planets. Mercury and Venus were of particular interest to him and he was very active in the BAA's section of that name.

For an incredible 60 years, from 1895 to 1955, Henry McEwen had been the director of that Mercury and Venus Section and Patrick had enjoyed a long correspondence with him, being one of the section's most enthusiastic observers. During 1955 McEwen became seriously ill (he was 91 years old) and as he was unable to answer correspondence he resigned as director. A few months later McEwen died and the BAA Council asked Patrick to become the new Section Director. After an enforced 1 year absence from the BAA Council table (due to the bye-laws) a Section Director post would enable Patrick to attend all of the Council meetings, without having to be elected or having to take a year off every few sessions. Patrick was already the Lunar Section secretary, but he felt that it was time to direct his energies into new projects and so he resigned his Lunar Section post. With Wilkins' resignation from the Lunar Section because of the ridicule of the O' Neill's Bridge saga, the Patrick-Wilkins team was no longer a BAA item and Patrick was dismayed by that embarrassing saga. Outside the BAA though Patrick was still a friend of Wilkins despite the fact that Wilkin's judgement appeared to be highly suspect and despite the fact that Wilkins' attractive daughter Eileen had not been attracted to Patrick. Patrick joined Wilkins in the short-lived International Lunar Society, which was outside the BAA, and they continued observing the Moon. Patrick and Percy Wilkins would still travel to Meudon too, for occasional looks at the Moon thru Parisian skies. A quote from Patrick's 10th Lunar Observing Notebook, for September 1956, follows:

> Observations this year were made with the 24 inch Cassegrain reflector as the 33 inch OG is dismounted for repairs. The main work done was of course on Domes. Unfortunately the weather conditions were not good. Work was possible on only the following nights:- Sep 12th, 14th, 15th, 16th....... The 24 inch is an excellent instrument but of course it can hardly be compared with the 33 in refractor. The 24 inch is a conventional Cassegrain.

As an aside, with the Royal Greenwich Observatory move to Herstmonceux Castle in Sussex being completed in 1957, there was no need for Patrick and Wilkins to travel to Paris to continue their lunar mapping quest with the giant refractors there. From 1957 and into the mid 1960s, Patrick would, whenever possible, travel on his ancient motorbike to use the 26 inch refractor at Herstmonceux, with or without Wilkins joining him. He would often arrive, late at night, wearing his old RAF flying jacket to keep out the cold, and as usual he resembled the living embodiment of the fictional character Biggles.

So colossal was Patrick's contribution to the BAA Lunar Section, until 1956, that his former workload was undertaken by a committee of people, appointed by the BAA Council. As with everything he did in the BAA, Patrick's energy and enthusiasm was boundless, but then he had no day job, no wife and no children. Unfortunately there was a tragic delay between Henry McEwen's death on May 6th 1955 and the appointment of Patrick as the new BAA Mercury and Venus Section Director, in January 1956. In the gap McEwen's executors cleared out his house and destroyed all of the BAA's Mercury and Venus Section observations, from 1890 to 1955! Sixty-five years of observations destroyed.......To put it mildly, this was not a good start for Patrick, but it was obviously not his fault. In subsequent years he would refer to this event as 'The Holocaust'! A term he used for more than one BAA disaster.

Patrick had always been interested in the so-called ‘phase anomaly’ of Venus; when the planet should appear looking like a half-moon it always appeared just a bit bigger, rather more than 50 % illuminated. As we saw earlier in this book, he was also fascinated by the alleged appearance of the ‘Ashen light’; when Venus appeared as a thin crescent observers sometimes reported that they could glimpse the dark side, possibly due to intense auroral activity on the planet (at least, that was one 1950s theory). Patrick generated enormous enthusiasm in studying these phenomena, and the Venus and Mercury Section went from strength to strength. Patrick was rapidly becoming known as a man who could revive people’s interest in astronomical topics and get them out of their beds and outdoors at the telescope eyepiece. In addition, BAA membership had strengthened in the 1950s and now stood at 2,300, more than triple the membership of the pre-war era.

Despite Patrick’s resignation as the Lunar Section secretary he was still fascinated by the Moon. Just after he had taken on his new role in 1955 he received some correspondence with Dr Dinsmore Alter, Director of Griffith Observatory at Mount Wilson, near Los Angeles, California. Using the 60 inch reflector there Alter had taken some photographs of the two large walled plains Alphonsus and Arzachel, which are members of the Ptolemaeus chain, and which lie near the centre of the Moon’s disk as seen from the Earth. Alter took pictures in both infra-red and in blue/violet light. He found, on several occasions, that part of the floor of the crater Alphonsus was blurred in the blue/violet filtered images. As infra-red light is better at penetrating haze or mist, the implication was that there was some kind of gaseous emission taking place in Alphonsus. This sounds fine, until you consider that the light has to travel through the Earth’s atmosphere, which somewhat confuses the issue! Blue/violet light is *always* scattered more than red light in our atmosphere; I do not know of any modern CCD imager who is unaware of this! Patrick wrote to a number of professional lunar observers, including one Soviet astronomer, Dr Nikolai Kozyrev, at the Crimean Astrophysical Observatory in the U.S.S.R. He suggested the professionals might like to keep a careful watch on the Alphonsus area. Three years later, Kozyrev would make a dramatic announcement in connection with Patrick’s appeal.

Patrick’s enthusiasm for lunar observing and sketching did not diminish after he resigned his Lunar Section post and he was still very much interested in features on the lunar limb; those tantalising glimpses of the lunar Far Side ‘libration features’ which could not be seen from the Earth. On the night of July 4th/5th 1955, at midnight GMT, Patrick had a great view of a limb crater which, in later years, as I mentioned much earlier, he would occasionally claim to have discovered. This was the crater Caramuel, later renamed as Einstein by the IAU. Patrick’s Lunar observing notebook entry reads as follows:

> 1955 July 5. 0h. Conditions fair, despite low altitude, and I was able to draw that elusive formation Caramuel – right on the terminator, under favourable libration. Several interesting facts came to light. The formation is a compound, and the interior crater is not dead central. From it a strong ridge runs N. and another S., causing, with the crater itself, a continuous line of shadow inside Caramuel. Outside is B (Vasco da Gama B) which is bright. North are two old rings in contact, the western and more conspicuous being Krafft A. The unnamed

formation X, E. of Vasco da Gama and S. of Caramuel, is deep, conspicuous, and too prominent for complete anonymity. It may be Rodes, probably it is. Krafft A is presumably "Armenter". On the whole, my early chart was not too inaccurate, but this observation does clear up a few points.

On January 16th 1956 the incoming Astronomer Royal, a man named Richard van der Riet Woolley, re-iterated his long held view that Space Travel was 'Utter Bilge', in the prestigious *Time* magazine. Needless to say, to members of the British Interplanetary Society, like Patrick, who dreamed of spaceflight in the near future, this was a devastating comment, especially if you planned earning money from writing fiction and factual books on the subject. The *Daily Express* newspaper decided to track down some more optimistic opinions and their reporter Donald Gomery interviewed the BIS secretary Leonard J. Carter at Vauxhall Bridge, and also Patrick at his East Grinstead home. This was a full 15 months before Patrick would start *The Sky at Night*, but he was already known to the press, following the success of *Guide to the Moon*. Mr Gomery reported his meeting with Patrick, in his newspaper column, on January 18th:

'I met Patrick Moore, author, astronomer, authority on the moon, and Interplanetary member.' Patrick's comments were reported as: "Orville Wright, back in 1903, could hardly have been expected to design a Comet airliner. Neither can we design a spaceship today. But man WILL get to the Moon."

Of course, just 21 months later, Sputnik would prove that the Astronomer Royal's comments were 'Utter Bilge' and Patrick's were spot on.

In 1956, an event which Patrick was later to call 'The Great Filter Battle' was to start. Patrick's longtime BAA friend J. Hedley Robinson had been invited by the prominent and controversial astronomer V. Axel Firsoff to make observations of the planet Mars, using coloured filters at the eyepiece. While involved in this work, Robinson decided to observe Venus through coloured filters too. He found that the phase of Venus looked smaller through a blue filter than through a red one. A number of prominent observers confirmed Robinson's findings, but others disputed them. The battle over who was correct would continue for many years. Indeed, it was only in the mid 1960s, when high quality photographs by another BAA member, Cdr Henry Hatfield, confirmed the effect, that the battle ended. The 1956 opposition of Mars, according to Patrick himself, marked the point at which he went from being a 'fair' planetary observer to a 'decent' planetary observer, after 8 years of regular planetary sketching. His eye, brain and hand were now as skilled in drawing at the eyepiece as all but the best planetary artists. He would remain hooked on Mars observing for the rest of his life. Patrick's planetary drawings were never beautiful renderings, but they were fairly accurate and loads of them were made at every planetary apparition.

During the 1956 May 30th BAA meeting, held at Burlington House in Piccadilly, Patrick was asked by the President, Dr R.L. (Reggie) Waterfield, to update members on recent observations of the tiny planet Mercury. During late April and early May of 1956 Mercury had been well placed in the evening sky and, as he was the Mercury & Venus Section Director, Patrick had collated the observations received. All of these observations were from his best BAA friends, young or middle-aged. Thus, the observers were

Patrick himself, a young Peter Cattermole and a young Brian Warner (these two using a 6.5-inch reflector and a 3-inch refractor), the eccentric 'water on the Moon' theorist V. Axel Firsoff, Bill (my wife has a 13-inch waist and there's a cat on my shoulder) Granger (much more on him to come later), and the highly experienced observer M.B.B. Heath. Patrick had used his 12.5-inch reflector and the aforementioned 6.5-inch reflector to study the tiny disc of the nearest planet to the Sun himself, just above his East Grinstead horizon. He observed Mercury ten times between April 21st and May 9th and Patrick reported: "I was unable to see any definite shadings, other than a general darkness in the area of Solitudo Hermae Trismegisti." In his report to the BAA Journal Patrick made sure his young apprentices, Cattermole and Warner, had their sketches reproduced alongside his own. Encouraging young observers would always be an absolute top priority for him.

The 1956 evening apparition of Venus also saw Patrick make his most sightings in one season of the enigmatic Ashen light, the glowing of the planet's dark side when it is a thin crescent. On every evening from May 18th to May 22nd he observed the phenomenon, and again on June 4th and 5th, as well as suspecting it on a number of other evenings.

The BAA meeting held on Wednesday June 27th 1956 at Burlington House was the occasion where the President Reggie Waterfield announced that 'the Council had reluctantly accepted the resignation from the Association of Dr H.P. Wilkins, who had directed the Lunar Section so ably for the past 10 years.' Mr Ewen Whitaker (born 1922) was appointed as the new Lunar Section Director. Whitaker was a British contemporary of Patrick's and, in truth, if anyone could claim to be a Moon-mapper, who assisted NASA prior to the Moon landings, it was him, not Patrick. In 1954 Whitaker drew and published the first ever accurate chart of the South Polar area of the Moon. After meeting Dr Gerard P. Kuiper, Director of Yerkes Observatory, at an International Astronomical Union Congress in Dublin in 1955, he was invited to join Kuiper's fledgling Lunar Project at Yerkes, to work on producing the best photographic atlas of the Moon available. Even as Wilkins' left the BAA, his life's work was being completely superseded by the photographic approach. Ewen Whitaker was, undoubtedly, the best man to take over the BAA Lunar Section. At that same BAA meeting on June 27th 1956, the date of Patrick's mother's 70th birthday, the 33 year old Patrick was certainly present (he spoke about radio noise from Venus!) but he made no comment about Wilkins' resignation. However, the octogenarian Robert Barker did pay a tribute. Barker was recorded as saying that Wilkins had built a strong Lunar Section and acted many times as an unofficial but successful BAA ambassador in other lands. It is interesting to note that another young British amateur astronomer, who would eventually become as well known as Patrick in the astronomical community, was at that meeting. A young Brian Marsden (1937–2010), not yet turned 19, was already a noted member of the BAA Computing Section and he delivered a paper on that day entitled 'Transits'. This was the result of calculations he had made to predict the dates (between 1930 and 1980) when an observer on any planet, from Neptune to Venus, might see an inner planet (from Uranus to Mercury) transiting the face of the Sun. In an era when there were no

PC's or electronic calculators this was quite an achievement from an 18 year old. Of course, Marsden would go on to become a world authority on comet and minor planet orbit calculation, as well as becoming the director of the Minor Planet Center at the Harvard-Smithsonian Center for Astrophysics.

In the years before Patrick would become famous he had been rather frustrated with events inside the BAA. His partnership with Wilkins had dissolved and they no longer ran the Lunar Section; Ewen Whitaker was now fully in control. Even though Patrick had become the Mercury & Venus Section Director this too had been dogged with problems, not least the loss of almost all of Henry McEwen's Section records. Then there had been the criticism of his early books by Henry King (who, horror of horrors, was even being tipped as the future BAA President). Also, as we have seen, there was that 'Great Filter Battle'. From 1956 to 1958 Patrick was happy to be elected President of the new Junior Astronomical Society (the JAS, now called the Society of Popular Astronomy or SPA) and spend less time immersed in BAA politics and feuds; but he would still attend all of the BAA meetings and continue being the meetings recorder. It was while he was installed in his new JAS Presidential role that he would become a household name on British television.

Chapter 6
The BBC

In 1955 the BBC producer Paul Johnstone had picked up a battered but recent book, obviously well-used, which had been left behind by someone in a flat where Johnstone was living; who that person was we will probably never know, but by leaving that book behind he created a monumental twist of fate for astronomers everywhere. It was that book *Suns, Myths and Men*, by Patrick Moore, illustrated by David Hardy, and given that scathing review by Henry King. I suspect that the title of Patrick's book was actually inspired by the popular book *Moons, Myths and Man*, written by H.S. Bellamy, which was published in 1936. As a historian and archaeologist Johnstone had little knowledge about astronomy, but he had not been reading the book for more than a few minutes when he felt sure that here was an ideal subject for television [he describes this in the Foreword of the first *Sky at Night* book]. It took Paul some time to persuade the BBC programme planners to agree to T.V. broadcasts on astronomy, but eventually a few programmes were transmitted. They were delivered by the aforementioned BBC radio 'voice of astronomy' and former BAA President, Dr J.G. Porter. However, it was quite obvious that Guy Porter was not happy in front of the TV camera, transmitting 'live' as almost everything was in those days. Neither was he happy with all the elaborate make-up needed in front of the low resolution, low contrast cameras of the 1950s. The resulting programmes had insufficient clarity and simplicity for the layman to follow and the broadcaster appeared more wooden, nervous and stilted than enthusiastic. Despite Porter's abilities on BBC radio, the programmes died a total death.

However, Paul Johnstone decided he was not about to give up. As he remarked, somewhat tongue-in-cheek: 'Having built a working model of scaffolding and elastic to illustrate the expanding universe theory, I had confirmed the visual possibilities of the subject'. He had also learnt the hard way that it needed a VERY special type of person in front of the camera to deliver such a subject on TV.

Some time later in 1955, Paul Johnstone was put in charge of a topical news item about the claimed sightings of Flying Saucers by the hoaxer George Adamski. As we have seen, in the 1950s Adamski had photographed ornate light fittings in his home and claimed they were alien spaceships. He had also hurled a variety of

M. Mobberley, *It Came From Outer Space Wearing an RAF Blazer!: A Fan's Biography of Sir Patrick Moore*, DOI 10.1007/978-3-319-00609-3_6,

circular and metallic objects, such as dinner plates, across fields and photographed them, claiming that they were alien spaceships too. But worst of all, he had boasted that he had been aboard an alien spaceship bound for Venus. In an act of breathtaking self-publicity Adamski told the BBC that he could make a Venusian spaceship appear over the Lime Grove TV studios! Patrick's good friend and, still, a Flying Saucer sympathiser, Desmond Leslie, was a friend of Adamski's and had originally made contact with Paul Johnstone and suggested a programme about Flying Saucers on the BBC. In 1954, Leslie's wife Agnes wrote an article in the *Daily Mail* in which she quoted Desmond as saying "if all goes well there will be Flying Saucers landing in London next year!" However, despite the interest from the public and the media in Flying Saucers, Johnstone insisted that an astronomer should appear on the programme to dismiss the claims.

Patrick, despite his friendship with Desmond Leslie and numerous other dubious UFO spotters, 'independent thinkers', parapsychologists and nutters (and even despite his alter ego of Cedric Allingham) was always keen on exposing UFO spotters as cranks. He would always publicly put forward his astronomers view, namely "it's all utter bunk". His book under the Allingham pseudonym had almost certainly been written partly to prove that "anyone can write this drivel", as well as to make money. He was a friend of Leslie because both were keen amateur astronomers and ex RAF staff, excited by the prospect of alien life; but Patrick was constantly telling Leslie that Adamski's stories were all made up. Thus, Desmond Leslie suggested Patrick's name as a sceptical astronomer who would appear on the planned BBC programme, to put forward the establishment view. Patrick agreed that he would be happy to appear as the 'voice of reason' on the proposed 1955 TV piece. He was shrewd enough to know that Flying Saucers attracted big publicity and so he could get a useful bit of free advertising for his astronomy books by appearing on the T.V. or radio. Initially, the BBC's Paul Johnstone was most reluctant to have someone suggested by Leslie to put the opposite view: after all, just how sympathetic might this amateur astronomer's view be? Johnstone wanted to use Dr Porter, but Desmond Leslie's friendly enthusiasm won him over and he agreed that he would meet Leslie's astronomer friend. So, Leslie played that crucial role in Patrick's first TV appearance. In addition, Guy Porter, who knew Patrick well from within the BAA, confirmed that Patrick would do a great job. In 1938 Porter had taken over as the BAA Computing Section Director, and the man he had inherited that post from had been none other than Major A.E. Levin, who 4 years earlier had proposed the 11 years old Patrick for BAA membership. It's a small world and various lucky connections were lining up in Patrick's favour in the 1950s.

The news item went ahead, and to no-ones surprise no Flying Saucers appeared over Lime Grove. As well as Desmond Leslie promoting the Flying Saucer case, Air Chief Marshall Sir Hugh Dowding backed the UFOs as real (remember, he believed in pixies and gnomes too!), and a couple of other Flying Saucer fruitcakes also appeared on the programme. Patrick delivered a brilliant performance in front of the TV cameras, as the only voice of sanity, and it had now dawned on Paul Johnstone that Patrick was actually the author of *Suns, Myths and Men*, the battered book left behind in his flat that had started off his idea for a BBC astronomy programme in the first place. It was also obvious that Patrick was such a natural performer that, at last, Paul had found the TV astronomer he had been looking for to replace Dr Porter himself. Also, as we have just seen, Porter knew Patrick's speaking abilities within the BAA only too well and was happy to reaffirm his backing of the young and boundlessly enthusiastic Mr Moore. Guy Porter just was not happy in front of a camera, only in front of a microphone. In the subsequent days Paul discussed the possibilities with an excited Patrick and put the case for a second trial astronomy series to the BBC board.

Whichever way you look at it, Patrick's BAA speaking performances had played a major role in his TV future. Not only was the BAA's Dr Porter aware of his oratory abilities, but so was Arthur C. Clarke, back in 1951, when his invitation to Patrick to speak in front of the British Interplanetary Society had led to his first astronomy book *Guide to the Moon*. Add the Desmond Leslie/Hugh Dowding connections into the equation, and Paul Johnstone having a copy of *Suns, Myths and Men*, and you then have all the breaks which made Patrick Moore a future TV star and a household name. So, you might ask, where was Arthur C. Clarke around the time that Patrick was on the verge of television fame? Well, being a keen scuba diver, in 1956 Arthur had decided to move to live in Ceylon (now Sri Lanka) and continue writing science fiction from there. He purchased one of the very first Questar 3.5-inch. Maksutov telescopes (which he absolutely loved) and from Ceylon he could see both celestial hemispheres of the sky, as well as scuba dive any time that he wished. Arthur would later purchase a fork mounted Celestron 14, which he used on many blissfully warm nights from the patio of his Ceylon home. Despite moving to the dark skies of Ceylon, Arthur's first love, as an organisation at least, would still be the British Interplanetary Society (BIS) and not the BAA. However, although Patrick had already acquired many astronomical roles, within the BAA and the JAS, he was happy to take on a demanding new post within the BIS itself. The society, unaware of his impending fame, asked him to become the editor of their new magazine *Spaceflight* in October 1956. Patrick accepted and would edit the new magazine for the next 3 years, after which he would hand the post over to a very worthy successor, the space travel enthusiast Kenneth Gatland (1924–1997).

Bizarrely, apart from Dr Porter, in 1956 the camp comedian Frankie Howerd (1917–1992) stood an outside chance of pipping Patrick to the post of BBC astronomy expert, despite the fact that he only had a passing knowledge of the subject. No, I'm not making this up! I was amazed when Patrick told me this tale and thought that he must be joking, but he was absolutely serious! The BBC of 1956 was reeling from the launch of Independent Television on September 22nd 1955 and less than 30 % of viewers were watching the BBC on some nights. Of course, the BBC was funded by a licence fee, not advertising revenue, but it was all a matter of pride. Howerd had worked with the BBC since December 1946 on the *Variety Bandbox* radio program and throughout the late 1940s and early 1950s had become a popular radio entertainer. He made his screen debut with the singer Petula Clark in the film *The Runaway Bus* and enjoyed scripts mainly written by a young Eric Sykes but with contributions from Spike Milligan, Galton and Simpson and Johnny Speight too. A three part BBC TV series called *The Howerd Crowd* had been popular in

1952, as had two other shows entitled *Nuts in May* and *The Frankie Howerd Show*. In 1956 Howerd was at a peak in his popularity, although the bubble would soon burst, but with BBC viewers switching to ITV en masse someone popular was needed to restore the ratings.

A junior BBC producer named Percy Lamb had worked with Frankie Howerd in films and suggested a new educational, but entertaining, series for the comedian where he explored space. The series would be entitled *Frankie Looks Up*! Each program would start with Howerd winking at the camera and quoting from Oscar Wilde: "We are all in the gutter but some of us are looking at the stars!" Presumably interspersed with Howerd's trade mark "Oooh no missus, Titter ye not!" But just to make doubly sure of the ratings figures there would be a co-presenter, namely the 'blonde bombshell sex kitten' Sabrina (born May 19th 1936) who was, in every way, Britain's answer to Marilyn Monroe and Jayne Mansfield combined, and more! Sabrina, born Norma Sykes, had the face of an angel combined with a jaw-dropping 42-17-36 figure. She frequently exploited her hourglass proportions with a plunging neckline. First introduced to TV by Arthur Askey when she was 18 she rarely said a word, she just posed, while the adult male viewers dribbled. When she was on TV, all 1950s men tuned to that channel! In Frankie Howerd's proposed *Frankie Looks Up*! show she would be called 'Cuzit from Cassiopeia' and wear a slinky silver space suit, no doubt with the front unzipped as far as 1950s TV would allow. Surely, the unknown Patrick could stand no chance whatsoever against that sort of competition?! This was not all though. The singer Alma Cogan was also booked to appear in the show. Dressed in an alien costume with waggling antennae she was due to sing "I'm going steady with Eddy from Earth!" and "I'm M,M,M,M,MAD about Martian men!". Surely it was time for Patrick to exit stage left. His career finished before it started? However, Paul Johnstone was a more senior producer to Percy Lamb and Paul's background was in archaeology and science, not comedy; and, ultimately it was a science program that the BBC wanted. There were plenty of other outlets for Howerd and Sabrina, more suited to their talents. Lamb was not about to give up though and even after he learned of the Paul Johnstone/Patrick Moore venture he arranged a rehearsal in front of the acting BBC Director-General Sir Henry Rawlinson. However Sabrina's charms failed to convince Sir Henry and even though he shared a meal with Alma Cogan and her lesser cleavage, after the show, it just had not worked. Frankie Howerd's double entendre gags, including the old favourite "let's have a good look at Uranus" just didn't work in a program that was supposed to be educational. With the strict BBC standards of the day the Howerd/Sabrina double act looked more pornographic than scientific. As 1957 started Sabrina starred as Virginia in *Blue Murder at St Trinian's*, a non-speaking role in which she was only required to sit up in bed wearing a nightdress while reading a book, despite sharing equal billing with the main star Alastair Sim and appearing on posters up and down the land dressed in a skimpy schoolgirl uniform! Meanwhile, Patrick was poised to become a star of a different sort, wearing a blazer, not a skimpy schoolboy uniform!

In fairness to Patrick, despite being unknown, he had a number of attributes which were especially well-suited to appearing on the BBC of the 1950s. Firstly, he looked the part. In those days the 34 years old Patrick cut a striking, if eccentric, figure;

not yet overweight, but tall (just over 6 foot) and with clean cut upper class looks and jet black hair, he fitted straight into the BBC mould. But, perhaps most importantly, he had an absolutely perfect BBC voice; clear, confident and rapidly speaking the Queen's English without hesitation, he could have been made for the BBC. This single factor and Patrick's enthusiasm and knowledge of his subject made him a dead cert winner for TV. He also looked straight at the camera as if he were looking at the viewer, unlike most TV scientists of that period. Not surprisingly, the BBC decided to give Paul Johnstone a second chance at bringing astronomy to a British TV audience. Of course, for Patrick, even a few months waiting for their decision must have been agonising. Patrick was summoned to an interview with the BBC to explore the possibilities of an astronomy programme in June 1956. He was asked to perform an audition, in front of the TV cameras, to see if he was 'any good' at broadcasting astronomy. Needless to say Patrick performed well, but the BBC was still not convinced there was any need for such a programme, whether it starred Patrick or even Frankie Howerd alongside a blonde bombshell. Nevertheless, Patrick was highly excited. In a more recent communication I received from Patrick's old friend Richard Baum, Richard recalled his own honeymoon in London at that time:

> Audrey and I married on July 21st 1956. Patrick gave us a sugar sifter which still holds an honoured place. Within minutes literally of our arrival in London at the start of our honeymoon, Patrick phoned and asked if he could come over to our hotel [The Strand Palace]. We said that was fine and at the meeting he confided to us his recent audition with the BBC for a projected series of six programmes on astronomy. We saw him again during our honeymoon when he, along with a number of young people (including if I recall right Brian Warner), accompanied us around the RGO.

Incidentally, I have said that the young Patrick was just over 6 foot tall, because that is how tall he appeared to me. I am 6 foot tall and, when I was a young man of 18 or 19 and Patrick was 53 or 54, he appeared roughly my height, or maybe a fraction taller. I have photographs of myself and Patrick standing together when I was 26 and he was 61 and we are the same height (if anything I look slightly taller by then). On the *Sky at Night*, Patrick actually once stated that he was "just over six feet tall" when he was standing next to a Voyager spacecraft, but the height quoted by the press, in tabloid newspaper interviews, often became exaggerated to six foot three or so! "The damage to my spine in the war lost me several inches of course" he would quip. In his seventies and eighties he definitely shrank quite a bit and he certainly cannot have been much over 5 foot 10 inches at the end of his life. Certainly, Patrick's Churchillian build, lack of a neck, and broad shoulders made him sometimes seem a lot taller than he actually was. Indeed, in Patrick's 1973 book 'Astronomical Telescopes and Observatories for Amateurs', published by David and Charles, his Selsey friend Reg Spry, who built Patrick's 5-inch refractor's run-off shed, describes Patrick (pages 226–227) as '6 ft 3in', which is clearly far too tall. Despite Patrick's editing of that book he did not correct that error! However, being a TV personality and occasional actor Patrick's details were available from entertainment agents who kept his data on file. Most of these agents list his height as 6 foot and half an inch (184 centimetres) which seems about right to me, at least when he was relatively young and in his prime; but I have digressed, so let's get back to the plot!

During the closing months of 1956 Paul Johnstone became the producer of a BBC programme called 'First Hand'. The first edition, in those monochrome 405 line days, was transmitted on November 20th, followed by a second, on the subject of the sinking of the Titanic, a week later. Paul decided that there was enough material available on Flying Saucers for a complete BBC show in his new series and, with Patrick having already waited 6 months since his audition, he, once more, invited him and his UFO friend Desmond Leslie to take part. Johnstone also invited the radio astronomer Dr R.C. Jennison, from the Jodrell Bank experimental station (as it was then) to join Patrick as a sceptic, and Air Chief Marshall Lord-Dowding also appeared again, promoting the UFO cause. This third programme in the 'First Hand' series was transmitted live on Tuesday December 4th 1956. The show also included filmed interviews with people, such as a tight-lipped veteran American pilot and UFO witness who was keen on the phrase "I'm not at liberty to say that Sir…." The programme was mentioned in a few newspapers the next day and the *Manchester Guardian* report was particularly interesting, stating: 'Patrick Moore, the leader of the opposition, had devilish eyebrows and the fanatical utterance of the born demolishing debater'. Regarding Desmond Leslie's argument, the newspaper also stated that: 'Mr Moore and his experts knocked large holes in it, although probably enough doubt remained for those who, as Mr Moore said, want to believe in Flying Saucers because they sound "so much more fun" than the orthodox explanations.'

The BBC did not make quick decisions in the 1950s, but Paul Johnstone kept prodding over the coming months and told Patrick there were encouraging signs regarding a BBC astronomy programme with him as the host. From October to December numerous letters passed between Paul and Patrick and they are still held in the BBC Archives. It is clear from reading these that Patrick was bluffing Paul that he was being headhunted by the ITA (ITV) for a similar programme to the proposed BBC 'Stars of the Month' show, as it was then envisaged, and so a quick decision by the BBC was required, following the 'First Hand' UFO programme. On the first Wednesday in January 1957 Patrick was attending a BAA meeting in London. After the meeting he made his way on the Underground to the British Rail Victoria station and then caught the train to East Grinstead. The weather, in Patrick's words, was "cold, damp and foggy – typical of the early months of 1957". As Patrick walked in the front door after the 2 hour train journey (to cover 30 miles!!) his excited mother handed him a telegram that read: 'You have been accepted for a television series. Please ring Paul Johnstone, at the BBC, as soon as possible'. A further communication from the contracts department would confirm he would be paid 25 Guineas per show. Not surprisingly Patrick was absolutely ecstatic, and their cat during that time, named Smudgie, was also very pleased. I must admit I have no experience with moggies so I shall just have to accept Patrick's word that his cat understood these things!

Early the next morning an excited Patrick phoned Paul Johnstone and then made his way back to East Grinstead railway station and thence to Victoria and the BBC's Lime Grove studios. Little could he guess at that stage that he would be travelling back and forth to the BBC in London, every month, for the rest of the century, and filming the show for the next 55 years.

The First Programmes

Paul and Patrick had a long discussion. The original attempt to broadcast astronomy on the BBC had not worked well, so what needed to be changed? Paul Johnstone felt that every television producer had a hidden sense of guilt and that, however good, amusing and interesting his programmes were, it was always a one-sided activity. The more people watch, the more people stop doing something creative and just start goggling at a TV. Even in the 1950s he felt that passive non-creative leisure was one of the failings of modern society. As an archaeologist, previously involved in making archaeology programmes, he wanted to encourage viewers to visit Stonehenge or to go on a dig. He was pleased to see that Patrick had similar views; he wanted to encourage people to go outside and view the night sky.

One of the fundamental issues was the choice of the programme's title. Initially, it was to be called *Star Map*, but just in time for inclusion in the *Radio Times* for the last week in April, Patrick and Paul decided to rename it *The Sky at Night*. The theme music was another issue. Holst's Planets suite was considered, but Patrick thought it was too obvious and too *astrological* (Holst's music was inspired by the planets' astrological characteristics). John Carter, the gramophone librarian of BBC Television, was asked to look out for a good and recognizable tune which had a feeling of 'the music of the spheres' about it. He came up with 'At the Castle Gate' from Sibelius' Pelleas et Melisande suite, which Patrick whole heartedly endorsed. The original set was the responsibility of Nancy Thomas, a skilled BBC programme director. She chose to make a distinctive giant planisphere as a backdrop for the programme. The illustrator of Patrick's 1954 book, *Suns, Myths and Men*, David Hardy (now 21), and another designer, Patricia Cullen, were chosen as the illustrators and then the programme was almost ready to go. The reader may recall I mentioned earlier that Patricia Cullen, who would illustrate many books (and even draw lunar charts) for Patrick, was married to Patrick's former teaching colleague and occasional author/translator Alex (Sandy) Helm. The talented amateur astronomer Leslie F. Ball was also hired to draw many of the illustrations in the early *Sky at Night* programmes and many figures in Patrick's books.

The final detail, by Johnstone and Moore, was to visit a bizarre establishment in London's Goldhawk Road, inhabited by an ex-Viennese polymath (a person of much and varied learning). The inhabitant was one Alfred Wurmser who, assisted by a Dane named Skat, and a massive black Alsatian ("the size of a pony" according to Patrick), presided over several cellars containing, arguably, the world's largest collection of assorted pieces of cardboard! Wurmser was a genius at turning out moving cardboard models in record-breaking times, for a variety of BBC programmes. Having an understanding of almost every aspect of science and technology, Wurmser had the gift of being able to see how the right animated cardboard figure could educate the viewer, at a stroke, and to produce the required model, to order, within hours! His massive dog Till was, according to Patrick: "under the delusion that he was a lapdog!"

No-one knew how the brand new programme would fare with the viewers and, initially, only three programmes, not six, were scheduled for April, May and June 1957.

So, with everything in place, *The Sky at Night* was ready to roll; but there was one dramatic development left and one that, arguably, would ensure *The Sky at Night* survived for more than three programmes. Some 5 months earlier, on November 8th 1956 to be precise, two astronomers at Uccle Observatory, in Belgium, had discovered a new comet heading for the inner solar system. The astronomers were called Arend and Roland so the comet was, unsurprisingly, named Arend-Roland. Measurements were made of the comet's position and the orbit was calculated. It would reach its closest position to the Sun (perihelion) on April 8th 1957 when it would be at its brightest and within 29 million miles (47 million kilometres) of the solar surface. Even more exciting was the fact that the comet was moving into the far north of the celestial sphere, so it would be easy to see from the UK. Further calculations suggested that the Earth would actually pass through the plane of the comet's orbit on the night of April 24th, the night scheduled for the transmission of the first *Sky at Night* and 106 years *to the day* since the birth of Patrick's first mentor, W.S. Franks! When the Earth passes through the plane of a bright comet's orbit, it is like looking through the edge of a sheet of paper. Suddenly you can see all the dust particles near to the comet, and a line or spike of material can often be observed in addition to the comet's tail. This spike, when visible, is called an anti-tail. Throughout the month of April comet Arend-Roland brightened until it was the most conspicuous object in the night sky (apart from the Moon). In fact, many considered it to be the best comet, visible in a dark sky (as opposed to twilight), since Halley in 1910. What a superb and lucky comet to usher in the new TV programme! Not surprisingly the topic of the first programme became Arend-Roland, with the original topic of a forthcoming lunar eclipse shunted to the end of the programme. With all this TV excitement going on it might be thought that Patrick would have been neglecting his contributions to amateur astronomy, but that was certainly not the case. Patrick's feet were firmly on the ground. Only days before the first *Sky at Night* Patrick gave a talk to the newly formed Guildford Astronomical Society (an offshoot of the Junior Astronomical Society) at Tangley Cottage, Wonersh. It was a lecture, with lantern slides, about observing the planets, and Patrick suggested to the society that they should invite local schools to send any budding junior astronomers to their meetings.

The first mention of the new astronomy programme was in *The Radio Times* that went on sale on April 19th, covering the next week's BBC schedule. It was a simple magazine in those days. It only had one BBC TV channel to list! Apart from the actual entry of *The Sky at Night* in the listing for the coming Wednesday, after Cyril Stapleton's Show Band Parade, before the 10.45 News and alongside adverts for Chivers Jelly and Tyne Brand Stuffed Pork Rolls, there was a preview and a constellation chart. The preview snippet was labelled 'Star Quest' and contained a comment by Patrick: 'People tend to think that astronomy is a difficult, expensive and unrewarding subject that has long become the prerogative of old men with long white beards. It is in fact none of these things and anyone can find interest and excitement in the night sky, if he knows what to look for'. The handy chart printed on the page for the coming Wednesday showed the Great Bear and Leo, with lines indicating how to identify the Pole Star, Arcturus, Spica and Jupiter. Underneath the star chart a caption read: 'This map should help you to follow Patrick Moore's talk.'

On Wednesday April 24th 1957 Patrick arrived at the BBC Lime Grove studio early on, for the rehearsals. Even when in his seventies and eighties Patrick would

NEVER use an autocue, and in the 1950s and 1960s he never used one either, despite being *live* for 15 minutes, on prime time BBC. How many other presenters can claim that? Well, none of the modern bunch that's for sure; they all read like dummies off the autocue or from printed cards, or enjoy the luxury of multiple retakes! Fifteen minutes is not much time to get your message across, and the producer suggested to the fast-speaking Patrick that he might even increase his delivery rate, as he could obviously speak fast without making errors. He also advised Patrick to dispense with the monocle in his right eye as 'monocles look old-fashioned'. Fifty years later Patrick's producer would be more likely to ask him to insert his monocle than remove it. It became his trademark. With the programme already being mentioned in that week's *Radio Times*, *before* it was made (like all of the live programmes of that era) Patrick had to deliver the goods. There would be no takes and re-takes. It had to be right first time, so exhaustive rehearsals before transmission were essential.

At 10.30 p.m. that night Patrick rememembered thinking: "My entire life depends on what I do during the next fifteen minutes." At precisely 10.30 p.m. the screen on the studio monitor began to glow and Patrick saw the words *The Sky at Night. A regular monthly programme presented by Patrick Moore*, appear on the screen. The stylus was lowered onto the Sibelius L.P. and the 'At the Castle Gate' theme music played out. The red light on the TV camera in front of him came on, and Patrick was off, his whole life changed for ever. No-one could have dreamed at that time that *The Sky at Night* would become the world's longest running TV programme with the same presenter and would survive for 722 episodes and for the next 55 years and 9 months, with him in charge. Patrick began the programme by explaining that astronomy was not just a hobby for white haired old gentlemen with beards and he said that in these monthly programmes he hoped to show people what they could actually see in the night sky each month. He described that bright comet Arend-Roland and explained that it would not return for thousands of years, so this was the viewers' only chance to see it. He also mentioned the comet's unique double tail: the main tail and its forward pointing anti-tail spike.

Undoubtedly, one of the reasons that the *Sky at Night* survived for so long was that it cost next to nothing to run. For the whole period of its existence, Patrick never signed a contract with the BBC! He had what he called a 'gentleman's agreement' with them. Essentially, the *Sky at Night* was a labour of love for Patrick. He had spent much of his early years giving talks and working for the BAA for no income and, even in the 1950s, he realised that TV was a fickle medium. One moment you were famous and the next you were a 'has-been', especially if you got greedy. Patrick knew that if his name became a household one he could earn a decent living, NOT from the BBC, but from selling books. His longevity and status in the world of astronomy was also far more important to him than making a 'fast buck'. Being a single man, and living at home with his Mum, he also had few outgoings. As long as the alter ego of Cedric Allingham never emerged to smear his name, he would be OK! He did not need a fat pay cheque. He just wanted the programme to survive and for his name to be known, and to encourage people, adults, or children, to observe the night sky.

Following the first programme, which seemed to go well, Paul Johnstone went nervously to his weekly BBC departmental progress meeting, to hear what his fellow colleagues (and bosses) thought of the new show. One particularly distinguished and articulate producer, later to become a fan of Patrick's, launched a scathing attack on the programme, ending with the emphatic pronouncement: "This is just the sort of thing we should not do". Paul Johnstone apparently won the exchange simply by stating, after this endless diatribe: "I don't agree" upon which all those present collapsed into laughter. Despite the timely intervention of the comet Arend-Roland it was by no means obvious that the programme would survive beyond its allocated three programme run! If it had not worked, but had been Paul Johnstone's second failure to deliver a successful astronomy format on BBC television, then that would have been that.

At home, in East Grinstead, Patrick also waited anxiously for feedback. He was especially concerned to know whether it had been too elementary or too advanced for the viewing audience. After a few days some reaction to the programme began to trickle in to the BBC in the form of dozens of letters. These were forwarded to Patrick who answered them all on his 1908 Woodstock typewriter. Most were very favourable, but as anyone who has ever been involved in astronomy will tell you, you will get plenty of crank mail! Patrick often recounted two letters in particular. One was from an old biddy who wanted to send a carrier pigeon to the Moon to talk to the aliens there. Patrick suggested sending a 'pigrot', a cross between a pigeon and a parrot, which could obviously talk to the aliens! Another crank letter was totally incomprehensible: 'Dear Mr Moore, Enjoyed watching your programme about Comets. On a similar subject, do you know where I can acquire an Army Tank. Have you got any?' "Mad as a coot" Patrick would invariably say of such people. I think he meant Mad as a hatter! After a short time Patrick developed a unique way of dealing with such letter writing cranks: simply put them in contact with each other. Thus he would tell a Flat Earth crank that he should seek more information from another crank, one who believed that the Earth's surface was the inside of a hollow sphere. The resulting letters must have been mind-boggling, but as Patrick often said: "Every Nut thinks every other Nut is a Nut". In other words, they could see the other character was a crank, but were convinced that their own mental state was perfectly fine!

The comet Arend-Roland, which ushered in *The Sky at Night*, had literally peaked in brightness and visibility during the week of the first programme. However, in the 1950s developing films and photographic prints took time, and Patrick was well aware that there could be some impressive photographs of the comet, taken by BAA members, which would be very suitable for the second programme. Totally unlike many modern twenty-first century TV astronomy presenters, the young Patrick was keen that *The Sky at Night* should showcase the work of BAA members. He was only too aware that he was just one amateur astronomer amongst thousands and in the field of astrophotography he was insignificant. These days it is common for vain TV presenters to 'hog' all the air time themselves and pretend other experts simply don't exist, but Patrick wanted as many of his friends and colleagues as possible to appear on *The Sky at Night* while it lasted. For Patrick this was as much about the BAA's observers getting onto television as anything else. On the afternoon of that very first programme the BAA had held its usual last Wednesday in the month meeting, and despite his date with destiny Patrick had made time to be there. A number of Arend-Roland photographs were on display, some by the comet expert and former President Reggie Waterfield, whose best picture had made the front page of *The Star* newspaper that week. Nevertheless, the one taken by Bill and Ethel Granger from Peterborough really caught his eye.

I will, unashamedly, keep mentioning the Grangers throughout this biography! They were the most entertaining eccentrics in the BAA throughout the 1950s and 1960s and so a bit of repetition will not go amiss. Apart from the fact that they invariably turned up at BAA meetings in a motorbike and sidecar, with big Bill plus the family cat shoehorned into the small carriage, and tiny Ethel astride the big motorbike, there was another jaw-dropping aspect to the couple. Ethel Granger was renowned for her 13-inch circumference waist. No, that is not a mis-print! Such a small waist was achieved by the traditional method of increasingly tighter corsets being applied over many years, and in the corsetry field Ethel was legendary. In addition, she was so be-decked with earrings that BAA members thought her ears would surely fall off! Bill Granger always carried a pocket full of earrings and jewelry insertion/extraction tools so she could change her jewelry throughout the day, whether they were out for a stroll, or on the motorbike! Around 1956 and 1957 Bill and Ethel were already hitting the headlines in Peterborough and beyond, mainly for Ethel's fame in the corsetry, lingerie and jewelry world, but also for their joint hobby of astronomy, which Ethel had become as interested in as Bill. Incidentally, Bill Granger, whose politics were distinctly communist, was a schoolteacher specialising in woodwork.

Fig. 6.1 Bill Granger (Shown here with his cat 'Treacle Pudding') was the first guest on *The Sky at Night*, appearing in May 1957 on the second programme (Photograph reproduced by kind permission of the Peterborough Telegraph)

So, to return to the plot, on the very day of *The Sky at Night* original episode, Patrick was impressed by Bill's photograph of comet Arend-Roland and asked him if he would like to appear on the second programme, thus being the very first guest on *The Sky at Night*? Well, how could Bill possibly refuse that offer, although surely he could not have dreamed Patrick would present it for another half century and more? Sadly, few programmes from the early years of *The Sky at Night* survive, but I managed to track down a description of Bill Granger's experience leading up to, and during, that second programme, amongst the surviving pages of his draft biography of Ethel. Bill's biography was never published, but the account of that period in 1957 is still intact. So, here are Bill's words about the experience, slightly edited for spelling and grammar, from the original draft manuscript:

> I had, in fact, exhibited some of my first photos of the comet, taken on the evening of 23rd April 1957, with a 21 minute exposure, with my 20 inch Aero Camera on Zenith Astronomical plate. The only other photo taken on the same date was by Dr Waterfield. This, as well as some planetary drawing I had done during the year, was put on show at the Exhibition Meeting of the British Astronomical Association, held in the rooms of the Royal Institution on April 24th, and these two were the first photos taken after perihelion return. It was then that Patrick had just started his new programme on T.V., "THE SKY AT NIGHT".
>
> When he saw my comet photograph he said, "I must have it on my next show." I said I was quite willing, and by then I might have something even better. He told me that he was going down, after the meeting, to put on his T.V. programme, and would make the arrangements for the next one. He said it would be nice if I could come up to be on the show, to talk about the photos. Ethel was with me that evening at the meeting. Meanwhile, at home, I continued with a long series of over 30 photos of the comet, right into June, when Comet Mrkos came along, and this time I was the first one to get a photo, on 10th August 1957.
>
> Patrick was as good as his word, for in a few days I had a letter from the B.B.C. asking if I was willing to come on the programme, "The Sky at Night", with Patrick Moore. Also they wanted two of 'my best comet photos and photos of the observatory'. These I packed up and sent off as requested, so they could prepare the necessary large prints for the programme. I didn't know whether we should both be on the programme or not, but I had expenses which allowed enough to cover the train fare for both of us, so I was determined she should come too.
>
> The official letter, dated 17th May, 1957 said:-
>
> Dear Sir, We invite you to prepare and deliver a talk in our television programme for broadcasting or for recording for subsequent broadcast reproduction .... etc. Title: "The Sky at Night" (to discuss Comet and provide photographs) to be arranged with the producer, Paul Johnstone. Studio H Lime Grove, London, Wednesday 22 May 1957.
>
> Time of Broadcast Wednesday 22 May 1957. 10–15 to 10-30 pm. Fee. £10-10-0 plus expenses, £3-6-2 (Rail fare and 1 nights subsistence).
>
> Of course I had to leave school early and the news appeared in the Peterborough Evening Telegraph of May 9th, with my picture, under which it says:
>
> Mr W.A. Granger (the children at the school used to talk of me as WAG) who will be seen on Television on Wednesday May 22nd. Mrs Granger, who is also an enthusiast at astronomy, will also be seen on T.V. They have accepted an invitation by Mr Patrick Moore to appear on his monthly feature, "The Night Sky". Mr Granger will take with him a photograph of the observatory he built at the back of his house in Priory Road, together with some of the 19 photographs which he has taken of the Arend Roland Comet. (My final total was over 30 photos in all.)
>
> I had previously been in more hot water [at school], because I had applied for regular permission to attend every monthly BAA meeting, as I was free every Wednesday afternoon. They soon stopped that by fixing up that I did not have my Wednesdays free, but I did get the concession that I should have permission if required for a special purpose. However, as this

did not take place until the evening, I just got off work early for the visit. So I left early, came home and fixed up Ethel, in case she should be needed, and left about 4-30 pm from Peterborough to King's Cross. She had on a lovely pair of high heels, her ears were decorated all round with diamante stones, and a nice pair of black pendant rings. She wore a well fitting dress in black velvet with a black leather studded belt, which set off her waist to perfection. We arrived at King's Cross about 6 pm and took a taxi to the Lime Street Studio, to arrive about 6.30 pm, to find that we were expected. In fact they expected us earlier, to have a preliminary run through; so as soon as we announced ourselves they said: "Take him upstairs to studio H. They are waiting." So upstairs we went, to the studio, which was prepared. It was a great empty barn-like building, in one corner of which was set up the necessary backcloth for the show, and all the other items required for the programme to be televised.

Pat was pleased to see me and the producer said "we expected you earlier, so we have had one preliminary run through, with a substitute for you", although why, if they wanted me earlier they did not say so, I cannot imagine, for we could have come up on an earlier train, because the school could not have refused permission. So they said, "We will now have another run through, to see how Mr Granger goes, and how it fits in with the time schedule." I had never seen inside a television studio before and, I must confess, when the actual show began, I had a few butterflies in my stomach and my mouth went dry, so I felt unable to say anything. But it was really only for a fleeting moment, then I got going, and I was O.K. I could see myself in the monitor screen in front, as I was talking. They had two of my comet photos and also a picture of the observatory, with Ethel outside by the observatory door, so she would be seen, but she was not on the live screen as I gathered they had not thought about it. Maybe if we had been in time for the first trial they might have fitted her in too?

While I was having the practice run through in front of the cameras, Ethel was standing just out of the picture at my side. When I felt more relaxed I could look around, and I could see the technicians behind me arguing amongst themselves. I could see by the way they were talking, putting their hands into a circle, it was her waist they were talking about, and I think they were laying bets on its size! They also pointed to their ears; she was not wearing a nose stud at the time, or no doubt that too would have come in for comment.

After a few run-throughs in this way we broke up, to go down for something to eat and drink for all of us: Ethel and I and Patrick. One of the female producers had to comment about Ethel's waist, and she said I would be very good on screen, and it could have been arranged for both of us to be on, but it had to go just with me as it was now all planned. The lady producer said she had never seen anything like Ethel's waist before, and could not imagine anything like it existed. What size was it she asked? When we told her it was under 14 inches she had a shock for she could hardly believe what even her eyes told her to be true. She said, "I am interested because they have just had a show on a collection of old costumes put on T.V. The smallest waist they could get to fit was for a 22 inch waisted dress, while the smallest waist in the collection was a 19½ inch one, which Ethel could have worn with ease." She was also interested in Ethel's earrings, so queried how they were fixed, and I told her she was brave and tough, as they were all pierced in place, to endure all that.

She also said that it was astonishing that Ethel could eat and drink normally and did not diet. After that drink and talk, Pat and I were bustled away to be made up, until I felt like a beauty queen, but I soon forgot the make-up. Then it was up to the studio, where the live broadcast was to take place. It was not scripted, it was not quite like any of the trials either, but we followed the plan fairly well to get it all in. Patrick was still new to the game and inclined to speak too quickly, and had to keep his pipe in hand to help him gain confidence. I must say that the butterflies returned, especially when I thought that thousands of people were probably watching me try to make a fool of myself, but I put the thought out of my mind and concentrated on what I was to say, and I felt fine again. It was very exciting and thrilling, so I felt on top of the world. The producer congratulated me and said that I had done very well and also that they might like to invite us again sometime.

> Pat took us back to the station, so we got home the same night, and I found I had forgotten to take off my make up in the excitement of the moment. Next day I had many friends, and people I did not even know, compliment me on the programme, so I became very well known. They said I came over very well, but why wasn't Ethel on too?
>
> Naturally the local papers, which had told the people about the broadcast, wanted to know all about it and how we got on, so I told them how it went off. I added as an afterthought that I could see the technicians arguing and laying bets, so I told the paper's reporters they were discussing Ethel's waist, and the [jewelry] collections, adding that Ethel had what was probably the smallest waist in the world.

So, that was Bill Granger's account of what it felt like to appear on the second *Sky at Night* as the programme's first guest. In the years to come another Peterborough schoolmaster would dominate the astronomical headlines, but we will come to George Alcock later.

As the new programme was being expertly presented and seemed to be of interest to at least a few viewers, namely the letter writing minority, and also because it was inexpensive to make, the BBC kept it going beyond the initial 3 month trial. The third programme was broadcast on June 20th and the fourth on July 25th. The amateur telescope and mirror making expert Henry Wildey, who Patrick often declared "made the finest mirrors you can buy" was chosen as the guest for the fifth programme, scheduled for August 19th. Remarkably, another naked eye northern hemisphere comet was discovered just a few months into the new programme's life, which can hardly have been a bad thing for Patrick! Comet Mrkos was officially discovered on August 1st, by the Czech astronomer Antonin Mrkos, but a Tamworth schoolboy, who was also a BAA member, Clive Hare, independently discovered it 2 days later and reported it as being 'similar in brightness to the star Pollux and with a two degree tail'. Clive would appear as a guest on the September 19th episode of *The Sky at Night*, which concentrated on that same Comet Mrkos. Another BAA member, namely W.A. Clark from Hull, spotted the comet on August 4th, without knowing of its discovery. Astronomy and comets were certainly getting a lot of publicity in 1957 and Patrick was informing the nation of all these developments.

Then there was another huge slice of luck. On October 4th 1957 the Russians launched the first artificial satellite, Sputnik 1. As well as being a colossal technical achievement for 1957 this was also a very disturbing one, which was a concern for the whole civilised world. If the Russians could put a nuclear warhead on a long range rocket, the world would become a much more worrying place! In addition, at Manchester University, Bernard Lovell had staked his reputation on the enormously ambitious (and deeply in debt) 250 foot diameter Jodrell Bank Radio Telescope; this was the only instrument that could track the new satellite, and it was British. Suddenly, space travel and Jodrell Bank were making the headlines and were topical, so the only BBC programme about astronomy and space was guaranteed a much longer lease of life.

One person who remembers that era very well, Jim Hysom, who was then serving on the BAA council with Patrick, told me once that it was the launch of Sputnik which made Patrick into a national institution and NOT the *Sky at Night*. When Sputnik was launched the *Sky at Night* had been running for 6 months and BBC bosses were seriously considering that it had run its course. Viewing figures were only modest: the public simply were not that interested in astronomy, despite the amazing appearance of two naked eye comets in recent months. When Sputnik 1 was launched the BBC had no space correspondent: there had never been a spacecraft before! So, when Sputnik went up Patrick was the only person the BBC could think of to contact. They phoned him at East Grinstead and asked him if he could appear on the BBC News. Patrick was delighted of course. Then they explained "It's in an hour's time and you'll be live". Patrick literally jumped on his motorbike Vesuvius, grabbed a tie and sped towards London. He got to the studio with minutes to spare and what the public saw was a man with his hair sticking vertically and his tie knot yanked almost over his shoulder. A distinctly nervous and sweaty Patrick spouted off what he could about rockets and Sputnik, live on the BBC1 News. He looked like a madman, but, to those who had never seen *The Sky at Night*, he was obviously a star. A crazed eccentric lunatic, yes, but they loved him. The real Patrick Moore era had begun! He was a man who could talk endlessly, live on television, without an autocue and, in one of the earliest *Sky at Night* programmes, he did not even falter when a fly buzzed into his mouth: he simply swallowed it in front of a million viewers and carried on! It, apparently, "buzzed all the damn way down" and his mother stated that although it was bad for him, it was so much worse for the fly. *The Sky at Night* following the launch of Sputnik, broadcast on October 19th, concentrated on The Sun and Patrick invited his good friend Colin Ronan along to talk about Solar Astronomy, but of course much of that programme was taken up with the Sputnik launch.

As an aside, the early *Sky at Night* must have been one of the few scientific TV programmes, then or now, to employ dancing girls! On many of the early programmes one or more would be dressed from head to toe in black, placed against a black background and directed to use white models of the stars, Moon, planets or comets. Advanced digital graphics were decades away, so this and the cardboard models were the only way of illustrating the Solar System, or the stars, in any animated or three dimensional form. Patrick was now famous and various magazines asked him to write for them. Almost as soon as *The Sky at Night* had started *Science & Fiction* magazine did an interview with him about his S.F. interests. He also started writing a regular weekly column in Arthur Mee's 1950s and 1960s *Childrens Newspaper*, as well as in *The Boys Book of Space*. These early articles would inspire a whole new generation of amateur and professional astronomers.

Sputnik 2 was launched on November 3rd 1957. On that occasion the satellite contained a dog, called Laika. The dog died a few days later, after its air supply gave out. A life-long campaigner against fox and deer hunting, Patrick made it clear to the press that he thought the Soviet launching of a dog, and its subsequent death, was cruel and unnecessary. A few years later, in his Penguin paperback 'Space in the Sixties', he made the following comment about the death of Laika: "Personally, I hold the minority view that it is ethically wrong to sacrifice any living creature unless it is quite definite that the results will be of major and lasting benefit to humanity……." Even in the late 1950s and early 1960s Patrick's views on fox hunting could make him unpopular, even in East Grinstead. He often related a tale of when he tried to thwart a Boxing Day hunt near to East

Grinstead and was told by the Hunt Master that a couple of his acquaintances were "going to duff Patrick over". Patrick claimed that it didn't happen simply because his unarmed combat skills were well known. Hmmmm, whatever you say Patrick!

The Sky at Night scheduled for November 16th 1957, the eighth programme, presented Patrick with something of a dilemma. The London Planetarium had just been completed and the projector had been demonstrated to a few dignitaries already. The Planetarium's first Director would be Dr Henry King, the man who had written the scathing review of Patrick's book *Suns, Myths and Men*. Patrick could not stand Dr King and the feeling was mutual, but with the Planetarium opening to the public in a few months it *had* to be mentioned. Patrick chose to invite two Planetarium experts onto the November programme, namely Reginald Edds, on the board of trustees of the new facility, and A.H. Degenhardt, an expert on the workings of the Zeiss projector installed in the Planetarium. Miraculously he managed to snub Henry King entirely and get away with it! However, it turned out that Edds was a close friend of King, who some months later proposed Edds as a Fellow of the Royal Astronomical Society. The Planetarium trustees had asked the now famous Patrick to write the visitors guide to the Planetarium, but a few months after the November *Sky at Night* Edds wrote back on behalf of the board claiming Patrick's final revisions (which he had not even made) had ruined it! Reginald Edds therefore entered Patrick's black book and while Degenhardt was happy to be invited to appear on *The Sky at Night* 13 months later, Edds would never appear on the programme again.

Despite now being a household name, Patrick was still very active at all the BAA's main meetings, indeed he was almost always the meetings recorder from the 1950s right through to the early 1980s; a tedious job at the best of times. As I have already mentioned, he was actually the President of the new Junior Astronomical Society when he started *The Sky at Night*. He was also still a very keen lunar observer and the director of the BAA's Mercury and Venus Section. Patrick had a close group of friends in the BAA, many of whom were lunar or Mercury and Venus observers and many of whom appeared on the *Sky at Night* in the late 1950s and 1960s. His best and most loyal friends would be rewarded with 15 minutes of fame. Patrick was never a person to try to mingle with the rich and famous, even if he would encounter them regularly in the BBC and as part of his considerable charity work. He always felt most at home with fellow amateur astronomers and BAA members and, of course, his Mum, but not with other famous people. What a refreshing difference to the fame-seeking TV presenters of today! The most notable exceptions to this, throughout his life, were perhaps the fellow author Arthur C. Clarke, comedians Michael Bentine and Jon Culshaw, and the rock guitarist Brian May, although all these were occasional amateur astronomers too. Most prominent in his list of amateur astronomy friends in the late 1950s and 1960s were Colin Ronan, Howard Miles, Henry Brinton, Richard Baum, Gordon Taylor, George Hole and J. Hedley Robinson, along with his youthful apprentices Peter Cattermole, Brian Warner and Iain Nicolson. Excluding his young trainees, only Hole, Taylor and Brinton lived in Sussex. In fact, Henry Brinton would move into the Old Mill

House at Selsey and encourage Patrick to move there too. They would be within walking distance of each other in 11 years time, from 1968. It was Patrick who actually encouraged Brinton to join the BAA and, with Patrick as his official proposer, he would be elected as a member in 1958.

As 1957 came to a close and 1958 commenced Patrick had his best ever view of the controversial Ashen light on the dark side of Venus, with the thin crescent sinking low in the western sky. According to Patrick's notebooks, which the author has studied at length, Patrick saw the phenomenon clearly with his 12½ inch reflector on five nights between December 29th and January 7th. I might add that the evening of January 3rd, the night this author was born, was a particularly good night. Patrick wrote that it was 'the best view of the Ashen light in my experience. It covered the whole circle, with a sharply defined edge, and appeared to be of a slightly brownish cast'. On January 8th *The Sky at Night* reached double figures; the tenth programme was about the brightest star in the sky, Sirius. The eleventh programme, transmitted on February 5th, was one Patrick had been planning for some time, ever since the first Sputnik had been tracked by Bernard Lovell's team at Manchester University. The studio guest was Roger Clifton Jennison, a member of Lovell's team and also the senior lecturer in Radio Astronomy at the Jodrell Bank Observatory. He had been a member of Patrick's anti-UFO team on that December 4th 1956 'First Hand' TV programme.

During February and March 1958 the BBC News and British newspapers would be dominated by the Munich air disaster in which seven of Manchester United's top football players were tragically killed on February 6th. However, for Patrick the show had to go on and his TV programme was now looking like it would survive for 12 months. With the March 2nd programme about the constellation of *The Plough* broadcast, and the April 2nd programme, discussing the planet Mercury (and an imminent penumbral lunar eclipse) also transmitted, he was almost there. The aim had been to put out 13 shows each year as, according to Patrick, this was the number of lunar months (29.5 days) in a year. Well, Patrick never was too good at maths; it actually works out at 12.4! Nevertheless, regardless of that mis-calculation, the April 2nd programme had been number 13, and for the next show 13 multiplied by a 4 week interval did indeed make 52 weeks, or 1 year.

So, the night of Wednesday April 30th 1958 marked the start of the second year of BBC *Sky at Night* broadcasts and so as an anniversary 'special' the BBC decided to do a live transmission, direct from the dome of Hampstead Observatory's 6 inch refractor. Patrick was very familiar with this observatory and the 1899 vintage Cooke telescope. It was, and still is, situated near the corner of Heath Street and Hampstead Grove, near Whitestone Pond, Hampstead, at the highest point in London. The Moon that night was 3 days from being full and Jupiter was on view too, low down in Virgo, but still accessible. Patrick was joined in the dome by the expert mirror maker Henry Wildey, who lived locally at 14 Savernake Road in Hampstead. Wildey was the BAA's Curator of Instruments and as mentioned earlier he had already appeared on the fifth *Sky at Night*. The cloud gods smiled on Patrick that evening and a direct image of the bright Moon was able to be transmitted 'live' on BBC 1. After the broadcast Patrick and Henry continued to observe Jupiter with the 6 inch refractor, and in his observing log book Patrick noted that a very dark, detached object, in Jupiter's North Temperate Zone (NTZ), first seen on April 20th, was observed by him and confirmed by Henry. Patrick made a sketch at 22.20 GMT using a power of 250× with the refractor. His log book also records:

> Most unfortunately, following the TV show I had to get the 1.45 am train to Nottingham and so I was unable to stay on at the Observatory to take any transits. Drawing had to be rather hurried, but I hope it is fairly accurate.

An abbreviation made just after that log book entry suggests that Patrick was heading for a meeting with the new (appointed in 1957) BAA Jupiter Section Director Bill Fox by taking that 1.45 a.m. train. Fox lived at Newark, near Nottingham. Three years and four months later another 'live' TV broadcast featuring the Moon would go disastrously wrong, but the Hampstead programme had been a great success.

Of course, after appearing every month for 1 year on prime time BBC TV and being on the BBC News, talking about Sputnik, Patrick was now receiving a substantial number of letters at his East Grinstead home. By the 1970s the volume was so great that he switched to postcards, but before then he always sent a letter in reply to each viewer's comment. Patrick was always courteous and encouraging, which must have been difficult considering the postal volumes involved. He would always address the letter to, for example, Mr John Smith Esq., and for a young person, being addressed as Esquire by a TV star was not only novel, but incredibly exciting.

In the early hours of April 4th 1958 there had been a very minor penumbral lunar eclipse, with the Moon very low down in the west-southwestern sky. Patrick had mentioned the eclipse on *The Sky at Night* broadcast on April 2nd and received quite a few letters from viewers who had tried to see it. One such person was a Mr Glasgow and it might be of interest if I reproduce Patrick's typed reply, as an example of the dozens of responses he was sending out every week, even in those earliest *Sky at Night* years. Note that, on all of Patrick's letters, he included his telephone number, just in case the viewer wanted to phone him. Can you imagine any TV celebrity doing that these days? Remarkable!

> From Patrick Moore FRAS
> Glencathara
> Worsted Lane
> East Grinstead
> Sussex
> Telephone East Grinstead 322
> 1958 April 13
>
> Dear Mr Glasgow,
> I am so sorry to have delayed answering your letter – I have been away for a day or two. Here, it snowed all through the penumbral eclipse. A slight limb darkening was recorded by two observers in Cornwall, using a 3 in. telescope, but it was very slight, as I said it would be; and I rather think that perhaps your 2 in. was not quite strong enough. Trouble is, one can never tell just how "dark" these small eclipses will be – and this seems to have been a bad one! Pity that we have no really good eclipse this year; I hope you will be more lucky next time. It was kind of you to send me a report – negative observations are just as valuable as positive ones.

Yours sincerely,
Patrick Moore.

With 1 year of *The Sky at Night* completed and the BBC mandarins having no plans to shelve the programme, as space rockets were still in the news, Patrick could breathe a sigh of relief. His name was now well enough known that his book sales were healthier than he could have dreamed of a year earlier and he just had to maintain the high broadcasting standard he had delivered during the programme's first year. On the May 28th *Sky at Night* he was joined by his friend Gilbert Sattertwaite and they discussed how amateur astronomers could observe the planet Jupiter.

Chapter 7
Encounters with Russians and George Alcock

Early in 1958 Patrick had received, completely out of the blue, a letter from the USSR Academy of Sciences in Moscow. Would he please send them all his published works on the lunar libration areas, together with all the charts he had drawn but never published? Patrick admitted that he didn't really appreciate why the Russians wanted his results, but he complied, sending his published papers as well as compiling charts, based on his recent observing notes. Patrick was not to know at that time that the Russians were planning on sending a probe past the Moon, sometime in the next year or so. They had reasoned that it would be useful to have at least some charts which showed the 9 % of the far side of the Moon that is occasionally visible from the Earth, if only to tie up any far side pictures they obtained. Patrick had no positive knowledge of these plans, but after the Sputnik successes sending a probe to the Moon was, surely, the next target in the Soviet's plan and probably in the plans of the US military scientists too, who had been left behind by their Space Race rivals. With this in mind, following the June 1958 *Sky at Night* (about the Hertzsprung-Russell stellar evolution model and why there were Giant stars and Dwarf stars) Patrick planned another programme with his friend Colin Ronan, this time about Moon Rockets and lunar craters. That programme, transmitted live, like the 16 *Sky at Night* programmes before it, was broadcast on July 23rd. Following the August programme, which was about the Andromeda galaxy, with Patrick joined once again by A.H. Degenhardt, Patrick managed a real journalistic coup by inviting the famous cosmologist Harlow Shapley onto *The Sky at Night* for the September edition. Shapley, then almost 73 years of age, had carried out groundbreaking work 40 years earlier, by estimating the size of our Milky Way galaxy from the study of Cepheid variable stars. He was the biggest name in professional astronomy that Patrick had managed to get onto the programme so far, but then that was only the nineteenth *Sky at Night* ever broadcast; there were more than 700 to follow! Shapley would make a welcome return visit to the programme just 8 months later.

M. Mobberley, *It Came From Outer Space Wearing an RAF Blazer!: A Fan's Biography of Sir Patrick Moore*, DOI 10.1007/978-3-319-00609-3_7,

Kozyrev

In November 1958, 3 years after Patrick wrote to professional astronomers to keep an eye on the crater Alphonsus, the Russian astronomer Nikolai Kozyrev announced that he had obtained a spectrogram, in the early hours of November 3rd, with the 50 inch (1.27 metre) Crimean reflector, showing hot carbon gas had been sent out 'causing a rise in temperature of, perhaps, 2,000 degrees'. He also claimed that the central peak 'appeared brighter and whiter than usual'. This was a dramatic and highly controversial announcement and Patrick realised that the next step was to see if there had been any permanent change in the appearance of the crater. During the following months a number of observers reported red patches near the site and these were alleged to be coloured material thrown out of the crater's central peak. However, as any lunar observer knows, the Earth's atmosphere acts like a prism and can throw a spectrum of light around any feature on the Moon, especially when it is low down in the sky! Brian Warner, that young friend of Patrick's, who was now an undergraduate astronomer at the University of London Observatory, claimed to have seen a 'bright red' patch in Alphonsus too, as did another friend, George Hole, but Patrick saw nothing at all.

As one might expect, Kozyrev's astonishing claims were discussed at the next BAA London meeting at Burlington House, on November 26th 1958, with Patrick's arch-enemy Henry King, now the BAA President, chairing proceedings. King opened the discussion by saying that he had been sent an envelope containing a newspaper cutting headed 'The Moon hits back' and then invited George Hole to say something about the 'reported volcano' in Alphonsus. George mentioned that he had experienced a fine run of five clear nights when he could examine the Moon before and after the Sun rose over Alphonsus. On November 19th, at 9.15 p.m., he had made a sketch showing a reddish/brown coloured area south of the central peak, which he considered abnormal to the region, along with the well-known dark patch under the east wall. His photographs, taken at the same time, seemed to confirm the same hint of colour to Mr Hole. Not surprisingly this claim of George Hole's was quickly challenged by that stickler for scientific accuracy Dr W.H. Steavenson, at the meeting. Steavenson was highly sceptical and thought it very unusual that Kozyrev's observation had been announced first by the Press, rather than through the usual astronomical channels. Patrick is recorded as saying the following: "I envy Mr Hole his five clear nights. I live only 30 miles from him, and I had only one possible observing night; even then the mist was so thick that I could do nothing of value. Of course, gaseous emissions from craters have been noted before, and I agree with Mr Hole that the dark patch in Alphonsus looks like the site of an old eruption. As I have always believed the craters to be igneous, I may of course be prejudiced! It seems that in this case the whole matter hinges upon the photographs which we understand Kozyrev to have taken. I may add that Dr Wilkins [no longer part of the BAA – M.M.] also has detected a patch of colour, and his description is in perfect agreement with Mr Hole's, which is excellent confirmation." Patrick also added the following comment: "On checking my old records, I found a note which

I made when observing in 1939: 'Alphonsus—central peak seems to be very hazy'. I added however that 'This may be due to imagination'. Whether the observation is significant I cannot say, but I cannot help wondering!"

After further sceptical points from Dr Steavenson any further discussion was curtailed by Henry King. However, while the BAA discussion might have been brought to a conclusion Patrick was determined that Kozrev's astounding claims should be the subject of an entire *Sky at Night* programme. The BAA Council had installed Patrick's friend Dr Gilbert Fielder as the new Lunar Section Director in October, so who better to have as a guest on the December 15th 1958 *Sky at Night*, which was entitled 'Volcanic Eruptions on the Moon'. In many ways the BAA had 'created' Patrick and so it is hardly surprising he chose a regular stream of BAA friends to join him on TV. Over the years many early guests on *The Sky at Night* would go on to great things as professional astronomers. As just one example take the case of a young Dr Roger Griffin, who appeared on the April 1959 edition talking about Binary Stars. Over the next four decades Griffin would establish himself as a legend in this field at Cambridge University.

On the subject of George Hole, Patrick was a regular visitor to Hole's observatory, as it had the largest working amateur telescope in the country: a 24 inch reflector. We will hear more of this telescope and its famous TV appearance shortly. Having leafed through nearly all of Patrick's observing logbooks it is clear that he was a regular user of the 24 inch. For example, one of Patrick's 1959 entries reads as follows:

> JULY 17th 1959.... Using George Hole's 24 inch reflector at Patcham.....taking photo's of Aristarchus, Herodotus and Billy...photo's taken at Gregorian focus but visual work was done at the Cassegrain focus....observing with George and Peter.

[Note: I think Patrick means the Nasmyth focus when he says 'Gregorian' in the above entry. Hole's 24-inch reflector had provision for Newtonian, Cassegrain and Nasmyth foci. The Cassegrain focus was used most often, but by using a third mirror, above the main parabolic primary, the Cassegrain beam could be diverted sideways through the tube wall and the declination axis fork bearing, to a Nasmyth focus, where it was easy and convenient to mount a heavy plate camera.]

Kozyrev would obtain another spectrograph, on October 23rd 1959, and he alleged that Alphonsus was a genuine volcano without a funnel! At the time, no-one knew what to make of all this, but the spectrograms were examined by other professional astronomers. Some claimed that his discrete 'emission bands' were an artefact of faulty guiding during the 30 minute spectrograph exposure. In fact there was no actual guidescope, but the observer could check the image drift by employing an optical arrangement on the spectroscope. The noted observer Gerard Kuiper also pointed out that Kozyrev's "emission bands" should have appeared as absorption bands against the backdrop of the sunlit central peak. In subsequent years Kozyrev claimed to have obtained spectra showing ionised Nitrogen in the atmosphere of Venus, ammonia surrounding the rings of Saturn and a hydrogen atmosphere around Mercury. None of his claims have survived the spaceprobe era, and these days it is clear to see that he was little more than a fame-seeking nutter with access to a 50 inch telescope! However,

Patrick was very proud of his Russian connections and treated Koyyrev with the same respect that he treated the other Russian astronomers he collaborated with. I guess this is not surprising really, when you consider that without the Soviet Space programme, and the *Sputnik* satellite, *The Sky at Night* may only have survived for 6 months.

Away from the controversial and crazy world of fictional lunar volcanoes Patrick was, as ever, always keen to observe anything unusual happening, every year, in the nighttime or daytime sky, and 1959 was no different. Patrick's friend Gordon Taylor, an expert in occultation predictions at the Royal Greenwich Observatory, Herstmonceux, had alerted the BAA, at the start of the year, to the fact that the bright planet Venus would occult the bright star Regulus in the daytime sky on July 7th 1959, at around 14:20 GMT. From southern England the star would disappear for around 8 minutes as it passed behind Venus. Locating Venus in a daytime sky, if your telescope does not have accurate setting circles, can be quite tricky and so Patrick chose to travel from East Grinstead to Henry Brinton's house, at his future home of Selsey, to observe the event with him. Gordon had been told by his astronomy colleagues at Herstmonceux Castle that he would be hurled into the Castle's moat if his predictions were wrong! Fortunately Gordon's calculations were spot-on and eight telescopes at Herstmonceux recorded the daytime occultation, which lasted 8 minutes. Thirty-six other observers around Britain also observed this Tuesday afternoon event, including Patrick and Henry, who noted a 0.8 second fade just before the star disappeared behind the planet. For once, a totally clear sky had coincided with an important event; Spode must have been away on holiday!

Luniks and Lunatics

Six months before that Regulus/Venus event, on the 2nd of January 1959, so only a year after Sputnik 1 had re-entered the Earth's atmosphere, the Russians had launched their Lunik 1 probe to the Moon. It had passed within 6,000 kilometres of the lunar surface 2 days later and proved that the Moon lacked a magnetic field. Communications were lost with the probe after 62 hours. As I have already mentioned, the December 1958 *Sky at Night* had been all about that alleged lunar volcano observed by Kozrev, but the Moon was still in the news 1 month later. Therefore, for the January 12th 1959 *Sky at Night*, Patrick was joined by his former mentor H. Percy Wilkins and also by that aforementioned friend with the 24-inch reflector, George Hole.

Eight months later, on September 12th, the Russians launched their second probe to the Moon, called (not surprisingly) Lunik 2. It crash landed on the Moon the next day, close to the crater Archimedes on the Mare Imbrium, after confirming that the Moon had no magnetic field. Only 3 weeks later, they tried again. On October 4th 1959 they launched Lunik 3. This time the probe went around the Moon, approached within 6,200 kilometres and sent back the historic first pictures of the lunar surface. With all these lunar events taking place Patrick might have been relieved that he had decided to give a trip to the October 2nd Total Solar Eclipse (passing over the Canary Islands and north Africa) a miss.

Fig. 7.1 Photograph of the participants at the meeting held at the Manor House, Crayford, on August 20th 1957. Patrick had been running *The Sky at Night* for 4 months at the time. Note the peculiar angle of Patrick's head, turning away as if snubbing those on the *right* (*his left*). Seated on the far *right* nearest the camera are Percy Wilkins' wife Hilda and her daughter Eileen, who was Mrs Coombes by then (Picture reproduced in the now extinct *International Lunar Society* magazine). N.B. Although Figures 7.1 and 7.2 are slightly out of chronological sequence I have included them here because they are rare pictures of Patrick with H.P. Wilkins, after Wilkins had left the BAA, but before he had appeared with Patrick on The Sky at Night

Fig. 7.2 An enlargement of the centre of the previous figure. From *left to right*: Mr Brewin, Mr Devin, Mr John Wall (inventor of the Crayford Eyepiece Mount!), Miss Francis, H. Percy Wilkins, Patrick Moore, The Mayor of Crayford (*with chain*), an unknown background face, and James (Doc) Q. Gant

There was another event that had occurred earlier in 1959 too; one that Patrick would prefer never to want to speak about and one which he felt, at the time, might have ended his early BBC career. Patrick was informed that the UFO fanatic George Adamski was actually coming back to London to give a number of TV, radio and lecture appearances. The BBC asked Patrick to appear on the *Panorama* programme and to openly debunk Adamski to his face. Patrick agreed to do this, assuming that this would be a formality. However, it was not as easy as he had hoped. Adamski had all his arguments very well thought out, and of course they both had a mutual friend in Desmond Leslie. Adamski must have known that Patrick was the alter-ego of the author Cedric Allingham and that Patrick's friend Wilkins had a very dubious UFO history too. Adamski got the better of Patrick on live TV on April 20th 1959 and so Patrick, 2 years into his TV career, seriously wondered if his BBC days were already over. He had expected to demolish Adamski's claims, but failed. However, Patrick's fears were unfounded, but the encounter with Adamski had shaken him up. Live TV can be a stressful business.

In the same month as his encounter with Adamski, Patrick had hit the headlines on an issue as far removed from Flying Saucers as possible. A lifelong animal lover and opponent of bloodsports, Patrick was revulsed by the fact that the RSPCA, an organisation supposedly devoted to preventing animal cruelty, had amongst its senior ranks some members of the upper classes and landed gentry who took part in fox hunting. To Patrick, and others, this was blatant hypocrisy and he was loathe to be a member of the association while this continued. However, early in 1959 the Crowborough and East Grinstead 'rebel' branch of the RSPCA made an offer to humanely gas fox dens on local farmers' land, for free, to prevent local fox hunting. On hearing this, Patrick joined the local 'rebel' RSPCA branch and became their main spokesman. Already known to BBC viewers, and a fine orator, he argued their case well and was quoted in newspaper reports on April 14th. The *Daily Express* report read as follows: '"Nearly all members of the RSPCA are against fox hunting, but there are a few officials at head office who like it, and are afraid to discuss it." Mr Moore said at his home in East Grinstead, Sussex, yesterday.' He added that the 'rebel' branch had been forced to withdraw its free gassing offer "after being bullied by head office". However, despite the bullying Patrick was determined to force an RSPCA vote on the matter, and he worked on this for the next 2 years, alongside all his other commitments.

As an aside, in that same year, the controversial founder of 'The Church of Scientology', L. Ron Hubbard, moved into the Saint Hill Manor at East Grinstead, not far from Patrick's home. Patrick always denied he had any contact with L. Ron Hubbard. However, 32 years later, in May 1991, Patrick's magazine *Astronomy Now* would feature a full page advert for an L. Ron Hubbard book. Patrick seemed genuinely horrified when, sitting next to me, he was shown the advert at a BAA council meeting; but surely, as the editor, he would have known it was in the magazine? But then again, knowing Patrick, maybe not!

When Lunik 2 was about to be launched, on September 12th 1959, Patrick's contact with the Russians meant that he knew that a prominent Soviet astronomer would be in London at the time of the September *Sky at Night*, coincidentally on the actual launch day of the probe. Patrick suggested to the *Sky at Night* director Paul Johnstone that they should invite the Russian along to the studio on this historic occasion. Very quickly a response came back from the Russian's hotel, stating that he would be delighted to be on the programme; this was looking like a BBC coup – a Soviet Astronomer on BBC 1, live, as a Russian probe headed for the Moon! At the allotted time *The Sky at Night* started, but the Russian had not turned up. Patrick had no problem filling in the details and 'treading water' until the guest appeared; after all, the subject was the Moon, something that Patrick knew everything about. Suddenly, from the corner of his eye, Patrick saw the studio manager arrive with the Russian. 'Great', thought Patrick, while he continued to broadcast 'live' to a million people. Then he noticed the studio manager holding a large sheet of paper just out of camera shot. It said 'HE DOES NOT SPEAK ANY ENGLISH!!' To many of today's broadcasters, used to being parrot-fed autocue material, this might have prompted a heart attack. But Patrick suppressed the horror he experienced at this news and simply welcomed the guest and pointed at a map of the Moon, stating the word 'Lunik'. The Russian replied and, in a string of incomprehensible Russian statements described the mission plan. Patrick had no clue what he was saying, but, with his knowledge of the Moon and Soviet rocket technology, did a convincing translation, live on BBC1! After a few more exchanges Patrick finished the programme on his own and breathed a sigh of relief. However, he knew that amongst the loyal followers of *The Sky at Night* there must surely be some viewers who understood Russian and he expected a barrage of mail about his translation skills (or lack of!) No letters were received, either due to apathy, or, because he had guessed right at what the Russian had been saying. Phew!

On the next night, September 13th 1959, Patrick was at the eyepiece of his 12.5 inch (317 millimetres) alt-az mounted reflector in Worsted Lane, East Grinstead. It was a clear night and the Russians had predicted that Lunik 2 would impact on the Mare Imbrium near to Archimedes at 21 hours 0 minutes GMT. The 250 foot Jodrell Bank radio telescope detected that, at 21 hours 2 minutes and 23 seconds, the signals from Lunik 2 ceased abruptly: in other words, it had hit the Moon. At precisely that time and without any knowledge of the Jodrell Bank result, Patrick suspected that he saw a tiny 'spark' on the lunar surface near to Archimedes. H.P. Wilkins, in Kent, (still a keen lunar observer despite the O'Neill's bridge affair and his resignation from the BAA) confirmed the event with his 15¼ inch (387 millimetres) reflector. However, despite this, Patrick was never sure that he, or Wilkins, had seen the impact. Observers all over Europe recorded flashes at similar times, but in different places; maybe they were just tricks of the eye? Shortly after the event Patrick received a telephone call from an elderly lady who claimed she had been watching the Moon through binoculars at the allotted time and had seen the Moon split in half by the impact! Patrick would later comment: "Astronomy attracts cranks like moths to a lamp".

In the late 1950s Patrick became involved with the nearby Croydon Astronomical Society (CAS) which had been founded as a JAS offshoot in May 1956. After giving his first talk to the thriving local amateur group he had invited anyone who was interested to come down to East Grinstead (only a short train ride away) and to look through his 12½ inch reflector. Croydon A.S. had some distinguished young members in its early years, many of whom, inspired by Patrick's enthusiasm, would go on to become well-known amateurs and professionals. These were people like Paul Murdin (later a professional astronomer), Mike Maunder and Rossie Atwell. Rossie would end up living near Patrick at Selsey in the latter years of her life. Some years later Patrick would have a falling out with some members of the society, as we shall see later. Mike Maunder, a member of Croydon AS in those days, recalls a story of when a few members of the society, visiting Patrick at East Grinstead, went to visit a Mormon Temple that had just been built not far from Patrick's home. Upon leaving the Temple Patrick was asked by an official if he wished to become a Mormon, at which he quickly splurted "No, I'm a very busy Druid" and promptly left, only pausing to bow to an Oak tree! From the late 1950s (even before Patrick was a TV star) Patrick's mother always had a warm welcome for any of Patrick's friends from Croydon A.S.

The Sky at Night was originally broadcast every 4 weeks, which worked out at 13 times a year, although this would become an average figure, with the gap between programmes ranging from 3 to 5 weeks. The programme of September 12th, with the aforementioned Russian guest, was the first of the occasional 'emergency' *Sky at Night*'s put out when anything dramatic happened. These were fairly rare, although they did happen even into the 1980s (such as when Halley's comet was recovered). In contrast, the programme scheduled for 7 weeks later, on October 26th 1959, just happened to be timed as pictures from Lunik 3, the next Russian Moon probe, might be released to the media. There was a good chance that, if the probe did not crash into the lunar surface, it might send back pictures from the far side. However, before the October 26th programme there was an encounter with George Alcock to deal with!

Alcock

Peterborough schoolteacher George Alcock was, without a shadow of a doubt, the greatest visual amateur astronomer to live under the cloudy skies of England. This is no exaggeration; he ultimately memorised the positions and brightnesses of *at least* 30,000 stars as seen through his binoculars and would ultimately discover five comets and five novae! He was born on August 28th 1912 in Peterborough and started submitting quality observations of meteors to the BAA Meteor Section Director, J.P. Manning-Prentice, in July 1931, 1 month before his 19th birthday. He was formally elected to the BAA 5 years later, on February 27th 1936, when he had raised sufficient funds to pay the annual subscription. Just over 14 months earlier the 11 years old Patrick's parents had paid the subscription fee for him with ease. From the 1930s through to the 1950s (apart from the war years), Alcock had been the BAA's most active meteor observer, and was more interested in observing than attending meetings. During the 1950s however, he felt that his meteor observations were being surpassed by radio work and he switched to hunting for comets in January 1953, just as Patrick's first serious book, *Guide to the Moon*, was being published. No-one in Britain had discovered a comet since 1894, when the legendary W.F. Denning had discovered his fifth, but George Alcock was a man with incredible determination and patience. On August 25th 1959, he discovered a new comet in the evening sky in the constellation of Corona Borealis. Most people would have been content with this place in history, but not George. He continued sweeping the skies in the days after his discovery and, in the dawn sky, on the morning of August 30th 1959, he discovered a second comet in the constellation of Cancer. To discover the first comet from Britain for 65 years was one thing, but to discover two, in a single week, from cloudy Peterborough, was absolutely incredible! George was suddenly famous. He was being chased by the BBC, by national and international newspapers and, by Patrick and *The Sky at Night.* Patrick wanted to bring a film crew to Alcock's house at Farcet, but when George heard that it would be Patrick interviewing him he decided to take another offer.

The *Sky at Night* was just over 2 years old at that point and, to George, Patrick was simply another BAA member, almost 11 years his junior and seemingly obsessed with making his voice heard and claiming there were strange things happening on the Moon. Patrick and George were total opposites, despite both being keen amateur astronomers. Patrick was very loud, confident and interested in bizarre notions like TLP, whereas George was quiet, modest and dedicated, as well as being a much loved teacher who knew the night sky far better than Patrick, or anyone else, and did not crave attention. Patrick's reputation had been known to George for more than a decade and he just rated him as a publicity seeker, like his mentor Wilkins. In later years Alcock's view of Patrick would mellow considerably, but in 1959 things were different. So, partially to annoy Patrick, George declined *The Sky at Night*'s initial offer and elected to be transported in a rickety old van, with his binoculars, to Birmingham, where he appeared on the popular *Tonight* programme. George was a great fan of Dr J.G. Porter's BBC radio broadcasts but Patrick's TV programme initially left him cold. He simply felt there were far more accomplished and less egotistical BAA members who were more worthy of presenting *The Sky at Night*. But, his rebellious journey to Birmingham, with his giant binoculars, in a rickety van, had not been pleasant either, and so, after some pressure from other BAA members, he agreed to appear on *The Sky at Night.*

The next convenient programme slot, following the *Sky at Night* 'special' on Lunik 2, was on October 5th and George reported to the Lime Grove Studios at 5.30 p.m., as requested. In the excellent biography of George, by Kay Williams (and in his own private auto-biography, in my possession, called "Beyond the Fringe") George states:

> I was led along a dimly lit tunnel and told to wait in a room that looked like a World War II bunker in Berlin – concrete walls, no windows, one unshaded light bulb and a couple of bare wooden forms without back rests. I sat there alone in the semi-darkness for two hours before the production team arrived to collect their victim.

Patrick, and his producer Paul Johnstone, were quite taken aback by George's account of how he had been treated, but the next 2 hours were spent rehearsing for the live transmission. George requested quite a few changes to the proposed format, which apparently did not go down to well. But the programme, transmitted live, as always, went very smoothly. After the programme a big limousine with a liveried chauffeur took him back to the railway station and the driver even found him a good seat on the train "to smooth my ruffled feathers" according to George! For many of his future discoveries George had even worse encounters with the press, and so ultimately he would leave his phone ringing, just to annoy them. George, in the following decades, used to show visitors to his house some classic press cuttings regarding his dealings on TV with Patrick. One of his favourites was the actual announcement of that October 1959 *Sky at Night* in the Radio Times:-

> 9.45 pm *The Sky at Night* with Patrick Moore – TOPICAL EVENTS
> Patrick Moore introduces George Alcock who recently discovered two new Planets. (!!!)

The Moon's Far Side

Anyway, from Patrick's perspective, although Alcock's feathers had been ruffled, as soon as one *Sky at Night* was over, the next was being planned, and even as Patrick had been talking to George, Lunik III had been en route to the Moon, and hopefully had taken some pictures. This, of course, was of great interest to Patrick as he had been mapping the lunar limb regions steadily since the end of the war and had provided the Russians with his best sketches. Planning for the October 26th programme was well advanced and no-one knew if pictures from the far side of the Moon would be released prior to the programme's transmission. The *Sky at Night* was not the only programme interested in broadcasting the pictures. The BBC News wanted to get their hands on them, as did the national newspapers. On the night of the 26th Colin Ronan, a good friend of Patrick's and the Director of the BAA's Historical Section, joined him for that *Sky at Night*, which was going to be about lunar observing history with, perhaps, some history being made by the programme. As had been the case with the September programme, nothing from the Soviets had arrived as *The Sky at Night* was being rehearsed, but even as Patrick prepared to start the programme a BBC motorbike courier was speeding through the streets of London and approaching the Lime Grove studios. Lunik 3 had actually passed by the Moon on October 7th, with the camera taking film based photographs at 04:30 GMT, but the Russians had taken 3 weeks to develop, decipher, interpret and release the pictures. In those days spaceprobes developed the film emulsions onboard, with a photodiode scanning the negatives for transmission. It had taken 11 days alone to develop and transmit the images! Eight days later the prints had arrived in London and were closing in on Patrick as *The Sky at Night* opened at 10.45 p.m. A few seconds into the programme Patrick heard a voice in his ear, from the producer Paul Johnstone: "Pictures of the Moon's far side coming up on your screen in 30 seconds. Scrap what we'd planned. Do it off the cuff. Good luck!"

Fig. 7.3 Patrick in *The Sky at Night* studio just after the Lunik 3 flyby of the Moon in October 1959 (Image by permission of the BBC)

To Patrick this must have seemed like fate taking control of events. For over 20 years he had been fascinated by the thought of what lay on the other side of the Moon and now here he was, in the hot seat with Colin Ronan, when a British TV audience would see the first pictures of the Moon's far side! He just hoped that he could make sense of the pictures. Fortunately, there was one feature on the disc that he spotted right away, as the pictures showed two-thirds of the far side and one third of the near side; this feature was the Mare Crisium, the lunar 'sea' he claimed he had given his first talk about, as a teenager, around 1937. In later years Patrick had to admit that the October 1959 *Sky at Night* was probably the one that stuck most in his memory. It has to be stressed that the first far side pictures were not brilliant. The Moon's far side is heavily cratered and the illumination was not favourable: most of the Moon looked bleached out and it was a miracle that Patrick was able to identify anything. But, even then, it was becoming obvious to Patrick that the days of the lunar mapper really were over and other lunar mysteries were soon to occupy his thoughts. The man Patrick had worked closely with, H.P. Wilkins, was "the last of the visual Moon mappers" according to Patrick, even if his map did look like it was drawn by a demented spider. As Lunik 3 took the first pictures of the far side, Wilkins was only a couple of months away from his retirement at the Ministry of Supply in London. Tragically, he died 3 months later, on January 23rd 1960, a few weeks after a major heart attack, following many years of cardiac problems. He had lived just long enough to see pictures from the Moon's far side. Patrick, as Wilkins' apprentice, was now, arguably, the last of the visual Moon mappers himself, even if in reality all of the visual work had already been totally superseded by Gerard Kuiper's photographic work of the last 10 years. Still, trying to peer around the Moon's edges had been fun! Patrick had derived much science fiction material from the Moon too. As Lunik 3 went past the Moon's far side he was putting the finishing touches to his latest novel *Captives of the Moon*. In passing it is worth adding that Wilkins' daughter Eileen donated her father's 300-inch map of the Moon to the National Maritime Museum in Greenwich and his wife donated a copy to the British Museum.

The Ark!

Now earning a regular but modest income from his new-found astronomical fame Patrick decided he could indulge, at last, in buying some kind of proper transportation. The 2 hour train journey from East Grinstead to London Victoria was absurd and for some time he had needed a car. After his motorbike accident of 1952 he had rarely used the bike for long journeys. He purchased a second hand Ford Prefect, registration GPN 924, for £140, and immediately installed a car compass on the dashboard. He had absolutely no time for maps: he was a former RAF navigator and so would simply pilot the car like a plane! Patrick's Ford Prefect was a model E493A, of which more than 190,000 examples were manufactured between 1949 and 1953. It had an 1,172 cubic centimetre 'straight four' side valve engine and a three speed gearbox. When new it should have had a maximum speed of 60 miles per hour, a 0–50 miles per hour acceleration time of around 23 seconds, and a fuel consumption of around 33 miles per gallon. Not exactly an Aston Martin! In addition, this was not a new car, and with the mileage Patrick added to it, those performance figures soon became even worse. Nevertheless, it just kept on chugging along.

Over the next 20 years Patrick would travel over three-quarters of a million miles in this car, mainly navigating by compass, and he affectionately named the Ford Prefect 'The Ark'. Three-quarters of a million miles is three times the distance to the Moon! By the late 1980s 'The Ark' was no longer the quickest form of transport; it may well have been the slowest. Patrick would often relate the story, in later years, that he found it had a maximum speed of 38 miles per hour on the motorway. On one occasion, while driving up Duncton Hill in Sussex, he was overtaken by a dog! In fact, a few years later, to allow for periods when the Ark was being serviced, Patrick bought an MG Magnette too (registration RPN 716).

The Mythical Morsgail Meteorite

On November 30th 1959 reports started to appear in various newspapers of a bizarre event on the Island of Lewis in the Hebrides. The reports were confused but involved suggestions of a meteorite strike on the scale of a mini-Tunguska (the Siberian 1908 event). There appeared to be a number of aspects to the story. One version suggested a postman, named Duncan Macdonald, had noticed that a local loch was suddenly 'missing'. The water was simply gone! Needless to say, aliens were cited in some of the more bizarre reports, along with lights in the sky. Other newspaper accounts claimed a meteorite had come down 2 miles south-west of Morsgail Lodge, which is itself 20 miles away from Stornoway, the capital of Lewis. It was stated that local people had seen the meteorite descend, and heard a massive explosion. It was also reported that various craters produced in the local bogland were still visible. According to Patrick *The Times* newspaper had asked him to travel up to Lewis at their expense to see what had actually happened. Patrick was already highly suspicious as some of the accounts cited previous meteorite falls in Stornaway (highly unlikely and Patrick could find none in BAA records) along with massive thunderstorms, which have nothing whatsoever to do with meteorites!

Now, East Grinstead and the Hebrides are a long way apart, especially when you own a Ford Prefect like 'The Ark', capable of a top speed of 40 miles per hour flat out; but, even so, Patrick decided to head north and convinced local Grammar School student Peter Cattermole (intent on studying Geology at University) to accompany him. After a solid day of travel they flew the final leg to Lewis from the Inverness airfield of Dalcross, and once on Stornaway they located Morsgail Lodge and a friendly Mrs Macdonald, who lived there. She showed the weary travellers where the 'meteorite' had supposedly landed, but admitted that she did not believe that anything had fallen from the sky. In Stornoway itself Patrick inquired about the so-called eye-witness reports and concluded that they had simply been dreamed up by the press. A few strange looking rocks proved to definitely not be meteorites. It was quite obvious that some form of local subsidence had caused the water from the missing Loch to drain out over several days. It was possible a small methane explosion had hurled one rock some distance, but even that was subject to speculation. No aliens were responsible and no miniature Tunguska event had occurred. So, Patrick's gut instincts about the reality of this media-created non-event were well founded, but it had been a jolly trip out! In the early 1960s Patrick wrote a weekly column entitled *Looking at the Sky with Patrick Moore* in *The Children's Newspaper*, a popular publication which was swallowed up by the children's educational magazine *Look and Learn* in 1965. If you dig out a copy from the week of 22nd September 1962 (almost 3 years after the Morsgail farce) he mentions the escapade on page three and he gave a BAA lecture about the adventure on January 2nd 1963.

TV Trivia

A couple of months after Wilkins' death, the *Sky at Night* almost didn't get transmitted. On the morning before a programme about meteorites, Patrick awoke at East Grinstead to find the entire south of the country shrouded in dense fog: a real pea-souper. There were no trains running and with the roads being gridlocked there was little chance of anyone outside London getting to the Lime Grove studios for that evening's live transmission. Patrick's only practical mode of transport in those conditions was his trusty 1929 motorbike, which he mounted and then set off for London, just over 30 miles away. After more than 4 hours Patrick arrived at the studio ("fog was so thick I couldn't see the handlebars") and was relieved that his guest, Dr Jo McCall, arrived shortly afterwards. However, with a skeleton studio staff, Patrick ended up producing the programme and holding up captions too! On that same programme, in his somewhat stressed state, Patrick inadvertently mentioned that many meteorites were probably thrown away as they often looked like slightly odd stones. In the days that followed the BBC were inundated with hundreds of pieces of assorted rubble sent by viewers, ranging from a few grams to over 10 kilograms in weight; needless to say, none were meteorites!

In 1960 the transmission time for *The Sky at Night* was far more 'prime time' than it would become during the 1990s and into the twenty-first century. Of course, in those days there were no domestic video recorders, as they did not arrive until the mid to late 1970s. There were no *Sky at Night* repeats either. If you missed a programme, that was it....tough luck! Your only option (from May 1959) was to read Patrick's monthly write-up in *The Listener* magazine. The start of the programme always featured a shot looking like a view of the night sky as seen through a house window, with the camera zooming in until the window frame disappeared and only the stars remained visible. Let us look at the BBC schedules for Wednesday January 20th 1960. Well, after the News at Six 'O Clock and the regional News summaries there was a topical programme called *Lookout* with Geoffrey Wheeler and the 'Roving Eye'. At 7.25 p.m. there was a brief 5 minutes News Summary followed by US cowboy adventure *Wells Fargo*. Following this excitement there were three more programmes of 35, 30 and 45 minutes duration, namely: *The Perry Como Music Hall*; *Sportsview* with Max Robertson and *The Amorous Prawn* with Evelyn Laye and Stanley Baxter. Then, for a mere 15 minutes, from 9.4 5 p.m., we find *The Sky at Night*, just preceding the Ten 'O Clock main evening News. Patrick was on immediately before the main News in mid-week. Wow!! The TV schedule finished with evening prayers at 10.50 p.m., after which the BBC closed down. There was one other TV channel, namely ITV; BBC2 was 4 years into the future. It is hardly surprising that Patrick was a household name.

On May 16th 1960, two science teachers at Kettering Grammar School, Geoffrey Perry and Derek Slater, attracted Patrick and the BBC's attention when their group (consisting of keen science students at the school) recorded transmissions from the Russian Sputnik 4 space probe. Patrick would keep regular tabs on the Kettering School's achievements from then on and frequently mentioned the school on his TV and radio broadcasts.

Another Telescope and More Russians

Around this time Patrick acquired a second major observatory at his East Grinstead home. Since 1951 he had mainly used his 12½ inch (317 millimetres) Newtonian reflector housed in a double-ended run-off shed. The telescope provided good views as it had a large mirror, but it was not equatorially mounted and had no clock drive, so following the stars and planets as the sky rotated was rather fraught, using crude altitude and azimuth slow motion controls. The low friction 'Dobsonian' style mounting had not been invented then. In the mid 1950s Patrick had acquired a fine 8.5 inch (216 millimetres) equatorial reflector of 59 inch focal length (just under f/7) and of 1860 vintage. It had a fine quality mirror made by the expert optician George With and a German equatorial mounting by Browning. The previous owners, the Royal Greenwich Observatory, had converted the drive to be electric and gave the telescope to Patrick for free! This heavy, equatorially mounted, With-Browning telescope should not be confused with his other 8½ inch Horne and Thornthwaite *altazimuth* Newtonian which he had used since 1950.

However, Patrick had no observatory for the heavy 8.5 inch, or for his smaller 6.5 inch reflector. This all changed in 1960 as the new 8.5 inch became permanently sited. At last, Patrick had a proper observatory and a telescope that would track the stars! The delay in regular usage of the telescope had come about because he had come to the conclusion that, for his second observatory, the design had to look good, because while his 12.5 inch reflector was hidden at the side of his mother's back lawn, this one was going to sit near the middle of the lawn, well clear of the trees and with paving slabs leading up to it. Patrick's two young second cousins, Brian and R.A. Gulley, kindly built him a unique and 'decorative' observatory over the course of a year. The total financial outlay for the observatory was a modest £40 and the design was unusual: a rotating octagonal wooden structure with glass windows, looking a bit like a rotating greenhouse. The lower section was, essentially, eight miniature fence panels held together in an octagonal formation. The structure was fully assembled, on a patio, by Patrick's second cousins, before being dismantled and re-assembled on his mother's lawn. I have seen pictures of the assembly process when looking through Patrick's photo albums. Patrick referred to this unusual building as his 'Merry-go-round Observatory'. Unfortunately, even with the roof slit and the windows fully opened, the glass caused the temperature inside to soar when it was sunny and scotched any hope of high resolution planetary work on such days, namely clear ones! In the Quarterly Journal of the Royal Astronomical Society, Vol. 1, p. 222, for September 1960, under the heading "Mr Patrick Moore's Observatory, East Grinstead, Sussex", Patrick wrote, regarding progress in the last year:

> Observations of the Moon, Venus and Jupiter were continued with the 8½ inch and 12½ inch reflectors. Lunar work was devoted largely to estimates of the depths of small craters by shadow determinations. It is hoped that the new 24-in reflector will be erected during 1960.

This brief note in the professional RAS Quarterly Journal was far more formal and restrained than the sort of typical TLP/Lunar vulcanism talk, paper or letter he would have been more likely to deliver to a BAA audience, or submit to the BAA Journal. The reference to a 24 inch instrument was related to the dream of Patrick's to continue with the late Percy Wilkins' plan to erect a giant reflector (Wilkins' referred to it as a 22 inch) at Bexleyheath, where he and Patrick could study the Moon with a giant aperture. Wilkins was now dead, but his mirror had been close to completion, and Patrick had acquired a lust for more aperture after using George Hole's giant 24 inch reflector at Patcham, near Brighton. Rather predictably (Patrick was not a handyman!) the 24 inch never appeared in Patrick's garden at East Grinstead. I dare say his mother might have drawn the line at such a monstrosity too! As with all Patrick's telescopes his friends were always welcome to use them and he and his mother's generosity were just as unbounded now he was on TV. Amateur astronomers, neighbours, schoolboys and scouts were all very welcome to be shown the night sky. One regular East Grinstead visitor in this era was a member of the BAA Computing Section, John Carl Vetterlein. John, or Carl as Patrick called him, published a number of BAA papers in the early 1960s, two of which were on the subject of the phase of Mercury and Venus, a topic which greatly interested Patrick.

On September 25th 1960 the United States launched the lunar orbiter probe Pioneer P-30 (also known as Atlas-Able 5A, or Pioneer Y) towards the Moon. The plan was to place a highly instrumented probe into lunar orbit, something which was obviously of much interest to Patrick. Sadly the mission failed shortly after launch and Patrick mentioned this at the start of the September *Sky at Night* programme, just 2 days later, when he introduced the programme. Coincidentally the planned subject of that *Sky at Night* was the Moon, with his guest being Gilbert Fielder, the Director of the Lunar Section of the BAA. Patrick opened the programme as follows, as always, word perfect and without an autocue:

> Good evening. Well, as you've all have heard by now, there's been another unsuccessful American Moonshot. This time the idea wasn't to land a vehicle actually on the surface of the Moon, but to put a rocket in a closed path round the Moon so that it could send us back useful information. Well, unfortunately it failed because there was some kind of a structural fault and the rocket fell back into our atmosphere and was burned away in the manner of a shooting star. Well, that's a great pity because a rocket round the Moon would be a really useful thing and send us back a great deal of information that we can't get in any other way. But I'm sure that an experiment of that kind will be made again in the near future and I think we've every hope of success.

In October 1960, Patrick accepted a long-standing official invitation to visit the USSR and deliver some lectures about his work in mapping the Moon, along with the work of H.P. Wilkins and the BAA. He flew from London on October 3rd, stayed there for 2 weeks, and had a most interesting trip. It is important to remember that at this time Russia was considered the enemy of the West and a nuclear exchange between the USA and the USSR was considered a serious possibility. Anyone involved in scientific defence work found travelling to the USSR was likely to be promptly arrested on spying charges! Exactly 12 months later the arms race between the USA and USSR reached its climax with a 50 megatonne atmospheric detonation trial by the Soviets over the Arctic Sea.

Patrick's first visit was to the USSR Academy of Sciences in Moscow. He almost caused an international incident by placing his fur hat on a bust of Lenin, but on seeing the horror-struck faces of his hosts, quickly removed it! On the 4th of October he addressed a 'House of Scientists' celebration on the third anniversary of the launch of Sputnik I, a prestigious role for a Western amateur astronomer. Of special interest to Patrick on the trip was the work of Professor Y.N. Lipski who showed him over 30 photographs of the far side of the Moon taken by Lunik 3. Only two photographs had been released to the media in the UK, so Patrick was very interested to study these new pictures. Professor Lipski had compiled charts of the far side and already identified over 500 craters. He had used the drawings of the lunar limb regions compiled by Wilkins and Moore to identify and correlate some of the regions photographed, which pleased Patrick greatly. Patrick would often claim, in future years, that his drawings of the Moon had been used by NASA. In fact, it was the Russians who used the Wilkins/Moore data because of their limb region mapping work.

Patrick was taken to a number of other fascinating places on his Russian excursion: to the Moscow planetarium; the Zvenigorod satellite-tracking station; and Leningrad's Pulkovo Observatory, containing the 26 inch refractor that Hitler had commissioned for Mussolini. Patrick was also taken to the Crimean Astrophysical Observatory near Simferopol in the southern USSR. This was mainly a solar observatory, but it also used the 50 inch reflector employed by Nikolai Kozyrev, who had claimed to have detected a hot carbon gas outburst from the crater Alphonsus in 1958. Patrick was especially keen to meet Kozyrev.

Fig. 7.4 Patrick with Dr Nikolai Kozyrev, at the Crimean Astrophysical Observatory in the U.S.S.R., in October 1960. Kozyrev claimed to have spectroscopically recorded a gas emission from Alphonsus, but most lunar authorities now regard him as a complete barking mad nutter (Photographed by an unknown Russian photographer and copied with Patrick's permission around 2002)

On his return from the trip he would report that "examination of the spectrograms taken on that occasion leaves no room for doubt that an outbreak (of carbon gas) did, in fact, occur". Well, that may have been Patrick's view, but today, the spectrograms are not taken seriously by anyone. Patrick was also taken to the giant optical workshop in Leningrad where the 236 inch (6 metre) Russian reflector was being constructed. This instrument, like the rocket programme, was another source of

Russian pride in 'beating the USA', although, as events turned out, although the 236 inch was eventually completed and was bigger than the Palomar 200 inch, the optical quality and observing site meant that the telescope made little impact on astronomy. Needless to say, when Patrick returned to the UK he broadcast his findings on the *Sky at Night*. That programme was transmitted on November 7th 1960. He also gave a lengthy talk about his experience in the Soviet Union at the next BAA meeting on November 30th, paying tribute to the interpreter Gregory Molyikov, who had accompanied him everywhere on the trip.

By the time of the 1961 January 4th BAA meeting there was great news. A new comet, Comet Candy, had been discovered by the BAA Comet Section Director Mike Candy, who had been testing a new eyepiece in his 5-inch comet sweeping refractor (built by Horace Dall) on Boxing Day. The comet was found in Cepheus and Patrick mentioned it during the next *Sky at Night*. It would appear that George Alcock was not the only British amateur astronomer who could discover comets, although Mike Candy was an amateur and a professional too, as he also worked at the Royal Greenwich Observatory at Herstmonceux.

Chapter 8
Some Classic 1960s Broadcasts

An Eclipse First

As we have already seen, Patrick had observed his first total solar eclipse from Sweden in 1954, but he was certainly keen to see another. On Patrick's return from the Soviet Union, on October 18th 1960, the BBC were already planning to implement his ambitious idea to transmit the February 15th 1961 total solar eclipse live from three continents in a *Sky at Night* special. In 1961 satellite TV did not exist, mainly because there would be no communications satellites until the middle of the decade! So, terrestrial transmitter links had to be used to communicate across Europe. In addition, trying to broadcast a total solar eclipse was fraught with problems, not least the weather; if it is cloudy, you do not see the eclipse, but then this was Patrick's reason for having three sites. Patrick's producer Paul Johnstone liked the idea and, in late 1960, he set up a conference, in Paris, between the major European TV networks. Patrick and Paul travelled to the conference to generate some enthusiasm. As the plan was to get broadcasts along the eclipse track, from southern France, Italy and Yugoslavia, it was essential to get the TV networks for these countries on board! Initially the Italian party was somewhat disinterested, but on the advice of the Danish conference chairman, Patrick got them drunk, after which they were much more sociable. As anyone who knew Patrick well would confirm, his ability to consume vast quantities of alcoholic spirit is legendary; he will always be the last man standing, even if he sways about a bit! According to Patrick: "They were so friendly after a few drinks that the Italian producer suggested timing the eclipse by releasing some lions into St Peter's Square and counting how many tourists were eaten during totality; unfortunately the BBC were not too keen on this idea!"

With Italian, French and Yugoslav parties all in agreement Patrick and Paul returned to London to plan the programme. Three of Patrick's best friends were roped in to help cover all of the locations. The astronomy historian Colin Ronan was dispatched to Florence in Italy, Hugh Butler (from the Royal Observatory at

M. Mobberley, *It Came From Outer Space Wearing an RAF Blazer!: A Fan's Biography of Sir Patrick Moore*, DOI 10.1007/978-3-319-00609-3_8,

Edinburgh) was stationed at St Michel in France and Patrick was dispatched to the frozen wastes of Mount Jastrebac in Yugoslavia! From an aircraft over cloudy England, Henry Brinton also commentated, although the eclipse was not total from here: only 91 % of the sun was covered from London. However, this was the best partial eclipse from the capital since 1927 (95 % covered from London) as although the 1954 eclipse was total from the Faroes and northern Shetlands, it had only been a 71 % partial eclipse from London. As the Moon's shadow approached St Michel the TV cameras caught the dramatic moment of totality well, but Hugh Butler's voice was never heard, due to a technical problem! Nevertheless, 15 minutes later Colin Ronan's sound and vision coverage from Florence came across perfectly; so now it was Patrick's turn! Only 5 minutes after the Moon's shadow departed from Florence, it was bearing down on Patrick, but Patrick's radio link was down, so he had no knowledge of what had happened so far! Despite being up a mountain, Patrick was clouded out, although breaks were appearing. Patrick was talking in broken French to a Belgian astronomer who was translating this into German. Fortunately, the Yugoslav TV director understood German and could translate the gist of Patrick's instructions into Serbo-Croat, which was the only language spoken by the cameramen. In addition to these minor problems it was cloudy and the Yugoslav director had artistic designs on the eclipse; designs he had not shared with Patrick! Our hero started his commentary in the gloom with no knowledge of whether he could be heard back in London. The darkness of totality rushed over them and they were still under cloud; however, a big clear gap had appeared and was moving rapidly towards the region where the eclipse was in the sky. At this point, the Yugoslav director implemented his master plan. To study how animals behaved in the darkness of totality, he floodlit a group of nearby Oxen with arc-lamps, so they were no longer in darkness at all! A horrified Patrick saw this unwelcome development on the monitor and gestured frantically at the director to get back to the Sun. Just in time the director agreed and most of the last minute of totality was captured and relayed back to London, including the spectacular diamond ring where the first sliver of the Sun re-emerges from behind the Moon. Patrick would find out later that it had all got through to London successfully, which was quite an achievement for a live TV broadcast of an eclipse in February 1961.

Seven days later, on February 22nd, Patrick, Henry Brinton and Colin Ronan gave an amusing account of their frantic BBC eclipse day at the next BAA meeting in London. As we shall note on numerous occasions in this book, Brinton lived at Selsey, just a short walk from where Patrick would live in 7 years' time. In the twenty-first century BBC camera crews would descend on Patrick's home every month to film *The Sky at Night*, but 1961 was 40 years earlier and Henry Brinton was the only really active astronomer based in the seaside town at the time. At that BAA meeting Brinton described how 'The Invasion of Selsey' had started a fortnight before the eclipse had even begun, and how on the day before the event a total of eight BBC TV vans had arrived; these contained many aerials and complex apparatus which were essential for the air-to-ground and ground-to-ground transmissions planned. Brinton described an array of 24 switches, all of which had to be in the right position for the transmission to work. He also described the BBC camera equipment, which used three consecutive 1 % transmission filters mounted in line, to transmit one millionth of the partially eclipsed Sun's light. He also said that inside the control van 'Bedlam reigned' during the transmission. Conditions on the ground from Selsey were so bad that nothing was seen, but Brinton had hitched a ride on the BBC aircraft which flew from Tangmere Aerodrome, a few miles east of Chichester. Henry Brinton thought the TV pictures had come through well, but the press, hoping for more entertainment, had been cruel, describing the pictures of the partially eclipsed Sun as resembling 'a squashed slice of melon in a mist'! Colin Ronan also described his BBC experience in Florence and then, in his inimitable style, Patrick described his day. He had helped the Oxen to pull an Ox-Sleigh full of BBC equipment through the snow on Mount Jastrebac and assisted in setting up a coelostat mirror and a 6 inch reflector. Patrick made it quite clear that after this BBC 'first', a monumental effort, and a technical success, he was most unhappy by the comments of a Mr Maurice Wigmore. Mr Wigmore was a television 'critic' who had written in the *Sunday Times* that the programme was 'almost completely uninteresting' Patrick stated that this was surely a remark which "will go down in posterity as the supreme example of a really ignorant comment."

But for Patrick there was no time for a rest. This had been a *Sky at Night* special and the next regular broadcast was due on February 23rd, just 8 days after the eclipse, which was the day after the BAA meeting. As Patrick had journeyed to the eclipse a dramatic announcement had been made in London by Professor Martin Ryle of Cambridge, allegedly totally disproving the so-called 'steady state' theories of messrs Hermann Bondi, Thomas Gold and Fred Hoyle. The scientific 'war' over the evolution of the Universe was about to intensify.

Big Bangs, Linné and a Cosmonaut

In the late 1950s and early 1960s radio astronomy had really taken off, principally due to the enormous financial and technical gamble of Bernard Lovell at Jodrell Bank, when the gigantic 250 foot radio telescope was built. Cosmologists were using radio telescopes to try to study how closely packed distant galaxies appeared at those frequencies. Although radio telescopes are huge, the wavelengths used in the early days of radio astronomy were so much bigger than the wavelengths of light that they could barely resolve fine detail with better resolution than the naked eye. However, in the 1960s various techniques were devised to improve the resolution of radio telescopes. One method was simply to move to finer wavelengths, though this requires a very smooth radio dish to work; the surface must be smoother than a quarter of the wavelength used. Another method is to use two radio telescopes a large distance apart to simulate a huge radio telescope. A third method was to actually use the Moon passing over radio sources to refine their precise position in space. At the start of 1961 some exciting results had been achieved by stretching the existing technology to the limit and by employing the two telescope (interferometry) technique. Professor Martin Ryle had used the technology to try to analyse some very weak radio sources which were associated with very distant galaxies (thought, at the time, to be eight billion light years away). By examining just how close these radio sources were in the sky, and assuming their density was typical for objects at that distance, either the steady-state or evolutionary ('Big Bang') theories would be favoured. Specifically, if the galaxies eight billion light years away were much more tightly packed than those in our local region, the evolutionary theory would be favoured, as a 'Big Bang' start to the Universe assumes matter is very tightly packed in the early stages. According to Ryle and his colleagues the new radio data favoured the Big Bang theory.

Incidentally, the term Big Bang was rarely used by its supporters in 1961, as it was a derogatory term coined by one of its main opponents, Fred Hoyle. Unfortunately, Patrick and Paul Johnstone were unable to get Professor Ryle for the programme, but a former President of the Royal Astronomical Society, Professor W.H. McCrea, was available instead and the new evidence was presented. Patrick was careful to stress that these were early days and although the new radio evidence looked like bad news for the steady-state theory, only time would tell. It might have been thought that Patrick's lack of any professional qualification in astronomy would have been a distinct disadvantage when presenting such a programme, but as he would prove on many occasions over the programme's huge lifespan, he had the rare ability of taming high-brow scientists to answer simple questions in a language the viewers could understand. The format obviously worked. No other TV programme would survive as long, with the same presenter, anywhere in the world. Even so, despite the programme's longevity, Patrick, unlike so many TV 'personalities', was never someone who thought that his fame would last for ever. So, during the early 1960s he continued churning out books on astronomy and books on science fiction too, while people still remembered who he was! He wrote four more novels from 1961 to 1964 alone, namely *Wanderer in Space* (1961), *Crater of Fear* (1962), *Invader from Space* (1963) *and Caverns of the Moon* (1964). Patrick would frequently turn his hand to absolutely any book for which he could clinch a book contract, even when it contained absolutely nothing about space travel or astronomy. For example, in 1961 his book *Conquest of the Air – The Story of the Wright Brothers* was published by Lutterworth.

Still a back garden amateur at heart, on the 23rd of March 1961, Patrick trained his 8.5 inch Newtonian on the first quarter Moon and was astounded to see the tiny crater Linné, on the Mare Serenitatis, actually looking like a crater. While this may sound rather obvious and strange, as I briefly mentioned in Chap. 3, on October 16th 1866 the astronomer Julius Schmidt had announced that the crater had disappeared. Where a crater had stood, all that remained was a small whitish patch. One famous observer and a pioneer of stellar spectroscopy, Angelo Secchi, used the powerful telescopes at the Vatican Observatory and wrote: *there is no doubt that a change has occurred*. No less a person than Sir John Herschel suggested that a moonquake had shaken down the walls of the old crater and the hollow had been filled with rising lava. Another leading Moon mapper (selenographer), Edmund Neison, wrote in January 1877: *A real physical change on the Moon's surface must have occurred at this point*. However, the great selenographer Madler wrote, in 1867, that he *found Linné shaped exactly, and with the same throw of shadow, as I remember to have seen it in 1831*. On that basis, there had not been a change. Indeed, I have already described how Patrick's friend Richard Baum initially located Madler's paper (presented at an 1868 meeting in Norwich) and described Patrick as 'flabbergasted' at the news. At a stroke it destroyed one of the pillars of his abiding belief that changes took place on the Moon.

However, Patrick's new observation revived his interest in the crater. Certainly, up to March 23rd 1961, Patrick had never seen Linné as a crater, just as a white patch. It could not have reverted back to its alleged pre-1866 appearance so it must simply be a crater that was highly susceptible to illumination angle. Nevertheless, Patrick thought it odd enough to write to his friend J. Hedley-Robinson in Teignmouth and they both promoted interest in the crater and observed its appearance many times over the next 10 years. Patrick was not keen to let any evidence of change on the lunar surface just die a death and be resigned to history. Once again we can find, if we look hard enough, the influence of Patrick's friend Robert Barker and mentor Percy Wilkins in this 'lunar change' obsession. As early as January 1954, while still deeply embroiled in the O'Neill's Bridge saga, Wilkins had stated that observations by himself, Patrick, Fauth, Thornton, Goodacre and Corder 'indicate a change of some sort in connection with Linné', adding, when he presented the Linné paper at the BAA meeting, that 'My paper is going to set the Thames on fire, I fear' (see the *BAA Journal* volume 64, page 86). However, in 1961 Wilkins was no longer around, but Patrick was happy to revive the Linné mystery following his own observations.

By April 1961 a full two years had elapsed since Patrick and his local 'rebel' RSPCA branch, at Crowborough and East Grinstead, had embarked on a course to force the RSPCA to hold a vote to condemn fox hunting. It seemed incredible to Patrick and his fox loving friends that senior figures of the landed gentry, in an animal cruelty prevention society, took part in hunting foxes, and were happy to see the furry animals ripped to shreds every Boxing Day, but that was the case. Despite various barriers being placed in the rebels' way, matters came to a head on April 18th, when a motion proposed by a former Tory M.P. for Brighton, Mr Howard Johnson, seconded by Patrick, was voted upon at an Extraordinary General Meeting at Central Hall in Westminster. Dr Robert Rattray, the RSPCA vice-chairman, and the Duke of Beaufort, the Master of the Queen's Horse, strongly opposed the Johnson/Moore proposal. A total of 876 votes were needed from the 1,460 members present to implement the rebel proposal. Much to Patrick's complete disgust, only 558 RSPCA members voted to ban fox hunting and so he resigned from the association. But, just 6 days earlier, a dramatic space travel first had occurred, which no doubt helped soften Patrick's disappointment just a bit, after a two years long cunning plan that had, ultimately, failed.

On April 12th 1961 science fiction became science fact. The Soviet Union launched Major Yuri Gagarin of the Soviet Air Force into space, aboard Vostok 1; it was a gigantic moment in history. Gagarin orbited the Earth and survived, which was a major relief to the Soviets and to all science fiction writers. Before any major step in technology there are always dire predictions of doom. In the 1800s it had been said that travelling in a train at a speed greater than 12 miles per hour would ultimately shake the human organs too much and result in death! In the 1950s there had been predictions that weightlessness itself would cause death, and any spacecraft going above the atmosphere would be destroyed by micro-meteorite bombardment and solar radiation; neither had proved to be the case. Suddenly the future of manned exploration of the Moon and the solar system looked possible and science fiction writers would have little to hold them back. A few months later Patrick would actually meet Yuri Gagarin, briefly, during a press conference held at Earls Court on July 11th, attended by 600 journalists! Tragically, Gagarin was destined to die in a Soviet fighter aircraft 7 years later, on March 27th 1968. His MiG-15UTI would crash near the town of Kirzhach, killing Gagarin and flight instructor Vladimir Seryogin. Although the *Sky at Night* briefly mentioned the achievements of manned spaceflight during the 1960s and 1970s the BBC covered the missions in more detail using special programmes and special space correspondents, such as Reginald Turnill. But, in the case of the Moon missions, Patrick still played a major part; after all, the Moon was his subject!

Of course, on *The Sky at Night* immediately following Yuri Gagarin's mission, Patrick's opening story was the Soviet success. As the opening credits on April 24th 1961 rolled past (THE SKY AT NIGHT, A regular monthly programme, Introduced by Patrick Moore) a clip was shown of Gagarin being greeted by Soviet dignitaries as Patrick commented:

> Good evening. Well, I expect you'll realise straight away who this is. Major Yuri Gagarin, just about to report Mission Accomplished to Mr Khrushchev. The first man in Space; the first spaceman in history. You know, if I'd come on the air in 1957 when we did the first of these *Sky at Night* programmes, and said that within five years I'd be showing you pictures of the first man to go round the Earth in orbit in a spaceship, well I think you'd have regarded me as mad! But, nevertheless, it has happened and the great question now is....what is going to happen next? Well, undoubtedly this flight of Gagarin's is one of the essential preliminaries towards man reaching the Moon. But of course that won't come yet-a-while. I think myself that the next step is going to be a soft landing of an un-manned instrumented probe upon the Moon's surface which will act as a real lunar transmitting station. But I've no doubt either that the first manned flight to the Moon will take place in the foreseeable future.

As I mentioned earlier, Patrick met Gagarin, briefly, during a press conference held at Earls Court, London, on July 11th 1961. With 600 other media people in attendance Patrick, for once, did not get an exclusive interview, and of course his conversation had to be directed through an interpreter. Patrick's questions were prepared in advance, but not recorded. However, he wrote up what he remembered of Yuri Gagarin's replies as soon as he returned to East Grinstead, based on some scrawled notes made a few minutes after the interview. I have seen Patrick's typewritten write up of the entire exchange and reproduce it below:

> Patrick: On due reflection, do you consider that the flight came up to expectations from a purely scientific point of view?
> Yuri: Most definitely. The technicians gained a vast amount of information which will make future flights very much easier and less dangerous.
> Patrick: You regarded your own venture as highly dangerous then?
> Yuri: Certainly not so risky as a wartime bomber raid, but of course it is always possible for faults to develop, and the whole experiment was something quite new. I may add that I had no personal misgivings and I had every confidence in my equipment and in the design of the Vostok.

Patrick: Did the Earth look much as it does from a high-flying aircraft?
Yuri: There is a subtle difference. The surface is well seen from a height of 200 miles or so, and I could observe the coastal lines, rivers, forests, clouds, and cloud-shadows. The sky was entirely black, while the Earth itself had a glorious blue halo. When I came out of the blue shadow a bright orange colour could be seen on the Earth's horizon, which then went through all the colours of the rainbow.
Patrick: Presumably you were much too busy making scientific records to do much casual 'gazing'?
Yuri: Very true. I was on a scientific mission not a sightseeing trip.
Patrick: What about zero gravity? I realize that you were not badly affected, but would you describe it as uncomfortable?
Yuri: No, it did not take me long to become used to it, and my working ability, both mentally and physically, was completely unimpaired. I do not personally believe that weightlessness will have any adverse effects, even over much longer periods.
Patrick: Were there any communication troubles?
Yuri: None at all. I was pleasantly surprised to find how easy it was to keep in touch.
Patrick: Do you believe that you yourself will land on the Moon?
Yuri: If I am selected there will be no difficulty.

However, unfortunately for Patrick the dawn of the manned spaceflight era prompted the International Astronomical Union's Commission 16 to propose that East and West be reversed on the Moon, to make life easier for any future lunar explorers. At the Commission's meeting on August 17th 1961, in Berkeley, the proposal was adopted much to Patrick's dismay. For almost 30 years Patrick had observed the Moon with Mare Crisium in the West and Aristarchus in the East. Now he had to mentally reverse all those directions. Needless to say Patrick strongly voiced his objections, but the IAU was not to be swayed. The Eastern Sea (Mare Orientale) was now sitting firmly on the Western limb of the Moon!

The Famous 50th Episode

With the success of the ambitious live filming of the February total solar eclipse still prominent in their minds, Patrick and *The Sky at Night* producer Paul Johnstone decided to try something similar for the 50th programme, on the 21st of August 1961, namely showing the Moon and planets 'live' on TV. Actually, this had been done before on the *Sky at Night*. As we have already seen, in the very early programmes the Moon had once been shown through the 6 inch (152 millimetres) refractor at Hampstead Observatory. Earlier in 1961 the planet Venus had been shown through George Hole's massive 24 inch (60 centimetres) reflector; this behemoth was sited at 44 Sanyhils Avenue, Ladies Mile, Patcham near Brighton. Hole, who I have already mentioned, was a telescope maker and amateur astronomer extraordinaire and had just taken on the role of 'Instruments and Observing Methods Section Director' in the BAA, so he was well known to Patrick. It was decided that George's 24 inch reflector could be used to view the Moon, Jupiter and Saturn 'Live' on TV for the 50th programme, starting at 10.40 p.m. It was also intended to show the glorious double star Albireo in Cygnus, as well as Mizar, the multiple star in the Great Bear. The 24 inch was ideal, as not only was it a large telescope (and TV cameras of that era were not very light-sensitive) it was also in the open air, so there was no dome to move about, and the outside broadcast cameras did not need to be crammed into an observatory. This whole idea was an ambitious plan, as most nights in the UK are cloudy, but at least August nights are usually fairly mild. To safeguard against cloud in the south of the country another camera was attached to the 36 inch reflector at Edinburgh observatory. The event was publicised in *The Radio Times* with a splendid sketch of Saturn being included to whet the appetite, alongside a photograph of Hole's monstrous telescope in his Patcham garden. The programme was entitled 'Celebrating with Saturn'.

As it turned out, George Hole's 24 inch at Brighton was clouded out (apart from a brief snatch of a ghostly Saturn) and the Edinburgh observatory was under mainly cloudy skies too. In addition, the low altitude of the planets at the time of transmission was a big problem for the Edinburgh telescope and only a few glimpses were seen from that instrument. So, with both sites predominantly cloudy, the onus was on Patrick to keep describing the sky and the cloud (there was little else to do) while George tried to swing the 24 inch monster onto any tiny patches of clear sky! This was to continue for 15 minutes! The resulting television (the archival footage famously survives today) is pure farce, but is also perhaps the most memorable piece of *Sky at Night* film ever shot. Of course, being night-time, the back garden of George Hole's garden was floodlit with arc-lamps, making lining up the telescope on a planet very tricky as both men were being dazzled. Most of the programme consisted of Patrick shouting "Look there's a gap….. no, it's gone again" and George mumbling "It's totally obscured Patrick, totally obscured". At one point George just misses a break in the clouds and says "Well, you can't move a 24 inch telescope any faster than that", prompting a somewhat stressed reply from Patrick: "No, I'm afraid YOU can't". The programme ended without a live glimpse of anything from Patcham being seen, except that ghostly view of Saturn. Rather predictably (astronomers call this Spode's law, everyone else calls it Sod's law) the sky started clearing just minutes after Patrick went off air, and 30 minutes later superb images of the Moon were being filmed!

Nevertheless, despite this embarrassing episode of failing to point a giant backyard telescope at the Moon and planets, in front of more than one million TV viewers, Patrick's lunar and planetary enthusiasm was undaunted. Nineteen days later, on September 9th, he was the leading light at the BAA's first Out-of-London meeting, held at the Herbert Art Gallery and Museum in Coventry. As well as having a lot to say about the current appearance of Jupiter and acting (as ever) as the meetings recorder, he had agreed with his friend Howard Miles to give a public lecture at the local technical college that evening. Patrick's talk, entitled 'Exploring the Moon by Telescope and Rocket', was, of course, packed out; just 4 years into his TV career he was already a huge crowd puller. Next month, at the October 25th BAA meeting in London, he delivered a major paper entitled 'Lunar Craters and the Moon's Crust', in which, not surprisingly, he repeated his entrenched and totally erroneous view that the craters of the Moon could not possibly have been caused by impacts.

Fig. 8.1 Patrick and George Hole in Hole's garden at 44 Sanyhils Avenue, Ladies Mile, Patcham (near Brighton) on August 21st 1961, during the fateful 50th anniversary programme when they were clouded out. Although George Hole is not shown clearly here it is interesting to compare his profile with that of Cedric Allingham. Bear in mind that in the early 1950s Hole had a moustache which was subsequently shaved off (Reproduced with the permission of the BBC Motion Gallery)

Fig. 8.2 Patrick at the BAA's first Provincial meeting, in Coventry, on September 9th 1961, standing alongside some of his best BAA mates of that era. The meeting was held in the City's Herbert Art Gallery and Museum. Patrick was very keen on having more BAA meetings out of London and this event was organised by his good friend Howard Miles. From *left to right*: W.M. Baxter (Solar Section), Patrick, Frank Hyde (Radio Astronomer), K. Fea (Artificial Satellites Section), Howard Miles (Artificial Satellites Section) and W.E. (Bill) Fox (Jupiter Section) (Photograph by kind permission of Alan Heath (future Saturn Section Director))

More TV, More Books, and a German Jaunt

At the end of 1961 Patrick secured another TV spot on BBC TV: a 15 minutes weekly programme for children entitled *Seeing Stars*, scheduled at 5.45 p.m., just before the 6 O'Clock News. Children's after-school TV on 1961 Mondays typically consisted of early editions of *Blue Peter* (with presenters Christoper Trace and Leila Williams), *Whirlybirds* (an Australian helicopter action programme) and Patrick's *Seeing Stars*. He would start this series again in the 1970s, when the Apollo Moon Landings inspired more interest in Space Travel. With Patrick also appearing once every 4 weeks on the *Sky at Night* at 9.45 p.m., before the 10 O' Clock News, his face was one of the most well known on British TV screens in the early 1960s. Patrick was also finding that he was in demand as a guest personality on various other topical programmes. By May 10th 1962 he was even a guest on the BBC TV series entitled 'Let's Imagine', which was hosted by the comedian Kenneth Horne, who invited celebrities to imagine future scenarios. Also appearing on that episode, with the 39 year old pipe smoking Patrick, was the science fiction author Brian Aldiss, who that same year had won the prestigious Hugo Award for his sci-fi work entitled *Hothouse*.

Fig. 8.3 The comedian Kenneth Horne (*left*), Patrick (*centre*) and science fiction writer Brian Aldiss (*right*) recording the *Lets Imagine* BBC TV programme on May 10th 1962 (By permission of the Press Association)

As well as writing four more science fiction novels himself between 1961 and 1964, Patrick's production of non-fiction literature for the amateur astronomer was prolific in the early and mid 1960s. These highly readable astronomy books would set new standards for the amateur astronomer and inspire generations of amateurs and professionals in the coming decades. In November 1961 his major book entitled 'Astronomy', published by Oldbourne and priced at 35 shillings, received brilliant reviews, even in national newspapers. The November 10th edition of the *Daily*

Express included a glowing tribute and described it as 'a Space Age textbook for the family'. The anonymous Express Science Reporter commented: 'Most books on astronomy are so technical that they seem as dry as moon-dust But now all the Space Age information which the average family will need has been gathered in one simply written and superbly illustrated book by Patrick Moore, the B.B.C. astronomer'. Such a review, in a newspaper read by millions, was worth its weight in gold. In December the Oldbourne publishers ran full column advertisements for its Christmas books in various newspapers, with Patrick's latest book taking pride of place.

In addition, during 1961, J.G. Porter, the BBC Radio voice of Astronomy, asked Patrick to become the Associate Editor of the new *Yearbook of Astronomy,* the first edition of which (for 1962) would appear at the end of 1961. For his contribution to astronomy, The Edinburgh Astronomical Society awarded Patrick its prestigious Lorimer Gold Medal in 1962. It would be the first of many awards that Patrick would receive in his lifetime. In the early 1960s, through Lutterworth Press, Patrick gradually developed a seven volume series entitled *The Amateur Astronomer's Library*. Patrick would write three of these books, edit a fourth, and delegate three others to fellow amateurs who specialized in Radio Astronomy (Frank Hyde), Solar Astronomy (William Baxter) and Lunar Photography (Cdr Henry Hatfield). The most popular books in the series were undoubtedly Patrick's *The Amateur Astronomer* and *Practical Amateur Astronomy*; the latter was edited by Patrick with each chapter being written by a leading amateur astronomer from the BAA. In many ways the 1963 volume *Practical Amateur Astronomy* was a similar, if more advanced version, of the 1953 BAA book 'Astronomy for Everyone'. In addition to these works Patrick had, since May 1959, been writing up every *Sky at Night* for the BBC *Listener* magazine and contributing a regular weekly astronomy column in Arthur Mee's *Childrens Newspaper*, as well as churning out Sci-Fi. All this work was carried out on his turn of the century typewriter too! In addition to the Lutterworth *Amateur Astronomer's Library*, numerous other smaller works were produced. In the 1960s, with Patrick's name guaranteeing to sell books and the USA/USSR space race generating much interest, publishers were constantly making a path to Patrick's door.

Perhaps the most inspirational work Patrick produced in the early 1960s was the small pocket book called *The Observer's Book of Astronomy*, which was first published, by Frederick Warne & Co., in 1962. Ask any amateur or professional astronomer born in the 1950s which was the first astronomy book to inspire them and the chances are it will be *The Observer's Book of Astronomy*. This was a diminutive book (145×92×20 millimetres in size) but was packed with essential data, would fit into a school blazer pocket, and the four groups of glossy illustrations (or plates as they were referred to) were fascinating. This was the book that started this author on a journey into amateur astronomy (I had the 1967 edition) and for me the pictures of amateur observatories and the colour illustrations by Leslie Ball were the features that really grabbed my attention. In addition, there were almost 60 constellation diagrams incorporated into the text pages. The '*Observers*' name was in fact a reference to the series of pocket books produced by Warne, which included titles on observing birds, butterflies, aircraft, cacti, lichen and fungi. Patrick was firmly into the swing of book writing by 1962 and, by his own admission, writing that incredibly popular and inspirational little *Observer's Book of Astronomy* took him just 1 month. It was an excellent title for an astronomy book and an excellent book all round. Patrick often encouraged other astronomers to write books too. One example of this was Thomas Rackham's 1959 book *Astronomical Photography at the Telescope*, dedicated by the author to Patrick for his encouragement.

In August 1962, Patrick answered several long-standing invitations to visit a whole host of observatories in both West and East Germany. On a wide-ranging tour, gaining information for himself, his books, and *The Sky at Night*, he visited the observatories in Frankfurt, Tautenburg, Potsdam, Treptow and the Wilhelm Foerster Observatory in West Berlin. He also visited the Jena Planetarium and the headquarters of Carl Zeiss. Three years later he would return there on official business, while working for the Armagh Observatory. While at the Treptow Observatory in East Berlin his hosts played Patrick an audio recording of Major Andrian Nikolayev aboard Vostok III. That spacecraft and Vostok IV (manned by Colonel Pavel Popovich) had been launched a few days earlier and so this was an historic space first: two spacecraft and two cosmonauts in orbit at the same time and while Patrick was en route to East Germany. Patrick wrote the trip up in an entertaining style in one of his log books and I have included a few of the more amusing extracts to give a flavour of his experience:

> Germany, August 1962
>
> Aug. 12 (Sunday). Brian drove me from Richmond to Victoria at the ungodly hour of 7.30. Met the party there, and departed for Ostend, via Dover. Weather tolerable; crossing calm. Having got to Ostend, we disembarked, and found our Minibus. We also have a driver, which lets me out. He is Belgian, and looks bovine to a degree. We drove off for Frankfurt via Liège, Aachen and Cologne. Apart from occasional stops, notably one at a café at 4 am, we went on till we got to Frankfurt at 7.30 am, by which time Ian and I, at least, were limp. We went to the Hotel Rebstock, and breakfasted off sausage and beer.

Under his entry for August 14th Patrick wrote:

> Left Frankfurt for Leipzig. A day of mishaps. Owing to our bovine driver (possibly a Martian?) we were late at Wartha, the frontier. We got there at 1.30, and were through by 4! We met Mr Haage, our East German guide, and got to Leipzig late. DDR beds are strongly reminiscent of boards, but we were tired enough to sleep on them. The driver is most odd. He speaks neither French, English nor German. He drives at a steady 10 km/p/h, changing down when going downhill and accelerating at corners. I am sure his home ground is a Martian desert.

On August 16th Patrick noted:

> Did some wandering in Leipzig – I managed to buy a Tchaikovsky record! –and then, guided by Mrs Langenhagen, drove off, bound for Berlin.

And, he added later:

> We got to Berlin at 8, and dined at the Hungarian restaurant. Subsequently we went off and looked at East Berlin, ending up at a restaurant (open-air) drinking coffee and Vodka. On getting to our hotel we entered a card-session (Ian, Mike, Richard and myself) until rather late!

Another entertaining diary entry by Patrick appeared on August 18th:

> Aug. 18 (Saturday) This is being written at the frontier Kontrol, 12.45 p.m., at Marienborn, on the Berlin-Hanover autobahn. Our efforts to talk German and semaphore have been

> hilarious, and I don't think the Germans appreciate them; the laughter in the forbidding Kontrol sounds out of place, but there was a lot of it! Moreover we have completely blocked a main gangway with our Minibus, and the Huns are trying to back lorries round us. Chaos reigns. All highly amusing.

The final comments for the trip read as follows:

> Aug 20. Smooth crossing; and the group finally broke up at Victoria at 3.30. I put Ian on his way- he's only a mile to go – and got back to East Grinstead at 5.30.
>
> So ends a weird trip.

What with endless book writing, foreign travel, a monthly television programme and observing every clear night it might be thought that Patrick's BAA and lecturing duties might suffer, but apparently not. He rarely turned down any request to speak for the BAA. So, when he was asked to be an astronomy course tutor at a Carnegie sponsored BAA event at Kingsgate College in Broadstairs, shortly after returning from Germany, he agreed immediately, along with two of his best mates Colin Ronan and Ernest Noon. Indeed, so big an attraction was Patrick by then that when the course was announced by the BAA, all 44 residential places were booked up within days, with 23 disappointed applicants having to be turned down.

February to May 1963 turned out to be a busy time for Patrick and for amateur astronomers in general. On February 6th the Swedish observer E. Dahlgren (and just afterwards the American observer Leslie Peltier) discovered the brightest northern hemisphere nova for 27 years in the constellation of Hercules, not that far from Alcock's new comet (see below). The nova was close to the bright star Vega in the constellation of Lyra and it reached magnitude 3.0. Indeed, it was so close to the Lyra/Hercules border that there was some debate in the International Astronomical Union about which constellation it was actually in. Nova Herculis 1963 faded to magnitude 6 by late March but it was an easy naked eye sight and Patrick made numerous magnitude estimates of the object. Just as the nova was fading from naked eye view, at 3 a.m. on the 19th of March, George Alcock at Peterborough discovered his third comet while sweeping the constellation of Cygnus with his 25 × 105 binoculars. Two months later it would flare up to naked eye visibility, reaching third magnitude. If that were not enough excitement, after a gap of 3½ years the Soviets announced that they were sending another Lunik probe (also called 'Luna' probes) to the Moon. The 1.4 ton Lunik 4 was launched on April 2nd 1963 from Tyuratum; its aim was to soft-land on the Moon and transmit from the lunar surface. Exciting stuff!

An extended *Sky at Night* special, lasting 90 minutes, was arranged for April 3rd, and after the success of the Lunik 3 broadcast in 1959, every possible arrangement was made. The BBC had set up a telephone link with Moscow, a radio link with Professor Bernard Lovell and Colin Ronan at Jodrell Bank, and cameras were attached, once again, to George Hole's 24 inch reflector at Patcham and the Royal Observatory's 36 inch reflector at Edinburgh (with Dr Peter Fellgett commentating). Both telescopes were pointed at the Moon. While the fiftieth anniversary programme had been a fiasco, this one turned out to be an even bigger one. It was raining in Edinburgh; there was solid cloud at Patcham; Jodrell Bank could not detect the probe; and no-one in Moscow knew a thing. A similar TV farce would await Patrick 23 years later when the Giotto probe went past comet Halley. Oh, and to cap it all, Lunik 4 did not soft land on the lunar surface: it missed the Moon by 5,000 miles! If there was one outstanding achievement it was, of course, Patrick, who managed to keep talking about the Moon and Russian space probes for the full 90 minutes! By comparison, the next scheduled 'normal' *Sky at Night*, on April 17th 1963, about the new nova, was a return to sanity!

Patrick's ability to speak live, with a faultless delivery, was becoming a favourite attraction to anyone who happened to be in the BBC studios in the early 1960s. Well known faces or up and coming actors would often peek around the set and watch Patrick in full flow. One such fan was the larger than life actor Brian Blessed, who played police constable 'Fancy Smith' in the *Z Cars* series of that era. He too would sometimes watch Patrick performing live, as so much television work, including dramas, was broadcast in those early days. Of course, in dramas there was always someone else to help if you forgot your lines, but with Patrick he was on his own for a minimum of 15 solid minutes every month. "Having the same amount of nerves as the average Rhinoceros" helps, Patrick would quip in later life.

Despite the second TV nightmare involving George Hole's enormous 24 inch reflector Patrick was always keen to use the giant instrument for looking at the Moon. It was, after all, double the aperture of his own 12½ inch reflector, Oscar. In the early 1960s Patrick often used to drive his Ford Prefect 'Ark' from East Grinstead, through the Ashdown Forest, and on to Patcham, near Brighton. One of Patrick's favourite stories was about the night he took a young student from East Grinstead Grammar School (Peter Cattermole, now Dr Peter Cattermole, a planetary geologist and regular future *Sky at Night* guest) on a trip to George Hole's observatory. On the way back, in the early hours, due to a spate of burglaries in the Ashdown Forest area, Patrick was stopped at a police checkpoint. They were investigating anyone who was driving through the area at such a ridiculous time. According to Patrick the discussion went something like this:

Policeman: "Would you mind telling me what you two have been doing at this time of night?"
Patrick: "We've been looking at the Moon".
Policeman: "Cmmmmonnn, you can do better than that, I want to know what you two have really been doing".
Patrick: "We have been down to Brighton, LOOKING AT THE MOON".
Policeman: "Don't give me that. I want to know exactly what you've been doing or you're in trouble".
Patrick: "HOW MANY MORE TIMES. WE HAVE BEEN OBSERVING THE MOON WITH A HUGE TELESCOPE, ALL NIGHT".
Policeman: "You honestly expect me to believe that Sir? Who do you think you are, Patrick Moore?"
Patrick: "YES!!!"

At this point the policeman shone his torch into Patrick's face, apologised profusely, and let them continue home!

Although, by July 1963, Patrick's alter ego of Cedric Allingham was well behind him (and certainly any belief in Flying Saucers and Little Green Men would definitely not have been compatible with his *Sky at Night* role) he still had quite a fascination with UFO sightings. Maybe he just wanted to expose the perpetrators as frauds, or maybe a small part of him really did want to believe there were a few genuinely unexplained events? Either way, when media reports of a mysterious crater appearing on the boundary between a potato field and a barley field at Manor Farm, Charlton in Wiltshire (near Shaftesbury, just over the county border in Dorset) appeared in the press, around July 16th, Patrick immediately drove up to the area to inspect it for himself. Flying Saucers had, apparently, been seen in the area. Was this going to be another farce like that of the Morsgail meteorite from almost 4 years earlier? Maybe, but at least this event was not in the Hebrides, so getting there was much easier. According to newspaper reports the crater was first discovered by local farmer Roy Blanchard. However, the 'UFO investigator' who went by the name of Leonard G. Cramp, claimed the crater was discovered by a chap called Reg Alexander. Regardless of the discoverer, as well as Patrick surveying the scene the Army bomb disposal squad were also called in, and questions were even asked in Parliament! I have examined Patrick's scrap book for this era and he took two photographs of the scene. One shows a very young beret wearing soldier at Manor Farm and Patrick's caption describes him as a 'Baffled Bomb Squad Soldier'! The Army claimed to have found magnetite at the site which scientists later said occurred naturally in that area. According to various reports the scene consisted of an 8 foot diameter crater, 1 foot deep, with a 3 foot deep hole in the centre. Four slots were seen radiating out from the crater, and the ground in a 12 foot radius was denuded, with the potato crop being obliterated. A lump of iron ore found in the crater by the Army team was pounced on by Patrick, who quickly declared that it looked like a meteorite, thereby staking his claim that this might be a major UK meteorite impact. However, the subsequent analysis of the rock eventually found it was naturally occurring iron stone, so Patrick had been wrong there.

Of course, the fruitcakes of the day soon made their voices known. An enigmatic character, calling himself 'Dr Randall', appeared on the scene and he claimed to be an 'Australian astro-physicist'! This mysterious chap was widely quoted in the press, despite the fact that his faked credentials were never checked. Dr Randall declared that, without a doubt, the crater was caused by a Flying Saucer weighing precisely 600 tons, with a crew of 50, and it originated (of course) from the planet Uranus! This account was printed verbatim in several newspapers! In the coming months an alternative and less hysterical view was put forward by an investigator called Alan Sharp. He originally thought that subsidence had cause the crater, but later concluded that a major lightning strike fitted all the evidence.

On the same day that Patrick investigated the Dorset crater he gave a public lecture at his local East Grinstead Grammar School (now Imberhorne School) on the subject of 'Life on Other Worlds'. He even brought some crater rock samples he had collected earlier in the day along to the lecture and told the pupils, parents, members of the public and teachers that they would be examined for radioactivity and signs of a meteoritic origin. As well as mentioning that the Moon was lifeless he informed the pupils that there were now (in 1963) some doubts about whether life could exist on Venus, but Mars had a far greater probability of hosting some form of life, in the form of vegetation. He added that he refused to believe that there was no other intelligent life out there in the Universe. According to the local newspaper report of Patrick's talk, he told the pupils to imagine that the Sun was 3 feet in diameter and inside the grammar school. On that scale, Patrick explained, the Earth would be several 100 feet away "in Halsford Park Road" and the nearest star, Proxima Centauri, would be in Siberia! As was so often the case with Patrick's talks to schools he waived any fee and the ticket proceeds paid for two East Grinstead Scouts to travel to the World Scout Jamboree in Greece, 1 month later. He added that he did not believe the recent Dorset crater, which he had investigated earlier that day, was caused by an alien spacecraft!

Anyway, for Patrick it was quickly back to reality and writing about real spacecraft, not illusory alien ones. A further book of his in this era, published later in 1963 by Penguin, was *Space in the Sixties*, all about the US and USSR space probes of the late 1950s and early 1960s. Patrick was particularly pleased to be writing for such a big name in the publishing business but was horrified when he saw the first prints of the book. He rang the publishers immediately insisting that all copies produced, but not yet sold, should have an erratum slip installed due to the gross errors describing the author on the front and back covers. One of the many mistakes was a howler, although Patrick never found it even remotely funny: *Patrick's hobbies include Chess, which he plays with a peculiar leg spin...........* The mind boggles!

In case the reader has forgotten, it is worth repeating that Patrick was indeed a very good Chess player and had even played for the Sussex county team in the 1950s. He often carried a pocket Chess set with him in his younger days. In the twenty-first century Patrick provided a cup for the Sussex County Junior Chess team and the relevant contest became known as the "Sir Patrick Moore Sussex Challenge Cup" defended by childrens' teams with contestants under 16, under 12 and under 10 years of age. Patrick was eventually made the Vice-President of the Junior Chess Team.

On the September 10th 1963 *Sky at Night* programme Patrick was delighted to welcome his old friend from the British Interplanetary Society, Arthur C. Clarke, into the studio. This was just 10 weeks prior to the assassination of John F. Kennedy. Clarke had, of course, been a prophet of communications satellites during the late 1930s, when still a very young man. Since 1954 Clarke had lived in Sri Lanka where he could see both celestial hemispheres and do plenty of scuba diving, so this was a rare trip to the UK. Patrick coaxed Clarke into some predictions on the programme and these were duly given as follows:

First circum-lunar manned spaceflight:
This could possibly be in 1964, but probably by 1968, and definitely by 1973.
First lunar landing:
1970, but could be earlier if Russian, as there were hints the Soviets wanted a Moon landing to coincide with the Russian 'October revolution' 50th anniversary (Nov 7th 1967). [Clarke actually gave the year of this anniversary incorrectly, as 1968, on the programme.]
First manned flight around Mars: 1983.
First man on Mars: 1988.
First base on Mars: 2000.

With the benefit of hindsight we can say that the first two predictions were pretty good, even if Clarke favoured the Russian's chances; but his other predictions were obviously not! A manned Mars mission has still not occurred, more than half a century later. Incidentally, at the same time, 1963, Patrick was predicting that the first manned lunar landing would not occur until 1980–1985!

Chapter 9
Back to the Lunar Section

During 1963 professional observations of the planet Venus had been transformed by the US spaceprobe Mariner 2, which had been launched on August 27th 1962 and had flown past the planet on December 14th of that year. All of the data had been analysed by the summer of 1963, just a few months prior to the assassination of John F. Kennedy on November 22nd and the first episode of the time travelling exploits of *Dr Who* on BBC 1, a day later. The contribution of Kennedy to the NASA space programme was huge as his famous speech to Congress of May 25th 1961, including the words "I believe that this nation should commit itself, to achieving the goal, before this decade is out, of landing a man on the Moon, and returning him, safely, to the Earth…." had fired the U.S. enthusiasm for the Apollo project. If anything, his shocking assassination, occurring just 30 months later, inspired the engineers, scientists and astronauts even further to achieve Kennedy's dream.

The Mariner 2 spacecraft had destroyed the popular Whipple-Menzel theory (that Venus was covered in an ocean of water) and had confirmed a very long rotation period, the lack of a magnetic field, and a scorchingly high surface temperature. Indeed, even the earliest results from the probe had forced Patrick to insert a last-minute postscript into his aforementioned *Space in the Sixties* book, at the proof stage, stating: 'On page 125 I commented that some of our views about Venus, at least, might prove to be wildly wrong. It now looks as though this is indeed the case!' It was looking increasingly like amateur observers had less of a role to play in observing the planet and so Patrick relinquished the BAA Mercury & Venus Section Directorship to a research bacteriologist and amateur astronomer, Dr F.L. Jackson, who had often been a guest on *The Sky at Night* when extraterrestrial life was being discussed. Dr Jackson was a scientist but did not have the infectious enthusiasm of Patrick and saw one of the main roles of the section in being educational. This approach almost killed the Venus & Mercury section off, but in 1965 Jackson would emigrate to Canada enabling Patrick's old friend J. Hedley-Robinson to restore the section's enthusiasm. Patrick would always place such zest higher than a strictly disciplined scientific approach, which is how he inspired generations of astronomers. Without a section directorship, Patrick did not have an automatic

M. Mobberley, *It Came From Outer Space Wearing an RAF Blazer!: A Fan's Biography of Sir Patrick Moore*, DOI 10.1007/978-3-319-00609-3_9,

seat on the BAA Council (though he would always be elected to the council by the voting members anyway) but, by chance, the BAA Lunar Section Directorship was about to become available. Patrick's old colleague H.P. Wilkins had resigned as Lunar Section Director in 1956 and handed over to Ewen Whitaker. Whitaker would be involved in the celebrated *Lunar Photographic Atlas* edited by Gerard Kuiper in the USA, and he had emigrated to the USA himself in 1958. At one time, as a young man, Whitaker had lived just a few miles from Wilkins and had occasionally cycled over to visit him.

Incidentally, Whitaker's excellent 1999 book *Mapping and Naming the Moon* is the authoritative work on the subject and his comments on Patrick's mentor Wilkins' map are worth repeating. He says: 'Unfortunately his [Wilkins] cartographic abilities did not match his infectious enthusiasm for the subject, and the smaller details in his crowded map are not trustworthy'. Plus, Whitaker adds: 'The addition of fictitious fine detail has so cluttered the map that it is virtually uninterpretable'. Interestingly, Patrick is not mentioned ONCE in Whitaker's book. Perhaps it is not surprising that Patrick had nothing to do with the BAA Lunar Section during Whitaker's reign, although Patrick actually gave Whitaker's book a good review some 40 years later. It is obvious that there was some friction between the two men, not least because Whitaker had thought that the era of sketching the lunar surface had been superseded by photography, even during the mid-1950s, when Wilkins and Patrick were still hard at it with the eye, pencil and sketchpad. In 2006 Whitaker and Richard Baum would join forces to set the record straight about who discovered the Mare Orientale and submit it for publication in Patrick's favourite publication, the BAA Journal. It must have been a painful paper for Patrick to swallow.

Anyway, let's get back to the plot: the Lunar Section vacancy. Whitaker's 1958 Lunar Section replacement was a good friend of Patrick, namely Gilbert Fielder. Gilbert himself resigned due to work commitments in 1962 and was succeeded by another friend of Patrick, Brian Warner, who himself emigrated in early 1964! This left the door open for Patrick to become the BAA's Lunar Section Director, and as always, regenerate enthusiasm in lunar observing.

Desert Island Discs

On Monday October 28th 1963 Patrick was the guest, or 'castaway', on the popular BBC radio programme *Desert Island Discs*. For anyone not familiar with this famous British institution, it was invented and presented by Roy Plomley (1914–1985) from January 1942 until Plomley's death in 1985. Following Plomley's demise the programme was hosted by Michael Parkinson and then by Sue Lawley and Kirsty Young. To be invited as a *Desert Island Disc* castaway you had to either be a household name, or someone of great achievement or importance. As the presenter of *The Sky at Night* for the past six and a half years Patrick was already a household name in the UK. Guests on *Desert Island Discs* were (and still are) invited to imagine which eight pieces of music (gramophone records in those days) they would take with them if marooned on a desert island; they are also permitted to select a book and a luxury item of their choosing. [No, a mayday radio or a speedboat are not allowed!] Examples of some other guests who were featured on the programme in 1963 may be of interest to the reader. On August 5th the James Bond author Ian Fleming was the castaway and on August 19th it was the racing driver Graham Hill's turn; on September 19th the BBC commentator Raymond Baxter was on the desert island and on October 7th, 3 weeks before Patrick, the popular actor Bernard Cribbins was marooned. It will come as no surprise that Patrick's choice of book was Norton's Star Atlas and his luxury item was (what else?) a telescope. All of Patrick's eight gramophone records were classical pieces; no *Beatles* or Elvis singles for him! His first gramophone record choice was also very predictable as it was simply *The Sky at Night* theme: Sibelius' 'At the Castle Gate' (from Pelleas and Melisande). Patrick also chose to take two Grieg records with him (Piano Concerto in A Minor and the Homage March from Sigurd Jorsalfar), along with Gilbert & Sullivan's Princess Ida Overture and the Dynamiden Waltz by Strauss. He also selected Chopin's Etude in C minor (Op. 10/12 'Revolutionary') and Tchaikovsky's Symphony No. 7 in E flat Major. Patrick declared his favourite desert island record (the penultimate played on that programme) to be the Bad'ner Mad'ln Waltz by Karel Komzák II. As Patrick's star turn on the famous BBC *Desert Island Discs* came to a close he could have had no idea that the next evening, in the USA, some dramatic lunar developments were about to be claimed by professional astronomers, viewing the Moon from a major observatory.

A TLP Sub-Section and a Dr Who Spoof

Patrick's first step on taking command of the BAA Lunar Section would be to create a Transient Lunar Phenomena (his term) or 'TLP' observer network. Remarkably, on October 29th/30th 1963, James Greenacre and Edward Barr, at the Lowell Observatory in Arizona, observed colour in the region of the very bright lunar crater Aristarchus. The observations were made while using the 24-inch (61 centimetres) Alvan Clark 'Lowell' refractor. That infamous telescope is the same instrument through which Percival Lowell sketched Martian canals! The description was of red and pink patches on the crater's floor and walls. Remarkably, these lunar observations, in media-hyped form, were the front page news in the Arizona Daily Sun newspaper for Thursday October 31st 1963. The headline blared out: MOON 'ERUPTIONS' SEEN HERE – LUNAR MAPPERS SPOT POSSIBLE VOLCANOES. There was even a photograph of Aristarchus on the front page! Equally remarkable was that the next month similar events were observed and this time they were confirmed by Dr Peter Boyce and Dr Kent Ford at the Perkins Observatory (located at Anderson Mesa). Boyce and Ford were using a 69 inch (175 centimetres) reflector on November 27th (Nov 28th GMT). A paper published around this time, by Dr Jack Green in the USA, seemed to indicate that strange glows on the lunar surface tended to occur when the Moon was closest to the Earth and under greatest gravitational stress. Patrick often stated that, but for bad luck, he might have been at Lowell on October 30th 1963 and seen the Greenacre/Barr TLP. In fact, he arrived at Lowell in the following month and spent many hours with James Greenacre, studying the Moon with the 24 inch Lowell refractor. To quote Patrick: "Had another TLP appeared I doubt whether we could have missed it, but Aristarchus remained stubbornly un-cooperative." Despite these comments though, the subsequent events, confirmed by Boyce and Ford, occurred around the time that Patrick *was* at Lowell.

A page or two ago I mentioned that the ever-popular science fiction series *Dr Who* was first transmitted in late 1963, on November 23rd in fact. This was the day after the Dallas assassination of President John F. Kennedy. *Dr Who* soon established itself as a cult programme and the first spoof based on the programme occurred only a month later, and Patrick himself was involved in that lark. It will come as no surprise to anyone who was a child in the 1960s to learn that the spoof was the brainchild of Patrick's close friend Michael Bentine (1922–1996) who, as well as being one of the infamous Goons in the early years of the Goon Show (with messrs Sellers, Milligan and Secombe), had a reputation for building all manner of models and amusing puppets. On Bentine's TV show *It's a Square World* he seemed to delight in blowing up parts of London, not least the BBC TV Centre! In the last programme of the sixth series, filmed on December 16th and December 20th 1963, Bentine used the actor Clive Dunn (later to be Corporal Jones in *Dad's Army*) to impersonate *Dr Who* in the form of 'Dr Fotheringown', using the real *Dr Who*'s (William Hartnell's) cape and wig! As Dr Fotheringown launched a massive rocket in the TV studio and propelled a model of the TV Centre into Earth Orbit, Patrick approached the microphone with a completely ad-lib, unscripted comment: "There are rumours that there is a form of intelligent life inhabiting this new and spectacular object in the night sky, but I, personally, do not believe this to be so". The episode was transmitted on December 31st 1963. As well as being 'a Square World' it is sometimes a small one too. Dr Fotheringown's real persona of Clive Dunn (1920–2012) had, as a 15 year old boy, acted with the comedian and amateur astronomer Will Hay in the 1935 film *Boys will be Boys*. How I bet the 12 year old Patrick and the older Patrick wished he could add that to his C.V.

When Patrick took over the BAA Lunar Section those enigmatic TLP were very much the hot topic of conversation. His new idea of a TLP network was that if anyone suspected a TLP was taking place they could phone others in the network for confirmation. A report from a single observer was just too subjective and too controversial. Patrick also began the practice of holding the Lunar Section meetings regularly in London, and in other large cities, on a regular basis. When Patrick was present at a meeting loads of BAA members were bound to turn up! Without the burdens of any family commitments and still being looked after by his 77 year old mother and a housekeeper, Patrick could still channel all his spare energy and enthusiasm into the BAA and amateur astronomy. He was not burdened with millstones like a day job, cooking, washing, cleaning, DIY, a mother-in-law or gardening.

Everything was done for him! Under Patrick's direction the TLP sub-section invented a 'Moonblink' apparatus, first proposed by the unique and eccentric V.A. Firsoff and then perfected by the multi-millionaire businessman Peter Sartory, who had made his fortune in the microscope industry.

The 'Moonblink' device consisted of a circular casing enclosing a disc containing a blue and a red filter. In addition, a 1¼ inch external diameter pipe on one side enabled insertion into a telescope drawtube and a 1¼ inch internal diameter pipe enabled insertion of an eyepiece into the other end. By twiddling a knob on the casing the red and blue filters could be rapidly interchanged making any coloured feature on the dull gray lunar landscape appear to blink. Somehow the term 'Moonblink' always seemed eminently suited to a man like Patrick, equipped with a monocle for his weak eye and with a tendency to half close it when the monocle was not inserted! Valdemar Axel Firsoff, who preferred to be called Axel or V. Axel Firsoff, and who proposed the 'Moonblink' device, was an eccentric and colourful character. Originally from Sweden, he lived for many years in Lochearnhead, Scotland, but then moved to Somerset. He was convinced there had been flowing water on the Moon. Patrick told me that, after moving to Somerset, Axel always slept with a loaded pistol under his pillow, because he was convinced that "the Scottish communists are out to assassinate me!"

Lunar Vulcanism

Apart from TLP, Patrick had another lunar obsession in the 1950s and 1960s; like TLP, it would ultimately go out of fashion and become an idea he should have dropped, rather than clung on to. This was the idea we have already come across in connection with H.P. Wilkins, namely that the lunar craters were formed, in the main, by volcanic activity on the Moon, rather than by asteroid impacts. Of course, TLP and Kozyrev's spectra of 1958 would be in keeping with an active Moon, or at least one venting gas.

A brief bit of history for the non-astronomer or non-geologist may be useful here. The volcanic crater theory requires internal heat from the Moon and used to be referred to as the endogenic theory, as opposed to exogenic for impacts. The first major paper suggesting that impacts were responsible for lunar craters was published by Gilbert in 1892. However, the most influential work in the twentieth century was the 1949 Chicago University Press book entitled *The face of the Moon* by Dr Ralph Baldwin of the USA. Baldwin, amongst other things, showed that the depth-diameter ratio for large lunar craters (they are relatively shallow compared to the small surviving impact craters on Earth) was in keeping with what would be expected for a large lunar impact. To the non-astronomer the thought that virtually every crater on the Moon is caused by an impact may seem bizarre, after all why don't we see them here on Earth? The crucial point here is that although there are hundreds of huge craters on the Moon they all occurred a long time ago, and so given time-scales of hundreds of millions of years, there would be a huge gap between major impacts, specifically, millions of years! Two-thirds of the Earth is water and our atmosphere, tectonic plates and vegetation do a remarkable job of covering up the smaller impact craters after only hundreds of thousands of years.

Supporters of the volcanic origin of lunar craters, including Patrick, had a number of key arguments. The first of these was that a number of lunar craters appear to lie on the Moon in long chains. This, they claimed, would be impossible unless asteroids fell in chains! In fact some of the crater chains they cite are, essentially, tricks of the eye, as stated by Baldwin; others are chains of craters caused by secondary ejecta flying from an initial impact. Another volcanic argument was that craters often appear in pairs and when a small and a large crater overlap, the wall of the smaller crater is never broken into; it is always the larger crater's wall that is broken by the smaller crater. The vulcanist's argument here is that it is impossible to believe that all the large asteroids impacted first and the smaller asteroids damaged them later; but as we have just seen, smaller craters in pairs or even threes and fours are quite easily formed by secondary impacts (rubble flying out of a bigger impact). Also, smaller craters are often totally obliterated by big impacts, whereas the reverse is not the case. However, it is true that volcanic activity has, indeed, played some role on the Moon, although not in the formation of the craters themselves. Another factor is that many craters smaller than 20 kilometre in diameter are not formed by an initial high energy impact, but, yet again, by less violent secondary impacts. The cratering of the Moon is not a simple matter and impact and vulcanism played major roles in the moulding of the surface. Nevertheless, the (almost) unanimous professional opinion now is that the major cratering of the Moon was caused entirely by impacts. This is not a view that Patrick was ever able to comfortably accept and at BAA and Lunar Section meetings well into the early 1990s he would make his view abundantly clear, often to great applause from his supporters. A standard Patrick talk would always contain this comment:

"The trouble with professional geologists is that they have never looked through an eyepiece; if they had they'd know the craters were all volcanoes. You know, some time ago, I showed the Moon to a bunch of schoolboys through my telescope, they all said that the craters looked like volcanoes and they do!"

Throughout the 1950s and 1960s Patrick would obsessively fly the vulcanism banner, publishing his opinions in amateur and professional magazines. If we go back to January 1963, we find, in a volume (197, page 273) of the prestigious Journal *Nature*, a letter from Patrick which was typical of his letters on this subject at that time, almost all of which mentioned the crater Thebit:

> Distribution and Origin of Lunar Craters
>
> The communication by McCall[1] leads me to make some further comments. It is certainly true that the non-random distribution of the large craters is a serious and probably fatal objection to the impact hypothesis so far as the major formations are concerned, though no doubt many small meteoric craters exist. The Mare Crisium seems to be nothing more than an exceptionally large crater-like formation, and is a member of the great western chain which also includes Petavius and Langrenus, in which case it seems that the other circular-type maria are basically similar. Moreover, in the case where one crater breaks into another, the wall of the broken formation is perfectly preserved up to the point of junction; Thebit is

a good case. This could not be so were the second, younger crater produced by a violent impact. I directed attention to this in 1956[2], and the same point has since been made independently by Kopal and others.

1. McCall, G. J. H., Nature, 196, 255 (1962).
2. Moore, P., Formacion de los cràteres lunares, Urania (Tarragona, 1956).
3. Moore, P., J. Brit. Astro. Assoc., 71, 362 (1961).

PATRICK MOORE
Glencathara, Worsted Lane, East Grinstead, Sussex.

On January 30th 1964 NASA launched its spacecraft Ranger VI toward the Moon. All such missions were of enormous interest to Patrick but, unfortunately, the power supply to the TV camera package failed shortly after launch. Nevertheless the spacecraft did impact the Moon as planned, on the eastern edge of the Mare Tranquillitatis, close to the crater Ross. The time of impact was 09:24:32 UTC on February 2nd. As soon as some clear nights arrived at East Grinstead, following the impact, Patrick was out studying that area with his 8½ and 12½ inch reflectors. Of course, the chance of anything that small making a visible mark on the lunar surface (it would have to leave a crater at least half a mile across to be seen in amateur telescopes) was negligible. Even so, Patrick wanted to be absolutely sure. At the next BAA meeting, on February 26th 1964, the President Ernest Beet invited Patrick to show his sketches and Patrick stated that he was confident that no trace of the impact could be seen with amateur sized telescopes.

Six days earlier, Patrick had been pointing a much smaller telescope at the Moon, from a road near to his future home of Selsey. His friend Gordon Taylor, who worked at the Royal Greenwich Observatory at Herstmonceux, had alerted him that the mountains on the southern limb of the Moon were due to occult a sixth magnitude star, as seen from the south coast of Sussex. Selsey was within the track where the star (SAO 93810) might be seen to wink in and out between the mountain peaks. Patrick's friend Henry Brinton, and his 12½-inch reflector at Selsey, were within the possible zone, so Patrick headed from East Grinstead to Selsey to stay with Henry on February 20th 1964. However, Gordon advised that there was still some uncertainty and it would be better if Henry stayed with his telescope, while Patrick took a portable instrument to the northern limit, a few miles up the road from Selsey. Patrick hauled a portable 4-inch refractor into his Ford Prefect and headed north, stopping at a convenient layby. As things turned out though neither Henry, nor Patrick, saw the star disappear, but it was a very close miss indeed. Gordon, and three colleagues, set up camp at Five Ashes, 50 miles further east, and witnessed the star disappear and reappear between four mountain peaks, just after 6:30 p.m. How frustrating; so near and yet so far!

One month later, at the March 25th BAA meeting, Patrick delivered a major lecture entitled 'Colour on the Moon' asking what could be learned from the recent Aristarchus TLP seen by Greenacre and Barr at Lowell Observatory and from many other reports of colour on the Moon. Concluding his talk, Patrick said that he would not really be happy until a coloured event was photographed, but he urged BAA Lunar Section members to keep a close watch on the craters Aristarchus, Alphonsus and Lichtenberg, in which colour had been suspected recently. Patrick had written to the Lowell Observatory in Flagstaff, Arizona, following the second major TLP of 1963 (that witnessed on November 27th/28th) and by the time of the March 1964 BAA meeting he had received more information. I unearthed a copy of the reply sent to Patrick, at East Grinstead, still stored within his somewhat chaotic Selsey filing system, even in the twenty-first century. The letter was dated January 7th 1964 and was from William D. Cannell, Chief of the Lunar Observation Section at Lowell Observatory. The letter read as follows:

> Observation on 27 November
>
> Dear Mr. Moore,
> Dr. Hall is presently in Chile, and therefore your letter has been given to me for an answer. I am surely a stranger to you more than you are to me, but perhaps we will become better acquainted when you visit New York next May. Mr. Greenacre has been invited to report on his recent observation of Lunar Color Phenomena at the Congress of Lunar Vulcanism, and I may go with him.
>
> I am enclosing a short description of the November 27 observations, which I think will give you the information you desire.
>
> The confirmatory observation with a large reflector was made on November 27 by Dr. Boyce and Dr. Ford with the 69-inch Perkins telescope. They were in the process of setting up a photo-electric spectro-scanner but were not able to put it into operation in time. However, they did observe visually with the guiding eyepiece at 900x with no filter. They observed a reddish glow in the shadow just outside the southern rim of Aristarchus. This was at approximately 6:45 p.m. (MST) near the end of the activity but exactly in the same place that the other four observers on the 24-inch refractor had seen it.
>
> We do not know what the phenomenon is, and at this time we are not venturing to guess. It is hoped that it will happen again when spectroscopic observations can be made. The area has been kept under close observation since November 27, but nothing has occurred.
>
> Sincerely yours,
> William D. Cannell, Chief Lunar Observation Section. Lowell Observatory. Flagstaff, Arizona.

Well, that letter from William Cannell did not tell Patrick much more than he already knew, but he was frustrated that a TLP spectrum had not been acquired. In May 1964 Patrick travelled to that International Lunar Conference in New York and delivered a lecture, presenting his views on why it was that what the lunar surface displays (overlapping craters, crater chains etc.) means that most of the crater forming must have been volcanic. Patrick described himself as an 'unrepentant vulcanist' at the conference and was supported by a small number of professional astronomers, most notably, Jo McCall of Australia and John Green of the USA. But, ultimately, the impact supporters would win: the evidence would become overwhelming, especially as spaceprobes journeyed into the solar system and found craters even on the smallest bodies, incapable of having internal heat sources. Anyway, although Patrick's views on TLP and Lunar vulcanism were, in hindsight, not the right horses to back, he invoked a new sense of urgency and excitement into the Lunar Section in 1964, as he did in everything he touched in the BAA. On that same trip he visited, once again, the Lowell Observatory in Flagstaff, Arizona and he managed to use the famous 24-inch refractor there to study various objects, including Venus.

While in America at that International Lunar Conference Patrick became aware that the whole scientific community was buzzing with the news of progress on the planned mission to land men on the lunar surface – the Apollo mission. As I stated earlier, if anything, the death of President John F. Kennedy had further galvanised the United States to land a man on the Moon 'before this decade is out'. Patrick mentioned the excitement about the project on the next *Sky at Night* programme, broadcast live on June 5th 1964:

> Well, I've just got back from America where there's a tremendous interest in what's called Project Apollo. Now, originally the Americans planned to get a man on the Moon by 1970 but, quite honestly, I tend to be a bit wary about this and I shall be a bit surprised if it happens before 1980 or so, though I may be completely wrong.

As it turned out Patrick was 11 years too pessimistic. Armstrong and Aldrin would be on the lunar surface only 5 years after he made those remarks. The cold war between the USSR and the USA would spur NASA on to achieve Kennedy's dream within his deadline.

Books, Quasar's and Moon Probes

In the summer of 1964, the BBC asked Patrick to compile the best of the past 7 years of *The Sky at Night* into book form. Patrick had been writing the programmes up for Maurice Ashley, Editor of the BBC's *The Listener* magazine since May 1959, but now they wanted to reproduce the best programmes into a book. The first *Sky at Night* book was published by Eyre & Spottiswoode at the end of 1964, priced 25 shillings (£1.25 in post 1971 decimal currency). Over the course of the next 46 years Patrick would produce a total of 13 *Sky at Night* books as well as three more associated in some way with the programme's major 25th, 30th and 40th anniversaries.

On July 24th 1964 Patrick was again joined by his good friend Colin Ronan in the *Sky at Night* studio, this time to discuss some enigmatic objects identified in previous years, namely the Quasi-Stellar Objects, sometimes called QSO's, but eventually called Quasars. These new objects were, arguably, the most exciting cosmological find of the 1960s. Radio astronomers had discovered powerful sources of energy in the sky which were not associated with normal galaxies. When spectra of these objects were obtained it turned out that they were very distant and amazingly powerful emitters. By the time of the *Sky at Night* programme a total of nine quasars had been identified. Even more remarkably these 'radio galaxies' appeared to vary in brightness on a timescale of a few days, sometimes doubling their luminosity in that period! It was hard to see how the object at the centre, thought to be 1,500 light years across, could vary its light output on a scale of days. It implied that most of the light must be coming from a central object which was only, at most, light days across. The quasar controversy would continue well into the 1980s and 1990s with some astronomers claiming that quasars had high red-shifts not because they were a long way away, but because they had been expelled at high speed from a parent galaxy. Nowadays, it is generally accepted that active galaxies have super-massive black holes at their centres which can generate huge amounts of energy as material swirls into them, but this was not appreciated in the 1960s.

On July 28th 1964 the U.S. rocket Ranger VII was launched toward the Moon. As with Lunik 4, some 15 months earlier, a *Sky at Night* 'special' was arranged. The aim of the probe was to take close range pictures of the Moon up to its impact with the lunar surface. It should be stressed that, in those days, the technology required to achieve this imaging was incredibly basic by today's standards. CCDs and digital video did not exist. Patrick appeared briefly on BBC TV after the *News Extra* programme on July 29th and said that in his view the main issues that would be solved by the probe would be:

1. Whether the lunar 'seas' (the smooth areas) were, in fact, deep dust drifts
2. Whether there were many small craterlets too small to be seen from Earth

At 13:25 hours GMT on July 31st, Ranger VII hit the Moon after having sent back a total of 4,316 images, at very close range, taken in the last 16 minutes before impact. This was at a rate of roughly four pictures per second, or about six times slower than a normal TV frame rate. At the start of this 16 minutes period the spacecraft was 1,300 miles from the Moon, but at the end it was smashed to pieces near the crater Guericke on the Mare Nubium, after a 5,000 miles per hour impact. This outstanding success ended almost 6 years of failures for the US space program. They had launched five Moon rockets between August 1958 and March 1959 (the first named Thor-Able I, the rest called Pioneers 1B, 2, 3 and 4) but, embarrassingly, only the last one escaped from the Earth's gravity! Five years before Ranger VII, on Patrick's 36th birthday (March 4th 1959), the U.S. had finally got Pioneer 4 within 37,000 miles of the Moon, only two months after the Soviet Lunik 1 probe had passed ten times closer. Of course, by July 1964, nearly 5 years had passed since the impressive Lunik 3 photography of the far side, and more than a year had elapsed since the failure of Lunik 4. After the Pioneer 4 mission the USA started the Ranger series, but Rangers I to VI had all been disasters: a mixture of rocket and imaging equipment failures between 1961 and 1964. At this stage the USA had spent hundreds of millions of dollars (a sum equivalent to many billions today) on their Moon probes with little to show for them. But they were confident that Ranger VII would be different.

Patrick was joined in the TV studio on the evening of July 31st 1964 by rocket expert Peter Stewart and questions posed by Patrick were answered by U.S. Jet Propulsion Laboratory (JPL) scientists at Pasadena. Remarkably, for 1964, a dozen lunar pictures were sent from Pasadena to the UK for the programme, only 10 hours after they were taken by the Ranger spacecraft. As a lifelong lunar observer this was an event almost as exciting as the farside pictures of Lunik 3 for Patrick. For the non-astronomical reader the significance of this success needs to be explained. Even the best professional Earth-based pictures of the Moon in the 1960s only revealed lunar features about half a mile across. No-one knew what the surface was like close up. The astronomer and steady-state supporter Thomas Gold had predicted, 9 years earlier, that the smoother lunar 'seas' or maria would be covered with fine dust and any space probe landing on this dusty surface would just sink into it, as if it were quicksand! This would be a major hazard for manned exploration if true. In addition, Patrick wanted to see evidence of vulcanism (volcanic crater forming activity) close up. During the programme the closest pictures available revealed craters as small as three feet in diameter, and only a foot or so deep. The craters were sharp and clear cut which would be impossible if Gold's dust theory was correct. The other obvious fact was that there were many small craterlets, sometimes in groups, on the surface. To Patrick this indicated secondary volcanic activity, but nowadays secondary impacts would be the consensus of opinion. Regardless of this, the July 31st 1964 programme was a great success. Once again, Patrick had presented new lunar data to the British TV audience almost 'as it happened' and the fiasco of the Lunik 4 program had not been repeated!

Shortly after that successful Moon programme Patrick was amazed at a letter he received in the post. It was from a lady called Miss Maud Nevill, who was the daughter of the great nineteenth century lunar observer Edmund Neison. If this sounds confusing it is because Edmund Neison's full name was Edmund Neville Neison Nevill. He only used the Neison part of his name when indulging in scientific matters! As a child of the 1930s, with a copy of Neison's 1876 book 'The Moon and the Conditions and Configurations of its Surface', Patrick had always assumed that anyone writing a book in 1876 must be long gone. However, Neison was only 27 when he published that book: he was born in 1849 and died in 1940. So, Patrick could have met him, especially as he lived at Eastbourne, just 30 miles from East Grinstead. As Patrick, in later life, just loved reeling off the names of famous people he had met, this realisation came as a major source of frustration. Nevertheless, after that July 1964 *Sky at Night*, during which Patrick had mentioned Neison, and his daughter had been watching, he was delighted that she had written to him. Needless to say, Patrick arranged a meeting with Miss Nevill and learned a great deal about her father from the encounter. Some time later (in April 1965) Patrick gave a lecture to the BAA about what he had learned from Miss Nevill. You just never know who might be watching when you talk about the Moon on national television.

Despite the fact that probes had now landed on the lunar surface, Patrick was still actively encouraging BAA Lunar Section members to observe the Moon and, especially, the limb regions. Patrick also created a regular monthly Lunar Section Circular at the start of 1965 to keep the hundreds of Lunar Section members informed. The duplicating and distribution of this circular was, without doubt, the most arduous task in the section and, perhaps, the BAA. One of the lunar limb areas that especially interested Patrick at this time was the crater which he and the late H.P. Wilkins had observed and Wilkins had called Caramuel (later re-named Einstein by the IAU). Towards the end of 1964 Patrick received a superb lunar photograph from a young amateur he had been encouraging, namely David Allen (later to become the director of the Anglo Australian Observatory at Siding Spring, Australia). David's photograph of the Caramuel region, taken with a 30 cm reflector, clearly showed that the region had been mis-represented in the chart of the area on the United States Rectified Lunar Atlas. At the 1965 Feb 24th BAA meeting at Burlington House in London, Patrick's paper on Caramuel convinced the audience that some valuable work in observing the limb regions, could still be of benefit.

Another development in 1964 was that Dr J.G. Porter, the original BBC voice of Astronomy, asked Patrick to take on the full editorship of the annual *Yearbook of Astronomy*. This was a guide for amateur astronomers on what was happening in the night sky over the course of the next year. Porter had produced the editions for 1962–1964, with Patrick as the associate editor, but he wanted Patrick to take the Yearbook on himself for subsequent editions and knew that with the name Patrick Moore on the book, it would survive. Patrick took the task onboard and would remain the editor for the next four decades and beyond, only employing help, in the form of associate editor Dr John Mason, from the 2002 edition onwards.The first 'all Patrick' Yearbook was published in December 1964.

The last month of 1964 would prove to be an important one for the BAA, Patrick and the Lunar Section. It was the month that the 43 year old Cdr. Henry Hatfield of the Royal Navy was elected to BAA membership. Two years older than Patrick, Henry Hatfield had not joined the BAA any earlier simply because he was, not surprisingly, at sea. However, from late 1964 Henry had a desk job at the Admiralty, rewriting their Hydrographic Surveying Manual. At last he could concentrate on building an observatory, because astronomy had interested him for some time. Based at Sevenoaks in Kent, Henry would become a firm friend of Patrick's in the coming years and decades. Like Patrick he had an upper class English voice, was very patriotic and he came from the same generation and a similar background. He was also another highly entertaining public speaker. In the coming decades Henry would serve as the BAA Treasurer, the BAA President, publish his own lunar atlas and be someone Patrick came to respect and admire greatly. For those readers who wonder what Cdr. Henry Hatfield was actually like, in middle age and into his seventies, well he always reminded me of the character 'Q', played by Desmond Llewelyn in the *James Bond* films; but maybe that's just my vivid imagination?

In 1965, Patrick's friends thought he might have suddenly become a film star, but only because of a movie called '*The Restless Ones*', which starred an actor of the same name. The film was produced by the US evangelist Billy Graham, and at one point Patrick (despite having no religious leanings) was mistaken for the preacher too! Patrick was opening a fete at his former childhood tutor's church at Coleman's Hatch, East Sussex, when a parishioner proudly told him he knew exactly who he was: Billy Graham! Well, O.K., there was some tiny resemblance perhaps! In fact, Patrick's only real film role (discounting the 1954 home movie with Desmond Leslie) would occur a year later when, in a 1966 film called 'The Master', he played himself, the TV astronomer Patrick Moore. 'The Master' (which also featured *Dad's Army* actor John Laurie) was a film about, wait for it, a 150 year old Shakespeare spouting telepath who is planning world domination when he meets two children on his secret island! Despite Patrick's fame as a TV astronomer he seems, perhaps, to have had aspirations towards the silver screen.

Chapter 10
The Move to Armagh

In early 1965 Patrick was invited to become the first director of the Armagh Planetarium in Northern Ireland, at the request of the Armagh Observatory director Dr Eric Lindsay. In the previous year (in April 1964 to be precise) Patrick had travelled to Armagh to film a *Sky at Night* about the research into photographs of the Magellanic clouds being undertaken at Armagh, and had enjoyed a pleasant few days in the company of Dr Lindsay and his wife and the observatory staff. Typical of Patrick, a decade earlier he had sent a 14 year old Armagh schoolboy and keen Armagh Observatory visitor, Pat Corvan, a telescope mirror making kit to get him started in astronomy; so Patrick's encouragement of newcomers to the hobby was well known in the area. The reader may recall that Dr Lindsay had taken the famous photograph of the comet Arend-Roland, from Armagh, which had been the opening caption on the very first *Sky at Night* programme, some 8 years prior to Patrick's job invitation. Lindsay was a man who Patrick liked. He was passionate about astronomy and often stated "I divide mankind into two classes: those who are interested in astronomy, and those who are not. And this division cuts across all races, nations and social groups…" That was the kind of comment Patrick could understand.

Patrick often said that he had previously declined an invitation to be the first director of the London Planetarium, 7 years earlier, back in 1958. That building had been constructed on the site of a cinema that had been destroyed in the blitz (not unlike the fate of the East Grinstead Whitehall Cinema) and was opened, in 1958, by the Duke of Edinburgh. As it turned out, the incoming BAA President in late 1958, Dr Henry King, had taken up the London directorship that Patrick claimed he had turned down. However, King was, undoubtedly, better qualified for the job; he had also known the Director of Madame Tussauds since 1955, when that establishment announced plans to build a planetarium. At that point Henry King had written a comprehensive paper in the April 1955 BAA Journal about Zeiss Planetarium projectors. It was clear that King had fixed his eye on the London Planetarium job for years, and although Patrick was already a household name in 1958, the last thing he had wanted then was another battle with his arch nemesis Henry King, or, horror of horrors, ending up as the Planetarium lecturer with Henry King as his boss, at the

M. Mobberley, *It Came From Outer Space Wearing an RAF Blazer!: A Fan's Biography of Sir Patrick Moore*, DOI 10.1007/978-3-319-00609-3_10,

same time as King had been the BAA President. That scenario, back in 1958, would surely have been a total nightmare from hell!

Just to recap here, to remind the reader: Patrick, in 1965 and since the mid 1950s, had hated Henry King and it was clear that the feeling was mutual. King had seriously criticised Patrick's book *Suns, Myths and Men*, and the criticism would never, EVER, be forgotten. He had also ripped apart Wilkins' book entitled *Mysteries of Time and Space,* in one of the most sarcastic reviews imaginable in the BAA Journal. If this was not enough, Wilkins' bizarre book entitled *Clouds, Rings and Crocodiles: By Spaceship round the planets* (a tour of the solar system) had also been destroyed in another scathing review by King and an Oxford astronomer, with King describing it as 'a mental clowning that has driven science from the scene' in his review in volume 65 of the 1955 *BAA Journal* (pages 253–255). So caustic were King's reviews that (as we saw earlier) Patrick and Wilkins had been permitted replies in the BAA Journal: a very rare event. King had, according to Patrick, not only criticised Wilkins' lunar work, but his own lunar work too. This would not have been too bad if King had been insignificant, but he was not. A few years later he had been elected as the BAA President (1958–1960) and King's own 1955 book, *The History of the Telescope*, remains a classic today and the sort of book Patrick simply did not have the patience to write, or research, in the 1950s or 1960s. Patrick's books were enthusiastic and exciting, but King's classic tome, while very dry, was the definitive and most meticulously researched book on the subject and has remained so for over half a century. In 1958 the newly famous Patrick had been quite happy to let King take up a post he had allegedly turned down and was not too confident of accepting. Patrick did not like having a boss, even if that boss was actually a group of trustees, and he certainly did not want piles of bureaucracy; he vastly preferred his freelance status. However, more important than all of that, was that he did not want any job which brought him anywhere near the sarcastic Dr Henry C. King, whom he suspected of constantly working against him, behind the scenes. As we saw earlier, when Patrick had agreed to writing a visitors guide to the London Planetarium, after it had opened, he had mysteriously received a letter from the board of trustees saying his 'revisions' had ruined the guide! Patrick had made no revisions at all and the guide that he wrote was published anyway. However, any mention of Patrick had been totally removed from the planetarium booklet. To Patrick this had "all the fine smells of rotting fish, trawled from the sea by Henry King……."

Mainly Projectors

However, 7 years later, in 1965, a more confident Patrick decided to accept the challenge of a planetarium directorship, although it was with a heavy heart that he left behind his beloved Sussex base *Glencathara* in Worsted Lane, East Grinstead. It had been his home for 36 years. As we have already seen, *Glencathara* was part of a trio of large linked houses, bound together in a will and a complex tenancy arrangement by Patrick's maternal grandmother, with the main proviso that her daughters should never be homeless. The house would not have passed to Patrick on his mother's death and arrangements had to be made with Patrick's aunts and cousins for the ending of the tenancy arrangement on the house. If he accepted a well paid astronomy job for a few years he might be able to buy a really big house outright and not be homeless, or find himself having to rent a property, when his mother died. Obtaining a large mortgage did not appeal to him at all; after all, he did not have a trade or a proper job and being in a huge financial debt to anyone was a very unpleasant prospect in Patrick's mind.

Although *Glencathara* was legally Patrick's mother's house, to inhabit until she died, there was no way he was leaving his 79 year old mother behind. She was coming to Armagh too! At Armagh, unlike at the London Planetarium, he would have full access to a decent telescope; the Armagh Observatory had a fine 10 inch Grubb refractor which was an ideal instrument for lunar and planetary work and a massive temptation for Patrick. With Saturn's rings soon presenting themselves edgewise on to the Earth, a huge refractor on a quality mounting would be a fantastic instrument to use. He would also be close to good friends, namely Dr and Mrs Lindsay, and he had already spotted a decent house on the market (The Armagh 'Judges House' at 4, St Mark's Place in the Mall) which he and his mother could just afford to move into. In fact, upon moving in, Patrick immediately moved his 8.5 inch With-Browning reflector and its decorative 'greenhouse merry-go-round dome' to Armagh too. The Planetarium director's salary was definitely an attraction. Patrick had not had a full-time salary since his teaching days and the years of his huge book sales were a few years ahead. Remember, he never had a contract with the BBC, just a 'gentleman's agreement'. Essentially, he was an author but had not sold any bestsellers, yet.

Fig. 10.1 Patrick and his 8½ inch With-Browning Newtonian inside its rotating dome in the 1960s. This instrument and the rotating glass observatory are now owned by Bruce Kingsley. (Copied with Patrick's permission around 2004)

Fig. 10.2 Patrick's 8½ inch With-Browning Newtonian's rotating observatory in the back garden of his mother's home *Glencathara* in Worsted Lane, East Grinstead, during the early 1960s. (Copied from a slide with Patrick's permission)

Patrick had been very active within the first East Grinstead Scout Group during the 1950s and early 1960s and had tutored many boy scouts towards obtaining their Astronomer Starman, Meteorologist, Archaeologist and Weatherman badges. How on Earth had he found time to do that?! Anyway, on announcing that he was leaving West Sussex for Armagh a camp fire in honour of Patrick was held at Hackenden on Saturday May 22nd 1965. Cubs and Scouts from all groups in East Grinstead district attended to say farewell and thank you for all the help and encouragement Patrick had supplied. The local Scout Group magazine reported that 'in addition [to his expert and enthusiastic instruction] Scouting had benefited immensely from his factual reporting of activity throughout the district'. The magazine added that, along with supplying his knowledge, these factual reports were his primary contributions, but the many other ways in which he had helped the Scout movement were too numerous and varied to list. As a token of their gratitude the members of the movement presented Patrick with an album containing the signatures of the Cubs, Scouts, Senior Scouts, Rovers and Scouters of the district and a photograph of the event was taken for Patrick's souvenir book. In later years Patrick admitted that despite decades of working with Scouts he still could not fathom how to erect a tent, something which greatly amused many of his friends. Just 4 days before this moving tribute from the Scouts Patrick had given his final public lecture before departing for Northern Ireland. He had been invited to give the Sir Alfred Herbert Annual Memorial Lecture at the Coventry Theatre, in Coventry, on May 18th. As usual, Patrick was happy to oblige, and talk about the Moon and planets to a packed house.

In June 1965, just after the giant radio telescope facility at Arecibo in Puerto Rico announced that Mercury's rotation period was 58½, not 89 days, Patrick started making plans to move all his belongings, all his telescopes, and his mother and cat Smudgie, to Armagh. He would be there for 3 years and it would be only the second period of his life when he lived away from Sussex (the first, of course, was the RAF period, especially the training period in Canada). Patrick's good friend Ian Baker masterminded the removal and transportation of all his belongings to Armagh, much to Patrick's relief; he would have been far too impatient to organise such a feat!

Of course, while at Armagh, Patrick still returned regularly to London for the *Sky at Night* planning meetings and broadcasts, as well as for the last Wednesday in the month (October to June) BAA meetings. Needless to say, he invariably managed to combine BBC planning meetings with the last Wednesday in the month BAA dates. Flights to London from Belfast Civil airport were quite regular, even in 1965. On a couple of occasions, when Patrick was travelling with his mother, he used the Belfast to Liverpool Ferry, but was appalled that it took 10 hours to make the short trip!

On July 15th 1965 the NASA space probe Mariner IV flew within 6,000 miles of the planet Mars and sent back the first close-range pictures of the Martian surface, showing unmistakeable craters. In 1892 the US astronomer E.E. Barnard had claimed to have glimpsed craters on Mars using the 36 inch Lick refractor. A similar claim had been made by the observer Mellish in 1917, using the 40 inch Yerkes refractor. However, this was the first real proof that craters existed. Despite this evidence, the idea that Barnard and Mellish actually saw Martian craters is, even today, highly controversial. For most of the past 100 years the more imaginative Earth-based observers had speculated that there were Martians and that the canals they had built were visible from the Earth. Mariner IV finally sounded the death knell for the canal theory; the pictures showed categorically that they did NOT exist and they were, like O'Neills Bridge, just tricks of the eye. Mariner IV also proved that the dark areas on Mars were not depressed sea beds and that the atmosphere was far too thin to support any form of animal life. The surface pressure was less than 1 % of that on the Earth. On the first *Sky at Night* programme since he moved to Armagh, Patrick discussed the preliminary Mariner IV findings and the electromagnetic environment between the Earth and Mars.

The next *Sky at Night*, broadcast on July 23rd 1965, marked a landmark for Patrick: it was his hundredth. He was joined by Desmond King-Hele of the Royal Aircraft Establishment at Farnborough. The subject was optimistic: communicating with extra-terrestrials using radio telescopes. Some five decades later, no aliens have been in touch and none have been detected.

By August 1965 Patrick was finally fully settled into his full-time post in Northern Ireland. His main task as the director of Armagh Planetarium was simple: construct a planetarium and publicise it! When Patrick arrived Armagh had an observatory, but not a planetarium. All of the Armagh Observatory staff would become firm friends over the coming years; these included the Director Eric Lindsay, the secretary Sheelagh Grew, and the other permanent staff, namely David Andrews, Margaret Cherry and Harry Grossie. A planetarium is, of course, a domed building onto which the appearance of the night sky is projected. The audience sits inside the dome, with the projector, and it looks just like the real night sky, except it is not freezing cold, everything is not dripping with dew, kamikaze moths are not flying towards your torch and there are no clouds! In addition you can display the sky as it appears at any time of the year, which is very useful for teaching purposes. Obviously, with thousands of stars visible to the naked eye from a dark site, a special kind of projector is required; not only special, but expensive! In the 1960s, software controlled fibre-optic projection systems had not been invented. So, the standard item of planetarium kit was the planetarium projector, a device looking like a giant two headed ant, with spheres at each end, crammed with a near-infinity of bulbs to depict stars of different brightnesses. These projectors were hand made, custom built, and very pricey! As the director, Patrick had to choose which model of projector to purchase. In 1926, when Patrick was three, the German optical experts Zeiss had built the first planetarium projector at Jena. But there were two other manufacturers in the mid 1960s, namely Spitz in the USA and Goto in Japan. In the autumn of 1965 Patrick embarked on a number of trips to suss out the competing systems and see them in action.

His first trip was to the Carl Zeiss facility in Jena in what had become East Germany. Of course, in 1965, the East–West cold war was at its coldest and Patrick was an ex RAF officer; but, as we saw earlier, he had been there before, in 1962 (remember the 'bovine' driver Patrick thought was a Martian). All went well with the trip to the planetarium and projector facilities at Jena and the knowledge gained confirmed Patrick's belief that a 40 foot (12 metre) diameter dome would be necessary for the Armagh facility. However, as with most things made by Zeiss, the projectors were of the highest quality, but also the highest price. Still, the trip had gone well and he had not been arrested as a spy! Unfortunately, Patrick stalled his jeep halfway across the 'Checkpoint Charlie' border crossing and found a near-infinity of rifle butts pointed at him as he struggled to re-start the jeep (so he told me….probably a degree of exaggeration there). Luckily, someone at the border-post spoke French and with his 'pigeon French with a Flemish accent' Patrick talked his way to West Germany and freedom! Phew!

Back at Armagh, the cost of the Zeiss projector was causing concern. The total budget for the Armagh Observatory was £70,000 and much of this was provided by the Ulster Government and the Armagh City and County Councils. Zeiss were probably the best projector makers, but what about the Spitz and Goto options? Although the Spitz projector was much cheaper, it looked, to quote Patrick "like a lavatory cistern". The other option, the Goto, was midway between the two and had a good reputation. Goto were renowned for quality telescopes as well. So Patrick set off for Japan, to assess the Goto projectors. Patrick was booked into a small hotel, on the outskirts of Tokyo, not far from the Goto factory. In traditional Japanese style any guests were expected to remove their shoes when entering the hotel and to dress in a kimono. Although Patrick had not yet attained the full 18 or 19 stone mass that he would be burdened with from the late 1970s onward he was still a big, bulky man and was still his full height. From the 1980s onward he gradually shrank by at least a couple of inches in height, increasingly exaggerated by a stoop and a stance reminiscent of Churchill. Patrick somehow acquired a 'normal fitting' (and that is by diminutive Japanese standard) kimono which he later described as looking 'indistinguishable from a ballet skirt on me'. His Japanese hosts frequently failed to maintain their normally disciplined composure at this sight at the meal table, where Patrick would attempt to sit cross-legged, on the floor, in his 'ballet skirt'. Maybe a Sumo costume would have been more appropriate? Another source of amusement was Patrick's massive feet. These were an impressive British size 13 'extra wide' fitting! When Patrick moved to Selsey he would ask a local boatyard to make his shoes. Incidentally, Patrick always wore slip-on shoes with no laces. He just could not abide the hassle of tying shoe laces, which "inevitably just come undone again", especially as simply shoe horning his gigantic plates of meat into his massive slip-ons could "never go wrong". Patrick often referred to his Japanese experience thus: "Someone once told me I gave every impression of being rather hastily constructed. Others have said I usually look like an unmade bed. They should have seen me in a kimono, cross-legged, wearing blue socks!"

At the first Japanese evening meal, with horror, Patrick spotted that one of his socks had a huge hole in it. This was quickly rectified for the next mealtime kimono appearance by using a blue biro to make the exposed foot flesh roughly the same colour as the sock! Patrick's kimono also revealed that he was seriously 'knock-kneed'. "There is no way I could stop a pig in a passage". Anyway, regardless of the humour, Patrick was impressed by the Japanese Goto system. It had similar capabilities to the Zeiss system but was considerably cheaper and, unlike the Spitz system, looked great too: Armagh would have a Goto projector.

While in Japan, Patrick wasted no time in visiting the Tokyo and Kwasan Observatories as well as various planetaria, in Tokyo and elsewhere. At Kwasan, the main telescopes were a 30 cm Cooke refractor and a 60 cm reflector. Patrick also made sure that he re-inforced collaboration between the Kwasan Observatory Director (Professor Miyamoto) and the BAA Lunar Section.

Following the official order of the Goto projector, Patrick insisted he was consulted in the design of the star projection system. He had long thought that many planetaria showed too many stars, which confused the view, and so insisted that the system should transmit 3,800 star images at most. Now that the projector decision had been finalised a 40 foot aluminium inner dome was ordered from Astrodome Inc. of Ohio, while the 50 foot glass fibre outer dome contract was awarded to a local Ulster company. One point Patrick insisted on was that the seats in the planetarium should be of the aircraft 'tip back' design, to enable customers to enjoy the simulated celestial view without developing a stiff neck.

Soon after Patrick arrived in Northern Ireland he was contacted by a 19 year-old amateur astronomer from Scotch Street in Armagh. The young man was Terry Moseley who would become Patrick's observing companion on many nights at the observatory. Terry, a psychology student at Queen's University, was delighted to hear that Patrick was now living in Armagh and wrote to him asking how he could build a decent telescope. Patrick wrote back immediately saying: 'I am the clumsiest man in three continents. I can't make you a telescope but I can tell you people who can'. Patrick invited Terry around and told him that he had now brought all three telescopes over from Sussex, namely his 3 inch refractor, his 8½ inch reflector, and his 12½ inch reflector. As Patrick explained, even he couldn't use three telescopes at once, so as he would be mainly using his largest telescope, Terry could use the 8½ inch. This sort of generosity was typical of Patrick. Of course, in practice, they would often end up using the 10 inch Armagh refractor together on those rare clear nights in Northern Ireland. On the evenings that were totally overcast, well, to quote Terry word for word: 'On cloudy nights, and there were a few, we wrote up our observations, worked on articles for magazines, played chess, and listened to his extensive collection of classical music, all supplemented by endless supplies of coffee and Marie biscuits, supplied by his mother, or his housekeeper.'

An Exciting Comet Eludes Patrick

While Patrick was considering the virtues of a Japanese planetarium projector, on September 18th 1965 two dedicated Japanese comet hunters, Ikeya and Seki discovered a comet that, technically, was set to become the brightest one of the twentieth century. The new discovery was ultimately named Ikeya-Seki 1965 VIII. Seki discovered six comets from 1961 to 1970 and his young rival Ikeya found five from 1963 to 1967. Ikeya would wait a staggering 35 years before discovering his own sixth comet, Ikeya-Zhang, in 2002, and he discovered a seventh (with Murakami) in 2010. England's own comet discoverer, George Alcock, would possibly have shared the discovery had September 18th not been his wife's birthday! However, 8 days later George was rewarded with his own comet discovery, which was his fourth. While sweeping in Hercules he spotted a faint fuzzy object not spotted by Ikeya, Seki, or the world's other amateur comet hunters. Like Alcock's previous three comets it bore his name and no-one elses; he was the sole discoverer.

At the time of its discovery the new Japanese comet was only eighth magnitude (so it was a binocular object, but certainly not naked eye). However, by the end of September, Ikeya-Seki was a naked eye object with a tail a third of a degree in length. In the weeks following its discovery the orbit of the new Japanese comet was calculated and it was found to have a marked similarity to that of the Great Comet of 1882. It was later deduced that both had belonged to the same original object, the source of the 'Kreutz' family of comets. A characteristic of Kreutz comets is that they are sungrazers, in other words they pass *very* close to the Sun's surface. This means they invariably brighten dramatically. Ikeya-Seki was set to pass within 470,000 kilometres of the solar surface on October 21st (the day of perihelion). While all this sounds very exciting, there are a few drawbacks. Firstly, comets that are only bright when they are close to the Sun have to be bright enough to be seen in strong twilight. Secondly, due to the speed such comets move around the Sun, they are only at their closest and brightest to the Sun for a few days and they may well break up due to solar heating. Thirdly, if it is cloudy you will not see the comet anyway, unless you can get in a plane and fly above the clouds! Nevertheless, by mid-October the new comet was second magnitude, with a 10 degree tail visible in twilight, and it was worth covering in a special emergency *Sky at Night* programme on the day of perihelion, October 21st. This particular *Sky at Night* had the distinction of being the only one of that non-digital TV era ever transmitted on BBC2, but not on BBC1. This was because the BBC1 schedules had already been set and could not be changed.

On the morning of October 21st, Patrick was driven from Armagh to Ballykelly in Northern Ireland, dressed in full flying regalia, reminiscent of his wartime RAF days. The sky was clear and the stars were out, as well as a crescent Moon. Accompanying Patrick was another member of the Armagh Observatory staff, Harry Grossie. The RAF Shackleton, piloted by Flight-Lieutenant Smith, took off at 4 a.m. and flew at an altitude of 6,000 feet until sunrise at 6:35 a.m. For over 2 hours Patrick searched for the comet through the cockpit windows, by eye and with binoculars, but the comet was not seen! The RAF Shackleton then flew straight to Belfast Civil Airport so that Patrick could catch his civilian flight to London where he was due to transmit the special *Sky at Night* dedicated to the 'spectacular new comet' a few hours later! Quite why Patrick failed to see the comet and chose that time to see it is still a mystery. Its head was dangerously less than one solar diameter from the Sun when he looked and he should have known this. Others saw it on the same day (though not from the UK), but it would not have been an easy observation. Looking near the Sun is a highly dangerous activity which can easily result in blindness, but the comet was seen with the naked eye by other observers as a line sticking out of the side of the Sun. Almost certainly the twilight must have blotted out the comet's tail on that day, at that time.

When Patrick arrived at the BBC studio at Lime Grove he was in a quandary. His hallmark was his endless enthusiasm; but how do you enthuse about a comet you have not seen? This was supposed to be the greatest comet since the Daylight comet and Halley's comet, both seen in 1910! Patrick managed to phone Lowell Observatory at Flagstaff, Arizona just before the live broadcast and learned that US astronomers had seen Ikeya-Seki from there and from Honolulu, though only a short tail had been observed. All he could do on the programme therefore was to report that the comet was bright, but its proximity to the Sun made seeing it almost impossible for the general public, at least for the next few days. Most of the programme had to be devoted to comets and historical great comets in general. As things turned out, Ikeya-Seki became a superb sight only a week later, displaying a head of magnitude – 2 and a 25 degree long tail. In hindsight, the programme would have been much better timed if it had been broadcast a week later and not when the comet was right on top of the Sun!

Chapter 11
Meteorites, Pills, Saturn and the Leonids

The Barwell Event

On Christmas Eve 1965, just before 4:15 p.m., a brilliant object flashed across central England in the twilight sky. No, this was not Father Christmas, it was a meteor and, as it flew over the mainly cloudy skies above Leicestershire, it broke up, depositing meteorites around the village of Barwell (mid-way between Coventry and Leicester). A few surprised onlookers who were outside saw 'an object half the size of the Moon with a golden-yellow tail' streaking behind the thin cloud. Others described the tail as 20 degrees long and looking like a firework. By good fortune the BAA's Howard Miles (who was a close friend of Patrick, director of the BAA's Artificial Satellite Section, an expert on meteorites and a regular guest on the *Sky at Night*) was living in Coventry at the time and saw the flash, from indoors, as the meteor broke up.

It should be explained here, for the non-astronomer, that although tiny meteors are visible on any clear, moonless night (you might, by chance, see one every 15 minutes or so from a typical country site) lumps of space debris that are big enough to leave a chunk of material on the ground (a meteorite) are very rare. Amazingly, a meteor smaller than a grain of sand will produce a very bright 'shooting star' as it fries up completely in the Earth's atmosphere; a meteor that survives the frying process will have to be rather bigger when it enters the atmosphere and may, briefly, be as bright as the Full Moon as it streaks through the sky.

Anyway, although Howard Miles was not outside when the meteor streaked through the dusk sky, he had the knowledge and presence of mind to realise what had happened and to gather all the facts about the event together, before the memory faded. His report to the BAA read as follows:

> At 4.12 pm I saw a very brilliant flash through the kitchen window and, at first, thought it must be lightning. In fact, I counted up to fifteen and came to the conclusion that it must have been a long way off. About one and a half minutes later there was a tremendous bang like a sonic boom, but not the double crack one associates with aircraft travelling supersonically. Both my wife and I dashed outside. Most of the sky was covered with cloud, but through a small gap we saw a puff of whitish-grey smoke expanding.

M. Mobberley, *It Came From Outer Space Wearing an RAF Blazer!: A Fan's Biography of Sir Patrick Moore*, DOI 10.1007/978-3-319-00609-3_11,

Howard spent many days travelling around the countryside between Coventry and Leicester where various chunks of the meteorite had landed. Remarkably, some witnesses had actually seen fragments of the object land. One piece hit a car bonnet and the owner, thinking it was thrown by schoolboys, threw it away! Another piece smashed through a window and embedded itself in a vase of flowers where it was found 3 weeks later! A large piece left a crater in a Leicestershire man's driveway and another piece cratered a nearby garden. In total 46 kilograms of meteoritic material was collected after the fall and there must have been much more left undiscovered.

Just 2 weeks later a 16 year old Miss Heather Couper wrote a letter to Patrick, as so many young people did over the years, asking for advice on whether a girl could have a career in astronomy. As always Patrick replied swiftly, on January 9th, advising Heather that she needed to be good at maths for a professional career in astronomy, but stating categorically: 'Let me reassure you on one point. Being a girl is no handicap at all!' Just 18 years and 9 months later Patrick would hand over the Presidency of the BAA from himself to Heather.

Patrick decided to devote an entire *Sky at Night* to the Barwell fireball event and it was broadcast on the 11th February 1966, just a week after the Russians soft-landed their Luna 9 probe on the Moon; another event of immense interest to him and which had prompted the week before's *Sky at Night* 'special'. The Barwell meteorite was a very rare event indeed, as although there must be quite a few meteorite falls in the UK and England every year, it is very unusual for fragments to be found. If enough eye witness accounts of the tracks of bright fireballs are available, amateur astronomers will mount expeditions to search for fragments, but they are almost unanimously doomed to failure. Only 23 authenticated meteorite falls have occurred in the UK between 1623 and 1991. Prior to the Barwell event, you have to go back to 1949 when 720 grams of meteoritic material were collected after a fall near Beddgelert in Wales. After Barwell there was the Northern Ireland Bovedy meteorite of April 25th 1969, most of which landed in the sea, and then 22 years later, the Glatton meteorite (Cambridgeshire) of 1991 May 5th, when a Mr Arthur Pettifor had a near-miss when a 760 gram chunk of meteorite landed at his feet. But the 46 kilograms collected at Barwell was only equalled by a fall in Limerick in 1813.

Following the February *Sky at Night* programme many more witnesses of the Barwell event contacted the BBC and the BAA, and so Patrick decided to see the Barwell area for himself on one of his return trips to the UK from Armagh. After his fruitless trips to Morsgail in the Hebrides and Charlton in Wiltshire, at last here was a real British meteorite fall and not something hyped out of nothing by the media. The next *Sky at Night* after the Barwell programme was devoted to the Luna 9 Moon landing results and was scheduled for March 11th, so Patrick decided to visit the site before his next broadcast, to see what he could find. On March 6th, 2 days after his 43rd birthday, Patrick and a small team of BAA friends set off for Leicestershire and Barwell for some meteorite hunting. By now over 2 months had elapsed since the fall and there had been the usual British mix of weather to muddy the impact area. Nevertheless, Patrick and his colleagues had high hopes that they might find more chunks of the Barwell meteorite. Patrick was joined on the trip by various BAA meteorite hunters including Howard Miles, Rossie Atwell, Ernest Beet, Keith Hindley and the aptly named Gladys Stone. Patrick often recounted the story that, while in a field at Barwell, prodding the ground with sticks, he spotted a small depression in the ground, prodded it, and unearthed a large stone. Much to his surprise, he would recall, it suddenly struck him that this was indeed a most unusual stone; in fact it turned out to be a sizeable meteorite, over 600 grams in weight! The Natural History Museum was, at the time, offering 10 shillings (50 pence in decimal coinage) per gram for Barwell fragments, so Patrick's find was worth £300 even at Xmas 1965. Other members of the party were amazed (astounded would not be too strong an expression) at Patrick's beginner's luck; only a handful of amateur astronomers ever discover a meteorite, but he had found one! Patrick said he always regarded it as a 'belated birthday present' and he donated it to the Armagh Planetarium, though they let him keep it when he left Armagh – as long as he left it to them in his will!

Fig. 11.1 Patrick and various BAA friends at the site of the Barwell meteorite in Leicestershire on March 6th 1966, more than 2 months after the Xmas Eve 1965 fireball event. Keith Hindley, Rossie Atwell and Gladys Stone (*holding camera*) are on the *left*, with Patrick and Howard Miles on the *right*. BAA Archive photograph by the late Ernest Beet (1904–1997)

However, there is another side to Patrick's remarkable Barwell meteorite fragment discovery. Patrick was well informed by Howard Miles and by *Sky at Night* viewers as to who had witnessed the meteorite fall and already collected fragments. One of the largest chunks was recovered by local Barwell lad David Radford: a young man in need of a new bicycle! Patrick seized the opportunity and offered David £23 for the brick sized meteorite (a fraction of it's real worth of at least £300, and it would sell for well over £10,000 these days). David gladly accepted Patrick's measly offer and purchased a new bicycle. Meanwhile, Patrick carefully slipped the meteor chunk into a furrow in the ploughed field where the BAA party was hunting and pronounced that he had found something very interesting!!!! I think the less said about this furtive episode the better!

More Moon; More TLP

On his return from Barwell to Armagh it was time to assess the latest Russian Moon probe results for the next TV programme. As we have already seen NASA managed to crash land their Ranger VII spacecraft on the Moon on July 31st 1964 after years of dismal and embarrassing failures. The probe had sent back some excellent close range pictures of the lunar surface. In 1965 even better results had been obtained from Rangers VIII and IX and the Russian Zond III probe had repeated the success of the 1959 Luna/Lunik III probe by imaging the Moon's far side. But Luna 9 was the first spacecraft to soft-land on the Moon. Despite the close range pictures of the lunar surface, and the obviously sharply defined small craters, Gold's 1955 dust theory still sent shivers down the spine when a manned lunar landing was contemplated. OK, the lunar surface looked solid enough, but what if it really was like quicksand? Only the real soft-landing of a spacecraft could close the door on this worry and Luna 9 was destined to be that spacecraft. On February 3rd 1966 the Soviet probe soft-landed in the Oceanus Procellarum (the Ocean of Storms), somewhere near the crater Hevel. Pictures were transmitted back to Earth and received by the Soviets and by Bernard Lovell's team at Jodrell Bank. From the pictures taken there was no evidence of the spacecraft sinking into the lunar surface and boulders in the distance seemed to have no desire to sink either. Gold's theory was, at last, disproved. Having said this, supporters of the theory still suggested that the Luna 9 landing site might not be typical! Of course, Patrick was also keen to look for any evidence of vulcanism in the pictures, as was his friend and colleague in vulcanism, Dr Gilbert Fielder, then of the University of London Observatory. But, like many of the results obtained then, and in later years, many scientists used the same early evidence to confirm their own beliefs. Patrick thought he could detect evidence of lava flows in the Luna 9 pictures, but really they were not of a quality that could be used to determine such things.

In 1966 the Moon was still Patrick's top observational priority and he would remain the BAA's Lunar Section Director until 1968, despite living in Northern Ireland. As with everything Patrick did in the BAA his term as Director was characterised by enthusiasm and energy. The Lunar section had enjoyed a regular periodical entitled *The Moon* which was edited by the experienced observer Harry Thornton. However, when Thornton resigned, there was inevitably a gap in communication. Obviously, there was no e-mail in the mid 1960s and, without a regular magazine, members could feel distinctly isolated. Patrick was keen for this not to happen and so ambitiously launched a monthly Lunar Section Circular (LSC) which was powered by his enthusiasm and the tireless efforts of Phil Ringsdore.

Patrick was not idle within his old Mercury and Venus Section either. Following the departure of the scientifically disciplined, but less-than-exciting, Director Jackson, Patrick's old friend J. Hedley Robinson had roped Patrick back onto the new BAA Mercury and Venus Section Committee in the spring of 1966. Active membership of the Section had dropped to single figures under Jackson, but with a committee containing Patrick and his friends Hedley Robinson, Richard Baum, Alan Heath and the 19 year old Bill Leatherbarrow, membership surged to almost 150 by the end of the decade. Patrick's lunar section pals Phil Ringsdore and Cdr Henry Hatfield, the celebrated lunar photographer, were also a valuable part of the team in this era of the 'Swinging Sixties'.

Patrick had also instilled a sense of purpose in the Transient Lunar Phenomena or TLP network he had created. As we have seen, the aim of this network was for independent observers, at different locations, and with different equipment, to independently verify suspected 'events' on the Moon. A single observer event just could not be taken seriously, even by other amateurs, let alone professionals. A number of members of the network possessed the red-blue filter wheel Moonblink devices (originally suggested by V. Axel Firsoff and designed by Peter Sartory) to enhance colour anomalies on the lunar surface, and since the October 1963 Aristarchus event the team members were poised to record any anomalies. Surprisingly, hardly any members of this team ever tried to photograph these so-called 'events'; indeed, when I joined the TLP network in 1980 I was essentially the only observer repeatedly photographing the Moon when alerts were announced. This would be my own early observing niche within Patrick's network.

Patrick scheduled a Lunar Section meeting at Burlington House in London for Saturday April 23rd 1966, despite having to fly over from Northern Ireland to attend and chair the meeting himself. Such was Patrick's boundless reserves of energy and enthusiasm that small hassles like that were ignored by him; as he often managed to arrange these events to coincide with live *Sky at Night* broadcasting days it was no problem. Seventy-one Lunar Section members attended the meeting including some very prominent observers and astronomical characters of that era. These included Cdr. Henry Hatfield, a 19 year old David Allen (later to be the Anglo-Australian Observatory Director), Leslie Rae, Jim Hysom, Horace Dall, Keith Abineri and Peter Sartory.

One week after the meeting, on the evening of April 30th 1966, Patrick was at Armagh when a TLP alert was received from the wealthy businessman Peter Sartory. Peter was observing the crater Gassendi with his 'Moonblink' filter device attached to an 8.5 inch (216 millimetres) reflecting telescope and had detected a faint orange-red glow in the region. It should be stressed that to avoid bias, the exact region of the TLP was rarely quoted; alerted observers were expected to find the TLP themselves. Gassendi is an impressive crater, over 100 kilometres in diameter. It lies on the border of the Mare Humorum and has a central mountain. The floor of the crater is especially interesting as it is crossed by a series of complex rilles. Various members of the lunar section were alerted to the event and Patrick and Terry Moseley, under rare clear skies at Armagh, were able to confirm it with the 10 inch (254 millimetres) refractor there. They could see a faint orange-red glow just outside the rim of the crater. Another red event was also seen inside Gassendi and the timings of Patrick and Terry seemed to tie up well with those recorded by Peter Sartory. Another lunar section member who witnessed the event was Phil Ringsdore, using another 8.5 inch telescope at Ewell in Surrey. One lunar day later, on May 27th, Terry Moseley observed another Gassendi event with the Armagh refractor and this too was confirmed by Patrick and Peter Sartory. Patrick was convinced that these phenomena were real and caused by some kind of venting from cracks in the lunar surface. To the sceptics he drew attention to the catalogue of suspected lunar events compiled by Barbara Middlehurst and her colleagues at the University of Arizona. This catalogue suggested that TLP were more likely to occur when the Moon was nearest to the Earth and under greatest stress and also in regions on the edges of the lunar 'seas' or maria. Patrick was excited by these recent observations and so *The Sky at Night* broadcast on July 8th 1966 was dedicated to TLP, with Peter Sartory as the main guest and a filmed interview with Barbara Middlehurst.

Incidentally, I once asked Patrick, in the 1980s, if he thought colour photography would have captured the April 30th Gassendi event and he thought not: "Frankly, the colour was just too subtle". So maybe I was wasting my time? However, film, and certainly CCDs, should stand a much better chance of revealing 'the truth' so I was not (and still am not) happy with Patrick's reasoning on that issue!

Now, on the subject of professional astronomer Miss Barbara Middlehurst (1915–1995) and TLP, well, there was far more to her agreeing with everything that Patrick said than most people realised. Barbara Middlehurst was Welsh (like Patrick's former mentor Wilkins and his wife and daughter) and, despite living in Arizona, often found herself in the UK attending BAA meetings. The reason was quite simple: she wanted Patrick as her 'toy boy'! In 1965, when Patrick and Barbara Middlehurst were 42 and 50 years old respectively, Patrick was keen to exploit Middlehurst's affection for him, but keep as far as possible from her in real life. More than one BAA member of that era remembers Patrick leaving BAA meetings as soon as possible when Barbara Middlehurst was around. "That damn woman's after me again" he'd hiss to the nearest person on the door, "After my body!" Then Patrick would be gone, in a flash, heading back home to his devoted mother. Having an adoring fan within the Lunar & Planetary Laboratory at Arizona was useful, but not when she was up close, personal and scarey!

For Saturday July 23rd 1966 Patrick had arranged with Dr Ron Maddison of Keele University to hold an Out-of-London Lunar Section meeting at Keele. A few months earlier Patrick had filmed Dr Maddison and Prof. Ingram at the University, during a *Sky at Night* programme devoted to Astronomy and Education. The July meeting was a huge success and the Keele venue would be used again for a BAA meeting later in the year. Remarkably, the meeting coincided with the 1966 Wembley World Cup Quarter Final match between England and Argentina (in which Hurst scored the winning goal for England after 78 minutes). However, most of Patrick's mates would always be cricket supporters, not football fans, and so more than 80 lunar observers attended. A short colour film about the NASA Apollo project was shown by Dr Ron Maddison at the start of the meeting, which certainly got members in the mood. Man was going to the Moon: no doubt about it! Subjects covered at the meeting ranged from lunar domes, Moon-blink devices, the eastern limb areas, photographic lunar maps, lunar photography and infra-red studies of the Moon. Everyone who went to that weekend meeting remembered it as one as the most memorable BAA meetings of their lives and the names of those present read like a 'Who's Who?' list of BAA lunar observers from Patrick's heyday. As well as Patrick and Ron Maddison others there included Harry Thornton, Leslie Rae, Keith Abineri, Henry Hatfield, Arthur Good, Phil Ringsdore, Bernard Doherty (father of the superb planetary artist Paul Doherty), Peter Sartory, a young Bill Leatherbarrow and Peter Lancaster Brown. After the formal meeting ended discussions continued in the Student Union Bar and then at Dr Maddison's house. Certainly it was a Lunar Section meeting that would long be remembered, but for Patrick, in a couple of days he would be flying back to Northern Ireland.

Pills and the Armagh Ghost

We have already seen that Patrick had an irregular heartbeat problem from childhood, and remarkably, up until a few days after his 83rd birthday, he did not have an artificial pacemaker fitted, and it was only then inserted at his doctor's absolute insistence. For all of his life Patrick had taken pills to keep his heart beating regularly. When he stopped taking them, if he over-exerted himself, he could be in serious trouble. Of course, Patrick being Patrick he found taking the pills an enormous hassle and relied on his mother to remind him. One such episode was recounted 40 years later by Patrick's young colleague at Armagh in the 1960s, Terry Moseley. This is Terry's account, in his own words:

> Then there was the time Patrick nearly died at my feet. OK, slight dramatic exaggeration.... he has a congenital heart problem affecting the 'pacemaker', and has to take a pill to regulate it. One winter night we were observing very late when Patrick said "I need to go outside." This was odd, so I followed him out. It was dark, so I couldn't see his face, but his voice was faint as he said "I forgot to take my beastly pill, I'm going to pass out. But don't worry, I'll wake up in a few minutes." With that he crumpled to the ground in front of me. OhMiGod! Even though it was freezing he was only wearing his RAF blazer (he didn't feel the cold!), The ground was frozen rock-hard with thick hoar frost on the grass. Minutes passed – it seemed like ten or more. What to do? He weighed about 17 stone, so I couldn't move him! I could have gone to the library and rung for an ambulance, but didn't want to leave him (although I would have had to do so when they arrived, as they would never have found us). So I took off my own coat and put it over him, and was basically running round like a headless chicken thinking of the headlines "Famous TV astronomer freezes to death: Colleague does nothing...." When he suddenly woke up and said "I told you I would wake up." Whew! So I got him into my car and took him home where his mother gave him a piece of her mind and a pill, and got him into bed. Next day he was right as rain!
>
> Then one time the sky suddenly cleared in time for us to attempt some important observation, so we rushed up in my little Morris Minor, over the ramp at the entrance to the Observatory, and when gravity and Patrick re-combined with the car seat, 'Boing!' went the spring in the seat – and on the front near-side suspension too!

I can't resist recounting another of Terry's stories at this point, about the Armagh Observatory ghost.... Again, in Terry's own words:

> One evening (at Armagh Observatory) I called for Patrick as usual to go observing, but he was expecting a call from the USA (a big event in those days, pre-booked through an operator!), so he told me to go on and he would come up later. So I was happily observing Jupiter at about 2 a.m. in total darkness and stillness, when I clearly heard footsteps coming along the path, right up to the door of the dome, and then stop. I assumed Patrick had stepped off the path on to the grass to check some binocular variables, or for aurorae etc., so I went on observing Jovian transits. When he didn't appear after about 10 minutes I went outside to look. There was no sign of him. I went in to the library building where there was a toilet and a little tea-room which we could use. He was not there. I went to where we parked; Patrick's car was not there. Could it have been Pat Corvan (he had no car)? No sign of him either. So who? Remember, that in those days there was no light pollution, and nobody who didn't know their way round the observatory grounds could have found their way to that dome in pitch darkness, let alone walking at a normal purposeful pace.
>
> Next day I asked Patrick. The call from the USA had been delayed and by the time it came it had been too late, so he hadn't gone observing at all. I asked Pat Corvan; it wasn't him. I asked Dr Lindsay and the other staff: David Andrews, Harry Grossie, etc.....It was not them either. It was only then that I began to get scared, and this was not helped by Patrick's response: "Oh well, it must have been Dr Davenport". He was the former Director who committed suicide at his desk because, according to Patrick's historical knowledge of the observatory: "His wife was an absolute fiend". Now, I do NOT believe in ghosts, but to this day I have NO explanation for what I unmistakeably heard!

Thanks to Terry for providing those highly entertaining accounts.

Saturn Edge-On

As well as developments on the Moon in 1966, and England winning the Football World Cup (of no interest to Patrick, who, as I said earlier, was a cricket man), an exciting series of events were taking place with respect to Saturn's rings; events which were of great interest to Patrick. Saturn and its ring plane are tilted with respect to its orbit around the Sun by just under 27 degrees (much as the Earth's own axis is tilted). Twice in its 29 year orbit the rings appear edgewise on as seen from the Sun, so that's roughly every 14 or 15 years. (Actually, it's every 13 years 9 months and 15 years 9 months because the orbit is not circular). The situation is further complicated by the fact that as the Earth goes around the Sun, in the year that the rings are almost edge on to the Sun, an observer on Earth is effectively bobbing to the left and right (or up and down, if you prefer) of the ring plane. So, in that critical year, the Sun will cross from shining on, say, the top side of Saturn's rings to shining on the underside, while an Earth based observer will get glimpses of both sides (which may or may not be illuminated) and a brief edge-on view too. Needless to say, as the 1966 event occurred some 7 years before Pioneer 11 reached Saturn much valuable data could be obtained by observing the appearance of the rings edge on. In 1966 there was little hard data as to how thin the rings were. We now know that they are only 100–200 *metres* thick, despite the visible rings being some

274,000 kilometres in diameter. In 1966 the Earth moved from the north to the south of the ring plane on April 2nd, but Saturn was too close to the Sun (from our viewpoint) to see anything. It was not until mid-May that the Armagh refractor caught a glimpse of the planet in the morning sky. Terry Moseley made a sketch on May 15th showing that the inner part of the rings were then visible. However, two further ring-plane crossings occurred on October 29th and December 17th. Between these two dates, separated by only 7 weeks, a very interesting situation occurred, as on Earth we could just glimpse the north face of the rings, but the Sun was shining on the south face; so what would be seen on these precise dates and in the period between?

Amateur astronomers worldwide were waiting in anticipation for the late autumn/winter of 1966 to see what would transpire! As with previous edgewise ring presentations those observers with clear skies at the *precise* time that the rings were edgewise on to the Earth, and with decent amateur telescopes, only lost sight of them for a few hours, despite their thin appearance. At all other times the rings were always visible, even when illuminated from the other side, but they sometimes appeared to have ‘condensations’ along them. Whether this is due to genuine differences in concentrations of material or just atmospheric seeing at the observer’s site is still a subject for discussion. One aspect Patrick was especially interested in was determining if the thin line of the rings ever appeared longer on one side of the planet than the other; this had been reported at previous apparitions but was puzzling. Patrick made numerous measurements at the eyepiece of the Armagh refractor and concluded that there was no difference in the ring length on opposite sides. Numerous other observers in the BAA would submit drawings of the rings during this period; some of the best were submitted by a talented 19 year old observer named Paul Doherty, who would become one of Patrick’s closest friends in the years to come and whose father was already a friend of Patrick’s.

As well as observing the rings themselves the edgewise presentation presents unique opportunities to study Saturn’s moons. Many readers of this book will be aware that the giant planet Jupiter has four large moons that can be seen even in binoculars. Saturn only has one giant moon, called Titan, but it also has a whole host of smaller moons, some seven of which were known to be visible in amateur telescopes in 1966. These seven are called, in order of brightness, Rhea, Tethys, Dione, Iapetus, Enceladus, Mimas and Hyperion. When the rings are edge on, the planetary equator is edge on too. This means that the satellites, travelling in the equatorial plane, can transit the globe of Saturn, as can their much more obvious shadows. However, in practice, only the giant Titan can easily be seen exhibiting these phenomena as it is 5,150 kilometres in diameter. The next largest moon, Rhea, is only a third of the size, at 1,530 kilometres across. From Earth, Rhea is only 0.22 arc-seconds in diameter (3,600 arc-seconds = 1 degree) so it looks like a star. Unfortunately, Titan orbits Saturn in just under 16 days and its shadow will only take a matter of hours to cross the planet. Add to this the fact that the planet was only at a decent altitude above the horizon for, maybe, 8 hours a day and the fact that most nights (especially at Armagh) are cloudy. Then take account of the fact that the series of events go on for less than a year, and it is not hard to see that witnessing a Titan transit or a shadow transit is quite a challenge. But Patrick was determined to see one! On the morning of August 11th 1966, using the Armagh refractor, Patrick's luck held out and he observed Titan's shadow crossing the disk, just below (north) of the ring plane. At the next opportunity (one orbit of Titan later), Paul Doherty had the luck and observed the same phenomenon on the evening of August 26th from Smallthorne, Stoke-on-Trent. Some observers (including this cloud-cursed author) still dream of witnessing this rare spectacle.

During August and September 1966, Patrick and his colleagues Terry Moseley and Pat Corvan (the same man who Patrick had sent a mirror making kit to when Pat was an enthusiastic teenager) were also keen observers of Jupiter whenever using the Armagh refractor. Despite Jupiter having been in solar conjunction (directly behind the Sun) on July 8th of that year, by the second week in August they were making numerous observations of the giant planet with the Armagh telescope. Being perched well above the ground, there were no horizon obstructions and seeing was often excellent. At the September 1966 BAA 'Out of London' meeting at Keele University, Patrick reported that they had already made hundreds of transit timings of features on Jupiter's disk in the dawn sky, suspected a disturbance near the Great Red Spot, and also seen a nebulous white patch in Saturn's southern hemisphere. Patrick's stamina for visual observing of the Moon and planets was truly staggering. One only had to leaf through the dozens of observing logbooks in the spare downstairs room at his Selsey home to see how phenomenal his observing output had been. On more than one occasion, when Jupiter was at a high declination from the UK, he managed to observe two transits of the Great Red Spot during a night, necessitating a winter observing session spanning 10 hours!! Although Patrick was not a regular observer of comets his logbooks record that he also observed the new, but faint, tenth magnitude comets Kilston (1966b) and Barbon (1966c) from Armagh during August and September 1966. These two faint fuzzy objects had been discovered during August by professional astronomers working at the Lick and Palomar observatories.

Returning to the subject of Saturn, when its rings are edge-on the dazzling glare from them is not present, so faint moons, that might not normally be visible, can be glimpsed. On various nights, between July and November 1966, Patrick made sketches of all eight visible moons, together with the occasional foreground stars which were sometimes seen.

In December the astronomer Audouin Dollfus, working at the Pic du Midi observatory in the Pyrenees, announced the discovery of another Saturnian satellite, later named Janus. Patrick quickly went back to his notebook and found that he had recorded it several times in the July-November period, when its orbit had just taken it clear of the left or right edge of the ring tip as seen from the Earth. On one especially good night (October 24th 1966) he had recorded Titan plus seven satellites at 21.10 hours, using a magnification of 400× with the Armagh refractor. This seemed perfectly normal at the time; Saturn had eight satellites just visible with a telescope of that size and he had seen them all. What he did not know was that the faint satellite Hyperion had not been recorded by him, but he had recorded the similar brightness Janus, just above (south of) the eastern ring tip. If he had checked his observing notes the next day he would have gone down in history as one of the few visual discoverers of a planetary satellite; but it was not to be! His French counterpart Dollfus, who he had been exchanging letters with for the past 20 years, had beaten him to it, although the Frenchman had used a much larger telescope at altitude. Only 2 weeks earlier (October 10th 1966) Patrick had broadcast a *Sky at Night* on the subject of 'Saturn without its rings' and the phenomena that might be glimpsed during this rare opportunity.

There is, however, a worrying postscript to Patrick's Janus story. While based at Armagh, Patrick submitted a paper to the *Irish Astronomical Journal* about his independent pre-discovery sighting of Janus. The editor of that important publication was the outstanding astrophysicist Ernst Julius Öpik (the grandfather of the future Liberal Democrat M.P. Lembit Öpik, who would be a guest on *The Sky at Night* some 24 years later). E.J. Öpik rejected Patrick's paper outright, citing that he had used the wrong comparison stars for his magnitude estimates and had not corrected for precession when identifying the star field. Many leading observers found it impossible to believe that such a faint satellite (magnitude 14.5) that close to Saturn could possibly be viewed with a 10-inch refractor at all; and Öpik was very familiar with the Armagh refractor. Even so, Patrick kept regularly repeating this pre-discovery claim for many years, at least until the next edge-on ring year of 1980, when it became clear just how incredible a claim it really was.

The *Sky at Night* had a new producer from the summer of 1966, namely Patricia (Pat) Owtram who took over from the founding producer Paul Johnstone when he was promoted in the BBC hierarchy, to run the prestigious programme *Chronicle*. Initially sceptical about having a woman producer Patrick would declare, in later years, that: "There are two types of woman producer: the rare excellent ones and the more typical awful ones. Pat Owtram came in the excellent category". However, to say that Pat Owtram took over from Paul Johnstone is not strictly true, despite the fact that she is often listed as Johnstone's immediate successor. In fact, during the period between 1964 and 1966, various temporary (or assistant) producers and directors had kept *The Sky at Night* going, according to Patrick. Perhaps the best remembered BBC *Sky at Night* backroom person from that era was Julia Cave, but other names include Frank Smith, Keith Clement and Julia Gaitskell, all of whom were involved with other 1960s BBC productions. Nevertheless, Pat Owtram became the longer term successor to Paul Johnstone.

Returning to the subject of the ringed planet, between July 1966 and January 1967 Patrick also made dozens of magnitude estimates of Saturn's moon Iapetus with the Armagh refractor to try to resolve the mystery surrounding its varying brightness. A star can be variable, but how can a moon possibly vary? The only solution was that it had to be bright on one face and dark on the other, with a rotation locked to the planet, and we must see opposite faces when it is on opposite sides of the planet (during its 79 day orbit). Patrick's brightest magnitude estimate was 8.7, but his faintest was 11.1, a dramatic tenfold difference in brightness. Of course, in years to come, space probes would indeed confirm that Iapetus had one dark hemisphere and one bright one, unlike any other major moon visible to amateur observers.

The Leonids

As well as the Saturnian system being edge-on there was another event looming in 1966, of equal importance to amateur astronomers, and potentially much more dramatic: the Leonid meteors.

Every 33 years or so the comet Tempel-Tuttle orbits the Sun. Behind the comet, and slightly ahead of it too, is a relatively dense concentration of dust particles which, if the Earth ploughs through them, results in a spectacular meteor shower. There are, of course, meteor showers throughout the year: the Perseids in August and the Geminids in December are the most active. But these showers are active over a period of days or weeks, slowly building up to peak activity on the maximum night. At most, a keen observer, from a typical site, might actually see one meteor per minute from such showers. Exciting? NO, not until you get the odd brilliant fireball, which you may get now and again, but only after you are frozen rigid in the case of the December Geminids. However, the Leonids are different; the dust particles are in narrow, dense ribbons of material, and every 33 years or so, if you are in the right place on Earth, at the right time, and if the Earth passes through a dense ribbon of dust, you might see meteors every second!

There had been dramatic meteor displays when Tempel-Tuttle had returned on previous occasions. In 1799 the great explorer Humboldt witnessed a meteor storm on November 13th of that year and, some 34 years later, an event of equal magnitude was observed on November 13th 1833. In 1866 it was Britain's turn and the amateur astronomer William Rutter Dawes was 'quite unable to count the thousands and thousands of meteors which seemed to flash through the sky'. However, in 1899 (the next predicted event) the Earth missed any significant ribbons of dust, and 1933 was equally dismal. What would happen in 1966? One vague indicator of likely activity is a slow build up in previous years of Leonid meteors; not a storm, just a sign that there are, maybe 20 or 30 meteors per hour, at the allotted times (in mid-November). Amateur meteor observers had detected encouraging signs from 1962 onwards, so Patrick decided to broadcast a *Sky at Night* on November 11th 1966, 6 days before the Earth was due to plough through the comet's orbit. The BAA meteor section director at the time was Harold Ridley and he was that month's guest on the programme. The Moon would be a thin crescent on the evening of the predicted maximum of November 16th/17th so moonlight would not swamp the meteors.

Patrick invited viewers to write to the BBC to apply for charts of the night sky and a staggering 11,000 were requested. Even more staggering, 9,000 were returned after the 17th. One problem with meteor watching is that the peak activity invariably occurs in the early hours, well after midnight. So, if the meteors did not behave there could be a lot of disgruntled and tired viewers. Needless to say, Patrick and Harold's enthusiasm inspired thousands of BBC viewers to get out of their warm beds, in the dawn hours of the 17th, to 'have a go'.

On the night of the 16th/17th Patrick was still based at Armagh, and predictably was under solid cloud. So he manned his telephone waiting for any reports of meteors to come in. Remarkably, skies over central and southern England were fairly clear, but it was becoming horribly apparent that nothing spectacular was happening. Even experienced observers were only seeing a couple of dozen meteors before dawn! This was starting to look like a disaster on the George Hole/Lunik 4 scale! However, *Sky at Night* viewers have always been fairly intelligent people and the majority realised that this sort of event was very hard to predict; indeed this had been explained fully on the programme. There were a few entertaining comments returned with the charts though. One classic letter stated:

> Watched from 12 o' clock to 5 o' clock in the morning. Meteors: from the sky – none; from the wife – plenty.

As it happened, there was a spectacular meteor storm, but it was 12 hours later than predicted and was well seen from California and Arizona, while it was midday in Britain.

Impressive Leonid meteor storms also occurred at the next return of comet Tempel-Tuttle, most notably in 1999 and 2001, but Patrick was clouded out for the former and not well enough to travel to the latter. It is worth mentioning at this point that after Patrick and his mother moved from East Grinstead to Armagh he regularly sent letters and postcards back to their former housekeeper, a lady called Mrs Dorothy Wheeler, who he affectionately called 'Mrs Weasel' due to her ability to ferret about in the nooks and crannies of their house! Patrick's letters were even addressed to 'Mrs Weasel' at Mrs Wheeler's address of 25 Upper Close, Forest Row, near East Grinstead. Mrs Wheeler's son, Peter, talked about the letters to the East Grinstead Courier and Observer in 2013. One of Patrick's letters, sent on 1966 July 5, makes it clear that all was not well in Armagh: 'This is a strange little country & one has to live here to realise the gaps between the various religions - quite unbelievable & all quite wrong - it all leaves one with a feeling of insecurity - it is a very great pity - & one wonders if the people will ever really grow up & Think!!'

Chapter 12
St Osyth, 2 British Novae, Serpents Everywhere, and an OBE

Hyde (and Jekyll?)

Shortly after the Leonid non-event, at least that's how it was from the UK, an unfortunate development occurred on the BAA council which caused Patrick considerable embarrassment. Patrick had been a friend, for some time, of the BAA member Frank Hyde, who lived at St Osyth near Clacton. Born in 1909, Hyde had been a guest on the *Sky at Night*, appearing in the June 1961 and May 1964 programmes and he was, arguably, Britain's leading amateur radio astronomer. Hyde was an electronics engineer with Crompton-Parkinson (later Marconi) in Chelmsford and, in 1950, had built a radio astronomy observatory in a World War II Martello tower at St Osyth. Like many of Patrick's friends, most notably Michael Bentine, he had an interest in the paranormal and psychic phenomena and also in airships like the R101 and the Hindenburg. He set up various antenna arrays in a field at St Osyth too. When Patrick started his *Amateur Astronomer's Library* series with Lutterworth Press in the early 1960s, Hyde had been happy to help and he wrote the second book in the series (*Radio Astronomy for Amateurs*) himself, as well as the Radio Astronomy chapter in Patrick's own book *Practical Amateur Astronomy*. Hyde's locals at St Osyth were convinced his antennas were turning their milk sour and at one point he used Patrick to reassure them that this was simply not possible!

In 1963, with Patrick's full and enthusiastic endorsement, the BAA council made Frank Hyde the new Editor of the BAA Journal. Up to Hyde's term as Editor there had been eight journals per year, but this schedule had created a few delays and Hyde reduced the frequency to bi-monthly. In 1964 the BAA council had decided to reprint a form of publication of the Computing Section known as a Memoir. A Memoir was usually a form of learned paper which was too large to fit into the bi-monthly Journal. Frank Hyde had been asked to get an estimate for the new Memoir from the printers of the BAA Journal, based at Bristol. However, unknown to the council at that time, Hyde had set up a grotty photocopying company in a caravan in a field near Clacton. The caravan was occupied by his young mistress too (he had

M. Mobberley, *It Came From Outer Space Wearing an RAF Blazer!: A Fan's Biography of Sir Patrick Moore*, DOI 10.1007/978-3-319-00609-3_12,

left his wife) and it doubled as a love nest! In due course Hyde presented two estimates for printing the Memoir; one from the Journal's printers and one from his own 'company'. The quote from his own 'company' was much lower and so he advised council that these second printers were conveniently in his local area (!), cheap, and highly professional. The BAA council accepted Hyde's recommendations and told him to instruct the new printers to proceed. Hyde was almost certainly hoping that he would eventually be able to funnel all of the BAA's printing needs (quite considerable) through his own company!

Unfortunately for him, due to technical problems (his cheap and nasty printing facilities), it took Hyde 2 years to photocopy all of the required Memoirs to any kind of 'acceptable' quality. They arrived at the BAA's registered office, which was also the Assistant Secretary's (Lydia Brown's) home at 303 Bath Road in Hounslow, just after the Leonid non-event; the Memoirs were available for inspection at the 1966 November BAA Council meeting. By this time Hyde had parted company with the then unpaid BAA Editor's post but was still on the BAA council, and the councillors had learned of his 'photocopying business'. After some debate, with a highly embarrassed Patrick involved, the council came to the decision that the price charged by the Clacton 'company' was too high for what had been delivered and the quality was virtually unacceptable. Initially, the council had wanted to pay Hyde the £150 he demanded but the new BAA Journal editor, Neville Goodman, was a man of principle and skilfully turned the council's view around. The council asked Hyde how he could justify the costs for such shoddy workmanship but, by letter, Hyde's 'company' (namely his mistress at the love nest) declined to provide any more information. 'Hyde and Jekyll' was a term being used by some for Hyde's alter-ego.

At a memorable meeting the BAA council, energised by Neville Goodman's moral stance, refused to pay the Clacton agency's bill and Frank Hyde's face "turned purple". According to various reports he banged the council table so violently with his clenched fist that councillor's cups of tea actually lifted off the surface and there was clear air between table and saucer! Patrick was not at all happy as he had already gone so far as to suggest Frank Hyde as a future BAA President! The actual new President, from late October 1966, was the Jupiter Section Director W.E. Fox. In early 1967 Hyde wasted no time in dispatching a writ for money owed to him directly to Mr Fox! The whole matter ended up in court the next year and, amazingly, Hyde won the battle, received 194 pounds and promptly resigned from the BAA. Once again, as with the Wilkins fiasco, Patrick had to distance himself from a BAA member who had once been a good friend. According to one former councillor, Patrick's relationship with Neville Goodman was never the same again and he referred to Goodman as 'a serpent' after the Hyde event. In reality, Patrick still bore a grudge against Goodman for allowing Henry King's scathing review of Patrick's *Suns, Myths and Men* into the BAA Journal more than a decade earlier. As an aside, 35 years later, Goodman left the BAA £300,000 in his will and he was a lifelong and tireless worker for the association. In Patrick's own biography '80 Not Out' there is a full size illustration of Patrick talking to Frank Hyde at a BAA meeting around 1955.

Serpents and the Polaris Club

In the 1950s and 1960s Patrick was especially keen on calling certain people 'serpents'. It seemed to be his favourite derogatory term. Basically, Patrick classed BAA council people into two categories: "a very dear friend" or "a serpent". If you had been a friend of Patrick's but then disagreed with him on a major point, at any time, you became "a serpent", unless you apologised, bought him a drink straight afterwards, or were very lucky. Patrick's views were pretty much black & white and not open to debate. While he would do anything for his fans, for children, and especially for sick children (his generosity was literally infinite for those categories), he had developed a deep and paranoid distrust of anyone who might be exploiting his fame or his kindness. People who criticised any of his books, or people who wrote competing astronomy books ("All copied from mine") were serpents, along with anyone who presented a rival TV astronomy show, however brief.

Anyone who had ever said anything negative about Patrick was also a serpent and their friends and relatives were all serpents too! The author, and former BAA President, Henry King, was the mother serpent, but UK authors James Muirden, Ian Ridpath and (by association) Robin Scagell were also classed as serpents too, as well as, eventually, Heather Couper and Nigel Henbest. In the mid 1960s Patrick referred to the Junior Astronomical Society (JAS), which became the Society for Popular Astronomy (SPA), as a 'nest of serpents', and yet he had been one of the first Presidents of that society only a few years earlier! However, he'd had a disagreement with a 'serpent' within the JAS and so that was that. He mainly stayed away from that society for over 40 years because of it, only returning at the 50th anniversary celebration, where he looked distinctly unhappy. When Patrick had taken over the BAA Lunar Section in 1964, and placed TLP and a volcanic origin for lunar craters firmly on the agenda, he classified a whole array of other people as 'serpents' too: namely, supporters of the impact origin of lunar craters.

As well as having nothing to do with the JAS he also had no more contact with the Croydon Astronomical Society, or at least with its leading members. Why? Well, a young and upcoming writer, who Patrick claimed in later life must have been the talented astronomy author Ian Ridpath, had allegedly written a highly amusing piece which found its way into the society's *Altair* magazine. In the article various lunar craters were allotted names according to certain famous characters in the astronomy world. Crater Moore was said to 'dominate all the smaller craters around it and had a deep white deposit on its shoulders'; Patrick's blazers were always coated in dandruff. Other jokes were directed at Patrick too, including an anagram of his name: 'Mike O' Craprot!' Patrick did not see the funny side of this at all (almost everyone else did) and so the whole of CAS were from that moment on classed as 'serpents', along with those TLP sceptics, lunar crater bombardment theorists, young up-and-coming astronomy graduates who were 'anti' lunar volcanos, and competing authors. Patrick seemed to have a real grudge against up-and-coming university educated authors who claimed to know more about geology than he did. He had a particular grudge against James Muirden (despite proposing him for BAA membership in October 1956!) because not only could Muirden write good astronomy books, he had (with John Larard) formed a breakaway group to the BAA known as *The Casual Astronomer* magazine, later renamed *The Astronomer* or 'TA'. The magazine survives today under the highly respected Guy Hurst.

In the late 1960s Patrick formed his own little 'inner sanctum' within the BAA. This was a circle of like minded non-serpentine friends whose role was to attend licensed premises now and again (not unlike his Halley's Comet Society of the mid 1980s) and try to furtively steer the BAA Council members his way. Patrick called this little clique 'The Polaris Club' and he and Colin Ronan (BAA Journal Editor) were the informal organisers. If you were considered to be of 'damn fine English stock', a BAA stalwart, a TLP believer, a cricket enthusiast, and had similar views to Patrick, he would write you a letter inviting you to join the club. The Polaris Club met after BAA Council and main meetings in the RAS library at Burlington House, in London's Piccadilly (where the BAA Office is located). Unfortunately, after the late 1970s, the RAS fire regulations made the Polaris club meetings impossible, but by then Patrick's mother's frail health was severely restricting him anyway. If you were a successful rival UK astronomy author, or a friend of one, it was very difficult to escape being classified as one of Patrick's 'serpents'! The Polaris Club was one of many short-duration little societies Patrick formed during his life. That aforementioned 'Halley's Comet Society' was another, as was the 'Anti-Centigrade Society' whose headed notepaper bore the banner 'Forever Fahrenheit'.

A Tragedy But No Shortage of Work

Throughout the mid 1960s the American and Russian space programs had been gaining pace and, fired by the speech of John F. Kennedy, the USA was well on its way to developing a viable spacecraft capable of landing men on the Moon before the Soviets. From 1961 to 1963 the USA had launched six one-man Mercury capsules into space. This was followed, from March 1965 to November 1966, by the ten two-man Gemini missions. At the start of 1967 minds were fixed on reaching the Moon by the decade's end, by developing the massive Saturn V launch vehicles and the three-man Apollo spacecraft. However, on January 27th 1967 (19 years, almost to the day, before the Challenger Shuttle disaster), a fire in the Apollo 1 spacecraft at Cape Kennedy killed astronauts Grissom, White and Chaffee and the first launch of a manned Apollo spacecraft into Earth orbit (on a Saturn IB launcher) was delayed by over 18 months. Less well publicised was the fact that, only a week after the fire killed those three astronauts, two technicians (Harmon and Bartley) died in another capsule in similar circumstances.

The Soviets were not without casualties too, although their more catastrophic failures were not revealed for many years later. At the end of April the Cosmonaut Komarov was killed when his spacecraft's parachute snarled on re-entry. Suddenly, there were doubts as to whether both the USA and the USSR had been just a bit too ambitious. Despite this, there had never been so much interest in astronomy and space research, or in Patrick's books. Early in 1967 Patrick learned with pride that he had been proposed for membership of the Lunar Commission of the prestigious International Astronomical Union (IAU), the governing body of professional astronomy (his membership would be confirmed in August). Around this time Patrick enjoyed a visit from his old friend Arthur C. Clarke. Over dinner the subject of suitable music for the film *2001: A Space Oddyssey* came into the conversation and Patrick suggested using the Strauss Waltz the Blue Danube, a favourite of his. The film producers liked the idea and the rest, as they say, was history! At least, that was Patrick's version of events.

Patrick's writing output during the 1960s was just as staggering as it had been in the mid to late 1950s. During this period it appeared that he had an infinite enthusiasm and energy where astronomy and space exploration were concerned. This energy came across in all the talks he gave too. Remarkably, the vast majority of the lecturing he did was for no financial benefit whatsoever! He was forever writing articles for the BAA Journal and the Lunar Section as well as co-ordinating the Lunar Section TLP network. He had boundless energy within the Jupiter and Mercury & Venus sections too, and on nearly every clear night he was at the telescope eyepiece, either with his back garden reflectors at Armagh, or with the 10 inch Armagh Observatory refractor. In addition to all this he was the Armagh Planetarium Director, meetings recorder at BAA meetings and presenter and planner of each month's *Sky at Night* programme. He was also becoming a regular voice on BBC radio whenever astronomy or space missions were the subject! Then there were the write-ups of *The Sky at Night* programmes for *The Listener* and planning the annual *Yearbook of Astronomy* content a year before publication. In early 1967 Patrick had not written a science fiction novel for 3 years, but that spring he started work on his 18th novel (19th if you include the spoof *Flying Saucers from Mars: The facts*) entitled *Planet of Fire*. This would be published in 1969.

As regards books of science fact, in the mid 1960s, as well as Patrick's *Amateur Astronomer's Library* series, published by Lutterworth, and his *Observers Book of Astronomy* and *Space in the Sixties*, he also rewrote his original *Guide to the Moon* as *Survey of the Moon*, which incorporated the latest space probe results. He then wrote 11 new major books entitled: *Guide to the Stars; The Planets* (a revision of *Guide to the Planets*); *Earth Satellite; The Planet Venus; Guide to Mars; The New Look of the Universe; Science and Fiction; Astronomy; The Sun; Basic Astronomy; The Sky at Night – Volume 1*

He also collaborated with a co-author he had known as a young man (and who would remain a life-long friend), Peter Cattermole, on a book entitled *The Craters of the Moon* and with the BAA Mercury & Venus Section Director (1963–1965) Dr Jackson on *Life in the Universe* and *Life on Mars*.

Whichever way you look at it, Patrick was already the author of over 80 astronomy and science fiction books by 1967, an average of more than five books per year since he started in 1952. However, the sales of these books (rarely more than 10 or 25,000 copies sold per title) would soon be totally eclipsed by the books he would write in a year or two's time.

Despite his full-time Armagh post, writing, and TV commitments, Patrick would observe whenever possible, and with whatever instruments came to hand. His Jupiter log for 1967 is packed with observations. As I mentioned earlier, sometimes he observed two Red Spot transits in one night, necessitating a ten hour session even in bone chilling weather. On January 28th/29th 1967 he used both of the Cambridge Observatory refractors, the 12 inch Northumberland and 8 inch Thorrowgood instruments. He wrote that:

> "The 12 inch does not appear to be as 'crisp' as the Armagh 10 inch, BUT conditions were poor….."

On February 13th 1967 Patrick had what he would later describe as one of his finest views of Jupiter and its Great Red Spot ever, using the Armagh refractor.

The Tenth Anniversary and Beyond

On April 28th 1967 *The Sky at Night* celebrated its tenth anniversary. Since April 1957 the programme had been transmitted once every 4 weeks, without a break, and there had also been various 'special' programmes. The tenth anniversary programme opened with the sad news of the death of the Russian cosmonaut Komarov, but the rest of the programme was a celebration of 10 years of the programme. In Patrick's own words, from his write up of the programme:

> Following the recording, many of the BBC producers and directors, floor managers and technicians who had been or are associated with the programme joined in an informal celebration – headed by Paul Johnstone, who created the *Sky at Night* and to whom I cannot sufficiently express my thanks. I will say only that this particular occasion ranks among my very happiest memories.

Shortly after the tenth anniversary programme Patrick travelled from Northern Ireland to Ireland itself, to the small town of Birr, in County Offaly, close to Athlone. In the mid 1840s the third Earl of Rosse set up a giant 72 inch (1.83 metre) telescope there, which for a time was the largest telescope on Earth. It was the first telescope to reveal the spiral form of the galaxy M51. This would be the first of many enjoyable visits to Birr, by Patrick. Being Ireland, no doubt there were plenty of other Patrick's in attendance too!

By 1967 Britain's greatest ever amateur astronomer, George Alcock, had bagged a total of four comets, discovered in 1959 (two), 1963 and 1965. But, since January 1961, George had undertaken the seemingly insane task of memorising all the Milky Way stars visible through his 15 × 80 binoculars, so that he could recognise novae as well as comets. Incredibly, after some 6 years of sweeping the sky with this aim, he had memorised some 30,000 stars in patterns! If a new nova appeared and he had a clear sky, he would have a good chance of spotting it. On July 8th 1967 George spotted a new 6th magnitude star in Delphinus. He knew the field so well that he was totally sure it was a nova; it was, and it was also the first nova discovered from England since Nova Herculis in 1934. No other nova had been discovered in Delphinus and this is still true today. Nova HR Del, as it was designated, quickly rose to fourth magnitude, so it was a naked eye object! It would rise again to a peak of magnitude 3.5 on December 13th 1967, fall, and then rise yet again to a magnitude 4.2 peak in May

1968. An extraordinary nova! On that historic July 8th night, George first telephoned Mike Candy at the Royal Observatory at Herstmonceux. Mike promised to try to get a photograph and a spectrograph with the 36-inch Yapp reflector. George then phoned various other BAA members, including Patrick at Armagh. By then it was about 1 or 2 am! Patrick had a good knowledge of the constellations, but it was insignificant compared to Alcock's. However, Patrick became one of the first observers to confirm the new object and, on the next night, Mike Candy's spectrum, from Herstmonceux, confirmed it as a nova.

Upon hearing of the official confirmation of Alcock's fifth discovery, Patrick, who had not long returned to Armagh from his trip to Birr, arranged to fly to London, meet up with a camera crew from the BBC, and travel to George Alcock's home at No. 55 Broadway in the village of Farcet near Peterborough. Patrick was aware that George had not enjoyed his trip to London for the October 1959 *Sky at Night*. He was also aware that George had a disabled wife whom he was loathe to leaving on her own. He also wanted a *Sky at Night* interview to be filmed on the almost sacred ground (to astronomers) that 55 Broadway, Farcet was fast becoming! The filming went well and Alcock photographed Patrick standing in his back garden, surveying the flat fenland as many other amateur astronomers have done from that spot. To an astronomer George Alcock's garden always would feel like holy ground. To Patrick's delight George agreed to join Patrick in the studio for the next programme on July 25th.

Patrick promised George that he would keep a close eye on his new nova whenever he had a clear night at Armagh. Patrick was always true to his word. Over the next 4 years, until 1971, despite all his other BBC and Moon-landing broadcast commitments, Patrick would make a total of 436 magnitude estimates of Alcock's nova. He observed it more than anyone else in the BAA! For the record, Patrick made magnitude estimates of HR Del on: 49 nights in 1967; 125 nights in 1968; 86 nights in 1969; 90 nights in 1970; 86 nights in 1971. Indeed, Patrick continued making visual observations of HR Del until the night of July 20th 1984 and until the naked eye nova of 1967 had faded to 12th magnitude. That final observation, 17 years after his first of this British nova, would be important to him, as it was his 700th magnitude estimate of Alcock's object. Despite his fame and his TV and book-writing commitments, Patrick was still one of the UK's most prolific observers during the 1960s, 1970s and 1980s, unlike so many vacuous so-called 'TV astronomers' in the twenty-first century.

IAU, IUAA, OBE and AAVSO!

The summer of 1967 certainly proved to be a very busy one for Patrick. As I mentioned earlier he had been proposed for membership of the IAU's Lunar Commission and so was invited to attend the 13th General Assembly of this professional governing body of world astronomy. The IAU meet every 3 years and this time the venue was Prague, Czechoslovakia; the conference spanned 10 days, from August 22nd to

31st. Patrick flew from Armagh to London and then joined his good friend, the professional Lunar geologist Gilbert Fielder. The two men then crossed the English Channel by ferry and drove to Prague for an entertaining week in a hotel with no lift doors, a surveillance bug in Patrick's room, but plenty of waltzes, which Patrick loved! It was during this conference that a few professionals suggested that Patrick should form an IUAA (International Union of Amateur Astronomers) and so he arranged a small meeting to gather support. In fact, Patrick and a Swedish amateur astronomer, Ulf Johansson, had written a letter to *Sky & Telescope* magazine in the previous year (published in May 1966) suggesting such an organisation, but the IAU Prague meeting finally provided the impetus to get things moving. At that meeting Patrick's friend Leif J. Robinson (an editor at *Sky & Telescope* and, later, Editor-in-Chief) acted as the secretary and recorded the minutes, while Dr Oburka of Czechoslovakia chaired the meeting. An international committee consisting of 19 members from 17 countries (there were two members from the USA and also the USSR) was formed by Patrick in the coming weeks, in his role as Provisional Chairman. In the years following the symposium Patrick and his international friends got the IUAA off the ground and successful meetings were held (as with the IAU itself) every 3 years. The first was at Bologna (Italy) in April 1969, followed by Malmo (Sweden) in 1972, Hamilton (Canada) in 1975, Dublin (Ireland) in 1978 and Brussels (Belgium) in 1981. However, during the 1980s the IUAA eventually floundered and to all intents and purposes it faded into insignificance. This appears to have come about for two main reasons. Firstly, the IUAA was set up to share astronomical information, not to create observing programs. Once it started doing the latter it was treading on the toes of established national bodies. Secondly, according to Patrick, "The Italians and Irish, who had done much of the groundwork, were just too different to work well together." This would be one of Patrick's few pet projects that did not survive, which he greatly regretted. Mind you, given his dislike of foreigners it was a miracle he founded it at all!

Three months after the Prague symposium, on the 10th of November 1967, Patrick was, once more, joined by a Russian who spoke no English on the *Sky at Night*; he was the Cosmonaut, Colonel Valeri Bykovsky. Fortunately, the BBC provided an interpreter this time and Patrick was able to discuss Bykovsky's experiences in space and the recent results from the Soviet Venera probes which had parachuted through the Venusian atmosphere.

Back in Northern Ireland, on November 17th 1967, Patrick and Terry Moseley, using the 10-inch Armagh Grubb refractor, observed a distinct red-blue blink with a 'Moonblink' filter device. Without the device Patrick could see nothing, but with it, between 18.36 and 18.50 GMT, there was a definite colour anomaly present below the interior south-west wall of the lunar crater Plato.

At the end of 1967 Patrick received a very pleasant surprise in the form of a totally unexpected letter. He would be receiving an OBE in the New Years Honours List. Needless to say, Patrick, and especially his mother, were absolutely delighted. The awards did not stop there either. Early in 1968 the council of the BAA nominated Patrick for their prestigious Goodacre Award, which is granted 'in recognition of a contribution to the progress of Astronomy over many years'. This was just as important to Patrick as the OBE, because above all he valued praise from his fellow amateur astronomers. The Goodacre Award was named after Walter Goodacre, the BAA Lunar Section Director from 1896 to 1938, the President from 1922 to 1924, and a real Moon mapper, well before the space rocket era. Goodacre had published a map on a scale of 60 inches to the Moon's diameter in 1910 and a popular book about the Moon in 1932. So, receiving the Goodacre medal at the Association's AGM would be very special indeed for Patrick.

On the 5th January 1968 Patrick broadcast a *Sky at Night* all about home-built observatories. For the programme he had filmed his own rotating greenhouse or 'merry-go-round' observatory, installed in the back garden of his Armagh home, as well as the observatory of Norman Starke. Mr Starke was a dental surgeon living at No. 8, The Avenue, Hatch End, in Pinner, Middlesex. This latter observatory was situated a stone's throw from Patrick's birthplace. Finally, Patrick also visited the 'Beehive' observatory of the lunar photographer Cdr Henry Hatfield, at Sevenoaks.

In the January 1968 issue of *The Astronomer*, an essential magazine for the 300 or so hard-core amateur observers in the UK, and now received worldwide, Patrick wrote an article entitled *The Future of Lunar Observation*. This was a rare departure for Patrick who had branded many of *The Astronomer's* supporters as 'serpents'. Nevertheless, the magazine was now well established and so on the subject of TLP, he wrote:

> So far as TLPs are concerned, I think that it is now recognised by all competent authorities that occasional red glows are seen. The trouble is that the observing of these is very difficult, and it is only too easy to be mislead….Anyone anxious to contribute to TLP studies must learn their way very thoroughly about the Moon, and he must realise that he will have to undertake many hours of fruitless searching before he has any hope of seeing anything unusual. Indeed, he may never do so…..

The 6th February 1968 would be a big day for Patrick, as the Armagh Planetarium was now complete and would be opening its doors for the first time. It would prove to be an immediate success and Patrick had already decided that the *Sky at Night* should film the view inside the dome. A week earlier *The Sky at Night* camera crew had arrived, but it was then that Patrick realised his huge mistake. While the inside of a planetarium is spectacular to the dark-adapted human eye, the stars are totally invisible to a TV camera, especially a TV camera of the 1960s. The cameras recorded NOTHING that was being projected onto the dome screen! In the event some photographs were taken of the projector and some time exposures were taken of the artificial night sky. When merged, with a bit of trick photography, the effect of being in a planetarium was re-produced, but Patrick was somewhat embarrassed at ever thinking the TV cameras could show the faint and tiny stars. The *Sky at Night*, concentrating more on the history of planetaria, was broadcast on March the 1st. Thirty-four years later, in July 2002, Patrick would finally use image intensifiers to show viewers what a planetarium display really looked like, in the South Downs planetarium at Chichester; a project where he played a major role in raising the funds.

In early 1968, while still at Armagh, Patrick checked through the final proofs of a NASA report he had been working on for some time, with colleagues at NASA (messrs Barbara Middlehurst, Miss J.M. Burley and Miss B.L. Welther). This was

NASA report TR r-277, the *Chronological Catalog of Reported Lunar Events*. It listed all 579 reported TLP up to 1967 October 19th, going back into the depths of history! It was a project that Patrick had been working on for many years and he was glad to have it published by such a respected authority as NASA, even if Miss Middlehurst's 'crush' on him had been exploited to the full!

In the early hours of Easter Monday, April 15th 1968, the tireless George Alcock, still living at his house in Farcet, Peterborough, discovered his second nova while sweeping the Milky Way with binoculars. The new star was magnitude 5.6 at discovery and was found in the constellation of Vulpecula, the Fox. It was very close to the position of the ancient nova discovered in 1670. Alcock's second nova was designated LV Vul and rose to magnitude 4.8 on April 21st. With Alcock's 1967 nova rising to another fourth magnitude peak at the same time, there were, for a couple of weeks at the end of April 1968, two naked eye novae in the April dawn sky, only 15 degrees apart, both discovered by an Englishman. George knew this and so did Patrick, but few other observers seemed to realise what an achievement it was. Patrick phoned George to congratulate him and to get all the details on its discovery for *The Sky at Night*. This remarkable sight of two British novae, easy to observe in binoculars, seems to have prompted Patrick to join the American Association of Variable Star Observers (AAVSO). He sent off his six dollars (membership and application fee) to Cambridge, Massachusetts, in the USA, a few weeks later, simply listing his vocation on the form as a 'Writer'.

Chapter 13
Selsey and the Race to the Moon

It was becoming increasingly obvious, throughout the early part of 1968, that Patrick's duties were stretching him in too many directions and, with a possible manned lunar landing only a year or two away, book commitments and broadcasting commitments were looming. The political situation in Northern Ireland was turning unpleasant too. Patrick had always enjoyed connections with Scout groups and youth groups wherever he had lived and Armagh was no exception; he was the badge-secretary of the Armagh Scout group and had served in a similar capacity when living in East Grinstead. However, he confided to friends in 1968 that: "Even the Protestant and Catholic Scouts hate each other in Armagh and things will soon get worse, mark my words." Patrick was getting tired of not only the politics of Northern Ireland but the religious extremes too. He had no interest in either. He admitted, years later, that it was through gritted teeth that he had meetings with hypocritical religious dignitaries who visited the Armagh Observatory and Planetarium. One such occasional visitor was Cardinal William John Conway, the Archbishop of Armagh and the Roman Catholic primate of all Ireland. In 2010, more than 30 years after the Cardinal's death (1977), a Police Ombudsman report would reveal that the same Cardinal was instrumental, during 1972, in transferring a local priest and IRA leader, Father James Chesney, across the border. This had taken place after IRA bombings in the small village of Claudy, in which Chesney (a priest!) was involved. Anyway, even though Patrick had no idea of such grim goings-on behind the scenes, he was becoming increasingly stressed by the political and religious extremes and tensions surrounding Armagh in 1968.

Over one particular incident Patrick decided to cause some trouble and have some fun. He was told that the religious authorities were insisting that children's swings should be locked up on the Sabbath to enforce a day of rest upon the youngsters. Speechless with this type of religious fanaticism he and his young apprentice Terry Moseley entertained themselves writing increasingly absurd letters to the local protestant newspaper. An example of one of these letters was a question from Patrick (under a pseudonym) as to whether, as a devoutly religious man, he should lock the swing in his budgie's cage up on the Sabbath too. After several of these

M. Mobberley, *It Came From Outer Space Wearing an RAF Blazer!: A Fan's Biography of Sir Patrick Moore*, DOI 10.1007/978-3-319-00609-3_13,

letters were published the editor started to get suspicious (no, surely not!) and the correspondence was stopped.

Humour aside though, Northern Ireland was becoming a very dangerous place, and as I mentioned earlier, Patrick was utterly tired of the politics and religious battles. Not only that, but now the Armagh Planetarium was up and running, he was the Planetarium Director, speaker, organiser, and even the toilet cleaner on occasions! In addition, his mother was keen on a return to Sussex as was their cat, which had never liked Northern Ireland and seemed to have pined for Sussex for 3 years.

Fig. 13.1 Patrick with the Goto projector he chose for Armagh Planetarium, after its installation. Picture taken on January 23rd 1968 (Photograph courtesy of the Belfast Telegraph)

According to Patrick the endless flights back to England to record and plan *The Sky at Night* were really wearing him out too, not least the airline food, which he hated. On one occasion he mistook an olive for a grape and his upper denture snapped clean in half, just hours before a *Sky at Night* broadcast. Despite various repair attempts with glue, sellotape and string at the BBC he had to speak, for the whole programme, through his teeth, with his jaw clenched, looking like he was seething with rage! That event was pretty much the last straw as far as Patrick was concerned. He wanted to return to England and Sussex and put a stop to the endless airline commuting.

There was another factor too. As I have mentioned a few times already Patrick regarded Dr Henry King as the most obnoxious man in his astronomical world. In the 1950s King had spared no venom on Patrick's literary work and King's role as the London Planetarium Director, BAA President, and more recently, BAA secretary, had meant it was almost impossible to truly enjoy a BAA meeting when the influential Dr King was present. In Armagh he was well away from Henry King and with firm friends, but the great news was that his arch nemesis had now left the UK to take up an appointment as the Director of the McLaughlin Planetarium in Ontario, Canada. Joy unbounded! Patrick could attend BAA meetings with no risk of his enemy being there....sheer bliss!

Fig. 13.2 Patrick, smoking in the background, alongside his young observing apprentice Terry Moseley, outside the Armagh Planetarium dome on April 5th 1968. Following Patrick's resignation Terry would briefly serve as the interim Planetarium Director before Thomas Rackham arrived (Photograph courtesy of the Belfast Telegraph)

So, without too much sadness, Patrick tendered his resignation at Armagh in April 1968. Although the Planetarium had then been fully functional for a couple of months the official opening ceremony had yet to take place and Patrick certainly wanted to be there; the date was set for Wednesday May 1st. The Northern Ireland Prime Minister, Captain Terence O'Neill, officially opened the Armagh Planetarium on that day and newspaper photographs showed Captain O'Neill, alongside Eric Lindsay, Mrs O'Neill, and Patrick, walking along the approach to the Planetarium Dome. The local newspaper, *The Ulster Gazette and Armagh Standard*, for May 2nd, headlined with 'P.M. OPENS CITY'S £70,000 PLANETARIUM'. Of course P.M. stood for Prime Minister, not Patrick Moore, but a front page picture showed Patrick explaining the projector to the Prime Minister.

Fig. 13.3 Patrick explains the operation of the Goto projector to the Northern Ireland Prime Minister, Captain Terence O'Neill, at the opening ceremony of the Armagh Planetarium on May 1st 1968. Photograph courtesy of the Belfast Telegraph

With Patrick's departure imminent, in late May he dismantled his 8½ inch (216 mm) With-Browning reflector and his 12½ inch (317 mm) too, ready for moving back to Sussex. This was, perhaps, a bit premature, as on June 14th the asteroid Icarus would pass within 13.7 million miles of the Earth and Patrick wanted to see it, and possibly, photograph it too. Icarus would be 13th magnitude and sweeping through Ursa Major, so it would be well placed for northern hemisphere observers, and surely visible with the Armagh refractor. Unfortunately, Patrick realised too late that the low northern sky was blocked by trees from the Armagh dome and, to frustrate things further, the Armagh Schmidt telescope was out of adjustment. So Icarus flew past and many amateurs saw it, but Patrick missed it. In late June, Patrick and his mother planned to move back to Sussex, just in time for his mother's 82nd birthday. He had, after all, achieved everything he had set out to do at Armagh. The planetarium was up and running and it was very popular. Patrick's young apprentice Terry Moseley was saddened at his departure. The 3 years he had spent observing with Patrick had been a joy. Conscious of Terry's sadness Patrick arranged for his good friend George Hole to make a 14½ inch telescope mirror for Terry and presented it to him on his departure; yet again an indication of Patrick's extraordinary generosity to his friends (and even complete strangers). Terry briefly became the acting Director of the new Armagh Planetarium until Patrick's successor, Dr Thomas Rackham, took over. In years to come various scurrilous theories would circulate, regarding why Patrick resigned and left Armagh in such a hurry. Some focused on the amount of his time he freely gave to helping children; but the truth is that when you are famous someone will always want to damage your reputation.

In addition, if you suggested that Catholic Scouts should be welcomed into a Protestant Scout group in that era, well, you would make serious enemies.

There was to be one house moving hiccup though, and it has affected more astronomers than Patrick over the years. On exchanging contracts for the sale of Patrick's house at Armagh, the new owner, a local Armagh veterinary surgeon, calmly informed Patrick that his observatory was now his property, as it stood on his land. The observatory had not been mentioned before, but as soon as the ink was dry on the contracts, the vet claimed it! When Patrick queried this, the flustered vet rapidly claimed it was absolutely nothing to do with him at all, but his wife "thought it would make a nice summer house!" This, of course, was the decorative, windowed, rotating, octagonal 'merry-go-round' observatory that Patrick's second cousins had built for him. Patrick, being Patrick, ignored the legal implications altogether and, with friends, dismantled the observatory in hours and had it, and the already dismantled 8.5 inch With-Browning reflector, shipped back to the UK. By the time the new owners arrived only the observatory's concrete base remained! In fact, Patrick was probably in the right. The wood and glass observatory was so heavy that it had never been fixed to the ground. It simply 'sat' on the concrete base, like a car sits in the driveway of a house. The new owners were not interested in a fight with a famous TV astronomer and the matter ended there. Patrick later stated that if moving it back to Sussex had been illegal: "I would still have moved it without a qualm." As in 1965, when Patrick had moved to Northern Ireland, it would be his friend Ian Baker who would mastermind the entire move back to Sussex, even driving Patrick's spare MG Magnette car onto the ferry on the big day, with Patrick and his mother following behind in the Ford Prefect 'Ark'.

With the move back to Sussex imminent and the 8.5-inch With-Browning dismantled Patrick decided it needed a better mirror. He had been using a 10-inch refractor for the past 3 years and felt his With-Browning lagged some way behind its performance, despite the reputation of 'With' as a fine mirror maker. So, he asked his good friend Henry Wildey, generally regarded as the finest mirror polisher in England, to make him the best mirror possible for when the With-Browning was reassembled. Wildey obliged and the future Sussex based 8.5 inch now had a Henry Wildey polished mirror of excellent quality. Strictly speaking it was now a Wildey/With-Browning Newtonian.

Farthings

Patrick was not moving back to East Grinstead though. 'Glencathara' was history. As a very young child he had liked living near the sea at Bognor, and so had his mother. Even then he had always resolved to return to the south coast one day and he knew from his many visits to his friend Henry Brinton (who had a fine, German Equatorial 12.5 inch Newtonian reflector, originally owned by Robert Barker) that the atmospheric seeing seemed to be especially good at Selsey Bill. This was where Brinton had lived, in the village's Old Mill House, since 1957. Patrick also liked

Brinton's rotating Newtonian top end and wanted a telescope with one of those. When Brinton had acquired Robert Barker's 12.5-inch Calver reflector in the late 1950s (following Barker's cycling accident) it had possessed a closed tube, but Brinton had modified it into an open 'skeleton' tube with the top end able to rotate. He had also slightly tilted the Calver telescope's 'astrographic' German mounting (which made it impossible for the tube to hit the pier) to adjust for Selsey's south coast latitude. Finally, Brinton made a run-off shed for the instrument, decorated with a mural depicting a traveller poking his head through the fabric of the celestial sphere, to gaze on the workings of the Universe! The mural was closely based on an engraving in Camille Flammarion's 1888 book *L'atmosphère: météorologie populaire*, described as showing a missionary of the Middle Ages who said he had found the point where the sky and the Earth touched; surely, a highly appropriate decoration for an astronomical facility.

Patrick was a regular visitor to Brinton's observatory in the late 1960s and being able to use Robert Barker's old telescope, a man who had believed in lunar changes and who had encouraged the young Patrick, even before the War years, was very special to him. The presence of Barker's telescope at Selsey, as well as the friendship of Henry Brinton, was like a magnet, drawing him to the small seaside town. It is interesting to remind ourselves, in passing, that Henry Brinton had written a few science fiction novels in his time, just like Patrick. Of course, in future years, Patrick himself would become a magnet for many other astronomers to move to the Selsey area.

Anyway, well before his return to Sussex, Patrick, via Brinton, had left his requirements with estate agents in Selsey and waited for details of suitable properties to arrive. Within a very short space of time a possible property came onto the market in West Street. This seemed like fate, as Major Levin, the BAA President who had originally signed Patrick's BAA application form in 1934, also used to live in West Street, just up the road, at a house called Elleray! In addition Brinton's own house, as I mentioned earlier, the one simply referred to as 'The Old Mill House', was on the corner of West Street and Clayton Road and was a short walk from the property that was up for sale. Being old friends it would be great if Henry and Patrick could live 5 minutes walk from each other.

Patrick flew back from Northern Ireland and stayed with Henry Brinton, at his house on the seashore in Selsey, for a couple of days. Brinton took Patrick for a walk to see the advertised house, but it was not quite what he had wanted; it just did not have the right feel to it and there was insufficient space for all his telescopes. He wanted a substantial back lawn, or at least, an area away from public view, where he could keep his present and future observatories. While surveying the property Patrick spotted a very interesting, if unkempt, massive ramshackle house with leaded windows, further up the street; so he went for a closer look. There was a big FOR SALE sign in the garden and the weeds were as high as Patrick; still, it looked interesting! The ramshackle lawn stretched 180 feet back from West Street on the extreme eastern edge of the property and the grounds were over 250 feet wide along the southerly end; there was over two-thirds of an acre of land in a sort of triangular shape, with the curved triangle's top being bounded by West Street. Investigations indicated that the house was not on the estate agent's list and that it had been unoccupied for 4 years. The previous owner had been a Mrs Rider but no-one had seen her, or anyone inside the property, since she moved out. However, after a few phone calls Patrick acquired the keys and managed to enter the large cobweb ridden six bedroom property, complete with dead birds on the floor!

Fig. 13.4 Patrick's house Farthings as it appeared around the time he purchased the property in 1968 (Photographer unknown. BAA Archive picture)

However, as soon as Patrick entered the building he felt this was the place he wanted to live. It was not quite on the sea front, but it was only 500 yards away, and it was big; with over two-thirds of an acre of ground there was plenty of space for observatories! Further investigations indicated that it was originally a thatched barn, as far back as 1671, and that it had been extended. In Patrick's own words: "I asked the agents if it had electricity, drains, woodworm, wet rot or dry rot. The answers were yes, yes, no, no, no, so I bought it!" At exactly £8,000 for two-thirds of an acre of land it was a bargain, but needed a lot of work. Nevertheless, 'Farthings' at 39 West Street, Selsey would be Patrick's home for the next 44 years. In fact, Patrick and his mother would live in the house in a state of semi-upheaval for several years, while the house was re-wired and oil-fired central heating installed. Re-thatching of the roof took until 1972 to complete. Many visitors to Patrick's house, over the next 44 years, assumed that *he* had named the house 'Farthings'. In other words, as it was an astronomer's house, with a garden full of telescopes for looking at distant objects, it was really named 'Far Things'. In fact it had been named Farthings for a long time before Patrick moved in. Locals told Patrick that the house was, long ago, owned by a Mr Penny, who had four sons, hence the name. But Patrick was never sure whether this was just a myth. In fact, a keen naturalist and amateur scientist named Henry George had lived at 39 West Street some 60 years before Patrick bought the property. So, as his period in Selsey overlapped that of the BAA President Major Levin it is possible that the man who proposed the 11 year old Patrick for BAA membership in 1934 had known a previous owner of Patrick's future property!

With all his impending Moon-landing commitments looming (assuming there were no more tragedies at Cape Kennedy) and with him moving house, Patrick also relinquished his Directorship of the BAA Lunar Section, handing it to his good friend at Keele University, Dr Ron Maddison. One would have thought Patrick had enough on his plate in the late 1960s, but while still at Armagh Planetarium he had invented a new quarterly publication and appointed himself the Editor. Called simply, *Planetarium*, and produced by 'Planet Publications', the magazine was available at three shillings and sixpence per issue. At the end of the 1960s and, briefly, into the 1970s the new magazine survived, just! It would be one of many ventures Patrick would make into owning his own astronomy magazine over the next 40 years. With all this on his plate it was hardly surprising that he had to relinquish the Lunar Section Directorship.

By the end of June 1968 Patrick was the owner of 39 West Street Selsey. They moved in just 2 days too late to have a combined celebration of moving in and of Mrs Moore's 82nd birthday, which had taken place on the 27th. After a week in the Irish Sea ferry system his belongings and telescopes were finally inside the property too, although not ready for action. On leaving Armagh Patrick had just over £8,000 to his name and the Selsey house had also cost him £8,000. He claimed that on the day he moved in he had only two shillings and sixpence left in his bank account, but those 3 years at Armagh, in a substantial salaried job, meant he did not need a mortgage. He had bought Farthings outright, just before property prices went through the roof and just before his period of highest income, which would be handy for all the renovation work. Brilliant timing! Patrick would never be interested in investments, savings or interest rates, but with lucky financial timing instinct like that he didn't need to be.

In the early hours of July 18th, at around 1 am GMT, Patrick received a phone call from Terry Moseley at the Armagh Observatory. He had been observing the crater Aristarchus with Pat Corvan, using the Armagh refractor, and they had observed a reddish glow and obscuration just south of, and the same size as, the 'cobra head' feature. Unfortunately Patrick's 8.5 and 12.5 inch telescopes were not ready for action, so he grabbed his childhood 3 inch refractor and zoomed outside. Fitting the moonblink filter wheel to the small refractor was a major hassle, it almost weighed as much as the telescope, and of course available light was at a premium! Nevertheless, Patrick reported a definite red-blue blink in Aristarchus, despite the small aperture used. If only his other telescopes had been available.

Siberia

In the midst of all the house-moving upheaval Patrick had another trip abroad to plan: it was to be his third total solar eclipse, this time in Siberia! On September 22nd 1968 the Moon's shadow would sweep across Mongolia, Siberia and the Arctic. The maximum duration anywhere on the track would only be 40 seconds, but after the successful TV broadcast of the 1961 eclipse the BBC were happy to fund Patrick to go off on another jaunt, despite having ignored four much longer ones in the interim. Unfortunately it soon became apparent that the practical difficulty of getting bulky TV cameras into Siberia was going to be a problem, so the BBC decided to fund Patrick to go alone. Then a second problem emerged, this time a bureaucratic one. For some unknown reason Patrick was refused a visa into Siberia, despite being an honorary member of the Astronomical Society of the USSR. After a few phone calls via academics in high places he was no nearer a solution so Patrick resigned himself to not going; he had plenty on his plate with moving to Selsey, not to mention an infinity of book writing and broadcasting duties. His friend, the new BAA Lunar Section Director Ron Maddison, was going, along with various other UK astronomers, so he asked them to bring back a photo for the *Sky at Night*. They had been granted visas for Siberia without any hassles, so maybe being a BBC broadcaster was the problem. It was all a great pity as this was, arguably, the first eclipse of the 'eclipse-chasing' era. It marked the point where astro-tourism really started as the event had been well advertised. Oh well; what could he do…….nothing!

Patrick had been having problems with his Ford Prefect car 'The Ark' while house hunting, so as he would not need it for a while, he booked it into a garage on September 18th. Late in the evening of September 19th he received a phone call from the Royal Society in London; the USSR Academy of Sciences in Moscow had arranged him a visa to Siberia after all, but he had to call in at the Soviet Embassy in London to collect it and get on a flight to Moscow by lunchtime the next day!! Patrick phoned the garage. "How's my car doing?" "It's propped up on bricks and the engine's on the hoist." Predictable! A few calculations indicated that he would, in practice, have to travel in the early hours to get to London, especially as roads in Sussex were flooded after local downpours and "British Rail were out of action – so nothing new there". His motorbike was in a bad way too and the MG Magnette had a flat battery, but with help from Henry Brinton and public services he got to the Soviet Embassy in the early morning of the 20th, picked up his visa, scrounged some travellers cheques from the BBC, and caught the plane for Moscow at midday. He packed his case in such a rush that it was full of equipment, but no spare clothing at all. He just had his trusty blazer….for Siberia! His mother hurriedly packed a jumper into the case insisting Patrick should not remove it to make way for more cameras. She imagined that, in Siberia, a jumper would definitely be required; but she also knew that Patrick NEVER admitted to feeling the cold. She even phoned the producer Pat Owtram to make sure he kept the jumper in his case.

While sitting on the plane Patrick studied his new 'Siberian Visa'. It did not seem to say anything that he recognised as Russian for Siberia, but he could only recognise the odd Russian phrase, so there was probably no need to worry. But when he arrived in Moscow and tried to get a flight to Siberia it transpired that his worst fears were realised. The visa was valid for Moscow; nothing more, nothing less! So, more hassle. Luckily Patrick got through to the Soviet Academy and his contacts there managed to rush through another Visa – phew! Patrick proceeded to fly to Siberia and then on to Kurgan. While in Kurgan he asked to be put in contact with Dr Maddison, at which the authorities assumed he was ill and wanted a doctor of medicine!

Well, Maddison and medicine do sound rather similar. But Patrick avoided being rushed to hospital....just! He eventually did meet up with his old friend Ron Maddison from Keele University as well as many other eclipse chasers, some with advanced equipment, at the Yurgamysh site only five hours before totality.

The eclipse site at Yurgamysh was just inside the Siberian border with Kazakhstan at longitude 65 East, latitude 55½N. A report of the eclipse was written up for *The Astronomer* magazine by Colin Hunt. He reported that the Russian border guards had confiscated fellow traveller Peter Lancaster Brown's copy of the *Observer* newspaper, but nothing else! The astronomer with the best idea (this author speaks from the experience of witnessing seven clear, one partially clear and two cloudy totalities) was a certain Dr Houtgast from the Netherlands. He had brought a substantial piece of equipment to the eclipse site: a large, comfortable armchair! Sick of fiddling with equipment at eclipses and getting stressed up, he was going to watch it. It sounds obvious, but many solar eclipse chasers do not see much of totality as they are too busy grappling with cameras! The skies were hazy, but clear, and Patrick and his friends witnessed just less than 40 seconds of totality. Patrick's personal project during totality was to sweep, near the Sun, for bright comets, which might not have been discovered when further from the Sun and dimmer. Eclipse comets had been discovered before, such as at the 1948 November 1st eclipse at Nairobi, but unfortunately Comet Moore was not to be discovered this time.

After the eclipse there was much celebration and Patrick proposed a toast to international co-operation between scientists and freedom from politicians. The scientists had certainly collaborated and they had even repaired Patrick's clothing, which had virtually disintegrated in the humid, mosquito-ridden, Siberian conditions! He had not needed a jumper after all, but a mosquito net! At this toast, Patrick told me that a normally mild-mannered Soviet astronomer bellowed, regarding politicians: "Nyet, ve do not vont zose BAR-STARDS here!" Patrick had survived the Siberian encounter but the climate and rigours of the journey had taken their toll on his blazer and shoes. The Russians had kindly staple-gunned his blazer pockets together and duck taped his size 13 extra-wide shoe uppers to his shoe soles. On his return Pat Owtram strongly hinted that he must get a new suit for TV. But Patrick had pre-empted her and already been to Moss Bros, the well known UK tailors of that era. He also called in at the local Selsey cobblers, to see what they could do to repair his size 13 'Extra-Wide fitting' shoes. According to Patrick they said: "It is beyond the Earth-based technology of the 1960s to repair your humungous Brogues Sir", so Patrick asked a craftsman at a local Selsey boatyard to carry out the repairs to his shoes! Anyway, despite the last minute panic, Patrick had seen his third total solar eclipse and he showed his photographs on the October *Sky at Night*.

A Medal and a Refractor

On Wednesday October 30th 1968 Patrick proudly received the BAA's Goodacre Award from the outgoing BAA President and Jupiter Section Director, Mr W.E. (Bill) Fox, to sustained applause from the packed audience at the AGM. As usual, the venue was the RAS building at Burlington House, Piccadilly. At that same meeting, Mr Fox would hand the Presidency to another friend of Patrick's, namely Gordon Taylor.

The Goodacre medal came with a small financial award too (the gift), and when Patrick became settled in Selsey he designated the money towards the cost of materials for constructing a small run-off roof observatory. This building was for his newly acquired 5 inch (12.5 centimetres) Cooke refractor. The observatory itself was designed and built by Patrick's friend and fellow Selsey resident Reg. Spry, quite a character in his own right and a keen BAA member during the 1970s. After his period at Armagh, using a decent refractor, Patrick decided that the late Will Hay had been right about refractors: Hay had always preferred his 6-inch refractor to his 12.5-inch reflector. Also, ever since the end of his childhood Brockhurst Observatory days, Patrick had yearned for a decent refractor of his own. Patrick once said to me: "If someone gave me a million pounds I'd go out and order the biggest refractor I could get. The Lowell 24-inch refractor must be my favourite telescope." With plans for a much bigger reflector at Selsey, his 5-inch Cooke refractor, plus the old 12.5 and 8.5-inch reflectors, he really would be spoilt for choice!

Patrick had in fact acquired his 5-inch Cooke refractor from the inventor of the Moonblink device, the Lunar Section member and millionaire businessman Peter Sartory. The elderly Sartory knew that Patrick wanted a good refractor of his own and so he offered his 5 inch f/12 Cooke to him, for the throwaway price of £10! In truth, Sartory was happy to give it to Patrick for free, but Patrick felt that he should pay something, so Sartory suggested that he would accept £10. Needless to say, Patrick snapped it up and the Goodacre gift money was donated to the refractor's run-off roof observatory 'materials fund'. Sartory's refractor sat atop a distinctive *Charles Frank* (of Glasgow) German Equatorial head and pier, with Frank's trademark dark grey hemispheres covering the Right Ascension and Declination wormwheels.

In fact, as well as the mounting, parts of the tube and the refractor's accessories and eyepieces, which Patrick inherited, had little to do with Cooke and more to do with *Charles Frank*. In the 1960s *Charles Frank* used components imported from *Swift* of Japan for its own refractors. These parts were very distinctive, with a brown/grey and cream colour contrast, and they were still that colour when Patrick acquired the 'scope. The biggest give-away that *Swift* components were utilised was the distinctive, brown/grey, three eyepiece *Swift* turret that used to sit on the rackmount when Patrick used the instrument in the late 1960s, 1970s and 1980s. This was a joy to use, as magnifications could be changed without fumbling for eyepieces in the dark. The finder, featuring a quality doublet achromat lens, was a pure grey and cream *Swift* unit too. Another 1960s *Swift* trademark was the refractor's two slow motion controls. These were of the long and bendy metal cable design seen on many Japanese refractors from that decade and they made positioning an object in the eyepiece a joy, as did the refractor's splendid 6-inch diameter engraved R.A. and Dec setting circles. So, the instrument Patrick purchased from Peter Sartory was not so much a Cooke refractor as a Cooke/*Swift*/*Charles Frank* refractor!

The new observatory was duly built by the aforementioned Reg. Spry who could easily turn any household junk into a 'Heath-Robinson' style telescope, or even an observatory. The large refractor, with a tube (including dew cap) some five and a half feet in length, had originally been supplied on a 'pillar and three vanes' pier with the words *Charles Frank* encased in wrought iron within the vanes. Such a free standing design could be, to quote Patrick, "as stable as a blancmange", but this pier was to be fixed to the ground. Rather than make a new pier, Reg. Spry simply bolted the existing three vanes to a huge, foot high, block of concrete, which raised the 'scope so that it was pretty much ideal for Patrick's stature and also for keeping the observatory roof under 8 feet in height. However, eventually Patrick had Spry's transparent plastic corrugated roof and walls replaced with wooden panels, mounted onto Spry's original existing wooden framework. The transparent structure had made the observatory overheat on sunny days, so that the refractor took far too long to cool down after sunset. Reg. Spry joined the BAA (as did this author) in late 1969, on Patrick's advice, and he lived just half a mile to the east of Patrick, at 60 Windsor Road.

Remarkably, Patrick actually had yet another refractor mounted in his garden during the 1970s, a 4-inch instrument (not to be confused with his childhood 3-inch refractor). This instrument was protected from the elements by an umbrella and was solely used for projecting the Sun's image to record sunspots. Not surprisingly, Patrick named this structure 'The Umbrella Mount'. Once the 5-inch refractor was up and running Patrick gave the 4-inch telescope to his local Chichester friend Roger Prout. Like the various other obstacles in Patrick's garden at night (including rain gauges and various other celestial contraptions) the Umbrella Mount was something else that any unwary visitor to Patrick's garden might horribly injure themselves upon during the night!

The Moon Draws Nearer

Despite the tragic Apollo capsule fire of January 1967 NASA had forged ahead with its plan for a manned lunar landing before, to quote John F. Kennedy 'this decade is out'. Although the pure oxygen supply for the Apollo spacecraft needed changing after the January '67 fire, the development of the massive Saturn V rocket itself was not directly affected by the tragedy; only the first launch of a manned Apollo capsule had been delayed. The three stage Saturn V Moon rocket was truly massive. It stood 365 feet high (111 metres) and weighed three thousand tons. Of course, most of this weight was just fuel. To get the Apollo Command and Service Modules, as well as the Lunar Module (some 50 tons in total), away from the Earth's gravity, and on to the Moon, required accelerating them to Earth escape velocity, some 25,000 miles per hour or 40,000 kilometres per hour (11 kilometres per second). This would take a colossal amount of energy, but the Saturn V provided it. Needless to say, the first launches of the Saturn V were unmanned.

On November 9th 1967 America resumed its place in the space race with the dramatic launch of the unmanned Saturn V rocket designated Apollo 4. Three unmanned (before the Apollo 1/AS 204 fire) Saturn 1B rockets (capable of *only* putting twenty tons into Earth orbit) had not been designated with Apollo names, merely with AS 201, 202 and 203 numbers, but the Apollo numbering system started again with the November 9th launch. In those heroic days the TV and press offices were rather closer than the 5 mile safety zone reserved for the later Shuttle launches. They were located in the Vehicle Assembly Building only 3.4 miles from launch pad 39A. The vibration from the 3,000 tons of thrust of the main engines caused the press stands to shake, and ceiling tiles in the CBS TV room fell down around the news anchor Walter Kronkite during the live launch transmission! NASA had not fully appreciated the effects of the audio onslaught 3.4 miles from the launch pad, and after Apollo 4 they installed a sound suppression system where hundreds of thousands of gallons of water from the ocean were dropped onto the launch pad, to dampen the noise during subsequent lift-offs. Further successful unmanned launches took place in January 1968 (the Apollo 5/Lunar Module test) and April 1968 (Apollo 6). The Saturn V rocket had some fuel flow problems on Apollo 6, but nothing disastrous. Finally, the work that had been delayed by the Apollo 1 fire, namely the manned testing of the Apollo spacecraft, was resumed in October 1968. This was the launch of Apollo 7, taking astronauts Schirra, Eisele and Cunningham and the Command, Service and Lunar Modules into Earth orbit, on a Saturn 1B rocket. Everything was 'GO' for the first flight around the Moon, and as it turned out, with Soviet pressure seemingly mounting, this would occur during the first manned Saturn V/Apollo mission, in December 1968.

Around this time the major incentive to the US Space program was indeed that apparent progress of the Soviet lunar program. As things transpired they simply did not have a rocket that was as powerful and reliable as the Saturn V, but they did have a space capsule to rival Apollo, namely the Zond craft. On September 15th 1968 the Zond 5 mission lifted off from the Russian launch site at Tyuratum and became the first spacecraft to circumnavigate the Moon and return to Earth, 3 months before Apollo 8. What is more, it contained living things: namely, tortoises, insects, plants and seeds! It was also the first Soviet splashdown. The craft came down in the Indian Ocean. Zond 6 repeated the success and was launched on November 10th 1968, with a return to Earth using a novel skip re-entry technique to reduce velocity. Disturbingly for NASA, the state controlled Soviet news agency TASS described the mission as 'an automatic test to finalise the functioning of the Soviet Union's manned lunar spacecraft'. The Americans were not in any mood to delay the first manned mission to the Moon!

For anybody reading this book who does not remember December 1968 (I was almost 11 years old) and the manned Apollo missions it is hard to describe the excitement that was in the air, both approaching Christmas 1968 and onwards through to the lunar landing in July 1969. Despite the deadline to "land a man on the Moon before this decade is out", men going to the Moon was still science-fiction. It was far and above anything that mankind had done before. Personally, I would say that only aliens landing on the Earth could generate the same degree of excitement and even apprehension. By the launch of Apollo 13 though, human beings being fickle creatures, it was old news. The human race had 'been there, done that, got the T shirt', unless you were an amateur astronomer or space enthusiast, in which case it was still incredibly exciting. Before 1968 Patrick was a TV astronomer with a similar status to a BBC science correspondent. Most people in the UK knew the name, but many probably could not remember exactly what he was about…..'just that space bloke on TV'. During the early Apollo Moon landings Patrick would suddenly become one of the most famous faces in the country, because everyone was watching the Moon missions from December 1968 and through 1969. The coverage would set Patrick up for life, as from that moment on everyone would know his name, opening doors to massive book sales and the status of a 1970s 'TV Personality.' In the decades to come Patrick would frequently be asked whether he would have liked to have gone to the Moon; he was, after all, only 7 or 8 years older than most of the Apollo astronauts. His answer would always be the same: "Oh, of course I would, but I'm afraid it would take a very massive rocket to launch me!" At around 17 stone in those days, this was something of an exaggeration! The Saturn V could get 50 tons to the Moon, so another 108 kilograms would make little difference. The 1968 Apollo 8 BBC coverage marked the first time that Patrick and BBC Apollo anchorman James Burke worked together. Patrick said that they developed a system: when James said the words "Patrick" he would speak about lunar matters continuously and when James deliberately blinked hard, Patrick would shut up! They were assisted by the BBC correspondents Reginald Turnill and David Wilson at Cape Kennedy and by Charles Wheeler at Mission Control in Houston.

Fig. 13.5 Patrick (*right*) with Cliff Michelmore (*centre*) and James Burke (*left*) during the Apollo 11 Moon Mission in July 1969 (Image by permission of the BBC)

Shortly after dawn, on December 21st 1968, the first men to leave Earth's gravitational pull were launched on the mighty Saturn V rocket, heading for the Moon. They were Frank Borman, James Lovell and William Anders. Patrick did not travel to Florida for the launch, but he did commentate live from Alexandra Palace. Patrick did not realise at the time, but quite a few amateur astronomers saw Apollo 8 en route to the Moon. From Sevenoaks, Cdr. Henry Hatfield even photographed it, and his picture was reproduced in the national newspapers.

By the afternoon (USA time) of December 23rd the crew of Apollo 8 were less than 50,000 kilometres from the Moon and the lunar gravity was stronger than that of the Earth. When the Apollo 8 spacecraft curved behind the Moon the crew would be out of radio contact with Earth for 32 minutes. During this time a lunar orbit insertion burn would have to take place, but its success would not be known until the Apollo spacecraft re-emerged from behind the Moon. As the spacecraft disappeared, Patrick and the mission controllers heard Jim Lovell say "We'll see you on the other side", just as a burst of static signalled the Moon getting in the way and the ground controller's spacecraft telemetry stopped. Patrick had been live on BBC1 when the first pictures from the Moon's far side had come through in 1959. Now, only 9 years later, he was 'on air' as men were actually travelling around that far side, taking pictures of the sights never seen by human eyes before. Thirty-two silent minutes ticked by on BBC1 and Patrick continued his commentary with the words "All we can do is wait. Listen for the signals in less than twenty seconds. This is one of the great moments in human history". And at that precise point in history, the BBC controllers switched BBC1 to transmitting a children's programme called *Jackanory*!

Patrick never did find out why that happened, but some minutes later he was reconnected to the TV audience, although by then telemetry with Apollo 8 had been recovered and mission controllers were chatting excitedly to the crew. As Apollo 8 rounded the Moon's Far Side on the ninth orbit, on Christmas Eve 1968, and as the Earth rose above the lunar landscape, William Anders, from the Apollo Command Module read from the book of Genesis:

> In the beginning God created the heaven and the Earth.
> And the Earth was without form, and void; and darkness was upon the face of the deep.
> And God said, Let there be light: and there was light.

It was a moving moment even for Patrick, who had no specific religious beliefs, but he often said, about that moment: "We must surely be here for some purpose, otherwise it is all just a waste of time."

NASA showed a live transmission from the spacecraft on Christmas Eve as it skimmed low over the Moon, and as BBC1 viewers joined BBC2 the sound of Bill Anders could be heard describing the Moon as: "a lonely, forbidding expanse of nothing – and it certainly would not appear to be a very inviting place to live or work". Now almost 72 hours into the mission and with both BBC channels joining to cover Apollo 8, James Burke turned to Patrick, saying "Well, Patrick Moore, what did *you* think of THAT?"

An excited Patrick, removing his glasses, replied at a speed only he could speak at on live TV: "Quite incredible. One thing we've got to bear in mind I think, they were magnificent pictures – I'm not going to say they show us more detail than we've got from the Orbiter's as they probably don't, but don't forget people were actually seeing them direct for the first time and this was bound to add to our knowledge." Via a satellite link to Jodrell Bank in Cheshire, Patrick's friend Bernard Lovell then joined in the excitement. Burke then asked Patrick (and Bernard Lovell) if they expected anything at all to come out of the flight that could not have been got from sending up an unmanned probe. Patrick replied: "I don't think the actual knowledge we are going to get is going to be of fundamental importance because I'm quite sure it's going to back up what we already know, but if we're going to get men to the Moon at all, if we are going to explore space, then this kind of operation has got to be carried out, and that, I think, is the fundamental point of all this and it is, quite probably, the greatest technological triumph ever made by man." Patrick then continued: "One thing I would like to make here if I may. There have been suggestions, I know, that this is just a publicity stunt. THIS IS PART OF SCIENTIFIC HISTORY JAMES, THIS IS PART OF OUR HERITAGE, and it would be criminal not to broadcast it as fully as we can".

A few days later, live on BBC1, Patrick was joined by James Burke, Cliff Michelmore and Michael Charlton for the return of the astronauts to Earth, which was flawless. After the successful splashdown of the Apollo 8 crew, less than 3 miles from the waiting aircraft carrier, a manned Moon landing was now looking inevitable for 1969. Patrick was contacted with some urgency, early in 1969, by the publishers George Philip and Mitchell Beazley. His name was now the name the public associated with the Apollo Moon missions. OK, there were other names: the BBC presenters Reginald Turnill and the aforementioned James Burke for example, but they were not prolific authors or experts on the subject of the Moon. The publishers Mitchell Beazley and Philip were convinced that there was a huge market for a *Moon Flight Atlas* to coincide with the first Moon landing and to explain, to the layman, how Apollo would get to the Moon, with simple explanations, maps of the Moon and lavish (for the time) illustrations. The Atlas, if printed with the same format and distributed through the same networks as Philip's World Atlases, could sell in very large quantities, if the release coincided with the first Moon landing (expected in late 1969). Patrick committed himself to the project with enthusiasm. Without a doubt this would be the biggest selling book he had ever written, if everything went to plan, and with workmen likely to be renovating his Selsey barn for years and turning it into a proper house, he needed the money!

Patrick quickly acquired all the space-probe photographs he wanted from NASA and all the amateur B&W earth-based photographs he wanted from his friend in Sevenoaks, Kent, the well-known lunar photographer, Cdr. Henry Hatfield. He also acquired one colour photograph from his neighbour in Selsey, Henry Brinton, taken with Robert Barker's old Newtonian. But despite all the Apollo fervour, he still had numerous observing projects and amateur astronomy commitments in and out of the BAA, not to mention voluntary work with the disabled, scout groups, boys clubs and charities.

Fig. 13.6 Henry Brinton's 12½ inch telescope and run-off shed at the Old Mill House in Selsey. Patrick was a frequent visitor to Henry's Observatory in the late 1960s and early 1970s (Copied with Patrick's permission in 2002)

A Man of Principle

Throughout his life, increasingly as he became more and more famous, Patrick would give talks up and down the country to various local astronomical societies and charities, especially British charities, children's charities and cancer charities. Even when the talk was not specifically in aid of a charity, or observatory fund-raising, he would always waive any fee for an astronomical society; he simply let them pocket the ticket money. I have heard cynics say "So what, he must have been loaded!" But he wasn't. He scraped along on the money he made from books and he had no material possessions to speak of. Unlike so many 'TV personalities' of today who deliberately let slip, or blatantly boast, about their charity work, Patrick was genuinely embarrassed if the subject was even brought up, often becoming tongue-tied if anyone praised his efforts. This was not false modesty: he was genuinely embarrassed by such praise. During the 1960s he helped raise a huge amount of money for the 'Mr Pastry Fund', named after the comic actor Richard Hearne's slapstick character of that era. Hearne (1908–1979) had been the President of *The Lord's Taverners* charity in 1963 and lived in Kent, just a short journey for Patrick when he had lived at East Grinstead. The 'Mr Pastry Fund' raised money for hundreds of hydrotherapy pools for children with polio and Patrick was one of its most active fund raisers. On a few occasions Patrick would reminisce with friends about driving 'sacks full' of donated money around to Richard Hearne's house in his Ford Prefect. Unlike with so many charities today the fund had no administration costs at all; the fund–raisers, such as Patrick and Richard, bore all the expenses. One hundred percent of the money collected went to the charity.

Those who knew Patrick best knew that his work for British charities over the years was equal to his contribution to astronomy. He would normally refuse to talk about such things, but a simple calculation, by anyone who knew him well, could prove that he raised many millions for charities, despite living very frugally himself. For example, Patrick would never spend money on a flashy new car, preferring to drive an old wreck into the ground and then buy a slightly healthier wreck! All his telescopes were second hand and his house was full of old furniture and fittings. In addition, while his health was relatively good, up to the late 1990s, he would rarely refuse a request to travel to a small society and give a talk. In 1969 he even gave a talk to Croydon Astronomical Society, which, as we have seen, he once described as a 'nest of serpents'! In the audience at that 1969 Croydon meeting was a 13 year old fan named Peter Fincham. Peter would go on to become the controller of BBC1 (briefly) from March 24th 2005 and the man who approved the budget for the *Sky at Night* during the programme's 50th anniversary year, some 38 years later, in 2007!

Another breathtakingly different side to Patrick, indicating that he really was a man of principle, was that he would never do advertising, or product endorsement, for personal gain. Indeed, for the vast majority of his life he would never do any form of advertising at all. During the 1970s, 1980s and 1990s he was frequently contacted by food manufacturers (usually soup and biscuits!) to advertise their products. Patrick thought that earning money like this was blatantly immoral, although he never criticised others for doing it. If he thought something was good, he would be happy to say so, but not for money; surely such behaviour was the hallmark of a very greedy, lazy and slimey person? He had a set of postcards pre-printed for him which read:

> Thankyou for your letter, but I do NOT do advertising.
> In fact, I would infinitely prefer to be found dead in a ditch.
> Yours, Ever, Patrick.

In the 1990s the offers became so lucrative that he did decide, on a few occasions to agree to do TV adverts where:

1. He was merely an actor, not directly endorsing the product, and
2. All the money went to charities

His 1990s *Abbey National* 'Tessa' and *Citroen* car adverts are examples of this, along with his 21st Century Post Office savings adverts.

Patrick undoubtedly boosted, through the roof, the attendances of national and local astronomy meetings around the country whenever he turned up to give invited talks about Space or the Moon landings. On February 12th 1969 he accepted one such invitation from the Peterborough Astronomical Society. Patrick drove his ageing 1953 vintage Ford Prefect to the village of Farcet, delivered his talk and stayed overnight with George Alcock and his wife Mary. Amongst amateur astronomers, George and Patrick were equally famous, as by then George had discovered four comets and two novae, using nothing more than binoculars.

More Moon, More Apollo

March 3rd 1969 saw yet another Apollo Saturn V launch from Cape Kennedy with astronauts McDivitt, Scott and Schweickart on board Apollo 9. This would not be a trip to the Moon; however, it would be a full test of the Command Module and Lunar Module in Earth orbit, with a full crew carrying out the docking and undocking manoeuvres between the two spacecraft. Again, the mission was successful, with the minor problem of Schweickart suffering from space sickness.

It was interesting that, around this time, there had never been more interest in looking at the Moon through a telescope, especially from Patrick's perspective as the BAA Lunar Section's TLP co-ordinator. Of course, this was mainly down to Patrick's fame and approachability, two things that are usually mutually exclusive, but not in Patrick's case. Patrick knew he was a magnet for the BAA and was quite happy to freely advertise his address and phone number in the Lunar Section Circular, for anyone to phone him if they thought something odd was happening on the Moon, and many people did, especially during the Apollo era. On March 27th however, Patrick himself saw something odd. He was observing with Lunar Section stalwart Phil Ringsdore using his 15 inch (38 centimetres) reflector based at 5 Rosedale Road, Stoneleigh near Epsom in Surrey. This was the same telescope that would, eventually, become Patrick's own; but more on that later. Terry Moseley from Armagh was also present. All three observers spotted a reddish area NNW of the central peak of Alphonsus, the same crater in which Kozyrev had claimed to have detected carbon gas emissions 10 years earlier. Attaching a BAA Lunar Section blink device confirmed the anomaly which was seen by all three experienced observers with excellent equipment. The next day Patrick managed to contact Kozyrev at the Crimea Observatory to alert him to the Alphonsus event. Remarkably, Koyrev's response to this, on the night of April 1st, a disturbingly poignant date, was to take spectrograms of the crater Aristarchus which he claimed showed, without a doubt, CN and N_2 emissions from a red spot on the inner western wall slope of that crater. I leave the reader to make what they will of such an instant response to Patrick's request, after years of inactivity (by Kozyrev), but I know what my opinion is!

From Friday April 11th to Sunday April 13th the BAA held its third residential week-end course on observational astronomy at King Alfred's Training College, Winchester. The first two courses, in 1967 and 1968, had been during Patrick's period at Armagh. He had always been a great supporter of the weekends, but in 1969 he was a major attraction himself; the Apollo publicity meant he was instantly recognized wherever he went. Patrick and his friend J. Hedley Robinson gave numerous talks at beginner and advanced level to the keen amateurs on the course. The course organizer was Alfred Curtis who paid tribute to the 'invaluable support' he had received from Patrick and Hedley. The course was an outstanding success, with participation up from 56 to 69 in the first 2 years, to 129 in 1969. Around this time, Patrick heard that he was to have yet another award bestowed upon him; this time it would be the Arturo Gold Medal of the Italian Astronomical Society.

On April 18th Patrick broadcast a *Sky at Night* on the subject of 'The Moon and the Earth' with guests Bert Foord (TV Weatherman), Henry Brinton (fellow Selsey astronomer) and Dr John Hutchinson (a well known psychiatrist). The contribution from Dr Hutchinson prompted a deluge of letters to Patrick on the subject of 'lunacy' and it was weeks before he cleared the resulting post from his desk! The pile of correspondence must still have been there in his study at Selsey 1 week later, on April 25th, when, just before 9.20 p.m., he nipped indoors to change the variable star charts he was using. It was a clear night at Selsey and he had already observed all the objects he could with the charts he had taken outside. Unfortunately, 9.20 p.m. was not the best time to nip inside; an incredibly bright blue-green fireball (a meteor as bright as the half Moon) shot through the evening sky while Patrick was indoors. The meteor travelled from SE to NW across the British Isles and a half kilogram meteorite ploughed through an asbestos roof in Sprucefield, County Antrim, while a bigger 5 kilogram chunk landed on a farm in Bovedy. By sheer bad luck Patrick had missed out on seeing the brightest meteor of his life by a matter of minutes. He would certainly have seen the flash and the trail as he was using his 8.5 inch Newtonian in that unusual observatory which had sides and a roof incorporating panes of glass. He was not a happy man when he heard from other astronomers in the BAA about what he had missed.

On May 18th 1969 yet another successful Saturn V launch propelled astronauts Tom Stafford, Eugene Cernan and John Young en route to the Moon. Apollo 10 would, like the Apollo 8 mission, take men around the Moon and back to the Earth, but this time two astronauts (Stafford and Cernan) would descend to within 10 miles of the lunar surface in the Lunar Module. They would do everything that a manned lunar landing mission would do, except actually land. A prime objective of the mission was, of course, to photograph the lunar surface in great detail, especially the planned landing site of Apollo 11 in the Sea of Tranquillity. Patrick was now fast becoming one of the most well-known faces on BBC 1 as the public became increasingly fascinated by the Moon missions. The BBC coverage of the Apollo 10 mission was comprehensive. The anchor man for all these broadcasts was, again, James Burke and the BBC's space correspondent was, invariably, Reginald Turnill, usually at the Cape Kennedy launch site or mission control in Houston. Patrick was, of course, the Moon expert, but was increasingly called upon whenever a precise amount of fill-in time was needed. Patrick was used to broadcasting live without any form of autocue and was in a different league to anyone else on the television when it came to talking for a specific duration. He already had 12 years of live TV broadcasting behind him. Patrick's infectious enthusiasm, and his eccentric but friendly manner, also made for compulsive viewing. The Apollo 10 capsule splashed down safely on May 26th, after a totally successful mission. The first manned Moon landing mission was set for July.

June 3rd 1969 marked the 200th anniversary of the observation of the transit of Venus, which had been observed, in 1769, by Captain Cook from Tahiti. Patrick thought that it would be a great idea to celebrate this on *The Sky at Night* by doing a broadcast from Tahiti itself (part of the island is actually called Point Venus). Strangely, the BBC thought Patrick's ambitious plan to take the whole *Sky at Night* team to Tahiti was rather expensive. Instead, they would send him to the less glamorous millionaire's playground (I jest!) of Much Hoole in Cheshire, where in 1639

Jeremiah Horrocks had made the first such observation. Oh well, I suppose you can't win them all! Mind you, that particular programme would stick in Patrick's memory. An hour before they were due to go on air they could not find *The Sky at Night* theme music record, always played with the opening credits. Remarkably they pressed a Henry VIII vintage harpsichord into service and Patrick played 'At the Castle Gate' on the instrument as the programme started! Thirty-five years later Patrick would eventually have his big transit day when the next Transit of Venus occurred, observed from his own back garden.

During the whole Apollo mission period there was no let up in Patrick's own scrutiny of the lunar surface for TLP. The Moon-mapping era was in the distant past but Patrick remained fascinated by the thought that the Moon was still, in some way, active. Observing alone at Selsey, on June 30th 1969, he detected an orange area over the SE wall of the brilliant crater Aristarchus, 'covering the wall and extending beyond'. His colour Moonblink device showed an anomaly, but outside of the 23.37–23.58 GMT time period it was, essentially, back to its normal appearance.

Chapter 14
Apollo 11

As far as his writing work was concerned Patrick's *Moon Flight Atlas* was now complete, except for one detail. The publishers wanted to include some facts about the first Apollo mission to the Moon in the atlas, and with the lunar landing now imminent it was decided to publish the atlas immediately after Apollo 11. The reasoning behind this was that a successful Moon landing would generate so much good publicity that the atlas would sell in droves; a disaster could result in a total rethink/shelving of the book. So, with everything complete and written up, Patrick and the world waited for Apollo 11. The launch date was scheduled for July 16th and touchdown on the Moon for the afternoon (in the USA) of July 20th. The idea was to get the maximum US viewing audience for the lunar landing, so touchdown and the astronauts leaving the capsule were scheduled for the afternoon through to the late evening on a Sunday.

Meanwhile, the Russians, who had started to look distinctly like also-rans ever since Apollo's 8 and 10 had upstaged their Zond 5 and 6 successes, decided to launch their Luna 15 probe to the Moon on July 13th. It was only 3 days before the proposed launch of Apollo 11. It had become obvious that they did not have anything to compare with the NASA Saturn V rocket and they had decided that the option of sending a single cosmonaut in a Zond probe (a much lighter payload when food, oxygen and water were considered) was too risky. They had one potential show-stopper left: they might still be able to grab a sample of Moon rock and bring it back to Earth before Apollo 11 returned.

Three days before Patrick's intensive Apollo 11 coverage was due to start the BBC dispatched him to Moscow with instructions to 'find out what's going on'. It should be said that, in all Patrick's dealings with the Soviet Union in the 1960s, he retained his lifelong philosophy of always backing anyone who had shown him support and kindness, as the Russians had. So Patrick always scoffed at the idea that Kozyrev's TLP spectra were anything other than proof of TLP (they were rarely taken seriously by anyone else), he always rubbished reports of cover-ups of horrendous Soviet rocket launch disasters (they happened) and he denied totally that Luna 15 was designed to steal the show from Apollo 11 (it was). On arriving in Moscow on July 14th,

M. Mobberley, *It Came From Outer Space Wearing an RAF Blazer!: A Fan's Biography of Sir Patrick Moore*, DOI 10.1007/978-3-319-00609-3_14,

Patrick found that the BBC's Soviet correspondents could shed little light on the Luna 15 mission. However, he had a stroke of luck. A conference of international astronomers was in progress at Moscow University and as a member of the International Astronomical Union he was able to wangle his way in. Once inside he found three Soviet astronomers, all speaking fluent English, who were able to categorically deny that the Luna 15 mission was anything other than a routine Luna mission, as it had been planned many months in advance. The launch timing, 3 days before that of Apollo 11, was 'purely coincidental'. Accompanied by a cameraman and a sound recordist, Patrick returned triumphantly to the BBC with an interview with the Soviet scientists, whereas all other journalists had failed. However, Patrick appeared to be the only person in the UK who believed the Soviets. But that was Patrick; he would always back his friends 100 %, but if they turned against him, he would NEVER forgive them, FOR LIFE! As events turned out, Luna 15 would be a failure anyway.

For the USA a successful Moon landing was far more prestigious than just beating the Soviets. It is worth remembering that there was so much dissent, rioting and unrest in the US at this time that *TIME* magazine had considered, in late 1968, running 'Dissent' as its cover story. The images from Apollo 8 of the blue serene globe of the Earth had made people much more aware of their spiritual side and of just how precious the Earth was. *TIME* magazine decided to change its tone; there was good news now, set against the riots and the Vietnam War body count, which had reached 40,000. The Kennedys were also in the news again. The killer of Bobby Kennedy had just been sentenced to death and Edward Kennedy was about to drive into the river Chappaquiddick, swim free, but fail to inform the authorities that 28 year-old Mary Jo Kopechne was trapped inside. In the UK, things were not much better in 1969: British troops were on the streets of Belfast for the first time (Patrick had left Northern Ireland at exactly the right time) and the full extent of the Kray twins East End reign of terror was revealed in their trial. They were sentenced to 30 years apiece. But, looking on the bright side, Concorde completed her successful maiden flight.

In the UK, the BBC and ITV coverage of the Apollo 11 mission would be intense, with broadcasts throughout the day, every day, whenever there was a live broadcast from the spacecraft, or whenever a mid-course correction was planned. Children's comics even published mission timetables! When Patrick was not on the radio or on TV he was living at home, with his mother, housekeeper (Mrs Denny) and cat in a state of semi-chaos (that's Patrick in a state of semi-chaos, not the cat). The roof was being re-thatched, the house re-wired and the whole house converted to oil-fired central heating; and, all the while, TV and newspaper journalists were phoning him up.

Lift Off!

On July 16th 1969, at 2.32 p.m. British Summer Time, the five main engines of another Saturn V rocket ignited and Apollo 11 was on its way to the Moon. Unlike the Space Shuttle launches of recent times the thrust to weight ratio for the Saturn

V's at lift-off was fairly close to one, so they seemed to take an eternity to lift off and clear the escape tower, enhancing the drama to feverish levels. To me, there has never been any more dramatic sight than a Saturn V slowly rising off the launch pad, amidst a sea of flame and smoke, the accompanying NASA commentary and the beep-beep of the communication system.

Your author well remembers that launch day. My parents had asked the village school headmaster if I could have the afternoon off to watch the launch live on BBC1. I was 11 and had even given a presentation at the school about the planned Moon mission. It was the school sports day, and I was not interested in sport, so was not involved in any way. Even as a timid and very shy 11 year old I thought that it was a reasonable request to make to the school headmaster. The miserable sod said no and I missed a moment in history. Of course, knowing what I do about the world, adults and teachers now, I would have taken the day off sick, and then booted the old git firmly in the gentleman's vegetables region at the next school football match. When you are young you assume teachers and adults have to be obeyed, but when you grow up you realise that many of them are just total hypocrites. I was resigned to watching the launch, later that night, on BBC1. There were no domestic video recorders in 1969 and our TV was a black & white model, though there were a few people with colour sets. To this day I can remember sitting on the sports field with a group of friends, counting down to the launch on my wristwatch and wishing desperately that I was near a TV set and that the headmaster was dead! What on Earth a teacher might gain from denying a pupil such a request, on such a momentous day, is beyond me. Indeed, it must rank as sheer incompetence for any teacher to schedule a sports day on such a day in history. My opinion of teachers has been very low ever since, unless that is they too are amateur astronomers. I told Patrick, a former teacher himself, about this incident and he was appalled. He said it wouldn't have happened if he had still been a teacher. He'd have made sure his entire class watched it on TV.

The Apollo 11 astronauts' names are etched into history: Neil Armstrong, Edwin (Buzz) Aldrin and Mike Collins. Armstrong and Aldrin were destined to land on the lunar surface, while Mike Collins would have the lonely wait in the Command Module, orbiting the Moon, while waiting for Armstrong and Aldrin to return. As we have seen, throughout the Apollo lunar missions Patrick and James Burke were the main BBC presenters with Burke co-ordinating the dialogue. Other regular studio performers were rocket expert Geoffrey Pardoe, Professor Geoffrey Eglinton, RAF medical expert Squadron-Leader Tony Nicholson, Dr Stuart Agrell and, occasionally, Gene Shoemaker the lunar geologist (in later years, he and his wife would become prolific comet discoverers). The BBC's Space Correspondent Reginald Turnill was, almost always, in mission control at Houston.

As the Apollo 11 spacecraft neared the Moon the Soviet Luna 15 probe was already in lunar orbit and there was considerable speculation as to whether there would be a conflict with the radio frequencies used. The Soviets (and Patrick) were adamant that this would be the last thing the USSR wanted and there were no problems experienced that could definitely be attributed to Luna 15; but there were certainly a few communication problems. In the *Radio Times* published on July 17th Patrick was given an entire page, dramatically entitled: 'What I believe they'll find'.

Unfortunately, this repeated the view that he believed the lunar craters were volcanic, an opinion that would, ultimately, prove to be totally flawed and very naïve.

Sunday July 20th loomed and BBC and ITV space correspondents were both 'on air' permanently from 6 p.m. Late July and early August are always the hottest times of the year in the UK and 1969 was no exception. Under the studio lights and the tension everyone was sweating profusely. This was made worse by the fact that behind Patrick was a 6 foot diameter papier-maché globe of the Moon, suspended by a single wire from the studio ceiling. With the air conditioning fans switched on, the globe swung violently; so the air conditioning was switched off! The news bulletins that evening were little different to the Apollo coverage; everything was about Apollo 11. The whole world was watching the TV and if you were British and hadn't heard of Patrick Moore before July 1969, you certainly had afterwards! Destiny had made July 20th 1969 Patrick's biggest moment, with nearly 20 million viewers watching BBC1. As the time ticked by, the critical hour approached: the landing of the Lunar Module on the lunar surface. The landing site chosen for Apollo 11 was in the Mare Tranquillitatis or Sea of Tranquillity, not far from the crater Maskelyne and the Hypatia rille. This was a relatively smooth area, but one easily recognisable from a Lunar module pilot's viewpoint. Patrick had already been describing the landing site in the Sea of Tranquillity, and what the Moon might look like from the astronauts' perspective, for nearly 3 hours on live TV, with his colleague James Burke scrabbling around for different things to say and different aspects of the mission to look into. Other experts came and went, talking about everything from how the Lunar Module works, to the effects of weightlessness. As I say, Patrick had been behind the desk for several hours now, but if all went well he would still be there at breakfast time! Live TV was something Patrick was used to, but not 12 continuous hours of live TV. Only 6 days earlier he had been flying to Moscow.

Descent to the Moon

From the UK, at the precise time of the landing, the Moon was at an altitude of 15 degrees in the SW evening sky, not far from the bright star Spica. Our nearest neighbour in space was approaching a half-phase, so the lunar terrain at the landing site would be illuminated, but at a low angle, so the landscape would stand out well and small craters could be avoided. No-one wanted the Lunar Module to tip over. In the orbit prior to the landing Armstrong and Aldrin had separated their Lunar Module from Mike Collins' Command Module and in the final orbit before the descent both craft would disappear behind the Moon. The Command Module would be flying around the Moon at an altitude of 60 nautical miles, but the Lunar Module would be below it, at a height of only 10 miles, as both spacecraft reappeared around the lunar limb. From the moment that radio contact was established between Earth and the Lunar Module, as it came around from the far side, there would only be an incredibly scant *30 minutes* to touchdown! This was because the spacecraft was moving over the lunar surface at some 3,700 miles per hour while in orbit and the landing site was barely 1,300 miles from the lunar limb, as seen from Earth. From the firing of the descent engine some 16 minutes after rounding the limb, there would be only 12½ minutes until the landing!

The Lunar Module onboard guidance computer, which would do almost everything except the final landing (which required human judgement and common sense) was remarkably advanced for 1969, but its memory capacity and operating speed were a fraction of that seen in today's PCs. Forget multi GHz processors; a few 100 kHz was state-of-the-art! In terms of memory the Lunar Module Guidance Computer had 36 K × 15 bits of program memory (think of this as fulfilling the role of an un-eraseable hard disk) and 2 K of RAM! Could *Microsoft Windows* land a spacecraft on the Moon with a 36 K hard disk and 2 K of RAM? No, it could not even talk to the mouse! In addition, each byte of RAM memory was often being used to store maybe half a dozen variables depending on which tasks were being handled at one time. If an alarm was generated by the computer, it would mean that the computer was overloaded. It would then reboot (Gulp!) but, unlike a modern PC, would come back on line almost instantly with the safety critical features still running and less important tasks (such as radar information) switched off. Needless to say, this reboot feature had been tested to exhaustion in the simulator on Earth, before the Moon landing. Unlike in a modern PC this system was foolproof. However, the alarms that were to come in the next few minutes caused some heart stopping moments at Mission Control, even if the commentators in the BBC studio had not got a clue what they actually signified.

So, imagine, if you will, the scene. It is a quarter to nine on a Sunday evening, British Summer Time (British Standard Time in that year). Some 600 million people are watching the Apollo TV coverage worldwide, 20 million of them are watching BBC1, with the vast majority having Black & White TVs. You have no idea if, in the next half hour, a huge event in world history is about to take place, or a tragedy. From Patrick's perspective, what you do know is that 20 million people are hanging on your every word because, back in 1956, you were on a programme about Flying Saucers, a comet called Arend-Roland came along and the Russians launched Sputnik 1, and also because of the belief of that first *Sky at Night* producer, Paul Johnstone, in your broadcasting abilities. Suddenly, a crackle of static announces that the Command Module (containing Collins) and then the Lunar Module (containing Armstrong and Aldrin) have both reappeared around the Moon's limb. Your colleague, James Burke, explains what is happening and invites TV viewers to listen to the commentary.

It is now 8.50 p.m. BST and everyone is listening to the NASA commentary and sometimes to the TV commentators, although most informed watchers would prefer it if the TV commentators would actually keep quiet! Your author can remember vividly sitting in the living room with his parents, transfixed by the TV commentary and hanging on every word from NASA, Patrick and James Burke. A big digital clock (with each digit on a metal plate, because LCD or LED displays had not been invented) in the BBC studio shows that Apollo 11 is now nearly 102 hours and 20 minutes into its mission, since the launch on July 16th. There still seem to be a few communications problems though. James Burke and Patrick are shown straining to make sense of the dialogue between mission control and the astronauts, interspersed with the standard NASA double beep and static. During this time the astronauts are also talking to Mike Collins, way above them in the Command Module, who seems to have a better radio link to Earth and can, when required, relay messages to them. Armstrong and Aldrin are also recognising landmarks on the lunar surface and relaying this back to Mission Control and they are adjusting the yaw axis of the spacecraft to try to improve the communication. Finally, they are going through the mission checklist, ready for the Descent Engine burn PDI (standing for Powered Descent Initiation), which will soon slow them down and lower them to the lunar surface. Their landing footprint is an oval, only 10 miles long by 3 miles wide. Ten miles is equivalent to a delay in firing the descent engine of only 10 seconds! Despite communication anomalies the checklist is completed and Armstrong and Aldrin are given a go for PDI; to descend the last 50,000 feet (just over 15 kilometres) to the lunar surface. The engine is fired just as the BBC studio clock is reaching 9.05 p.m. At this point they are just 250 nautical miles from the landing site, from a horizontal perspective. The next 12½ minutes are just constant exchanges of altitude and descent information between Charlie Duke, the capsule communicator in Houston (CAPCOM) and Armstrong and Aldrin.

At 9.10 p.m. BST Armstrong reports the first Lunar Module Computer Program Alarm, with just a hint of urgency in his voice. He qualifies it as a type "1202". All over the world, TV viewers ears prick at the word "Alarm"...fingernails get even more bitten! Within 30 seconds a 26 year old software expert in Mission Control, named Steve Bales, has worked out that the onboard computer is simply being overloaded by a flood of radar data, but is otherwise OK. Charlie Duke tells the crew "We're Go on that alarm". Everyone breathes a sigh of relief and the descent continues. A few minutes later they are only 3,000 feet above the lunar surface and descending at 70 feet per second when Aldrin reports a 1201 alarm. Almost instantly, Duke replies with "We're Go. Same Type. We're Go". Coming up to 9.15 p.m. and with the Lunar Module only 600 feet above the lunar surface, descending at 19 feet per second, Armstrong takes over manually for the actual landing, pitching the Lunar Module so that it is then virtually upright and ready to land. For the next 2 minutes Aldrin just calls out height and descent velocity from the radar.

Tranquillity Base

The last 30 seconds of dialogue will permanently send shivers down the spine and moisten the eyes of anyone who was watching the TV that night:

> Aldrin: "40 feet, down two and a half. Picking up some dust. 30 feet, two and a half down. Four forward. Four forward. Drifting to the right a little. 20 feet, down a half."
> Duke: "30 seconds". (This meant 30 seconds to the point where you have to land in 20 more seconds, or abort because of lack of fuel).
> Aldrin: "Drifting forward just a little bit; that's good. Contact light." (This meant that the 1.7 metre long probes on the Lunar Module footpads had touched the Moon!)

Aldrin: "Okay, Engine Stop. ACA out of Detent".
Armstrong: "Out of Detent. Auto."
Aldrin: "Mode Control, both Auto. Descent Engine Command Override Off, Engine Arm Off. 413 is in."
Duke: "We copy you down, Eagle."
Armstrong: "Houston, Tranquillity Base here. The Eagle has landed."
Duke: "Roger Tranquillity, we copy you on the ground. You got a bunch of guys about to turn blue. We're breathing again. Thanks a lot."

It was 9.17 p.m. BST in the UK, on July 20th 1969, and it was, arguably, the biggest ever moment in history. Patrick, everyone in the studio, and at least 600 million TV viewers on Earth, breathed a huge sigh of relief. Patrick would later admit that, up to the moment that the Lunar Module had landed, he was still worried about Gold's lunar dust theory of the 1950s. What if the Lunar Module really did sink without trace? What if the unmanned probes that had previously landed on the Moon had just been lucky? They had not; the whole of the lunar surface was rigid. But for Patrick the night had only just begun. Assuming the post-landing checks looked OK and the spacecraft was not tilting, Armstrong and Aldrin would be stepping out onto the lunar surface in about 6 hours time, hopefully before people went to bed in the USA. For the British viewer this would mean a very long night, well into the early hours. Again, there was no point setting the video recorder: they did not exist; anyway, people wanted to see this live! I well remember, as if it was yesterday, my parents and I waiting to hear when the Moonwalk would take place, then going to bed, and my father setting the alarm for 3 a.m., when we would all get up again. The whole thing, from my perspective as an 11 year old child, mad keen on space and astronomy, seemed totally unreal. Here, in front of our eyes, science fiction was becoming science fact. If aliens had just landed it could not have been more dramatic!

In the BBC TV studio Patrick told James Burke that he did not believe the Apollo 11 astronauts would take their planned sleep period. He was sure they would just prepare to go out and they would have to stay 'on air'. Burke informed the programme controllers of this, but they were sceptical. However, Burke had total faith in Patrick's judgement and he bet the producer a bottle of champagne that Armstrong and Aldrin would go out without any sleep. Patrick, of course, was right and so the presenters Burke and Moore won that bottle of champagne.

One Small Step

At 3.00 a.m. my parents and I got back out of bed and trudged down to the living room again. As the TV set warmed up (the valves took ages in those days), I distinctly remember Patrick's BBC tones being the first thing I heard. All was apparently still going well. It is incredible to think that Patrick had then been 'on air' for 9 hours and still had things to say! Armstrong and Aldrin had, as Patrick predicted, opted to scrub the 4 hours rest period scheduled in the flight plan and had just gone straight ahead with the meticulous checklist leading to their planned 135 minutes moonwalk. How could anyone be expected to rest when they were sitting on the surface of the Moon? In the cramped confines of the Lunar Module two astronauts kitting up with their bulky spacesuits and backpacks, and then depressurising the spacecraft, took much longer than expected. Nevertheless, by just after 3.30 a.m. BST the spacesuits were on and working and the Lunar Module was depressurised. It was time to open the hatch. In years to come there would be much speculation as to who should have stepped onto the Moon first. Armstrong was the mission commander, but traditionally the seat occupied by Aldrin, the Lunar Module pilot, would have meant he was the first man out. Technically, Aldrin was the most highly qualified member of the crew, with both a B.S. from the U.S. Military Academy in 1951 and a Ph.D. in astronautics from M.I.T. in 1963. But Armstrong, despite being a very quiet and uncommunicative man, was deemed to be the coolest under pressure. So Armstrong would be the first man to step onto the Moon, but strangely the only decent photographs of a man on the Moon, during the Apollo 11 mission, were those taken of Aldrin, by Armstrong!

Patrick met both Neil Armstrong and Buzz Aldrin and Buzz became a good friend in the years after Apollo. Patrick always stressed that Aldrin avoided taking pictures of Armstrong because he was too busy "that's all there was to it". But there were other opinions. However, in the years to come, Aldrin was undoubtedly the best publiciser of the Apollo missions, and of NASA, whereas Armstrong would avoid publicity. He was a shy and modest man and refused to sign autographs for huge fees, like some of the other astronauts. Aldrin was also a vocal supporter of Patrick and his achievements. Anyway, we have digressed. At around 3.50 a.m. BST on July 21st 1969, UK viewers saw the first live TV pictures from the camera on the leg of the Lunar Module. They were highly contrasty (soot and whitewash) pictures. The sky was jet black and the Moon was bleached white and the pictures were, initially, upside-down. At this point I turned myself upside down in my parents' living room chair, to try and make sense of the picture, but slowly the shape of Neil Armstrong was seen coming down the Lunar Module ladder.

After a few attempts by Armstrong to see how easy it was to haul himself back from the Lunar Module foot pad to the ladder, when fully kitted up, he was ready. So, at 3.56 a.m. BST, he stepped onto the Moon with the immortal words "That's one small step for man: one giant leap for mankind". As had been the case only hours earlier, the faces in the BBC studio were ones of immense pride, with people barely able to contain their emotions; what a moment in history! As soon as Armstrong had stepped onto the surface James Burke said: "Timed for us, this morning at 3:56" and Patrick's first words, looking at the stark black & white TV pictures were: "There we have a daylight scene, but on the airless Moon of course, a black sky, because on the Moon the sky is always black".

For Patrick it must have been an especially indescribable moment. He had been observing the Moon since he was a child, had been on air when the first pictures from the far side arrived in 1959, and now, aged 46, he was watching a man walk on the surface of the Moon, while almost 30 million British viewers listened to his commentary! Minutes later James Burke said "Well, this is a moment, if there ever was a moment, for Patrick Moore.........Patrick...." Patrick said "Well, I really feel absolutely overcome. I've been living with this idea all my life and now that it's actually happened I can hardly believe it. No admiration can be too great for those magnificent men who brought this strange spidery module down on the Moon, and this obviously is a moment that humanity is never going to forget...."

There were, of course, more moments of drama to come and Patrick would virtually live in Studio 7 at the BBC TV centre for another 3 days, covering the heart-stopping lift-off from the Moon (one engine, no back up and a felt tip pen was used to flip a broken switch!), then the docking with the Command Module, the return to Earth and, finally, re-entry and splashdown. While on the lunar surface they received a live telephone call from President Nixon and unveiled a plaque which read: "Here men from the planet Earth first set foot upon the Moon, July 1969 A.D. We came in peace for all mankind". Twenty-two kilograms of Moon rock were returned to the Earth.

So, in all this time, what had happened to the Soviet Luna 15 probe? Despite Patrick's denials, the obvious intent of the Soviets was to steal Apollo 11's glory, by returning moonrock to the Earth first. In fact, at the time of the Apollo 11 lunar landing, Luna 15 had already been orbiting the Moon for 3 days. Two hours before the Apollo 11 Lunar Module was due to lift off from the Moon, Luna 15, on its 52nd lunar orbit, began to descend towards the Mare Crisium, a region only a few hundred miles from the Apollo 11 landing site; however, the landing phase went wrong, and 4 minutes into the descent the spacecraft crash-landed.

Apollo 11 splashed down on Thursday July 24th at 17:50 BST, in the Pacific Ocean, just 15 miles from the recovery ship the USS Hornet, and the BBC coverage showed the three astronauts being taken in 'Biological Isolation Garments' (NASA were worried about importing germs from the Moon) from the helicopter to a quarantine trailer. One of Patrick's final comments was: "I doubt whether historians of the far future will remember 1066, 1914 or 1939, but they will all remember the 21st of July 1969."

After more than 8 days of almost continuous TV coverage, following his breakneck trip to Moscow, Patrick was exhausted, as well as being, arguably, the most famous face in Britain at that time.

Chapter 15
A Bestselling Author, Nutters and More Moon Landings

Bestselling Books

With the Apollo 11 flight being a 100 % success, Patrick's *Moon Flight Atlas* had the green light. Within days of the mission ending the publishers told him they were planning a print run as soon as he could supply some extra data on Apollo 11 and as soon as they received publicity photographs from NASA. Just 6 weeks later the initial 50,000 copies of Patrick's *Moon Flight Atlas* rolled off the presses. The first editions were in English, but, eventually, it was published in nine different languages. Patrick had never had this many copies of a single book printed before, but this was only the start! Even before the first print run the back orders exceeded 50,000! Patrick had pessimistically thought that it might only sell 10,000 copies when the idea was first mooted, as that was typical for most of the books he had written so far, but Mitchell Beazley had been right. This was a blockbuster! By the launch of Apollo 12, on November 14th, the *Moon Flight Atlas* had been reprinted three times; there were now 200,000 copies in circulation and they were still in demand! Patrick was now a bestselling author; the money was rolling in! But Patrick was still not a man to take things for granted. He would still continue to use his 1908 typewriter and 1953 Ford Prefect car. He knew that fame and money were fickle things and he wanted to maintain his way of life for ever; buying a flashy car and spending money was not the way to do this and simply not his style. In addition, the work needed to renovate *Farthings* was going to be costly. It needed a completely new central heating system for a start, as the wood burning stove in the kitchen, which had heated the water up to 1969, frequently had mini-explosions which filled the kitchen area with black smoke. The serious money was arriving at just the right time!

Ultimately the *Moon Flight Atlas* sold three quarters of a million copies, which must have netted Patrick a nice amount of money and given him a financial security he could only have dreamed of when he moved from Armagh to Selsey. Remember, he claimed he only had two shillings and sixpence in his bank account!

M. Mobberley, *It Came From Outer Space Wearing an RAF Blazer!: A Fan's Biography of Sir Patrick Moore*, DOI 10.1007/978-3-319-00609-3_15,

However, when I hinted at this to Patrick he denied that these huge book sales made him a rich man. "Martin, you have to remember what damage Harold Wilson inflicted. Under his super tax rates I was paying 83 % tax on most of that. Heath wasn't any better...WORSE, he wanted us to get into bed with the Frogs and the Hun. Maggie was the only leader who really lowered taxes and kept Brussels noses out of our business." Strangely, Patrick sometimes said he quite liked Harold Wilson (!), but I think that was just because he despised Ted Heath's pro-European stance even more. Yet, in many ways, the octogenarian bachelor and musician Heath was quite similar to Patrick in his older years! As well as the 83 % super taxes of that era, the 1970s would be a period of high UK inflation too, with the pound buying less than a third, in 1980, than it did in 1970; and Patrick had no interest in investments and finance whatsoever. Even so, his huge book sales during that era meant that he now had no financial worries.

Only days after Patrick had finished his Apollo 11 coverage he was off to Pasadena in California to review the results from the Mars probes Mariner's 6 and 7. A few newspapers showed Patrick enjoying his trip to the USA and wearing a Stetson hat while smoking his pipe. The Mariner results showed a world very much like the Moon, rather than a place for life; it was heavily cratered. One of the principal investigators told Patrick: "It's all very fine. We work out an experiment which is every bit as hard as reaching the Moon. We pull it off and we manage to get pictures of superb quality – much better than we'd dared to hope. And what do we see? Darned Craters!"

The success of Patrick's *Moon Flight Atlas* had also given the green light to another Mitchell Beazley/George Philip project too, the *Atlas of the Universe*; an ambitious title for a truly stupendous book. While huge, glossy, lavishly illustrated books are commonplace enough nowadays, they were almost unheard of in 1970 when this monster was published. The *Atlas of the Universe* would follow the same standard as the *Moon Flight Atlas*, namely quality paper, big pages and the best colour photographs and illustrations, but in a much bigger way. The pages would be a massive 38 × 28 centimetres in size and there would be over 270 of them. With the paper used being almost as thick as card, this would make the book more than 3 centimetres thick and a real strain to pick up! The cover was so thick it could have been made from plywood! The *Atlas of the Universe* was, essentially, five atlases in one: an Atlas of Observation & Exploration of Space; an Atlas of the Earth from Space; an Atlas of the Moon and its exploration; an Atlas of the Solar System; and an Atlas of the Stars. With a foreword by Sir Bernard Lovell of Jodrell Bank and an epilogue by Dr Tom Paine, the NASA Administrator, this was a prestigious book indeed. Also, at 10 pounds, in 1970 it was very expensive; but it sold in huge numbers. It appeared that, just after Apollo 11, the public could not get enough of space and astronomy. Patrick's *Atlas of the Universe* would ultimately sell over 250,000 copies! None of his other books, before or after would come anywhere near these huge Atlas sales. Indeed, it would not be until

Stephen Hawking's *Brief History of Time* that these sales figures were surpassed by a British astronomer.

Away from Apollo Moon landings and huge coffee table books, the monthly *Sky at Night* programs continued unabated. On September 9th 1969, Patrick invited Professor Samuel Tolansky to explain, in laymans terms, the theory of relativity on the programme. The explanation was so good that the BBC was deluged with protests from schoolboys whose parents had refused allowing them up as late as 11 p.m. to watch it. There were, of course, no VCR's in 1969. The BBC bowed to the demand and repeated it on a Saturday afternoon!

For the October *Sky at Night* Patrick decided to devote a programme to amateur variable star observers. He visited the observatory of J.S. Glasby, the Director of the BAA's variable star section, who owned a 13 inch reflector at Stevenston, Ayrshire, and then to the observatory of teenager Julian Jones, at Astley Abbotts, near Bridgnorth in Shropshire, who was a keen young observer.

Fig. 15.1 Patrick and most of the BAA Council for the 1968/1969 BAA Council year. *Front row*, *left* to *right*: Rev. Cameron Dinwoodie; Ernest Beet; Gordon Taylor (BAA President); E.P. Duggan; David G Hinds (on *lower step*); Gilbert Satterthwaite; William A. Granger (with camera, whose wife had a 13 inch waist and was in the Guinness Book of Records!). *Second row*, *left* to *right*: Cdr Henry Hatfield (legendary lunar photographer); Neville Goodman; Colin Ronan (Journal Editor); Vince Barocas; A.W. Vince. *Third row*, *left* to *right*: Howard Miles (above/next to Cdr Hatfield); Alfred Curtis (with glasses, the Winchester Weekend organizer); William Baxter (Solar Section Director); A.K. Appleton; W.E. Fox (Jupiter Section Director). *Fourth row*, *left* to *right*: Horace Dall (short with glasses, an optical genius); J. Leslie White; Jim Hysom (black hair, mirror maker for Astronomical Equipment Ltd.). *Fifth row*, *left* to *right*: E.H. Collinson (Mars Director); Gladys Stone; Keith Hindley (Meteor Director); J.S. Glasby (Variable Star Section Director); Patrick Moore; Ron Maddison (Lunar Section Director); Stan Milbourn (Comet Section Director) (BAA archive photograph by the late Cdr. Henry Hatfield (1921–2010), taken on 1969 April 30th)

Fig. 15.2 Patrick's Ford Prefect 'The Ark' and his MG Magnette in his side driveway at Selsey in 1969 (Photographer unknown. Copied with Patrick's permission)

Independent Thinkers

Despite Patrick's obvious abilities at broadcasting science fact he was still fascinated by the people he referred to as "Independent Thinkers", or when off-air, "cranks", "nuts", and people who were "not sailing with a full crew". It might be thought that after probes had been sent to the Moon, and certainly after men had *walked* on the Moon, the likes of *The Flat Earth Society* would have thrown in the towel, but not a chance! In 1969 Patrick spent many weeks filming extraordinary interviews with a variety of people for a BBC 2 programme entitled 'One Pair of Eyes'. In this programme he met Flat Earth guru Samuel Shenton, a man who was still 100 % convinced that the Earth was as flat as a pancake; he was the secretary of *The Flat Earth Society*, and could easily answer all of Patrick's questions. For example:

> Patrick: "Aircraft can circumnavigate the Earth. Fly east, or west, and eventually you will come back to your starting point; this couldn't happen with a flat world."
> Shenton: "Of course it could. The Earth is shaped like a gramophone record and all you have done is to complete a circuit round the central North Pole."

Another guest on the programme was Mr John Bradbury, a chiropodist from Ashton-under-Lyme and the inventor of a telescope which, he claimed, enabled him to observe the stony metallic casing of the Universe. Mr Bradbury's Earth was not flat: it had a flat top and a convex bottom; the North Pole lies in the middle and so all directions (according to Mr Bradbury) are south! In the Bradbury Universe the

Moon was 220 miles away, slightly convex and made of carbon. The Moon's changing phases were due to the carbon Moon accumulating plasticine phosphorus circulating above the liquid-air zone! Needless to say, Patrick managed to put Messrs Bradbury and Shenton in touch. The resulting correspondence must have been fascinating! Mr Bradbury was obviously a deep thinker as after much research and deliberation he confidently declared the following: "Light is nothing more than darkness, lit up"; surely, a man of great wisdom!

Patrick also visited the Reverend P.H. Francis at Stoughton Vicarage in Sussex; a mathematics graduate of Cambridge University, he believed that the Sun was cold and that "the notion of an incandescent Sun is childish, and not worthy of a grown-up person". During the interview, this expert on mathematics, heat and cold, suggested the vicarage drawing-room fire needed livening up and so poured a *gallon* of paraffin on the coals. Fortunately the Sussex fire brigade responded quickly and the vicarage was just saved. Patrick also visited the Aetherius Society's HQ in London's Fulham Road, where a Mr Robertson informed him that they were just being invaded by Fish-Men from the other side of the Galaxy. He also confirmed to Patrick that the beings living on Saturn were perfectly spherical; in fact, they were "complete balls".

The grand finale of the TV programme was provided by Mr Bernard Byron of Romford, England. Mr Byron was, by his own admission, one of the few Earthmen privileged to be able to speak, and write, languages of other planets. He was fluent in Venusian, Plutonian and Krugerian (a planet going around the star Kruger 60!). The sight and sound of Mr Byron (who seemed to lack a competent dentist on his home planet) speaking in his alien tongues, while Patrick, hair standing on end, watched in fascination, was one of the programmes enduring moments!

Apollo 12

Unfortunately, Patrick had to return to the real world. Nevertheless, in 1969 it was a very exciting world to live in. On November 14th Patrick and James Burke were back in BBC Studio 7 for the Apollo 12 mission, the second manned mission to the Moon. With them in the studio were Geoffrey Pardoe and ex-NASA astronaut Curt Michel. As the clock marked 4.22 p.m. in the UK, Apollo 12 thundered off the launch pad at Cape Kennedy, bound for the Moon's Oceanus Procellarum, or Ocean of Storms. This was an apt destination, for at the time of the launch Cape Kennedy was in the middle of a heavy rain storm. Although the rain was intense, the winds (crucial to the launch) were light and there were no reports of thunderstorms in the area. It was 11.22 a.m. in Florida. Aboard the Apollo spacecraft were the astronauts Pete Conrad, Dick Gordon and Alan Bean. All was going well until, suddenly, 30 seconds into the launch, a bolt of lightning forked out of the clouds, struck the massive Saturn V rocket and then continued down the vapour trail, hitting the launch pad at the Cape too! That it was a lightning strike was confirmed later, but the first the crew knew about it was that rows of warning lights came on in the capsule; the ground controllers just saw garbled telemetry coming back. Fortunately the spacecraft was not damaged and a flight controller called John Aaron advised the crew to switch the signal conditioning equipment to the auxiliary power supply, which restored the telemetry and saved the mission. A 2 hour check of all the systems, while in Earth orbit, revealed the spacecraft to be OK.

The landing site for Apollo 12 was the same as that for the NASA moon probe Surveyor 3, which had soft landed there 31 months earlier. Incredibly, after travelling all the way to the Moon, the Apollo 12 Lunar Module landed only 180 metres from the spacecraft. The astronauts brought back the TV camera and other parts of the Surveyor 3 spacecraft for analysis. Unfortunately for Patrick and the BBC team the Apollo 12 Lunar Module TV camera was pointed at the Sun almost as soon as it was switched on, so there were no pictures. The whole of the lunar surface TV commentary had to be done 'blind', which was a big disappointment as the colour equipment carried by Apollo 12 was far superior to that on Apollo 11. Nevertheless, this was another big success for NASA and 34 kilograms of moonrock was collected, while six scientific experiments were deployed. It was also another big success for Patrick, as without pictures his live TV ad-libbing skills were called into service even more regularly. As the astronauts prepared to leave lunar orbit they deliberately crashed the lunar module upper stage into the Moon. The Apollo 12 seismometer, only 46 miles from the impact, detected the event, but surprisingly the effects reverberated for 45 minutes. One scientist said that the Moon was "ringing like a bell".

The year 1969 had been truly momentous; not one, but two manned spacecraft had landed on the Moon, four astronauts had walked on the lunar surface, and moonrock had been returned to the Earth. In addition, two NASA Mariner probes had flown past the planet Mars. To celebrate all of this a special half-hour long *Sky at Night* was transmitted on Christmas Eve and much of it was recorded at Patrick's Selsey home, rather than live from London. It was a very memorable programme for all amateur astronomers of that era, mainly because Patrick was shown in his garden. He demonstrated his 12½-inch reflector and its run-off shed and also his unique, octagonal, merry-go-round observatory, housing his smaller 8½-inch reflector. The reader will recall that this latter observatory, with its glass windows, was rescued from a last minute house moving crisis when Patrick had left Armagh during the previous year. In addition, that 1969 Christmas Eve programme showed Patrick entering his *Farthings* study door, sauntering across the bookshelf lined room, and sitting in his favourite rocking chair. There was also a guest appearance by Eric Ilett, the man who made all the animated models for *The Sky at Night*, in that era before computer graphics. Eric was filmed standing in Patrick's study, while demonstrating a hand operated model which he had constructed; this model showed how the Mariner 6 and 7 spacecraft had travelled in a 'transfer orbit' to arrive at Mars that summer. All in all that Christmas Eve *Sky at Night* was quite a classic episode and also a fascinating glimpse into Patrick's observatory-strewn garden and his book-strewn study.

In fact 1969 was the last full year that the *Sky at Night* itself was transmitted live. In the late 1960s colour TV cameras and TV sets were introduced, and by 1970 recordings were almost as good in quality as live coverage. So, to take advantage of all the new techniques and to permit re-takes, live TV was phased out for many programmes, especially for dramas like the police series *Z Cars*, which were traumatic to flawlessly act out without errors. Another factor was the lateness of *The Sky at Night*. Recording a programme before overtime pay hours was cheaper in studio salaries than showing it live at 11 p.m. So, from 1970 onward, Patrick's main live appearances would be during the Apollo missions' coverage.

Life Goes On

Needless to say, away from the Apollo studio Patrick's writing work, BAA duties and planning *The Sky at Night*, continued unabated. The first *Sky at Night* of 1970, a 'special' was about a potentially spectacular comet, discovered by the Japanese observers Tago, Sato and Kosaka in October 1969. Patrick was hoping this really would be a comet to remember, like Arend-Roland, which had ushered in the *Sky at Night* in 1957. He certainly did not want a repeat of his 1965 programme when he had flown above Northern Ireland at the wrong time, failing to spot Ikeya-Seki. Unfortunately, Tago-Sato-Kosaka, while being a good comet for suitably equipped amateurs to find, was a disappointment to the viewing public. Ironically, a few months later, in April 1970, a really obvious comet, called Bennett, would appear and peak at the time of the Apollo 13 mission. Patrick succumbed to the flu in January 1970 and had to miss the first conference on the analysis of the Moon rocks brought back from the Apollo 11 and 12 missions; this was a major frustration for him, although he was able to gather enough information from other delegates to present the major findings on *The Sky at Night.* These findings were that the Moon is about as old as the Earth, namely between 4,500 and 5,000 million years old, there has never been liquid water on the Moon, and there was evidence of very weak lunar magnetism. Also, there was no evidence of life ever existing on the Moon. In addition, no meteorites had been collected and the rocks were essentially basalt. Needless to say, Patrick saw these last two points as evidence that the craters were volcanic, but life is not that simple!

Some Variable Stars

During the late 1960s Patrick had become increasingly interested in making magnitude estimates of bright variable stars, especially those easily visible in binoculars. This had also been a favourite pursuit of his during the War Years, when he was away from home. Undoubtedly, this revived interest had partly arisen because of his interest in George Alcock's two bright novae, namely HR Del in July 1967 and LV Vul in April 1968. Both had become easy binocular targets, indeed technically they were naked eye objects. Patrick had become one of the BAA's most active observers of such bright variables in the late 1960s and, during late 1969, and full-time from

1970, the BAA's Variable Star Section Director, a Mr J.S. Glasby, employed Patrick as the BAA's Binocular Variables coordinator. Patrick quickly drew up a short list of interesting objects within binocular range and published the list in the BAA's February 1970 Journal. For the observational astronomers reading this book, Patrick's list was: VZ Camelopardalis; μ Cephei; AR Cephei; SS Cephei; W Cephei; CH Cygni; P Cygni; EU Delphini; U Delphini; RY Draconis; V Eridani; R Lyrae; W Orionis; RY Ursae Majoris; VW Ursae Majoris. The naked eye targets β Pegasi, α Herculis, ρ Persei, α Cassiopeia, γ Cassiopeia and η Geminorum were also added. A further paper on binocular variables, by Patrick and Michael Ring, was approved for publication in the April Journal.

Patrick agreed to dispatch all the necessary charts for these stars to interested observers and to handle the incoming observations. There were no suitable charts for EU and U Delphini in the BAA collection, so Patrick prepared them himself in early 1970. As with everything he did within the BAA, through his enthusiasm and his legendary status, the Binocular Variables sub-section became highly popular and efficiently organised.

Many of Patrick's magnitude estimates of variable stars, prior to 1967, are only easily accessible through his observing logbooks, which I have studied in detail. However, the BAA records of his observations from 1967 until the late 1980s are easier to research in the association's archives. To be strictly accurate that archive does list a few magnitude estimates by Patrick in the 1990s too: they are all of Betelgeuse, the last one being made on the 30th December 1994. The BAA records list a total of 5,612 magnitude estimates submitted by Patrick during that 27 year period, with the most for a single star (as I mentioned some time ago) being his 700 observations of Alcock's nova HR Del. Those 5,612 magnitude estimates refer to 48 individual stars and are mainly of binocular brightness variables. Some, such as those of Betelgeuse, are indeed naked eye estimates, but he also observed a few very faint variable stars through his three Newtonians. One of the faintest stars studied by Patrick, during the 1970s, was AY Lyrae, which he recorded at magnitudes between 12.2 and 15.5. TZ Per was another, and his almost 200 estimates for that star, from 1967 to 1984, ranged between 11.4 and 15.3. After his 700 estimates of HR Del, his next highest totals in the 1967–1994 era are for the stars SS Cyg (467), R CrB (439), SU Tau (369), U Cyg (304), W Cyg (285), T CrB (267), U Ori (265), RU Peg (252), R Leo (217), R Cyg (206), TZ Per (198), Gamma Cas (170) and Chi Cyg (153).

For completeness I should add that Patrick also submitted many of the same observations of variable stars to the AAVSO (American Association of Variable Star Observers). That database lists his joining date as 20th May 1968, with 2,183 observations being submitted, covering the period 1954 (14 years before he joined the AAVSO) to 1990.

By February 1970 Patrick was fully recovered from his January flu, so when the Japanese observer Honda discovered a bright new nova in Serpens on February 13th, he was keen to observe it. The nova peaked at magnitude 4.5 by the end of the month and during the February 25th BAA meeting at 23 Savile Row in London (only the second meeting held at that venue) Patrick urged members to observe it. On the instruction of the Variable Star Section Director, Patrick created a chart of the region and distributed copies to interested observers. In later years I always used to find it amusing that the BAA London meetings venue, from the 1970s through to the early twenty-first century, was in Savile Row, the home of the traditional men's bespoke tailoring empire. The contrast with this image and Patrick, who usually resembled an unmade bed, even in his blazer and tie, could not have been greater. After BAA council meetings the councillors would invariably exit the RAS building at Burlington House in Piccadilly (opposite Fortnum & Mason's) and turn right, then right again, into the exclusive shopping tunnel known as Burlington Arcade. They would then turn right along Burlington Gardens and then left along the entire length of Savile Row, before arriving at number 23, also known as the Civil Service Scientific Societies meeting venue. Maybe it was just my warped imagination, but as the geeky bunch of BAA councillors, led by the huge form of Patrick, wearing a blazer that hadn't been dry-cleaned in decades, shuffled along Savile Row to each BAA meeting, I had visions of the resident tailors all gaping in horror, and wishing that we would just walk past their premises as quickly as possible, so as not to give Gent's outfitting a bad reputation!

On March 7th 1970 a total solar eclipse would take place across a narrow track down the eastern side of the USA, through northern Florida/southern Georgia, across the Gulf of Mexico and through Mexico itself. Occurring in the USA, this was a major event which inspired many future eclipse chasers. Needless to say, having been turned down regarding his proposal to do a *Sky at Night* from Tahiti in 1969, Patrick had high hopes that the BBC could fund him to travel, perhaps to sunny Mexico, for this eclipse. Patrick had, as we have seen, viewed total solar eclipses from Sweden, Yugoslavia and Siberia, so maybe this would be his fourth. Sadly, the BBC again declined, but suggested doing a live *Sky at Night* 'special' from the Lime Grove studio, with TV pictures beamed in from Mexico and Georgia, via a US TV network. Patrick was joined in the studio by Dr Vinicio Barocas, who was the Director of the Jeremiah Horrocks and Wilfred Hall Observatories in Preston, and also a solar expert. Because the eclipse would take place in the early afternoon in the USA, the programme was broadcast from 17.55 BST in the UK. Totality occurred at 18.32 BST and excellent pictures were transmitted from Mexico to London (drifting clouds caused problems in Georgia). It was another successful *Sky at Night* eclipse, but from Patrick's perspective it would have been so much better if he could have been there, in Mexico.

With yet another NASA Moon mission looming, namely Apollo 13, *The Sky at Night* for late March 1970 covered the issue of photographing the lunar surface. Patrick's friend, and one of the most regular visitors on the *Sky at Night* over the whole of the programme's history, H.J.P. (Douglas!) Arnold discussed the subject of lunar photography. Little did they imagine what traumatic events would unfold over the coming weeks.

However, before the launch of Apollo 13 it was time for Patrick to, once more, be the great attractor at the BAA's annual Winchester Weekend residential course. The fourth of these events was held, as always, at King Alfred's College in Winchester (April 3rd to 5th) and Patrick was joined by his best mates Hedley Robinson, Ron Maddison and Alfred Curtis as the main lecturers. A major attraction at the weekend was Comet Bennett, which was peaking in brightness at around first magnitude in the dawn sky. With the Moon being new on April 6th the sky was Moon-free from the time the comet rose (just before 2 a.m. BST) until twilight started to wreck the morning sky after 4.45 a.m. Friday night/Saturday morning was rather cloudy, but the Saturday night/Sunday morning period was nice and clear. I recall the late Charles Wise, a well known BAA member and eventual Winchester Weekend organiser himself, telling me that he had awoken to see the magnificent Comet Bennett rising above the horizon at Winchester, and so thrown stones at Patrick's bedroom window, to raise him from his slumbers after a good session at the bar the night before! The Winchester residential blocks had keys to allow access to each block, as well as each room, and Charles was in a different block to Patrick, but he knew which room Patrick was in, as seen from the outside. Within minutes of being woken Patrick was outdoors, avidly observing the comet with a variety of home made telescopes that had been brought to the meeting. Not surprisingly, by 4.30 a.m., he was joined by a lot of other course attendees, who had also braved the chill dawn air to see one of the best comets of the twentieth century alongside Patrick. Unlike him, they were shivering, despite being wrapped up. Patrick, as always, was just wearing his blazer and totally oblivious to the temperature!

Houston, We've Gotta Problem

Apollo 13 blasted off at 13:13 hours Central Daylight Time on April 11th 1970 from Cape Kennedy. It was early evening in the UK and, once again, Patrick and James Burke were in the BBC's Apollo studio. The events of the next few days are well known of course, with there even being a major film made about the mission. After the overwhelming successes of Apollo 8 and 10 (the non-landing Moon missions) and then the Moon landings of Apollo's 11 and 12, public interest in the Moon missions was waning. Human beings are fickle creatures and despite the awesome technical achievement of landing a man on the Moon, the viewing public would never be impressed by the same story. They needed something different: well, they certainly got it with Apollo 13! There was a minor worry at launch, when the central engine of five on the second stage had failed, necessitating a longer burn of the other engines, and of the third stage; but, at least the rocket had not been struck by lightning on launch. Aboard Apollo 11 were astronauts Jim Lovell (one of the Apollo 8 crew), Fred Haise and Jack Swigert. Swigert, the only bachelor on the mission, had replaced Tom Mattingley at the last moment due to fears that Mattingley might have contracted German measles. At 2 days and nearly 8 hours into the mission, at just after 9 p.m. Central Daylight Time on April 13th, the crew of Apollo 13 had been doing a TV commentary and trying to locate Comet Bennett through the capsule window. They had been asked to 'stir up the cryo tanks' to resolve an oxygen tank pressure anomaly. What they did not know at that time was that a heater thermostat switch had been left in the wrong position when the spacecraft was tested on the ground. In addition, faulty wiring insulation meant that only a spark was needed to set off an explosion in the cryogenic oxygen tank. That spark was provided by the 'stir up the cryo tank' instruction.

At 55 hours and 55 minutes into the mission an explosion ripped through the service module of Apollo 13 and Swigert announced the problem, followed by Lovell himself, uttering the immortal understatement "Houston we've gotta problem". By an extraordinary coincidence, only months earlier, a Hollywood film called 'Marooned' had been released which depicted a similar situation developing in Earth orbit. [Aged 12, this author had seen that film when it was released, but as it was completely upstaged by Apollo 13, I have never seen it repeated on TV since.]

There was a complete reversal in public opinion on Apollo 13 at this point. From the 'been there-done that' attitude at launch it was now 'will they live or die in space?' The horror of three astronauts, potentially dying, watched by their friends, families and the world, was turning Apollo 13 into a real life drama unprecedented by anything else. Suddenly the viewing audiences were back at Apollo 11 levels and, once more, Patrick, a grim-faced Patrick this time, was on BBC TV throughout the day. As I have said, the drama of Apollo 13 has been well documented, but for me I will never forget the splashdown, covered live on TV. An estimated 30 million BBC1 viewers watched a tense studio, containing James Burke, Patrick, Cliff Michelmore and Geoffrey Pardoe. They were nervously fiddling with their pencils and glasses and biting their fingernails (and in the case of Michelmore, he appeared to be trying to swallow his entire hand) as they waited to see if the Apollo 13 Command Module survived re-entry; waiting for the first radio message from the crew after the radio blackout period. James Burke had his fingers theatrically crossed for the whole re-entry period. When the Command and Service Modules had separated in Earth orbit, the astronauts had seen the considerable damage to the side of the Command Module. But had the explosion damaged the heat shield? If it had, the crew would surely be fried. As the seconds ticked by after the theoretical end of the radio blackout period the NASA commentator reported that one of the aircraft in the area had reacquired the signal with the spacecraft. Burke was heard to murmur "But is it true?" Then, the cameras zoomed in, onto the sight of the Command Module and three parachutes descending through the clouds. Burke shouted "There they are". Patrick shouted "There they are" a split second later, followed by "They've made it" and deafening applause from Mission Control in Houston. It was suddenly all beaming smiles in the BBC studio.

Before the re-entry, the NASA head, Thomas Paine, said that even if the astronauts returned safely he would regard Apollo 13 as a "qualified failure". Patrick's view was, however, entirely different. He stated many times that he regarded the mission as "a human triumph, for the technical team at Mission Control, and above all for Lovell, Haise and Swigert, who showed us all the qualities that are great in mankind."

In early May, Patrick was off on his travels again, this time for a series of lectures in South Africa, which was one of his favourite destinations under white apartheid rule, but certainly not afterwards! While there he was fully aware that the planet Mercury was predicted to cross the face of the Sun on May 9th. Never one to miss an observational opportunity, Patrick arranged to use the fine 8 inch (20 centimetres) refractor at the Port Elizabeth Observatory. The refractor was fully equipped with an appropriate filter – essential for safe observing of the Sun; anything less than the correct filter can result in instant blindness. Mercury would start to cross the Sun's disk at 4.20 a.m. GMT and leave it, 8 hours later, at 12.12 GMT, thus the whole of the transit would be visible from the UK and all but the start would be visible from South Africa. Patrick was in position with the telescope from sunrise till midday, but sadly the clouds over Port Elizabeth did not clear. Frustratingly, skies were actually better over much of the UK, and many observers there saw the transit.

As well as Patrick's monthly appearances on *The Sky at Night*, and his Apollo broadcasts, in early 1970 Patrick was asked to present a series of eight programmes for children on BBC 1, entitled *Seeing Stars*. Patrick had presented an almost identical series in the early 1960s, with the same title. Interest in astronomy had never been higher, indeed the BAA's membership had risen from 2,700 in 1960 to 4,000 in 1975; it would peak at 6,000 in the mid 1970s. Patrick's publishers for his *Moon Flight Atlas* and imminent *Atlas of the Universe*, Mitchell Beazley, lost no time in collaborating with the BBC on a large format paperback booklet to accompany the *Seeing Stars* TV series, using many of the diagrams already designed for the *Atlas of the Universe*. The series was well presented, but like so many other astronomy programmes over the years, it only survived the first run of programmes. While children might be interested in space travel and aliens, few have the stamina to become fully fledged amateur astronomers, at least not in a country as cold and as cloudy as the UK! One of the most notable features of the new programme was that Patrick made himself available to a live audience of children, answering any astronomy questions they wanted to put to him. However, he was made to use an autocue on that *Seeing Stars* series, simply to force him to speak more slowly, so that children could keep up! He absolutely hated having to read words "like a standard BBC parrot!"

Also in early 1970, the BBC started a children's quiz programme hosted by the pop group *The Scaffold*, who were famous for their hits *Thank U very much* and *Lily the Pink*. Band member Mike McGear (real name Peter Michael McCartney) was the brother of Paul McCartney. As well as the quiz format the show had short sketches and musical items and eventually used *Dr Who* assistant Zoe (Wendy Padbury) as a co-host. Patrick appeared as a guest astronomer/entertainer on the children's programme twice during 1970.

Chapter 16
Mainly BAA and TLP

Despite Patrick now being one of the most famous people in the UK he would always do whatever donkey-work was needed to help out in the BAA and he was forever promoting the association, amateur astronomy and back-garden observers who were close personal friends. To this end, for *The Sky at Night* at the end of May 1970, Patrick travelled to visit his old friend Frank Acfield (1905–1975), at 61 Great Lime Road, Forest Hall, near Newcastle-on-Tyne. Aged 65 at the time of that *Sky at Night*, Acfield had been a BAA member since 1937 and owned a well-used observatory, housing a magnificent 10-inch f/9 Newtonian reflector, in his back garden. The observatory had been built in 1949 and so had seen 21 years of regular use, with all the observations being sent to the BAA. Frank had often provided Patrick with pictures of comets and asteroids for his books, as he was a keen astrophotographer and had even built darkroom facilities within his observatory dome. In fact, during the 1950s and 1960s, Acfield had been the Astronomical Correspondent of the *Newcastle Evening Chronicle*, often using photographs he had taken on the previous night in the next evening's edition, especially if something like a lunar eclipse had occurred. Frank had also taken some particularly pleasing pictures of Comet Bennet in April 1970, which appeared on *The Sky at Night*. Patrick used the May programme to emphasise the good work that could be done with even a modest pair of binoculars, as well as with larger instruments. Standing next to Frank and his impressive observatory dome, Patrick commented: "This very neat and efficient looking dome contains a 10-inch reflecting telescope, and just to make sure we all know where we are the latitude and longitude is given on the door." The camera focused on the plaque reading LAT 55 01 35 N, LONG 01 33 58 W, affixed to the observatory door, next to Frank's maximum-minimum thermometer. Frank replied "The Royal Greenwich Observatory gave it to me in order to pinpoint my position if I sent in any observations". Patrick asked "Do you find that the dome is easy to turn?" "Yes", said Frank "once you get it going Patrick it's quite easy; look here, you see, just you have a try". Patrick moved the dome about back and forth, easily with his hand. "Yes, well that's pretty easy, there's no difficulty there", Patrick says. All very simple stuff, but for most keen amateur astronomers, seeing Patrick in the

M. Mobberley, *It Came From Outer Space Wearing an RAF Blazer!: A Fan's Biography of Sir Patrick Moore*, DOI 10.1007/978-3-319-00609-3_16,

back garden of an observer just like them made the whole hobby seem all the more attractive, and who knows, maybe he would visit their observatory one day too? Patrick absolutely loved Frank Acfield's 8 foot long Newtonian and especially its splendid fork mounting, which a local Newcastle man, Tom Whittome, had built for Frank, at no cost!

A month later, during the June 24th 1970 meeting of the BAA, at their new London meeting venue at 23 Savile Row, Patrick outlined the aims of the new 'Binocular Variable Star Sub-Section' and answered many questions regarding suitable binocular equipment. Just as the observations of binocular variables were beginning to steadily flow through Patrick's mailbox, George Alcock, still at Farcet in Peterborough, used his own binoculars to discover his third nova, on July 31st. While sweeping through the constellation of Scutum he thought that a magnitude 6.9 star, with a slightly fainter star to the north, was not one he had seen before (of the 30,000 he had memorised). The next night, August 1st, it had faded to magnitude 7.8, confirming its nature. George had bagged another nova and provided another binocular variable for Patrick to study.

The end of August 1970 was a very hectic time for Patrick because the next IAU (International Astronomical Union) meeting was being held just down the road from Selsey, at Brighton on the South Coast. The IAU was, and still is, the governing body in world astronomy. It meets every 3 years and had not held a meeting in Britain since 1925. Needless to say, Patrick offered to help out with the meeting and was also keen to hear about the latest developments in lunar results from Apollo's 11 and 12, as well as getting some footage for the *Sky at Night*. During the week he was dragged off into endless meetings with the press, as despite being an *amateur* astronomer he was the best communicator, and also a familiar face to all TV viewers. Of particular interest to Patrick was the announcement by Dr Maurice Ewing, at that IAU meeting, that the Moon definitely suffered moonquakes. The seismometers placed on the Moon by Apollo 11 and 12 had been sending data back for 13 and 9 months respectively and the 'quakes were definitely associated with the time of lunar perigee (when the Moon is closest to the Earth). Dr Ewing paid tribute to the amateur astronomers who had been patrolling the Moon (largely under Patrick's guidance) for TLP. There now was categoric proof that the Moon was not totally dead; at least in a moonquake sense!

After the Apollo 13 near-disaster in April the next Apollo landing was delayed until February 1971. NASA decided to scrap the proposed Apollo 14 landing site near the crater Censorinus and divert Apollo 14 to the proposed Apollo 13 Fra Mauro area landing site; however, this was many months away. In the meantime, the USSR decided to take full advantage of the situation with their Luna 16 Moon probe, their first Moon mission since the failure of Luna 15 in July 1969. Luna 16 lifted off from Tyuratum on September 12th 1970 and safely soft-landed in the Sea of Fertility on September 20th. An onboard drill penetrated the lunar surface and 100 grams of moonrock was collected. The spacecraft then blasted off from the Moon and was safely recovered in Kazakhstan on September 24th; quite an achievement. Luna 16 would be swiftly followed by an even more ambitious mission, Luna 17, which would be launched in November. Patrick's contacts at the Soviet Academy kept him fully informed of developments.

Patrick was still very active in the BAA Lunar Section too, despite no longer being the section director. In September 1970 the section was still being directed by Patrick's friend Dr Ron Maddison of Keele University. The Lunar Section secretary Phil Ringsdore was another friend and one, much like Patrick, who was a tireless source of energy and enthusiasm, but remained a man of incredible modesty. Ringsdore was an excellent organiser too and a man who was universally liked. Patrick's 1965 invention, the monthly Lunar Section Circular, was still going strong, but the Lunar Section was extremely popular in those days and distributing the monthly circular involved duplicating up to 500 copies (typically 12 pages per circular) per month and posting 500 envelopes: a colossal chore for Ringsdore! After 1969 there was a feeling that maybe the days of the amateur lunar observer were over. After all, man had now been there and rock had been brought back; what more could be done from the Earth? However, Patrick's view was, and always would be, just the opposite. The Apollo 11 astronauts, on July 19th 1969 (from 1845 to 1847 UT) had described the NW wall of the crater Aristarchus as 'luminous' and 'more active' when viewed from lunar orbit, through binoculars. This, to Patrick, was reason enough that TLP were still on his agenda.

There were other activities within the lunar section that were still, arguably, valuable science too. Howard Miles was co-ordinating a project called 'Moonhole' whereby the depths of lunar craters were established by measuring the internal crater shadows when the lunar terminator (the night/day boundary) was near the crater in question. In addition, Harold Hill was trying to co-ordinate observers to map the South Polar regions which had not been mapped by any space probes. David Allen (destined to be the future Director of the Anglo Australian Observatory) and J. Hiscott had a programme running to investigate infra-red photography of the Moon and Cdr. Henry Hatfield was encouraging all to photograph the lunar surface. Henry had produced his own photographic lunar atlas in 1968 as part of Patrick's *Amateur Astronomer's Library*, published by Lutterworth Press. The TLP network would always be most associated with Patrick, although in 1970 the TLP co-ordinator was Phil Ringsdore.

Patrick and the Grangers!

September 1970 was a month this author remembers very well. I was 12 years old and had been a Lunar Section member (and a BAA member) for just under a year. On September 26th my father took me along to my first Lunar Section meeting in London, at the Royal Geological Society lecture rooms at Burlington House in Piccadilly. This was the first time that I had met Patrick, who seemed to dominate the meeting with his presence, even though he wasn't an official speaker that day. Being a famous face on TV, and having a voice like a loudspeaker, rather singled him out at the meeting! He was literally buzzing with enthusiasm and energy in those days and was remarkably quick on his feet for a man of his size. Many other adults seemed almost in a coma by comparison, but there were certainly plenty of other notable characters in that Burlington House room.

The first BAA member to ever speak to me, at that very meeting, was a certain Mr William (Bill) Granger from Peterborough, a very good friend of Patrick's. He nearly always had his cat 'Treacle Pudding' with him (usually on his shoulder, like a parrot) and that day was no exception! The careful reader will be well aware that I have mentioned the Grangers before, namely when Bill appeared on that second ever edition of *The Sky at Night*, 13 years earlier, in 1957. Bill was Patrick's first TV guest on the programme. I am unashamedly repeating some background on them here because, frankly, they were so unusual that I feel compelled to, and the reader may have forgotten who they were.

Bill's wife Ethel was extraordinary, though I did not know it at the time. She was (and still is) in the *Guinness Book of Records* for having the narrowest recorded waist measurement of any adult female; it measured an amazing 13 inches in circumference between 1929 and 1939 and was not much more in middle age! Bill's wife always wore huge amounts of jewellery and sported a beehive hairdo. BAA members of that era always regarded the dramatic arrival of Bill and Ethel Granger at the association's provincial meetings as something never to be forgotten. They travelled everywhere in a motorbike and sidecar combination, but with the diminutive wasp waisted Ethel, bedecked with jewellery and goggles, actually driving the huge motorbike, while the large frame of Bill was shoehorned into the sidecar, along with their moggy Treacle Pudding. Extraordinary! So, you thought Patrick was the most eccentric astronomer in the UK? Think again! Anyway, at that Lunar Section meeting, where I first met Patrick, I also spoke to Bill Granger too. The most amazing thing was that Bill and Ethel were highly intelligent people, despite their eccentricities.

The speakers I remember from that day in September 1970 were V.A. (Valdemar Axel) Firsoff (talking about the possibility of water once having flowed on the Moon), Miss Cicely M. Botley and Mr Keith Abineri. Just like the Grangers, messrs Firsoff and Botley were highly unique individuals too! Perhaps even more amazing than seeing Patrick 'in the flesh' was the fact that a sample of Moonrock, brought back by Apollo 12 10 months earlier, was sitting in a glass jar on the table in the middle of the lecture theatre. It looked exactly like a grey sugar cube and had obviously been cut to that shape by the University scientists who had brought it along. The sample was completely unguarded and members were asked not to remove it from the jar and touch it! I remember being absolutely terrified of going anywhere near it, just in case I fell over and damaged it in any way! But it was incredible to see. Of course, these days someone would pinch it within minutes of it arriving and put it on eBay! Anyway, that's enough of my memories…back to the plot!

In the September 1970 BAA Lunar Section Circular, Patrick's friend J. Hedley Robinson cautioned against seeing things on the Moon that weren't really there, citing the 1953 farce over O'Neill's bridge, some 17 years earlier, as a good example. Unfortunately, in the October issue, Mr M. Whippey followed this up, specifically stating that he had recently observed the region "where the bridge had been reported by O'Neill and later confirmed by the late H.P. Wilkins and Mr Patrick Moore". Patrick was obviously not at all happy by the association between himself and O'Neill's bridge, or by linking TLP to O'Neill's bridge. The November 1970 Lunar

Section Circular contained various items by Patrick, regarding TLP. On the subject of Mr Whippey's comments, Patrick wrote:

> Will there never be an end to the mis-quotes about this? When the observations were published, in the 1950s, the Press seized upon them. I was asked if I had seen the 'bridge'. I replied, dryly: 'Well, I've seen the area'. Of course, I was then reported as having seen a vast, artificial arch straddling the shore of the Mare Crisium......'

On the front page of the same Lunar Circular there was a big announcement that Kozyrev (who had claimed to have spectroscopically proven that the crater Alphonsus was 'active' in 1958) had received the 'Pioneers of Astronautics Gold Medal' for his study of vulcanism on the Moon. Also, on the inside front page of the same circular, Patrick emphasised the findings of Dr Ewing in the USA, who had stated at the Brighton IAU conference that (in Patrick's words): "There was absolutely no doubt about the correlation between moonquakes, perigee and TLP, which is what has been suspected for some time".

In early November 1970 two events of special interest to Patrick occurred in the same week. Firstly, on November 10th, the USSR launched their first wheeled vehicle to the Moon, Lunokhod 1, on Luna 17. Seven days later the probe soft-landed in a shallow crater in the north-western Mare Imbrium (Sea of Rains). A ramp was lowered and the wheeled vehicle Lunokhod 1 trundled onto the lunar surface. Lunokhod 1 would be an outstanding success; it operated for 11 months and travelled over 10 kilometres around the Moon. Over 20,000 photographs were taken. In addition, soil analyses were undertaken at 500 locations and chemical analyses at 25. The second event in early November (the 14th) was that the first man to set foot on the Moon, Neil Armstrong, was in London, and Patrick had arranged to meet him and film a special *Sky at Night* around the interview. According to Patrick, Armstrong was relieved to get down to talking some science, rather than answering the press-pack's predictable "how do you justify the cost of the Moon landings" questions. However, the surviving footage of the interview shows an enthusiastic Patrick, hair sticking up at crazy angles, confronting a positively terrified Armstrong! Apparently, although considered to be the calmest NASA astronaut under pressure, Armstrong was more terrified of interviews and public speaking than of anything else. A lot of the interview concerned the colours of the lunar surface seen by the astronauts at various Sun angles and the deceptive distances of near and far features on the lunar surface. Armstrong thought that there would undoubtedly be manned lunar bases in his lifetime; who would have believed then, that after 1972 no-one would revisit the Moon that century? The interview was broadcast on November 16th, only 2 days before the next 'regular' *Sky at Night*, which was on the subject of infra-red astronomy.

Back in the BAA Lunar Section, Patrick was becoming increasingly frustrated by the apparent lack of interest in observers scouring the surface for colour anomalies. While there had been a flurry of TLP reports around the time of Apollo 11, there now seemed to be a wave of TLP apathy present. In November 1970 there had been no credible TLP reports sent to Phil Ringsdore, or to Patrick, since December 28th 1969, when K. Kilburn had observed an 'event' in the southeast of the crater Aristarchus. Nearly a year had elapsed with no TLP! While the sceptics would say that this was good news, as TLP did not exist, for Patrick it was bad news. In a letter published in the December Lunar Circular Patrick urged members to participate in the TLP network and send in their reports of TLP. However, in the same circular, the Director Ron Maddison stressed that observers could not report TLP if none had occurred. There seemed to be a definite difference of emphasis between Patrick and Ron. Patrick wrote a letter, which was published in the January 1971 Lunar Section Circular; it read as follows:

> Do let me clear up one point. Of course I agree absolutely with the Director when he says that TLP reports cannot be made if there are no TLPs! Also, I do believe that the dearth of them during the past 18 months or so is genuine. But when I inferred that interest in the Section seemed to have fallen off, I based my view on the fact that I had virtually no negative reports. When a TLP is reported, the vital thing is to get confirmation. It was in this way that we obtained the conclusive observations of Gassendi – how long ago that seems! We then set up a network, and collected a great many reports, almost all of which were (predictably) negative. They enabled me to weed out one or two reports of activity which proved to be due to observational or instrumentational error. It also provided a list of people who were doing the work, and who could be contacted if anything positive came in. But this seems to have ceased now, and I have no idea who is or is not continuing with the programme – which was what prompted my comment in my letter. I greatly appreciate the Director's help and encouragement; please, take heed in particular of the last paragraph of his comments last month!

However, even before Patrick's letter was actually printed in the January 1971 circular, the Lunar Section director Ron Maddison had tendered his resignation to the BAA council, on the grounds of work commitments. The situation was further confused by a postal strike in the UK and the change, in February, to decimal coinage, something which Patrick opposed totally. "There was nothing wrong with proper money, pounds, shillings and pence". In the UK, the whole of the 1970s would be dominated politically by postal strikes, car workers' strikes, rail strikes and coal miners' strikes. Because of the postal strike the January Lunar Section circular was not distributed until mid-March, along with the March circular; the latter contained the resignation announcement. The BAA council appointed Phil Ringsdore as Acting Director but did not want to appoint a new Director in mid-term (BAA Sessions run from the end of October to the following end of October). However, everyone knew that Patrick was the automatic choice, as Phil Ringsdore was just too modest a man to accept the directorship and everyone in the country, and the BAA, associated Patrick with the Moon. As we have seen, Patrick was the Lunar Section Secretary under H.P. Wilkins from 1946 to 1956 and a previous Director from 1964 to 1968; his resuming of the Lunar Section directorship was inevitable!

While a precise decision on the future Lunar Section directorship hung in the air, there was no doubt that Ringsdore & Moore would make a brilliant team. Both had a total dedication to the observation of the Moon from the Earth, even though they were both totally different people. Phil Ringsdore was a quiet, unassuming man, flattered to be working alongside Patrick, while Patrick was a brilliant speaker, motivator and writer, not to mention a TV star. In addition, Phil and Patrick were two of the four BAA vice-presidents, so it was unlikely that the Council would decide anything that they were not happy with.

In the early evening of January 13th, Patrick rolled back the double-ended run-off shed at Selsey, containing his 12½ inch (318 millimetres) reflector and trained his telescope on Saturn. The ringed planet was on the borders of Taurus and Aries, 2 months past opposition (opposition meaning opposite the Sun, so highest up at midnight and best placed) but still nicely on view in the evening sky and with the rings wide open. From 18:00 to 19:00 GMT the stability of the Earth's atmosphere was exceptional and Patrick enjoyed what he described as his best ever view of the planet. While he could see detail on the globe, what really amazed him was the detail in the rings. In particular, the elusive so-called 'Encke' division appeared to be visible all around the edge of ring A. As all Saturn observers will know, the main division in Saturn's rings, between ring A and ring B, is called the Cassini division. However, a much narrower division exists at the edge of ring A and can only be seen in exceptionally stable atmospheric conditions. Such conditions are rare and unpredictable, but often coincide with a large high pressure region over the country and misty, or even foggy, weather. Observing sites near to the sea are often favoured for good seeing and Selsey was no exception; but, even by Selsey standards, the night of Jan 13th 1971 was exceptional. Patrick did not believe he should really be able to see Encke's division all around the planet, but this was before space probes had been to Saturn (Pioneer 11 was 2 years away from its arrival). Patrick thought that the observation was an illusion, caused by the exceptional stability. Nevertheless, the observation inspired him and he urged members of the Saturn section, then directed by his friend Gilbert Satterthwaite, to see what they could glimpse of the elusive Encke division.

On January 16th 1971 Patrick left Selsey early to attend the next London Lunar Section meeting (where the resignation of Ron Maddison was announced) and to attend a celebration at the Dorchester Hotel in London. It was the seventieth birthday party of one of the BAA's legendary figures, a man named Horace Dall (1901–1986). Horace was an extraordinary man. By profession he was an engineer in Luton, making instruments which measured the flow of liquids and sewage. But in his spare time he was an amateur astronomer, optician and explorer extraordinaire. Dall had joined the BAA in 1925, when Patrick was only 2 years old. In 1937 he had built a 15½ inch (39 centimetres) reflector on top of the highest hill in Luton. By amateur standards this was gigantic for those times. He invented the Dall-Kirkham Cassegrain reflector design in 1929, when aged 28, and made microscopes of record-breaking power, by using exotic glasses. Remarkably, Dall had built a folding 15 centimetres reflector that could fit inside his jacket pocket. Add to all these achievements the fact that he was the best lunar and planetary photographer in the country and the first man to cycle across central Iceland, and you had a real one-off character. His house, at 166 Stockingstone Road in Luton, had a camera obscura telescope in the roof. Dall could project the outside world onto a white surface in his attic using the instrument and even zoom in on the local cricket match taking place a mile away, something that fascinated Patrick who was a regular visitor. At the Dorchester event 150 guests, mainly from the BAA, turned up and Patrick was the master of ceremonies. Patrick had been the best man at Horace Dall's second wedding in 1970, which had been held at Caxton Hall in Westminster. Dall's first wife had died in 1964. He had met his second wife, Helena, in a Café in a Patagonian rain forest in 1968! Both were addicted foreign travellers and both had been in their 1960s and recently bereaved; an extraordinary meeting of like minds. In his toast to Horace, at his seventieth birthday party, Patrick said that "No-one had ever gone to Horace for help and come away empty-handed".

As an aside to all this, 13 years later, in 1984, on a visit I made to Horace Dall's house at 166 Stockingstone Road, the subject of Patrick's bachelor status came up and the fact that he had been Dall's best man. I was somewhat taken aback when the usually quiet Horace described Patrick as: "The World's Biggest Mother's Boy.... I have never, EVER, in my life, met a bigger Mother's Boy". I was particularly shocked as Patrick and Horace had been firm friends since the 1950s. I pointed out that Patrick claimed he had planned to marry a girl who was killed in the War. Dall shook his head and said that Patrick would never, ever, have done anything that would have taken him away from "His Mummy" and then he added "...especially after his father died just after the War." I was also told that Horace's wife Helena had compiled some considerable notes on her husband's remarkable life, and at one stage the astronomy author James Muirden was seen as the obvious choice to assemble them into a Horace Dall biography. Patrick got to hear of this plan and was appalled. Patrick disliked Muirden (he disliked nearly all rival authors unless they were his protégés!) and, behind Muirden's back, told everyone that Muirden stole data from his books. In fact, nothing could have been further from the truth. Outraged that Muirden might end up writing Dall's biography, Patrick insisted that only he could write Horace Dall's biography. He took Helena Dall's notes home to prevent her giving them to Muirden. However, Patrick was not a meticulous researcher. Patrick wrote all of his books off the top of his head. Sifting through data patiently was not his style at all. After several months Patrick admitted that he could not do it because "I've too much else going on........." The book was never written.

Apollo 14

At the end of January 1971, it was time for Patrick to re-enter the BBC Apollo studio for another 10 days: Apollo 14 was now on the launch pad. The Saturn V rocket lifted off from Cape Kennedy on January 31st 1971 just as Patrick's revised and enlarged Moon *Flight Atlas* was appearing in the shops, with a special additional feature on the dramatic Apollo 13 mission. So far this was the seventh reprint of the Atlas, with sales of over half a million! Interest in the Apollo missions had, ironically, been boosted by the near disaster of Apollo 13. Everyone prayed that this time there would be no onboard drama and a successful landing in the Fra Mauro region would finally take place! Of course, in February 1971, no-one realised that the Apollo series would stop in three more missions' time. There were still six more Apollo missions on the drawing board, right up to Apollo 20, targeted to land close to the crater Copernicus in 1973/1974. For the whole of the Apollo 14 mission Patrick was back with his Apollo broadcasting colleagues. From January 31st to

February 9th, he would eat, sleep and drink Apollo 14. The mission's landing site was, like those of Apollo 11 and 12, close to the lunar equator and it was only a few hundred miles east of the Apollo 12 site. The Fra Mauro region was an old walled plain some 80 kilometres in diameter within the Mare Nubium (Sea of Clouds). As for all of the Apollo landing sites, Patrick was very familiar with the region as seen through the telescope, at least, down to a resolution of a mile or two.

The mission would not be without its moments of drama! The astronauts on the Apollo 14 flight were Commander Alan Shepard (who Patrick reminded everyone was "the same age as me, he's 47"), Command Module Pilot Stuart Roosa, and Lunar Module Pilot Ed Mitchell. Alan Shepard was already an American hero as he had become the first American in space during the Mercury 3 mission, on May 5th 1961. However, that flight had only lasted 15 minutes and the other Apollo 14 astronauts had never been in space before. So it was an interesting crew! In addition, Shepard had undergone an ear operation, so some questioned his fitness. For the first time ever, a Saturn V launch was delayed on the launch pad. The T-10 minutes point was held for 40 minutes while a rain squall passed overhead. NASA did not want another Saturn V rocket to be hit by lightning. Three hours after the launch and over 7,000 miles from the Earth, Stuart Roosa needed six attempts to dock with the Lunar Module; the docking latches refused to lock. Eventually they did work, but there was some anxiety about whether the mission should continue. On this mission the command module (Kitty Hawk) was put into such a low orbit around the Moon that it was only 15 kilometres above the lunar surface at its low point. This would save Lunar Module fuel when descending for the landing.

The landing itself was not short of added drama. Mission Control in Houston received an 'Abort' warning just as the final descent to the lunar surface was about to start. However, Ed Mitchell (so the story goes) thumped the control panel and the 'Abort' light went off! Mission Control decided the landing was still on. After considerable hovering over the landing site, Shepard finally brought the Lunar Module Antares down only 27 metres from the Fra Mauro target landing position. The only other problems encountered during the two EVA's (Extra-Vehicular Activities, namely walking on the Moon) were the astronauts' breathlessness on the walk to 'Cone Crater', 1.6 kilometres away and 120 metres uphill. Shepard's heart rate reached 150, Mitchell's reached 130. They turned back just short of the crater rim, where they would have had a fantastic view of the inside of the 40 metres deep crater. But, in all other respects, the EVA's were a great success. A total of 44 kilograms of moonrock was returned to the Earth when the command module splashed down on the evening of February 9th. In the studio, Patrick, James Burke, and the BBC space team, had once again captured the viewing public's imagination; men were, yet again, walking on the surface of the Moon. Also, for the first time ever, really high quality TV pictures were being relayed back to the Earth.

Writing in the March 1971 Lunar Section Circular, Phil Ringsdore announced the situation regarding the resignation of Ron Maddison and also suggested that a provincial (outside London) meeting should be held in either Preston, Coventry or Chester; he invited comments from members living in those areas. Within a few days of the circular going out he received a petition, signed by 82 Lunar Section members, requesting that the next section meeting be held in Chester. This was not really a surprise, as Richard Baum was an old friend of Patrick's, lived in Chester, and there were many other keen lunar observers in the area. Phil and Patrick set the date for Saturday June 5th. In the meantime, Patrick was at the telescope whenever he could be, scouring the lunar surface for any signs of TLP and observing planets and variable stars. As if to usher in a new era of TLP enthusiasm from the Lunar Section both Phil Ringsdore and Hedley Robinson reported seeing coloured anomalies on the lunar surface on March 2nd and 3rd. Ringsdore's suspect was an orange-pink glow on the central mountains of Theophilus while Hedley Robinson's was a reddening south of the crater Madler.

At the end of March, Patrick addressed the BAA monthly meeting and reported that the variable star SU Tauri was experiencing a deep minimum. Three weeks later, on the nights of April 17th, 19th and 20th, he had failed to find the star in twilight with his 12½ inch (318 millimetres) reflector and concluded that it had faded below magnitude 13. These observations were reported to the BAA at the April 28th meeting, where Patrick added that Alcock's amazing 1967 nova, HR Del, had now faded to below magnitude 9.7 and SS Cyg had been estimated at magnitude 8.7, by himself, on April 27th at 4 a.m.

Another event, on April 19th, of special interest to Patrick, was the launch of the huge Soviet Space Station, Salyut. On April 23rd, Cosmonauts Shatalov, Yeliseyev and Rukavishnikov flew to the space station to examine it, but mysteriously returned to Earth only a day later. The Soviets might not have put a man on the Moon, but they certainly still had the capacity to achieve space 'spectaculars'.

Flying the TLP Banner

In the May 1971 Lunar Section Circular Patrick was, once again, trying to drum up interest in searching for TLP. In an article entitled *TLP's A Personal Note* Patrick wrote:

> Some years ago, the Lunar Section had a 'network' of observers who were making close searches for TLP's. This was during my period of Directorship. Inevitably one's ideas are coloured by one's personal interests; I have always been fascinated by the idea of mild activity on the Moon (I was writing about it even before the war, and certainly I called a lot of criticism on myself during the 1940s and early 1950s, when the Moon was officially classed as totally inert). Therefore, we had some criticism that the Lunar Section was too TLP-minded. There may well have been something in this. Nevertheless, the research proved to be very successful, as we all know, and I think it had a great deal to do with the official change of attitude with regard to the phenomena. Incidentally, the very term TLP comes from the Lunar Section – it was our first name for the reddish phenomena, and it has now become at least generally accepted everywhere.
>
> Since those days, much research has been undertaken. The setting-up of seismometers on the Moon has yielded new data. As we had expected, there is a definite link between TLPs, the time of perigee, and moonquakes – as had been forecast earlier by Barbara Middlehurst, Jack Green and myself. At the last IAU Meeting in Brighton, last autumn,

Dr Ewing came over from America to present an official report; and he was very kind in his praise of the work which had been done on TLPs by our Lunar Section.

The catalogue, drawn up by Barbara Middlehurst and her colleagues in Arizona and myself in Britain, is complete to October 1967. No doubt some of the reports are faulty; but the general trends are clear. At present I am bringing the catalogue up to date, and I hope to have this completed within a few months. I had intended to defer it until after the end of the Apollo series, but I have been officially requested by NASA to expedite it. Mrs Cameron and myself are the official 'collectors' of TLP information for NASA.

During the past few years it seems that the BAA Lunar Section network has ceased to operate. This is a great pity. We all know that even though there is much lunar observation left to be done from Earth, some aspects have been removed; most of the limb cartography, for instance, is not complete. Though I may be laying myself open to the charge of being TLP-mad, I suggest that TLP work is some of the most important research still open to the amateur Lunar observer, always provided that it is carried through with the greatest care (a faulty observation is disastrous). And I believe that the time has come to re-establish our network as quickly as possible.

I have been discussing this with Phil Ringsdore, who has been deeply involved in the project all the way through (he was, of course, one of the observers of the famous TLP in Gassendi way back in 1965 [it was actually 1966 – MM]). In his capacity as Acting Director, he has given me full backing in the appeal to get the network started up again and has asked me to resume as TLP Co-ordinator. I would suggest that an initial step is to publish in the Circular, a list of all those who are ready to take part, together with their instrumental equipment and whether or not they have a Moon-blink device.

I do regard this as urgent and of course we have, in the Circulars, a magnificent medium of communication! So if you are interested, please write to me – Patrick Moore, 'Farthings' 39, West Street, Selsey, Sussex.

Immediately after this, in the same circular, Patrick added a few words about the enigmatic crater Linné, reported as having changed in 1866. The article certainly seems to indicate that Patrick was still semi-obsessed with the idea of change on the lunar surface. Patrick wrote:

I wonder if anyone was looking at Linné on the evening of May 1? I began a blink survey at 2130, which proved to be negative. During it, I looked at Linné, and was struck by its appearance. It was a very obvious small crater with walls raised high above the outer Mare and with the impression of being on a mound. The exterior shadow of the wall was also obvious. It was very like other craters on the Mare Serenitatis, though of course larger than the others apart from Bessel.

The white nimbus was very dim, and I might have overlooked it had I not been expecting it; it was not regular, but was enclosed by the familiar low ridges which surround Linné. I made a sketch at 2147 with x360 on my 12½ inch reflector, and another with the same result, at 2200 with x330 on my 5 inch refractor. Seeing was 3 on the Antoniadi scale, with considerable shimmer.

It may sound rather old-fashioned now to discuss the age-old Linné question …but have we solved it completely as yet? I am not sure! – Patrick Moore

Patrick was also out observing 2 weeks later, in the early hours of May 14th. One of his best BAA friends, the orbital expert Gordon Taylor, had alerted the association to a possible occultation of the fifth magnitude star Beta Scorpii C (SAO 159683) by Jupiter's giant moon Io. This was before any spaceprobes had visited the planet and its moons and so timing the period when a star passed behind the moon would help to calculate Io's diameter. Exciting stuff! Unfortunately, at around 0152 GMT on that May 14th morning, Jupiter and Io were just 17 degrees above Patrick's south-southwest horizon and only his 8.5-inch reflector could easily view that low. However, remarkably the sky at Selsey was clear, but sadly no occultation occurred from his observatory. Patrick watched excitedly in the pre-dawn twilight, at high magnifications, as the two objects (looking the same brightness to him) closed in, but Io drifted roughly 1 arcsecond north of Beta Scorpii C and the light from the star did not dim at all. Patrick later found out that the experienced observer Jean Dragesco, based in Central Africa, had seen the star disappear behind Io for a full 5 minutes.

Two weeks later, on May 26th, Patrick attended the 1971 BAA Exhibition meeting, staged at the Royal Commonwealth Society in London's Northumberland Avenue. At that meeting the South African discoverer Jack Bennett, who had found both Comet Bennett 1969i and an eleventh magnitude supernova in Messier 83, was awarded the Merlin Medal by the President, Dr Barocas. It was quite an emotional meeting for Patrick because the BAA's recently retired Assistant Secretary (for almost 40 years), Miss Lydia Brown, had been gravely ill, but she had managed to attend the meeting. Patrick was delighted to be able to speak to Miss Brown and also to Jack Bennett, whose comet he had very much enjoyed observing. Patrick had apparently been involved, behind the scenes, in Lydia Brown receiving the MBE from the Queen 2 months later, for her services to the BAA, but sadly Miss Brown passed away in October 1971, just 3 months after her trip to Buckingham Palace.

Fig. 16.1 Patrick with the comet and supernova discoverer Jack Bennett at the BAA Exhibition meeting on May 26th 1971 (Image by kind permission of John C. McConnell)

After his May Lunar Section Circular comments Patrick continued plugging the TLP theme on June 5th at the Guildhall in Chester, where, not surprisingly, he was the first speaker at the next BAA Lunar Section Meeting. The meeting was opened by Mr L. Coburn, President of the Chester Astronomical Society. Following Patrick's talk, a Dr Mills addressed the meeting with his own ideas on the possible origins of TLP. Dr Mills was of the opinion that TLP were caused by the occasional release of accumulated gases in response to tidal stresses. He also thought that the separation of particulate material, at these times, could lead to an electrostatic discharge on the Moon, similar to lightning. Certainly, these theories were of great interest to Patrick who had many questions for this speaker. Following the Lunar Section meeting, Patrick gave a public lecture about Mars at the Guildhall, to a packed audience. He then went to Richard Baum's home at Boughton where he spent the evening. Richard Baum recalled that very night in an e-mail to this author:

> After the major lecture in Chester about Mars, Patrick wrote the foreword to my book, 'The Planets'. We got back from the lecture around 2230, a number of colleagues stayed talking until after midnight, and then at 0245 hrs Patrick suddenly said "I like your book very much and will do the foreword now". I was aghast. After a hard night of talking, and leaving on the 1000 train that very morning, I did not think he could do it in so short a space of time, and was rather reluctant for the sake of his health for him to even attempt it at that time. Still he asked to borrow my typewriter and one sheet of paper. We closed the sitting room door and all chaos broke loose. One finger typing plus enthusiasm equals intensity. Fifteen minutes later he came into the other room and asked "will this do?" What he wrote then is as it stands in the book with the exception of a minor error.

Three days later, on June 8th, that month's *Sky at Night* programme was broadcast. The subject was 'Orbiting Space Stations'. As we have seen, on April 19th 1971 the Salyut space station was placed in Earth orbit and visited, briefly, by three cosmonauts. At the start of June (the 6th) a further set of cosmonauts flew to the station on Soyuz 11, and it was announced that they intended to stay on board for longer than the previous Soviet space duration record of 19 days. Thus, it all seemed like a good subject for a *Sky at Night* programme.

However, on June 29th, tragedy struck. After 23 days of weightlessness, the three cosmonauts transferred back to Soyuz 11 and re-entered the Earth's atmosphere. When the spacecraft hatch was opened the cosmonauts were dead in their seats. Apparently, a pressure loss and subsequent oxygen deprivation had been the cause, though at the time there was speculation that 23 days of weightlessness had taken its toll. Certainly, Patrick thought this might be a factor. He quickly arranged a sombre *Sky at Night* programme on the day after the tragedy, to discuss the possible cause and the implication on space travel. Patrick had very good contacts within the USSR; indeed that same year he had been made an honorary member of the Astronomic & Geodetic Society of the Republic, an honour he treasured. So, he was greatly saddened by the loss of life. Despite all the recent Apollo successes and the averted tragedy of Apollo 13, it was still obvious that space travel was a dangerous game. The scheduled *Sky at Night*, already recorded days earlier, was broadcast as planned on the next night (July 1st) and it was a much happier affair: a visit to the back garden observatories of Patrick and Henry Brinton at Selsey, and to Henry Hatfield, the lunar photographer, at Sevenoaks.

In July 1971 Patrick completed what was, essentially, a new book entitled *Guide to the Planets*. Admittedly, he had originally written a book with this same title in 1954, to match his *Guide to the Moon*, but as Patrick himself said, the original was: 'published at a time when the idea of travelling to the Moon was officially regarded as ridiculous'. The original *Guide to the Planets* was revised several times and then published, in 1962, as *The Planets*. However, 9 years on and with many space probe results to hand, it was time for a complete re-write. The new manuscript was completed for Lutterworth Press in July and would be in bookshops before the end of the year.

In the summer of 1971 Patrick was visited by the journalist John Sandilands (1931–2004) at his house in Selsey. Sandilands wrote an entertaining account of a day spent with the TV astronomer, and also of his journey, with Patrick and his mother, to a local hotel restaurant. The article was published in the now defunct (1965–1975) British magazine called *Nova*. Patrick's car stank of petrol fumes en route and the back seat was littered with yellowing astronomical journals, both of which his 84 year old mother mentioned, suggesting the fume-filled car should be taken to the local garage for investigation. Of course, in 1971 no-one had heard of political correctness, and so Patrick wasted no time talking about subjects which, in the twenty-first century, might have seen him arrested! "There are certain things that make me rather cross (pronounced *crawss*). Weirdies, Beardies, Commies, Lefties. You've only to switch on the television to see these long haired gentlemen making nuisances of themselves on public money. If they don't like it here let them go somewhere else. Both Mother and I are rather keen on Enoch Powell. Enoch for P.M. we say, don't we Mother?" Then, lowering his voice: "I've nothing against our coloured brethren. Splendid chaps. But I may say that if I went to somebody else's country I'd make a certain effort to behave myself! On the other hand I suppose I'm a bit of a socialist in some respects. Dead against blood sports, you see. Absolutely dead against!" John Sandilands day spent with Patrick and his mother would stand him in good stead 3 years later, because when Thames Television decided Patrick was a suitable victim for the programme *This is Your Life*, they commissioned Sandilands to write the entire script.

For the July 21st *Sky at Night*, viewers were treated to the extraordinary sight of Patrick bowling a cricket ball at Holmewood House preparatory school, near Tunbridge Wells in Kent. This was the school where Patrick had taught in the early 1950s. One of Patrick's cricketing colleagues, Robert Bairamian, was the headmaster and he invited Patrick to use the boys at the school to demonstrate the concept of parallax: as the Earth moves around the Sun, the nearer stars shift against the background of distant stars. Patrick's description of his bowling style was: "Medium-paced leg-spin featuring a long, leaping run-up and windmill action". My own opinion is that Patrick's bowling action resembled a kangaroo suffering from substance abuse. After the programme Patrick received many letters about his bowling action, but none about parallax! Even cricketing friends of Patrick admitted that his bowling action was a secret weapon; one friend put it thus: "The opposing batsmen fall about so much laughing at Patrick that he bowls them out!" Mind you, compared to his attempts at playing golf, his cricketing techniques were pretty conventional. A Selsey friend once took Patrick for a round at the local golf club. Patrick completed the 18 holes with a score of 231, including a 43 on the third hole.

Apollo 15

Five days later, and only 6 months after the Apollo 14 mission, it was time for Patrick, once more, to take his seat in the BBC Apollo studio. Apollo 15 was the first of three so-called 'J Series' missions designed to maximise the collection of scientific data. This time the crew consisted of Dave Scott, Al Worden and James Irwin. The launch, at 14.34 BST in the UK, on July 26th, was the most powerful ever, with the Saturn V rocket developing an estimated 3.556 million kilograms of thrust. This time the destination was the area known as Hadley Rille, in the mountainous Lunar Apennines; the most rugged landing site yet chosen and the first to be well away (26°N) from the lunar equator. Hadley Rille is roughly 1.6 kilometres wide and nearly 400 metres deep and is visible through amateur telescopes. On a good night the region is so distinctive through an Earth based telescope that a skilled observer can easily see the main features visited by the astronauts. This was especially relevant for Apollo 15 because they had brought a 'car' with them, so their range was much greater than on previous missions, and amateur astronomers could visualise the scale of their excursions compared to the view through the telescope. The car was certainly not a Cadillac! It was called the LRV (Lunar Roving Vehicle) or just 'The Lunar Rover' and it was stowed in a triangular shaped segment of the lunar descent stage. The LRV added 180 kilograms to the Lunar Module's weight. The BBC built an exact replica of the Lunar Rover and much footage was taken of James Burke and Patrick going for a spin, in and out of the studio, in the mock-up! The summer of 1971 was a season that the British male population would remember very well. It was the peak of the 'Hot Pants' era, when young ladies wore outrageously tight and skimpy shorts in broad daylight. In keeping with those blissfully non-politically correct times Patrick and James Burke were filmed in the mock-up Lunar Rover, while being pulled along a disused airfield by their 'slave', the BBC TV's 'Summer Colour Girl' Christina Lepora; and yes, she was wearing a suitably skimpy pair of Hot Pants!

The real LRV, built by Boeing, was fed by a 36 volt battery system powering four quarter-horsepower motors: one per wheel. With two fully laden astronauts and equipment the whole LRV weighed about 550 kilograms, but in the Moon's 1/6th gravity this would have been equivalent to less than 100 kilograms. Patrick was also given his own rather different vehicle, to use on the Moon should he ever get there! This was the Emmet Lunacycle, a bicycle built for Patrick by the amateur inventor Roland Emmet, incorporating various novel features. The accessories included an astro-cat (to ensure you always landed on the Moon the right way up), a protective meteorite umbrella, and a cheese comparator for testing the lunar surface. Patrick gave a serious commentary on the virtues of the Emmet Lunacycle and was so convincing that a *London Evening Standard* reporter phoned him at Selsey to ask if it really was going to be used on the Moon! Patrick was quite happy to cycle around the Sussex countryside looking totally eccentric, while mounted on the Lunacycle; he was never someone to take himself too seriously. As if to emphasise this, during the Apollo mission he appeared on the BBC 2 *Late Night Line-up* programme, lightly disguised and with a fishbowl on his head. He claimed to be a Martian and insisted that life could not exist on the Earth, because the atmosphere was too thick and there was too much water! Actually, riding the Lunacycle was probably a far safer exercise for Patrick than riding his own deathtrap of a bicycle, on which the brakes had never worked. For getting around Selsey Patrick often preferred to just leap onto his bike, and he cycled like he talked...manically! He would quite often dash out of the house on a whim, to visit friends, without telling his mother. If the phone rang and Patrick didn't pick it up, his mother would first look into his study, and then look on the peg where he hung his bicycle trouser clips. If they were missing he had obviously dashed out somewhere, wearing the same clothes as ever: a blazer, shirt and tie!

While on the subject of weird forms of transport in the Selsey region, it was in 1971 that a local resident and, like Patrick, a supporter of the local RAF association, came to him with an idea for raising funds for that organisation. The resident was Selsey gift shop owner George Abel and his idea was to have a Selsey Birdman contest, where a prize of £1,000 (later increased to £3,000) would be given to anyone who could fly 50 yards over the sea, taking off from the lifeboat ramp. Of course, this was just the sort of crazy idea that Patrick loved and so with his full backing and energy the first Selsey Birdman contest took place in the summer of 1971. By 1972 the BBC were even televising the event from Selsey, with Patrick and Anne Nightingale as the reporters, watching while 17 flapping and gliding men attempted to become human birds. An amusing cartoon in the *Radio Times* depicted a wide-eyed Patrick, arms strapped to planks, plummeting seaward, as a watching stork laughed its head off! In 1976 the organisers were told that they could no longer use the Selsey lifeboat ramp, but by 1978 the event had moved to nearby Bognor, and Patrick continued to support the event, raising funds for the RAFA for more than 30 years. In recent times similar copycat events have been held at other locations, even in China, and for much larger prize money. Typically, Selsey and Bognor residents attempting to fly have dressed up as all kinds of celebrities, including the Pope, Dr Who, a skateboarding cow, an Apollo rocket, a vampire, and Mary Poppins; and all because George Abel had a chat with Patrick about raising funds for the RAFA in 1971.

In the July 1971 Lunar Section Circular, BAA members were asked to monitor the Moon for evidence of the impacts of the third stage of the Apollo 15 Saturn V rocket and the Lunar Module, when they were crashed into the Moon. This double crash landing was a regular procedure adopted by NASA and it served two purposes: the impacts were used to calibrate the Apollo seismometers on the lunar surface and it was much tidier than leaving space junk in lunar orbit. By the time

Apollo 15 was launched it was known that the NASA budget cuts had meant the scrapping of Apollo's 18, 19 and 20, so space fans were realising it was time to savour the remaining Apollo missions.

Apollo 15 was a spectacular scientific success. A small satellite was launched from the spacecraft to monitor the Moon's gravitational anomalies and instruments in the spacecraft's service module 'SIMBAY' took photographs and spectrographs of the Moon. Unlike previous landings in flatter areas, the Lunar Module landed with a 10 degree tilt because one footpad was in a 1.5 metre diameter crater. This caused Scott and Irwin quite a few problems, as it meant the hatch door had a tendency to swing shut. The astronauts described the ride in the Lunar Rover as 'a real bucking bronco' in 1/6th Earth gravity and they had to wear their seat belts all of the time. Nevertheless, Patrick, James Burke, and the viewers, had never seen such entertainment on an Apollo mission, because the Lunar Rover had its own TV camera and umbrella-like antenna, which relayed quality pictures back to the Earth. Scott and Irwin drove a total of 28 kilometres around the Hadley Rille area in their Lunar Rover and collected nearly 80 kilograms of moonrock. Perhaps the most dramatic moment though was the Lunar Module lift off. For the first time, because of the Lunar Rover's camera, viewers on Earth could watch the top half of the Lunar Module lift-off from the Moon on its way to rejoin the Command Module. This was always a heart-stopping moment as the ascent engine had no back-up. If it did not fire the astronauts were stranded on the Moon. Fortunately it did work, and despite the Lunar Module's 10 degree tilt, the rendezvous with their colleague Al Worden, in lunar orbit, was flawless.

Although there were some worries about the seals on the Command Module hatch and about the Command Module colliding with the Lunar Module when it was released, the Lunar Module finally impacted on the Lunar surface at 07:20 GMT on August 3rd, some 100 kilometres west of the landing site; this proved close enough to provide some nice vibrations for the seismometer left by the astronauts. The Saturn V third stage had already impacted with the Moon on July 29th at 21:00 GMT (at approx. 3 degrees South, 8 degrees West). No-one in the BAA Lunar Section reported seeing any evidence of the impacts.

On the way back to the Earth, Al Worden did a space walk to retrieve scientific data from the instrument bay on the Apollo service module. It was the first ever spacewalk carried out for a definite purpose. Unfortunately, the Lunar Rover camera failed not long after it was left on the Moon, so scientists were deprived of seeing the Earth eclipse the Sun from the Hadley base on August 6th. More dramatic though, was the failure of one of the three parachutes after re-entry on August 7th. Fortunately the splashdown was successful and NASA had achieved another extraordinary success.

With another Apollo mission completed, Patrick's life could, at last, return to normal, and so he was able to get back to what he enjoyed doing best, namely looking through the eyepiece of a telescope and mixing with his biggest fans and best friends.

Fig. 16.2 Patrick in 1971, aged 48, with his 5 inch refractor at Selsey, which he purchased for a token £10 from the millionaire businessman Peter Sartory in 1968. The observatory materials were purchased using his Goodacre Award money from the BAA. The mounting and various other components were made by Charles Frank of Glasgow. (Photograph by Beverly LeBarrow)

I am grateful to Damian Peach for pointing out an especially interesting set of observations that Patrick made of the planet Jupiter, just after the Apollo 15 mission. In August 1971 the Giant Planet had a declination close to −19 degrees and was sinking rapidly into the murk as soon as the Sun set from Selsey. Nevertheless, Patrick must have been keen to observe it given the amount of time he had spent in the BBC TV studios in recent weeks. One especially interesting set of observations were made by Patrick on August 16th, between 19:40 and 20:10 U.T. His logbook shows he was down the road at Henry Brinton's observatory, using Henry's 12.5-inch reflector; possibly because Jupiter was just too low to reach from Patrick's own garden. His notes read: 'Henry Brinton's reflector with Henry, Martin, Mark, John, Peter! SEB very active, RSH [Red Spot Hollow] noticeable. Alas, seeing worsened quickly as Jupiter descended.' The sky would still have been a strong twilight hue by the end of the session and Jupiter would have been a mere 16 degrees above the horizon. That is serious observing dedication!

Throughout early 1971 Patrick had been working on his extension to NASA Technical Report R-277: the chronological catalogue of reported lunar events (TLP) from 1540 A.D. (!) to 1967 October 19th. Patrick's extension would cover events from November 1967 to May 4th 1971, the last event to be reported prior to Patrick submitting his catalogue extension to NASA and to the BAA. The original report covered 579 events; Patrick's extension added 134 new events, through to event 713. The extension was heavily biased towards the periods when Apollo's 8 and 10–13 were orbiting the Moon, simply because more people were looking at the

Moon then; but nevertheless, the extension was a major piece of work. It was published in the August 1971 BAA Journal.

At the start of September Patrick made a brief visit to Moscow to present a paper at an International Geology meeting. The theme of the paper was that different areas on the Moon produce different sorts of TLP. Patrick's theory was that most of the red glows were in the Imbrium/Humorum region, while most of the colourless obscurations are around the Mare Serenitatis or Crisium regions. On his return Patrick described the Russians as 'Most interested in the whole TLP problem'. Later in September, Patrick was accompanied by Gilbert Satterthwaite and Commander Derek Howse (head of Astronomy at the National Maritime Museum) when he used the historic Airy transit instrument at the old Royal Observatory in Greenwich Park, to make the first scientific measurement with the telescope since 1954. Patrick had broadcast a *Sky at Night* about the observatory in February and resolved then to use the historic instrument once more. Shortly after using the transit instrument, Patrick heard news of a major dust storm developing on Mars: more about that later!

On Tuesday October 19th 1971 Patrick drove down to the Royal Greenwich Observatory at Herstmonceux in Sussex to witness a moment in history. The historic 28 inch (71 centimetres) refractor there was being removed and re-installed in its old home at Greenwich Park, as a visitor attraction at the National Maritime Museum. The instrument had been installed at Greenwich in 1893, but because of light pollution the telescopes were moved to Herstmonceux in the mid 1950s. However, the big refractor was largely a visual telescope and so was of little use at Herstmonceux. In addition, a modern telescope, unwanted by the government of the Congo (!) had become available for Herstmonceux, so the 28 inch was going home. As Patrick watched the giant refractor and its 30 foot (9 metres) tube being winched out of the dome a gale blew up and so the operation had to be postponed. Patrick made a *Sky at Night* programme about the move, featuring Cdr. Derek Howse, Gilbert Satterthwaite, John Millman and the retiring Astronomer Royal, Sir Richard Woolley. It would be transmitted 2 months later, on December 7th. As Patrick stated on the program, entitled *A telescope comes home*: "The next morning there was a 70 MPH gale and when I arrived at Herstmonceux the wind was really tremendous, and it was just as much as I could do to open the door to get inside the great dome". The program showed the massive Patrick, dressed in a brown suit and trousers, striding purposefully up to the dome, with everything flapping in the gale, and then pushing with all his strength to open the glass door into the observatory. Of course, with the slit opened to enable the telescope to be winched out the pressure inside the dome was substantial, especially when trying to open a door inwards. However, on October 21st, the removal teams, Patrick, and the TV crew, reassembled and so the operation was completed successfully. Two days after that, on the following Saturday, the telescope was returned to its old home in Greenwich Park and it is still there today.

Chapter 17
Back in Charge of the Lunar Section

The last Wednesday in October has always traditionally been the date of the BAA AGM, and 1971 was no exception. Before the AGM the council meets to discuss the state of the BAA and to resolve any issues, financial or otherwise. Prominent on the agenda for Wednesday the 27th October 1971 was the choice of a new director for the Lunar Section. Phil Ringsdore had then been acting director for 10 months, but did not want to be the director. There were three main reasons for this. Firstly, Ringsdore was a modest, backroom man; a tireless worker, but one who preferred to keep out of the spotlight. Secondly, he had enough work to do, producing the monthly circular and distributing it to nearly 500 observers. Thirdly, he and almost every other council member thought that Patrick was the best figurehead for the Lunar Section and the others knew that Phil Ringsdore would never accept the directorship. It must have been an interesting council meeting with Patrick and Phil Ringsdore, both vice-presidents of the BAA, insisting that the other would make a fine director, but everyone knowing that, ultimately, Patrick would give in to the inevitability that he should take on that role. Patrick accepted the post with pride and also with reservations, and on the front of the November Lunar Section circular he wrote the following:

> During the past session Phil Ringsdore has been acting Director of the Lunar Section. He has carried out these duties with his usual efficiency and energy; nobody could possibly have done better. It would have been a source of pleasure to the whole Section (and to others) if he had been willing to become permanent Director, but he was adamant that he did not wish to do so, despite considerable pressure being put on him (not least by myself!) When I was asked to reassume the Directorship I had considerable misgivings about it. For one thing, I feel that in most cases ex-Directors should not come back; and I also feel that I count as an old-fashioned Moon-mapper of the pre Space Age era. I have accepted the post because the Section Committee and the Association's Council wanted me to do so. I can only say that I will do my best, and that I will step down at any time should the Section wish it. But it is not fair to leave Phil Ringsdore 'holding the fort' any longer. My first act was to ask him to remain as Assistant Director, and I am glad to say that he has agreed to do so. There is a certain recurrence of events here. Virtually my first act in being appointed Director many years ago, when Orbiters lay in the future, was to ask him to become Section Secretary.

M. Mobberley, *It Came From Outer Space Wearing an RAF Blazer!: A Fan's Biography of Sir Patrick Moore*, DOI 10.1007/978-3-319-00609-3_17,

A Dusty Mars

At the same time as Patrick was settling in again as the new BAA Lunar Section Director, three space probes were approaching the planet Mars and there was much excitement about what they would find. These were not the first space probes sent to the red planet. In 1965, NASA's Mariner 4 obtained close-range images from a distance of 9,800 kilometres and, 4 years later, Mariners 6 and 7 were also successful, both passed within 3,500 kilometres of the Martian surface. However, Mariner 9 was due to enter Martian orbit on November 13th 1971 and was expected to take thousands of high resolution images of the Martian surface. In addition, the Soviet probes Mars 2 and Mars 3 were due to arrive at the same time. Apart from the competition between the USA and the USSR, Mars was making a close approach to the Earth in the autumn of 1971, so this was the optimum time to send probes to the red planet. Patrick had not overlooked the public interest in Mars at its closest and his childrens book entitled *Mars: The Red World*, illustrated by David Hardy, had just been published to coincide with the planet's approach.

Despite the low altitude of Mars from the UK (typically 17 degrees above the south horizon at best, when at its closest in August 1971) Patrick had been observing the planet regularly with his 12½ inch (318 millimetres) reflector throughout the summer and the autumn. On September 23rd Charles Capen, at the Lowell Observatory in Arizona, had reported a bright yellow cloud in the Martian region of Noachis. During October the dust cloud extended all around the planet, obscuring even such prominent features as the obvious Syrtis Major, much to the surprise of all who were observing Mars! The BAA observer Jean Dragesco, based at Yaoundé in Cameroon, submitted some excellent drawings of the planet and Patrick plotted the progress of the storm, from Dragesco's sketches, for the BAA Mars Section Director Edward Collinson. Patrick, and other UK observers, were able to follow the dust storm from their own back gardens, even if the planet was horrendously blurred and little higher than the tree tops!

However, shortly after the dust storm started officials at NASA phoned Patrick at Selsey and paid him to travel to South Africa and use the giant 26½-inch Innes refractor at Johannesburg to study the dust storm's progress over several weeks. NASA wanted to know when the storm was likely to clear and because the planet rotates with a period only slightly longer than that of the Earth only one side of Mars could be observed from the USA at any time, unless you waited 2 or 3 weeks for the 40 minute Earth-Mars 'day difference' to accumulate. Patrick was delighted, as the red planet was almost overhead from that latitude, but he was frustrated too, as much of the globe was virtually featureless during the storm.

Despite the fascination of such a planet-enveloping dust storm, the events on Mars were a problem. Patrick had scheduled a *Sky at Night* for November 17th as there should have been be a wealth of high resolution Mariner 9 images to show by then. But, just as the spacecraft had closed in, the planet had (almost as if to conceal its secrets) shrouded itself in the aforementioned dust. Patrick decided the programme should go ahead; there was fascination in the dust storm itself, and with guests

Gilbert Fielder (Patrick's lunar geologist friend) and Arthur Cross (a NASA Mars mapping artist) they found plenty to talk about. Unusually for Patrick, just after the November *Sky at Night* he was taken ill with "Something that may have been pneumonia". Fortunately, there was enough material from the Greenwich 28 inch refractor move to make the majority of a *Sky at Night* programme, leaving just a short interview with the outgoing Astronomer Royal to film; so Patrick could spend the end of November and early December in bed, looked after by his octogenarian mother and their housekeeper.

The Dundee TLP and Goodbye to a Friend

On December 6th 1971 Patrick's newly inspired TLP Network was put to its first big test when he was phoned from Dundee, by Morgan Findlay. Morgan had been observing the Moon (roughly 3 days past full) with the 10 inch (25 centimetres) Dundee refractor when, at 9.35 p.m., he noticed that the eastern floor of the crater Theophilus had a 'very distinct bright red-orange' colouring. At 9.40 p.m., messrs Ford, Taylor, Robbie and Bolger, also at Dundee, confirmed the appearance, but by 10.45 p.m. the feature was 'virtually nonexistent'. Patrick reported that he was clouded out at Selsey, but he phoned one of his many astronomer friends: the highly experienced observer, Richard Baum, at Chester. Baum, observing with a 4.5 inch (114 millimetres) refractor, recorded a brownish red patch in Theophilus, which he described as 'distinct but faint' and 'reminds one of the penumbral shadow during a lunar eclipse'. Baum's view was confirmed by fellow Chester observers Wilson and Kemp. Phil Ringsdore managed to cram the event as a last minute STOP PRESS item into the December Lunar Section Circular.

Patrick's elation at this TLP success was quickly flattened by the news that his friend Bill Baxter, the BAA's keenest solar observer at that time, had died on December 9th, aged 75. Baxter was famous in the amateur astronomy world for acquiring exquisite photographs of sunspots from his back garden observatory at 164 Gunnersbury Avenue at Acton, in London. Almost 11 years earlier, on January 30th 1961, Baxter had been Patrick's guest on *The Sky at Night* when sunspots were the topic being discussed. In the 1960s, Baxter had written Volume III of Patrick's *Amateur Astronomer's Library*, entitled *The Sun and the Amateur Astronomer*. The book was now out of print, but was still in heavy demand. Patrick had encouraged Baxter to revise his book and, despite other commitments (he was the BAA's Solar Section Director), he agreed to take it on, completing the first draft of the manuscript in September 1971. But, even in July, he had started to feel unwell, complaining of annoying chest pains. These became acute as he completed his manuscript, but he soldiered on; 1971 was, of course, before the time of routine heart by-pass operations and coronary stents. A few days before Bill Baxter died, Patrick, unwell himself, phoned him to see how he was feeling. Little did he realise he would never speak to him again and that Baxter would not live to see his revised book in print. Patrick wrote a tribute to Bill Baxter for the next Lunar Section circular.

However, Baxter's death would ultimately cause a 30 year rift between Harold Hill, regarded by many as the greatest living lunar observer, and Patrick. Aware that his cardiac problems were serious and that he needed to rest, Baxter had entrusted Hill with instructions for the final revisions required to complete his solar book. Hill gave the dying Baxter his solemn word that these revisions would be completed and Harold Hill was a solid, cast iron, man of his word: obsessively so. Hill had even been known to stand by postboxes in his village, with a watch, and admonish the postman if he was just 1 minute late collecting the mail! Strangely though, for such an intense man, Hill was a Flying Saucer enthusiast and a member of the Wigan Flying Saucer Club! Upon Baxter's death, Harold Hill contacted Patrick about the revisions and Patrick agreed they would be incorporated into the book, even though the author was now deceased. However, when the book was finally published, with Patrick as the series editor and the link to the publishers, none of Baxter's final revisions had been incorporated at all! Harold Hill was appalled. He had given his word to a dying man and told Patrick of his agreement, and yet here was Baxter's book, without those essential revisions! Baxter must surely be spinning like a whirling Dervish in his grave?

According to Richard Baum, who knew Patrick, Hill and Baxter well, Hill was invited, by Patrick, to review the published Baxter book, but Hill was sure that Patrick would not allow any publication of what he would have to say about the missing revisions. Hill was right; the critical review never appeared (Patrick blocked it) and the rift between Harold Hill, probably the greatest visual lunar observer of his generation, and Patrick, the man the public *thought* was the greatest ever lunar observer, existed up to Hill's own death in 2005. The simple fact was that Patrick wrote and edited books and articles and did *everything* at a blinding speed; he had no patience at all for revisions, proof checking, or anything that involved hassle. Getting permission from people to use their photographs was something he had no time for either. If he simply wrote a book, off the top of his head, and there were a few dozen errors, he could churn it out in a few weeks. If he triple checked everything, did some research and asked for the photo permissions, the hassle would take extra months of work. So, he ignored all such things! It gave him far more time to use his telescopes and Mars was the planet of the moment in late 1971.

So, despite the planet Mars' diameter shrinking to less than 9 arc-seconds in December 1971, Patrick followed the dust storm developments from his back garden and declared the storm 'largely cleared' by late December, despite not being fully recovered from his illness. As far as the space probes were concerned, Mariner 9 turned out to be a great success, returning 7,000 images when the dust storm finally cleared. The Soviet probe Mars 2 left a USSR pennant on the planet, but Mars 3 stopped transmitting 20 seconds after it successfully landed on Mars: a big disappointment. Even as Patrick's new *Guide to the Planets* book was appearing in bookshops he was preparing an updated edition incorporating Mariner 9's results.

Some time before the Mars 3 probe's failure Patrick had been delighted to be filmed for the 1971 Morecambe & Wise Christmas show! Although the appearance was very brief, starring as a 'top hat and tails' male guard of honour for Glenda Jackson, in a spoof of the Fred Astaire/Rita Hayworth number *You were never lovelier*, any appearance on that show was highly prized. Alongside Patrick, as the other top hat and tails performers were Cliff Michelmore, Frank Bough, Eddie Waring (just before Patrick in the line up), Michael Parkinson and, finally, Robert Dougal. All these extras had to say were the words "lovelier" and pull a grimace as they glided off past the camera. Patrick's bizarre facial expression to the TV camera was especially memorable! Being on the Morecambe & Wise Christmas show ranked as one of the greatest honours that could be bestowed upon an Englishman in the 1970s! Indeed, for many, the highlight of Christmas each year *was* the Morecambe & Wise Christmas Show.

Everything that Patrick did in amateur astronomy was done with total enthusiasm; his second term as the Lunar Section Director was no exception. Patrick was always keen to have many meetings, both in London and at 'out-of-town' or 'provincial' venues. The first Lunar Section meeting of 1972 was held on January 29th at the Royal Geological Society Rooms (opposite the RAS/BAA offices, at Burlington House, Piccadilly) and over 90 members (including this author, aged 14) attended. With Patrick organizing and participating in a meeting things would never be boring and the meeting of January 29th was far from that! Patrick had arranged for Mr Donald Francke to bring along a 16 mm format projector and a copy of the 1950s Sci-Fi classic 'Destination Moon' as well as a copy of a short newsreel of the Apollo 11 lunar landing. I vividly recall a member asking Patrick: "We'll need an entertainment license to show these films, surely?" To which Patrick replied: "No, we just need a working projector and mains electricity, unless there's a policeman in the room!" The main speakers at the meeting, as well as Patrick, were: Keith Abineri, speaking about how to start observing the Moon; Howard Miles, speaking about measuring crater depths (project Moonhole); and finally, Mr H.K. Robin, speaking about his method of timing lunar occultations of stars. Patrick also announced that a new Section Handbook was being prepared and he hoped to have the booklet commercially printed in the very near future.

Many readers of this book will never have experienced Patrick, in his prime, giving a talk. It was a sight to behold. Patrick's voice was always incredibly loud, but clear: like listening to a 1950s BBC newsreel, amplified through a loudspeaker. But Patrick never used a microphone; he just did not need one! The pace was rapid too, typically 200 words a minute. Unlike almost every other astronomy speaker, there were never any moments of hesitation or boredom. It was pure enthusiasm every time!

By February 1972 it was becoming obvious to Patrick that November's *Sky at Night* programme, the one partially thwarted by the global Martian dust storm, would have to be updated with another *Sky at Night*. The 7,000 pictures returned by Mariner 9, after the dust storms had cleared, were fascinating and so another programme was inevitable. Patrick's friend Dr Peter Cattermole, one of the most regular studio guests over the whole course of *The Sky at Night*, was invited to film the revised programme at the BBC TV Centre just prior to the March 1st broadcast. However, the curse of Mars had more tricks up its sleeve to thwart Patrick. Anyone older than about ten in the early 1970s will remember the numerous strikes in the UK at that time and the resulting hyper-inflation of around 25 % a year.

The early months of 1972 were in the era of Ted Heath and the miners' strikes and there was simply not enough energy available to keep the whole UK going! So there were power cuts almost every day in February. It was very likely that the BBC studio would be unavailable for the *Sky at Night* recording, so it was agreed to film the programme well in advance (rather than a day in advance) and just transmit the film on March 1st. That might have solved the problem, but no; Peter Cattermole phoned Patrick the day before the filming: he had the flu and could not make it!! So Patrick reviewed the results from Mariner 9 himself, in his study at Selsey, surrounded by camera equipment. Predictably a power cut occurred in the middle of filming, but portable generators were used to get past that particular hurdle. The programme was transmitted on March 1st, as planned.

On the evening of Sunday March 19th 1972 the Moon was predicted to drift across a number of bright stars in the famous Pleiades (Seven Sisters) star cluster. The Moon was a pleasant (in other words, not dazzling) 4 day old crescent in the western sky, so the stars would be occulted by the dark limb and not be swamped by moonlight. The planets Mars and Saturn were also in the western evening sky. Typical of Patrick, and his mother, he had invited a number of his friends around to his house at Selsey to view the occultation. These included a number of BAA members, namely: Laurence Anslow (positioned next to Patrick and using his childhood 75 millimetres refractor), Reg Spry from Selsey (using Patrick's Wildey/With-Browning 216 millimetres reflector), John Miller from Bognor (using Patrick's 125 millimetres Cooke refractor) and Roger Prout, also from the local area, using a 100 millimetres reflector. Patrick was stationed at the eyepiece of his own 12½ inch (318 millimetres) alt-azimuth reflector. Just before 19.45 GMT, Patrick and Laurence were waiting for the star SAO 076152 to disappear behind the invisible lunar limb. Simultaneously, they both exclaimed "It's fading!" Patrick estimated that it took maybe 0.3–0.5 seconds to fade away. This should not happen, unless the star is a double or multiple star system, as the Moon travels across the star rapidly, at roughly 0.5 seconds of arc per second of time, and stars do not show discs, they are point sources. The other three observers on Patrick's lawn were preparing themselves for the occultation of the bright Pleiades star Merope, so did not see the fade. Shortly afterwards Patrick learned that two Swiss observers, Brandli and Germann, had also seen the fade, as had Frank Tobin in north London. Brandli had estimated the fade as 0.7 seconds in duration and thought the star had turned red as it faded. He speculated that the fade might be due to a cloud of gas or dust above the lunar limb. Needless to say, this was of great interest to Patrick who contacted L.V. Morrison at the Royal Greenwich Observatory at Herstmonceux; he confirmed that the star was *not* a known double star. Patrick decided to instigate a new project, called 'Project Fade', within the Lunar Section, to investigate these phenomena. Subsequently, Morrison received a report of a photoelectric occultation of the star from the USA indicating that it was actually a close double star, so there was no need for a lunar cloud of gas to explain the observation.

By the early 1970s Patrick had taken a strong dislike to modern psychiatrists, not least because some of them had attempted to analyse him: a man approaching 50, devoted to his telescopes and his mother, and without a wife or a lady friend in sight.

The fact that the barking mad solar system theorist and author Immanuel Velikovsky (1895–1979) had once been a 'respected' psychiatrist pretty much confirmed Patrick's views about the whole bunch of them.

When a prominent and controversial psychiatrist, Professor Hans Eysenck, wrote in the magazine *New Scientist* in 1972, and Patrick's friend Henry Brinton, of the Old Mill House in Selsey, took him on in the letters column, Patrick decided to wade in too. His letter was published in the March 30th edition, p. 710:

> Sir, – The discussion between Mr Brinton (Letters, 24 February, p. 449) and Professor Eysenck (Letters, 16 March, p. 615) is very interesting, but it does seem to miss the main point – which is that psychiatry has ceased to be a science, and has become a cult. Modern psychiatry is as foolish as astrology, and much more harmful. Its effects on education are particularly unfortunate – and anyone who doubts this need only reflect that twenty years of "education by psychiatrists" has resulted in the "comprehensive jungle", to say nothing of establishments such as the London School of Economics.
>
> This being so, playing with words – the favourite habit of psychiatrists – is rather meaningless. I suggest that what we need today is less psychiatry and more common sense.
>
> Patrick Moore

Moving back to astronomy (phew!), The *Sky at Night* for April 12th 1972 was the 15th anniversary edition of the programme. Patrick chose to do a review of what had happened in astronomy since 1957 and invited three astronomers to join him in the review: two professionals and one amateur. The astronomers were Professor Anthony Hewish, a leading radio astronomer, Professor James Ring, an Infra-Red astronomer, and Cdr. Henry Hatfield, the well-known lunar photographer from Sevenoaks in Kent. Who would have thought then that the programme would still be being transmitted, fronted by Patrick, well into the next century!

Apollo 16

On April 16th 1972 the penultimate Apollo mission lifted off from the launch pad at Cape Kennedy, bound for the Descartes area of the Moon. The time was 17.54 GMT in the UK and Patrick was back in the BBC Apollo Studio. Onboard Apollo 16 were the astronauts John Young, Tom Mattingly and Charlie Duke. Young had been an astronaut on two Gemini missions and had already been within 10 kilometres of the lunar surface on the Apollo 10 mission. Nine years later he would be the first Space Shuttle pilot. Mattingly had been lucky enough to be pulled out of the traumatic Apollo 13 mission 2 years earlier with suspected German measles; this was his second chance. Charlie Duke had been the capsule communicator at Mission Control on the historic Apollo 11 mission and it was his words that everyone still remembers: "Roger Tranquility, we copy you on the ground. You got a bunch of guys about to turn blue. We're breathing again. Thanks a lot!"

The third stage Saturn IV-B was expected to hit the unlit part of the Moon at about 0100 GMT on April 20th and, once the astronauts had no further use for it, the Lunar Module would follow suit at sometime between the evenings of April 23rd and 25th. Once again, the aim was to crash the Lunar Module within 100 kilometres of the landing site to calibrate the seismometers. The landing site was at 9 degrees South and 15.5 degrees East. Patrick arranged for Phil Ringsdore to phone any positive observations of the impacts to him, in the BBC TV Studio.

A major scare caused much discussion in the BBC studio during the 12th lunar orbit. The Lunar Module (Casper) and the Command Module (Orion) had already separated ready for Orion's lunar landing when Mattingly detected a problem with the Command/Service Module backup engine firing system. Only minutes from starting the lunar descent, engineers on Earth were debating whether to scrub the landing; if there was a problem with the Command/Service Module (as in Apollo 13) the Lunar Module would be needed as a life raft. The BBC slot reserved for the coverage would have to be filled with waffle. As Patrick was almost the only team member not in the USA for the landing he was in charge. But even Patrick found it hard to talk about a situation for which there was no information. He must have said "We simply don't know" dozens of times. The TV farce made the *Daily Mirror* the next day. On the 15th lunar orbit Houston approved the landing and Orion subsequently touched down, many hours later than planned, on the Cayley Plains of the Descartes landing site at 02.23 GMT, on April 21st. They were about 300 metres from the target site, but only 3 metres from a 7 metre deep crater! Patrick pointed out to BBC viewers that John Young had become the first man to orbit the Moon on two different missions as he had also been on the Apollo 10 mission. (Jim Lovell had been to the Moon twice, on Apollo's 8 and 13, but not actually orbited it on the doomed Apollo 13 mission; they just swung round the back and returned to Earth). Only one other astronaut, Eugene Cernan, would equal Young's record by going back to the Moon and stepping on its surface on the second visit, but that would be on the next mission.

The astronauts carried out three EVA's on the lunar surface, and again had the use of a Lunar Rover. This vehicle was duplicated, once more, by a copy in the BBC studio. Unfortunately the normally flawless John Young managed to break the heatflow experiment cable, but Duke obtained a superb 3 metre deep core sample. The TV image quality, especially on the second EVA, was excellent, but frustratingly for Patrick, there was a distinct lack of volcanic geological features. Due to the lunar landing delay a revised lunar take-off time was declared by NASA, which reduced the third EVA to less than 6 hours. Nevertheless, the mission was record-breaking in every respect, with nearly 100 kilograms of moonrock being collected in 20 hours of work on the lunar surface. There was yet another scare when the Lunar Module was separated from the Command Module. A switch was left in the wrong position causing the Lunar Module to tumble dangerously. Houston told the astronauts in the Command Module to take evasive action and, much to the disappointment of the BAA's project Moonglow team, the Lunar Module was not crashed into the lunar surface after all. Once again, nothing was seen impacting the Moon, even in the case of the third stage of the Saturn V rocket. There was much discussion in the BBC studio about the diet the astronauts were on; it was a special solid food/orange juice mix, laced with potassium, which was effective in eliminating the previous problem of irregular heartbeats (something which Patrick had suffered from all his life). The crew splashed down safely on April 27th and Patrick's life could, once again, return to normal.

Yet More Books

After the Apollo 16 mission Patrick was able to finish off a number of major new book projects he had been working on. The first was a colourful collaboration with David Hardy, the space artist who had first worked with Patrick in 1954 on his book *Suns, Myths and Men*, and had provided the illustrations for many of the early *Sky at Night* programmes. Their joint book, *Challenge of the Stars*, was a gaze into the future exploration of the Solar System, lavishly illustrated with Hardy's superb paintings. Hardy had only been 18 when he illustrated Patrick's first book; he was 36 now. It would be almost a quarter of a century before Hardy and Moore were on such friendly terms again, as we saw earlier in the book. (To recap: Patrick thought David had stolen some book contracts from him, but he had been paranoid and wrong!) The new book's foreword had been written by Patrick's old friend Arthur C. Clarke, by then of course world famous as the author of *2001: A Space Odyssey*.

Fig. 17.1 The world-renowned Space Artist David A. Hardy played a crucial role in Patrick's early career, providing the critical illustrations for his book *Suns, Myths and Men* in 1954. Here he is shown with Patrick in 1972, discussing their joint book *Challenge of the Stars* (Image by kind permission of David A. Hardy)

The second book project nearing completion was completely different. Since Patrick's 1969 BBC broadcast *One Pair of Eyes*, on the subject of barking mad cranks, more politely titled 'Independent Thinkers', Patrick had been slowly collecting enough material to form a book. By the spring of 1972 the book was complete and was published, by David & Charles, later that year. If you want to know everything there is to know about the deluded nutters that were around in the late 1960s and early 1970s, this book is for you. Patrick decided to call it *Can you speak Venusian?* It is a very entertaining read! At about the same time he also completed the *Sky at Night* Volume 4, a summary of many of the programmes broadcast between 1970 and 1972. Finally, in collaboration with Lawrence Clarke (the son of his childhood friend Pat Clarke) who drew the diagrams, Patrick produced a charming little Corgi mini-book entitled *How to Recognise the Stars*. This was a pocket-sized paperback and the constellation diagrams were so well thought out that no-one reading it could fail to learn their way around the night sky. This was a bargain at 20 pence and one of my favourite little books of all time. The front and back covers showed a photograph of Patrick at the helm of his 5 inch Cooke refractor, a picture that, for me, spoke a thousand words. The photo was taken by a certain Beverly Lebarrow. I used to think that whoever *she* was, she had captured Patrick, and the Cooke refractor, perfectly. However, rumour has it that Beverly Lebarrow is just another name for Beverley Goodway (1943–2012), a male photographer, who for 30 years from 1973 to 2003 took *The Sun* newspaper's topless *Page 3* model pictures. Hmmm, Patrick was, perhaps, a rather different photogenic challenge!

Lunar Section Projects and BAA Meetings

With all the aforementioned projects cleared from his desk, promoting the BAA Lunar Section and promoting TLP was back at the top of Patrick's agenda. For some time Patrick had been formulating a theory about different classes of TLP, and he put his ideas forward in a paper, submitted to the BAA Journal, in early 1972. In that paper he wrote:

> I have re-plotted the TLP sites, trying to be more selective than previously; sites which depend upon one report only have been ignored except in one or two special cases. I have also distinguished between red events and other events – the latter category including local obscurations, colourless TLPs, etc. The results are of interest. The red events are commonest in the second quadrant (N.W., astronautical; classically N.E.) and, to a lesser extent, in the third quadrant (astronautical S.W.). There is an obvious association with the border of the Mare Imbrium and the ill-defined boundary between the Oceanus Procellarum and its neighbours. There are very few 'red sites' in the first and fourth quadrants (eastern hemisphere, astronautical); but the Maria Crisium and Serenitatis peripheries are equally clearly linked with non-red TLPs. So too, to a less obvious extent, are Maria Tranquillitatis, Vaporum, Nectaris and Foecunditatis.
>
> In short, there is a definite tendency for different areas of the Moon to produce different classes of TLPs. This certainly lends added support to the gaseous emission theory favoured by McCall, but in this paper I propose only to report my results rather than theorize about them.

Patrick was also forging ahead with the Lunar Section Handbook and with a whole host of Lunar Section Meetings around the country. During the same period, all the other BAA Observing Sections either had no meetings at all, or annual meetings. The Lunar Section was by far the most active and had far more members. Patrick drew up plans to have another Lunar Section meeting in London on May 20th, one in Dundee on June 10th and another one in Hull, on September 23rd. At the March BAA meeting the council had given approval for Patrick to have the Lunar Section Handbook professionally printed, even though the cost of £140 for 500 copies was considered excessive. Patrick was confident that most Lunar Section members would order a copy by post and so the council had agreed to proceed. Patrick was very vocal in his thanks to the BAA President Barocas and to the Treasurer, Mr Vince, but made it clear that he thought the Royal Mail, a year after the national postal strike, was, if anything, worse than ever, as well as being more and more expensive. He vented his feelings on this to anyone who would listen and the *Daily Express* for Tuesday 23rd May 1972 carried a story headlined 'Patrick Moore threw a rocket at the Post Office'. The trigger to this had been that with the post being so slow Patrick had decided to see if he could successfully send a Telegram in an emergency, such as if he spotted a TLP, or a new comet, or a nova. So, as an experiment, he sent a Telegram from his house to a friend at Highleigh, just 4 miles away. The friend received the Telegram 21 hours later! Patrick calculated that his Telegram must have been travelling at 335 yards per hour and so he contacted his right-wing friend Ross McWhirter, co-editor with his twin brother Norris of *The Guinness Book of Records*, claiming a world record for the Telegram's slow speed. An embarassed Post Office spokesman quickly admitted that they had failed miserably and that Patrick's fee would be returned immediately.

The London Lunar Section meeting at the Royal Geological Society on Saturday May 20th was a great success, boosted by the fortuitous presence of Dr Donald H. Menzel from the USA, who was able to show the latest Mariner 9 Mars images to those present. While not strictly a lunar matter, it was of great interest to members to see the degree of cratering on Mars, with respect to the Moon. Patrick had acquired the proof copy of the Lunar Section Handbook just before the meeting and proudly showed it to those present, urging all members to part with the 30 pence necessary to buy a copy. Another speaker at the meeting was Mr Harry Ford of the Mills Observatory Dundee, where confirmation of the December 6th TLP had been obtained. Mr Ford described the 10 inch refractor in use at Dundee, explaining that it had once been owned by Walter Goodacre, the Lunar Section Director between 1896 and 1938.

A few days later Patrick was speaking, by phone, to Dr Joseph Brady of the Livermore Observatories in the USA, about Brady's amazing claim that the tenth planet was in Cassiopeia and might even be as bright as magnitude 14. Patrick also got the opinion of Dr Clyde Tombaugh, the discoverer of Pluto, and a *Sky at Night* was broadcast on the subject of Planet Ten on May 24th.

Only a week later and Patrick was at another BAA meeting: the annual Exhibition meeting on Wednesday May 31st. This was held at the Royal Commonwealth Society in London's Craven Street. The Exhibition meeting is the annual meeting where the BAA observing sections show the results of their observations during the last year, using posters and exhibits, but it always includes a formal 'Ordinary Meeting'. Patrick and Phil Ringsdore exhibited numerous displays for the Lunar Section with Patrick providing a map and a catalogue of the TLP regions and an unusual Russian lunar globe.

The highlight of the Exhibition meeting was the award, to the incredible George Alcock, of the BAA 'Merlin' award for the second time: an unprecedented event. Alcock had, by 1972, discovered four comets, three novae and also memorised 30,000 stars: an amazing achievement. Standing in for the President, the Vice President Gordon Taylor told the audience that, after being the first recipient of the Merlin medal in 1961, Alcock might have rested on his laurels, if he had been a lesser man. But Alcock had now notched up an incredible 3,000 hours of searching the night sky! To loud and prolonged applause Alcock accepted the award, saying that he thought it should go to his wife, as she had to reconcile herself that if it was clear, observing was an absolute necessity! Later in the same meeting Mr Hedley Robinson, talking on the subject of the ghostly Ashen light on Venus, that is, the controversial glowing of the dark side, said that Patrick had reported seeing the phenomenon recently. Patrick, however, said he was not confident that the effect was real as he had not on this occasion used a curved occulting bar to blot out the light from the crescent. The Jupiter Section Director Bill Fox, also speaking at the Exhibition meeting, drew attention to Patrick's drawing of the giant planet, made two nights earlier, which clearly showed that there was no evidence of the Great Red Spot fading.

Ten days later and Patrick was in Dundee, this time chairing the June 10th Provincial BAA Lunar Section meeting, which was a great success. The meeting was opened by Mr James Mullholland, Dundee's Parks Convener, who welcomed the Lunar Section to the city. He said: "If John Mills, who donated the money for the Observatory to be built, was alive today, he would be proud that the Lunar Section of the B.A.A. was meeting in his Observatory." Patrick commented that the 250 millimetre Cooke refractor at Dundee was once owned by Walter Goodacre who was famous for his fine lunar maps. Patrick presented his copy of Walter Goodacre's original, unique, observing notebook to the Mills Observatory, stating that the book should be kept with the telescope Goodacre used. Needless to say, the December 6th 1971 TLP observed from Dundee and Chester, was a major talking point. Shortly after the Dundee meeting, Patrick announced that he had found someone to co-ordinate the occultation activities of the Lunar Section: a Mr Ken Gayner of Bristol.

For the June 1972 *Sky at Night* Patrick had decided that a programme about Stonehenge and the Summer Solstice would be in order. There were two reasons for this. Firstly, Professor Gerald Hawkins, who had theorised that Stonehenge was an ancient eclipse calculator, was in the UK. Secondly, the *Sky at Night* was due to be transmitted on mid-summers day, June 21st. Patrick, with Professor Hawkins and the BBC film crew, filmed sunrise over the famous Stonehenge 'heel stone' on the day before transmission. Skies were clear and they enjoyed a good view of sunrise. The next day, the Druids were not so lucky: it rained! Patrick often mentioned that he'd once heard an American visitor at Stonehenge comment that the monument was built much too close to the freeway!

Shortly afterwards, on July 1st 1972, Patrick appeared on an early BBC 1 Parkinson programme; it was programme number 5 of Parkinson's second series. The film star Tony Curtis was also on the show. It was a very strange contrast of styles, with Patrick looking distinctly out of place at times, despite having already had his own TV series for 15 years. Also on that particular show was the glamour model Vivian Neves! Patrick could always produce eccentric humour on such occasions though. Michael Parkinson and Patrick had one thing very much in common. They were very keen cricket fans. Patrick's strength was bowling and when Parky asked him about his worst performance Patrick was only too happy to elaborate. He admitted that 11 runs were once scored by the opponents from just one of his balls: "Wretched man hit the ball into the outfield, it went into a rabbit hole….Fielder forgot to shout lost ball and by the time it was found they'd run 11!"

In the July Lunar Section Circular Patrick commented on the success of the London and Dundee meetings and was pleased to confirm the appointment of the new occultations co-ordinator, Mr Ken Gayner, of Bristol. Meanwhile, Patrick's plans were also well advanced for the September 23rd Lunar Section meeting in Hull as were his plans for a paper he was delivering at the September 30th BAA Bristol meeting, on the subject of Martian Dust Storms. It was all GO in Patrick's world, with no time for a rest.

In August, Patrick spent a few days on the island of Hven, near to Malmö in Denmark. Hven had been the observatory home of the famous astronomer Tycho Brahe and Patrick wanted to do research for a forthcoming *Sky at Night* about the astronomer.

The Lunar Section meeting in Hull was the first for quite a while held in collaboration with one of Britain's senior universities and it was a major success, but for the first time since Patrick had instigated the Lunar Circulars the October issue could not be produced. Phil Ringsdore, the backbone of the BAA Lunar Section, had returned from a fortnight's holiday in Jersey to find he had a buyer for his home at Stoneleigh, near Epsom. At last he could move to where he wanted to be, namely Burnham-on-Sea in Somerset. The November Circular would have to be a joint October-November one, but it was a miracle that only 1 month of disruption was experienced.

I must digress at this point to tell a really amusing anecdote, told to me by one of the UK's top lunar photographers and imagers, Mike Brown of York. Mike knew all of the leading lunar section people like Patrick, Phil Ringsdore, Peter Sartory, Henry Hatfield and Ron Maddison in the 1960s and 1970s. To quote Mike precisely:

> A Lunar Section meeting was held at Hull University, can't remember the date, but what sticks in my mind was the superb buffet laid on for us. I was talking to Patrick, who'd got a huge cream cake on his plate and he gesticulated at something I said and the cake fell on the floor - completely unseen by Patrick. When I told him what he'd done he said "Oh Dear" and slyly pushed the cake under the table with his foot where it was hidden by the droop of the tablecloth!

The Famous 15 Inch Newtonian

Due to his ill health Phil Ringsdore would have to give up his large 15 inch reflector upon moving to Burnham-on-Sea. A number of amateur astronomers in the Ewell area were told that Phil was selling his impressive telescope, not least members of Phil's local astronomical society. Indeed, some of them even made tentative bids for the instrument. However, much to the frustration of many, Patrick had a word with Phil and outbid them.

It seems incredible, but Patrick had never owned a large, equatorially mounted telescope until that year of 1972, when he was 49. He had owned his altaz 12.5 inch reflector (named 'Oscar') since 1949, but had only used it from 1951 onwards. He had also used an 8.5 inch Horne & Thornthwaite altazimuth Newtonian from 1950 (until the 12.5 inch was fully operational) and his 8.5 inch With/Wildey-Browning equatorially mounted reflector (formerly owned by Greenwich Observatory) since 1960. But he had *never* owned a decent, driven reflector of large aperture! In truth, Patrick's connection with the Ringsdore telescope went back far before it was owned by Phil. In the mid 1960s the wealthy business man, Peter K. Sartory, had originally built that 15 inch reflector. Sartory had retired and sold his microscope and precision optics manufacturing company some years earlier and honestly admitted he had more money than he knew what to do with. Indeed, as we saw earlier, he virtually gave Patrick his own 5 inch Cooke refractor, for a token £10, just a few years later!

Sartory was a keen amateur astronomer, and highly active in Patrick's BAA TLP network. As I mentioned previously, he had manufactured the first 'Moonblink' red-blue filter holder for the Lunar Section. Sartory decided to equip his magnificent Surrey Tudor mansion, *Mellow End*, situated in 4 acres of grounds off the Littleworth Road at Seale, Farnham, with a really impressive telescope in a brick-walled dome. He manufactured an octagonal mahogany tube for this instrument and acquired a 15 inch f/6 mirror for it. However, that 15 inch telescope, in its early life, was slightly different to the one Patrick eventually acquired from Phil Ringsdore. Sartory had mounted his telescope on a German equatorial mounting, one where the telescope weight is counterbalanced across the declination axis with a huge set of counterweights. He also had a 6 inch Maksutov mounted piggyback on the tube for solar observing. When Peter Sartory became too frail for regular observing he had originally approached Patrick with a view to him owning it. However, at that time (1967) Patrick was still at Armagh in Northern Ireland and a new huge telescope would not have been convenient. So, Patrick suggested that his good friend Phil Ringsdore, living just 25 miles west-northwest of Sartory's observatory, would be a good recipient. In fact, Patrick, Phil and Peter were all good friends who often met up at Peter's RAC club in London.

Anyway, the 15 inch telescope transferred from Peter Sartory to Phil Ringsdore in its original 'German' specification in 1967, but 5 years later, in 1972, Patrick wanted it on a low-slung Fork mounting, where the telescope is more user-friendly. However, a huge Fork needs manufacturing to specifically fit the tube width. Patrick had long admired the design of his old Selsey friend Henry Brinton's telescope. That was on a German mount, but one where the metal pillar was inclined to be parallel to the polar axis, thus the tube never hit the pillar (strictly speaking this design is called a German Astrographic Mounting). The really great feature of Brinton's telescope though was its rotating top end; a very useful facility for blissful user comfort at the Newtonian eyepiece and one that was quite common on Victorian and turn of the century (twentieth) Newtonians. Even Will Hay's 12½-inch f/7 Newtonian had employed a rotating top end. However, Patrick really wanted a Fork mount, mainly for its convenience, but partly because he hated anything with German connotations, even if two of his existing telescopes were already on German mountings! Also, with a Fork you never have to swap the tube to the other side (to avoid hitting the plinth) just as you move through the south meridian, where the Moon and planets tend to be at their best. Another influence on the choice of a Fork mount had been Frank Acfield's splendid 10-inch f/9 Newtonian at 61 Great Lime Road, Forest Hall, near Newcastle-on-Tyne. In the 2 years since Patrick's visit to Frank (for the May 1970 *Sky at Night*) he had wished he owned a similar long focus Fork mounted reflector, within a dome. Also, coincidentally, Patrick's 15-inch f/6 reflector was the same focal length as Frank's splendid telescope. I might add, for the non-astronomer, that all mounts with the tube at one end and the counterweights at the other end of the declination shaft are called 'German' after the design of the German Dorpat refractor of 1824. Patrick's 5 inch refractor and his 8.5 inch reflector were both on German mountings, but the 15 inch was to be the main instrument at Selsey and Patrick wanted it to be his ultimate 'dream' telescope.

At this point I will digress yet again; this time into the history of that 15 inch mirror. Patrick always stated that the original mirror maker was unknown and that Sartory had never said who made it. However, Patrick was very confused about the mirror's history and also said that his friend George Hole had re-figured the surface at some point. But there is a separate account of where the mirror originally came from. Some time earlier an employee of the company *Optical Surfaces* (a company well known to many 1960s BAA members, and employing BAA Council member Jim Hysom) had seriously damaged an 18 inch mirror blank in their workshop. I was told that an apprentice had left a mirror grinding and polishing machine unattended and it had trashed the mirror rim, leaving damage even 1 inch in from the edge. The apprentice was eventually relieved of his post, but the mirror was salvaged by a senior employee, named John Mortleman, who ground the wrecked 18 inch blank down to a 15 inch disk and made a decent telescope mirror out of it. However, it was a mirror without a purchaser. Around this time Patrick's friend George Hole (remember the 50th *Sky at Night* and Hole's 24 inch reflector farce?) acquired this 15 inch mirror from Optical Surfaces Ltd and reworked it slightly, declaring it to be "as fine as a mirror could be". That 15 inch f/6 mirror went into Sartory's telescope and so became the heart of Patrick's biggest telescope. This story rings true for me, because Patrick's 15 inch mirror was ludicrously thick for its diameter. Even in the 1960s, when mirrors were thicker, you rarely saw one with a diameter to thickness ratio less than 6:1. Indeed, this was the standard historical 'rule-of thumb' ratio. Patrick's 3 inch thick mirror (5:1) corresponds perfectly with an original diameter of 18-inches. Its final artisan, George Hole, was both a mirror and a telescope maker in his rather badly organised company, and Patrick and his Selsey friend Henry Brinton had used their own funds to keep Hole afloat on several occasions.

In late 1972, with the 15 inch telescope in his possession, Patrick asked Dudley Fuller, the owner of Fullerscopes (later Broadhurst Clarkson & Fuller, initially of London's Farringdon Road) to make him an especially low-slung Fork mounting for Sartory's mahogany tubed telescope. He also asked Dudley to get his engineers to make the top end of the 15 inch tube rotate, like Henry Brinton's tube did; the feature that Patrick had always admired. This would have to be engineered very carefully if the telescope was not to lose collimation. The perfect 'sweet spot' of an f/6 Newtonian is just under 5 millimetres in diameter at the focal plane and that tiny region is where the light from the Moon and planets is undistorted and where the eyepiece axis needs positioning. Even a few millimetres of misalignment will make the images less than perfect at high magnifications. Fuller's folder on that 15 inch and the engineering work carried out on it still exists, and he told me about the work in a telephone call on March 20th 2005. Fullerscopes' engineers measured the focal length of the 41 pound (19 kilogram), 3 inch thick (76 millimetres) mirror at just less than 91 inches, making it between f/6.1 and f/6.0. My own measurements made it very close to f/6.0. Patrick always claimed it was f/6.2, as the mirror had '93' scrawled on the back, which he had assumed to be the guesstimated focal length in inches; apparently not! They also determined that the 2.5 inch minor axis secondary mirror was barely large enough to collect the light, so they re-positioned that secondary mirror for optimum performance.

Although the 14 inch diameter, 359 tooth, sidereal drive wheel and its associated worm system (incorporating a smaller piggyback 6 inch wheel and worm for fast slewing) were Fullerscopes' design, the fork itself had some history behind it. That big Fork was originally owned by Horace Dall for one of his pioneering Dall-Kirkham Cassegrains (a 14 inch model). The Dall instrument had, apparently, looked like three tea chests bolted together. The huge triangular base, the polar axis, and the sidereal drive worm and wheel system, mounted beneath Patrick's Fork, are virtually identical in construction to that used on the huge 18-inch f/7 Newtonian at the Charterhouse Observatory in the Mendips hills in Somerset. That instrument was also constructed by Fullerscopes in the early 1970s, but has a German equatorial declination shaft atop the head, rather than a Fork. Having used both instruments I can truthfully say that Patrick's Fork mounted Newtonian, with its rotating top end, was at least 100× more user-friendly than the giant Charterhouse telescope!

After the telescope engineering work was completed the Fork mounted instrument would eventually be erected in a major new observatory at Selsey. This, of course, would be that unique 16 foot diameter, cylindrical, green 'Oil Drum dome' or 'Pill box' at the south end of Patrick's garden. This unusual structure was built, to Patrick's specification, by the local building and engineering firm of 'Males' at Selsey. Whenever it was possible Patrick would employ local people to carry out his projects, whether it was building or publishing. The new observatory was specifically designed to allow plenty of room for bulky TV cameras, cables, producers and cameramen to fit inside, and for them to be able to work within the dome while filming on *The Sky at Night*. Patrick was not a man to have the patience to apply for a planning application of any sort and so the dome height was set at 9½ feet, just a few inches below the height needed for planning permission (nowadays set at 3 metres or 9 feet 10 inches). The observatory was also sited more than 16 feet from the house, to avoid it being classed as an extension. The last thing Patrick wanted was some 'jobsworth' bureaucratic pen-pusher telling him that his new observatory was illegal the second it was constructed. As Patrick would eventually write a book called 'Bureaucrats: how to annoy them' (as well as 'The Twitmarsh Files') he was only too aware of how these sad people operated!

A week after the 1972 BAA Lunar Section meeting in Hull, with his new observatory nearing completion, Patrick delivered a controversial paper to a BAA audience at Bristol. Patrick was no stranger to controversy of course; the whole subject of TLP research was highly speculative, yet Patrick had based his observing reputation upon it and on volcanic activity being the main crater forming mechanism on the Moon. Patrick had been studying the frequency of Martian dust storms following the global event at the end of 1971. He had noticed that there seemed to be major Martian dust storms at perihelic oppositions, namely those years when Mars is at its very closest and best-placed for observation from the Earth's southern hemisphere (it is always too low down for the best views from the far northern hemisphere at such oppositions). An opposition of Mars is simply the period when Mars is closest to the Earth at that particular apparition, in other words the Earth is 'overtaking' Mars and the red planet is between 56 and 99 million kilometres behind the Earth. A perihelic opposition is one where Mars is, typically, only about 60 million kilometres away. Patrick also speculated as to whether the major dust-storms could be caused by dust ejected by *recent* volcanic activity on Mars and whether the close proximity of the Earth (if 60 million kilometres can ever be considered close) might influence Martian vulcanism! Needless to say Patrick's theory was questioned by those at the meeting, in particular by his friend Hedley Robinson. In a subsequent write-up for the December Journal Patrick wrote: 'I am well aware that this idea is highly speculative, and that it may prove to be completely wrong. I propose it merely as a tentative hypothesis which may possibly merit more detailed investigation.'

Apollo 17

By the end of the year the last NASA moon mission, Apollo 17, was looming and in October 1972 Patrick received instructions from the BBC that, for this Apollo, he would not be in the London studio, but in Mission Control at Houston. However, he would be at Cape Kennedy for the launch; an exciting prospect! Initially, Patrick's friend Miss Rossie Atwell, who had recently joined the Lunar Section committee, had managed to book the Swedenborg Society rooms in London for a December 16th meeting. However, it was doubtful whether Patrick would be back by then, so the meeting was scrapped. The visit to the USA would involve two separate trips as the BBC wanted Patrick to film various interviews while he was in the USA, and as a lunar observer Patrick had invitations to various technical conferences on lunar geology. Patrick was much happier on short trips abroad; two long ones in quick succession were things to dread. Apart from the hundred plus letters a week he answered, he was constantly book writing and he loved his Selsey home, mother, cat, housekeeper and telescopes, as opposed to living out of a suitcase in various hotel rooms. Patrick departed to the USA for the first trip on November 11th, more than 3 weeks before the launch of Apollo 17. He would just have to finish his latest book, *Guide to Comets*, and sort out the Lunar Section business after Apollo 17 ended. After a brief return to Sussex in late November, during which he was able to deliver a talk to the Royal Aircraft Establishment at Farnborough, and see their observatory with its 6-inch Coudé refractor, Patrick departed for Florida again at the start of December. This time he would cover the Apollo 17 launch and then move to Houston, in Texas, for the rest of the mission. Accompanying Patrick was James Burke and the Apollo production team.

On the early evening of December 6th 1972, Apollo 17 was ready for lift-off. This would be a night-time spectacle; the first time a Saturn V had been launched in darkness. Onboard the spacecraft were two astronauts, and a geologist trained as an astronaut. The latter was Dr Harrison (Jack) Schmitt who would accompany Eugene Cernan to the Moon's surface. The Command Module pilot was Ron Evans. Various technical problems during the countdown pushed the launch well into the night, but at 11.33 p.m. Central Standard Time, 3 hours late, Apollo 17 turned night into day for southern Florida as over 3,000 tons of rocket and fuel lifted off the launch pad. It was Patrick's only experience of a Saturn V launch; he had always been in the studio before, but this was altogether different. Patrick simply said that it was impossible to give an adequate description of the scene. The sound of the rocket launch took almost 20 seconds to reach Patrick, as the press were kept 4 miles away. But, when it arrived, it was like 'being hit by a wall'. Later, Patrick admitted that he was 'acutely aware that he was watching the start of the last part of the most ambitious project ever undertaken by mankind'. Apollo 17 was heading for the Taurus-Littrow area of the Moon, on the edge of the Mare Serenitatis. The Commander, Eugene Cernan, like John Young and Jim Lovell, had been to the Moon before; in fact, on the same Apollo 10 flight as John Young. So, (as I mentioned earlier) like John Young, he would orbit the Moon on two separate missions and walk on the Moon on the second one. Apollo 17's Lunar Module, named Challenger, would land in an 11 kilometre wide valley, sandwiched between 1,400 metre high mountains on the one side and 2,100 metre high mountains on the other side. This was an ambitious mission, but as Patrick was well aware, the NASA astronauts had originally wanted a grander finale to Apollo. They had suggested a landing *inside* the crater Tycho, or even on the far side of the Moon. Unfortunately, NASA ruled these out for fuel and communications reasons and not for safety reasons related to the terrain.

The Taurus-Littrow region was selected because it was well away from other lunar landing sites and was thought to contain a mixture of young and old features. The lunar landing went perfectly and Challenger came down within a few hundred feet of the target. Cernan and Schmitt carried out three EVA's and the heat-flow experiment, damaged on Apollo 16, was finally carried out. There was much excitement on the second EVA when the geologist Schmitt spotted some orange soil near a crater nicknamed 'Shorty'. The orange colour might have indicated the historical presence of moisture, but this did not, ultimately, turn out to be the case. Once again, the Lunar Rover camera showed the Lunar Module ascent engine firing to return Cernan and Schmitt to their colleague, Evans, in the orbiting Command Module. The next day, December 15th, Patrick and James Burke presented a special extended summary of Apollo 17 highlights on BBC1, at 1.45 p.m.; it was entitled 'The last Apollo: Farewell Moon'.

Apollo 17 splashed down on the 19th December 1972; it was the end of an era, for Patrick and for mankind. He would never again enjoy quite so much TV coverage and the BAA's membership, then standing at a record 6,000 members, would never be as high again. The Moon landings, and Patrick on TV, had marked the golden years of the BAA, at least, as far as membership was concerned. With Patrick at the very height of his fame a founding member of the Bonzo Dog Doo-Dah Band, Roger Ruskin Spear, even devoted a track to him, entitled 'Patrick Moore', on his solo album *Electric Shocks*.

Chapter 18
A Telescope, a Comet and the Monte Umbe

As the last men returned from the Moon, Patrick returned to Selsey, and was keen to use his latest and biggest telescope, that 15 inch (38 centimetres) Newtonian. The new observatory was up and running! He was also keen to show his new toy to the world and so the 1973 February 7th *Sky at Night* would be about amateur observers and observatories. The BBC hooked up a TV camera to the 15 inch and *Sky at Night* viewers were able to see Saturn as it appeared through Patrick's telescope.

Fig. 18.1 Patrick's famous 15 inch f/6 Newtonian, formerly owned by Peter K. Sartory and, for a time, by Phil Ringsdore. Patrick had Dudley Fuller's engineers mount the telescope on a Fork and they added, at his specific request, a rotating top end. The instrument modifications were completed in 1972, although this picture was taken in the 1980s (Courtesy of the BAA/John Mason)

M. Mobberley, *It Came From Outer Space Wearing an RAF Blazer!: A Fan's Biography of Sir Patrick Moore*, DOI 10.1007/978-3-319-00609-3_18,

This was the telescope Patrick had always wanted: a large, but easy to use reflector, with a wooden tube in a proper observatory. As I have already mentioned, the telescope was, according to Patrick, f/6.2, giving a focal length of some 93 inches. On that Feb 1973 programme he described it as "fifteen and a half inches"; but I have measured every dimension of that telescope and I made the mirror 15 inches *exactly* and a touch over f/6.0. It was a big instrument, but the low slung fork design ensured that the eyepiece was never more than about 7 feet above the observatory floor when looking at the Moon and planets. In addition, the telescope had that vital feature Patrick had always wanted, ever since he had seen it on Henry Brinton's 12½ inch reflector: a rotating top end, so the eyepiece could be conveniently swivelled around for easy viewing. It cannot have been perfect for telescope collimation, but it did make life easier. Comfort is a major factor for the visual observer. I realise that some readers, who are telescope fanatics, may be interested in all the critical dimensions of Patrick's 15 inch f/6 Newtonian, so here they are.

Tube: Octagonal, 84 inches long and 18 inches wide between octagon flats. The Fork 'Dec. axis bearing' fixing points were positioned 23.5 inches along the tube from the primary mirror end of the tube and that end of the tube was very heavily counterweighted. When pointing at the zenith the primary mirror end was 12 inches off the floor and the eyepiece was roughly 95 inches above the floor. Internal diameter of rotating top end ring = 15.5 inches. Primary mirror surface to secondary mirror surface distance = 76 inches. Secondary mirror surface to focal plane = 14 inches. Fork: Dec axis mid-point just under 3 feet above the observatory floor. Fork tines are 28 inches in length, separated by 21 inches and are 2 inches thick. One 8 inch diameter declination circle was fitted, but no R.A. circle. When you know the sky as well as Patrick, and the instrument is polar aligned, all you need is a good declination circle and a decent finder telescope.

After the cancellation of the December 16th 1972 Lunar Section meeting, Patrick's friend Rossie Atwell managed to reschedule the meeting for February 24th 1973, at London's Swedenborg Society rooms at 20 Bloomsbury Way. Patrick arranged for the previous Lunar Section director, Ron Maddison, to address the meeting. Phil Ringsdore was still the assistant director at this time, but living in Somerset made communication somewhat fraught. Patrick was often quite vocal in his opinion of Britain's industrial disputes, trade unions, British Rail, British Leyland (the car makers) and the GPO. On this latter subject he wrote, in the February 1973 Lunar Section Circular: 'Some of the delays and troubles have been caused by the post. I have only just realised how bad the service has become. It takes a long time for a letter to get from Selsey to Burnham – if indeed it arrives at all'.

In addition to the distance between Patrick and Phil Ringsdore, and the postal problems, it had been known to Patrick, for some time, that Phil was now a very ill man. Phil had put a brave face on things for many months, but it was obvious that he might not be around for much longer. Despite his illness, Phil Ringsdore continued editing and producing the Lunar Section Circular, but it was clear that this could not continue for ever. Patrick proposed to the BAA council, in February, that Phil Ringsdore should be awarded the Association's highest honour: the prestigious

Goodacre Award. The council agreed, and Patrick just hoped that Phil would be able to travel to London to receive the medal and gift in October.

The February Lunar Section meeting was the first one for many years that Phil Ringsdore had been unable to attend and Patrick voiced how much Phil was missed in the April circular. There was also some very sad news in that an old friend of Patrick's, Professor Samuel Tolansky, had died suddenly. Patrick had arranged for Professor Tolansky to speak at the next Lunar Section meeting and he was due to appear on the April *Sky at Night*; but it was not to be.

Barely had the February Lunar Section meeting ended and Patrick was planning the next one. It seems incredible in the twenty-first century, when BAA Section meetings occur, at most, annually, that the BAA Lunar Section was able to hold meetings every few months all around the country. However, Patrick was a big crowd puller and local astronomical societies were always keen to host Lunar Section meetings. The next such meeting was scheduled for May 5th, at the Norwich Astronomical Society venue.

Fig. 18.2 Patrick and friends at the 1973 Winchester weekend. Between the refractor and the reflector are J. Hedley Robinson, Alfred Curtis, Cdr. Henry Hatfield and Patrick. The short bearded gentleman on the *extreme left* of the picture is Harry Ford (From the BAA 'Second Fifty Years' Memoir. Photographer unknown, but the picture may have been taken by Cdr. Hatfield on a time delay)

Kohoutek

On March 7th 1973 Dr Lubos Kohoutek of Hamburg Observatory discovered a new comet on a photographic plate. It was his third discovery and was very faint: only magnitude 16 in fact. Even more interesting was the fact that it did not seem to be moving much against the background stars. This could mean it was heading straight for us, or more likely, it was a long way away. The latter proved to be the case and it transpired that it was 700 million kilometres from the Sun, as far out as Jupiter in fact. To be that far out and that bright it must be a very active comet. In addition, the orbit showed that it would pass only 20 million kilometres from the Sun on December 28th, in 9 months time. The comet was brightening rapidly and all the indications were that it could be the comet of the century, reaching perhaps magnitude −15, and that's brighter than the Full Moon! Of course, it would be almost on top of the Sun, in the daylight sky, but even so, this was exciting! After Patrick's disappointment with failing to spot Ikeya-Seki over Ireland in October 1965 he was very cautious in his expectations for the comet, but the press were not. Very soon the tabloid journalists were doing their favourite trick of taking the scientists' remarks out of context and hyping the best bits to infinity. Comet Kohoutek was being branded 'The Comet of the Century' a full 9 months before it would be at its best. Patrick decided to wait and see what might develop. If the signs were good later in the year he would broadcast a *Sky at Night* about Kohoutek's comet.

Meanwhile, after Patrick's controversial talk and paper on the subject of possible Martian vulcanism causing dust storms, there was, not surprisingly, quite a bit of criticism coming his way. In the April 1973 BAA Journal the renowned astromathematician Jean Meeus took Patrick's theory apart, piece by piece, in a long letter to the Journal. Meeus cited the lack of hot-spot evidence from Mariner 9 as the main obstacle to Patrick's theory, but also pointed out that tidal effects on Mars, caused by the Earth, will be three million times less significant than any effects on our Moon. In the same Journal, Patrick supplied a rather vague reply to Mr Meeus, stressing that "All I tried to do was to point out the coincidences."

Patrick's letter concluded with:

> I wish I could explain all these coincidences [between Earth, Mercury and Venus]; but I cannot. That of Martian dust-storms and perihelic oppositions may be another. I hope Mr Meeus will be more successful in finding an explanation; and I am very glad he has made the points which he did in his letter.

The May 5th 1973 Lunar Section meeting, held at Norwich, was one which this author, aged 15, attended. For various reasons, it sticks in my mind more than any other meeting of that era. I think this is because the Norwich Astronomical Society was so enthusiastic about the construction of their massive 30 inch telescope. As a 15 year old, with only a little observing experience through a 60 millimetre refractor, I naively thought that bigger was always better. Over the last four decades of looking through a telescope, experience has taught me the following: that while big is good, ease of use and having a telescope in your own garden, ready to go at a moments notice, is far better than size. Large society telescopes are rarely used, night after night, by observers. Nevertheless, they do occasionally inspire beginners, who will queue up, in anticipation, for a peek through the eyepiece. Patrick insisted that the meeting was chaired, not by himself, but by Frank Harvey, Chairman of Norwich Astronomical Society. The society's Reverend Cyril Blount then stressed the educational aspects of the society's work and explained how government funds might be obtained for amateur astronomy, if you knew how to apply. Mr Brian Mitchell then described how the society was grinding the 30 inch mirror for the new telescope, including an amusing account of how the weight of the mirror blank in his car boot meant his car headlights were illuminating the tops of trees and low flying aircraft as the blank was driven to Norwich! All this was of great interest to this author as I was grinding an 8½ inch mirror at the time. The meeting wound down with talks about orange lunar soil, by Keith Abineri, as well as lunar photography, by Henry Hatfield.

Shortly after the Norwich meeting Patrick received the news he had been dreading. Phil Ringsdore phoned him to say that he really was most unwell now and he would have to relinquish the editorship of the Lunar Section Circular. Patrick was both depressed at Phil's health deterioration and daunted by the prospect of taking the Lunar Section Circular on himself. Editing a small monthly magazine and dispatching it to nearly 500 members worldwide was a tedious task. The magazine was produced by a stencil controlled duplicator, whose handle had to be turned once for every page printed! For 500 subscribers this translated to, typically, 6,000 handle turns. Then they had to be collated and stapled, shoved in envelopes and posted! A thankless task, especially if added to Patrick's writing and broadcasting commitments and the hundred or so postcards a week he typed in reply to *Sky at Night* viewers and amateur astronomers' queries. Remember, there was no e-mail in the 1970s. Nevertheless, Patrick decided to take over Phil's roles. Fortunately, Phil had said that he could carry on until October, the end of the BAA year.

The fate of the Lunar Section Circular hung in the balance in June 1973, but fortunately Patrick did have something really exciting to look forward to, namely the so-called 'Eclipse of the Century'.

A Monster Eclipse

As we have already seen, prior to 1973 Patrick had already witnessed three total solar eclipses, namely: Sweden in 1954, Yugoslavia in 1961, and Siberia in 1968. In 1970 the BBC had kept him in the London studio for the popular USA/Mexico eclipse, but in 1973 events would be different. On average there are around 70 total solar eclipses per century; roughly one every 17 months, or so. However, the length of these can vary from a split second (when the Sun and the Moon are exactly the same size) to a theoretical maximum of 7 minutes and 31 seconds. A really long eclipse results if the Sun appears small (and we are furthest from it around July 4th) and the Moon appears large (it is closest to us every 4 weeks). Another factor is simply where you are situated on the Earth. The Earth's equatorial diameter is wider than the poles, so an eclipse passing over the equator means the observer is lifted up slightly, making the Moon appear fractionally larger. Basically, when all these factors come together, you get a really long eclipse. In the twentieth century, the longest total solar eclipse occurred in the Philippines, on June 20th 1955; that monster lasted 7 minutes and 8 seconds.

Eighteen years and 10 days later (the *Saros* interval) on June 30th 1973, the Moon's shadow would pass over Africa, with the maximum eclipse duration being 7 minutes and 4 seconds, just fractionally shorter than the 1955 event. The track would start in Guyana, then cross the Atlantic Ocean and the Cape Verde Islands, before arrival at Africa. The track would then pass over various African countries (Mauritania, Mali, Niger, Chad, Sudan and Kenya) but none of them had particularly good roads or facilities suitable for the BBC to set up a base. Patrick, and his BBC *Sky at Night* producer Patricia Owtram, decided that the best plan would be to observe the eclipse from the ship already organised by the BAA. Fortunately, the BAA had set up an eclipse committee as early as 1970, to investigate the best way to get members to this 1973 spectacle. They had, in collaboration with *Transolar Travel* of Bebington, Cheshire, decided to charter the *Monte Umbe*, a ship sailing from Liverpool for the eclipse. Incidentally, *Transolar Travel* was run by BAA member David McGee; the company's astronomy trips eventually transferred to his brother Brian McGee, under whom it still ran the eclipse trips and then scuba diving holidays, under the name *Explorers*. In the 1980s, and up to 1994, Patrick was roped in as the company's eclipse expert, which was obviously a huge attraction. Brian sold the entire company to *Thomson* in 2008. In 2009 Brian became the front man for a new astronomy holiday company called *Astro Expeditions* and then launched another company for astronomy holidays called *Astro Trails*.

Originally, the solar observer Bill Baxter had been the BAA eclipse committee chairman, but following his death in December 1971 Howard Miles was put in charge. Howard had considered a number of sites from which BAA members might view the eclipse and was keen to involve *Transolar Travel*, as many BAA members had already journeyed with them to watch the lift-offs of the Apollo missions to the Moon, as well as tours of the US Observatories. Howard's committee looked into observing from Lake Rudolf in Kenya, but the cost was considered prohibitive. Inland Mauritania was also looked at, but there were few roads and even fewer facilities available. The Cape Verde islands were also considered, but the cloud risk was high. Eventually, observation from a ship, lying off the Mauritanian coast, was resolved as the best solution. Although the maximum 7 minute eclipse duration would occur over Mali and Niger, it would still be about 6 minutes long from a ship off Mauritania. Of course, the ship would have to be positioned accurately and this was well before the days of the GPS system. In particular, Gordon Taylor, of the BAA and the Royal Greenwich Observatory, wanted to make accurate timings during totality and he needed the ship positioned within 50 metres. The problem was solved by Patrick's lunar photography friend Cdr. Henry Hatfield, whose navy connections enabled him to get *HMS Endeavour* to lay buoys at the precise place where the ship should be on eclipse day. Howard Miles was a meticulous planner, and in April 1973 he had travelled up to Liverpool to meet the captain of the good ship *Monte Umbe* and to finalise the arrangements. Captain Vincente Millerave's first comments were that *HMS Endeavour's* buoys appeared to be above a reef! However, it ultimately transpired that the charts of that region of the sea were not in good agreement: the French maps showed a reef, the Spanish ones did not!

The one remaining problem for Patrick was how to get the film of the eclipse back to London quickly. The *Sky at Night* would not be broadcast for another 11 days, but James Burke, Patrick's Apollo broadcasting colleague, was providing a summary from London on the BBC that evening and they wanted the films flown back from Mauritania as soon as totality had occurred. Patrick decided to check out the BBC's contacts in Mauritania before he went. His producer gave him the phone number of the BBC's 'man in Nouadhibou', that town being the second largest settlement in the country. This BBC correspondent had been 'on the payroll' (according to Patrick) since the end of the second World War. Well, maybe he had, but Patrick could find no trace of him and no-one who had spoken to him in living memory! In the end a local pilot was contracted to fly the BBC eclipse film out of Mauritania on eclipse day. One wonders where the salary of the BBC's 'man in Nouadhibou' had been going for over a quarter of a century!

On June 22nd 1973 Patrick and all his BAA friends set sail from Liverpool on the *Monte Umbe*. The ship stopped at Las Palmas on Gran Canaria, in the Canary islands, and travellers were able to visit the NASA tracking station there as well as the solar observatory. Patrick and the film crew also visited a French Concorde on the Las Palmas runway, which had been specially modified to film the eclipse from 55,000 feet altitude. Because Concorde could travel at twice the speed of sound it could just about keep up with the Moon's shadow (no other commercial airline could then, or can now). This meant that instead of the eclipse lasting 7 minutes it could be extended to 90 minutes for anyone onboard the plane, the duration being limited only by Concorde's need to divert from the shadow to land and refuel at a suitable airport.

After visiting Concorde, Patrick decided to get a taxi into Las Palmas itself to phone his mother on her 87th birthday (June 27th). Satellite phones and cell phones, along with GPS, were many years in the future. The trip was a success and with the phone call completed he got back into the taxi and told the driver to head straight for the Las Palmas dock, where the *Monte Umbe* was moored, "Pronto". The Spanish taxi driver then promptly put his foot to the floor and accelerated straight into another Spanish taxi! Suddenly, Patrick and the driver were surrounded by semi-hysterical jabbering and gesticulating policemen who were obviously intent on extracting a huge written witness statement from him with only 2 hours till the ship set sail for the eclipse track! Patrick could see utter disaster looming, especially because, as he put it, "I was dealing with hysterical foreigners". None of the police spoke English, so Patrick tried his 'French with a Flemish accent'. This seemed to confuse them even more! Patrick bided his time until the point where all the police were arguing with each other and with the two taxi drivers, and until a big crowd had gathered. He then tried to make himself look small (not easy for an 18 stone, 6 foot man) and gradually sidled around the nearest corner, out of sight, whilst whistling to himself. He then walked very briskly away from the scene, sweating profusely, and hailed the first taxi that appeared, praying that it wouldn't collide with the existing wreckage just around the corner. Trying to look as small as possible he crouched down in the replacement taxi, much to the bafflement of the driver, and prayed that they would get to the docks without further incident. His luck held, and on arriving at the dockside he stuffed a wad of cash into the taxi driver's greasy mitt (far more than the journey cost) and headed, like a Rhinoceros on heat, straight for the Monte Umbe gangplank. Once onboard ship he locked himself in his cabin for 2 hours, until the Canary Islands disappeared into the distance. Phew! There was no way he was going to allow himself to be locked inside a Spanish prison cell while the ship steamed towards the eclipse.

The *Monte Umbe* continued onwards to a point 24 miles off the Mauritanian coast, namely the place where the Moon's shadow would pass through on June 30th. The first notable thing on arrival at the eclipse position was that someone had pinched the buoys that were so carefully placed by HMS *Endeavour*! Gordon Taylor and Henry Hatfield were, presumably, somewhat peeved at this development and may even have said something like "Oh dear, how annoying". One added bonus for Patrick though was that he did see the famous 'Green Flash' on the evening before the eclipse, in other words the last rays of light from the Sun as it sets beneath the Ocean. Watching the Green Flash at such times is a favourite pre-occupation of many cruise ship eclipse chasers, even if it infuriates the restaurant staff when sunset occurs at meal time!

The last hours before the eclipse were spent with Patrick filming interviews with the amateur astronomers on board. The legendary optical designer and planetary photographer Horace Dall was on the trip, along with his wife Helena. Horace was planning to balance his camera on his nose during the period of totality. Cdr Hatfield had designed a gimbal mounting to counteract any rough swaying of the ship during the eclipse. The former Lunar Section Director Ron Maddison, and his son Chris, were also onboard. According to Howard Miles there were two elderly passengers on the *Monte Umbe* who had no idea this was an eclipse cruise, or even what an eclipse was! When they put their deckchairs down on the deck area that was specifically reserved for telescopes, minutes before the start of the eclipse, they were promptly moved! The BBC was not the only media company on the voyage either as Alan Sidi productions recorded an entire film about the trip entitled 'A line to the Sun', which was released a few months after the cruise.

As the eclipse proceeded and more and more of the Sun's disc was eaten away by the Moon, the sea became very calm, which was a great relief to all the photographers onboard. However, unfortunately the skies became very hazy too, and it became obvious that the full extent of the Sun's corona would not be seen. Nevertheless, it was with great drama and excitement that, at just before 10.40 GMT (9.40 a.m. local time) on June 30th, the Moon's 250 kilometre diameter shadow swept over the *Monte Umbe*. The sky turned turquoise and Venus could easily be seen, with Patrick claiming to have spotted Saturn too. With 6 minutes of totality to savour it was a completely different experience to Patrick's 40 seconds at his Siberian eclipse, some 5 years earlier. The haze did render the outer corona invisible, but it was still a spectacular experience. As the shadow streaked over Africa, the Concorde followed it, and the onboard experiments worked flawlessly.

Immediately after totality, Patrick realised that he had dropped his small hand camera, and bending down to pick it up, nearly everyone onboard heard the massive rip as his trousers split! He rapidly sat down near to a BAA colleague of mine, Laurence Anslow, who, as a young man, was on that famous eclipse trip. "Laurence, Disaster Has Struck!" Patrick wailed, as he explained what had happened. Apparently, after the eclipse, there was much more interest shown in the photo's of Patrick's split trousers than in the eclipse photographs! Fortunately, a fellow traveller called Ed Turco saved the day, by using considerable amounts of masking tape to repair the ripped trousers. Another event from that famous trip that everyone remembers is the totally non-technical Patrick trying to take a group photo of his friends and focus the picture. After moving further and further back the camera still refused to focus and Patrick manically twisted and twisted what he assumed was the focusing ring until, suddenly, the pricey lens fell clean off, with a sickening thud, onto the deck of the ship.

With totality over, the Captain of the ship quickly headed for the port and airport town of Nouadhibou, on the northern border of Mauritania with the Western Sahara, so that the films for the BBC could be flown back to London for that night's eclipse summary, co-ordinated by James Burke. Patrick took a taxi to the airport, handed the films over to the pilot and watched him take off. They were shown on BBC 1 that evening. The astro-tourists were then invited to spend 'the afternoon at leisure' in Nouadhibou. This was not exactly an exciting prospect as there was only a foul smelling fertilizer factory, plus sand, dirt and flies to see. Needless to say, there was no sign either of the BBC's enigmatic 'Man in Noadhibou', or the money he had accumulated on the BBC payroll for a quarter of a century!

Patrick described Nouadhibou as the most wretched, foul-smelling dump he had ever been to, and in a matter of minutes he composed a music-hall ditty about the place, which he performed manically as part of the traditional last evening concert on the ship. Anyone witnessing this rendering has never forgotten it. It went down in eclipse history. The lyrics are reproduced below, in their entirety, and a few audio tapes of Patrick in full cry still change hands at great cost:

We walked into the desert sun,
The day had only just begun,
We'd heard such glowing rumours of the place.

We looked for sheikhs on Arab steeds,
And women dressed in gorgeous beads,
But when we saw what we had got to face –

(Chorus) Boo, boo, Nouadhibou,
We're glad you're far away.
True, true, Nouadhibou.
You told us we could stay.
But pooh, pooh, Nouadhibou.
It's time to say 'Good-day';
So toodle-oo, Nouadhibou.
Shall we come back? Nay, nay!

We watched with glazed and goggling eyes
As multitudes of desert flies
Surrounded us and nibbled at our coats.
We dodged the crowd of market boys

Who tried to sell us junk and toys,
And cursed us as we ran back to our boats:

(Chorus) Boo, boo, Nouadhibou,
We're glad you're far away.
True, true, Nouadhibou.
You told us we could stay.
But pooh, pooh, Nouadhibou.
It's time to say 'Good-day';
So toodle-oo, Nouadhibou.
Shall we come back? Nay, nay!

And as we leave this sandy land
It isn't hard to understand
That we will come here never, never more.

One visit here was quite enough,
And Mauritania now can stuff
Its sand back whence it came in days of yore:

(Chorus) Boo, boo, Nouadhibou,
We're glad you're far away.
True, true, Nouadhibou.
You told us we could stay.
But pooh, pooh, Nouadhibou.
It's time to say 'Good-day';
So toodle-oo, Nouadhibou.
Shall we come back? Nay, nay!

What could possibly top that for entertainment? Well, arguably only the BAA's eccentric Mrs Ethel Granger, famed for her 13 inch waist, who won the Monte Umbe fancy dress competition dressed as, what else, a wasp!

Apart from a stopover at Tenerife and an exploration of the Teide volcano, that was the end of Patrick's 1973 Eclipse. The Monte Umbe docked in Liverpool on July 8th and the *Sky at Night* of July 11th showed all the footage obtained on the ship. In fact, so much footage was taken that even the following *Sky at Night* programme, broadcast on August 1st, was essentially an interview onboard ship, filmed with Henry Hatfield, about Navigational Astronomy.

Back Home to Lunar Section Duties

On Patrick's return to Selsey, after the eclipse, it was back to business as usual and, of course, the need to find a replacement for Phil Ringsdore, as Lunar Circulars Editor. Unfortunately, during Patrick's eclipse trip, Phil's condition had worsened and he was no longer able to carry on beyond the August circular. In that circular, Patrick paid tribute to Phil's tireless work over recent years and also announced, what BAA councillors already knew, that Phil would be awarded the Association's highest honour: the Goodacre Award. In a short 'goodbye' from Phil in the August circular, he described this time as both the saddest and proudest of his life: sad to be stepping down as Editor, but proud to be receiving the Goodacre medal. Patrick insisted that Phil retain his post of Lunar Section Assistant Director, even if he was too ill to take part in running the Section. Phil doubted if he would be well enough to receive the Goodacre Medal in person though. He was now quite frail and travelling from Somerset to London was a real risk.

At the end of July Patrick travelled down to Somerset to see Phil and described him simply as ‘not well’. However, there was a sliver of good news. Patrick had received an offer to take on the Lunar Section Circular production, from a local astronomer called Geoffrey Buss, of Stroud Green, Bognor Regis; he was just up the road from Patrick which was an ideal situation. Patrick was also able to offload his TLP co-ordinator role to the duo of Morgan Findlay and David Taylor at Dundee Observatory. Patrick was still the driving force behind TLP studies, but collating the TLP observations and running the alert network took a lot of spare time, which Patrick rarely had.

For the *Sky at Night* programmes in late August and September, Patrick once again managed to persuade the BBC to come to him. The Selsey link for the August programme was, at best, tenuous, but Patrick was proud of his Selsey home near the sea, and of his garden full of telescopes, especially the biggest and newest instrument. Venus was the subject and was in the evening sky, and visible through that 15 inch f/6 Newtonian. Plus, why not make craters on Selsey beach and talk about whether they might look like the craters under the Venusian clouds? At one point Patrick recalled to me that while making that programme he leapt off a groyne and caught his toe on a half-buried anchor.

On September 15th, Patrick was back at Norwich, this time for a BAA Provincial meeting. Following the successful Norwich Lunar Section meeting in May, Patrick had urged the BAA Council to have a full-blown meeting at Norwich, as the local astronomical society had organised the Lunar Section Meeting so well; the council were in full agreement. This author attended that meeting, not only to see Patrick, but also to try to speak to the expert telescope maker Horace Dall, as I had a number of queries about problems I was having when polishing my own 220 millimetres mirror. As it happened, I did get to speak to Dall, but my problems were not ones he had ever suffered! Nevertheless, he gave me some sage advice.

Travelling with Patrick, from Sussex, to that meeting were the new Lunar Section Editor, Geoffrey Buss and fellow Selsey resident, Reg Spry, who was a keen observatory builder; if you can call a rotating tool shed an observatory! Spry’s expertise was in making telescopes from junk. At the meeting itself, in response to a comment from a Mr Wadbrook along the lines of “Why isn’t astronomy taught in schools?” Patrick was very vocal. Patrick stated that it would be: “A great mistake to teach astronomy in state schools; the quality of instruction would inevitably be bad, and enthusiasm would be killed”. Patrick felt that it would be much better to: “Defer anything of the kind until the present eccentric trends in psychiatric education have run their course, and educational sanity has returned.”

It was back to Selsey beach in late September, for a classic *Sky at Night*. Everyone knows that the Moon looks much bigger when it is on the horizon, rather than when it is overhead; but why? This effect is called the Moon Illusion and is thought to be caused by the brain seeing the Moon against familiar distant objects (trees, houses etc.) and allotting it more importance. Another factor is that a low-down Moon, just above the horizon, is much dimmer and so its features can be more easily seen, as there is less glare. Patrick's guest on the Moon Illusion programme was Professor Richard Gregory from Bristol University, who had a few ideas on this subject as he was an acknowledged expert in optical illusions and visual perception. As an aside, I might add that Prof. Gregory's father, Christopher Clive Langton Gregory, had been a Director of the University of London Observatory at Mill Hill and had known the stage and screen comedian and amateur astronomer Will Hay. Once again, Patrick strolled along Selsey beach, this time at night, and with Professor Gregory and himself inviting members of the public to guess what size of pebble would cover up the Moon. During the programme, every species of moth decided to commit suicide by hurling itself at the TV lights. In addition, every time Patrick groped for suitable pebbles he found himself sinking into a vile-smelling brand of seaweed. "The stench was very reminiscent of Nouadhibou", he once told me. The seaweed was of a variety which he had "Never seen before, and would never sea again" on Selsey beach.

In late September of 1973, as Patrick was extracting himself from the evil smelling Selsey seaweed, the Japanese astronomer Seki was the first to photograph the over-hyped comet Kohoutek as it escaped from the Sun's glare into the September morning sky. It was now just over 3 months to perihelion (the comet's closest approach to the Sun) but it was only eleventh magnitude. It was clear that Kohoutek would not be the 'daylight comet' that the press had hyped up, but it would still be a nice object for amateur astronomers. Patrick's pessimism had been well-founded: this was not 'The Comet of the Century' after all!

During filming for the October 1973 *Sky at Night* the BBC briefly thought Patrick had plunged to his grisly death, over a cliff, on the slopes of the Mt. Teide volcanic caldera in Tenerife. While recording with Dr Ron Maddison, Patrick's straw hat blew off, and determined to re-capture it, Patrick chased it out of sight. When he did not come back the film crew, briefly, feared the worst, but minutes later Patrick re-emerged triumphantly, straw hat in hand. Why were they inside a volcanic caldera anyway? Well, Patrick and Ron were discussing the highest volcano in the solar system, Olympus Mons, which is located on Mars.

By early October preparations were now well advanced for the next out-of-town BAA Lunar Section meeting, to be hosted by Torbay Astronomical Society on November 3rd. Patrick had announced the details in the October Lunar Section Circular:

'The meeting will be held at the Abbey Lawn Hotel, Scarborough Road, Torquay, at 2.30 p.m. It will last until 6 p.m., and at 7.30 p.m. there will be a public lecture – but as this will be given by myself, I can hardly recommend it! I will myself be staying at the Quay Tor bed and breakfast (£1.50 for a single room).' Despite being a TV celebrity and a household name it would never be a five star hotel for Patrick.

Shortly after concluding the Torbay meeting arrangements, Patrick received a brief letter from Mr Ken Gayner, the Lunar Occultation sub-section co-ordinator who he had installed back in June 1972. The letter informed Patrick that Mr Gayner was 'resigning due to personal circumstances'. Patrick was obviously disappointed by this news but hoped to appoint a successor and transfer all of Mr Gayner's records to the new co-ordinator. However, Patrick, and everyone else, never heard from Mr Gayner beyond that point. At some point in October 1973 he disappeared from planet Earth and no-one saw him ever again! Patrick endeavoured to contact him over the course of the next 4 months, trying to trace him with the help of the Bristol Astronomical Society, but to no avail. Sixteen months of valuable occultation data were lost and, years later, Patrick concluded at a meeting that: "He was abducted by aliens who wanted the occultation data: nothing else fits the bill".

Before the Torbay Lunar Section meeting there was the BAA AGM to attend. The AGM was always held on the last Wednesday in October and 1973 was no exception. Patrick received two phone calls, which delighted him, just prior to the AGM. One was from Phil Ringsdore at Burnham. He was feeling slightly better and was pretty sure that he could attend the BAA AGM to collect his medal in person. The second phone call was from no less a person than Arthur C. Clarke. He was flying from Sri Lanka to London in a few days and would like to see Patrick if possible. Needless to say, Patrick invited Arthur to spend the night before the AGM at Selsey. What Patrick was totally unaware of at this time was that Patrick's mother Gertrude and a select band of his friends and TV colleagues were all in on a secret. In recent weeks, Eamonn Andrews, the TV presenter, had contacted many of them, as Thames TV was planning a surprise *This is Your Life* TV programme for Patrick early in 1974.

On Wednesday October 31st, Neville Goodman, the BAA President, opened the BAA AGM. Sitting in the front row were Patrick, Phil Ringsdore and Arthur C. Clarke. Patrick was always in the front row as he was the meetings recorder. This was the moment Patrick had hoped for when he had proposed the Goodacre Award for Phil, due to his tireless work for the Lunar Section. Following his summary of the BAA year, the President announced that it was his 'proud and pleasant' duty to present the Goodacre and Lydia Brown medals to Mr P.A. Ringsdore and Mr A.W. Vince. In presenting the association's senior award to Phil Ringsdore, the President said:

> Mr Ringsdore joined the Lunar Section a decade and a half ago and has been one of the mainstays of that extremely active Section ever since, all the time as Section Secretary and part of the time as Acting Director – virtually full-time occupations. Perhaps his principal task, daily taking many hours and earning him friends all over the world, has been the editing of the Section Circular. But his work for astronomy has gone beyond his official work in the Association and the hours he has spent as an observer in his own observatory. He founded the Ewell Astronomical Society, which I believe to be the largest local society in the country, with a membership approaching twelve hundred – a fruitful source of members for our Association. He has introduced innumerable people to astronomy and was the prime mover in establishing a fine observatory in Ewell with a 254mm reflector. Mr Ringsdore, we are all delighted to see you here this evening, and I am pleased to offer you the Association's thanks and good wishes and to present to you the Walter Goodacre Medal and Gift.

On accepting the medal, an obviously frail Phil Ringsdore simply said that this was a very proud moment for him and he had enjoyed every minute of his work for the Association. The award was greeted with loud and prolonged applause.

Following the second presentation, to Mr Vince, and the Presidential Address (on the subject of using the stars to navigate when hill-walking) there was the usual short ‘Ordinary Meeting’ during which Arthur C. Clarke showed some recent photographs he had taken of the giant Arecibo radio telescope, in Puerto Rico.

Three days later, on November 3rd, Patrick was booking into his Quay Tor bed and breakfast room in Torquay, ready for the next Lunar Section meeting, hosted by Torbay Astronomical Society. Patrick had always had close links with Torbay, following his 30 year friendship with J. Hedley Robinson at Teignmouth and now with his son, Donald Hedley Robinson too. Indeed, his close links with Torquay Boys Grammar School directly resulted in one of Patrick’s protégé’s from that school, Chris Lintott, becoming the *Sky at Night* co-presenter some 30 years later.

On November 10th Patrick was out of bed early and positioned by his 5 inch refractor at Selsey, hoping to catch a glimpse of the transit of Mercury across the face of the Sun, which he planned to observe by safely projecting the light from the refractor’s eyepiece onto a piece of white card. This was the first transit of Mercury since 1970 and there would not be another until 1986. Mercury would cross the Sun’s disc, observable from the UK, between 7.48 a.m. and 1.17 p.m. Unfortunately November 10th proved to be cloudy, whereas the days before had been “brilliantly fine” in the whole Bognor-Chichester area, according to Patrick.

Throughout October and November Patrick had endeavoured to spot Comet Kohoutek with his 15 inch reflector, but success only came when he found it, at magnitude 7, on November 17th. Patrick could only detect a trace of a tail but, by the end of November, it was magnitude 6, and Patrick estimated the tail as a quarter of a degree long. Also in late November, Patrick was a speaker at the Sunderland Arts Centre, in a festival of Science Fiction and Space Exploration entitled ‘Beyond this Horizon’. Other speakers at that festival were the Black Hole expert Professor John Taylor and the acclaimed Science Fiction writer Brian Aldiss.

On December 3rd 1973, the space probe Pioneer 10 flew past Jupiter and returned spectacular close-range images. Patrick filmed a *Sky at Night* with Dr Garry Hunt to cover the results. But the next *Sky at Night* would have to be about comet Kohoutek, which was rapidly being labelled by the press as a ‘flop’.

Playing the Fool!

On December 8th the BBC transmitted the thirtieth episode of the comedy series called *The Goodies*, featuring Graeme Garden, Tim Brooke-Taylor and Bill Oddie, who were constantly involved in far-fetched schemes to ‘save the world’. Episode 30 was entitled *Invasion of the Moon Creatures* but was perhaps better remembered by its alternate name of *The Planet of the Rabbits!* The storyline involved Graeme Garden sending rabbits to the Moon, which evolved and converted Tim-Brooke Taylor and Bill Oddie into rabbits too. Patrick appeared in the programme, lightly disguised as a giant moon-dwelling rabbit, hell-bent on taking over the world! Patrick never took himself too seriously and was always happy to take part in almost any television programme, however bizarre. Even when he was made to look like a total eccentric it did not bother him at all. All three of the cast members of *The*

Goodies starred in a long-running BBC radio series called *I'm Sorry I'll Read That Again*, alongside comedians John Cleese, David Hatch and Jo Kendall. That series came to an end in December 1973, but one of its enduring highlights (for this fan of Patrick at least) was when *Monty Python* star John Cleese used to occasionally come out with a totally manic impersonation of Patrick, much to the amusement of the other cast members and the live studio audience.

In the 1970s barely a day went by without Patrick appearing on television or on the radio. Between 1973 and 1981 he even appeared five times as a guest, or expert, on the *Generation Game*, hosted either by Bruce Forsyth or Larry Grayson. In a 1974 TV programme the magician David Nixon (the first partner of the famous fox puppet Basil Brush) recalled how Patrick and himself had recently appeared on the BBC radio panel game programme called 'Fair Deal', in which Nixon was the card dealing question master and panel chairman. Patrick, the cartoonist Willie Rushton, and the comedienne June Whitfield, were the regular panellists. Patrick turned up for the programme with a bottle of wine. He suggested the team had "a quickie" before the show went out on the radio, in front of a live studio audience. Anyway, Patrick grabbed the wine bottle between his massive thighs, inserted the corkscrew, applied even more pressure from his thighs (so the bottle wouldn't slip) and........ the bottle exploded. The lower half of Patrick was saturated in glass fragments and wine and the show was minutes away from going live on air. When the show started, David introduced Patrick to the studio audience. Patrick walked onto the set backwards, so it would not look like he had suffered a serious exploding bladder problem, and he spent the whole programme behind the desk with sodden, steaming trousers. It was very fortunate for Patrick that it was not a TV programme!

Panel game producers were always keen to have Patrick in a team during the 1970s, whether the show was on TV or on the radio. He could always be relied upon to cause amusement. Some of these programmes were never seen by the majority of Patrick's fans as they were aired on the regional BBC South network. One such show was called *Slightly Quizzical*, a programme in which panellists set questions for their colleagues about regional knowledge. Patrick's best remembered question for his colleagues on the programme was: "Why was a one time American Wild West showman using a butcher's scale roped to a tree in Farnborough?" The answer was that the man 'Colonel Cody', a self taught aviator, is said to have tied the scale to his aeroplane to measure the torque of the engine. He made the first recognised powered flight in Britain on October 16th 1908, covering 1,390 feet in 29 seconds. Maybe someone at RAE Farnborough had told Patrick this story when he had given a talk there in November 1972? The astronomical society that now operates the small observatory near to where Cody made his flight is aptly named the Cody Astronomical Society or CAS.

Chapter 19
Kohoutek Flops and This Is Your Life

Following the Japanese observer Seki's recovery of comet Kohoutek, at eleventh magnitude in the September dawn twilight, many amateur astronomers, as well as Patrick, had been following the comet as it became more visible in the night sky. It was obvious that Kohoutek had not lived up to press expectations. It had obviously been abnormally bright when a long distance from the Sun, a characteristic of a comet which was on its first ever visit to the solar system. Nevertheless, for amateur astronomers with binoculars, telescopes and photographic equipment it held every prospect of being a fine sight, possibly the best comet since Bennett's of 1970. Patrick flew to Hamburg Observatory to interview its discoverer, Dr Lubos Kohoutek, for *The Sky at Night*. Although working at Hamburg, Kohoutek was a Czech, and Czechoslovakia was renowned for comet discoverers. Between 1946 and 1959, eighty new comets were discovered by Czech observers, using photographic patrols, and visual patrols with huge (ex WW II) binoculars. Despite the fact that Kohoutek was clearly not going to be 'The Comet of the Century', as it had been labelled by the tabloid press, Patrick was very much in demand during the final weeks of 1973, for his advice on what people might see in the night sky. He made his very first appearance on the BBC children's programme *Blue Peter* in December, explaining to Peter Purves where youngsters should look to spot comet Kohoutek.

Patrick was both amused and irritated to see that, as always in these circumstances, the comet had brought the cranks out of the woodwork. A religious statement put out by a Mr Moses David of the 'Children of God' suggested that the apparition heralded the end of 'Fascist America and its new Nazi Emperor'. In the London *Times*, Patrick's friend Bernard Levin wrote an article accusing Patrick of *stealing* comet Kohoutek! In case the reader is wondering, yes that is the same Bernard Levin that Patrick's other friend, Desmond Leslie, had punched on live TV in 1963! The best naked eye views of the comet were probably obtained by the Skylab astronauts Carr, Gibson and Pogue. The *Sky at Night* about the comet was broadcast on December 28th 1973, when Kohoutek was closest to the Sun. In the first few weeks of January it was expected to be at its best, emerging into the evening sky, so the programme was timed to encourage people to search for it.

M. Mobberley, *It Came From Outer Space Wearing an RAF Blazer!: A Fan's Biography of Sir Patrick Moore*, DOI 10.1007/978-3-319-00609-3_19,

With the usual cloudy Christmas skies over the UK (and over Ireland) Patrick was especially pleased to get an early New Year report of Kohoutek's magnitude from the Northern Ireland astronomer John C. McConnell, who he knew from his Armagh years. The newspapers had been asking Patrick how bright the comet was, but no-one else Patrick knew had seen it. John, a friend, and a local pilot, Captain Mike Woodgate, had flown from Belfast International airport in a light aircraft to escape the cloud. Over Lough Neagh the comet was spotted, very low down in the west-southwest dusk twilight, in 10×50 binoculars and at fourth magnitude; this information was relayed to Patrick who passed it straight to the London *Times* newspaper. McConnell's magnitude information appeared on the front page the next day! One week later, on January 10th 1974, Patrick, accompanied by one of his astronomy acolytes, schoolboy David Lowne, flew on a *Transolar Tours* early evening flight above the Irish Sea where he secured his best binocular view of Kohoutek from 33,000 feet. The brilliant planets Venus and Jupiter, together with the comet, were a pleasing spectacle, but Kohoutek was still only fourth magnitude, although it did have a 'delicate' 10 degree tail. The public would remember Kohoutek as a flop and the integrity of astronomers was certainly damaged by the press hype, nevertheless it became a good comet for well-equipped amateur astronomers in the weeks to come.

The year 1974 did not start well in the UK. Apart from the hyper-inflation and strikes of the 1970s, the world oil shortages made travel somewhat fraught. This was a threat to Patrick's plans for Lunar Section meetings around the country. There was no point planning a meeting if there was no way of getting there. Nevertheless, Geoff Buss and the South Downs Astronomical Society hosted a good Lunar Section meeting on January 19th, at Chichester, only a few miles from Patrick's home. One of the main speakers was Charles Cross, who drew the NASA Moon and Mars maps and had collaborated with Patrick on various projects. Unfortunately, as well as the UK's energy crisis there was a national paper shortage (there was no paper recycling in those days) and so the January Lunar Section Circular would have to be combined with the February one.

This Is Your Life

On February 6th 1974, two days before the splashdown of the Skylab three crew, Patrick was finally collared by Eamonn Andrews, disguised in a space suit, and subjected to a *This is Your Life* experience. Patrick was sitting in a scale model of the NASA Lunar Roving vehicle when Eamonn bagged him. Totally oblivious to what was about to happen, Patrick, led by James Burke, had said hello to the mysterious astronaut as he positioned himself in the Lunar Rover. Even when Eamonn got the red book out it did not register with Patrick that this was anything other than a joke. Then it suddenly dawned on him that this was for real! "You swine" he laughed at Eamonn (and he said more than that 'off camera'). Guests on the programme included: James Burke; Patrick's mother Gertrude; magician David Nixon; his childhood tutor (the 84 year old Reverend John Missen); his lifelong friend Pat Clarke and Pat's son Lawrence; Patrick's former RAF Wellington crew colleague Flt. Lt. Guest Dempster; a schoolteacher from his Holmewood days, Mary Trotter;

the co-founder (with Patrick) of the Shoreham Light Opera company, Ray Lowrie; Patrick's first BBC producer Paul Johnstone; the entire Selsey Cricket Club; and, a video recorded message from astronaut Dave Scott.

Patrick was very keen to correct people when they gave the wrong information on that programme. When James Burke told Eamonn that Patrick had predicted the Moon landings when he was 15 and had been the youngest British Interplanetary Society member, Patrick quickly corrected him that it was the British Astronomical Association, not the BIS, and he stressed that plenty of others had made that same prediction about the Moon landings. He also exploded when Eamonn stated that Patrick's opera, *Perseus and Andromeda*, was just about to be released by the Shoreham Light Opera company in *Essex*. "SUSSEX, SUSSEX, SUSSEX", Patrick bellowed! Of great interest to this author is that there was not the merest hint of Patrick's wartime fiancée in the programme, or his alleged piano accompaniment with Einstein. These were stories that Patrick only really promoted after his mother died; presumably there was then no-one to scold him for "telling fibs", and Patrick loved telling stories and adding a degree of embellishment!

Strangely, there were no amateur astronomers on that programme, and only a few people from Patrick's distant past. It was as if Patrick's life had solely been associated with being on TV. The programme was transmitted a week later, on February 13th 1974, and was the 116th *This is Your Life*, immediately following the one on the comedy actor John Alderton (from *Please Sir!*) and just preceding the one on the *Orlando* actor Sam Kydd.

As I mentioned much earlier in this book, around this period, with Patrick's fame at its very peak, various half-baked and sordid theories emerged, generated by faceless people and psychiatrists, about why such a man would devote so much spare time to helping teenagers and scout groups in astronomy, and why he did not have a wife, and lived with his Mum! Baseless rumours even circulated about why he had left Armagh so quickly in 1968! Well, the answers were simple: he was incredibly generous, shy with women, and Northern Ireland was about to become a war zone! Patrick was always visiting sick children, schools, and scout groups at this time, and with no children himself, and having had a sickly childhood, he was always very generous towards children, just as W.S. Franks had been with him. In many ways these malicious rumours were inevitable for such a famous, eccentric, but totally naïve TV celebrity, who spent so much time with young people on dark, clear nights. The theories were ludicrous, but his unmarried status did him no favours here, and it was shortly afterwards that the story of a wartime fiancée, killed by the Luftwaffe, emerged. It is easy to see why: Patrick wanted to convince people he was 'normal', but unmarried, simply because of a tragedy long ago.

Farewell to Two Old Friends

With signs of the UK energy crisis easing in February, Patrick confirmed arrangements for a Lunar Section meeting in Manchester on March 9th. He also appointed Mr Geoffrey Amery as Occultations Co-ordinator (Mr Gayner was still with his alien abuctors). Patrick also had big plans for trying to obtain a spectrum of a TLP. Only Kozyrev's controversial TLP spectra had ever been 'obtained', but Ron Maddison at Keele thought it might be possible to secure spectra with the new 24-inch (60 centimetres) Keele reflector, which was nearing completion. Getting such spectra would obviously be a major breakthrough and it was a prospect that certainly excited Patrick. It might seem strange that a University department would wish to collaborate at all with the BAA Lunar Section, especially on a subject as controversial as TLP. However, without the Lunar Section there would have been no new 24-inch reflector at all. One of the section's most experienced observers, the elderly Harry Thornton, was a friend of Patrick and a very wealthy man too. In the early 1970s, with encouragement from Patrick, he donated £2,000 to Keele University so they could proceed with the 24-inch reflector. The BAA Telescope Making Section Director, Jim Hysom of Astronomical Equipment Ltd., would provide the optics, and a local company, William Boulton, would provide the rest of the engineering. So, Keele University owed a debt of thanks to Harry Thornton, Patrick, and the BAA Lunar Section.

Meanwhile, in the Lunar Section Circular, Patrick urged members to observe the occultation of Saturn on the night of March 2nd/3rd 1974. As it turned out the skies over much of England south of London were cloudy, so Patrick missed the event, but many other observers viewed or photographed Saturn disappearing and emerging from behind the Moon. One of the successful observers of this event was the charismatic Bill Granger of Peterborough (the character with the cat on his shoulder and the wife named Ethel with her 13 inch waist) who obtained some excellent photographic slides of the Saturn occultation in freezing conditions. However, tragically, Bill died suddenly the next day. It was the day before Patrick's 51st birthday.

Barely had the March Manchester Lunar Section meeting finished when Patrick confirmed another Dundee meeting, scheduled for May 4th at the Mills Observatory. Also, as we have already seen, Patrick was a keen supporter of the BAA Winchester Weekends and so from Friday April 5th to Sunday April 7th he was back at King Alfred's College. He was one of the four course mainstays, the others being Alfred Curtis, Henry Hatfield, and Hedley Robinson. Two weeks later and Patrick drove down to Exeter with Reg Spry for Hedley Robinson's own meeting on April 20th. Hedley, at that time, was the BAA Mercury & Venus Section director and Patrick was acting as the Mercury & Venus Section secretary. Needless to say, Patrick gave another public lecture while in Exeter. Four days later Patrick was at the monthly BAA London meeting and just 10 days after that he was at the other end of the country for his own Lunar Section Dundee meeting. He certainly got about in the 1970s! This Scottish meeting attracted a number of leading professional astronomers and geologists.

On the subject of Reg Spry, who was a great friend of Patrick's, and living in Selsey regularly cadged a lift to meetings in his car, Patrick once told me:

> He was a great chap, but I gave him lifts to save his life. Reg was the world's worst driver. I never once was in the car with him driving when he didn't mount the kerb at some point; usually every hundred yards. I saved countless lives by giving Reg a lift!

I was also told a hilarious story by Ken Kennedy, a Dundee Astronomical Society member, that hails back to those 1970s Dundee meetings: "I well remember little Reg Spry being bundled unceremoniously into the boot of a big Volvo outside my house, as there wasn't enough room in the car for him on the way back [from the meeting] to the hotel!"

Fig. 19.1 Patrick using an 8.5 inch Newtonian in the observatory of Ken Kennedy at Dundee on May 3rd 1974. A young Dundee Astronomical Society member (M. Fodor) is observing with Patrick. By kind permission of Ken Kennedy

Fig. 19.2 Patrick (*right*), Harry Ford (*centre*) and John McConnell (*left*) at the May 4th 1974 BAA Lunar Section meeting held at Dundee Astronomical Society. By kind permission of Ken Kennedy

After the Dundee meeting, Patrick could catch his breath for a few days before heading back to the TV studios for an interview with Carl Sagan, the US TV Astronomer and 'exo-biologist'. The resulting *Sky at Night* was entitled 'Life Beyond the Earth'.

Although Phil Ringsdore had been too unwell to do any serious observing from Burnham, he had told Patrick that the skies there were less light polluted than at his old Stoneleigh home and maybe he could manage to use a small reflector now and again. On hearing this, Patrick had informed the Lunar Section members that he would like to start a fund for a small telescope for Phil, and very quickly a sum of 35 pounds was collected. Patrick went to his friend Dudley Fuller, owner of Fullerscopes (now Broadhurst, Clarkson & Fuller) in London's Farringdon Road and Fuller offered him a 4 inch (100 millimetres) reflector at cost price, namely 35 pounds. Patrick collected the telescope in May and decided to take it down to Somerset at the earliest opportunity, in early June, when Phil returned from a holiday in Jersey.

May 29th was the date of the annual BAA Exhibition meeting at the Royal Commonwealth Society in Northumberland Avenue, London. The late Bill Granger's wife had travelled down from Peterborough with the developed Saturn occultation slides that her husband had taken on the night of March 2nd, just before he died. Ethel Granger was famous in her own right of course. We saw much earlier that the record breaking narrow waist of her younger days earned her a permanent place in the *Guinness Book of Records* and won her the *Monte Umbe* fancy dress competition, when she went as a wasp! Patrick was more than happy to show Bill Granger's slides at the meeting and he said: "The Association has lost a great character, a great amateur astronomer and one who would be remembered with affection by everybody." The BAA had certainly lost one of its most colourful characters and the man who had been the first ever guest on *The Sky at Night*.

In the Exhibition hall itself a large exhibit had been placed by the Lunar Section, including some fine lunar photographs by messrs. McBain, Foley, Hatfield, Doherty and Pennell as well as some lunar drawings by Patrick and Paul Doherty. Patrick had known Paul's father Bernard, and Paul would become an even closer friend to Patrick in the coming years. Paul was quickly becoming established as the finest planetary artist the BAA had ever known.

At the start of June 1974, Patrick received the news he had been dreading: Phil Ringsdore had died, while on holiday in Jersey. Patrick wrote a tribute to him for the Lunar Section Circular and an obituary for the BAA Journal. It was only 8 months since he had accepted the Goodacre Award at the BAA AGM. In the Lunar Section Circular Patrick wrote:

> It was as an organiser that he was at his best, and there are very many people, both amateur astronomers, professionals and layman friends, who owe him much. I do not think I have ever heard an unkind word said about him; I do not believe that there could have been anything unkind to say. He was the friend of everyone, and we all respected him as well as valued his friendship. He will be sadly missed, and all of us will extend our deepest sympathy to his wife and son. But though Phil is no longer with us, he will not be forgotten; there are many people who will never forget him – and who, like myself, are proud to have known him.

Sadly, Phil never found out that Patrick had organised the collection of donations for a telescope that he could use at Burnham. The small reflector was, at Patrick's suggestion, kept within the section, named 'The Ringsdore Telescope', and made available to young astronomy enthusiasts. Some years later, I think it must have been around 1980, Patrick was appearing on the popular Saturday morning BBC TV childrens programme *Noel Edmonds Multi-Coloured Swap-Shop*. As always, he was answering questions on astronomy from a small audience of children, and from children who were phoning in to the programme. Patrick was taken by surprise when Phil Ringsdore's granddaughter phoned in, asking Patrick if he remembered her granddad. Patrick was delighted and wasted no time telling her, and the viewers, what a great friend Phil Ringsdore had been.

On July 4th 1974, Patrick received the first of over a dozen honorary doctorates that he would be awarded during his lifetime. That first one was from Lancaster University, but he eventually received them from Leicester, Keele, Sheffield-Hallam, Glamorgan, Hatfield Polytechnic, Liverpool John-Moores, Dublin, and Birmingham, to name but a few!

A few days later, Patrick flew to the USA to interview the inventor of the V2 rocket developed at Peenemünde during the war, Wernher von Braun. At one stage, while Patrick was in Bomber Command, the RAF bombed the Peenemünde facility. Patrick certainly was not on that raid "But in theory, I could have been", and if that really had been true, he could have killed the eventual designer of NASA's first rockets, thus scuppering the entire Moon program! Wernher von Braun thought that a manned lunar base would be set up before the end of the twentieth century and men would walk on Mars before 2024. When speaking privately about Wernher von Braun, Patrick would often tell visitors to his house that the inventor of the V2 had said to him "My role voz to make ze rockets go up. Vare zay came back down again voz anuzzer man's problem……."

Returning to the UK, Patrick was pleased to learn that arrangements for the next Lunar Section meeting in London were finalised for August 3rd, at the Conway Hall; Rossie Atwell had, as always, organised things flawlessly. On the day of the meeting, Lunar Section members stood silently in memory of Phil Ringsdore. The meeting was well attended and the well-known observer and mirror maker Tom Cave, from the USA, was particularly welcome. He had known Patrick from the 1940s and even met him during the War Years. The meeting was dominated by TLP talks; Patrick was determined to increase the coverage of the lunar surface, and hopefully obtain a spectra. The previous director, Dr Ron Maddison of Keele University, stressed the importance of spectroscopic confirmation of TLP, and Dr Allan Mills of Leicester University was, apparently, designing a spectrograph for this purpose.

In August, Patrick and his 88 year old mother, and no doubt the family cat (a moggie named Beno during that era) celebrated the publication of his mother's first and only book; it was entitled *Mrs Moore in Space*. The book featured her talents as an artist, not as an author, and contained her remarkable paintings of alien life-forms or 'bogeys' (friendly and cute little green men) on other worlds.

Chapter 20
Not Enough Hours in the Day

Following the death of Phil Ringsdore, Patrick appointed Harry Ford of Dundee as the assistant director of the Lunar Section, with Morgan Findlay, also of Dundee, joining the committee. All seemed to be going smoothly in the Lunar Section, but then disaster struck! Geoff Buss, who had taken over as the Lunar Section Circular editor a year earlier, had lost his typist and so found juggling a day job, his local Bognor Astro-Society commitments, and the Lunar Section Circular, just too much. Something had to give, and unfortunately it had to be the Lunar Section Circular. In a letter to the Lunar Section Committee, inside the combined August/September Circular, he implored them to dissuade Patrick from taking over the Circular production. Geoff wrote: 'He is already grossly overworked' and 'Take it from me, he works a 30 hour day'.

Of course, as well as Patrick's writing and BBC work, plus his speaking and charity commitments, he was running the Lunar Section. Things were getting fraught and Patrick could, perhaps, have done without a trip to Birr Castle in Southern Ireland to re-visit the extraordinary historic observatory there, which *The Sky at Night* had first visited in 1967. The 72 inch telescope tube had now been restored and a museum had been set up to commemorate the achievements of the third Earl of Rosse, who built the world's largest telescope there in 1845.

Another event had been organised well in advance too, and this was one that Patrick could not possibly avoid. The clock manufacturer *Westclox*, who had a factory on the Vale of Leven Industrial Estate in Dumbarton, Scotland, had up to 1968 been a division of the US Company *General Time*, who had worked on the Apollo spacecraft timing systems. Business was booming for *Westclox* and the quartz crystal era was just starting too, so to celebrate this the US parent company invited Apollo 11 astronaut Neil Armstrong to attend a Space Seminar on 'Time Keeping' at the factory. Armstrong had genealogical connections with Scotland, namely Armstrong clan roots tracing back to the town of Langholm. Indeed, 2 years earlier, Armstrong had actually accepted an invitation to visit Langholme to be made a freeman of the town, much to the surprise of the residents.

M. Mobberley, *It Came From Outer Space Wearing an RAF Blazer!: A Fan's Biography of Sir Patrick Moore*, DOI 10.1007/978-3-319-00609-3_20,

Nevertheless, Armstrong was a shy and private man who hated the limelight, so the *Westclox* organisers suggested that Patrick should also be invited to accompany the 'First Man on the Moon' and deal with speaking to the press: Patrick was never tongue-tied! Of course, Patrick was absolutely delighted to accept this proposal and so when Neil Armstrong flew into Heathrow on October 9th 1974, Patrick joined the *British Airways* Trident charter flight (BA 351) from Heathrow to Glasgow with Neil Armstrong onboard. [*British Airways* was only 7 months old at that time, having formed from the merger of the *British Overseas Airways Corporation* (BOAC) and *British European Airways* (BEA) in March 1974]. Armstrong really hated newspaper, radio and TV interviews, and so Patrick acted as his escort and publicity buffer, to answer any press matters and accompany Neil, while dealing expertly with the media. After landing at Glasgow (where Armstrong agreed to publicity pictures with a Glasgow Police band piper and the *General Time* Vice-President John Santos) and then touring the Westclox factory the two men had lunch at the Dumbuck Hotel. They were subsequently driven around to give a short talk to local schoolchildren, followed by a reception at the Lomond Castle Hotel, the day before the seminar. It was at that reception that Patrick's favourite picture of him and Neil sharing a drink together (with Patrick looking extremely pleased to be photographed alongside Neil) was taken. It was a highly memorable couple of days for Patrick, and the contrast in styles worked well, but, returning from Scotland, Patrick must have wondered what on Earth to do without a Lunar Section Circulars editor.

Fig. 20.1 The well-known picture of Patrick with Neil Armstrong, taken at the Lomond Castle Hotel in Glasgow on October 10th 1974. Photographer unknown

Mason and Atwell Save the Circular

The disruption to the Lunar Section Circular was immediate and the October 1974 edition never appeared. Patrick was definitely NOT the man to do the job of Circular production. He was happy with his 1908 Woodstock typewriter and did not want to use a modern typewriter, but his ancient machine could not cut the stencils for the duplicator! At this point a young undergraduate student and amateur astronomer from Bognor, named John Mason, who would become a lifelong friend of Patrick's, a regular guest on *The Sky at Night*, and a future BAA President, stepped in to help with using the Lunar Section duplicator. Six years earlier Patrick had invited the teenage schoolboy John Mason, along with one of John's friends, to his *Farthings* home, shortly after moving to Selsey. John and his friend had written to Patrick, asking to look through his telescopes, but had not expected an instant reply. As soon as Patrick had lifted John to the eyepiece of his 12.5-inch Newtonian, back in November 1968, and showed him Saturn, John had been hooked.

John would become a younger version of Patrick in the coming years, making many TV appearances and becoming a BAA stalwart. Like Patrick, John had a remarkable ability for delivering astronomy talks in front of an enthralled audience; having learnt from the master himself he was never lost for words when giving a lecture, or when a camera was thrust in his face. For anyone of my generation John Mason in the 1970s, 1980s and 1990s, was Patrick Mk II or 'Son of Patrick'! After a night observing at Patrick's, John would often be driven back to Chichester (the nearest large town) in the infamous Ford Prefect called the Ark, at its maximum speed of well under 40 miles per hour. On a number of occasions John has recounted how a typical magical evening at Patrick's would commence in the mid to late 1970s. On arrival at Farthings it was not uncommon to hear the sound of Patrick's piano playing as you approached the front door (sometimes it was the Xylophone), invariably opened by the housekeeper, or Patrick's mother. Upon entering the house three requests would be issued by Patrick, in verbal machine-gun fashion, with his pipe, unlit, in his hand:

1. "John, Please can you get this lighter working. It doesn't work. Must've run out of fuel". Followed by:
2. "John, none of these staples work. Can you put some new staples in the stapler". And then:
3. "John, the wretched light bulb over there has blown. Do you think you can find some way of fixing it?"

And finally, after these simple chores (but not simple for the impatient Patrick) were completed, Patrick would utter his famous cry, very familiar to anyone entering Farthings: "Have a drink!" It is quite possible that Patrick gave more people alcohol, in his home, or at a bar, than any other man in history! His generosity with alcohol simply knew no bounds! Frequently he would simply tell people "Iremonger's rules apply in this house". This would mean nothing to most people. Asked to clarify this statement Patrick would simply rattle off the following: "My old friend Colonel Edmund Iremonger's rules apply in this house; in other words, if you want a drink, help yourself!" Patrick never mentioned the mysterious Colonel Iremonger in any other context except in the consumption of alcohol! However, when I pressed him on this issue in 2005 he explained that the Colonel had been an independent candidate in the 1979 General Election, standing for Chichester with the policy "Put Britain First"; more on this later in the book.

But I have digressed; in 1974, with a Lunar Section Circular production crisis on Patrick's hands, the undergraduate student John Mason stepped into the breach, took control of the duplicator, and saved the day. Another local friend, Roger Prout, also offered his services. Meanwhile, Rossie Atwell finalised arrangements for a Lunar Section meeting in Ipswich on December 7th. But even with this help, as always, Patrick was working a 25 hour day, constantly in demand on TV, on the radio, in the BAA, and by local astronomical societies. He even appeared on his second Michael Parkinson show on November 2nd, with his mother Gertrude also being interviewed; maybe after she had appeared on *This is Your Life* she had developed a liking for TV work!? Oscar Peterson and Ronnie Barker were the other guests on that programme.

So, at this particularly headless chicken stage of his life, Patrick could not have welcomed the resumption of the criticism regarding his Martian Dust Storm theory, by the Belgian astronomical mathematics specialist Jean Meeus. Mr Meeus submitted a letter to the BAA in late 1974 which began: "Contrary to the prediction of Patrick Moore, there was a dust storm on Mars in 1973." Mr Meeus continued: "This supports the hypothesis that the obscurations of Mars are not due to vulcanism but to dust raised by winds, and that such storms occur almost each Martian year near the time the planet reaches its perihelion." In the December BAA Journal Patrick offered a brief reply: "Yes, there was a dust storm in 1973; I observed it myself! It is interesting that the dust-storm of 1971 was particularly severe; and for anything comparable we must go back to the dust-storms of 1909 and 1911. It seems, therefore, that a particularly severe dust-storm is apt to be followed by another at the next opposition, whereas an ordinary perihelic dust-storm is not."

Despite the ongoing hassles with Lunar Circular editing, production and dispatch, a successful Lunar Section meeting was held, at the Orwell Park School Ipswich, on December 7th 1974. This author remembers that meeting well, as I vividly recall talking to Cdr. Hatfield about his fine lunar photographs and showing people pictures of my newly completed home-made 8.5 inch reflector. Patrick opened the meeting by saying that, so far as he knew, this was the first time that the BAA had ever held a meeting in Ipswich. He said that he was glad to see so many members present and he expressed his appreciation of the excellent organization by Mr Roy Cheesman and other members of the Orwell Astronomical Society. The society had the use of a fine 10-inch (254 millimetres) refractor, originally owned by Colonel Tomline, and very similar to the one Patrick had used at Armagh between 1965 and 1968. Mr Cheesman explained the history of the telescope. While inspecting the giant refractor Patrick jokingly asked if it was for sale as he would like to add it to his collection at Selsey, but Henry Hatfield said that whatever Patrick offered he would outbid him!

Patrick asked the audience if anyone present would be prepared to take written notes of the meeting. As almost always happened in these situations no-one offered to do the job, and so as he did at so many BAA meetings, Patrick scribbled down and wrote up the entire afternoon's proceedings. The other speakers at the Orwell meeting included Lunar Section enthusiasts Henry Hatfield, Arthur Good and Malcolm Ellis and also, from Selsey, Reg Spry. Reg, who had again travelled up from Selsey with Patrick (to spare the lives of himself and any other motorists on the road that day) described his unique 'rotating tool-shed observatory'. At 5:30 p.m. Patrick adjourned the meeting and Roy Cheesman, David Bearcroft and others took Patrick and a few Lunar Section stalwarts to the 'Crown and Anchor' Hotel in Ipswich. Apparently the other diners at the restaurant spent the entire evening just staring at the large table of astronomers, astounded that the legendary Patrick Moore was in their midst! Once again, after the meeting, Patrick gave a public lecture entitled 'Into Space'. This was held at the Copleston School nearby, which was packed to bursting point with 530 people crammed into the school hall just to listen to Patrick's 70 minute talk, delivered at over 200 words per minute without a break. After his lecture Patrick signed 200 autographs and then had coffee with the headmistress Miss Beeson. By 9:30 p.m., and after a very long day of travelling, meeting recording, meeting chairing, public lecturing and socialising, Patrick was still keen to do some observing with the 10-inch Tomline refractor! However, clouds intervened and so Patrick headed back on the almost 200 mile trip to Selsey, with Reg Spry and Rossie Atwell as his companions (although Rossie did not yet live in Selsey at that time). In the 1970s Patrick's stamina and enthusiasm were truly without bounds!

On January 2nd 1975 Patrick attended the BAA's Christmas meeting at the 23 Savile Row London venue and detailed the work of the Lunar Section and, in particular, the search for TLP. The main theme of the meeting was a talk about the evolution of the atmospheres of the Terrestrial Planets, by Professor A.J. Meadows of Leicester University, but the close passage of the asteroid Eros to the Earth in the next few weeks was also covered by Gordon Taylor, who had discussed this very subject with Patrick on the November *Sky at Night*. Throughout the 1950s, 1960s, 1970s, and even into the early 1980s, Patrick would record almost all of the BAA meetings. There were no word processors in those days and he used to write the meeting reports up, literally in real time, as the meeting progressed. OK, they were a scrawl, but still legible. Every BAA meetings recorder since Patrick, including me, took the rough notes home and knocked them into shape, but Patrick's notes were just about in finished form when he left the meeting. A fanatical half hour of typing would complete the write-up! Unusually for those times, Patrick possessed a pen with a light on the end, one that he could check his notes with in the dim illumination of a lecture theatre, when the lights went down. He always wrote meeting notes up with a pencil, and edited them with a rubber as he went along in the semi-darkness. However, Patrick pressed so hard on the paper that the pencil tip often snapped! Some time ago I had an e-mail from a BAA member, Kevin Bailey, who often used to sit next to Patrick when he was recording the meetings in the early 1970s. Kevin, a poet, had worked out a clever strategy to get to know Patrick:

> In the early '70s I used to get to the Scientific Society's lecture theatre in Savile Row early (for BAA meetings) because PM was meetings recorder then and I knew where he sat… and if I got there early, I got to sit next to him. I kept a supply of pencils in my jacket – he was always breaking pencils while taking the minutes… thus we got to talk.

On January 13th 1975 Patrick left the UK for a 19 day trip to South Africa where he was recording a series of programmes for the South African Broadcasting Company (SABC). Due to the disruption this would cause to the Lunar Section Circular, Patrick was particularly grateful that John Mason, once again, stepped in and undertook the duplicating work himself. Writing in the January Lunar Section Circular Patrick described himself as "John Mason's assistant". Another helper, Andrew Gannon, was dispatching the overseas Lunar Circulars, but a permanent assistant with this responsibility was needed. The foreign tour went well and Patrick recorded a number of TV programmes and interviews, but he would need to go back to South Africa in March, for an even longer 26 day commitment.

I have actually seen some of the programmes that Patrick recorded for the South African Broadcasting Company and the format was very similar to *The Sky at Night*, with Patrick in a studio, or on location, alongside individual astronomers. Perhaps the most memorable interview he carried out (as remembered by South African amateur astronomers) was when he visited the comet and supernova discoverer Jack Bennett, at his home in Pretoria. Patrick, dressed in a huge white blazer and fawn coloured trousers, followed Jack onto his back lawn, obviously expecting to see an observatory there. Then Jack disappears through an archway leading to a garden building (probably an outside lavatory!) with Patrick looking baffled and his body language expressing 'Where on Earth has he gone?' Jack returns sometime later, carrying his highly portable comet sweeping refractor and its tripod. Patrick, hands thumped firmly onto his hips, swivels around to the camera and shrieks "Well, why not? That's one place to keep a telescope!"

Anyway, Patrick returned from South Africa to Selsey on February 1st. To avoid disruption to the *Sky at Night* (missing a programme was Patrick's biggest dread) much of the research and filming for the next programme, about the La Palma Northern Hemisphere Observatory, had already been completed. All that was needed was for the RGO (Royal Greenwich Observatory) director designate, Professor Graham Smith, to join Patrick in the studio. The programme was broadcast on February 3rd.

As soon as Patrick was back in Selsey, producing the February Lunar Section Circular was the top priority. John Mason threw himself into the duplicating and stapling and Rossie Atwell took on much of the onerous task of putting the four hundred-plus circulars into envelopes and posting them! The overseas circular distributor, Andrew Gannon, then reported that the GPO had lost all of the overseas circulars posted to him, from Patrick; a disaster that would need sorting out in the coming weeks. Rossie Atwell had also stepped in and arranged the next London meeting, scheduled for May.

The TLP Return

At the end of February 1975, after a very long gap, there was a flurry of TLP claims. Firstly, on February 22nd, Lunar Section member Peter Foley alerted lunar observers to a bright patch near the crater Prinz. At the same time observers in Cambridge, and Lawrence Fitton, observed a blue appearance in the crater Aristarchus, although Fitton regarded it as an optical effect and not a real event on the Moon. On the next night, Peter Foley and Patrick, observing independently, thought that Aristarchus was brighter than normal at around 18:40 GMT. Geoff Amery, in Reading, and Reg. Spry, in Selsey, also recorded abnormalities. By 20:51 GMT, Patrick, Reg Spry and Peter Foley thought the north wall of Aristarchus looked blurred. The alleged activity would continue on February 24th, 25th and 26th, with Patrick, Peter Foley, Geoff Amery, Howard Miles and others reporting various strange appearances in and around Aristarchus, as well as in Gassendi and Plato. A few years later, from the early 1980s, this author was a member of the TLP network. While I could invariably confirm the suspicions of others when observing colour anomalies, there was always considerable uncertainty about whether what I was really looking at was simply caused by a combination of variable atmospheric seeing and spurious colour aberrations within the Earth's atmosphere; in other words, nothing to do with the Moon at all! Nevertheless, this kind of independently 'confirmed' TLP activity (I have heard it called 'independently confirmed mass hallucination' before now) was something that excited Patrick. He was convinced that things did, occasionally, happen on the Moon and most of his TLP network observers and fans were disturbingly keen to confirm his views. In the March 1975 Lunar Section Circular Patrick delayed production so that he could include 'a preliminary account of the exciting events in Aristarchus'. He also presented a full account of the Aristarchus 'events', together with Peter Foley, at the BAA London meeting on February 26th.

Patrick had to depart for his second long trip to South Africa on the day after his 52nd birthday; he flew off on March 5th and warned Lunar Section members to expect disruption to the circular. While in South Africa Patrick was able to fly over the Vredefort Ring, a 140 kilometre diameter, highly eroded, faint remnant of a two billion year old crater. Its crater-like origin was only determined in 1961, when aerial photographs of it were examined. Patrick made a leisurely study of it in a helicopter and had discussions with the University Department studying it. The majority of geologists regard it as an impact structure, but it goes without saying that to Patrick there was 'absolutely no doubt it is a volcanic structure'. Patrick also visited the major South African Observatories while he was there. After a series of South African TV programme recordings, interviews and lectures, Patrick returned to the UK on March 31st, just in time to see himself in his second appearance in '*The Goodies*' BBC Comedy series. This episode, the 43rd in the '*Goodies*' series, was entitled 'Lighthouse Keeping Loonies', in which the comedy trio of Graeme Garden, Bill Oddie and Tim Brooke-Taylor were employed as lighthouse keepers but had run out of oil for the lamp. Obviously, they dig for some, find some, and the resulting conflagaration sends the lighthouse into Earth orbit: cue Patrick, who announces that a new comet (the lighthouse) is visible in the night sky! With his trusty childhood refractor Patrick describes how you can see it by finding the Sun and then going 'up a bit', down a bit', 'right a bit', in the manner of the 'Golden Shot' TV programme which was popular at the time.

Patrick allowed Lunar Section member Lawrence Fitton to criticise the Aristarchus TLP observations in the April 1975 Lunar Circular. In a well thought out series of arguments, covering five pages, Fitton effectively rubbished the observations, or at least concluded that the 'Aristarchus event of 1975 February 22nd to 27th was spurious and due to atmospheric effects'. This was a brave and damning conclusion, especially when considering that most Lunar Section members regarded Patrick as their 'hero'; aged just seventeen, I know I did!

On April 4th 1975 Patrick set off, once again, for the annual BAA Winchester Weekend course. This was the first Winchester Weekend that this author attended: they are still going strong more than 38 years later! Once again, the Winchester Residential weekend Course on Observational Astronomy, as it was then called, consisted of Patrick and his mates, and it was a huge draw, pulling in 200 participants, mostly there just to see and hear Patrick. The course organiser was still Alfred Curtis, but Lunar Section stalwarts Henry Hatfield, Harry Ford, Hedley Robinson and Patrick himself formed the tutorial panel. All the weekend's lectures, eleven in all, were given by these five observers, apart from one excellent talk by Dick Chambers on Star Atlases and Catalogues. Patrick had a co-lecturer, David Wright, for one of his talks on occultations. Patrick's other talks were on lunar observing and variable star observing.

Fig. 20.2 Patrick and friends at the 1975 Winchester weekend. From *left* to *right* in the *front row* are Patrick, Alfred Curtis, J. Hedley Robinson and Cdr. Henry Hatfield. BAA archive photograph by the late Donald Fletcher Hedley Robinson (1935–2007)

The weekend was chaired by the BAA President, Howard Miles. I well remember queuing for the Winchester lunch outside the college canteen; you could always hear Patrick from a hundred yards away, and frequently he could be heard voicing his low opinion of the then Prime Minister, Harold Wilson (whose taxes were preventing Patrick becoming a wealthy man), or of British Rail, the GPO (Post Office), or British Leyland (the car makers). It was a time of hyper-inflation in the UK, reaching more than 25 %, and also a time of never-ending Trade Union strikes. Patrick was keen to see the Conservatives back in office, but hopefully not with the Europe-loving Ted Heath in charge, or any of the Tory landed gentry who hunted foxes and indulged in other bloodsports!

Needless to say, Lawrence Fitton's criticism of the February Aristarchus event did not go unanswered amongst fellow Lunar Section observers. Peter Foley, Chris Lord and Dr James Reading had alternate views in the May Lunar Section Circular. On May 17th the Lunar Section met at the Swedenborg Society HQ in London. Messrs Fitton, Foley and others were locked in considerable TLP discussions in which there were serious disagreements, but Patrick described the discussions as 'strong but friendly'. Much of the meeting was spent arguing about colour anomalies on the Moon, caused by, according to Lawrence Fitton, the Moon's altitude and the weather conditions. Miss Atwell commented that most of the men at the meeting were seeing the colour red! Patrick chaired the meeting and John Mason acted as the meeting recorder. Patrick was particularly pleased that Harry Ford had travelled down from Dundee and Mr F.M. Flinsch, from America, was also present. As well as arguing about colour, Reg. Spry, who had travelled up to the meeting from Selsey with Patrick, demonstrated how his crater-extinction filter wheel worked, and David Jewitt (later to become a notable professional astronomer who recovered comet Halley) spoke about how the device could be calibrated.

Eleven days later and the BAA's annual Exhibition Meeting saw Patrick arranging the Lunar Section display on May 28th. This consisted of numerous drawings and photographs by members, that same crater extinction filter device made by Reg Spry, and some 'Moonblink' filters and paraphernalia from Patrick and Peter Foley. There was always a brief 'Ordinary Meeting' at the BAA's Exhibition Meeting and in those days, technically, members could still be proposed for the BAA council elections at that meeting; there was just time to get their name on the ballot paper and into the August Journal. Patrick liked the BAA council to contain a lot of his old mates and as few 'serpents' as possible. Many of his closest friends were in powerful positions on the council and that was how he liked it. The 1975 President was his best mate Howard Miles, and one of the Vice Presidents was Alfred Curtis, another staunch ally. One of the secretaries was the dedicated lunar photographer Henry Hatfield, and also present on the council were his best mates Reg. Spry from Selsey, and Iain Nicolson. The council had that morning decided, rather prematurely, on the final list for the council ballot, forgetting that the rules did allow additional names to be put forward that afternoon. Mike Maunder, in later years a very close ally of Patrick and a co-author of two of his books, was still classed as a Croydon AS 'serpent' in 1975, and so Patrick was concerned when his name was proposed at the Exhibition meeting, after the Council had prematurely finalised the ballot.

The President explained that Mr Maunder's name could not be put forward for this year's ballot as the ballot list had been finalised, and Patrick bellowed his strong agreement. By all accounts, Mike Maunder gave as good as he got to Patrick (no mean feat!) and the meeting was remembered by many for the public exchange between Mike and Patrick. However, in an e-mail to this author, Mike thinks the row was exaggerated, even if it *was* down to Patrick. To quote Mike precisely:

> 'I think the affair was down to Patrick though, and arose from the black-balling I'd got from the Altair' [i.e. the Croydon AS magazine article which had some fun at Patrick's expense – M.M.] 'I'd been nominated for Council and in those days members could do that at the Exhibition meeting (then end of May, part of regulars). Forget exact details but the lists were somehow closed such that I was prevented from getting that nomination accepted that day. I was livid and said so. The rules were never gerrymandered like that again, so it did do some good in the end and I got elected next time round – Mike.'

Sadly, the row had reinforced Patrick's view that Croydon AS was a 'nest of serpents', but 12 years later Patrick and Mike would become firm allies on the same side, during the biggest row the BAA would ever see!

Less than 2 weeks after that exhibition meeting, Patrick was presenting papers in front of a distinguished audience of lunar scientists at the Royal Society Lunar Conference in London. Patrick and NASA researcher Barbara Middlehurst presented a number of papers about visual observations of transient lunar phenomena. At the BAA meeting on June 25th, Patrick and Barbara continued the debate, which largely centred on the speculative link between Moonquakes (detected by the Apollo seismometers) and TLP detected visually by amateur observers.

Despite Patrick's immersion in all things connected with astronomy he still voiced his opinions strongly whenever the subject of animal cruelty arose. In 1975 the Labour Government, under Harold Wilson, passed a bill to ban Hare Coursing, which delighted Patrick, at least until the House of Lords threw the bill out! Nevertheless, in June it had all looked very promising, and Patrick had written a letter to the *Daily Express*, which was published on June 18th:

> May I stress that, contrary to a widespread impression, cruel sports are opposed not only by Labour Party members, but by all decent members of other parties as well. It is most unfortunate that a few brainless and despicable Conservative MPs opposing the Hare Coursing Bill have created the impression that all Tories are cruel sport enthusiasts. I am decidedly Right Wing but nobody could be more opposed to blood sports than I am. Mercifully, the first battle has been won. Now we must press on for legislation to outlaw the equally filthy hunting of otters, stags, and foxes.
>
> Dr. PATRICK MOORE. O.B.E., F.R.A.S.
> Selsey. Sussex.

Two weeks after Patrick's public criticism of the landed gentry's participation in blood sports he was mingling in the company of royalty, but the subject matter was a very happy one this time. The new William Boulton Observatory, at Keele University, was declared open on July 1st 1975 by Her Royal Highness Princess Margaret and Patrick was one of the invited dignitaries. The reader will recall that the observatory's 24-inch reflector was made possible by a £2,000 donation from Patrick's friend, the elderly lunar observer Harry Thornton. Maybe now, at last, the big telescope could be used to obtain spectra of TLP, under the leadership of Ron

Maddison? Disaster almost struck at the opening ceremony though because Patrick, being Patrick, simply turned up in his normal blazer, unaware that this was a dinner jacket and bow tie event, as royalty was present. Some of the photographs taken that day show Patrick talking to Ron Maddison, with an extra hand groping around near Patrick's groin, and with both men looking highly amused! I asked Bill Leatherbarrow (a lunar section committee member from that era) about this mystery and he said that the extra hand was that of a tailor, trying to frantically fit a spare dinner jacket to Patrick before the opening ceremony started!

Fig. 20.3 Patrick with Ron Maddison (*left*) at Keele University on July 1st 1975, during the royal opening of the University's William Boulton Observatory. Ron was the BAA Lunar Section Director from 1968 to 1971 and Patrick hoped to obtain spectroscopic confirmation of TLP using Keele's brand new 24-inch reflector. Note the extra hand between Patrick and Ron's legs! A tailor was frantically trying to fit a spare dinner jacket to Patrick's girth! With thanks to Ron Maddison and by kind permission of Keele University Library

Barely had the summer recess from the main BAA astronomy meetings started when Patrick found himself organising another BAA Lunar Section meeting in London. A group of amateurs from the New York Amateur Astronomers association contacted the BAA to see if there were any summer meetings which they could attend, as they were going to be in London in early August. There were none, so Patrick decided to create a Lunar Section meeting to be held at the Swedenborg Hall, just off Bloomsbury Square, in London. Once again, Rossie Atwell and John Mason saved the day, with Rossie organising the meeting and John producing the Lunar Section Circular. The meeting took place on August 7th. This meeting became infamous as there was a problem with fully blacking out the room to enable Patrick's slides to be seen. Patrick was adamant that if the room could not be blacked out it would be a disaster and he dispatched Lunar Section member Peter Gill to go out onto the roof of the building (!) to see if a drape could be suspended from there to block the sunlight streaming in. Well, Mr Gill's feet promptly plunged through the fragile paper thin roof (although he survived) and the BAA was soon presented with a hefty repair bill for the damage!

Nine days later Patrick had a scarey TV encounter with a *Dalek*, although as it turned out, maybe the programme's host was the more terrifying alien lifeform! He appeared on the BBC Television programme *Jim'll Fix It* on August 16th, a series in which the disc jockey and *Top of the Pops* presenter Jimmy Savile made childrens' dreams come true, along with their nightmares it later transpired! Patrick had also appeared on *Jim'll Fix It* 5 weeks earlier. During the programme a *Dalek* suddenly trundled into the studio, which prompted some interesting facial grimaces and eyebrow raising from Patrick; when the *Dalek*, with reference to Patrick, asked "AND WHO IS THIS?" Patrick enquired as to whether it had ever had a meteorite rammed up it's antenna?! Patrick and Jimmy Savile (who was just 3 years Patrick's junior) seemed, at that time, to have an enormous amount in common. Both became unique British eccentrics and household names from the late 1950s onward (in Savile's case on Radio Luxembourg to start with), both were lifelong bachelors absolutely devoted to their mothers, both were eventually knighted and both raised enormous amounts of money for charity. However, a year after Savile's death in 2011, hundreds of disturbing revelations about the disc jockey emerged, prompting a police investigation and a BBC inquiry. It was obvious that the two men had virtually nothing in common after all….

Chapter 21
A Naked Eye Nova and a Director Resigns

On the *Sky at Night* programme Patrick always aimed to be very topical. Although the days of the programme being transmitted live had ended in 1969, it was still recorded, where possible, on the day before transmission. The broadcast day, in the 1970s, was usually a Wednesday, but sometimes it was a Tuesday or a Thursday, and the scheduled time seemed to be 11:37 p.m., 11:38 p.m., 11:42 p.m., or 11:43 p.m. Clearly, rounding late night times up to the nearest 10 minutes was not fashionable in the BBC of the 1970s, and as video recorders were mainly found only in libraries and Universities, programmes after midnight would simply not be watched by anyone with a day job. The August 1975 programme was broadcast on the 28th of the month. The subject was 'The Galilean Satellites', the collective name for the four largest moons of Jupiter. The guest was a regular on the programme, Dr Garry Hunt. When the programme was transmitted there was nothing dramatic happening in the night sky, but this was soon to change! Britain's most extraordinary observer, George Alcock, then a discoverer of three novae, had been outside on the night of August 28th/29th 1975, sweeping the summer Milky Way constellations with his powerful binoculars. Spotting nothing new, he retired to bed in the early hours of the 29th, convinced there was nothing bright to be discovered. The next day, still August 29th, at around noon in the UK, the Japanese nova hunter Osada discovered a bright new star of third magnitude in the constellation of Cygnus. He was swiftly followed by many other Japanese amateurs, including the legendary Minoru Honda, and as night fell over the UK many British amateurs, including Patrick, spotted the new star.

Alcock was devastated. He had memorised the positions of 30,000 stars, visible through his binoculars, and yet had not spotted this naked eye newcomer which even casual observers had spotted the next night. Alcock thought he would surely be a laughing stock! It had now been 5 years and over 2,000 hours of patrolling since Alcock's third nova discovery and he had surely missed a sitting duck! Fortunately, photographs of the Cygnus region, taken in the USA at the same time he had been patrolling, proved that the nova had been too faint to discover during Alcock's August 28th/29th patrol. Alcock's relief was tremendous; he decided to carry on. In the meantime, on the evening of August 29th, Patrick had strolled out onto the

M. Mobberley, *It Came From Outer Space Wearing an RAF Blazer!: A Fan's Biography of Sir Patrick Moore*, DOI 10.1007/978-3-319-00609-3_21,

lawn of Farthings, totally unaware of the new nova discovery and spotted it instantly, as a second magnitude star, totally distorting the familiar shape of Cygnus: it was the brightest nova for some 33 years. Patrick phoned the Royal Greenwich Observatory; only to find out they knew all about it. They reckoned he was probably the fiftieth person to discover the nova, but he may well have been the 500th! Either way, the August *Sky at Night* had been transmitted the day before and so news of the nova would have to wait until the next programme. In fact, Patrick and the BBC did a special *Sky at Night* programme about Nova Cygni 1975 (also known as V1500 Cygni) on October 22nd. It proved to be a very fast fader and was well below naked eye magnitude by September 7th, despite having peaked at magnitude 1.8 a week earlier! Patrick made 45 magnitude estimates of the nova between August 29th 1975 and February 11th 1976, by which time his observing notes recorded that it had faded to magnitude 10.4.

The *Sky at Night* programme inherited a new female producer, Patricia (Pat) Wood, in 1975. She was the third producer and would work with Patrick on the *Sky at Night* for the next 6 years, but had already directed many programmes during Patricia (Pat) Owtram's 9 years (1966–1975) in charge. What with Patrick as the presenter there did seem to be a lot of Pats working on *The Sky at Night*!

September 25th 1975 saw Patrick attending a major BAA Provincial meeting at Dundee, the sort of 'out of London' event that Patrick's friend, the BAA President Howard Miles, had championed for years. Sadly, Howard was unable to attend the meeting, but it was a great success and Patrick's evening lecture, entitled 'Mars', had the Dundee University lecture theatre packed to capacity. There was a strong Lunar Section and TLP flavour to the BAA meeting, which was hardly surprising with Harry Ford, the Lunar Section TLP co-ordinator, being based at Dundee, and with Patrick in attendance.

On October 21st 1975 Patrick made his first ever appearance on the humorous BBC Radio game show called 'Just a Minute'. The aim of the show was to speak for 'Just a Minute' on some allotted subject, without hesitation, or repetition, and without deviating from the stated subject. The first subject this time was "How I carry on". Nicholas Parsons was the chairman and the other guests were Kenneth Williams, Clement Freud and Peter Jones. Patrick was quick on the buzzer when Clement Freud repeated the word 'quietly', to which the camp comedian Kenneth Williams exclaimed, in a voice only he could produce: "Yerrrsss. He said that before. You're quite right. That is absolooootely right. You're brilliant!" Patrick then tried to speak continuously for the next 30 seconds with the line: "Well I don't actually carry on very often. I do my best, but it all rather depends on what you mean by carry on. One can carry on by going into space, by not going into the void. One can in fact carry on by looking around one and seeing this magnificent audience who have come here especially to hear us make this discourse. We also can look around at our panel and, and…" But that second "and" saw Clement get his revenge for hesitation by Patrick. And so it continued. Patrick would be a guest on the radio programme on many more occasions.

Despite all of the other BAA sections holding only one or two (or zero!) meetings a year, Patrick's appetite for Lunar Section meetings was endless. A meeting in

Reading was arranged for November 1st, and one in Chichester, hosted by the South Downs Astronomical Society, organised for December 6th. The November meeting had a distinctly occultation biased theme, with for once barely a mention of TLP. On November 12th Patrick flew out to South Africa to attend an astronomy conference and to present a television series on South African TV. Patrick returned to the UK on December 5th, his absence requiring the November and December Lunar Section Circulars to be combined. Despite only arriving back in the UK one day earlier, Patrick was in fine form at the December 6th meeting. Despite his workaholic schedule he even found the time during 1975 to perform in his own opera 'Perseus and Andromeda' at the Selsey Amateur Dramatics Society!

Never afraid of making a fool of himself, Patrick appeared in his third BBC 'Goodies' comedy sketch for the 1975 Christmas special (aired on December 21st): 'Goodies Rule O.K.' In this episode, regular Goodies Graeme Garden, Tim Brooke-Taylor and Bill Oddie were joined by Patrick and numerous other BBC 'personalities', including disc jockeys Terry Wogan and Tony Blackburn, as well as the Rugby and *It's a Knockout* commentator Eddie Waring.

Speaking of *It's a Knockout*, I well remember an episode of the BBC's *It's a Celebrity Knockout* show, broadcast in the late 1970s, where Patrick appeared as part of *The Lords Taverners* team, raising money for charity. On this occasion, Patrick and his team members were tasked with carrying water buckets on a bouncy inflatable surface, to fill up a barrel. All of the team were soaked to the skin, including Patrick, who did not seem to have come prepared for a drenching: everyone else was in very sporty kit, but Patrick appeared to be dressed for an astronomy meeting! He had come in his standard attire, just removing his blazer and tie for the messy bits. After a total soaking, his trousers, already pushed to the limit trying to contain his peak weight 19 stone girth, tried to fall towards the floor, revealing far, FAR too much of his considerable buttocks. The commentator, Stuart Hall, was paralytic, but Patrick then seemed to disappear for 15 minutes. I began to wonder if he had gone home in a sulk! When Patrick re-emerged he had a completely new set of black trousers (completely water-sodden) tied around his waist with inch-thick ship's anchor quality rope. He was asked where he had found them. "Ohh, I just rummaged around in the grounds and found a perfectly good pair of boilermaker's slacks" was his reply, as if nothing could be more normal. To this day, I am still pondering why boiler makers have specially made trousers of that girth that they throw into the undergrowth!! I'll have more to say about *It's a Celebrity Knockout* later.

On the subject of Patrick's precise weight I do recall a *Sky at Night* program from the 1980s where Patrick mentioned on TV that he was "around the region of 18 stone".

However, I was once told by one of Patrick's producers, Pieter Morpurgo, that during the 1980s Patrick used to tell people: "I am NOT overweight. I am only 19 stone. I am certainly not 20 stone, and if I was 20 stone, well, yes, I would be slightly overweight. But my bathroom scales don't lie". However, Pieter added that Patrick's bathroom scales only went up to 19 stone and so however much beyond that point he went, he would still only weigh 19 stone!

During the latter stages of 1975 it had become obvious to Patrick that his numerous writing and broadcasting commitments, as well as his running of the BAA Lunar

Section, was just too much. Patrick never gave anything other than 100 %, and compared to other BAA Section Directors, his running of the Lunar Section was more like 1,000 %. In addition, his mother, now 89, was in need of more of Patrick's attention, even with the help of a full-time housekeeper. Reluctantly, Patrick concluded that he would have to resign his Directorship of the BAA Lunar Section. He just could not give 100 % any more. At the BAA Council meeting on January 7th 1976 he submitted his resignation to the BAA Council, at their meeting in the Council room of the Royal Astronomical Society at Burlington House, Piccadilly. Patrick had already discussed this with the Lunar Section committee and sounded out his successor, namely, Harry Ford, the Lunar Section assistant director. Patrick gave a slightly different version of his reasons for resigning in the February 1976 BAA Lunar Section Circular:

> With only one rather short break, I have been Director of the Lunar Section for about fifteen years; I was the Section Secretary before that – which takes us back to 1946 and the time of Wilkins! I am quite sure that it is wrong for anyone to retain the Directorship of an observing Section for too long, particularly when (as now) there are excellent people to take on and infuse new ideas into the Section. The appointment of a new Director is, of course, entirely the decision of the BAA Council, but it is fair to say that the recommendation of the retiring Director is generally followed; and if this is the case now, I am sure that the members of the Section will whole-heartedly approve. I will make an announcement as soon as possible. Meanwhile, I must make it clear that I am in no way withdrawing from the Section. Far from it! I will continue observing as energetically as I can, and I will serve the new Director in any capacity he wishes (if any). I am confident that all members will appreciate, and will approve of, my reasons for stepping down now after a long spell in office. If the Council follows my recommendations, the change-over will take place on 1st March, which means that I will attend my last Council Meeting at the end of February and will take the Chair for the Section meeting in London on 21 February. Obviously I will have some final comments to make, but these can be left over for the present – and I will only say that, as usual, this Circular would have been hard to produce but for the untiring efforts of the usual team (Miss Atwell and Messrs. Mason, Spry, Prout and Gannon).

It was the end of a BAA era, with Patrick departing as the Lunar Section Director. In truth, the Lunar Section would never be the same again. Harry Ford would be a good and popular director but no-one could really step into Patrick's size 13 shoes; no-one else had the charisma or the work-rate. Even in his final weeks in office, Patrick was organising the precise details of the next Lunar Section London meeting for February 21st. At the end of that meeting Patrick formally handed the Lunar Section over to Harry Ford, to loud and prolonged applause.

But even if Patrick would be having less to do with the BAA Lunar Section it was a rare week when he was not on the radio, in the newspapers, or on another TV show, in addition to his own monthly *Sky at Night*. In a hysterical Morecambe & Wise TV sketch, broadcast on January 14th 1976, Patrick introduced Eric & Ernie as two leading astronomers, namely: Dr Ernest Bavistock Peregrine Cecil Wise, from the Department of Astronomy, University of Harrow and Mr Eric St John Faraday Waverley Morecambe, assistant caretaker at Milverton Street Infants School! After pointedly asking Eric Morecambe how big his reflector was (to much amusement from the audience) Patrick said ..... "And yet you know gentlemen there is one place from which you can not only see Mars & Venus but all the other heavenly bodies". Cue Eric Morecambe, who said "And where is that may I ask?" Cue Patrick, who positively shrieked: "I'm delighted you asked me that because I am going to show you". At which point he pranced off to the piano and gave forth a rendering of "The stars at night are shining bright [Bang, Bang, Bang, Bang] deep in the heart of Texas". The bangs were provided by Ernie Wise slamming the piano lid. Extraordinary!

The Glasby Affair

Patrick's disappearance from the BAA Council table conveniently spared him some embarrassment too. During the 1950s he had been in a very tricky position regarding his good friend Percy Wilkins' belief in O'Neill's bridge on the Moon. Then, in the 1960s, he had been embarrassed by his good friend Frank Hyde's dodgy printing activities from a caravan Hyde shared with his mistress. Both Wilkins and Hyde had appeared on *The Sky at Night*. As the 1960s changed into the 1970s, Patrick had become increasingly friendly with the BAA Variable Star Section director John S. Glasby (1928–2011). We have already seen that Glasby had appeared on the October 1969 *Sky at Night*, demonstrating a 13-inch reflector from his observatory at Nirvana, 7 Kilwinning Road, Steventson in Ayrshire, Scotland. Glasby had appointed Patrick as the Variable Star Section's binocular co-ordinator. Physically, Glasby was a formidable presence, looking even larger than Patrick himself! Standing a solid 6′ 2″ and weighing in at around 240 pounds he was the textbook 'brick out-building' of a man, resembling either a rugby player, or the sort of debt collector with a lantern jaw that you didn't argue with!

As well as his BAA role and his professional life as a research chemist and a mathematician, Glasby was a highly prolific author of all manner of strange works of fiction, all of them published under a variety of pseudonyms, but never under his own name! These works, some short stories, but many complete novels as well, ranged from the genre of science fiction to war stories (100 of these!), spy stories, crime and adventure tales, romance novels and novels with a cowboy western flavour! His pseudonyms, many under the 'Badger Books' label, included Victor LaSalle, Karl Zeigfreid, John E. Muller, A.J. Merak, Chuck Adams, Tex Bradley, Manning K. Robertson, D.K. Jennings and many others. None of these works of fiction were bestsellers and most were less than exciting, but he could certainly churn them out at an amazing rate. He probably produced 300 articles and novellas in total! In truth, he had little choice regarding the rate he had to write at, because Glasby and another author, the reverend Lionel Fanthorpe, practically provided Badger Books with most of their material, and the Badger deadlines for specific stories were often just a few days long, sometimes just a single weekend! The reason for such manic deadlines was the way Badger Books operated. The money-spinning formula was that a pulp fiction artist would be asked to provide a scarey piece of artwork, to grab the potential buyer's attention. Once the artwork was completed the author was required to spin a yarn around the cover artwork in days to ensure a quick cashflow into the company. Crazy!

However, Badger Books went out of business in 1967, but this was just a minor hiccup in Glasby's writing career. From 1968, encouraged by Patrick, Glasby turned his attention instantly from pulp fiction to writing science books, but now under his real name; well, there's a turn up! His book *Variable Stars*, published in 1968, three years into his term as the BAA Variable Star Section Director, was regarded as a useful guide for the variable star observer of that era, who had very little else as a reference. He would eventually churn out nine other science books, including four more on variable stars and astronomy: an almost Patrick-like production rate.

Glasby resigned as BAA VSS director in 1971, to be succeeded by the excellent director John Isles. However, Patrick had encouraged Glasby to write a book on the class of variable stars known as 'Dwarf Novae'. Glasby completed the book in 1970, but over the next 5 years it attracted much unwelcome publicity; he was accused of fabricating some of his own alleged observations, ones that were totally impossible to make from UK latitudes. Worse, some of the observations he cited were claimed to be 'from the BAA VSS archives', which they were certainly not. It soon became apparent that Glasby's sidelines of writing science-fiction and science-fact, two totally different disciplines, had become horribly blurred. If he was short of data, no problem, just make it up! As the incumbent BAA President for the October 1974 to October 1976 period, Howard Miles decided to set up a committee to investigate the case against Glasby, and to stop any damage to the BAA's reputation from the book that Patrick had encouraged him to write. A number of specific points amongst dozens of anomalies in Glasby's book were investigated, namely:

1. The author claimed to observe the star CV Aqr when it was very close to the Sun
2. Light curves in the book, for the stars X Leo, UV Per, SU Uma, SW Uma and UU Aquilae were attributed to come from the BAA, but no such BAA records existed
3. A light curve of the star FQ Sco, based on the author's observations, showed variations between magnitude 12.5 and 15.6 in a period of 24 days, but the maximum theoretical altitude of this star from Glasby's observatory was 2 degrees!

The BAA Council committee found that all of these points were valid and so there was considerable egg on Glasby's face, although the official Council statement did not call Glasby a liar, simply that there were 'certain inconsistencies and errors in the book…..it appears that some of these are printing errors'. Patrick must have been relieved he was not a serving council member when the investigation produced its findings! Glasby had been a good friend and a guest on *The Sky at Night*, but now his entire reputation was the cause of much speculation. However, after resigning as the BAA VSS Director in 1971, and after the Council investigation of 1975/1976, Glasby had nothing more to do with the BAA, and Patrick had absolutely nothing more to do with John S. Glasby!

Comet West

At the start of November 1975, the noted professional astronomer Richard West, of the European Southern Observatory, discovered a promising comet on photographic plates exposed 3 months earlier, in August. The comet was due to reach perihelion (closest approach to the Sun) on February 26th 1976 at a very small solar distance of only 29 million kilometres. This was a very exciting prospect as an active comet travelling this close to the Sun would become very hot and therefore very bright. However, only 2 years after the 'flop' of Kohoutek, astronomers and the media were being very cautious. There was still considerable egg on the faces of the scientific community after comet Kohoutek failed to become the daylight object which the press had 'hyped' it up to be. However, this time the comet did live up to expectations, although UK observers had to battle with considerable cloud in the pre-dawn March mornings when Comet West had peaked. Even the hardiest UK observers, such as Paul Doherty and Guy Hurst, only had one awesome view of the comet, when it was an easy naked eye object with a curved dust tail, dominating the pre-dawn sky. Unfortunately for Patrick, when the comet peaked he saw nothing of it. He contracted influenza and was confined to bed for much of March. He was, as he put it 'Hors de combat!' For a normally fit workaholic like Patrick this was a total disaster! Still not fully recovered by the time of the annual BAA Winchester residential weekend (April 9–11th that year) and with a backlog of BBC and writing commitments, Patrick had to pull out of the weekend, to his immense disappointment and the dismay of those attending. To make things worse, the course organiser Alfred Curtis, a good friend of Patrick's, would die 2 months later, much to everyone in the BAA's sadness.

However, Patrick did see Comet West with his 15 inch reflector, before and after its peak. His last view was on May 25th 1976 at 23 hours GMT, using that big Newtonian at 200×. By this time its close encounter with the Sun had caused it to break up and the nucleus was obviously multiple, as seen through Patrick's eyepiece. Patrick was also thwarted on April 8th when observers had been alerted that the bright (3rd magnitude) star Epsilon Geminorum would be occulted by the planet Mars as seen from southwest England. Unfortunately, Mars was only at an altitude of 9 degrees, and as Patrick observed the star and planet converging, they both sank below the dome wall of his observatory, just 10 minutes before the occultation. Patrick swiftly carted his small childhood 3 inch refractor outside the dome and followed the occultation, but through such a small aperture all he could determine was that he was 'pretty sure' the occultation did take place.

At the end of April 1976 Patrick travelled to the spectacular Greek island of Thera (also known as Santorini) with 30 other British eclipse chasers, to view the solar eclipse of April 29th. This was not a total solar eclipse, but an annular eclipse, where the Moon passes directly over the Sun, but is slightly smaller, so that a ring of light surrounds the black disc of the Moon. Patrick chose to ride a mule, rather than to walk, up the 800 steps from the harbour to the town. Patrick had enjoyed riding on donkeys as a small boy at Bognor, but in those days the donkeys were heavier than him! The sight of the 18 stone Patrick, atop a mule of only fractionally larger apparent size, caused much amusement! The comments of the mule are not on record, just the very unhappy expression on its face, as it trudged up a steep 800 step winding road with Patrick on its back. This author walked down those steps in 2006 and the mules were sweating profusely even with a light payload, so I can only presume that Patrick's mule had to be put out of its misery by the time they reached the top. Possibly, for such a passionate animal lover, this escapade was a bit out of character for Patrick?

Fig. 21.1 Patrick, wearing a white blazer and a tie, riding a mule up the 800 steps on the island of Santorini. Whether the mule had to be put down after the experience is unknown. This event occurred during his annular eclipse trip in April 1976. His good friend Rossie Atwell is riding the mule on the *left* side of the picture. Photograph by kind permission of John C McConnell

Two weeks later, on May 13th 1976, a partial eclipse of the Moon took place from the UK, although it was very low down on the eastern horizon. Patrick's main telescopes at Selsey were obscured by trees but he again managed to employ his childhood refractor to observe the event. His report appeared in the July Lunar Section Circular, as follows:

> The Moon was, of course, very low and I could only use my portable 3 inch refractor. The shadow was chocolate in hue, and I had the impression that it would have been a dark eclipse. After the umbra had gone the penumbra was virtually imperceptible. No colours were seen at any time. There was no cloud, and only a little haze; conditions were in fact tolerable, and photographs were successfully taken on Fujichrome, ASA 100.

The 27th of June 1976 was a big day in Patrick's life: it was his mother's 90th birthday. Since his return from the RAF in 1945, Patrick had lived with his mother at East Grinstead, Armagh and Selsey. Indeed, apart from the RAF years and a few trips abroad, he had lived with his mother for his whole life, namely 53 years.

Patrick organised a big party at the nearby Seal Hotel in Selsey, and the weather was very kind on the day. Although Patrick had no wife, children or brothers and sisters he did have plenty of friends and quite a few cousins and second cousins, all of whom attended Gertrude Moore's 90th birthday party. Patrick and his mother had recently taken on a new full time housekeeper, Mrs Hester Woodward, known affectionately as 'Woody'. With Patrick often away from home, and a frantically busy man, a full-time totally reliable housekeeper was essential, especially with his mother becoming increasingly frail.

Around the time of his mother's ninetieth birthday, Patrick had a phone call from a Miss Philippa Savery, a resident of the city of Bath; the city from which William Herschel discovered the planet Uranus. The news was dreadful! The Bath authorities had plans to demolish Herschel's old home at 19 New King Street, the very address from which Uranus had been discovered! Patrick was horrified and drove straight down to Bath and inspected the house. It was dilapidated, but after Patrick drummed up publicity the Herschel Society was formed and the house restored into an excellent museum to celebrate Herschel's life. The house and museum survive today, in no small part due to Patrick's involvement.

Chapter 22
An End to Lunar Section Chores

Despite Patrick handing over the Lunar Section Directorship to Harry Ford, and Geoff Amery being appointed as the assistant director, Patrick was still heavily involved in Lunar Section activities throughout most of 1976. He was, after all, still the Lunar Section circulars editor which brought with it the tedious duty of duplicating hundreds of copies each month (typically half a dozen double-sided A4 sheets), placing them all, folded, in hundreds of envelopes, and posting them off. Many years ago, in a conversation between this author and John Mason, John recalled a number of occasions where he had accompanied Patrick on a night-time drive through Selsey, searching for post boxes which could swallow up several hundred letters in one go. One post box was rarely large enough! Patrick was not a patient person and when people sent stamped addressed envelopes which were not really big enough for the circular Patrick would not attempt different folding techniques. He would simply place the bulging envelope on the floor, jump on it with his massive weight and size 13 extra-wide shoes and then, because the envelope would split, wrap a light-year length of selotape around the ripped envelope!

However, with the Director Harry Ford living in Dundee, Patrick living in Selsey, and the GPO postal service becoming ever worse, Patrick was only too happy to hand over all of the Lunar Section editing tasks to Harry at the Dundee end. This had been a colossal amount of voluntary effort by Patrick who, as a self employed author, always on call to the BBC, and a tireless charity worker, had zero spare time. The October 1976 Lunar Circular would be the last that Patrick ever produced and Patrick wrote a lengthy 'goodbye' on page 1:

> The time has come for me to say farewell as editor, I am very sad at severing my official link, though, needless to say, I will continue to be active in the Section as I have been for the past forty years or so!
>
> The decision by the Director is absolutely right. To have a Director at one end of the British Isles and an editor at the other simply does not work, mainly because the postal service today is so slow. It often happens that the Director wants some notice inserted at the last minute – and if it is too long to convey by telephone, there ensues about a week's delay. The situation has not occurred before. During much of my Directorship, Phil Ringsdore was editor, and he was within half an hour's drive of my home; even when he went to

M. Mobberley, *It Came From Outer Space Wearing an RAF Blazer!: A Fan's Biography of Sir Patrick Moore*, DOI 10.1007/978-3-319-00609-3_22,

> Somerset, the distance was not great, and of course when Geoff Buss took over he was very close to me in Selsey. But since I retired as Director, the writing has been on the wall; and now the decision has been made, things should be more co-ordinated. Latterly, I am afraid that the Circular has been an 'independent publication', and this was not the intention of anyone – least of all myself.'
>
> I now bow out with a final expression of thanks to those who have helped me so nobly during my spell as Editor (for much of which I was also Director). Miss Atwell and Andrew Gannon have done wonders with the distribution; Reg Spry has always been on hand in an amazing number of ways; Roger Prout and Mark Savill have also been of assistance. But I must make special mention of John Mason, who has been a tower of strength, and who has done most of the actual duplication on my elderly machine. Without him the situation would have been wellnigh impossible.
>
> Now under the new regime, the Director will be on hand to control everything. I hand over with the greatest confidence, and I will always be ready to help in any way I can. So, as Editor – farewell, and thank you all.

This final severing of all Patrick's Lunar Section duties must have been a huge relief as he still had a near infinity of speaking and writing commitments, not to mention his monthly *Sky at Night* TV programme and charity work. His output of books was still as prolific as ever with his title *The Next Fifty Years in Space* being published even as he resigned his editorship. In addition, 1976 had seen the completion of a total re-write of his original, first science book, *Guide to the Moon*, originally published in 1953. Various upgrades had appeared; the last, in 1963, being renamed *Survey of the Moon*. But this new edition was a total re-write and the original title was restored. It had been the straw that almost broke the camel's back, when combined with Patrick's other chores and media appearances.

Patrick could turn up on almost any TV or radio show during the 1970s, in a serious or humorous role. On April Fool's day 1976 he announced, on BBC Radio, that at 9.47 a.m. Pluto would pass behind Jupiter and that this would reduce the Earth's gravitational pull. He invited listeners to jump at that minute, upon which they would start to float! Although humorous, there was an alternative agenda to this prank. It was a dig at the book entitled *The Jupiter Effect*, written by John Gribbin and Stephen Plagemann, which Patrick described as "pseudo-scientific claptrap even worse than astrology". Amazingly, following Patrick's April Fool, the BBC received numerous phone calls from listeners, confirming that they had actually experienced the floating sensation! By the mid 1970s, after almost 20 years on television, it was hard to think of anyone who was more famous than Patrick. He would never have the wealth of a rock star, but he was just as well-known as the biggest names in the pop industry; and they knew who he was too. On the *Rolling Stones* 1976 European Tour their new guitarist, Ronnie Wood, was always booked into hotels under the comedy codename of 'Patrick Moore', with the request that 'he needs a large bed'; hilarious!

Another TLP and an Observing Rota

A few weeks after relinquishing his tedious Lunar Section editorial chores, Patrick was observing the Moon with his 15 inch reflector on October 4th 1976, when he spotted an unusual appearance to the lunar crater Gassendi. At 8.45 p.m. GMT

Patrick had started observing the Moon with a ×400 eyepiece inserted. At 8.52 p.m. he had reached the region of the crater Gassendi and had detected a pinkish colour extending south from the smaller crater Gassendi A, into the main crater Gassendi itself. Using the red/blue lunar section 'Moonblink' filter he got a positive blink when switching between filters. This was a highly coloured event. Patrick made a series of phone calls, firstly to the observatory at Dundee (no answer!) and then, more successfully, to Peter Foley in Dartford, Kent and Hedley Robinson in Teignmouth. Each of these last two observers confirmed various red anomalies related to Gassendi. A fourth phone call, to Lawrence Fitton, got through, but Fitton was clouded out. Unusually for Patrick, who was a visual observer, he then took some photographs of the area, but upon resuming his visual inspection, at 9.30 p.m., there was no trace of the pink colouration. However, by 11.00 p.m. GMT Patrick noted that the southern part of the middle member of Gassendi's central peak complex was red. To quote Patrick's report: "This was the most startling colour I have seen for many years". Numerous telephone calls ensued throughout the night between Patrick, Peter Foley and Hedley Robinson, and their accounts of the event, especially those of Moore and Foley, are virtually identical. A subsequent analysis of the observations by Lunar Section member Lawrence Fitton, who had developed a balance of probabilities 'technique' for weeding out colour anomalies caused by the Earth's atmosphere, determined this phenomenon to probably be a real lunar event, possibly caused by 'the emission of light by a luminous gas'. There would be much to discuss at the forthcoming Lunar Section meeting in Coventry on October 30th.

Unbeknown to Patrick, the Lunar Section committee had decided that his recent years as the Lunar Section Director and editor could not go unrewarded: a presentation was planned for the Coventry meeting. Harry Ford had secretly informed Lunar Section members, via an insert in the Circular, that a collection was being made for Patrick's 'retirement', and from all over the world a total of £32 was received. With the assistance of his new wife, Lynne, Harry decided to purchase a quartz crystal clock 'which matches the décor of the Moore household' along with six etched Brandy glasses. The presentation, at the Coventry meeting, was a closely guarded secret and took Patrick completely by surprise. Needless to say, rapturous applause accompanied the presentation. Patrick sent the following letter to the Lunar Section, for publication in the circular:

> When I came to the Section meeting in Coventry, I was completely unaware of the presentation to be made, and for once in my life I was lost for words! I can only say 'thank you'. The clock is magnificent – it has the place of honour in my study, where it will always remain, and at last I can tell the proper time (my cuckoo clock, a present from my mother on my sixth birthday, is hardly a precision instrument). And the glasses will be in regular use…I would stress though that the Section is a team effort and always has been. I have tried my best to play a part and though I am no longer the Director I will continue to do so to the best of what ability I have. But I do want you all to know how much I appreciated the presentation. It was a very proud moment for me, and I can only say again – Thank you. Very best Ever Patrick.

Needless to say, after the events of October 4th, the TLP sub-section were keen to keep the crater Gassendi under close scrutiny, and on the night of November 3rd/4th more red glows were observed by messrs Peter Foley, Howard Miles, Geoff Amery and J. Hedley-Robinson. However, Patrick observed no further events in

Gassendi on that date. But, despite having retired from his Lunar Section duties, Patrick was still keen to improve the section's coverage of the lunar surface so that future events, in Gassendi or elsewhere, could be detected. Patrick put together some thoughts on the matter, which were eventually published in the January 1977 BAA Lunar Section Circular:

> From time to time there have been comments on the effectiveness of our TLP Network. I was considering this again recently, when carrying out a routine Blink Survey. It is quite true that if an event does occur, it may well pass unnoticed by any of our members, particularly if it takes place in a formation not noted for TLP activity. For instance, no TLP has ever been recorded in, say, Stofler. This is one of the formations which I check quite often; but if a TLP did occur there, what are the chances of my being observing it at the vital moment?' One has to admit that they are slight; and so far as I am aware, other Network members do not survey it at all regularly. (I give Stofler only as an example, not because I have any reason to suspect it of any activity.) There are very few formations which are watched sufficiently regularly to hold out any hope, apart from sheer chance; and even with these there are wide gaps in the monitoring. I agree that this is a problem which can never be properly solved, partly because of the weather but mainly because nobody can sit up all night concentrating on a few craters in the hope of seeing something anomalous. But I do think that we should go back to an idea I suggested some time ago – if it can be organised and maintained. I suggested that for a few set formations, we could ask for our leading TLP observers to be on watch at definite times: for instance –
>
> Aristarchus:
> Observer A, 20 to 21 hours on Feb 10th.
> Observer B, 21 to 22 hours on Feb 10th.
> Observer C, 22 to 23 hours on Feb 10th.
> And so on.
>
> Other formations could be similarly 'issued out'. I appreciate that we may be asking for trouble in the shape of observational bias toward 'suspect' formations, but I think we are now past the stage of having to prove that TLP exist; we now need to find out how they behave. Obviously we require spectra, and when Dr Maddison and his team at Keele are ready we have a good chance of securing something; our task is to alert Keele when a suitable TLP appears. But at the moment our monitoring is random, or virtually so, and the chances of a TLP escaping detection are obviously increased. I do see the difficulty of any systematic monitoring rota, but I believe that we must try to put some kind of rota into operation. At any rate, I make the suggestion for what it is worth.

Although a few Lunar Section members were keen on Patrick's plan, it was never a success. Few members, especially those with wives and families, could stick to such a rigid system; fewer still would be cloud free at the appropriate times. None had Patrick's single-mindedness to stick to a timetable 'Come Hell or High Water'.

During Patrick's life he lost an extraordinary number of his friends to cancer and, unknown to all but his closest friends, he donated huge amounts of money, from giving public lectures and other events, to cancer charities. As 1976 ended Patrick found himself writing yet another obituary, one for his Canadian TV astronomy counterpart Kenneth Chilton. Ken was only 38 when he died, but like Patrick he was an astronomy popularizer, and he was also the serving President of the International Union of Amateur Astronomers.

Around this time Patrick had been experimenting seriously with wine-making: in particular, rice wine. To keep his fermenting brew at the right temperature he agreed with his mother that he would store it at the top of the airing cupboard: the warmest place in the house. His mother was worried about it leaking, but reluctantly agreed. One night, while Patrick and his mother were soundly asleep, there was a colossal explosion which, according to Patrick "rocked the house". "It sounded like the last trump! I thought the third World War had started". Patrick had got something badly wrong in the wine making process. The vat of potential wine had exploded dramatically, swamping his entire stock of underwear. His mother was not amused!

On the subject of not being amused, Patrick was appalled when he heard that his guest on the October 28th *Sky at Night*, Dr Percy Seymour of the National Maritime Museum, had apparently used the programme appearance to springboard himself to a new role, as.......an Astrologer! Many years later I asked Patrick about this bizarre turn of events. "I wasn't at all happy Martin and I should have known better. The first time a non-British guest appeared on the programme. I should have stuck to my instincts about foriegners!"

At the very start of 1977, on January 7th in fact, Huddersfield Postman John Hosty discovered a nova in Sagitta, very low down in the evening sky. It was so low down that, according to Patrick, he had asked the keen variable star observer, Storm Dunlop, to check whether the new star existed. Storm told Patrick that he had failed to find the object in his binoculars, although at only 5 or 6 degrees altitude this was hardly surprising! Patrick took Storm at his word, without taking the altitude into account, and on the January *Sky at Night* he announced to one million viewers that there had been reports of a nova, "But frankly, I don't think it is really there". Patrick would end up with egg on his face over that. It was a real discovery, by a dedicated (if eccentric) British amateur, and Patrick had rubbished it. Patrick, when speaking privately to others, unfairly blamed Storm for the mistake, and in future years they would have a big falling out over other issues.

Chapter 23
20 Years on TV and a Fall in the Bath!

As 1977 started Britain was at the height of the Punk Rock Era and Johnny Rotten, Sid Vicious et al., alias *The Sex Pistols*, were going strong. The group even released a record to 'celebrate' the Queen's Silver Jubilee entitled "God Save the Queen", although as the next line was "She ain't no human being" it was hardly meant to be complimentary! But Patrick's thoughts were firmly rooted, as ever, on astronomy. He was, as always, constantly in demand, on TV and on the radio and by newspaper columnists. Patrick had appeared on his third Michael Parkinson show on New Year's Eve, alongside fellow eccentrics and scientists Magnus Pyke and David Bellamy. Roland Emett, the designer of Patrick's lunar bicycle, The Lunacycle, was also featured. Parkinson kept trying to convince Patrick that the reason people watched *The Sky at Night* was because they were fascinated by him and his delivery, and not by the subject itself. Patrick was having none of this though, even though Parky was clearly correct. So Parky turned to the studio audience for help, shouting "Am I not right…is it not the performance as much as what he says….." The audience shouted back a resounding "YES" and Patrick looked somewhat embarrassed.

During the show Magnus Pyke agreed with Patrick that astrology was bunk and could be harmful. The grumpy *Daily Express* columnist George Gale took issue with that, and in his newspaper column stated that although he was not a believer in astrology he felt astrology was not harmful. Gale also stated that it was not unreasonable that people born at the same time and place might share certain characteristics. Well, this prompted a swift response from Patrick, and his favourite astrological phrase "There's one born every minute" was directed at Mr Gale in the *Daily Express* letters column during January 1977. Patrick issued a challenge to George, to carry out a scientific test to prove or disprove astrology. When Gale had not replied within a few days Patrick sent him a card, stating:

> 'I have, as yet, had no reply to my letter or my challenge. Don't tell me you mean to chicken out?'

The exchange of words went on for several weeks, but in the meantime there was a big TV birthday looming for Patrick, which was far more important than discussing astrology.

M. Mobberley, *It Came From Outer Space Wearing an RAF Blazer!: A Fan's Biography of Sir Patrick Moore*, DOI 10.1007/978-3-319-00609-3_23,

As well as the Queen's Silver Jubilee, in a few months time *The Sky at Night* would celebrate its own anniversary of 20 years on television: an unbroken run of one programme every 4 weeks, with a mixture of leading professional astronomers and Patrick's amateur BAA buddies appearing each month. A 20 year anniversary demanded a special programme and the BBC decided to fund a 40 minute anniversary special with an audience of invited guests. The special programme was filmed on April 13th 1977 and was followed by a party afterwards, in the sixth floor suite at the BBC TV Centre in London. It was transmitted at 10.25 p.m., exactly 1 week later. The guests paying tribute to Patrick and attending the party were numerous and eminent, and leading professional and amateur astronomers were present. These included Sir Bernard Lovell, Sir Martin Ryle, Professor Graham Smith, The Earl of Rosse, Dr Anthony Hewish, Dr Jocelyn Burnell, Dr Kenneth Pounds, Dr Garry Hunt, Professor Geoffrey Eglinton, Commander Henry Hatfield and Dr Ron Maddison. Patrick was particularly proud of the tributes from professional astronomers; he was, after all, merely an amateur, with no formal degree and no scientific training; he just had an incredible ability for speaking about his subject in an infectious and accurate manner, with an autocue nowhere in sight. Complementing his 20 years on TV he learned around that time that he would be the recipient of the prestigious Jackson-Gwilt Medal of the Royal Astronomical Society, an award bestowed on few amateur astronomers.

The 20th anniversary programme certainly had its humorous moments, not least, a 'Two Ronnies' sketch, broadcast a few months earlier, in which Ronnie Barker pretended to be Patrick Moore's brother, deputising for Patrick. In the sketch, Patrick's Orrery, often seen on *The Sky at Night*, fell apart and rolled across the floor during a 'fake' *Sky at Night*. Barker said he was standing in for his brother Patrick because "He had to go and show his telescope to the local townswomens' guild.....if they like it they are going to knit him a cover for it......" That famous clip from the 50th *Sky at Night*, with Patrick's friend George Hole and his massive 24 inch telescope being 'totally obscured' was also shown, amid much laughter! Needless to say, numerous TV impersonators loved to mimic Patrick. As well as Ronnie Barker, the impressionists and comedians Mike Yarwood and Spike Milligan had great fun impersonating the fast-talking, monocle-wearing Moore as did Jon Culshaw some 30 years later.

A Serious Injury

But it would soon be down to earth with a bump, after the passing of the 20th anniversary – literally! At the start of May 1977 disaster struck when Patrick trod on a bar of soap in the bath at Selsey, cartwheeled (Patrick's own words) and knocked himself unconscious by headbutting the bathroom tiles. It must have looked like a scene from a Laurel and Hardy film but it was deadly serious. His mother, hearing the bang, phoned for an ambulance and Patrick was taken to hospital. He suffered severe concussion of the spine and was drifting in and out of consciousness for 2 weeks. [One crazy media report mixed the facts up and declared that he had been unconscious in the bathroom for 2 weeks, before his mother realised that he was missing and not just having a long soak!] When the *Daily Express* columnist Jean Rook visited him a year later, and saw the chaos he and his mother lived in (including a scene where Patrick dropped potatoes all over the floor), he said about the fall "It was one of those silly accidents you think could never happen to you". Rook's view was rather different: "I didn't like to say it could NOT happen to anyone else".

Worse news was to follow though, as when he was fully conscious after the fall he appeared to have virtually no movement in his hands. According to Patrick an RAF wartime spinal injury, caused by a crash landing, had been aggravated by the fall. This was a serious situation, but Patrick was adamant that the next *Sky at Night* had to take place and he had to be on the programme. Patrick had an unbroken 20 years behind him and had no intention of missing even a single episode of the programme. The BBC agreed to an emergency plan with Patrick. For the next few months he would be unable to travel. Indeed, he was due to have a spinal operation on May 25th and the success of that was impossible to predict. Patrick's friend Paul Doherty had already agreed to travel down to Selsey to help him and his 91 year old mother (and their housekeeper) to cope and maybe snatch a peek through Patrick's 15 inch reflector too. So, the BBC agreed to film the next few programmes from Patrick's Selsey study.

The first programme, on May 24th, would be about sunspots and Dr Ron Maddison would travel down from Keele to Selsey to present most of the information. After Patrick's operation and recuperation, Gordon Taylor and Dr Garry Hunt would also travel to Selsey, to talk about the newly discovered rings of Uranus. Finally, a programme scheduled for July 1977, about the UK Infra-red telescope on Hawaii, would involve a BBC camera crew travelling to Hawaii without Patrick, but with Patrick fronting the programme, assisted by Professors Ring and Reddish. To compound Patrick's misery during the period while he was recovering from his spinal operation, on June 1st he heard that his dear Selsey friend Henry Brinton had died. Henry was 76, and as we have already seen, when Patrick had been leaving Armagh in 1968 it was Henry who had convinced Patrick to move to the charming seaside town of Selsey. Henry had suffered a series of crippling strokes in recent years, and his death was not totally unexpected, but it must have been a bitter blow to Patrick, whose own health appeared to be hanging by a thread. After his health had started to decline, in 1976 Henry Brinton had donated his 12.5-inch reflector to Hatfield Polytechnic (now The University of Hertfordshire) where Patrick's friend Iain Nicolson was a lecturer. A few years later the Old Mill House where Brinton had lived, just a few hundred yards from Patrick on the junction of West Street and Clayton Road, was sold to property developers, and according to Patrick "Demolished and turned into boxes." Patrick added that "The same developers came to me wanting to buy Farthings…..I told them to get stuffed!" [Note: Brinton's 12.5-inch reflector, originally the Calver reflector owned by Robert Barker, was moved back from The University of Hertfordshire's Bayfordbury Observatory, to West Sussex, in the twenty-first century. The move was due to the telescope's replacement with a modern 16-inch Schmidt-Cassegrain. Brinton's old Newtonian now resides in an observatory at Slindon College, near Chichester, opened in 2009.]

While Patrick was lying in his hospital bed in June 1977 it may have been some consolation to him that the BAA had finally published his paper regarding *The Linné controversy*, discussing his verdict that nothing structural ever did change on the Moon these days after all. Interestingly, this paper contained the only reference (that I know of) where Patrick, unprompted, specifically mentioned that he disagreed with his old mentor Percy Wilkins. To quote: 'More recently another former Director of our Section, H.P. Wilkins, suggested that the post 1905 observations were discordant enough to indicate further structural changes, but I do not for a moment think that this can be so'.

Thankfully, Patrick's spine did recover, although problems would resurface more than 20 years later. Just as important to Patrick though was that his unbroken record on *The Sky at Night* was intact. He had survived to live another day, unlike Elvis Presley, who died on August 16th, just as Patrick was becoming mobile once more.

Canals

Although Patrick's first love would always be the Moon, the planet Mars had often run a close second. Early in his writing career he had, as we have seen, written a number of science fiction books about Mars, including one under a pseudonym, presented as science fact! When Patrick was a young man there were still those who claimed that the Martian canals, hyped by the turn-of-the century wealthy American Percival Lowell, were both real and artificial. Indeed, the Guzman prize of 1900 offered 100,000 Francs to anyone who could communicate with beings from another star system, and the planet Mars was specifically excluded as communicating with Martians was deemed to be 'too easy'! Certainly, under mediocre seeing conditions, with modest telescope apertures, linear features on Mars had been widely observed, but rarely was the Earth's atmosphere stable enough to allow photography to disprove their existence. In many ways, prior to the 1960s and the first planetary space-probes, amateur astronomy was a far more romantic hobby. Taking sharp photographs from the Earth was so badly hampered by the Earth's turbulent atmosphere that a talented visual observer, with a 20 centimetre aperture telescope, could sketch just as much detail on the planets as the largest telescopes on Earth could photograph. All telescopes were simply limited by the Earth's atmosphere. Mars had features which might be interpreted as vegetation and a temperature which might support life, and some observers were reporting canal-like lines across the planet.

Unfortunately, where visual observations were concerned, it was always a case of one man's view versus another's and the lack of space-probe images of Mars, before the 1960s, meant that amateurs could claim they had seen Martian canals and no-one could prove them wrong. The arrival of Mariner 9 at Mars, in 1965, changed all that and the Viking spacecraft, which landed on Mars in 1976, really did put the nails in the coffin! It was quite obvious by 1977 that the canals never had existed, but Patrick felt obliged to publish a paper on this topic in the Journal of the BAA. Patrick's approach was to do something which, remarkably, had not been done properly before: to superimpose the best Martian 'canal' map onto the best modern

Mars map. Patrick chose the Flagstaff map of Mars, compiled mainly from Mariner 9 images, and the historical Flammarion/Antoniadi 'canal' map from Flammarion's book (La planete Mars et ses conditions d'habitabilité Volume 2) published in 1909. He then superimposed the 1909 canals on the Flagstaff map. There was no correlation at all! Indeed, in his summary, Patrick wrote: "Nobody has ever seen a canal on Mars; there are no canals to see".

However, Patrick's verdict was not received without criticism. The BAA Mars Section Director E.H. Collinson, writing in the October 1977 Journal, disagreed with Patrick's view that the giant Vallis Marineris (Mariner Valley) could not be identified on the Martian canal maps. Collinson pointed out that the Vallis Marineris corresponded with the canal feature Coprates on the IAU map and with the feature Agathodaemon on the 1909 Antoniadi map. It was also shown as a fine line on Lowell's maps. Collinson, a respected authority on the subject of Mars, also suggested that other broad streaks, although not corresponding with any physical features, might be explained as albedo effects, which might not appear on the Mariner photographs owing to the large scale of such photographs and consequent loss of contrast. Patrick conceded ground with respect to the Coprates-Vallis Marineris issue, but emphasised in his letter of reply that he was merely trying to show that there was no physical basis for Lowellian canals.

Patrick was, finally, healthy enough in October to plan a rare out-of-the-studio *Sky at Night* by visiting one of his closest friends, Paul Doherty at Stoke-on-Trent. In many ways this was a reward for Paul's unwavering support and kindness during Patrick's recent spinal injury period. But, in addition, Paul had just completed a superb 16½-inch (419 millimetres) Newtonian for observing the planets. Unusually, Paul's giant telescope was simply housed in a huge, hinged, wooden box, which protected the massive instrument from the elements.

Fig. 23.1 Patrick with his dear friend Paul Doherty in 1977, filming Paul's 16½-inch reflector at Rode Heath, near Stoke-on-Trent. A scan sent to the author by Paul Doherty in the 1990s. Taken from an upstairs bedroom window. Patrick had known Paul's father Bernard and was godfather to Paul's son Chris. Tragically, Paul would die from Cancer in 1997, aged just 50

To recap, Patrick had been a friend of Paul's father, Bernard Doherty, and Paul's son Chris would become one of Patrick's godsons. Paul was, undoubtedly, one of the UK's finest planetary observers, and certainly the best artist. At the time he was a door-to-door insurance agent but he would later become a freelance astronomy artist. He was one of the nicest people one could ever meet, but his life was cursed in many ways. His daughter was born with a life-threatening heart condition and his son, many years later, was almost killed when he was struck by a car. Paul's 419 millimetre Newtonian would be badly damaged in a freak gale and Paul himself (along with many of Patrick's friends) died of cancer, at the tragically young age of 50, in 1997. The giant fork of Paul's 419 millimetre Newtonian was made from the steel remnants of a bed frame and he said that one of the biggest problems with building the instrument was eliminating rust, something that Patrick had experienced quite a few problems with, living so close to the sea. The giant mirror in Paul's telescope had been ground and polished by Patrick's friend George Hole (the same man who owned the 24-inch telescope in the farcical 50th anniversary *Sky at Night* program).

Just after the memorable *Sky at Night* with Paul, Charles Kowal of the Palomar Observatory in California, was examining some photographic plates and made a remarkable discovery. He discovered what the press called 'The Mini-Planet', later renamed Chiron. This object was unique at the time, as it orbited the Sun predominantly between the orbits of Saturn and Uranus. Patrick's next *Sky at Night*, broadcast in November, was devoted to Chiron, and Patrick's great friend Peter Cattermole was the invited guest.

On November 29th 1977 Patrick appeared on his fourth BBC 'Goodies' programme with Graeme Garden, Tim Brooke-Taylor and Bill Oddie. This time Patrick appeared as a Punk Rocker!

The December 1977 *Sky at Night* was an unusual departure into science fiction. With the film *Star Wars* sweeping the world, Patrick decided to run a programme entitled *Suns, Worlds and Story-Tellers*. His invited guests, no less a duo than the comedian Michael Bentine and the Science Fiction author Arthur C. Clarke! Bentine would appear on another classic *Sky at Night* 2 years later. Patrick's final mention on BBC 1 in 1977 was on the *Morecambe & Wise Christmas Show* of that year; a show that attracted 28 million viewers. Although he did not actually appear on that programme, the last that the famous duo filmed for the BBC before switching to ITV, Patrick was still mentioned. In the opening credits to a hilarious sketch about Cyrano de Bergerac, the non-existent 'Ballet Sequence' was listed as being performed by 'The Patrick Moore Dancers' – Hilarious!

Chapter 24
Back in the TLP Driving Seat

As 1977 drew to a close it became increasingly obvious that Harry Ford, and his team at Dundee, who had been running the BAA Lunar Section, producing the circular and co-ordinating the TLP sub-section for the last year, were finding the tasks too onerous when added to their family and work commitments. This was not the first time in the BAA that even a full team of people had been unable to match one-man Patrick's work output! Remember, 18 months earlier Patrick and his Selsey helpers had been running everything in the Lunar Section. The big advantage, of course, was that Patrick had no rigid clock-based work commitment and no bosses, wife, day job, children, or mother-in-law! There were no DIY chores in his house either. If something went wrong he would get a friend to bodge it back into working order, rather than attempt it himself, or pay a local workman. He simply worked flat out for 18 hours a day on the love of his life: astronomy! O.K., he had an elderly mother, but he also had a housekeeper too.

Ironically, one of the reasons all of the Lunar Section activities had shifted to Dundee in 1976 was because Patrick's southern team had 'shouldered the burden for too long' and 'it was impractical to have Section co-ordination split across the country'. However, at the end of 1977, Harry Ford resigned as the Lunar Section Director and Geoff Amery, based in Reading, took over the reigns. By this time Patrick's spine was fully back in working order, but he had been 'a crock' for much of the year. In the Lunar Section Geoff Amery had been the assistant director and the occultations co-ordinator and so was well known and highly respected. He was a man with no enemies too: an essential pre-requisite. Fortunately, Peter Foley offered to take over the onerous task of producing the monthly Lunar Section Circulars, using that horrific stencil based duplicating machine! To the relief of all, Patrick agreed to resume as the Lunar Section TLP co-ordinator. There had never been any doubt that he was the 'father' of TLP studies (if that is an appropriate term for so dubious a phenomenon) despite the fact that he had relinquished his TLP post in mid-1973 to the Dundee team.

A Lunar Section meeting in London was arranged, as soon as Christmas was out of the way, for January 14th 1978, at the Hawkstone Hall in Lambeth. This venue,

M. Mobberley, *It Came From Outer Space Wearing an RAF Blazer!: A Fan's Biography of Sir Patrick Moore*, DOI 10.1007/978-3-319-00609-3_24,

sought out by Rossie Atwell, would become a popular, if grotty, meeting place for Lunar Section and BAA Exhibition meetings throughout the 1980s, because it was very easy to get to and it was dirt cheap! Patrick, Henry Hatfield and Arthur Good were booked as the main speakers. If Patrick was listed as a speaker, the attendance would double. It was that simple!

In the January 1978 Lunar Section Circular Patrick wrote the following:

> It is very sad that Harry Ford has found it impossible to continue as Director of the Section. However, we are not losing him, and we will all be glad that he is to remain as Assistant Director, as he was during the last part of my own term of office. As a past Director, Secretary, Editor and Co-ordinator of the Lunar Section, let me say how glad I am that so worthy a successor has been found in Geoffrey Amery. He will more than maintain the great traditions of the Sections. Since the Dundee team also wishes to relinquish the co-ordination of TLP work (though again, they will, I am glad to say, continue to play a major role observationally), I have been asked to resume as co-ordinator and this I will willingly do. This note is more or less a preliminary announcement, because for various reasons I have been out of touch for most of 1977; I had a long illness which effectively stopped my observing and neither have I been able to check the records. What I am anxious to do, as soon as possible, is to re-establish the TLP Network that we used to have and which proved to be so valuable. This involves knowing just how many members are ready to co-operate. Therefore, will anyone who will join the Network please write to me, and give me the following details:
>
> Name. Full address. Telephone number, if any. Telescope and other equipment (such as a Moon-blink). Number of years experience in lunar observation. Latest hour of contact by telephone.
>
> The latter is important. I remember that some years ago I sent out an alert for a TLP in Gassendi (which was subsequently well confirmed). The time was well after midnight, and I had to check through my list to see who did not mind being telephoned in the early hours of the morning! I will guarantee that nobody is disturbed after their personal 'limiting hour', but please do not forget to notify me. I will not pretend that observers with telescopes below 6 in aperture (for reflectors) or 4 in (for refractors) can do very much, but I would be the last to put a limit upon membership of the Network; after all, we may get a really violent T.L.P at any time. Many schemes are afoot. We hope to link up with Dr Maddison at Keele, so that when we see a really prominent T.L.P we can alert him in time for he and his team to obtain a spectrogram. The link with solar activity may turn out to be crucial and this, of course, is where Commander Hatfield is a key figure. Altogether there is much to be done – and do not forget that the BAA Lunar Section has always pioneered T.L.P. work. Finally: I want to update the official list (which at present ends in 1971). Would members who have observed T.L.P since then be kind enough to send me brief details? I may well have most of them, but in a list of this sort which will have world-wide distribution, it is essential to double-check where possible. I look forward to hearing from you. Patrick Moore, Farthings, West Street, Selsey, Sussex. Telephone: Selsey (024361) 3668.

By February, 15 Lunar Section members had either joined or re-joined the TLP network; rather less than Patrick had hoped for, but it was a start. However, considering the Lunar Section had several hundred members it must have been a bit demoralising. Indeed, Patrick described the situation as 'rather disappointing'. In the February circular he stated: 'Without wishing to appear single-minded, I do feel that TLP observation is of particular importance at the moment, if only because our knowledge of the phenomena is still so incomplete'.

Patrick's guest on the February 1978 *Sky at Night* was the 28 years old Greenwich Caird Planetarium lecturer Heather Couper, some 12 years after she had written to

Patrick for careers advice. She was a young, attractive, and bubbly character and her specialist subject was the nearest galaxies. She would quickly become a popular 'TV Astronomer' herself and, for a time, become a sort of female rival to Patrick. Heather had joined the BAA in 1974 and had attended many BAA London meetings in the previous year, where Patrick was invariably to be found. Ultimately, she would follow Patrick as the BAA President, after which things started getting just a bit less friendly in the BAA. After appearing on her first *Sky at Night* Heather suddenly started appearing regularly on TV and she also joined (undoubtedly after encouragement from Patrick) the BAA Lunar Section.

Patrick worked hard over the spring months of 1978 to regenerate interest in the TLP sub-section, despite first his mother and then himself becoming ill. As she was now 92, Patrick could take no risks with his mother's health, and so things became rather fraught. Patrick was the TLP co-ordinator, and as such, placed himself on 'alert' status all night. He was happy for anyone to phone him *at any hour of the night* if something unusual was suspected on the lunar surface. However, with his mother ill, calls after midnight where becoming somewhat undesirable, especially if the phone never stopped ringing. Patrick, never a man happy with anything technical, decided to take the plunge and buy a new-fangled gadget: an ansafone! At last, he could listen to whatever crank was phoning him, before picking up the receiver! In addition, when he was away from home, giving talks, the BBC or book publishers could easily get hold of him without disturbing his mother or their housekeeper Woody. Remember, in 1978 mobile phones did not exist. When you left home you were out of contact. But despite various hassles, Patrick had built the TLP network team up into a force of 25 observers by April.

On April 29th the Lunar Section met in the Stockbridge Hall at Chichester, not far from Patrick's Selsey home. Needless to say, Patrick was the main speaker and his emphasis was not only on the subject of TLP, but the possible correlation between TLP and solar activity. In addition, Patrick gave a short account of some of the ways in which lunar craters were NOT formed, and described published theories ranging from ice-hollows to coral atolls and atomic bomb craters. Mr K. Manning, Chairman of the South Downs Astronomical Society, pointed out that there had been a new revelation by a Mr Leonard, in a book, proving that artefacts existed on the Moon. Patrick agreed that this was most remarkable; he had read the book *Somebody Else is on the Moon* and felt that it proved his own thesis: "There's one born every minute!"

Patrick also helped prod the Lunar Section Circular into another era. In the May 1978 issue he wrote a detailed article about the lunar crater Plato, complete with a reasonable quality (by photocopying standards) image of the crater itself. This may seem trivial today, but although drawings had been crudely reproduced before, in the stencil generated circular, reasonable photographs had never been reproduced. Indeed, the only way they could be reproduced was by inserting a separately funded photocopied page. Patrick and Peter Foley pioneered this 'event'. It is hard to believe, in these Internet days, that as recently as the 1970s and 1980s distributing quality pictures, amongst members of an amateur astronomy community, was so difficult.

By May 1978 Harry Ford had resigned totally from the Lunar Section committee. He was replaced as the assistant director by Peter Foley, who was already churning out the Lunar Section Circulars and becoming ever more involved in TLP observing, especially the monitoring of Aristarchus, the brightest crater on the Moon. Away from all this Patrick appeared in a small cameo role in a BBC2 *Comedy Playhouse* programme on May 22nd. An unusual production, it was entitled *Curriculee Curricula*, and was shot entirely on the Birmingham University campus. It starred Scottish comedian Chic Murray as a boiler repair man and *Mastermind's* Magnus Magnusson was the narrator. I guess Patrick just happened to be on the campus, giving a lecture while it was being filmed, and offered a minute or so of his time, for a suitable donation to charity!

On May 31st 1978 Patrick attended the monthly BAA London meeting and presented a paper on his increasing interest in the possible link between Lunar TLP and solar activity. Patrick explained that attempts had been made to correlate TLP frequency with 'all manner' of other phenomena, sometimes with a degree of success, as in the case of perigee and moonquakes. The real trouble, Patrick said, was that TLP reports were uneven in number. There had been a peak of reports at the time of the first Apollo missions, but this had obviously been spurious, as it was due to the fact that more observers were watching the Moon, and some of these observers lacked experience. Patrick showed a graph which brought this point out. He then went on to discuss the possible correlation with solar activity and said that while he had originally doubted the reality of a connection he was starting to have second thoughts. Now that the Sun was approaching its peak in activity this would be an excellent time to make some tests and there was full collaboration with the BAA Solar Section, whose members were being extremely helpful. Replying to a question from the audience, Patrick said that the minimum useful telescope aperture for TLP work was a matter for debate. He thought that keen-eyed observers could use a 150 millimetre reflector to good advantage, though he himself was 'frankly unhappy' with any reflector of less than 315 millimetre aperture; at his observatory he always used his 380 millimetre reflector.

By the summer of 1978 Patrick's winter appeal for 'all observers who had seen a TLP since 1971 to contact me' had borne fruit, and Patrick wrote up his findings as an extension to his ongoing NASA report TR R-277. That report, originally published in 1968 by Patrick and NASA's Barbara Middlehurst, had already been extended twice. This latest extension took the catalogue forward from April 30th 1971 to April 23rd 1978. A total of 45 new events were listed. At the same time Patrick received what he described as a 'splendid new catalogue' of TLP from Mrs Winifred Cameron in the USA.

After a very poorly attended June Lunar Section meeting in London the Lunar Section committee took stock of the situation. The June meeting had advertised its main theme as TLP, but the turnout had been just a handful of people, from over 200 members. At least, there were over 200 people who received the monthly circulars. Patrick was most disappointed when he heard of the turnout and it can be no coincidence that, a few months later, in September, he arranged a special *Sky at Night* on the subject of TLP. The Lunar Section committee, after much deliberation, decided to organise a provincial meeting at Bristol University on November 11th. It was, in hindsight, quite obvious to see why the June meeting was poorly attended! It was the start of the holiday season and Patrick had not been in attendance. He had been preoccupied with his forthcoming performance on the Xylophone with the Worthing Sinfonia, culminating in his 'Penguin Parade' composition, and there was even a plan to release a 'Moore Music' Long Playing record at some point! He was Mr TLP; so not being at that Lunar Section meeting was a disaster for the section. When he was coming to a meeting, he was invariably listed as a speaker. Because of the low June attendance, plans for a meeting at Keele University were postponed. Lunar Section stalwart John Pedler organised the Bristol meeting and Patrick agreed to give an extra talk, a public lecture entitled 'Into Space' in the evening. Predictably, that meeting would be well attended and highly successful.

At the start of July Patrick and his mother were hosts to a woman once described as 'Britain's bitchiest, best known, loved and loathed woman journalist', the *Daily Express* columnist Jean Rook. However, considering Rook's reputation the interview came out fairly well. The full interview appeared on pages 16 and 17 of the July 10th edition, with the largest headline being reserved for Gertrude Moore's comment: "He was the untidiest, oddest little devil as a boy and he hasn't changed. Life with him is a bit strange…"

On Sunday July 16th 1978 Patrick took part in the first of four *It's a Celebrity Knockout* competitions he participated in, during which famous personalities got drenched with water and surmounted obstacle courses, in the traditional BBC *It's a Knockout* gameshow style. The show was broadcast a week later and one of the highlights was Shaw Taylor, presenter of the crime prevention show *Police 5*, tipping a gallon of water over a T-shirt and shorts clad Patrick, while he was speaking to Eddie Waring and trying to prevent his shorts falling off. Earlier in the book I mentioned one of these shows in which Patrick's enormous buttocks were briefly revealed after his gigantic trousers became totally sodden! Patrick took part in these events because he was a leading light in *The Lords Taverners*, the UK's leading charity for recreational cricket, youth cricket and disabled sportsmen. *The Lord's Taverners* was founded in 1950 by a group of actors, led by Martin Boddey, who used to enjoy a pint watching cricket from the Tavern pub at the Lord's Cricket Ground. Patrick supported the charity for the whole of the period he was on *The Sky at Night*. Patrick appeared in four of these summer *It's a Celebrity Knockout* programmes in 1978, 1979, 1980 and 1981 and was always highly entertaining, if not particularly quick or agile! In each event several dozen TV personalities took part, but the astronomer Patrick was always a leading light, even with comedians like Eric Morecambe present. Morecambe was *The Lord's Taverners*' President in 1978 and the 'Goon' Harry Secombe followed him into that role. All but one of the four events that Patrick took part in were filmed at the Fulham Football Club ground at Craven Cottage, the exception being the 1979 competition which was held at the Queens Park Rangers Football Club ground at Loftus Road, not far from Shepherd's Bush or the BBC TV Centre. It was always a case of *The Lord's Taverners* versus some other team and the other teams were named, in order: Celebrities (78), Celebrities (79), The Rest of the World (80) and The Dodgers (81). In 1979 the

Lords Taverners drew with the Celebrities, but they won easily on the other three occasions where Patrick was involved.

In August 1978 the international amateur astronomy organisation founded by Patrick 11 years earlier, the IUAA, was still surviving, although the next decade would see it flounder. However, its fourth triennial meeting, held from August 14th to 19th, was attended by 81 participants from 20 countries, and being held in Dublin, Patrick was able to participate. The former BAA President (1970–1972) Dr Vin. Barocas was the acting IUAA President at that time. Patrick gave a public lecture at the conference entitled ‘The New Mars’, which was very well attended.

The October 1978 episode of *The Sky at Night* was both memorable and controversial. Patrick travelled to Cardiff where he discussed the theory that life was brought to Earth by comets, with Sir Fred Hoyle and Professor Chandra Wickramasinghe. Hoyle was certainly a very clever man, especially in his younger days, but his controversial biological excursions had made many people question whether he was sailing with a full crew.

On November 2nd Patrick was invited by the Duke of Edinburgh to a Foyles Bookshop Literature Luncheon, at the Dorchester Hotel in London. The event was staged to celebrate the Duke publishing a collection of his ‘Environmental Revolution’ speeches. Various literary dignitaries and celebrities were invited and media pictures showed Patrick and the artist David Shepherd deep in conversation with the Duke.

Two days later, on the fourth of November, Patrick had the pleasant duty of officially opening the completed dome and 20 inch telescope of the Hampshire Astronomical Group at their Clanfield reservoir site, not all that far from his home at Selsey. This had been the culmination of 6 years of hard work by the group’s members.

On the November *Sky at Night* Dr Simon Mitton of Cambridge University guested on a programme about the Crab Nebula. A few months later, in April and May 1979, Simon would suddenly have his own six part ITV programme entitled ‘Beyond the Moon’.

Chapter 25
Nursing Mother

In December 1978 Patrick resigned from the Lunar Section committee and as the TLP sub-section co-ordinator. His mother's deteriorating frail state of health meant that, increasingly, he was devoting more of his time to looking after her; she was 92. When combined with planning each month's *Sky at Night*, book-writing, and public lectures, it was all just too much. Patrick was, after all, no longer a spring chicken! At 55 years of age and weighing 18–19 stone, with a dodgy heartbeat, a fragile spine, a 'gammy' knee and the first signs of arthritis, his own health was at risk if his breakneck pace of life continued. For the next 2 years Gertrude Moore would require round-the-clock nursing, and her condition caused Patrick much anguish. Indeed, I would venture to suggest that after this time Patrick's character was never quite the same. It was rather rare to see a truly happy and excitable Patrick after the end of the 1970s.

The Moore's housekeeper Woody would organise the nursing day shift, allowing Patrick time to work. Patrick would organise the night shift. Occasionally, Patrick would hire expensive full-time nurses to help too. As far as Patrick's mother was concerned, no expense was too great. For the next 2 years, and for the first time in his freelance career, astronomy would have to take a back seat. Of course, *The Sky at Night* would continue, but book-writing, public-speaking, playing cricket, amateur dramatics, writing music, BAA work, and observing would be severely curtailed. Patrick allowed himself a few local diversions though. On December 28th 1978 he honoured his promise to be at the opening of the 55th Annual Hastings International Chess Congress; local newspapers showed him taking on the young Chess expert Nigel Short (aged 14) at the Congress. Even so, BAA and Lunar Section duties were severely curtailed.

Peter Foley would take over the running of the TLP sub-section, but Patrick would still retain an interest and get to meetings whenever he could. Fortunately, Patrick had never been a person who threw his money around on material possessions, despite being incredibly generous to friends and charities; so, turning down work was not a disaster, even though he was employing a full-time housekeeper and private nursing help. He had enough in cash reserves to keep him going. For the whole of 1979 and

M. Mobberley, *It Came From Outer Space Wearing an RAF Blazer!: A Fan's Biography of Sir Patrick Moore*, DOI 10.1007/978-3-319-00609-3_25,

1980 Patrick was tied increasingly to his Selsey home and his mother's condition was paramount. However, *The Sky at Night* continued unabated. Television is a fickle business, and if Patrick missed even one programme *The Sky at Night* might sink without trace, and with it the free publicity that guaranteed good book sales and a steady income. However, one good piece of news during that period was that Patrick would have yet another international award to hang from the wooden beams in his study: the Astronomical Society of the Pacific announced that they were awarding him their prestigious Roberts-Klumpke Medal.

The United Country Party and Guinness Books

With Patrick spending more time at home he decided, in January 1979, to involve himself more in local Sussex politics. Patrick had always retained very strong political views, which were decidedly racist by today's standards. He was not simply worried by a person's skin colour: if they weren't British they weren't the best stock, it was as brutal as that. In particular, Patrick was concerned about the increasing number of immigrants and the fact that Britain was becoming more and more part of the European Union. Ted Heath had let the country down, and after the winter of discontent so had the Labour leader Jim Callaghan. We wanted less to do with Europe, not more, and he did not initially think that the young Margaret Thatcher would be able to reverse things. Patrick's own father had died in 1947, after 30 years of ill health caused by a German gas bomb thrown into his trench in WWI. In addition, many of East Grinstead's residents had been killed by Germans in the second World War and many of Patrick's RAF colleagues had been shot down over Germany. Two World Wars had been fought to keep Germans out and keep England for the English, yet now it seemed to Patrick that a more subtle type of invasion was taking place: an invasion of asylum seekers, seeking benefits and a free ride at the tax payer's expense.

Patrick was usually a Tory voter, but the Conservatives had let him down over Ted Heath's pro-Europe stance. As a young man Patrick had been a Liberal. Most of the Selsey Cricket club were sympathetic to his pro-English views and so they decided to form their own, rather 'tongue-in-cheek' party: The United Country Party or UCP. As Patrick was the club's devious spin bowler, he was made the chairman of the party and the future Foreign Secretary. As Selsey was deemed to be such a special place they decided you would need a passport to get in, and in practice only good white English stock would be allowed in automatically. It was hard to say where the boundary between out and out racism ended and hilarity began with the UCP, but there were some amusing parallels with the Raving Loony Party, which Patrick would eventually join too! For example, the United Country Party would have a 200 mile fishing limit, so Selsey fishermen would have unlimited access to fish in the river Seine in Paris if they so desired. Mysteriously, the Cricket Club Chairman, an ex-military man named Colonel Edmund Iremonger, did not get elected at the May 1979 general election in nearby Chichester, much to the party's dismay. Iremonger had an impressive military record and the reader of this book may recall that Patrick's 'Iremonger rules' (namely, help yourself to my booze) applied in his house Farthings too. Patrick eventually ended being the chair of the UCP when it was absorbed by the New Britain and Keep Britain United parties on September 16th 1980. In front of press photographers, Patrick, Col. Iremonger, Dennis Delderfield and E. J. Clark announced they were joining forces. In later years Patrick would be a staunch supporter of UKIP, the U.K. Independence Party. This was the only party that really represented his views on a national level, although it has to be said that the British National Party came pretty close! Incidentally, according to the Space Artist David Hardy, Patrick told him (regarding forming a new party) that: "I'm thinking of starting the Politically Incorrect School of Sociology – and the acronym says it all."

Throughout the early 1970s Patrick had worked with the twin brothers Ross and Norris McWhirter, compiling the section on Astronomy for the *Guinness Book of Records*, which the twins had founded in 1955. Patrick always got on famously with the McWhirters as they spoke his language, were ultra-patriotic, and very right wing in their political views. In fact, it was as a result of these strict views, especially regarding Irish terrorists, that Ross was assassinated by the provisional IRA in November 1975, an event which greatly shocked Patrick. However, the surviving twin, Norris, continued to edit the *Guinness Book of Records* following his twin brother's murder, and on Patrick's advice, more than 20 years later, he too would join UKIP. Norris also continued to co-host the BBC TV programme *Record Breakers*, alongside the entertainer Roy Castle, and when Astronomy was the subject, Patrick, of course, would also appear on the show. I well remember when Clyde Tombaugh discovering Pluto was the 'record' being mentioned, and Patrick's appearance on that very episode. In the late seventies the *Guinness* publishing company diversified into producing more detailed record books on a variety of subjects. Patrick was the obvious candidate to compile the *Guinness Book of Astronomy: Facts and Feats*, and Norris asked him to produce the book sometime in 1978. The first edition was published in early 1979. It was an absolute mine of information containing 288 pages crammed full with astronomical data. Although not a best seller in the class of his *Moon Flight Atlas* or *Atlas of the Universe* it would go on to sell over 60,000 copies and was reprinted and revised six times. Twenty-one years later, in 2000, it would transfer to the *Institute of Physics Publishing* as *The Data Book of Astronomy*: a 529 page bookshelf bursting monster!

With Patrick devoting more time at home to his frail mother any big writing projects, such as the Guinness one, were welcome, as they were very compatible with spending day after day indoors. Patrick was a workaholic and could never stay idle, so he always had to direct his considerable energies into something. One plan he came up with in this period was another French-to-English translation. During 1949–1951 he had translated works by Gérard De Vaucouleurs and Audouin Dollfus, but now he had plans for a much bigger translation: converting Camille Flammarion's massive 1892 two-volume work, *La planète Mars et ses conditions d'habitabilité*, into English. Patrick completed the translation as a rough draft around 1980 and had four copies made. These ended up with Lowell Observatory, the BAA, the RAS, and with the author Bill Sheehan. In the late 1980s Bill tried to get the work published for Patrick, and Patrick signed away the rights to Bill in 2004. However, during Patrick's lifetime the work would, sadly, never be printed, largely due to publishers getting 'cold feet'.

In the week of Patrick's 56th birthday (March 4th 1979) Patrick was invited onto the BBC Radio 4 programme 'Midweek' as the birthday guest of the week. The programme was hosted by Sue Lawley and Desmond Wilcox. Desmond was the head of the BBC department that produced *The Sky at Night* in the mid 1970s, so knew Patrick well. During the programme Patrick admitted that he gave "every impression of being somewhat hastily constructed" and being absolutely useless when it came to being practical. "When it comes to using my hands I am, without any shadow of a doubt, the world's worst. A little while ago I was faced with the knotty problem of driving a nail into a hard piece of wood. Well, I bent the nail into a hoop, I split the wood down the middle and I knocked the head off the hammer. Luckily, the paper boy then came in and did it for me!" Miss Lawley pointed out that *Style* magazine had recently voted Patrick 'The world's most inelegant man whose level of elegance quotient is astronomically low'. Patrick queried whether this magazine was a photocopied publication and said he was "sure the magazine was enjoyed by *both* its readers". Then the subject of conversation turned to astrology, as opposed to astronomy. Patrick was sure that the subject proved one thing "There's one born every minute". Patrick was presented with some hot buttered toast as a birthday present and he insisted that, despite his weight and his bulk, he ate very little.

The next month Patrick appeared on the Lennie and Jerry TV show in a comedy sketch with another TV scientist, Magnus Pyke. The two of them sang and danced around to the tune of 'Tiptoe through the Tulips'! Magnus tried hard to match Patrick's natural eccentricity but it was obvious that he was trying too hard. Patrick was just being his normal self.

At the end of April, Patrick received the sad news of the death of Bernard Doherty, the father of Paul and a lunar section stalwart of the 1960s. Bernard had submitted many excellent drawings of lunar domes, all of which had been passed on to his son Paul, now widely regarded as the BAA's leading planetary artist.

The month of June 1979 saw Patrick taking a brief break from nursing his mother, and honouring a promise he had made long ago, to give a talk to the personnel of RAF Akrotiri (BFPO 57) on the Mediterranean island of Cyprus. His talk, delivered under a clear night sky to 400 people, was appropriately entitled 'The Sky at Night'. On the morning after Patrick's talk, and before his flight home, Group Captain Paul Grey, the Station Commander, gave him a guided tour of the RAF base and allowed Patrick to sit inside the cockpit of one of 19(F) Squadron's Phantom jets. He was made an honorary member of the Squadron and was even presented with a Squadron tie.

Despite Patrick's pre-occupation with his mother's frail condition and his resignation from the Lunar Section committee he agreed to participate fully in the Lunar Section's proposed Keele symposium over the weekend of July 7th and 8th. This was planned to celebrate 10 years since the Apollo 11 Moon landing of July 1969, and of course the big hope was that the Keele 24 inch telescope would soon be ready to take spectrographs of the lunar surface, whenever the TLP sub-section issued an alert. The planned symposium would be a residential weekend course, with leading lunar professionals, as well as Patrick and Henry Hatfield, giving some amateur astronomy themed lectures.

Unfortunately, Patrick's mother's health, a filming session for his fifth and sixth appearance on *The Goodies* BBC Comedy shows (shown 7 months later) and the need for an 'emergency' *Sky at Night* caused him to pull out at the last minute, which was a great disappointment to all. A few members were heard to grumble that *The Goodies* were more important to Patrick than the Lunar Section. Nevertheless, the Keele symposium was a great success with over 150 participants attending. The whole event was masterminded by Lunar Section members Peter Gill and Rossie Atwell. Dr Maddison of Keele, a former Lunar Section Director himself, provided the group with excellent facilities. Even the BAA President, J. Leslie White, attended the meeting. However, Patrick's mother was now so frail that she was essentially bedridden and Patrick was now housebound, except for essential trips. Apart from visits to the BBC in London, mainly to plan and film the monthly *Sky at Night*, he was virtually grounded at Selsey, unless he paid for nursing cover for his mother.

So what was the reason for the aforementioned emergency *Sky at Night*? Well, a sudden deterioration in the orbit of the Skylab space station, due to greater-than-expected atmospheric drag, meant that this huge spacecraft was going to burn up in the Earth's atmosphere just after the time of the Keele symposium. The BBC wanted Patrick to get some experts together and quickly broadcast a programme before Skylab fell to Earth. Patrick managed to get hold of the UK orbit expert Desmond King-Hele, and his US counterpart Dr John Granger. The programme was broadcast live on BBC1 on Tuesday July 10th and Desmond King-Hele's predictions of where and when Skylab would re-enter proved to be spot on. It burned up on the following day, in the Australian desert, just East of Perth. Two weeks later Patrick managed to get nursing care arranged for his mother so he could take part in his second *It's a Celebrity Knockout* event, raising money for his favourite *Lord's Taverners* charity. Fortunately, this year it was held, on July 29th, at the Queens Park Rangers football ground, not far from the BBC TV Centre.

1979 was a year of planetary encounters with NASA spaceprobes and these were covered fully by Patrick on *The Sky at Night*. In April and August the Voyager 1 and 2 probes rendezvous' with Jupiter were described. In September the Pioneer 11 probe passed Saturn. All of the findings were covered by Patrick and his guest Garry Hunt on the *Sky at Night*. However, Patrick was less than impressed about a show The London Planetarium was running, concerned with the much hyped 'planetary alignment' due in a few years time. The show had been based on a book I mentioned earlier, published in 1974, and called 'The Jupiter Effect –The planets as triggers of devastating earthquakes'. This bestseller had been written by John Gribbin and Stephen Plagemann and even had a foreword by Isaac Asimov! The authors claimed that the alignment (if that is the right term) of seven planets within a 95 degrees arc, during 1982, would cause enough tidal stress to cause a major earthquake in Los Angeles. Patrick had already declared the book to be bunk but was mortified upon hearing that the London Planetarium was devoting a show to the concept. In September he wrote a stern letter to *New Scientist* (published on p. 907 on the 20th) stating: 'I hope that this programme will be taken off immediately as it is harming the reputation of the Planetarium. And let us hear no more of this nonsense about the 'planetary alignment.' It happens every 170 years or so; nothing spectacular will be seen in the sky; and in the opinion of almost everyone, it can affect nobody and nothing.' Following Patrick's continued protests the show was dropped.

However, there was a much more pleasant task for Patrick that month. September 19th saw Patrick opening the Dyson Gallery of the Old Royal Observatory at Greenwich. Dyson was the ninth Astronomer Royal and a new history gallery at the Royal Observatory Museum was being named after him. Despite the presence of the then Director of the Royal Greenwich Observatory, Professor Graham Smith, and of descendants of William Herschel, Patrick was the man invited to do the honours and declare the new gallery open.

An Unwelcome Book Review

A book review of Patrick's revised publication *Basic Astronomy*, written by the future BAA papers secretary Storm Dunlop, caused Patrick to comment in the October 1979 BAA Journal. Dunlop was a well known variable star observer, a tireless worker for the BAA Variable Star Section and a BAA councillor too. Living at East Wittering, near Chichester, he was almost a neighbour of Patrick's. In his review, Mr Dunlop made the following comments, after pointing out a few technical errors:

> However, my major criticism of this book lies not with the author, who has done a good job, but with the publisher for the way in which it has been produced. One might even go so far to say that it is nearly a classic example of what should be avoided. Some of the pages look most peculiar, as it is possible to see where the text has been changed, as the different typeface is quite obvious. Page 109 has two sentences in the new type, which are immediately followed by four lines in the old which repeat them word for word. Blocks of text slope up and down where they have been cobbled together, and the text on page 14 is fully 20 mm deeper than that on page 15 opposite! The paper is of poor quality and quite unsuitable for half-tones, which have consequently reproduced very badly. The positioning of the pictures is also poor, with the Pleiades and the spiral galaxy NGC 2841 in the chapter on the planets, and the photographs illustrating galaxian types are at least 30 pages too early. The four photographs facing page 9 are far too small, with the photograph of the surface of Venus having an actual picture area of 49x12mm, while that of Mercury is nearly all black. This very basic design and the rather crude line diagrams must lessen the book's appeal to any intending purchasers. The lack of an index is also a considerable disadvantage.

Storm Dunlop had made some valid and inescapable points in his review, but with Patrick's spare time at a premium due to his mother's frailty, he had not had the patience to properly check the book proofs. He hated such hassles as proof checking at the best of times, and certainly Patrick had even less patience for such things as he got older. In addition, when you publicly criticised Patrick you made an enemy for life and were never, ever, forgiven. Storm had become an official Patrick Moore

'Serpent'! The Council allowed Patrick a letter of reply to the reviewer, in the same issue of the BAA Journal. This luxury was afforded to very few authors in the BAA's history. Patrick replied with the following, rather sarcastic, letter:

> Dear Sir,
> I noted Mr Dunlop's comments in his review (page 619). I quite agree that the spectrum of the Sun is G2 not G0, though I should be interested to know whether Mr Dunlop could tell the difference between a G2 and a G0!
>
> So far as presentation is concerned, Mr Dunlop cannot be expected to realise that to make a book attractive is also to make it more expensive. This, alas, is one of the facts of life appreciated only by those experienced in authorship and publishing!
>
> Yours faithfully,
> PATRICK MOORE
> Farthings, 39 West Street, Selsey, West Sussex.

In fact, in the coming decades, Dunlop went on to become an author of many good astronomy books as well as translating many more into English. Maybe Patrick's sarcastic comments spurred Dunlop on to prove he could produce high quality books?

Despite the nursing requirements of Patrick's mother he continued, whenever possible, to give talks in aid of his favourite charity, Cancer Research. Saturday October 13th saw him giving a talk at this author's old school: the King Edward VI Upper School in Bury St Edmunds. The Bury Cancer Research Group Chairman, Mr Hilary Hammond, announced that Patrick was donating all the ticket sale proceeds to the charity, to much applause from the 400-strong audience. With tickets selling for 4 pounds each this was a £1,600 donation, at a time when the average wage in Britain was about £500 a month. No doubt the organisers thought this was a one-off gesture, but Patrick did this all the time. The amount raised by Patrick for charity over his lifetime must have been truly staggering.

Also during October, that month's *Sky at Night* featured another amateur observatory, that of Peter Gill. As well as being an active member of the BAA Lunar Section, Peter carried out solar observations from a small fourth floor flat, a mile or so from Victoria station. With a modest 3 inch refractor he projected the Sun's disc onto a wall and observed the Sun safely by the 'projection' method. Patrick was a lifelong advocate of that method. In the 1970s cheap Japanese refractors often came with grotty 'sun filters' which screwed into the eyepiece. They were highly dangerous as they let some infrared radiation through to the eye and often cracked in the heat. Patrick's oft-repeated advice was as follows: "Take your dark sun-cap. Hire a boat. Row at least a quarter of a mile out to sea. Hold the sun-cap out at arm's length. Release it!" Perhaps one reason for Patrick's vocal concern about such filters was because, in 1979, a Kent schoolboy was awarded damages against the local County Council for injuries to his eyesight. Patrick wrote a letter to the BAA Journal emphasising his views:

> Dear Sir,
> A schoolboy of Napier Road School in Gillingham has been awarded damages against Kent County Council. During a geography lesson, a teacher allowed him to study the eclipse of the Sun through filters, direct; with the inevitable result that his retina was permanently damaged.

Over the years I have been accused of alarmism for my constant warnings about this sort of practice. One silly book (not by me!) even advocated 'sweeping around' for Venus with binoculars even before Sunset. May I, therefore, repeat my warning yet again? There is only one rule for direct solar viewing: don't, as the luckless schoolboy now knows to his cost. Of course his teacher was to blame for sheer stupidity, but that will not restore the sight of the boy's eye.

Yours faithfully,
PATRICK MOORE,
Farthings, West Street, Selsey, West Sussex

A happy event took place on November 17th 1979. The Selsey Cricket Club presented a delighted Patrick with a unique marble and silver plated ornament inscribed:

Presented to Dr Patrick Moore O.B.E. by Selsey Cricket Club in appreciation of every way he has helped the club. 17-11-79.

The December 1979 *Sky at Night* was a classic and definitely one of the favourite *Sky at Nights* ever made. Filmed in Patrick's study at Selsey, Patrick and Michael Bentine discussed the possibility of extra-terrestrial life. Like many of Patrick's friends, Bentine was interested in areas on the fringes of normal science, like spiritualism and the paranormal and even Flying Saucers, but he was not quite so eccentric that he could be labelled as a crank. Patrick and Michael discussed some of the early ideas about fictional spaceflight, the anti-gravity compound 'Cavorite', and the idea of shooting men in missiles to the Moon. As the programme ended, two of Bentine's puppet creations, aliens in this case, were seen discussing whether there was any possibility of intelligent life on Earth, as they wandered outside Patrick's study window. Surely this was the most entertaining *Sky at Night* since the 15th anniversary farce of searching for cloud gaps with George Hole's massive back-garden telescope. In the write-up of that programme, Patrick stated that he knew both the Flying Saucer travellers George Adamski and Cedric Allingham. As we saw much earlier in the book, it was not difficult for Patrick to have known Allingham: he *was* Allingham!

As if this *Sky at Night* classic with TV Goon Bentine wasn't enough, Patrick was on prime time BBC1 for the 1979 Christmas Day edition of the quiz show Blankety-Blank, hosted by Terry Wogan. For the record, the other 'celebrity' guests on that programme were David Jason, Kenny Everett, Sandra Dickinson, David 'Diddy' Hamilton, Roy Kinnear, Liza Goddard, Beryl Reid, Shirley Anne Field, Wendy Craig, Lennie Bennett and Lorraine Chase. The show had been recorded some 2 weeks earlier, on December 10th. Publicity photographs showed Patrick, standing between David Hamilton and Beryl Reid, gazing skyward, as a youthful Terry Wogan, complete with star-shaped magic wand, flew through the air above them on metal wires. Various guests, notably Messrs Jason, Everett and Hamilton, proceeded to pelt the airborne Wogan with artificial snowballs as he soared overhead, just like the traditional Christmas Fairy! As always on these shows Patrick looked embarrassingly out of place, but no doubt he was paid well for his appearance; and with the escalating nursing costs for his mother, this was just as well.

Back in the world of amateur astronomy a new year dawned, and in January 1980 Patrick was saddened to hear of the death of Ernest Noon, a BAA member for 47 years and the co-founder, with Eric Turner, Patrick and Richard Baum, of the Junior

Astronomical Society. Ernest Noon was 93 and had been one of Patrick's oldest friends, so this was yet another sad loss. Patrick was finding out that as you get well into your fifties, many of your older acquaintances and mentors start to rapidly fade away.

At the start of February 1980 Patrick appeared in two more *Goodies* comedy programmes on BBC TV, his fifth and sixth appearances on that programme. Playing the fool never bothered him: it was easy money! In mid-February 1980 Patrick made special arrangements for nursing staff to look after his mother while he flew out to a one day Pluto conference in Las Cruces, New Mexico, to celebrate the fiftieth anniversary of the discovery of Pluto by Clyde Tombaugh. The one day conference was scheduled for February 18th, exactly 50 years to the day since the announcement of the discovery of the ninth planet. Patrick was especially keen to attend, as he had been honoured that Tombaugh had agreed to collaborate with him on a book about the discovery: *Out of the Darkness the Planet Pluto*. Patrick had been coming up to his seventh birthday when Tombaugh had discovered Pluto, so this was an opportunity the schoolboy Patrick could surely never have expected. At a ceremony, attended by Patrick and *The Sky at Night* team at New Mexico State University, Tombaugh was awarded the University's highest honour: the Regent's Medal. He was then surprised by an announcement by Henry Giclas of Lowell Observatory, that minor planet 1604, formerly 1931 FH, one of the many asteroids found in Tombaugh's original search, had been officially named 'Tombaugh'. Replying, Clyde said that "At last I now have a piece of real estate that no-one can touch!" Patrick and *The Sky at Night* film crew also made other visits to astronomical sites in Arizona, including the Multiple Mirror Telescope, Meteor Crater, and the National US Observatory at Kitt Peak, each of which provided enough material for specific *Sky at Night* programmes during the spring and summer months.

The spring months of 1980 offered a treat for Saturn observers: another chance to view Saturn's rings when presented edge on. The last time this phenomenon had occurred Patrick had been based at Armagh, and he had made many valuable observations; so had his good friend Paul Doherty, then aged only 19. Fourteen years later both Patrick and Paul were keen to grab as many observations of the edge-on rings as possible, and so for the February *Sky at Night* Patrick invited Paul as the guest on the programme, to describe precisely what might be seen through a decent amateur telescope. Patrick had also managed to take a peek at Saturn using the Flagstaff 24 inch refractor while he was there in February. One month later the March *Sky at Night* showed Patrick's visit to the Pluto conference in New Mexico. The sequence of Saturnian events had actually started in October 1979, but Saturn had then been too close to the Sun to observe and the peak 'edge-on' ring passage would not occur until March 12th. In addition, the rings would be edge-on to the Sun (as opposed to the Earth) on March 3rd. Patrick and Paul Doherty (then Assistant Director of the BAA Saturn Section) made many observations of the edge-on rings. Paul's observations were described by Patrick as outstanding, but there would be trouble when the Saturn Section report was submitted to the BAA Council; read on!

On the 4th of January 1980, Patrick, while observing with Dr John Mason, hoped to observe the asteroid 48 Doris occult a 7th magnitude star in the constellation of

Orion. For once conditions were excellent, but from Selsey at least the asteroid did not occult the star; it missed it by over 1 second of arc, as photographs later revealed. However, with asteroid occultations, especially in the 1980s, a negative result was almost as valuable as a positive one. Three months later, in the early hours of June 15th, the minor planet 59 Elpis was scheduled to occult a 9th magnitude star at 02 hour 17 minutes UT. Again, Selsey was very close to the track. Unfortunately, the cloud arrived at Selsey just before the predicted event and nothing was seen by Patrick. At a BAA meeting, some 10 days later, Gordon Taylor of the BAA and the Royal Greenwich Observatory, reported that only two observers were clear, but neither had seen the star wink out.

In the May 1980 *Sky at Night* Patrick's guest was, once again, the bubbly Heather Couper, speaking on the subject of 'Journey to the Centre of the Galaxy' and Black Holes. At the end of the programme, Patrick and Heather were, courtesy of rather crude BBC special effects, to be seen spiralling down into a black hole. Interestingly, when the eighth volume of *The Sky at Night* books was published in 1985, Patrick chose to use a special Paul Doherty painting of a Black Hole on the cover, and during Heather Couper's term as BAA President, Patrick donated a copy of the book to the BAA library, inscribed 'To Heather, in a Black Hole.......Patrick'. The May 1980 *Sky at Night* programme would mark Heather's second and final *Sky at Night* appearance, but it did enable her to springboard to other TV projects during the coming years. By the start of 1981 she was working with Terence Murtagh, who was doing Patrick's old job at the Armagh Planetarium, on a Yorkshire TV programme called *Heavens Above!*, broadcast for children on Mondays at 4.45 p.m. With Patrick in his late fifties I am sure that a lot of presenters of that era must have been seeing themselves as his replacement when he retired. Hmmmm, not a hope! Heather Couper's co-presenter, as Armagh Planetarium Director, had actually presented a regular astronomy programme on Ulster TV, from 1973, called *Look Up with Terence Murtagh*. A few UK regions had also shown the Ulster show, most noticeably London Weekend Television on Sunday afternoons. However, none of these other fleeting shows, or their presenters, could ever compare to Patrick and *The Sky at Night* in the affections of the viewers. Apart from his established popularity, Patrick had a secret weapon: he did the programme for love, not money. In addition, he was a real amateur astronomer, who made observations every clear night, unlike many twenty-first century TV presenters, who observe their own egos and nothing else!

The Ashen Light

At the end of May 1980 Patrick had an opportunity to use his 15 inch reflector to look at the planet Venus, which was then a crescent in the evening sky. It was a memorable night for Patrick and for Paul Doherty, and Patrick communicated the following to the BAA's Venus & Mercury Section and to the Lunar Section:

> We have had evidence of lunar activity lately, though unfortunately I have not seen it myself. However, I have made an observation which may, or may not, be relevant. On 26th

May, at 20.20, using x130 and x400 on my 15 inch reflector, I hid the crescent of Venus with an occulting bar and saw the Ashen light clearly, extending for more than half the diameter from cusp to cusp. On the following night, 27th May, from 20.20, using x72 to x400, I again saw the Ashen Light. With or without an occulting bar, it was more evident than I have ever seen it – and I began observing in 1934! It is hardly an exaggeration to say that it was like the earthshine on the Moon, and moreover there was appreciable limb brightening along the night hemisphere. The entire disk was visible, and it was so plain that I am in absolutely no doubt that it was real; it was seen whether or not the crescent was hidden by the occulting bar. Seeing was I on the Antoniadi scale, which is rare indeed (it had been I to II on May 26). Paul Doherty, who was with me, agreed entirely with my observations on both nights. There have been suggestions that the Ashen Light is affected by solar activity; the Sun is active – and we know that there has been reported TLP activity. Is there a connection? I am inclined to believe that there may be.

PATRICK MOORE

Some years ago, while at Patrick's house, I examined his Venus Observing Log for that day, May 27th. The log records the following:

My best ever view of the Ashen Light. With or without occulting bar a full circle, dimly luminous, with a sharp edge slightly brighter than the interior. Quite unmistakeable and even striking. Paul Doherty and Woody [Patrick's housekeeper!] agreed. Best in white and green, less striking, but still visible in yellow. Also glimpsed in red which enhanced the bright patches on the crescent above the one shading and near the S. cusp. My best view of Venus in many years.

Many experienced observers, including this author, have striven to see the Ashen light on the dark side of Venus, and there have been many attempts at explaining the phenomenon. As most observers have never seen it when the bright crescent is occulted the usual opinion is that it is just an illusion: a contrast effect between the dark night sky and the bright crescent, or simply stray light reflections inside the telescope. This is strengthened by the fact that it has never been photographed or CCD imaged (except in the infra-red). On the other hand, one would certainly not expect observers of the calibre of Patrick and Paul Doherty to be fooled! Some have suggested that the phenomenon is caused by the illumination of Venus' dark side by light from the bright Earth. But when you do the maths, this illumination source is incredibly weak at the distance of Venus. Others have suggested it may be due to massive electrical storms in the Venusian atmosphere, light scattering, or the heat of the Venusian terrain. Either way, it is a mystery, and like TLP, just the sort of visual observing 'amateur at the eyepiece' romantic enigma that Patrick thrived on!

Despite Patrick's mother's deteriorating condition throughout the summer and autumn of 1980 he did manage to get some further observing in and also to attend a few astronomy meetings. He was also able to take part in his third *It's a Celebrity Knockout* contest for *The Lord's Taverners* at the end of June, held at the Fulham football ground. One of those astronomy meetings was the Lunar Section Meeting at Birmingham's Aston University, held on Saturday October 4th. It was an interesting date to hold a meeting as there would be a double dawn sky lunar occultation for many on the following day, the Sunday morning, involving the bright star Regulus in Leo and the planet Venus. Regulus would be occulted for all UK observers with a clear sky, but Venus would have a near miss from locations south of Birmingham, a disappearance from locations north of Birmingham and a graze along the Moon's south limb at Birmingham's latitude. The Lunar Section Director, Geoff Amery, was thwarted by a last-minute family 'mini-crisis' and could not attend, but Patrick did make the trip. The meeting was chaired by the Birmingham Astronomical Society Chairman, Dr John Penney, and by Alan Wells, the Lunar Section occultation co-ordinator. On the occultation theme, Mr Gordon Taylor drew members' attention to an unusual occultation of a bright star by Venus in 13 months time. Patrick, with respect to the imminent Venus occultation, reminded members of an observation by the BAA's W.H. Pickering, as long ago as 1892. Pickering had observed the planet Jupiter being occulted by the Moon and had seen a black line join the two just before the planet went behind the Moon's disc. This had been regarded as evidence of a thin lunar atmosphere by Pickering, although later spaceprobe data did not support this idea. At the conclusion of the Lunar Section meeting, Patrick brought proceedings to a humorous end with a brief look at two 'independent thinkers' of the type that appeared in his book *Can you Speak Venusian*? One of these characters had pronounced "All light is green, except green which is red" which brought howls of laughter from the audience. The other 'thinker' concluded that "Light is darkness lit up", which brought the meeting to a suitably hysterical end. Members then headed home, for perhaps some sleep, and maybe a pre-dawn rise to see the double occultation event. A few others headed for Cannock Chase, near Birmingham, to witness the Regulus/Venus occultation in the dawn sky.

Chapter 26
A British Comet Marks the End of an Era

The months leading up to Christmas 1980 were an absolute nightmare for Patrick, as his mother's health deteriorated rapidly and it soon became obvious that the end of the happy 33 year era of Patrick living solely with his doting mother (after his father's death) was drawing near. Patrick had to turn down almost all media appearances during that period as he wanted to be at his mother's side when she finally declined and passed away. After he had filmed the early December *Sky at Night* and been recorded for the Boxing Day edition of the humorous BBC quiz show *Blankety Blank* he was not leaving his mother's bedside if at all possible. Patrick was determined that his mother should stay at home and not be moved permanently into a hospital or a hospice, but the workload on himself (in the night-time) and his housekeeper Woody was becoming intolerable. Friends in the Selsey area and from further afield arrived at the house to help Patrick and Woody out, but it was becoming a full-time job, even with hired help.

While all this chaos was taking place and peaking over the Christmas 1980 period, a remarkable event occurred in the UK skies over Northamptonshire. Walgrave amateur astronomer Roy Panther, who had been sweeping the skies for 33 years, the same length of time as Patrick had been living just with his mother, came across a faint fuzzy patch in the night sky. On Christmas Day 1980, 20 years (−1 day) after BAA member Mike Candy's UK Christmas comet discovery, Roy had swept up his own comet. Comet Panther 1981 II (C/1980 Y2) was very close to the double star Epsilon Lyrae at discovery and Roy had found it using his 20 centimetre altazimuth Newtonian. After more than 600 hours sweeping, on 699 clear nights, he was due for a discovery! The comet would reach perihelion (its closest point to the Sun) on 1981 January 27th and it passed extremely close to the pole star, Polaris. Living only a few miles from the editor of *The Astronomer* magazine, Guy Hurst, who was also based at Northampton, Roy sent his discovery message to Guy via Taxi on Christmas Day! Roy was a reclusive character who had deliberately not had a telephone installed at his home. Guy had just moved house too and his phone was not working either! Nevertheless, Roy and Guy did manage to phone George Alcock and alert BAA astronomers Michael Hendrie, Harold Ridley, Stan Milbourn and

M. Mobberley, *It Came From Outer Space Wearing an RAF Blazer!: A Fan's Biography of Sir Patrick Moore*, DOI 10.1007/978-3-319-00609-3_26,

Brian Manning. Hendrie and Manning managed to secure photographs of the comet. A British comet discovery was a rare event indeed, but Patrick had little time to celebrate or to appear on TV. His beloved mother was dying and it was obvious that she had little time left.

Some say that Great Comets mark the deaths of Kings and Queens. Well, comet Panther was not a spectacular comet but it was a comet discovered in England: an even rarer class of comet. As comet Panther peaked, on January 7th 1981, Gertrude Moore died, and a major part of Patrick's life ended for ever. At 94, she had, undeniably, had a good innings, but Patrick was totally devastated. Friends such as Iain Nicolson and Peter Cattermole (who Patrick had known as young students from the 1950s and 1960s) and Patrick's housekeeper Woody, were a tower of strength, but Patrick was totally distraught. He was inconsolable in the days following his mother's death. Although not a religious man, in the years to come he would tell close friends that he believed it was only a temporary parting and she would always be there for him. However, while this was a very sad event, Patrick was once more free to pursue his hobby and his life in astronomy. Indeed, he wanted to immerse himself totally in his hobby and his astronomical friends, it was all he felt he had left now. He had always been a workaholic, and work plus Woody would have to keep him sane in the coming weeks and months. Christmas time had always been a happy time in previous years at Farthings, with Patrick's mother sending out many hand painted cards to their best friends. However, Christmas 1980 had been so grim that it was, as far as Patrick was concerned, the last Chrismas he would acknowledge. It could never be the same.

During Patrick's mother's long life she had often been interviewed by the press about her son. Amongst the comments attributed to her there are probably three which most stick in this author's mind. Firstly: "He was the untidiest, oddest little devil as a boy and he hasn't changed. Life with him is a bit strange, but not bad when you stop worrying about what could happen next". Secondly: "Patrick has done so well because he speaks in a language you can understand. So many scientists think they are talking only to other scientists." Thirdly: "He used to talk at normal speed until he started *The Sky at Night*. But he has to get such a lot of material into a small space of time that I think he got into the habit of speaking fast. Now he does it all the time."

On the day Patrick's mother died, the BAA held their first meeting of 1981 and Roy Panther, normally a reclusive character, addressed the meeting to describe his Christmas night discovery of Comet Panther, 1980u. At the same meeting (held at 23 Savile Row, London) Patrick's friend John Mason described other remarkable events occurring on Christmas Night 1980: two brilliant meteors and the re-entry of the Cosmos 749 rocket! Patrick was the BAA meetings recorder, but they had to manage without him on January 7th. However, despite his grief and the funeral arrangements, Patrick did attend his good friend Horace Dall's 80th birthday party 9 days later. One of Patrick's closest friends over many years had been Edna Rosa Atwell, known as Rossie to all in the BAA and the BAA Lunar Section. She had, for many years, lived in Croydon and had been the salaried deputy assistant secretary in the BAA's London Office, from 1970 to 1978, and the assistant secretary (the BAA office boss) from 1978 to 1981. Rossie had been a backbone of the BAA since the 1950s, and arguably Patrick's closest female friend. She was 2 years older than Patrick, and like him had never been married. Immediately following Patrick's mother's death, Rossie resigned from the BAA job and moved south to West Wittering, a few miles to the west of Patrick at Selsey. With his mother's death leaving a huge void in his life, and his only other female company being his housekeeper "Woody", Rossie became a very welcome weekly visitor to Patrick's house in her retirement and in his bereavement period.

A few years later Rossie actually moved into the same house as Patrick's elderly friend Reg. Spry (born in 1902) at 60 Windsor Road in Selsey, to be within walking distance of Farthings. When Reg. Spry died she moved half a mile further north and into a flat just off the Selsey high street, at 56 St Peter's Crescent. Rossie herself would die, aged 80, in April 2002. In many ways Rossie was as close as Patrick ever came to having a female companion after his mother's death. Indeed, she even appeared with him on a BBC TV programme in the mid 1990s, where they were sampling some cuisine together, as a couple. As a committee member of the Selsey Manor Women's Institute she soon became as well known to village residents as Patrick himself.

On January 18th 1981, another UK astronomical discovery was made…..or was it? Amateur astronomer Dave Branchett, a well known 'Deep Sky' and variable star observer, was sweeping the constellation of Scutum in the January dawn sky when he came across what looked like a new eighth magnitude nova in the field of his 80 millimetre binoculars. An exposure was made, by professional astronomers at the Royal Greenwich Observatory at Herstmonceux, and there appeared to be an object (much fainter than eighth magnitude) on the plate. However, no other observers, professional, or amateur, could confirm the discovery. Nevertheless, an official IAU (International Astronomical Union) circular was issued to the world's professionals. A subsequent examination of the RGO plate seemed to indicate that the new 'object' might have been a photographic flaw, but the episode remains a mystery today, and Branchett was an experienced observer.

On the next *Sky at Night*, at the end of January 1981, Patrick started the programme by briefly introducing Roy Panther and Dave Branchett and congratulating them both on their discoveries. This was in total contrast to his *Sky at Night* dismissal of John Hosty's nova discovery in January 1977. Maybe he felt he could make amends by having both discoverers on the programme? It was a subdued, tired looking and bereaved Patrick on that *Sky at Night*; still a man obviously in a state of shock, disbelief, and sadness.

On January 24th, Patrick attended his first astronomy meeting since the death of his mother. This was a BAA Lunar section meeting held at the Hawkstone Hall in Lambeth, in southeast London. I remember that meeting well as I had just installed my brand new 14 inch telescope and was keen to show Patrick my photographs of the new telescope and observatory. Although I had met Patrick at various Lunar Section meetings since 1970, when I had only been 12, and exchanged many letters with him, I felt this was the first occasion that he had actually recognised me as a serious, contributing, Lunar Section member. I was a force to be reckoned with: I had a telescope as good as Patrick's! I was also very keen to become a leading member of the TLP sub-section. My aim: to try to photograph the lunar surface at as high a resolution as possible, whenever TLPs were reported. This was a niche that, strangely, no-one seemed to be filling and I wanted to fill it.

Following the director Geoff Amery's introductory comments, Patrick, despite his recent loss, launched into a lively and entertaining history lesson on the whole subject of TLP. He hinted that the section should try to undertake spectrographic analysis to try to confirm that gases were being released from beneath the lunar crust and he stressed that there would be much prestige from achieving success in this area. During the tea break I managed to grab a few minutes with Patrick. He was surrounded by Lunar Section members, especially younger members, who were offering their sympathies on his mother's death. Patrick, his eyes very watery, was obviously genuinely touched by their response. I always found talking to Patrick a daunting experience in my early twenties. Perhaps it was that severe Churchillian scowl, but even when I got to know him better, it sometimes did not seem much easier. He was always a man wrapped up in his own thoughts, self absorbed, and his opinions were always fixed and unmovable. This always gave Patrick a rather unfriendly appearance, which was quite opposite to his real character. He simply never had time for small talk or debate, but if you agreed wholeheartedly with him on life's essential issues, you had a friend for life. He loosened up considerably with some alcohol inside him though!

I proudly showed Patrick the photographs of my new telescope at that January 1981 meeting and told him that I intended to use it for the photography of TLP. He was very enthusiastic and his encouragement was just what I wanted: for the next 10 years I would be a keen lunar photographer, until I temporarily lost interest in the Moon. At that meeting I also showed Patrick my first pictures of the Moon, through my newly installed 14 inch telescope, taken a month earlier. I was proudest of my picture of the Mare Crisium and was fascinated to learn that Patrick's first BAA talk had been submitted on the subject of small craters on that Mare, when Patrick was only 14 years old. I left that meeting with a sense of destiny and purpose; it was my longest chat with Patrick to date, as a fellow lunar observer!

The spring of 1981 saw Patrick heavily involved in various celebrations of the 200th anniversary of the discovery of the planet Uranus by William Herschel. For almost 5 years, since the phone call from Bath resident Philippa Savery, Patrick had been heavily involved in saving and restoring Herschel's house at 19 New King Street, Bath. Thanks to the generosity of Dr and Mrs Leslie Hilliard, Treasurer and President of the Herschel Society, campaigning by Patrick, and the hard work of local residents, Herschel's house had been purchased and restored during the past 5 years. It had been decided that the Herschel house and museum should be opened on March 13th 1981, 200 years to the day since Herschel discovered the planet from that address. Patrick, together with the BAA President Commander Derek Howse, and the Mayor of Bath, attended the opening ceremony alongside various other invited guests and astronomers. The great-great-granddaughter of Sir William, Miss Caroline Herschel, was also present. At a public lecture in Bath, following that opening ceremony, Patrick spoke about Herschel's achievements to a packed house of 400 people, and he donated the proceeds to the Herschel Society. Four weeks later, the University of Bath celebrated the discovery of Uranus by hosting an International Astronomical Union (IAU) Colloquium entitled 'Uranus and the Outer Planets'. Of particular interest to Patrick was a reception on April 14th, held by the Mayor of Bath, followed by a concert including some of Herschel's music. Unfortunately, by a particularly bad timing coincidence, Patrick missed both: the first ever flight of the Space Shuttle occurred at this time and he was on TV. However, on the 15th, Patrick was able to attend an excellent evening lecture by his good friend Sir Bernard Lovell, entitled *The Life and Times of William Herschel*, which was punctuated with appropriate music.

Chapter 27
Musings on Planet 10; Life After Mother

Although Patrick was putting on a brave face regarding the loss of his mother he was still in a state of numb shock for the whole of 1981. Farthings had been a happy home for both of them for more than 12 years and although his housekeeper Woody was a great source of comfort, as were his closest friends and neighbours, the house seemed horribly empty without the mother he had known for almost 58 years. For 33 of those years, since his father had died, she had been his best friend and sole surviving parent. With no brothers or sisters and no wife or children, it was a massive shock to overcome. When returning from an astronomy meeting, a trip abroad, the BBC TV Centre, or just walking in from using the telescopes, his mother's friendly face would never welcome him indoors again and there were no siblings to share family memories of her with either.

Fortunately, Patrick was famous and so he had many friends (genuine ones and plenty of hangers-on) and so, with their support, he survived. However, he always believed her spirit was still around and that they would meet again. He also told me, after my own mother died more than 25 years later, that his mother was still "always there for him" and not just her memory either. Although Patrick had no religious leanings he strongly believed we were on Earth for a purpose "otherwise what's the point?" But he did not believe anyone knew what that purpose was. A few months after his mother died he received a copy of the brand new book *The Door Marked Summer* from his good friend the comedian Michael Bentine. As I have mentioned earlier, Bentine was strongly into researching paranormal behaviour and spiritualism and seems to have convinced Patrick that there was something in it. Bentine had researched the subject and was convinced some genuine mediums could talk to the departed. *The Door Marked Summer* contained much of his thoughts on paranormal activity. Bentine was the President of the 'Association for the Scientific Study of Anomalous Phenomena' in the final years of his life. Patrick had always been fascinated by this sort of stuff anyway, not least because the astronomer and author Camille Flammarion (1842–1925) had been a strong believer in spiritualism, believing it could be analysed by scientific methods. Patrick claimed he had discussed this with Flammarion's second wife, also an astronomer, Gabrielle Renaudot Flammarion (1877–1962).

M. Mobberley, *It Came From Outer Space Wearing an RAF Blazer!: A Fan's Biography of Sir Patrick Moore*, DOI 10.1007/978-3-319-00609-3_27,

Incidentally, while mentioning Madame Flammarion, I might add that Patrick used to claim that she had told him what a lovely sight Donati's Comet of 1858 had been; thus he could claim that he knew someone who had seen Donati's Comet, one of the most beautiful comets of all time. Clearly, as she was born in 1877, this was 19 years too late (!) and Flammarion's first wife had died in 1919, four years before Patrick was born. Of course, Gabrielle Renaudot Flammarion would have been told by her astronomer husband, 35 years her senior, that Donati's comet had been stunning, but she was just too young to go back that far in time. Patrick often liked to say that he had known, or met, someone famous, born in the 1800s, whose life overlapped with his, but the Donati story is a story too far! Anyway, I have digressed, as spiritualism was the subject.

Patrick was convinced that the Armagh Observatory possessed a ghost, "a former Director who killed himself because his wife was an absolute fiend" and after hearing creaks and groans in Farthings he became convinced that his ancient house had one too. He even named the ghost 'Claude'. Patrick made a few enquiries and someone "skilled in these matters" (!) came to Farthings and told Patrick that the house was smack bang on a Ley Line and that, yes, there *was* a ghost and that there was a significant presence under the floorboards too, where Patrick claimed "something had been found!" This 'evidence' was backed up by Patrick being told that the house next door had once had a priest hole/tunnel that ran across West Street.

On rare occasions, with people he trusted, Patrick would hint at his spiritualist leanings, but it was not anything he ever committed to print, especially given his scathing views about astrology. Of course, the death of Patrick's mother not only involved sadness, but bureaucracy too. There were lots of forms to fill in and as he was the executor there was no escape from these. As the reader will appreciate, Patrick loathed this sort of red tape with a passion. To add to the misery, another bureaucrat, Mr K. Whitmarsh, had recently demanded that Patrick pay £10 to the Southern Gas Board for repairs to his central heating system, despite the fact that the Farthings system was oil-fired and had needed no recent repairs. A further five letters arrived from Mr Whitmarsh, either duplicating or totally retracting the previous letter! Shortly afterwards Patrick's next book, *Bureaucrats: How to Annoy Them*, was published under his R.T. Fishall pseudonym. Amongst other gems it advised people to smear a thin layer of candle grease on those parts of a form marked 'For Official Use Only', to prevent the 'jobsworth' Civil Servant from writing anything in the space! "It's very useful when dealing with the Inland Revenue" according to Patrick. A few years later Patrick would follow this up with another R.T. Fishall production, this time actually named in honour of the Gas Board official: *The Twitmarsh Files*!

Anyway, ultimately, the best way to lift his bereavement and bureaucracy induced depression in 1981 was simply to throw himself back into astronomy and one avenue he hurled himself into concerned the ninth planet, as it was still regarded then, namely Pluto.

Since his trip to meet Clyde Tombaugh in 1980 for the fiftieth anniversary of the discovery of Pluto and with the 200th anniversary of the discovery of Uranus on his mind, Patrick had spent much of 1980 mulling over the possibility of a tenth planet and where it might be found. There was a mystery regarding the discovery of Pluto which goes as follows. After the discovery of Uranus by Herschel in 1781, and of Neptune in 1846, astronomers were understandably keen to discover any ninth planet that was in the outer solar system. The kudos associated with discovering a planet was enormous. The only way to discover a ninth planet would be to photograph the entire ecliptic path around the sky (essentially the so-called constellations of the astrological Zodiac, which mark the plane of the planetary orbits) or, to try to calculate where the next planet was from any gravitational pull it might produce on the existing outer planets. In fact, when Neptune was discovered, it did not fully account for the discrepancies in the motion of Uranus around the Sun, prompting astronomers to speculate that there might be a big planet much further out than Neptune. Astronomers and mathematicians set to work; in particular, Percival Lowell and W.H. Pickering made predictions. When Clyde Tombaugh actually discovered Pluto (or Lowell's Planet X as some called it) in 1930, it was quite close to where Lowell had predicted it to be and it's orbital period of 248 years was not too far off the 282 years predicted by Lowell. However, as the years went by it became clear that Pluto did not have the bulk to perturb Uranus or Neptune gravitationally, despite the fact that it appeared to have been tracked down by those very perturbations!

Patrick had been mulling over all these matters since his collaboration with Tombaugh on the book *Out of the darkness: the planet Pluto*. He consolidated his thoughts in a paper, submitted to the BAA Journal in 1981. Patrick's reasoning went along these lines, taken from his 1981 paper, entitled 'Some thoughts on Planet 'X'':

> This is where I depart from science and enter the realm of speculation. I am quite prepared to be laughed at, particularly as I am no mathematician. I merely want to put forward an idea, with the full knowledge that it may be completely out of court. The essential basis of my argument is that Lowell's calculations had real validity. (If not, we are back to 'Square One'.) His Planet 'X' was much further out than Pluto is; as we have seen, its predicted period was 282 years. So far as I can see, the only assumption which is not quite illogical is that in 1930, when Pluto was located, Pluto and Planet 'X' were aligned. In fact, Planet 'X' was in the same region of the sky as Pluto, but escaped detection partly because it was beyond the limiting magnitude of Clyde Tombaugh's search, and partly because once Pluto came to light it was tacitly assumed that the hunt had been brought to a successful conclusion. Now, assume that the real 'X' does have a period of 282 years, and that the orbital inclination is only a few degrees. (For the known planets, excluding Pluto, remember that the inclination is 7° for Mercury and less than 4° for all the rest.) Pluto turned up near the star δ Geminorum, whose right ascension is 7 h 19 m. Assume that at this time the RA of Planet 'X' was also 7 h 19 m. In the 50 years which have elapsed since then, it will have increased its RA by 4 h 15 m. That makes its present position RA 11 h 34 m. Of course, this is carrying 'accuracy' to a ridiculous extent, so let us round things off to 11½ hours. If the inclination is very slight, this gives us a position roughly between σ and τ Leonis. But, all in all, it seems that the distance and period are likely to be greater than this. If we extend the Uranus/Neptune sequence, we arrive at a period of 324 years. In this case we have an RA increase of about 3 h 40 m, which, again on the assumption of slight inclination, gives a present RA of 10 h 59 m – let us say 11 h, close to χ Leonis. You will probably say that this is mere guesswork and, of course, you will be absolutely right. Had I the access to a powerful Schmidt, capable of reaching down to (say) magnitude 21, and had I unlimited time, I would be tempted to make a very close survey of the sky round χ Leonis. But the chances

> of anyone doing so on the basis of a calculation which could have been made by anyone who can add and subtract are virtually nil, and I would be the last to demur. All I will say is that if Planet 'X' is ever found, and did, in fact, lie somewhere near χ Leonis in 1981, I will give myself a modest pat on the back.

Needless to say, no-one took up Patrick's search suggestion and Planet 'X' was never discovered, although smaller Kuiper belt or Trans-Neptunian objects were found in abundance. To be honest, only Patrick could have got away with such a speculative and simple paper in the BAA Journal. Believe me, I should know, I was the BAA Papers Secretary from 1990 to 1997 and it was an established fact that Patrick expected his papers to be accepted by the referees without a single word of criticism. If there was even a single word uttered against one of his papers, he would withdraw it, instantly, like a spoilt schoolboy! However, Patrick and the council knew, that whenever members received their Journals, if it contained a paper by Patrick, that was the one the members would read. So, many of Patrick's papers were published, even though they were often simply stating the obvious, albeit in a unique and highly entertaining way. Maths and orbits were not Patrick's strong point, but enthusiasm was.

On April 17th 1981 over 500 amateurs attended the Hampshire Astronomical Group's National Astronomy Week, an event originated by Hampshire amateur astronomer Robin Gorman. Of course, 500 people turned up because they knew Patrick was coming! The event was held at the Queen Elizabeth Country Park near Petersfield. Patrick was delighted to meet a young girl, Deborah, who was a distant descendant of William Herschel (the actual discoverer of Uranus) at that event.

In the spring of 1981 Patrick had been absent from the BAA Council table ever since he resigned as the Lunar Section Director in January 1976: a period of more than 5 years. This was unprecedented for Patrick. Admittedly, at that time, the BAA bye-laws forbade anyone serving for more than four consecutive years as a councillor, without taking a 1 year break, although section directors, treasurers, secretaries and editors were immune from this law. However, Patrick had not served the BAA Council in any capacity since January 1976, simply because of his mother's deteriorating health. Needless to say, the council were keen to see Patrick resume his seat at the table and Patrick was keen to return. Patrick's name was added to the final ballot paper list in April 1981. The ballot paper would be distributed, as always, with the August BAA Journal and the votes would be counted at the end of October. The new council would take their seats at the meeting in November, held on the last Wednesday in that month. Perhaps it should be pointed out that the BAA council elections took place every year, but no-one had ever contested the posts of President, Treasurer, or the secretaries (up to this time) and even the ten councillors were generally elected out of, at most, a pool of 14 candidates. The new President was, by tradition, proposed by the outgoing President and the decision was always rubber stamped by the Council; the bye-laws prevented a President serving more than 2 years in succession. In fact, since the formation of the BAA in 1890, all of the Presidents had been re-elected after year one and so had served for the two consecutive years maximum term; none had served a second 2 years term as President, although there was nothing to prevent this happening. This was largely because the

President was fully expected to serve the full 2 years and because there were always good candidates willing to be the next President.

For the Council Session of 1981/82 fourteen candidate councillors featured on the ballot paper. These included a mixture of aspiring astro-media people, such as Patrick's friend John Mason and the duo of Heather Couper and Nigel Henbest. The President, with one more term to serve, was Cdr H.D. Howse of the National Maritime Museum and the business and papers secretaries were Neville Goodman and Storm Dunlop. No-one had any doubt that Patrick would be one of the ten successful councillors when the votes were counted.

We have already noted that Patrick had not seen eye to eye with Storm Dunlop's accurate review of his very scruffily produced book *Basic Astronomy*, published in 1979. Well, in October 1980, Storm had been elected to succeed Cdr. Henry Hatfield as the BAA Papers Secretary. He was therefore responsible for recommending the acceptance of papers for publication in the Journal, subject to the two referees' approval and that of the council. Shortly after the change of Papers Secretary, the Saturn Section report for the 'edge-on rings' apparition of 1980 was submitted to the BAA Journal by the BAA Saturn Section Director Alan Heath. Alan was, without doubt, the hardest working Saturn Section Director the BAA had ever seen, producing detailed reports every year, for the BAA Journal. Patrick's good friend Paul Doherty had submitted by far the greatest number of quality observations of the apparition. Paul's eyesight and artistic ability were extraordinary and he had been able to observe the finest details in the edge-on rings. Paul, the Assistant Saturn Section Director, was certainly looking forward to seeing his sketches and paintings in the Journal, as was Patrick. Paul told this author that he had been looking forward to one of his best sketches appearing, perhaps, on a whole BAA Journal page, to "show all the detail that was visible". The Saturn Section report, as submitted to the council, was a thorough one and a *very* long one. Unfortunately, the Papers Secretary and the council said that it was *too* long: Paul's observations would have to be severely curtailed and reproduced to the same 'postage stamp' size as all the other sketches. In a telephone conversation with this author, some years later, Paul said that in the opinion of one critical person in the BAA refereeing process, "The observations by Paul Doherty show details that cannot *possibly* be seen through an amateur telescope". Paul and Patrick were distraught. If Patrick had been on the council already, he might have been able to alter things, but he was not and he had crossed swords with the new BAA Papers Secretary on other occasions. Despite pleadings from Patrick, Paul Doherty would resign shortly afterwards. His best paintings of the edge-on rings were reproduced instead in the American publication *Sky & Telescope*. The BAA had lost its finest planetary artist for no good reason. The final Saturn Section paper was put in the Journal queue (published in 1981 December) with a note from the Director Alan Heath:

> It is regretted that this report has had to be shortened on instruction of Council due to the length of the original. A copy of the full report is retained in the Section files. It is hoped that observers will continue to contribute their valuable work to the Section as the reduction of the report was due to reasons of economy only.

For the BAA Exhibition meeting of May 30th 1981 the Lunar Section put on their traditional display of lunar photographs, maps and drawings at the Hawkstone Hall in Lambeth. A lunar photometer and combined Moonblink and Crater Extinction Filter Device were also displayed. Roy Panther, who had discovered Comet Panther back in January, was awarded the Merlin Medal by the BAA Vice-President Leslie White (the President, Cdr Howse, was overseas). For the first time the Lunar Section organised a photographic competition at the Exhibition meeting, which excited this author considerably! I had taken some pleasing colour photographs of the Moon in December 1980 and thought that one of them, featuring the Moon's Mare Crisium region, stood a small chance of winning. The judges for the competition were the Lunar Section photographic co-ordinator John Pedler, the TLP co-ordinator Peter Foley and the lunar photographer Henry Hatfield. Patrick also cast his beady eye over the entries. I remember that day very well. I especially remember drifting around the Exhibition Meeting stands, looking at all the section displays and then sneaking the odd glance to see if the judges had made a decision yet! Early in the afternoon of May 30th the Lunar Section Director Geoff Amery approached me. "Congratulations Martin", he said, "you've won the Lunar Section photographic competition". I was ecstatic, as there had been some good photographs on display, although some had, admittedly, been taken from countries much nearer to the equator, where the Moon is always riding higher in the night sky. Geoff told me that the prize was any book that was on sale at the meeting. I chose Patrick's book *Out of the darkness: the planet Pluto* and Geoff signed it 'Congratulations on your excellent photographic work – Keep it up' and Patrick congratulated me too. For this author, that was a day to remember. Geoff and Patrick's congratulations on my work at that meeting were a spur for me to carry on with my lunar photography.

Two weeks later, on June 12th, Patrick was out observing the Moon with his 15 inch reflector and decided to check out the crater Plato at 21.00 hours GMT. Plato had been a source of many TLP alerts over the years. Patrick was immediately struck by the appearance of Plato's south wall. It appeared indistinct, as if it were obscured by something. The obscuration extended over the south part of the crater floor too and there was no sign of Plato's central craterlet, which, in the prevailing conditions, Patrick had expected to see. At 21.10 Patrick phoned the TLP co-ordinator Peter Foley who confirmed the appearance, describing the obscuration as 'dense'. Peter phoned me, Hedley Robinson and Gerald North and we all saw Plato's south inner wall as looking 'indistinct'. I took some colour photographs between 21.45 and 21.55 UT and black and white photographs between 22.10 and 23.27 GMT. Patrick's view was that the obscuration had almost gone by 21.50 UT, but Peter Foley was convinced that my colour photographs around that time did show the obscuration. Certainly, my photographs seemed to show the south wall looking blurred, but to me the atmospheric seeing could quite easily have been the culprit and not events on the Moon. Still, I had achieved something that night: I had taken photographs of an alleged TLP, which had been my major aim ever since acquiring my 14 inch reflector in 1980 and ever since joining the Lunar Section in 1970! When Patrick saw my photographs he was of the opinion that they did seem to show the appearance he had seen, but that it had been a lot more obvious visually (and his most significant observations had been made more than half an hour before my photographs).

Later in June, Patrick was invited to attend the Astronomical Society of the Pacific's awards ceremony and banquet in Seattle. With his mother's suffering finally over, 1981 would be a year when Patrick frequently travelled to the USA, filming for the *Sky at Night* and for its 25th anniversary in one year's time. The ASP Seattle event was not a *Sky at Night* trip, but it was a very grand affair. The banquet took place during a boat cruise of the city's bays and waterways, while 150 ASP members dined on Pacific Coast Salmon. After that the ASP awards were presented by President Halton C. Arp in a converted paddle-boat (called the Showboat Theatre) owned by the University of Washington. The BAA had been invited to suggest worthy recipients for the ASP's first ever International Amateur Achievement Award and, not surprisingly, George Alcock had been nominated for his visual discovery of four comets and four novae between 1959 and 1976. Bizarrely, the BAA council asked George if he would kindly write his own nomination (!), but George asked his old friend J.P. Manning-Prentice to submit the citation to the BAA's Neville Goodman instead. To the delight of the BAA Council, George won the ASP International Achievement Award and was invited to the Seattle function. However, George was a retired Peterborough schoolmaster with a sick wife to care for. He did not have the time or the wherewithal to fly to Seattle and collect the award and its $250 cheque. So Patrick agreed to accept the award on his behalf and present it to George on his return. Patrick was also being honoured at the same meeting, with the distinguished Klumpke-Roberts award, presented to him by Halton Arp himself. Arp would, a decade or so later, become the most controversial figure in cosmology.

Following Patrick's return to the UK the next BAA London meeting would be the October AGM. It was suggested that George could travel to London to accept the award at the AGM, but because of a lot of dithering by the BAA Council and because of George Alcock's invalid wife and his teaching commitments, the presentation never took place. A disgraceful situation then ensued in which Patrick kept the award plaque in his custody and vowed to 'present it to George somehow' but kept putting it off and off and off! Tragically, another year would pass before the BAA was able to present the ASP award to a quietly fuming George Alcock. One theory about this farce suggested that it had dragged on for so long that Patrick decided another 6 or 8 months would make no difference and he, Patrick, could present the award to George personally, if he was the BAA President (as was on the cards). The first opportunity when this could occur would be in November 1982, some 17 months after Patrick received the award on George's behalf! In 1991 I filmed a video interview with George Alcock. He became so agitated at the mention of the ASP award and the delay caused by Patrick that he asked me to turn the video camera off, in case he said something he would seriously regret.

On July 19th 1981 Patrick took part in his fourth and final appearance on the BBC's *It's a Celebrity Knockout* program, once again held at the Fulham Football Ground at Loftus road and with Patrick, as always, raising money for his favourite charity, *The Lord's Taverners*. At 58, and carrying the weight Patrick did, it was remarkable that he was still able to compete at all, in an outdoor activity where getting soaked to the skin was an essential part of the entertainment! The programme was transmitted on August 31st.

In September 1981 Patrick's old mate Professor Sir Bernard Lovell retired as Director of the Nuffield Radio Astronomy Laboratories at Jodrell Bank in Cheshire. Patrick and Sir Bernard's astronomical fame had started around the same time, in 1957, and both owed a huge debt to the Russian Sputnik 1 spaceprobe of October 1st that same year. Jodrell Bank had been able to famously track the carrier rocket of Sputnik 1 and due to Lord Nuffield's personal cheque, the Manchester University facility had avoided bankruptcy. Also, the start of the Space Race had guaranteed the survival of *The Sky at Night* beyond its first 6 months. Patrick devoted a special *Sky at Night* to Sir Bernard's career.

On September 12th 1981 the BAA held a provincial meeting at the Merseyside County Museum in Liverpool. Patrick gave the major afternoon talk, about Voyager 2's images of Jupiter and Saturn, to a packed audience.

On September 13th the very first BBC 'voice of Astronomy', Dr J.G. Porter, died. He was 80 years old. Guy Porter's Radio 4 talks on 'The Night Sky' ran, every month, from 1948 to 1961. He had been a huge influence on Patrick and his radio programmes had been a forerunner to *The Sky at Night*. In addition he had played a small commending role in Patrick getting *The Sky at Night* job, especially as he had vastly preferred to stick to radio presenting, allowing Patrick to clinch that dream position. Porter had been the BAA President from 1948 to 1950 and had known the comedian Will Hay who died during Porter's presidency. Patrick said that the BAA had never enjoyed a better President and he was quick to offer his services, by writing Porter's obituary for the BAA.

At the start of October 1981 Patrick set off on a trip to Colombia. A new BAA member, originating from Haverhill in Suffolk, named Kevin Marshall, had been working as an English teacher in Colombia and was trying to stir up interest in amateur astronomy in that country. Kevin invited Patrick to be the Colombian Astronomical Association's 'Honorary President' and to participate in their first 'National Encounter' of amateur astronomy in Colombia, from October 10th to 12th. Unfortunately, poor weather meant that no observing was possible during the conference weekend, but a musical concert replaced the observing session and Patrick treated the fascinated audience to a solo musical performance of his, by playing……. the spoons! Needless to say, Patrick also delivered an entertaining talk, about the cratering of the lunar surface.

In October the ballot results for the 1981–1982 BAA council election were counted and to nobody's surprise Patrick was returned to the council table. Barely had he taken his seat when people started clamouring for him to be the next BAA President. It was truly remarkable that he had been such a prominent BAA figure for so long, without holding the ultimate post. By the end of 1981 he had been a BAA member for 47 years and had been the Director of both the Lunar Section and the Mercury & Venus Section. He had also been the recipient of the association's senior award, the Goodacre Medal, as well as being the face of astronomy, as far as the public were concerned. Indeed, he was also the face of xylophone music at that time. Only a month later, on November 23rd 1981, Patrick was at the Drury Lane Theatre in London, playing the xylophone in front of the Queen at the Royal Variety Performance, watched by 15 million viewers on TV! His only comment to close friends about this was "If only mother could have lived to see it".

So why had Patrick not been the BAA President already? Well, since Patrick's earliest days on the BAA Council he had recorded the London meetings, for 30 years in fact, and you cannot easily be the meetings recorder and the President at the same time. Whenever Patrick's name was suggested as the next President, throughout the 1960s and 1970s, he had always shaken his head, shrugged his shoulders and said "But who would record the meetings?" OK, that is a pretty feeble reason! In truth, the President needed to be a patient man, not bored by debate, and not opinionated. Patrick was just the opposite. The Presidency also involved much bureaucracy and a mandatory appearance at every council and main meeting. With Patrick's hectic schedule in the Apollo and post-Apollo era this just would not have been possible. Then, of course, his mothers increasing frailty had kept him away from the council table from 1976 to 1981. But now things were different. It had always been the tradition for the outgoing President to name his successor after some consultation with the senior BAA officers. Cdr Derek Howse, the President from October 1980 to 1982, thought Patrick was the best man for the job, now that he had finally returned to the council table. However, it was, according to Patrick, Heather Couper's insistence that finally swayed him. Heather was very keen that he should become the President and in Patrick's own words:

"I said if she would be President after me, I'd do it. In fact I first said this at Horace Dall's 80th Birthday Party, just after Mother died. Heather, Derek and Peter Foley were witnesses."

Barely a few months after resuming his council seat, Patrick, much to the council's joy, accepted Cdr. Howse's nomination and had his name marked down as the President's choice to be his successor. To be brutally honest there were not that many other obvious contenders at the time. Horace Dall, the legendary optician and designer of the Dall-Kirkham telescope system in the 1920s, had always refused the offer to be President. Despite Horace's legendary reputation he was often virtually silent at council meetings and was a very shy man. Other senior council figures, such as Gordon Taylor, Neville Goodman, Howard Miles, Harold Ridley and Leslie White had already served as the President. The lunar and solar photographer Cdr. Henry Hatfield was, realistically, the only other Presidential contender, with the exception of Colin Ronan, but Henry was another naval Commander, keen for Patrick to be the next President, and Colin was the editor of the BAA Journal, which excluded him from the post. In all honesty Cdr. Derek Howse, the existing President, had been a rather unusual choice as a President as he was not a noted BAA observer, simply a man of good character who had strong links with the National Maritime Museum and Greenwich. With virtually everyone on the council in the early 1980s being a fan of Patrick's he was always destined to be the President. Despite his opinionated views on many issues he was a hero figure, an inspiration and hardly ever took himself too seriously. Yes, he had enemies, but he also had very few council opponents in that immediate post-Apollo era. A Patrick presidency would also be fantastic publicity for the BAA, whose membership had started to decline rapidly after the last of the NASA Moon landings.

The second week of 1982 started with a good dose of snow over the southern UK. This was a major blow for BAA Lunar Section observers, as a potentially excellent lunar eclipse would take place on the evening of January 9th. As luck would have it, this author was just north of the snow cloud line and saw the whole event! No-one further south saw anything of the eclipse. I was delighted when Patrick showed my best picture of the eclipsed Moon on the January *Sky at Night*! For me, it marked the start of 30 years of Patrick showing my pictures on TV. [I should like to add that the BBC used my pictures, over 30 years, for free!]

On January 24th 1982, unknown to Patrick, the astronomer Ted Bowell in the USA, based at the Anderson Mesa station, discovered a new asteroid, provisionally designated 1982 BR. A few years later, when the orbit was fully determined, it would be named 2602 Moore, after Patrick. I would venture to suggest that Patrick must have felt rather chuffed that 'his' asteroid was given a lower number than the one eventually named after his friend Arthur C. Clarke (4923 Clarke), which had actually been discovered in 1981, but took longer to be numbered due to the time taken to fully define the orbital parameters.

In the February BAA Lunar Section Circular Patrick wrote of his sadness on hearing of the death of V. Axel Firsoff, a well-known lecturer on controversial matters (for example, he had theorised that there had been liquid water flowing on the Moon). Patrick's obsession with the idea that the major lunar craters were volcanic in origin (highly controversial) made him and Firsoff occasional allies, flying in the face of accepted lunar geology.

In April 1982, Patrick's beloved *Sky at Night* programme would pass another historic milestone: its 25th anniversary. In recent years Patrick had starred in all sorts of TV roles. In TV charity shows he had sung *Yes Sir that's my Baby* on TV but as follows: "Yes, Sir, That's my Baby Nowwwwwwwwwwwwwwwwwwwwww-wahh" sounding like a baying hound at the conclusion and accompanied by dancing girls! He had also sung *The stars at night are shining bright, deep in the heart of Texas* on TV and appeared as a TV 'personality' in quiz shows such as *Blankety-Blank*, *The Generation Game* and *It's a Celebrity Knockout.* On the 1981 Royal Variety Performance he had entertained the Queen with a performance on the xylophone! He was also a regular guest alongside the likes of Richard Baker and Joyce Grenfell on *Face the Music*, the BBC classical music quiz, and barely a month passed without him being a guest on *Noel Edmond's multi-coloured Swap Shop* – a children's Saturday morning programme. On this latter programme he was occasionally asked to judge newly released modern pop records: a task he hated. "Sounds like a Tom cat being put through the mincer" was his typical reaction. "And your score, Patrick?" he would be asked. "I'll be generous and give it zero" was his usual reply, scowling behind his monocle.

Patrick was an established British institution and the BBC had planned a special programme for Sunday 25th April 1982: a celebration of 25 years of *The Sky at Night* in which Patrick travelled the world visiting major observatories. To coincide with the anniversary Patrick had published his latest book, *The Unfolding Universe*. Patrick claimed that he had written all 80,000 words of it in 2 weeks, on his 1908 Woodstock typewriter! In the year leading up to the anniversary programme Patrick had been jetting all over the place. He visited the Canary Islands, a mine in South Dakota, the Kennedy Space Center, Mount Palomar, Kitt Peak Observatory in Arizona, the Anglo-Australian Observatory at Siding Spring, New South Wales: you name it, he filmed there.

The *Sky at Night* had acquired a new and highly regarded producer, Pieter Morpurgo, in 1981 and never had Patrick travelled so far and wide for his TV programme. Morpurgo would work with Patrick for a memorable 17 years (1981–1998) and, for Patrick, "Pieter was the best". The BBC had rarely allowed such travel expenditure on *The Sky at Night* (according to one source, £4,000 per programme was the budget at the start of Morpurgo's reign) but so much international observatory material was filmed it could be used over much of the anniversary years' programmes. In addition, the BBC had never before had a programme running for a quarter of a century with the same presenter, so splashing out was fully justified. There was also a big 25th anniversary party, held at the BBC TV Centre in London at the end of April 1982, which many astronomers, past guests and producers attended. An enormous cake (plus a tiny plastic telescope adornment) had been made in the form of two huge '2' and '5' numbers, each of which was more than a foot across. The number '2' was iced with the words 'Congratulations Patrick' and the number '5' with the words 'The Sky at Night'.

Pieter Morpurgo had actually been *The Sky at Night* director before being its producer and was originally told that he would just be "looking after the program for a month" until a "proper" producer was found! As things turned out he became the programme's best producer and his 17 year collaboration with Patrick would be a very happy time for both men. Pieter told me, in 2008, that the only thing that irritated him about Patrick was that the great man referred to Pieter's other excellent BBC programs, like *Crimewatch*, as "Your minor programs!" Pieter now has an asteroid named after him, following some string pulling by Patrick, and the object formerly called 1991 PM1 is now designated 5521 Morpurgo. Incidentally, Pieter's younger brother is the best-selling children's author Michael Morpurgo; it would seem that media excellence must be rooted somewhere in those brothers' DNA!

During May 1982 Patrick removed the mirror from his 15 inch reflector for re-aluminising. While it was removed he would use his 12.5 inch reflector or smaller instruments. Before the 15 inch mirror was re-installed, while observing the Moon on July 3rd with the 12.5 inch instrument, Patrick received a TLP phone alert from Peter Foley. The Lunar Section Director Geoff Amery had reported seeing 'an indistinctness' on the north-eastern wall of the crater Plato and had alerted Peter. Patrick's observing notes from that night read as follows:

> A distinct brownish hue to a stretch of the N. wall of Plato. Of course I took all the usual precautions to guard against illusion and am confident that the effect was real and confined to that area. I did not note any loss of definition either along the wall or to the north of it. 22.10 GMT. There was a perceptible blink in the TLP area. By 22.16 the length of the affected wall seemed to have diminished, and things remained static until 22.35 when the

colouration decreased temporarily – to return at 22.40 and somewhat increased by 22.45. Sadly, the Moon then retired behind my chimney.

I had also been alerted to this event and, from my observatory near Bury St Edmunds, I observed the crater Plato too. I recorded red, yellow and blue on the south wall and red/yellow with a blue tinge on the north wall, but regarded this to just be a result of spurious colour caused by the Moon's low altitude in the atmosphere. Patrick also recorded Aristarchus as exceptionally brilliant that night. I disagreed, seeing Aristarchus as, if anything, looking dimmer than normal! Interestingly, Patrick also reported anomalies in the craters Aristarchus and Grimaldi A, only 6 days later, on July 9th, describing Aristarchus as 'bluish'. This was highly unusual for Patrick as his eyesight had been measured across the visual spectrum by Cdr. Henry Hatfield and Patrick agreed with the findings: he was definitely more red-sensitive than the average person and less blue-sensitive.

In August Patrick was able to attend the International Astronomical Union's 17th General Assembly. IAU assemblies are held every 3 years throughout the world, but Patrick had not been able to attend one since the 1970 Brighton gathering, largely due to his mother's declining health. However, for the 1982 August 17th – 26th event, the location chosen was the city of Patras in Greece, with the venue being the local University. Patrick, as we have seen, had a pretty low opinion of foreigners, but he had a soft spot for Greeks (as well as Scandinavians). So, accompanied by John Mason, he flew to Athens and then headed for Patras. Patrick spent an entertaining 10 days meeting old and new IAU acquaintances, drinking a huge amount of the local wine, and having a very amusing time indeed.

Back in the UK, a BAA Lunar Section meeting was arranged at the Hawkstone Hall in Lambeth, for September 18th 1982. I was one of the speakers at that meeting, giving an account of my photography of the January lunar eclipse. It was quite a special meeting because an old friend of the Lunar Section, Dr Ewen Whitaker, would be there. Ewen's association with the Moon went back as far as Patrick's and the two men's views on the formation of the lunar craters were totally at odds. Ewen had spent his life as a professional lunar and planetary scientist, whereas Patrick was an amateur. Ewen had been a member of the BAA Council from 1954 to 1956 and the BAA Lunar Section Director from 1956 to 1958, a period in which Patrick had been notably absent, after the O'Neill's bridge affair. But Ewen had emigrated to America in the 1950s and enjoyed a very successful career at the Lunar and Planetary Observatory in Tucson, Arizona. The BAA Council had decided to present its senior award, the Goodacre Medal, to Ewen in 1982. Fortunately, Ewen was visiting the UK and so speaking at the Lunar Section meeting and receiving the award at the same time was possible. The BAA President, Cdr Derek Howse, presented Ewen with the Goodacre medal at the meeting, to much applause from the audience. Ewen gave a talk which Patrick must have hated: how the craters on the Moon were formed by meteor impacts and NOT by vulcanism. Certainly, Patrick's supporters, Henry Hatfield and Peter Foley, gave Ewen quite a grilling when it came to questions! Patrick doubted whether any artificial meteor crater simulations could produce craters with a central peak, but Ewen claimed this was perfectly possible. Patrick was generous in his praise of Ewen Whitaker, but to this member of the audience there was something of a history of discord between the two men: a suspicion that older members of the BAA Council have confirmed to me over the years. As far as Patrick was concerned, the Moon looked like a body littered with volcanic craters and it was as simple as that: an opinion fixed rigid and unmovable until Patrick reached 80 years of age! To the professional astronomer, patient, systematic research proved that a volcanic origin was simply not the case: the lunar craters were caused by meteoritic bombardment over aeons.

Patrick often claimed his drawings of the Moon had helped NASA plan their Apollo Moon landings, but even in the 1950s Patrick had been backing the wrong horse. It was Ewen, who had emigrated to America in that era, and embraced the new photographic techniques, who had really played a huge part in mapping the future Apollo landing sites, not Patrick. However, Patrick was a household name, whereas Ewen was not, so most people have never heard of Ewen Whitaker, but everyone has heard of Patrick Moore. Even in astronomy, life can be very unfair at times, but then, having met Ewen Whitaker, he did not strike me as a man who craved attention, just one who sought scientific and historical accuracy; a very modest and patient man indeed.

Chapter 28
Halley Recovered as Patrick becomes President

The BAA and its Lunar Section had inspired a number of young amateurs to become professional astronomers over the years. One of these amateurs had been a keen TLP sub-section observer in the 1970s and a regular contributor to the Lunar Section Circular. His name was David Jewitt and the start of the 1980s saw him emigrate from the UK to the USA to work on the 200 inch telescope at Mount Palomar. In the early 1980s professional astronomers were just starting to use CCDs (Charge-Coupled Devices) to image the night sky, rather than using photographic film. Early CCDs were ten times more sensitive than film, but were tiny in size, necessitating a very accurate positioning of the telescope. David Jewitt was fortunate enough to be using one of the first CCD detectors on the world's largest telescope. Working with his colleagues, Danielson and Dressler, Jewitt, in October 1982, was involved in an ambitious project to be the first to detect Halley's Comet returning to the inner solar system. On October 16th they struck gold when a series of CCD images showed the famous object, at magnitude 24.3, only 8 arc-seconds away from its predicted position. The recovery hit the headlines and the BBC produced an 'emergency' *Sky at Night* to cover the success. On the live programme, Patrick, even more excited than usual, spoke on the telephone to David in California, congratulating him on his success. A BAA Lunar Section member had recovered the most famous comet in history!

Little more than a week after the announcement of the recovery of Comet Halley, Patrick became the President of the BAA. Typical of Patrick, he remained BAA meetings recorder even for the October 27th meeting, when he took the chair. The next meeting saw his old mate Howard Miles take over as meetings recorder. The format of BAA Annual General Meetings is that they always occur on the last Wednesday in October and they are followed by a short 'Ordinary Meeting'. Every second year the outgoing President vacates the chair at the end of the AGM and the incoming President chairs the ordinary meeting immediately afterwards. Thus, Cdr. Derek Howse vacated the chair, at 23 Savile Row, on October 27th, and then, amid much applause, Patrick became the President. On becoming seated Patrick said that when he joined the Association, exactly 48 years ago, he had no thought that he might one day be elected as President. He regarded it as a tremendous honour,

M. Mobberley, *It Came From Outer Space Wearing an RAF Blazer!: A Fan's Biography of Sir Patrick Moore*, DOI 10.1007/978-3-319-00609-3_28,

particularly since the BAA had always been a big part of his life. He might, he said, be a bad President; but if so, it would not be through lack of effort and he promised that he would do his very best. Then, with the Papers Secretary Storm Dunlop seated to his left and the Business Secretary Neville Goodman seated to his right, he adjourned the AGM and opened the Ordinary Meeting. Patrick had, at last, been installed as the President, in the organisation that had done so much to form his character since the age of 11. As might have been expected, the first speaker Patrick asked to address a BAA meeting under his chairmanship was one of his oldest friends, the aforementioned Howard Miles; his subject was one close to Patrick's heart: *Twenty-Five Years of the Space Age*.

Despite Patrick and Storm Dunlop sitting together on the meeting podium and at the council table, and despite them living a few miles apart, they could hardly be described as bosom friends. They had travelled to BAA Council meetings from Chichester railway station on many Wednesdays, but on more than one occasion Patrick told me he would sit in a carriage as far removed from Storm as possible! Storm had, after all, criticised one of his books in the Journal and advised him that Nova Sagitta 1977 did not exist, when it did! This would not have been too bad if Patrick hadn't broadcast its non-existence on TV. In addition, Patrick held Storm and others fully responsible for Paul Doherty's resignation. Patrick's views on all these matters were highly unfair, but Patrick had made his mind up and like a spoilt schoolboy, that was that! Storm was a tireless worker for the BAA and, as Papers Secretary, was doing the sort of tedious chore that Patrick would never have done.

Alcock gets the Plaque

At the next BAA meeting, Patrick would, at long last, be able to present the UK's most revered amateur astronomer, George Alcock, with his plaque from the Astronomical Society of the Pacific. The plaque was for their International Amateur Achievement award: George being the first recipient. We have already seen that Patrick was happy to receive the plaque, on George's behalf, at a lavish awards ceremony, banquet and river cruise in Seattle, some 17 months earlier. However, Patrick did not communicate anything to George for many months after the Seattle ceremony. In fact nobody told George anything about what had happened in Seattle. It was only when George was sent the December 1981 Volume 10, Number 6 ASP Journal 'Mercury' that he even knew that the Seattle award had been presented to him, via Patrick, in his absence! Eventually, Patrick promised George that he would present the plaque too him, at a BAA meeting. However, George wanted the plaque immediately on Patrick's return to the UK.

Whatever Patrick's reasons for not getting the plaque to George (and many thought it was simply Patrick's desire to present it to George in public, when he, Patrick, was the President) George was livid. Normally a very placid man, as I mentioned a few pages ago, when I filmed a video of George's life in 1991, he demanded that I stopped the tape when we got to this point in his life, so he could tell the three of us present exactly what had occurred! The ASP award was George's award, NOT Patrick's (remember, he'd received the Klumpke-Roberts award at the same ceremony) and it was because of George's devotion to his crippled wife and because of his limited funds that Patrick had represented him at the ceremony in Seattle. To George it was clear cut: Patrick could not be bothered with the hassle of packing the plaque off to him and he wanted the prestige of presenting the award when he, Patrick, was the President. From a non-astronomer's viewpoint this theory might seem daft, but Patrick's standing amongst his peers was important to him and amongst amateur astronomers Patrick was a famous TV astronomer, a great writer and a speaker, but he was simply not in George Alcock's observing league. Yes, Patrick observed the Moon, planets, bright variable stars and bright comets on every clear night, but the patience exhibited by Alcock was staggering. In addition, Patrick loved the limelight, whereas Alcock hated it. For Patrick to present George with an award was an honour for Patrick, not for George! George had memorised 30,000 stars in patterns and had, by 1982, discovered four comets and four novae. Patrick had made countless observations at the eyepiece but had discovered nothing: he was famous, a prolific book author and a brilliant motivator, writer and broadcaster, but George was a God and a living legend, amongst even the most critical of amateur astronomers *worldwide*.

Anyway, George agreed to travel to the BAA meeting of Wednesday November 24th and, after a lot of irrelevant (to George) announcements about BAA telescopes, Comet Austin 1982g and some waffle about noctilucent clouds, Patrick finally completed the task he had started 17 months earlier! Patrick stood up (as always, he bellowed, never using a microphone) and said that he now had a very pleasant function to perform. He said that the Astronomical Society of the Pacific had recently (!!!) awarded to one of the Association's members, Mr George Alcock, their International Amateur Achievement Award for his distinguished work in the field of comets and novae. He then invited George to come forward and accept the award. "I've waited a long time for this" said George, looking Patrick squarely in the eye, as Patrick made sure he grabbed George's hand in a firm handshake for the cameras. Years later, George said "I could have had a real moan at him there and then, but I was so glad to just snatch that plaque off him. I thought I'd just be a good boy after I'd got it".

George told the audience that he was delighted to eventually receive the award and that he had, that morning, made his 1,079th comet sweep since his last comet discovery. Patrick said that this showed that George was still active and he wished him all success for the future. For Patrick's second BAA meeting as the President he then, once again, invited his old mate Howard Miles to be the main speaker: that month's talk was on *The Meteoritic Craters of Australia*.

As 1982 drew to a close Patrick was delighted to hear that the British Pipesmokers' Council were installing him as 'Pipesmoker of the Year' for 1983. The previous year's holder had been the disc jockey Dave Lee Travis and the next year's recipient would be the boxer Henry Cooper. The award ran from 1964 to 2003, after which it was banned due to regulations preventing all promotion of tobacco! The very last recipient would be the actor and writer Stephen Fry. On January 28th 1983 Patrick

Fig. 28.1 Patrick, at long last, presents George Alcock with the International Amateur Achievement award of the Astronomical Society of the Pacific on Wednesday Nov 24th 1982 at 23 Savile Row, London. On the extreme *left* of the picture Storm Dunlop is seen applauding the award (Photograph by kind permission of Robin Scagell)

posed with a special pipe, supposedly shaped like a telescope, but looking more like a fountain pen attached to a fish eye lens, to amuse the press photographers after he had officially received the title. On accepting the award Patrick told his fellow pipe smokers that: "I regard two classes of people as being beyond the pale: weight-watchers and those who have just given up smoking."

The Torricelli B Event

On January 29th 1983 the BAA Lunar Section was involved in one of the most celebrated TLP events in its history. This author remembers the events well: I should do, I was one of the main witnesses, although to this day I am still not clear what we all saw. An alert was issued by the TLP co-ordinator Peter Foley, who had been making routine measurements of the brightness of the small crater Censorinus when he became aware, at 8.35 p.m., that 'lying to the southwest was an unbelievably bright point'. Peter Foley originally misidentified the crater as Censorinus B: it was, in fact, Torricelli B. Peter reported that the crater seemed to have 'an incredible halo around the inner rim, intensely coloured, brilliant electric-blue'. He quickly phoned around other members of the TLP alert network. The Moon was 1 day past Full and at some 30 degrees altitude, in the East, when I trained my 36 cm Cassegrain telescope on the region, at 9.25 p.m., following Peter Foley's call. I had no idea where the crater Censorinus B was, except that it must be close to Censorinus, but I had various maps and charts available. However, no charts were necessary, because as soon as I had trained the telescope on Censorinus the source of the TLP alert was obvious. A tiny crater in that region appeared to be 'illuminated' by an almost ultraviolet light. I described it as 'most unlunar and eyecatching'. The nearest I could come to describing it was that the colour was like that seen in ultraviolet sterilizing cupboards in the medical world. The deep violet colour seemed to extend outside the rim at times. From my maps the crater was labelled as B, but it was hard to agree that it was Censorinus B. I thought it must be Torricelli B and, indeed, it was.

Interestingly, not everyone saw this appearance. My observation was almost identical to that of Peter Foley and we both saw starlike flashes near the region. Marie and Jeremy Cook, two other TLP sub-section observers, also saw the deep purple coloration. However, Patrick himself, who did not start observing until 10.20 p.m., with his alt-azimuth 31 cm reflector, could see no anomaly at all. Neither could Geoff Amery. However, Patrick was well known to have poor sensitivity at the violet end of the visual spectrum. Over the 10 years that I spent as a member of the TLP sub-section, this single event was the most dramatic one that I witnessed. Indeed, I would say it was the ONLY one, out of dozens of alerts, which made me think, perhaps, that TLP were real. Undoubtedly the crater was ridiculously bright and electric-blue that night, especially when one considers its size. However, in subsequent years I have looked at the region at the same lunar phase and detected that electric-blue colouration every time, but never quite to the same intensity. I am led to conclude that this event was caused by a specific critical illumination angle on that night, which, for a period of an hour or so, made the crater look very odd indeed, especially to someone with a blue bias to their vision. The flashes may well simply have been a result of straining so hard, at a neck-cricking angle, to see more detail. In retrospect, I should have attempted photography, despite the freezing weather, but when there was cloud about and you were responding to an alert, the priority was always to see the thing first and phone the co-ordinator back with a Yes or No. In addition, this event was in a tiny crater. I seriously doubted whether photography would have revealed the colour of the tiny, but intense, violet glow: it would probably have been drowned out. This, was, of course, well before the era of high resolution digital imaging by amateurs.

Despite Patrick's failure to witness anything odd about Torricelli B on January 29th 1983 he was most excited about it and enthused greatly about the event at the February 12th Lunar Section meeting in London, congratulating me on my observations. At that same meeting I was approached by Geoff Amery (Lunar Section director) and John Pedler on another matter. John was stepping down as the Lunar Section's photographic co-ordinator and the Lunar Section committee wanted me to take his place. I was greatly flattered! I would be invited to all the Lunar Section committee meetings and I would be serving on the same committee as Patrick, my lifelong hero! I was delighted. For the next 8 years I submitted a voluntary monthly article about lunar photography to the Lunar Section Circular. At the March 30th BAA meeting, Patrick, as President, introduced himself as a speaker and wasted no time enthusing about the events of January 29th.

On April 8th, after a gap of quite a few years, Patrick attended the annual BAA Winchester weekend course. He gave the opening lecture on the Friday evening entitled 'Naked Eye Astronomy'. Patrick had been what he himself would have described as "A stalwart, a backbone and a mainstay" (had he been describing anyone else in the BAA) of the Winchester weekends and it was a great pleasure for everyone to see him at Winchester in 1983.

Immediately after Winchester, Patrick set off on a 5 week trip down south as a guest of the Royal Astronomical Society of New Zealand (RASNZ). His aim, as BAA President, was to increase collaboration between UK and New Zealand astronomers, especially between the BAA and RASNZ. Patrick visited various observatories on the north and south islands, spending some time at the southern Mount John Observatory, which was equipped with two 600 mm reflectors. This 5 week trip meant he would have to miss the April BAA meeting (an unusual occurrence for an incumbent President). Some time ago the comet discoverer Rodney Austin kindly sent me an account of Patrick's visit to New Zealand and I have reproduced his account here:

> I vividly remember the first time Patrick came to New Zealand in 1983. He did a complete lecture tour through the country. On the evening he stayed in New Plymouth I met him at the meeting when he arrived, with the New Plymouth Observatory Director, in a Ford Capri. I still can't believe how he actually got into the car! Luckily John Walmsley is a fairly small guy. After the meeting I had the pleasure of taking Patrick out to the then Tikorangi Observatory in my little Vauxhall Chevette Station Wagon. On the way home we had to divert as the main road was blocked by a heavy transporter carrying a component of the Motunui Synthetic Fuel plant which was being built near the observatory (Patrick was most impressed by the action), so we headed around through Inglewood. It started to rain, and the rain got heavier – and heavier, until the road was under about 5 cm of water. Patrick was stunned and told me he had never seen rain so heavy. It is relatively common in the area, but I have to say it is still the heaviest I've ever actually driven in. A couple of weeks later I attended the RASNZ Annual Conference in Wellington where Patrick was the principal guest. I got to show him Comet IRAS-Araki-Alcock through one of the Carter Observatory telescopes.

Alcock Bags Comet No. 5, and a Farce

On the evening of May 3rd 1983, the living legend George Alcock had just finished putting his crippled wife Mary to bed, at their house in Farcet, Peterborough, when he stopped by the landing window at the top of the stairs. As always, George had his 15 × 80 mm binoculars to hand. George had not discovered a comet since his fourth one in 1965. Because of local light pollution, and his unique memorising of the Milky Way star fields, George had concentrated on novae since the late 1960s, bagging, as we have seen, nova discoveries in 1967, 1968, 1970 and 1976. As George had entered his seventies he had taken to searching for novae from indoors, as the house had recently been fitted with high quality double glazing, which allowed binocular viewing. Wandering around the various rooms of number 55 Broadway, in that small village of Farcet, armed with binoculars, was as natural to George as watching TV was to most people! As George raised his binoculars that night, crouching on the landing to get a good view over the top of the next door roof, he spotted a large fuzzy patch almost instantly. It was a sixth magnitude comet, (technically, a naked-eye object from a very dark site) and he had spotted it within seconds! George called out to his wife in the bedroom that he had bagged a comet. She took some convincing! Even by George's standards this was an extraordinary discovery, through a double-glazed window, but then Mike Candy's Boxing Day comet, 23 years earlier, was made in similar circumstances.

George knew, at a glance, where the comet was in the sky and immediately phoned Guy Hurst of *The Astronomer* magazine who was able to confirm the discovery. As it turned out, the comet had been imaged by the Infra-Red Astronomy Satellite (IRAS), although the IRAS team had not been sure what it was. In addition, a Japanese amateur astronomer, Araki, had made an independent discovery only hours before George. The comet was officially named Iras-Araki-Alcock, but it was George's extraordinary bagging of his fifth comet that would always be remembered. Iras-Araki-Alcock proved to be an extraordinary comet, passing within 5 million kilometres of the Earth and growing to over a degree in size. Its motion across the sky, at closest approach, was too rapid for many 1980s amateurs' telescope drives to cope with!

The Sky at Night, being typically recorded a day before transmission, was usually able to alert the general public to bright discoveries, but not this time. The early May programme, on the quickest Pulsar, had been recorded before Patrick set off for New Zealand. Such was the comet's rapid progress across the sky that it was gone from northern skies by the time of the late May programme (on the subject of the South Dakota Homestake mine) and all Patrick could do was apologise and show a photograph of the comet taken by Ron Arbour. A much larger embarrassment was to follow though. The BAA paper circulars service, usually a bit on the slow side, had totally failed the membership in the case of IRAS-Araki-Alcock. The Circulars editor had been away during early May, Patrick had only just returned from New Zealand, and a BAA circular (693) about the comet was not issued for 2 weeks! By this time the comet had gone south and was invisible from the UK. The only people who saw the comet from Britain were people who had been alerted to the news by telephone, were dedicated comet observers, or who subscribed to *The Astronomer* magazine alert service, run by Guy Hurst. Their paper circular was issued the day after the discovery. There would be no such thing as E-Mail for several years to come. One of the best comets for many years, co-discovered by an Englishman, and the closest comet to the Earth since 1770, had been missed by most BAA members, due to a farcical BAA cock-up during Patrick's Presidency! The fact that *The Sky at Night* had failed to announce the comet too was a double-blow for him.

The date of May 28th saw Patrick presiding over the BAA Exhibition meeting at the Hawkstone Hall in Lambeth, southeast London. He apologised profusely to those present for the problem with the belated comet circular and said that he had taken steps to ensure such a disaster never, ever, happened again! The BAA Comet Section Director, Michael Hendrie, showed pictures of IRAS-Araki-Alcock at the meeting, taken by himself. Also, a young up and coming astronomer, Rob McNaught, showed an image he had taken with a fish-eye lens, of the comet. McNaught would go on to become the world's most successful comet discoverer. Contrasting with the disaster of the circulars, Patrick did have a very pleasant duty to perform at that meeting. He awarded the association's Lydia Brown Medal to the recently retired BAA Treasurer (1973–1982), Cyril Belchem.

On June 2nd 1983 the hawk-eyed planetary observer Paul Doherty was staying at Patrick's home in Selsey. Skies were clear that night and so Patrick and Paul went out to observe Saturn (Paul's favourite object) with Patrick's 15 inch (38 centimetres) reflector. Although Saturn was at minus 8 degrees declination and only transiting at an altitude of 30 degrees, they had an exquisite view of the ringed planet at around 21:30 GMT: the atmosphere was exceptionally stable. Paul made detailed sketches at the eyepiece and, over the coming days, produced one of his finest paintings of the ringed planet. He made a spare copy for Patrick who, from that moment on, displayed the painting in his study as a memory of that night. Patrick was still very conscious of Paul's disappointment at his edge-on ring sketches not being published in the Journal and so he suggested to Colin Ronan, the BAA Journal Editor, that the painting should be used as a future Journal cover. Patrick was still unhappy about the way Paul and the Saturn Section report had been treated and wanted Paul to rejoin the BAA as soon as possible. Colin Ronan placed the painting on the Journal cover one year later, much to Patrick's delight.

An Eclipse, a Full Diary and a TLP Critic

On the 6th of June 1983 Patrick set off on a trip to see his fifth total solar eclipse. Because of his mother's increasing frailty throughout the 1970s, and various other commitments, it had now been 10 years since Patrick had witnessed his last totality, from the deck of the good ship *Monte Umbe*, moored off Africa. The path of the June 11th eclipse began in the south-western part of the Indian Ocean, crossed south-eastern Indonesia and the southern part of New Guinea and ended among the islands of the Republic of Vanatu (the New Hebrides) in the Pacific Ocean. The maximum duration of the total phase was 5 minutes 11 seconds which made this a very desirable eclipse to see. The optimum place from which to view the eclipse was thought to be Java, from a 'probability of clear skies' viewpoint and also from a tourism aspect. Various stories abound about Patrick disappearing with some Jakarta 'Ladies of the Night' on that trip, and only reappearing at the hotel the next morning. Patrick's only comment was "I was simply teaching them star recognition!" I think perhaps the less we say about that sordid tale the better! Despite almost total cloud cover on the day of the eclipse the clouds did break (possibly due to various prayers to statues of Buddha made at that time) and the eclipse was seen by all on the *Explorers Tours* trip. Patrick took a number of slides, even though his camera jammed in the 38 degrees Centigrade heat and 99 % humidity conditions. He had now seen his fifth total solar eclipse: a very healthy five out of five success ratio.

At the June 29th 1983 BAA meeting Patrick invited Heather Couper, now well known to all amateur astronomers for her TV appearances, to address the meeting.

Heather briefly announced that she was quitting her job as the lecturer at the Greenwich Planetarium to work freelance on writing and broadcasting. She also gave details of a new television series and a new radio series, both dealing with astronomy, in which she was deeply involved. Patrick wished her every success in her new roles, but it must have been obvious to him that here was, potentially, his only serious 'TV Astronomer' rival. Heather had, of course, been spring-boarded to TV fame by appearing on Patrick's *Sky at Night* programme and she was now, because of Patrick's "I'll do it if you will" challenge, the next President in waiting, because BAA Presidents always recommend their successors. Shortly after that June meeting Patrick set off for California to do some filming himself for the October *Sky at Night* on the subject of the Mount Wilson Observatory.

By the end of July Patrick was back in Sussex and steadily ploughing through his endless calendar of TV, radio, newspaper, lecturing, BAA, charity and writing commitments. On the 27th he officially launched the tiny (6 feet 9 inches long) yacht 'Big C', owned by the yachtsman Tom McNally, in which Tom hoped to sail the Atlantic! Two days later, on the 29th, Patrick shook hands with a robot called Denby after he performed the opening ceremony at the 'Living with Robots' exhibition, held at the British Engineerium in Hove.

During Patrick's presidency he was keen to hold as many so-called 'Out-of-Town' or 'Provincial' meetings as possible, up and down the country. Patrick was well aware of his popularity and the fact that he was a huge crowd puller wherever he went. Normally, BAA meetings outside London were sparsely attended (typically 40–70 people would turn up) simply because BAA members were usually reluctant to travel more than 20 or 30 miles to a meeting. As the overwhelming bulk of BAA members lived in the southeast, and because it was so easy to get into central London, the BAA meetings in the capital typically attracted well over 100 members in those pre-Internet days (sometimes 200 in the 1970s and 1980s). However, if Patrick was attending and holding an evening public lecture you could always expect over 100 people to turn up, wherever you went in the UK. Patrick was well aware of his pulling power and so, as a previous BAA President Howard Miles had done, many years earlier, Patrick was keen to promote these Provincial Meetings and to get council members out of London to meet the people.

The BAA Council organised a joint meeting with Loughton Astronomical Society on August 4th 1983. Loughton was in Essex and so not that far from London, but it was one of the country's biggest local astronomical societies. Patrick was accompanied to this out-of-town meeting by his protégé Dr John Mason, the BAA comet section director Michael Hendrie, the assistant secretary Eddie Watson-Jones and Rossie Atwell. A very healthy 165 people turned up at the meeting. Patrick seemed to spend most of the time at Loughton signing copies of his books and even scraps of paper and, not surprisingly, he gave a rousing talk on Transient Lunar Phenomena. John Mason, again unsurprisingly, gave a talk on meteors and fireballs and Michael Hendrie spoke about comets, including the recent IRAS-Araki-Alcock fly-past and the returning comet Halley. As always at these events Patrick tried to encourage observers to join the BAA and this led to a classic Patrick moment. Patrick pointed out that at 18 pounds a BAA membership was superb value for money and there was no reason NOT to join. This brought a comment from a man in the audience who complained that he had a family, a mortgage and a car and just could not find the spare cash to pay the BAA subscription. Patrick bellowed back an instant reply "You should get your priorities right!" Needless to say, the audience fell about with prolonged laughter and it was some time before order was restored! A superb report of the meeting was written up by the Loughton Society's Steve Ringwood, well known for his humorous writing.

Fig. 28.2 Patrick and his Ford Cortina Estate in 1983, photographed by Brian Moody. The Cortina, despite its condition, was his main method of transportation at that time!

August 1983 also saw a new BAA innovation highly promoted by Patrick. Ever since the IRAS-Araki-Alcock BAA Circular fiasco Patrick had been aware of criticism regarding the rapid communication of astronomical events and news. The BAA Journal had a typical lead time of 2 months and the BAA Circulars, while quicker, mainly dealt with the brightest new discoveries. Something else was needed and so Patrick championed the creation of the 'BAA Newsletter'. This was a small insert in the main BAA Journal designed to be a friendly, chatty, less formal medium with which to communicate imminent news and events to members. Remember, in 1983, even basic e-mail was some 4 years in the future. The BAA Newsletter was very popular for the next 15 years, until the Internet rendered it redundant.

For the weekend of September 16th to 18th the BAA arranged their main autumn meeting at the University of Exeter. Once again, Patrick agreed to give a public lecture on the Saturday, with all profits going to the BAA. The weekend included a guided tour of the Norman Lockyer Observatory at Sidmouth, Exeter Cathedral, Exeter Maritime Museum and Plymouth Planetarium.

Patrick's non-stop schedule, representing the BAA at as many astronomical functions as possible, continued on October 7th, when he formally opened an exhibition commemorating the bi-centenary of the birth of Dr John Lee (1783–1866) of Hartwell, Buckinghamshire, a noted local astronomer of the region. The exhibition was held at the County Museum in Aylesbury. As if all this BAA travelling was not enough, Patrick was occasionally asked to take part in BBC Open University programmes and, around this time, the BBC decided that it would be good if he could demonstrate relative planetary motion, by piloting a barge on the Winchfield canal between Basingstoke and Farnborough! Fortunately Patrick survived this escapade without falling into the canal, because, as he had never learned to swim, it could have resulted in the demise of an incumbent BAA President by drowning. Yet even with all these bizarre type of commitments, none of which he ever seemed to turn down, it was a rare week during his Presidency that Patrick was not, at some point, on BBC TV or on the radio. During 1983 Patrick seemed to be the most regular guest on the BBC quiz show *Blankety-Blank*, on which he had appeared since the show's creation in 1979. This was the last series before Terry Wogan handed over to Les Dawson as the series host. Even in the final months of the year Patrick was on the programme three times, twice in October, and also on the special Christmas Day show.

On October 23rd 1983, the Lunar Section's TLP co-ordinator Peter Foley was observing the Moon with his 30 cm Newtonian when he noted, at 6.50 p.m., a large brilliant extension to the usual bright point on the east wall of the lunar crater Aristarchus. Foley issued a TLP alert and Patrick was the first person that he was able to reach. Patrick had a couple of visitors at Farthings that night, namely, the author and lecturer Iain Nicolson and Lawrence Clark, who had illustrated some of Patrick's books. Patrick's big reflectors were unable to reach the Moon due to trees being in the way. However, he was able to train his 5 inch Cooke refractor on the Moon and immediately noted that the crater was 'brilliant' and 'unmistakeably abnormal'. He reported that Iain saw a very bright starlike point on the eastern edge of the crater. Some other observers also reported a similar view that night. Around 8 p.m. I managed to join in the alert and trained my 36 cm Cassegrain on Aristarchus. My own observing log recorded 'Spurious colour a total mess around Aristarchus, nothing abnormal seen'. However, despite the turbulent atmospheric seeing on that night I did take some photographs of Aristarchus which Peter Foley said showed the abnormal appearance. However, I was far from convinced, despite Patrick and Peter's views.

While we are recalling the controversial subject of TLP during the closing months of 1983 it is worth including a relevant disagreement here between Patrick and the man most amateur astronomers regarded as the most experienced lunar observer in the U.K., namely Harold Hill (1920–2005). The diligent reader may recall that some 12 years earlier, due to Patrick's editorial impatience excluding the essential last minute revisions to the late Bill Baxter's solar book, a rift had developed between Harold Hill (Baxter's friend) and Patrick. The two men had not been on speaking terms since 1971. This rift would last for the remainder of Harold Hill's life. However, this rift did not prevent Harold Hill having a dig at TLP in the September 1983 issue of 'The Astronomer' magazine (page 71) following a paper and letter on the subject from Denis Buczynski, which had appeared in the previous issue. Hill was a TLP sceptic and no-one could claim to have more experience in lunar observation, not even Patrick. Amongst other remarks in the September magazine, Hill stated: 'Having observed the Moon for more years than I care to remember in pursuance of varied programmes it may seem strange that not once during that time have I unquestionably recorded a TLP – even onwards from 1958 when the Russian Kozyrev announced an eruption in the formation Alphonsus – an event which seemed to trigger off the TLP vogue in earnest – and was to be followed by an absolute plethora of such sightings'. These were pretty damning comments from such an experienced observer and he sent them to Patrick out of 'courtesy'. However, Patrick had a somewhat infuriating standard method of dealing with such scepticism, namely by pretending any serious criticism was not damning at all, but minor. He would often start any counter attack by simply saying that he thought the criticism was absolutely valid and he welcomed it, but then he would follow this up with a paragraph that showed just the opposite! Patrick's reply eventually appeared in the November 1983 edition of 'The Astronomer', on page 106, and it read as follows:

> Mr Hill has been kind enough to send me a copy of his notes. He is, of course, absolutely correct. Speaking as one who has been involved in TLP work from the very beginning (in the days when all such reports were treated with official scepticism!) I am quite sure that many of the reports are spurious. The observation of these elusive phenomena is extremely difficult and needs not only experience but great caution. In the catalogues which I have published, I have given very little weight to any report which has not been confirmed by at least two observers at different observing stations. I must also say that, in my view, some of the lists issued have not been nearly critical enough. I well remember a splendid TLP that I observed in Alphonsus some time ago. Unfortunately, there were similar glows in at least four other craters! This sort of thing can be checked, but it is not easy to weed out all the spurious reports. But, despite all this, I am quite sure that the remaining reports are genuine. Before the Apollo landings, Barbara Middlehurst and I collaborated in a paper linking TLP areas with moonquake sites, and we turned out to be correct. Also we have one good spectrum (Kozyrev's) and we are hoping to obtain more, which is really the only way to put any particular TLP beyond doubt. The BAA Lunar Section has a good group of experienced observers and we are doing our best to make reports as careful and as objective as we can. Meanwhile, I hope that everyone will heed Mr Hill's warning!
>
> Patrick Moore, Farthings, West Street, Selsey, Sussex.

Meetings, Meetings and More Meetings

The 1983 BAA AGM, held on October 26th, saw Patrick celebrating his first year as President and, as is the President's duty, delivering the annual report of the Association and his own 'Presidential Address'. For his lecture Patrick chose the topic of 'The Mapping of Mars.' In his first year as the BAA President, Patrick had not made any radical changes. However, he had increasingly involved the local societies affiliated to the BAA into the association's activities, and introduced additional observational material into the Journal. Also, it was Patrick who had introduced the BAA Saturday meetings, rather than the traditional and difficult to attend 'last Wednesday in the month' meetings. As a start, experimental November, February and April Saturday meetings were introduced. Patrick had also proposed that far more introductory 'review' articles should be published in the BAA Journal, specifically, easily readable summaries on all categories of object in the night sky and how they could be observed. The council agreed that such articles would be very welcome, but as no-one seemed able to muster the energy to write them, Patrick wrote nearly all of the planetary articles on his own.

At that 1983 AGM, Mr Leslie White and Mr Harold Ridley proposed the vote of thanks to Patrick for his Presidential summary and his lecture on the subject of Mars. Following the AGM a short 'Ordinary Meeting' was held in which Patrick introduced the main speaker as Dr David Whitehouse, who would later become the BBC Science Editor and, in the late 1990s, the BBC News Online web editor. Dr Whitehouse spoke on the subject of the next Space Shuttle 'Spacelab' mission. Needless to say, Patrick's reign as President was endorsed by the Council and he continued, as had all Presidents before him, since 1890, into his second year of office.

November 12th 1983 was a big day during Patrick's term of office. It was the first 'extra' BAA Saturday meeting. Patrick was not a President for half measures and he also did not want the traditional 'last Wednesday of the month' BAA meetings to evaporate. Since the end of the war Patrick had made every effort to attend all of the BAA meetings and they had always been held, from October to May every year, on the last Wednesday in each month, except December. A special Christmas meeting was usually held, as soon as possible after January 1st, in later years. Patrick knew that with members' family commitments Wednesdays were very difficult times for meetings, even when they started in the early evening. However, scrapping Wednesdays for Saturdays had never been an option for him. He wanted additional, not alternative, Saturday meetings. For November 1983, Council approved two meetings: one on Saturday November 12th and one on Wednesday November 30th. Both meetings were a success, even though most of the attendees appeared to be the same people, namely dedicated BAA members and fans of Patrick living in the London area.

December 3rd saw yet another Patrick-inspired meeting at a big, affiliated, local society; there really was no holding back Patrick's enthusiasm for spreading the astronomical word far and wide and promoting the BAA during his term as President. This time the venue was Dundee: specifically, the Mills Observatory, long associated with the Lunar Section. Once again, Patrick gave a talk on TLP and a public lecture (in the evening) entitled "Into Space". Although the attendance was only 70, less than half that of the Loughton meeting, the attendees were enthusiastic. Also representing the BAA at this event were Ron Livesey and Dave Gavine of the BAA Aurora Section and Neil Bone of the Meteor Section.

On December 20th another keen project of Patrick's, the first 'BAA sponsored' Public Lecture at the London Planetarium, attracted 150 people; they enjoyed a talk about Pulsars from Professor F. Graham Smith, the Astronomer Royal for England.

The first BAA meeting of 1984, on January 7th, saw Patrick introducing the Astronomer Royal for Scotland, Malcolm Longair, who addressed the audience on the subject of 'The most distant galaxies'. Severe weather at the end of January forced a rapid rescheduling of the main speaker at the next meeting but it went ahead nonetheless.

February 11th saw the second of Patrick's extra Saturday meetings to be held in London and Patrick had introduced another innovation: it was a joint meeting between the JAS, that's the Junior Astronomical Society (known as the Society for Popular Astronomy or SPA these days) and the BAA. Only Patrick could have clinched such a meeting because there had always been an SPA desire to remain separate from the BAA, ever since they formed, as a BAA spin-off, in the 1950s. As Howard Miles and Henry Hatfield, who often gave a brief resumé of events in the night sky, were not able to make the Saturday meeting, Patrick gave the monthly prospects himself, before introducing the main speakers.

On February 14th Patrick was out observing the Moon with his 15 inch reflector when he became aware that there was no detail on the floor of the lunar crater Plato. As all lunar observers will be aware, the crater Plato has a number of tiny craterlets on its floor, most of which are well under 2 miles in diameter. Thus, in poor seeing, these craterlets simply vanish. Patrick was one of the most experienced lunar observers in the world, but always seemed to regard the disappearance of the craterlets as 'mysterious', suspecting that there was some obscuring medium at work. This was his suspicion on February 14th and he also reported that the east wall of Plato was 'badly defined, hazy and obscure' with the northeast wall appearing dim. Patrick also reported that there was 'something odd' about the lunar mountain Pico which was 'very bright and gave the impression of a crater'. He phoned Peter Foley who issued a section telephone alert. As with almost all of these TLP alerts, the situation was badly confused by the Earth's turbulent atmosphere and although others saw similar sights to Patrick (or were influenced by what Patrick had seen) there was no clear-cut verdict. Patrick described his observations at the February 29th meeting and added that it was a long time since the BAA had met on a leap day and he had been informed that this would not recur until 2012. Henry Hatfield reported there had been considerable activity on the Sun in late February and wondered whether this could be linked to TLP activity.

On March 22nd 1984 Patrick's former friend Frank Hyde, briefly the editor of the BAA Journal, died. As the reader of this book will recall, Hyde was a keen amateur radio astronomer and appeared on early *Sky at Night* programmes. Hyde, like Patrick's friend Michael Bentine, had an interest in parascience but when he tried to covertly award a BAA printing contract to a company he operated from a Clacton caravan, with his mistress, his days on the BAA Council were numbered. Patrick had (in public) disassociated himself from Hyde for many years, but on Hyde's death he handed the BAA editor Colin Ronan a very brief obituary. This mentioned nothing of Hyde's argument with the Council but merely described him as someone 'whose talks older members would remember' along with his 'pleasant personality.'

The next BAA London meeting, on March 28th, saw the poorest attendance of any meeting during Patrick's tenure; but this was no fault of Patrick's. Only 64 members turned up due to a strike affecting all London Transport services! However, Patrick was delighted to announce that the council had voted in favour of awarding two of his closest friends, Howard Miles and Paul Doherty, the association's Goodacre and Steavenson awards respectively. Patrick had pushed hard for both these awards and he very much wanted to present the medals himself. This was especially true in the case of the Steavenson award as Patrick was still annoyed that Paul's Saturn sketches, from the 1980 apparition, had been forced out of the BAA Saturn Section report. Patrick also knew that in the month following the medal presentation to Paul, namely in June, Paul's magnificent Saturn painting, made while using Patrick's telescope, would be appearing on the Journal cover.

Patrick repeated the news of Paul and Howard's medals to a much larger audience during the third of his extra Saturday meetings on April 14th. He emphasised that he was looking forward to presenting the medals at the Exhibition meeting. Once again, with Howard Miles and Henry Hatfield absent, Patrick started the April 14th meeting with an enthusiastic look around the April night sky. This meeting was an important one for Patrick as Peter Foley also gave a major talk on the subject of TLP, Patrick's number one obsession.

Also in April 1984, Patrick was privileged to be allowed to take part in a commissioning run with the Isaac Newton Telescope at the Los Muchachos Observatory on La Palma in the Canary Islands. The test target was M57, the ring nebula in Lyra, and the resulting 'first light' run of the newly re-sited telescope (it had formerly been based at Herstmonceux in Sussex) was shown on the June *Sky at Night*.

At the April 25th meeting of the Association Patrick announced that George Alcock had donated £250 to the BAA. This was a generous donation indeed as George was not a wealthy man and had often experienced much envy from certain leading BAA members, not to mention the fiasco surrounding Patrick's delay in presenting him with his ASP plaque! The amount, in pounds, was the same as the amount in dollars he had received with the ASP award.

In the spring of 1984, Patrick was entering the final 6 months of his Presidency and it was time for the BAA Council to endorse the outgoing President's choice of successor, namely Heather Couper. Council were very happy to do this. Heather was fast becoming a TV astronomer in her own right and with Halley's comet returning during her Presidency, the association needed someone who looked enthusiastic and extrovert on TV and was a good populariser and presenter of astronomy. Heather fitted the bill. She would, at 35, be the second youngest ever BAA President (second only to Dr W.H. Steavenson) and the first woman President in the Association's 94 year history. Her name officially went down as the Presidential candidate on the 1984 ballot paper.

I well remember the BAA Exhibition meeting of Saturday May 19th 1984. Patrick was in excellent form and even more enthusiastic than usual. This was a day he had long awaited. He could, at last, present two BAA medals which he had championed, to his good friends, Howard Miles and Paul Doherty. As with most Exhibition meetings the main bulk of the day was informal: members simply drifted around the exhibits, chatting to friends, old and new. The Exhibition meetings were held, throughout most of the 1980s, in the dinghy 'Hawkstone Hall' in Lambeth, which was a grotty venue, but cheap and easy to get to. The short 'formal' part of the meeting was invariably held in the Hawkstone Hall basement, which always seemed to be unbearably hot and humid and devoid of sufficient air to breathe, at every Exhibition Meeting! Patrick opened the formal part of the meeting by saying that one of the more pleasant of the Presidential duties was the presentation of the

Association's medals and awards, all of which were most thoroughly well earned. Patrick explained that the Association's highest honour, the Goodacre Medal and Gift, was this year to be awarded to Howard Miles who, Patrick positively bellowed the words: "Is a Stalwart, a Backbone and a Mainstay of the Association". Patrick explained that Howard had been a past President of the BAA, Director of the Artificial Satellites Section and "One of the great pioneers of artificial satellite observing since the early days of the 1950s".

Howard accepted the award modestly and recounted some amusing tales regarding letters he had received from BAA members over the years. One especially bizarre letter went as follows: "I am interested in satellite observing. Please send me details of the sky at night as seen from Camaza, Gateway to the West, with the rise and fall of the tide, Spitfire, seagulls, white cliffs of Dover, scintillation of starlight radiation"! The audience fell about at this and Patrick joined in. Patrick bellowed out his story as follows, along with facial expressions that cannot be adequately conveyed on a printed page: "I remember getting a letter from a member who said he had followed my advice and ground and ground a glass blank to make a telescope mirror. He said that he now found he had his radius of curvature *inside* the focal length. What should I do next?" Patrick wrote back: "Put it on the mantelpiece and use it as a vase". Patrick pulled an extraordinary grimace at this point, his eyebrows joining in the performance. It could surely have won him the World Gurning Championship outright! By this time the BAA audience were writhing with laughter. With Howard's award now dealt with, Patrick moved on to the second presentation. Patrick said that the Steavenson award for observers had been granted this year to Paul Doherty. On handing Paul the award, Patrick praised Paul's artistry and said that when he saw his drawings of Saturn or Jupiter it made him positively green with envy!

Although there was no BAA meeting on May 30th, there was the usual council meeting and, by a fortunate coincidence, just after the council meeting, there was a partial solar eclipse in the UK, visible from London. Even more fortunate was that the Sun was visible! Patrick had arranged for two telescopes to be made available from Dudley Fuller, the proprietor of Fullerscopes (later Broadhurst Clarkson & Fuller) in London's Farringdon Road. These were a 21 cm reflector and an impressive 12.5 cm refractor, both equatorially mounted. BAA and JAS member Robin Scagell brought along his 15 cm reflector as well, and the telescopes were used to project the Sun's disk onto card as the eclipse progressed. The press and TV had all been invited along by the association's public relations man, Nigel Henbest (the co-habiting writing partner of Heather Couper). In addition, Heather had announced the gathering on the BBC radio programme *Woman's Hour* on the previous day and Patrick had also mentioned the BAA's presence in Green Park on Radio 4. Heather commented live on London's LBC radio from Green Park, as well as on TV. The BBC weather forecaster Jack Scott presented that day's weather forecast from near the BAA gathering too. A substantial crowd of over 500 people gathered to see what was going on and to meet the one-and-only Patrick. It was all good publicity for the BAA.

Three days later, on Saturday June 2nd, Patrick set off on a long trek north to Lancashire and a provincial BAA Lunar Section meeting, organised by the well-known Lancastrian butcher and astronomer Denis Buczynski. As a Lunar Section committee member, like Patrick, I attended that meeting myself, heading north from Suffolk in my Ford Cortina, which, sadly, had no air-conditioning. It was a sweltering weekend! The meeting was an excellent one, but only 50 people attended: all but four were locals of the Lancaster region. The four exceptions were Patrick, Paul Doherty, me, and a keen amateur with northern roots, Andrew Elliott.

The venue was the Headway Hotel, Marine Road, Morecambe. Denis had assembled an array of speakers well-known for their lunar expertise. These included a local professional geologist Dr Lionel Wilson, a local visual observer Peter Wade, Patrick himself, and an up and coming expert on the history of astronomy, Dr Allan Chapman, now a living legend in astronomical public speaking. That 1984 meeting marked the first time I had seen Allan Chapman and the first time I had come across anyone who could match Patrick's speaking abilities. Allan, then and now, could deliver a 40 minute talk, without pausing or hesitating, or showing any slides, and without boring the audience. Allan always shows his slides at the end, as if to prove that he does not need any visual prompts whatsoever. In addition, he speaks and dresses like a Victorian Gentleman, complete with gold pocket watch on a chain! Patrick was the final speaker and his talk was about the lunar crater Thebit, a crater which, to Patrick, epitomised the view that the lunar craters were formed by volcanic activity, because it obeyed the rule that smaller craters broke into the walls of larger craters and so the big ones must have formed first. [Readers with good memories will recall that I mentioned all this much earlier in the book.] You rarely see larger craters with complete rims, cutting smaller craters in half, therefore, in Patrick's world, the smaller craters formed last as volcanic activity subsided. In fact, this reasoning was simplistic and totally flawed and, even in 1984, it was fast becoming an obsolete and naïve view. Nevertheless, Patrick was always amongst fans and disciples at Lunar Section meetings and few would publicly disagree with him!

I still have an audio tape of that meeting which brings back good memories of Patrick in times gone by. As he took his turn to speak he was keen to say how good the previous talks had been, but that now things were going downhill because it was his turn! A few excerpts from my recording may help to illustrate Patrick's speaking style, when at his peak. He had a certain set of stock phrases which never failed to amuse fellow amateurs and, even now, bring back vivid memories, to myself and others, of his popular public talks. They were pitched at just the right level, in that they praised Patrick's fellow amateurs, but mocked his own efforts, and challenged the professional view that lunar craters were of impact origin. Indeed, they made Patrick's fellow amateurs feel like they were unsung heroes who were the only people who really understood the Moon. Why? Because amateurs actually looked through telescopes at the Moon and professionals did not, as they mainly sat at desks shuffling stacks of paper, applying for research grants and spouting hot air; that was Patrick's main thrust in his lunar talks, time and time again.

Patrick started:

"Well, Ladies and Gentlemen, we've just had some fine talks and they say after the Lord Mayor's show comes the dustcart. I will do my best. I am slightly outrageous at times and I'm quite aware that in some respects my views are not quite the same as those of many other people."

Patrick started by showing a sketch of the Thebit region of the Moon sketched by Leslie Ball. “One of our great lunar artists” Patrick explained. “I’m not an artist; I am, in fact, the most inartistic person in three continents.” He continued with one of his favourite quips: “Thebit is near the region on the Moon called the Straight Wall and it’s called that because it is NOT straight and it is NOT a wall”. This always got a good laugh!

“The wall of the crater Thebit is broken into by the smaller crater Thebit A and that in turn has its wall broken by an even smaller crater, Thebit B”. Patrick continued: “On the Moon, it’s always smaller into larger, smaller into larger. Back in about 1955 or 1956 we counted how many cases there were of smaller craters breaking into larger craters’ walls: it was 99.9999 % of all cases. Of course, in those days south was always at the top and east was east and west was west. Then we had the battle for IAU Commission 16 in which I led the case for the defence, but failed. So, after that everything had north at the top and east was labelled west and west labelled east. I always still mark my slides with south at the top, except for the one I’m showing here, which I got the wrong way up”. This was another standard Patrick laugh raiser!

“When we look at craters like Thebit we see the wall of the bigger crater which is broken into by the smaller crater, is perfect right up to the small crater, but if craters were caused by impact, the larger crater walls would be brought down for miles around. Then, of course, we have crater chains. It would take some very educated meteors to fall like that. Back in the 1940s, Baldwin claimed that, by chance, there would be about seven crater central peaks which had, themselves, been hit by meteors, just by chance. But, when I went to Meudon and studied the craters with central peaks with the 33 inch refractor, I counted seventy-four craters on central peaks. THE BOMB CAN’T LAND IN THE BUCKET EVERY TIME! I also discovered the rille at the bottom of the Alpine valley from Meudon and people said I was crazy, but there it is, taken from space.”

Then Patrick moved on to the dust theory of Thomas Gold. “In the 1950s, Thomas Gold thought that with all this alleged meteoritic impacting on the Moon, the surface would be covered with a deep covering of fine dust. It caused quite a flap, because people thought a spacecraft might sink without trace. But we now know there is no risk. It just goes to prove that astronomers might be brilliant in the fields of mathematics and astrophysics, but when it comes to the Moon they really should go outside and look at it through a telescope.” This always got a loud murmur of approval from a BAA Lunar Section audience and the Lancashire audience that day was no different.

“Then there is the grid system”, Patrick continued, “Mark all the craters more than 50 miles in diameter on the Moon and you see that the ones facing Earth fall in chains running north to south down the Moon. Professionals say it’s a trick of the eye, but it’s not. Turn a chart of the Moon on its side and you still see those crater chains running from left to right.” Patrick then moved on to cover the far side of the Moon and one of his favourite memories which always brought a big laugh. “When the first pictures from the Moon’s Far Side first came in, in 1959, I was doing a live *Sky at Night*. I heard the producer’s voice in my ear ‘Pictures of the Moon’s Far Side coming up in 1 minute. Scrap what we planned. Do it off the cuff. Good Luck’.” He also threw in, for good measure, the comment that he had met the Russian father of space travel Tsiolkovski (where and when he never stated) and that the man had made "Hitler and Stalin look distinctly left wing", plus the fact that he had once shown a slide of part of the Moon to a lunar conference and then admitted it was a photo he had taken of boiling coffee! Patrick could never resist including the alleged discovery by him and Wilkins of the Moon's Mare Orientale, or *Eastern Sea,* with the quip "thanks to the IAU, our eastern Sea is now on the west of the Moon." Patrick rounded his entertaining talk off with a slide of what looked a bit like a crater with a broken wall, explaining that it was a sand crater built by himself and Peter Cattermole on the beach at Bognor!

It was a classic Patrick lunar talk and many there would have heard the jokes before, but Patrick, of course, got the loudest applause of the day! During that meeting I was invited to join Patrick and Dr Lionel Wilson at the meal table. This was the first time (but certainly not the last) that I had ever eaten a meal with Patrick and witnessed his eating habits. Patrick never ate any greens or vegetables of any kind on any occasion that I shared a meal with him! Mind you, in later life, to disprove people's theories about his eating habits, he regularly produced a 'seafood salad', made with local fish from the Selsey beach fish merchant 'Honest Bill'; but, whenever I ate with him, his plate was always piled high with mountains of meat and with potatoes: at least a dozen large potatoes! He always went up for a second helping and then always said something along the lines of "Despite my size and despite my bulk, I have the appetite of an anorexic sparrow. I only ever have one big meal this size a day". I said nothing, but politely raised my eyebrows. I could have mentioned that the 'one meal' I had just seen him eat was more than most people could eat over 2 days, let alone one…but I kept quiet! Patrick also absolutely hated brown bread, which he admitted was an "obstinate childhood grudge" adding "I will simply not have brown bread anywhere near my plate or near my palate".

As with so many meetings that Patrick attended, the June 1984 meeting in Lancashire was a success, largely because Patrick knew how to get amateurs on his side, if they weren't already converted. On June 21st Patrick travelled even further north, to officially open the newly re-furbished Mills Observatory at Dundee. The next item on Patrick's BAA Presidential agenda was the last main BAA London meeting before the summer break: that of June 27th. The main speaker at that meeting was no less a person than Dr Brian Marsden (1937–2010) of the Minor Planet Center in the USA. Marsden's BAA affiliations went right back to the 1950s when he had been a major player in the Computing Section. Maths had always been a weakness for Patrick, but he had been good friends with all those in the BAA Computing Section since the 1930s, when he was just a boy. In particular he had got on famously with the Section's Directors: Major A.E. Levin, Guy Porter, Cameron Dinwoodie and Gordon Taylor.

Despite the BAA summer break, it did not stop Patrick from attending another of his championed BAA affiliated society meetings in Liverpool on July 7th. Once again he gave a rousing lecture in the afternoon entitled simply "Mars". Also around that time, while in the north of England, he performed another duty, not normally associated with BAA Presidents: he was chosen by Chester Zoo to be the celebrity who opened their brand new Penguin pool!

Patrick's penultimate meeting as the BAA President was the September 14th to 16th Out-of-London BAA gathering held at the University of Durham. Professor Arnold Wolfendale officially welcomed Patrick and the Association to Durham and then handed over to Patrick who chaired the Saturday meeting. As with all of the BAA's provincial meetings a tour of local places of interest was held on the Sunday. A visit to the parish of Tow Law was arranged in this case, as the local vicar, the Reverend Thomas Espin (vicar from 1888 to 1934), had been a noted observer in this area. The current incumbent, the Reverend Andrew Driver, had arranged an exhibit to celebrate the 50 years since Espin's death and Patrick thanked him for welcoming the Association.

October 13th 1984 marked the 100th anniversary of Greenwich Mean Time and the Greenwich meridian being adopted as the International Time and longitude reference standards. Of course, being so patriotic and anti-foreigner, as well as being convinced that "The English are the best", this was a celebration that Patrick was keen to be a part of, especially as he was still the BAA President. He filmed a special *Sky at Night* about the anniversary and, despite hating water, he even waded through a Greenwich pond which lay along the meridian!

On October 31st 1984 Patrick chaired his final BAA Council meeting and Annual General Meeting as President. Like the 46 Presidents before him, stretching back to 1890, Patrick had survived the 2-year term of office and was now ready to hand the chairing of the association to his chosen successor; but before that, he had to deliver the council report to the AGM audience and present his second and final Presidential Address. Patrick had been a President full of energy and with no time for bureaucracy and waffle. As we have already seen, his real strength had been in his promoting of additional Saturday meetings and other meetings flung far and wide across the UK. He had been especially keen in going out to meetings of the larger BAA-affiliated societies. During Patrick's term as President a new observing section, under the Directorship of Andy Hollis, had been formed: the Minor Planets Section was devoted to the study of the asteroids. In addition, Patrick had continued to encourage and write many 'Review' articles for the BAA Journal: concise summaries, aimed at the beginner, covering different planets and other targets in the night sky.

Following Patrick's 'State of the BAA' summary he gave his second Presidential Address entitled "The Mapping of Venus". At the end of Patrick's lecture there was prolonged applause from the audience. Patrick's predecessor as President, Cdr Derek Howse, proposed a vote of thanks to Patrick and he said that the association was not only fortunate in having Patrick as its President, but also for all his past work. Patrick had been the director of the Mercury & Venus Section and the Lunar Section and, prior to becoming President, had served for 30 years as the BAA meetings recorder! Cdr Howse said that Patrick had needed a great deal of persuading to become President, but he had at last been convinced after it had been pointed out how absurd it was that he should not stand, after having served the BAA so well over so many decades. Mr Leslie White, from the audience, seconded the vote of thanks and commented on Patrick's extraordinary enthusiasm and energy. Mr White said that some years ago he had been invited to the opening of a school planetarium and on the programme the school organisers had written: 'We had hoped for the planetarium to be opened by Patrick Moore, but as he was unable to attend we have had to make do with the Astronomer Royal!'

In his final official act as President, Patrick then asked the retiring Business Secretary, Neville Goodman, to read the result of the annual BAA council ballot. Patrick then thanked Mr Goodman for all his work and presented him with a gift: a pair of binoculars from the association. Patrick then said that it was almost exactly 50 years since he had joined the association and so it was a suitable occasion to step down in favour of someone younger. He then invited Heather Couper to take her place in the President's chair, on the Savile Row stage, to prolonged applause from the audience. Heather said that she was aware of the great honour it was to be elected as BAA President and she hoped that if any members thought that their opinion was not being heard they should not be afraid to approach her, or any member of the council. Heather then adjourned the AGM, opened the short 'Ordinary Meeting' and Patrick's magnificent term as President became part of BAA history, legend and folklore.

Just 10 days later Patrick had another unique and memorable experience in his life: he was riding on the back of a giant cardboard cut-out, while waving at the London crowds! This was part of the London Lord Mayor's show, held on November 10th. Because of the Greenwich Mean Time centenary the National Maritime Museum had been allocated a float within the annual Lord Mayor's Show. A lorry trailer had been decorated as a Greenwich Observatory Park building profile, complete with Onion Dome. Patrick, alongside astronomers and staff associated with the Museum and with Greenwich, manned the float. Everyone, including Patrick, wore white T-shirts with a big stripe down the middle and either side of the stripe the words 'West' and 'East' were printed. Patrick's prop was a cardboard refractor spyglass, 3 feet long, which he wielded dramatically as the float trundled through the streets of London.

Chapter 29
Fifty Years in the BAA and Halley

To be perfectly accurate, Patrick had been elected as a member of the BAA on November 28th 1934 and, not surprisingly, the BAA Council were keen to celebrate his golden anniversary in November 1984. He had been 11 when he joined and, at 61, he still had the childlike enthusiasm for astronomy that most lose when work and family commitments swamp their hobby. By a fortunate coincidence the BAA's last Wednesday in the month November meeting would fall on the 28th in 1984, so a precise celebration of Patrick's 50 years of membership would be possible. Those councillors who knew Patrick well checked that he would be attending the late November meeting (he rarely missed BAA meetings) but they managed to keep news of the celebration a secret.

On that day the new President Heather Couper invited Patrick's friend, and the BAA Journal editor, Colin Ronan, to address the meeting. Colin was a similar age to Patrick, had known him since the War Years, and was a noted author on the history of astronomy. He described Patrick as "Probably the most famous amateur astronomer in the world" and said that he was probably best known for *The Sky at Night* television programme, which would soon have been running for 30 years. Colin said that Patrick's enthusiasm had stimulated many people to take an interest in astronomy, many of whom were now highly respected professionals. Patrick had brought this enthusiasm into all he did for the association, being first the director of the Mercury and Venus Section and also the director of the Lunar Section and, for the last 2 years, the President. Colin explained that Patrick's main interests were planetary and that in his lunar work he had managed to get professional astronomers interested in TLP: no mean achievement. He went on to say that he thought Patrick was an indefatigable observer and, like Sir William Herschel, he ignored the cold. On one occasion, whilst observing with Patrick at East Grinstead, Colin remembered that there was snow on the ground and, despite wearing an overcoat, he (Colin) was unbearably cold and his feet were freezing. However, Patrick, standing beside him, was wearing his normal blazer and just a scarf as acknowledgement of the weather! The audience were most amused by this story and burst into spontaneous applause as Patrick ascended the Savile Row stage at Colin's request.

M. Mobberley, *It Came From Outer Space Wearing an RAF Blazer!: A Fan's Biography of Sir Patrick Moore*, DOI 10.1007/978-3-319-00609-3_29,

Patrick said that he had absolutely no idea that a tribute was planned. He also added that he was delighted that *The Sky at Night* had inspired others to achieve great things and, to much applause, Heather Couper then presented Patrick with a card and a decanter engraved with the BAA's logo. Two weeks later, at the BAA's second public lecture at the London Planetarium, Patrick gave a rousing talk entitled 'Exploring the Planets' to a packed house.

Mainly Lunar Stuff

As 1984 came to a close there were great hopes within the BAA Lunar Section that a major step forward in TLP research was about to be made. Patrick was a person who could open doors in the astronomy world and his influence had managed to open a couple of significant 'observatory doors' as his Presidency came to a close. During the 1970s there had been high hopes that the 60 cm reflector at the University of Keele could be used to take spectrographs of the lunar surface during TLP alerts. If gas was seeping out from the lunar crust it might be possible, with a large enough instrument, to obtain a spectrum which identified the gas. Sadly, Patrick's friend Dr Ron Maddison, along with the team at Keele, had never quite perfected being able to take lunar spectrographs (it can hardly have been a priority for the University). A decade had drifted by without any progress on this front. In addition, it was felt that with small craters like Torricelli B a larger aperture was required just to scrutinise the region.

The new doors that were opened in this respect were in the form of access to two large telescopes: the 28 inch refractor at Greenwich and the 30 inch Coudé reflector, plus spectrograph, at Herstmonceux. Professor Stuart Malin had granted access to the 28 inch refractor and Professor Alex Boksenberg had, after a specific request from Patrick, made available the 30 inch reflector. The initial plan was that the Greenwich refractor would be used by myself and by Peter Foley and that the Herstmonceux reflector would be used by another lunar section member, Gerald North, who was then based only a few miles from the observatory, at Bexhill. January 31st 1985 saw the first attempt by Peter Foley and I to use the 28 inch refractor at Greenwich for TLP region photography, but we were thwarted by cloud and nothing was achieved. We were assisted that night by Carole Stott from the observatory and by my Lunar Section colleague Nick James. Three days earlier Gerald North had experienced his first trial run in obtaining spectra of the lunar surface, from Herstmonceux. Patrick, Peter Foley, Gerald North and I all had high hopes that the use of these giant telescopes would bear much TLP fruit, but they never did, although Gerald certainly became an expert at obtaining lunar spectra in the coming months and years. The simple fact is that a telescope in your own back garden is 100 times more useful than a large, unfamiliar, 'brute' that you have to spend an hour of travelling to get to. By the time you arrive, it has clouded over.

I feel I must relate a bizarre personal story at this point, which serves to illustrate how the whole subject of TLP was right on the edge of amateur astronomy and possibly nearer to pseudo-science like UFOs, the Loch Ness monster and the Yeti in the 1980s. In early 1985 I drove to the TLP co-ordinator Peter Foley's house 'Tree Trunks' at Nettlestead Green, near Maidstone in Kent. The plan was to wait there to see what the weather was doing and, if it was favourable, to dash up to Greenwich to use the 28 inch refractor to study the TLP 'hotspots' with a much larger aperture. If Patrick was the Father of TLP and preoccupied with them, Peter was the Godfather! He was totally obsessed with them and seemed to think there was some lunar 'activity' every night. I suspect he was such a big fan of Patrick's that he just never wanted to disappoint him. Anyway, I finally got to Peter's house, after having to change a punctured wheel en route, but it turned out to be cloudy and so I stayed up with Peter late into the night talking about TLP, lunar photography and Patrick.

Peter Foley had a manner with those younger than him that you simply dared not cross him. If you tried proving him wrong on scientific facts he suddenly became deaf in one ear or changed the subject! Foley's enigmatic persona was very similar (to my mind at least) to that played years later by the so-called 'Cancer Man' or 'Cigarette Smoking Man' in the X-Files and as there are many similarities between TLPs and UFOs perhaps this is not too surprising! Like that X-Files character Foley was always smoking (usually a cigar); he always spoke in a deep voice and he was always economical with words, many of which were unfathomable. Ask him a searching question about what he was implying was happening in the lunar crater in question and, in a cloud of cigar smoke, he would growl out words like "bow-shock wave" or "piezoelectric discharge" and then become deaf to further questions. Ultimately this attitude would be his downfall. He was a hard working TLP co-ordinator, Lunar Section Circulars editor and, ultimately, BAA Lunar Section Director, but he just would not listen to the views of younger observers. Peter, like Patrick, could exaggerate and tell tall tales, so I do not know who is most to blame for the following yarns. As Peter seemed to believe everything Patrick said, maybe the first yarn was all Patrick's doing; I know not. As the night wore on, and we did the subject of TLP to death, Peter made me promise that I would not repeat what he was about to tell me; I was intrigued!

Fig. 29.1 Patrick with Peter Foley, in the mid 1980s, at a Lunar Section meeting in the National Maritime Museum at Greenwich. Jeremy Cook is just visible in the background on the *left* hand side of the picture (Photograph by the author)

First, Peter related the story that quite a few years earlier (during the late 1970s) Patrick had been contacted by an acquaintance in the Soviet Union who worked in the Soviet space program. Patrick was invited, James Bond style, to meet this Russian in Green Park in London, on a bench at an allotted time. The meeting took place, and a newspaper concealing a thick folder of photographs and notes was passed to Patrick. On Patrick's return to Selsey he opened the folder. Inside were a number of high quality photographs, taken in 1968 allegedly, by a cosmonaut in lunar orbit (on a Zond probe); in other words, a manned Soviet lunar orbit mission before the historic flight of the NASA astronauts around the Moon in Apollo 8. Now, I had heard that tortoises and other small animals had been flown around the Moon in Zond probes during 1968, but not a man. Peter Foley told me that the pictures taken by the cosmonauts on the mission were of such a high quality that there was no doubt the camera had been only a few miles above the lunar surface and angled by the hand of a man. He also said that he had the best Plato pictures in his office at Nettlestead Green, where we were talking. I said I would very much like to see them. After much dramatic reluctance (faked or real, I do not know) Foley went and got the pictures of Plato. Yes, they were very high resolution pictures of Plato's surface indeed but, frankly, I had seen many just as good from later Apollo missions and I simply could not agree with Peter that they had to be taken by a man. They could easily have been high resolution Soviet unmanned images.

Rather disappointed by my scepticism, and still insisting the 'Zond' pictures were genuine, Peter Foley then said that he was going to show me some pictures of the most dramatic TLP to date, photographed by him. Of course, if I disagreed with *this* claim I would be calling him a liar to his face! In a highly dramatic manner Foley opened a draw in his office desk and pulled out a buff envelope marked 'Top Secret'. He then showed me a sequence of his photographs (taken with his 30 cm Newtonian) of the large lunar crater Ptolemaeus. On the floor of Ptolemaeus lies a smaller crater Ptolemaeus B. Over a period of a couple of hours (according to Foley) the floor of Ptolemaeus B had filled with a sharp black circle, increasing until it filled the crater, and then it slowly receded. The black feature looked most un-lunar and, frankly, I thought it was simply a fake. However, I was a guest in Foley's house and did not want to sleep on the lawn that night, so I simply expressed my view that it was "remarkable". However I did query, when the entire BAA TLP sub-section was trying to prove these things and get photographic and spectroscopic proof, why we simply did not stop and push this sequence of pictures to the forefront. Foley claimed that both he and Patrick had tried and Patrick had pushed NASA to analyse the pictures. He claimed that every test had been carried out on the pictures and NASA were sure it was something happening on the Moon but that they had become very uncooperative and now refused to discuss it further. The whole subject was "a hot potato" and "Top Secret" and "Someone high up knows what is going on, but if we push this one they will make it look like it's a fake, so we will put this one back in the drawer, and not mention another word about it Martin"!

I mentioned these pictures to Patrick many years later and he claimed not to know anything about them and then rapidly changed the subject. To this day I still do not know what to make of it all, but I am tempted to think that both Patrick and

Peter Foley had a strong desire to spin an entertaining yarn when they craved attention! Peter Foley had a big falling out with me and others in the Lunar Section 6 years later, largely due to him simply never listening to the views of younger people. By profession he was a senior accountant for an electronics company called Burndept, based in Dartford, Kent. He died in 2008. I ended up writing Peter Foley's obituary for the BAA Journal, but left the bizarre story above out of it, as it would have embarrassed Patrick, and the Editor would have censored the crazy stuff anyway!

On March 2nd 1985 the Lunar Section held another London meeting at the Hawkstone Hall in Lambeth in southeast London and Patrick was the main speaker. Patrick chose to speak about the libration zones of the Moon, that is, those regions near the edge of the disc that can only be observed well under favourable circumstances. Patrick used some excellent slides by Cdr. Henry Hatfield to illustrate his talk, but he did emphasise that, ever since 1959, when the first probe photographed the far side of the Moon, trying to map the libration regions from the Earth had become a redundant activity, but there was still work worth doing at the South Pole, where space probes had not flown past. Following Patrick's talk, I gave a brief demonstration of some software I had written for a Commodore 64 home computer, to tell me whether the Moon would be clear of obstructions on my observatory horizon. Before these Lunar Section meetings there was, nearly always, a committee meeting at which the Lunar Section Director (Geoff Amery) and the co-ordinators would discuss developments. I found out at these committee meetings just how paranoid Patrick was about other authors copying his ideas. Essentially, if there were any other books about amateur astronomy that had just been published, they were "Just copies of my books!" There were a few exceptions to this view though, namely if the books had been encouraged by Patrick, or where they were written by his very closest friends.

I remember at one committee meeting I happened to have brought along a very good new book by James Muirden (*How to use an Astronomical Telescope*) and I had placed it next to me on the committee room table. Patrick looked horrified and asked if he could see it. The book had plenty of tables of data in it, regarding the planets, their moons, future eclipses and other information. Patrick barked: "All these tables, they're just straight copies out of my books". I could not agree with Patrick there as I had most of his books and, anyway, the tables of data were common knowledge and could have come from NASA or any number of sources. "No", I said rather nervously, "These tables are compiled from data he could have got anywhere, they are just standard facts and figures about the solar system". "No Martin" said Patrick firmly. "Like everyone else, he just copies everything straight out of my books, in fact this whole book is just word-for-word my *Guide to the Planets*". Well, I owned Patrick's *Guide to the Planets* and it was totally different, but I decided to keep the peace and said nothing more!!

During the mid-1980s Patrick seemed to appear on BBC TV or BBC Radio every few days. His quick brain, sense of humour, and his remarkable memory for amusing stories, seemed to make him a natural guest on science, quiz and game shows. His voice was so clear and distinctive on the radio that he was unmistakeable for anyone else. In the 1970s (as I mentioned earlier in the book) Patrick had regularly appeared on a BBC radio programme called 'Fair Deal' alongside magician David Nixon, cartoonist Willie Rushton and comedienne June Whitfield. Nixon had died in 1978 and that programme title had not been resurrected. However, in the 1980s the format was recreated with another magician in charge of 'dealing the cards', namely Paul Daniels. The new show was revived with the title 'Dealing with Daniels'. It revolved around choosing a question card, based on standard playing cards, but not choosing the same card as another player had already chosen; it may sound complicated, but needless to say it always descended into hilarity with so many entertainers involved. Despite a regular supply of other entertaining panellists on the show, such as John Junkin, Anneka Rice, Duggie Brown, Sarah Greene, Barry Cryer, Lorraine Chase and Matthew Kelly, Patrick became the most regular panellist, as well as being the only survivor from the original 'Fair Deal' series. It was an amusing and witty show and Patrick could be just as entertaining as the professional comedians alongside him; unlike them however he could, just as easily, switch to serious science broadcasting in an instant.

On May 12th 1985 Patrick broadcast a fascinating programme on Radio 4 about the remarkable achievements of Geoffrey Perry, Derek Slater and their team at Kettering Grammar School who had been monitoring Russian space missions, from their radio transmissions, since 1960. Two weeks later, on Monday May 27th 1985 (a bank holiday in the UK) Patrick was live on BBC Radio 1, as part of the Radio 1 'Road Show' alongside D.J.'s Steve Wright and Janice Long. The Road Show was being performed at Herstmonceux Castle, home of the Royal Greenwich observatory. Patrick entertained the thousand-strong crowd with a xylophone rendering of 'The Entertainer' to cries of "We want Patrick!"

Three days later, on May 30th, Patrick was, as usual, at the eyepiece of his 15 inch reflector at Selsey. Paul Doherty was with him and they were looking at the Moon (of course!) At 20.10 GMT Patrick noticed what he described as an 'anomalous brightening' on the west wall of the crater Aristarchus, which was just coming into sunlight. By 20.36 he thought the north wall was 'definitely pinkish' and 'there was a certain lack of local definition at the northwest wall (5 o'clock)'. The effect was a little less pronounced at 20.44 when he described the colour as 'rusty' and 'the west wall not so anomalously bright as it had been.' By 21.03 GMT Patrick thought the rustiness on the crater's north wall had shifted to the northeast wall (7 o'clock). At an early stage Patrick issued an alert to Peter Foley who phoned other section members. Gerald North, Marie Cook, Howard Miles, Peter Madej and Peter Foley all reported a similar appearance. Most significant was that spurious colour elsewhere, due to the Earth's atmosphere, was NOT reported. This was a very important event as many experienced observers simultaneously reported Aristarchus as looking odd. In the June 1985 Journal Patrick's entertaining paper relating to 'crank' theories of how the lunar craters might be formed, was published. Of course, there were many who thought Patrick's own volcanic and TLP theories should have been included in the paper!

At the start of August 1985, Patrick, Heather Couper and Nigel Henbest embarked on a joint trip to Colombia at the invitation of Kevin Marshall, President of the

Colombian Astronomical Association. As we have already seen Patrick had attended the first Colombian 'National Encounter' in Astronomy back in October 1981. But now he was back, with the BAA President and incoming BAA Editor accompanying him. Neither Patrick or Heather or Nigel spoke a word of Spanish, so every word they spoke had to be translated sentence by sentence via interpreters! The Colombian Association now boasted 200 members, but according to Patrick the country was permanently buried under cloud and he had never seen a single star on either of his visits there!

Around this time Patrick completed the eighth volume of his *Sky at Night* books. This new volume covered the period from 1980 to 1984 and, as I mentioned earlier in this book, Patrick commissioned Paul Doherty to paint the cover: an artist's impression of a black hole that might be at the centre of our galaxy. This was the subject of the *Sky at Night* Patrick had presented with Heather Couper in May 1980.

Halley and a Broken Shoulder

During 1985 much of Patrick's TV and lecture time was devoted to the imminent return of Halley's Comet to the inner solar system. Needless to say the media was tempted to hype the event up, but in reality it would be a very poor return of the comet as it was never coming anywhere near the Earth. Despite this, it was an exciting time for amateur astronomers as you only get one chance to see a comet that returns every 76 years or so.

A friend of Patrick's called Brian Harpur formed the Halley's Comet Society. Apart from arranging one charity function this society served no other function at all, except to meet on licensed premises. This suited Patrick perfectly; no-one ever drank Patrick under the table!

Patrick had collaborated with Heather Couper to produce "The Halley's Comet Pop-up Book" and earlier, in 1984, had teamed up with John Mason on a more scientific book entitled simply "Halley's Comet". There was added excitement because the European Space Agency was sending a probe to pass very close to Halley and, hopefully, return the first ever pictures of a comet's nucleus. In the UK a National Astronomy Week was arranged for November 9th to 16th to celebrate the return of the comet, at a point when it would be easily visible in small amateur telescopes. Although a few amateurs in the USA and Japan had photographed Halley at the end of 1984 and early in 1985, there was quite a competition in the UK to become the first to photograph and see the comet when it emerged from the dawn twilight in August 1985. From an astrophotography viewpoint the situation was confused by the fact that the comet was just brightening to the verge of 'UK photographability' in early 1985 but then passed behind the Sun and would be unobservable in a dark sky until August, when it should easily be within the range of amateur UK photographers. So, who would be the first to photograph it had little to do with who could go the deepest and a lot to do with who would make the effort to get out of bed in the pre-dawn August skies and be lucky with a lack of clouds! As it turned out the UK astro-photographers Alan Young and Ron Arbour both obtained indisputable images on the morning of August 15th and they made a brief joint appearance on the BBC's Breakfast Time programme, along with Heather Couper, who was rapidly becoming established as a younger rival to Patrick as Britain's 'TV Astronomer'; her Channel 4 series 'The Planets' was broadcast that same year. I might add that the infamous David Icke was the sports presenter on that breakfast programme at the time. He would later claim, according to the press, to be the Son of God, then the Son of the 'Godhead' and the man who realised that the Earth was already controlled by shape-shifting lizards from the constellation of Draco!

On the night after Young and Arbour's success, according to Patrick, namely on August 16th, he and John Mason managed to photograph Halley with Patrick's 15 inch reflector. I obtained a photograph on the 25th August and on the 18th September and, confusingly, my photograph of September 18th was deemed to be the official 'third UK photograph' by the BAA Comet Section, as it showed a 'definite image at the right place'. However, by September, the comet was also becoming a real possibility for visual observers using large amateur telescopes. Once more, Patrick and John Mason, using Patrick's 15 inch Newtonian, were keen to be the first to grab Halley; but with their eyes this time. They reported seeing the comet, at around magnitude 13, on September 11th and Patrick was sure they were the first UK observers. However, the comet observer Jonathan Shanklin had already logged a possible sighting himself, with a 13 inch instrument, as early as August 15th, and he claimed a definite sighting on August 27th. So, as it turned out, Patrick was certainly one of the first UK observers to witness Halley's return, but not the very first. Of course, with a photograph you have cast iron proof, but with a visual observation you always have to trust the word of the observer.

Patrick knew the pioneering astro-photographer Ron Arbour very well and following Ron and Alan Young's first UK photographs of Halley, Patrick filmed a memorable *Sky at Night* programme at Ron's Hampshire Observatory, in September 1985. In fact, the main thrust of that *Sky at Night* had been Ron's completion of his 16 inch computerised Supernova Patrol telescope, a telescope that was 10 years ahead of its time, both from its automation capability and its remarkably accurate friction roller drive. Ironically, the 16 inch never did discover any supernovae photographically, but Ron did, using mainly a commercial Schmidt-Cassegrain telescope and various CCD cameras. It would be 21 years later that Ron's revised 16 inch reflector would bag its very first supernova with a CCD camera.

I cannot help mentioning a personal event at this point. On October 5th 1985 I managed to couple an experimental CCD camera to my 14 inch Cassegrain telescope and video-taped the lunar surface at high resolution. When I sent the tape to the TLP co-ordinator Peter Foley he was amazed at the detail that was visible on the recording. Patrick was keen to see the tape too, as this meant that the Lunar Section now had a way of video-taping (in Black & White) events on the lunar surface. Patrick's reaction was one of great excitement when he viewed the tape: "Really must have Martin on the *Sky at Night*" he told a friend. Well, I did eventually appear on the programme……13 years later!

Sadly, the autumn of 1985 would mark a turning point in Patrick's opinion of Heather Couper and Nigel Henbest. This was mainly due to an article that appeared in the *Mail on Sunday* 'YOU' colour supplement, written by Angela Levin, entitled "At Home With The Stargazers", which not only appalled Patrick, but shocked many of the older BAA Council members too, many of whom had distinctly Victorian codes of conduct. The interview, with President Heather and partner Nigel, at their then home in Greenwich, made them look eccentric as a minimum, and publicity seeking too. However, what really made many senior BAA figures question whether Heather should be the President at all was the comment, at the bottom of page 35, that "It's much more fun to have sexual involvement outside the relationship" and a picture of her on page 36, in the bath, holding an inflatable Space Shuttle. This interview, and Heather's choice of the next BAA President, would, Patrick told me, ensure that she would never appear on *The Sky at Night* again. However, Patrick was very wary of anyone who might look like they were capable of stealing his total dominance as Britain's favourite TV astronomer. It was a role he loved and would protect to the death, especially after presenting *The Sky at Night* for 28 years. In reality, his paranoia was unjustified, as he was so well loved by the British public that no-one could ever replace him. He was also, unlike all his rivals, a real back-garden observer.

With comet Halley brightening rapidly during the closing months of 1985, and many amateur astronomers training their telescopes on the returning celestial visitor, Patrick was frequently talking about it, on *The Sky at Night* and on many other programmes too. He even managed to mention it on the BBC *Children in Need* fundraising programme, broadcast on November 22nd on BBC1. Patrick had appeared on the earliest editions of this annual event in 1980 and 1981 and would also be involved in the 1990 edition. Standing next to Terry Wogan on the 1985 show he acted as a link to introduce the arrival of *Dr Who's Tardis*, along with a total of four *Dr Who* actors and numerous assistants of the famous time traveller. The *Dr Who* series would have been running for 22 years on BBC1 on the following day, only 6½ years less than Patrick's *Sky at Night* programme. With the subject being Time and Space, Terry Wogan introduced Patrick as "The extremely spacious and timeless Patrick Moore", at which point Patrick looked him straight in the eye, hands thumped on hips and, with monocle firmly in place, queried Terry's use of the word spacious. As Terry phrased it: "I know, you can turn on me like a wounded Rhinoceros!" Terry raised the issue of observing Halley's comet (which he dared admit to Patrick he was less than excited about) and Patrick said that he hoped to see it the next time around too, even though "I will be 139 years old." Patrick added that he thought that many of the *Children in Need* they were helping that night would actually see the comet on its next visit. At this point Patrick spotted *Dr Who's Tardis* materialising in the studio and, when he pushed the door of the police box open, out came another Patrick, the second *Dr Who*, Patrick Troughton. He was followed by numerous assistants and UNIT personnel, along with the third Doctor Jon Pertwee, the fifth Doctor, Peter Davison, and the sixth and current Doctor, Colin Baker. Amongst the 16 non-Doctors who emerged in front of Patrick and Terry Wogan were the series' much loved UNIT character Brigadier Lethbridge-Stewart (Nicholas Courtney), Captain Yates (Richard Franklin), Sergeant Benton (John Levene), Sarah Jane Smith (Elisabeth Sladen), Liz Shaw (Caroline John) and Perpugilliam Brown (Nicola Bryant). The fourth Doctor, Tom Baker, was notably absent. Cheques for *Children in Need* (donated from a *Dr Who* Exhibition and the *Dr Who* appreciation society) were presented by Colin Baker and Patrick Troughton, amid much heckling from Jon Pertwee, at which point Terry Wogan left Patrick's side and ran off into the Tardis! Sadly, Patrick Troughton would die just 16 months later, while in Georgia, USA, where he was a star at a Science Fiction convention.

Throughout the late autumn and winter months of 1985 Patrick was keen to get as many glimpses of Comet Halley as he could. The comet brightened rapidly throughout November and was an easy object in any amateur telescope, although it could never be classed as a naked eye comet like Hale-Bopp, which would come along 12 years later, in 1997. Sadly, the weather in the UK was not kind, especially during December. Because the comet was heading south and would be at its best in the southern hemisphere in March 1986, this meant that the last week of December 1985 and the first of January 1986 were the last possible opportunities to see Halley well from the UK, before its peak. Patrick was determined not to miss a single observing opportunity, whatever the weather: if there was a gap in the clouds he would go for it! Unfortunately, disaster struck for Patrick on the night of December 29th/30th. He had been out observing Halley and other objects, in freezing conditions, and had also been attempting to take some photographs. The rain that had fallen earlier in the day had frozen and, as Patrick straightened up after leaving his observatory by its cramped door at around 2 a.m., and walked on the short path back to his study door, he slipped on the ice, fell badly, and broke his left shoulder. When interviewed by the Press he said that what really annoyed him about the injury was that he would probably not be able to play cricket for the Selsey Team or for *The Lords Taverners* that summer.

Voyager 2, Challenger and Giotto

On January 24th 1986 the NASA spaceprobe Voyager 2 gave astronomers their first close-up view of the planet Uranus and its faint rings and Moons. Patrick, despite his shoulder injury, was at the Jet Propulsion Laboratory in Pasadena, California as the spacecraft went past the planet and he later said that the tension and excitement were greater than for either the Jupiter or the Saturn encounters. Why? Well, Jupiter and Saturn could be studied well from the Earth, but Uranus was simply so far away that it was virtually featureless; astronomers had never seen it this close up. Patrick had travelled to Pasadena with Garry Hunt but, as if his recent shoulder injury was not enough of a pain, he caught a virus on the way out and completely lost his voice. Patrick, without his voice! Surely, a show stopper? Actually, no! The producer, Pieter Morpurgo, filmed a virtually silent Patrick mouthing his commentary like a goldfish, along with Garry Hunt's actual commentary and, on return to London,

Patrick dubbed his (still croaky) voice over the top. Another *Sky at Night* had been salvaged! In fact, I was told by Pieter in 2008 that a similar trick was used in 1987, when Patrick caught a chill on the top of Mauna Kea in Hawaii and had to go down to a lower altitude. In that case they filmed astronomers answering questions on the mountain and then edited in Patrick asking questions to thin air some days later!

However, 4 days after the Uranus flyby, Patrick, fellow astronomers and the civilised world, were shocked when the Space Shuttle Challenger exploded shortly after take off on January 28th 1986. Ice damaging the rubber solid rocket motor seals had been the culprit. Not only was this an appalling human tragedy, but there had been plans for the Shuttle crew to observe Halley during the mission. Shuttle missions were grounded indefinitely. For once, Patrick was totally unaware of a crucial moment in the history of space travel when questioned by the press. It had occurred while he had been en route back to Heathrow airport from the USA and the first he knew of the tragedy was when a reporter spotted him emerging in the arrivals hall after the long flight home. At the same time as the Challenger crew had, to quote Ronald Regan, "Slipped the surly bonds of Earth and touched the face of God", Patrick had also been airborne in US Airspace.

As Comet Halley passed its closest point to the Sun, on February 9th, excitement was mounting for the flyby of the European Giotto probe on the evening of March 13th The BBC arranged a joint *Horizon/Sky at Night* programme to cover the live encounter, but it ended up being nothing more than a farce, as Patrick had confidently predicted. He would have vastly preferred the event to have been covered by *The Sky at Night* and a few selected guests, but the BBC *Sky at Night* budget did not, sadly, stretch that far. Patrick arrived at Darmstadt in Germany a week before the encounter but he was not the man in charge this time: the BBC *Horizon* people were in control. Patrick's hand-picked guests, for his brief appearances on air, were the professional comet experts Fred Whipple, Jan Oort and Roald Sagdeev. The *Horizon* team of 'experts' were a bizarre mixed bag of scientist-celebrities, none of whom seemed to know much about comets! Patrick's old Apollo Moon-landing anchorman colleague, James Burke, was called in to co-ordinate the studio guest debate, but he was not a comet expert either and so was totally out of his depth. In fact, the only people who were speaking any sense at all were Patrick and his three guests (in brief flashes) and a teenage amateur astronomer called Lee Sprotes, who had appeared on *The Sky at Night* in the previous year.

A few hours earlier things had seemed relatively under control. That evening's episode of the BBC Science programme *Tomorrows World* had mentioned the Giotto encounter and the producer had obviously seen sense and so included a live link to Patrick, at Darmstadt, within the show. The *Tomorrows World* presenter Judith Hann was the anchor for the link and announced "We've got Patrick Moore waiting in the studio in Germany and I know that he's hoping soon to bring us those live pictures, or later on television tonight. So, Patrick, how close are we to getting live pictures?" As always with Patrick he was off the starting blocks barely before Judith Hann had stopped speaking. "Well, we are getting pictures alright and this is a picture of the kind that Giotto is actually sending back now……" Patrick continued, without pausing, at 250 words per minute for the next 90 seconds, explaining precisely what the situation was, with no more prompting required from Judith Hann. The *Tomorrows World* part had, at least, gone smoothly.

However, a few hours later, as the Giotto probe closed in on Halley's nucleus, it became increasingly obvious that the *Horizon* programme controllers had given zero consideration to what the cameras would actually transmit. Later on, it became clear that if the astronomers Fred Hoyle and Chandra Wickramasinghe had been listened to (they had advised that Halley's nucleus would be dark) the returning images could have been interpreted far more accurately. From the perspective of the amateur astronomer watching the programme it was a case of 'please cut back to Patrick now, cut this waffle, cut the debate'; but, strangely, that is not how the 'head in the clouds' *Horizon* programme producers decided to play it. All the public could see, at first, was a diagram indicating how many particles were hitting the Giotto dust particle detector as the probe closed in. Then, within the final half hour of close approach, we started seeing more pictures from the onboard camera; but those pictures made no sense whatsoever and they had been converted into a gaudy pseudo-colour, where a strange colour represented a brightness level. The viewer had no idea which colour indicated bright, or which indicated dark. Even as the spacecraft achieved closest approach, no-one on TV, not even Patrick, had a clue what on Earth we were all looking at!! It was a total fiasco and Patrick, for the first time in his life, was tongue-tied when the director cut back to him. While trying to voice his view that the Giotto spacecraft had in fact survived the encounter, as well as contain his anger that the programme was becoming the disaster he had confidently predicted, he momentarily looked like a slobbering rabid dog, unable to control his jaw, dentures or tongue. Patrick repeatedly burbled "It has, it has, it has, it has, it has, it has in fact survived!" followed by such an extraordinary facial grimace that I thought he had either been entered into the World Gurning Championships or his dentures were going to fly out! Undoubtedly being part of such a massive astronomical farce had rattled him and he admitted, some time later, that he was very annoyed by the way the Giotto programme had turned out. After the programme it was widely reported that the Prime Minister Margaret Thatcher was not impressed either and she had voiced the view that if that was the best the European Space Agency could do they would not get any more UK funding. Both Patrick and Pieter Morpurgo were embarrassed by the program that resulted. The only good thing had been the fantastic cover artwork for that week's *Radio Times* 'Halley – the closest encounter', painted by Patrick's friend Paul Doherty, who was delighted to have secured the commission for a *Radio Times* cover.

Nevertheless, despite Patrick's intense annoyance that the BBC had not accepted his original advice, namely that working with the *Horizon* people, "Who know nothing about comets", was doomed to failure, he did agree to playing a part in a BBC licence fee promotion trailer for the corporation. The trailer first appeared in the spring of 1986 and featured a bearded John Cleese, sitting in a pub, and having a good gripe about the TV license fee. "What's the BBC ever given us for 58 quid?" Cleese moans. It quickly becomes obvious that the pub is full to the brim with the BBC's biggest stars of the 1980s, who all have plenty of answers. Those having a pint in the pub included David Attenborough, the snooker player Steve Davis, numerous TV newsreaders, David Dimbleby, Peter Snow, Alan Whicker, Bob Geldof, David Jason, the Two Ronnies (Barker and Corbett) and many others. They all remind John Cleese that the BBC makes lots of programmes that he seems to have forgotten about. The camera cuts to the *Blue Peter* presenters Peter Duncan, Janet Ellis and Simon Groom (plus Labrador), who all say to Cleese "there's children's telly too", after which the camera swings onto Patrick who is, of course, battling with the pub's Space Invaders machine! Patrick turns around, in a gap from destroying alien spacecraft, adding "…and Science…", before resuming his battle. Maybe this was Patrick's shortest ever TV appearance, but teaming him up with the Space Invaders machine was a stroke of genius.

With the excitement and the farce of the Giotto programme behind him, Patrick was now free to plan his next foreign excursion. Four weeks later, he was off, with some 40 other travellers, to observe Halley from Australia, with *Explorers Travel Club*. Actually, Halley had peaked in brightness in early March, but he secured a good view of it from Australia, from April 11th to the 23rd. Patrick's first point of contact in the Outback, and therefore near to some very dark and clear observing sites, was Alice Springs. There had been no appreciable rain there for 5 years, until Patrick arrived, when there was a downpour. This was somewhat frustrating for a comet observer! Remarkably, the Australian national newspapers had Patrick dominating their front pages with headlines such as "Famous British Astronomer brings Rain to Alice!"

Once again, on that holiday, Patrick's trousers split, just as they had on the 1973 Total Solar Eclipse trip. This time both seams on the inside of his trousers had split open, close to the groin. Of course, he had not brought a spare pair along; after all, he was only going to be abroad for 2 weeks! Some women were found who could repair his trousers and Patrick seemed quite happy to address an attentive audience of his fans, while the two women frantically repaired his trouser crotch area with sharp needles, even while Patrick was still wearing them. Extraordinary! Towards the end of that trip, after a clothing shop had been located, Patrick obviously decided that his old trousers were only useful as a swimming costume. With everyone getting into the hotel pool, Patrick joined in, at the shallow end, to cool down, but embarrassed by the terrible state of his left leg (injured during the war, it was usually a green colour) he waded through the pool like a whale, but dressed in those ruined black trousers and a shirt. It was his *It's a Celebrity Knockout* appearance all over again! Patrick's final view of Halley at its peak was from Bali, on April 24th, during a total lunar eclipse. It had been an interesting cometary apparition, even if Halley had been very poorly positioned compared to its 1910 return and even though Patrick had broken his shoulder too. The trauma of the live Giotto TV farce would, surely, fade with time, after a few years or so. As Halley peaked the Cockney pub sing-a-long duo of Chas & Dave even released a record entitled 'Halley's Comet', or perhaps that should read 'Alley's Committ! The lyrics included the lines: "You'll be a bit stiff when 'e next comes round… Pushin' up daisy's I'll be bound…. So all 'ave a knee's up Muvver Brown…. 'Alley's Comet's 'ere……!" Clearly, Patrick must have found this an amusing little ditty as Chas & Dave visited him at Selsey the following year.

In early May Patrick heard news that one of his oldest friends, the master optician Horace Dall, had died: he had suffered a fatal stroke while repairing a friend's microscope objective. Dall had been a BAA member since 1925 and was the inventor, in the late 1920s, of the Dall-Kirkham Cassegrain telescope (the maths for which was worked out by the American, Alan Kirkham, a few years later). When Patrick had joined the BAA in 1934, aged 11, Dall was already a respected member. Patrick, as we saw earlier in this book, had been the best man at Dall's second wedding (Horace, a widower, had married Helena in 1970, aged 69) and had been the master of ceremonies at Dall's 70th birthday party. Horace had undergone surgery a few years earlier on a suspected carconegenic growth in his throat and, coincidentally, Patrick was admitted to the King Edward VII hospital at Midhurst, a few weeks after Horace's death, to have throat and thyroid surgery himself. Dall had told Patrick that he had no intention whatsoever of dying: he appears to have got it wrong!

Throughout the early summer of 1986 Patrick was fighting a battle not only with Thyroid problems but with the Science and Engineering Research Council (SERC), who intended closing the Royal Greenwich Observatory at Herstmonceux and moving most of the staff, but few of the facilities, to Cambridge. At the May 28th BAA meeting Patrick addressed the audience and showed that SERC's reasons were totally flawed. Patrick explained that the move would lead to the observatory losing its identity, but with enough public pressure it could be blocked. Patrick then called upon Mr Gordon Taylor, a retired member of the RGO staff, to give his views. Gordon explained that he had worked at the RGO for many years and the only growth area there had been in the administration department! He agreed with Patrick and felt that the SERC administrators had not understood the role of the RGO. He added that the observatory had moved from Greenwich, in London, to Herstmonceux to obtain better weather conditions and that it would not be cost-effective to move it again. The BAA audience, with a show of hands, revealed they were 100 % behind Patrick on this one, but it was to no avail. Despite another meeting arranged by Patrick, on June 6th, to which many journalists were invited, the RGO move took place, and by 1998 it had been entirely dismantled by bureaucracy.

Another irritation at this time was that during July and August there was a brief media flurry of interest in Patrick's alter ego, from the 1950s, of Cedric Allingham. To recap, this was the pseudonym that Patrick had written under when writing a book entitled *Flying Saucer from Mars*, a UFO encounter spoof novel which he had always denied having anything to do with. An article in the July edition of the UFO magazine *Magonia* was quickly followed up by short pieces in *The Star* newspaper (on July 28th) and, more disturbingly, in the respected *New Scientist* magazine (on August 14th). The *Magonia* allegations and *The Star* story, both claiming that he was Allingham, were relatively minor irritations; after all, one was a cranky UFO magazine and the other was a tabloid newspaper. However, *New Scientist* magazine was a completely different matter and Heather Couper's friend and business partner, Nigel Henbest, was the magazine's astronomy consultant, as well as being the editor of the *BAA Journal*. The *New Scientist* piece was featured in its Feedback column, which was part of the regular Forum section edited by Martin Alder, but the article alleging that Patrick was Cedric Allingham had no author listed; clearly the culprit wanted to remain anonymous. "There's a plot against me, a plot to muddy my name", I heard Patrick say at a BAA Lunar Section meeting, followed by his standard line: "This has all the fine smells of rotting fish!" Some weeks later a Radio Sussex interviewer also raised the matter of Allingham with Patrick. Not surprisingly he emphatically denied that he was Cedric Allingham, live, on air, and said he was going to sue for libel; but, of course, he did nothing of the sort! It would be a few years before Allingham would rear his head again.

From September 12th to 26th Patrick could forget about the Allingham allegations as he found himself a nice little number as a tour guide on board HMS Canberra, a P&O cruise ship which sailed the length of the Mediterranean on, in this case, an astronomy cruise. Patrick was on hand to point out the stars by night and to give talks during the day, as the ship travelled to such fascinating ports as Istanbul, Izmir and Athens.

Trouble Brewing on the Council

During the final year of Heather Couper's Presidency she had to present the name of her successor to the BAA council, to be added to the ballot paper. If the council strongly disagreed the President might have to think again, but this had never happened, at least, not in living memory. Heather had gained much publicity for the BAA as she was a young, friendly and extrovert character and just the sort of person TV producers love to have on their programmes; in this respect she was a very unusual astronomer indeed. Most BAA members were middle-aged, serious, scientific types and 95 % were male. In much the same way that Patrick, in the 1950s, had energised a BAA society of mainly elderly greybeards with his exciting, enthusiastic and youthful style, Heather had shown a less stuffy approach to the BAA of the 1980s.

Patrick had chosen Heather to succeed him and to be the BAA President during the era of Halley's Comet; now it was her turn to choose a successor. A number of candidates were plausible choices. There was Patrick's lifelong friend, the newly-retired BAA editor and historian Colin Ronan for example, although he had retired due to taking onboard a new and time-consuming professional commitment. There was the business secretary Michael Maunder too, although he had only held that post for a year. Another candidate, and one who Patrick would have fully approved of, was Cdr. Henry Hatfield, the lunar, solar and planetary photographer. However, the candidate Heather chose was Storm Dunlop. He had been the association's papers secretary for 6 years and was very active in the Variable Star Section. Storm had been most disappointed when Heather and the outgoing editor, Colin Ronan, had proposed Nigel Henbest (Heather's former boyfriend, now live-in business partner) to be the new BAA Journal Editor. As the unpaid BAA Papers Secretary and a self-employed translator/writer, Storm could have done with the steady income of the BAA Editor's job. Whether this was a factor when Heather nominated Storm as her successor, no-one knows, but he did at least have 6 years service as

BAA papers secretary behind him, a length of service which was usually sufficient to justify being made a President! I should know as they made me the President after 7 years in that tedious role!

Patrick was, privately, dead against Storm being made the President. For a start, Storm had openly criticised his 1979 book *Basic Astronomy* in the Journal and Patrick would hold that against him for life. If the book had been good, it would not have mattered, but *Basic Astronomy* was a grotty effort, unworthy of Patrick, so Storm's critique had struck a nerve. It had struck a nerve even more so because it had been Patrick's devotion to his dying mother that had caused his literary standards to slip. Patrick had also held Storm responsible for the omission of Paul Doherty's Saturn paintings from the Saturn Section report for the 1980 edge-on rings apparition. As we have seen, Paul had been most upset by this: his best ever set of paintings were simply ignored by the BAA. Patrick bore grudges for eternity, but he also stayed loyal to his kindest friends for eternity. If Paul thought Storm was responsible for the problem (in reality it was the whole council's decision) then Patrick was right behind him. Then there had been the denial on TV, by Patrick, that Nova Sagitta 1977 existed, allegedly on Storm's advice: this had resulted in a lot of egg on Patrick's face…. So, Patrick was totally against Storm becoming the President for a variety of reasons. A few other council members, such as Henry Hatfield, had minor difficulties with Storm too, citing an unusual, somewhat distant and aloof manner, which most BAA councillors were perfectly tolerant of, but others simply could not stand. However, those others were, in the main, friends of Patrick. Colin Ronan, a lifelong ally of Patrick's, and the former BAA Journal Editor, had told Patrick he had been 'traumatised' by a late night phone call from Storm's mother, lambasting him over the council's decision to give the Editor's job to Nigel Henbest, instead of to Storm. It all just added to Patrick's dissatisfaction with Heather's choice of successor.

Anyway, there was little that could be done about the matter. Patrick would have much preferred Henry Hatfield to be Heather's choice of President, but it was not to be. Patrick knew full well that his decision to make Heather the President had led to this state of affairs and this was starting to play on his mind. However, the majority of the councillors felt that 6 years as the papers secretary, plus Storm's tireless work for the BAA Variable Star Section, more than justified him being proposed as the President, especially as he had missed out on the Editor's job. A number of councillors have told me since that they wondered if Heather chose Storm simply because he was totally different to her. She was a chirpy, excitable, extrovert party-goer and was enthusiastic when appearing in public. He was a dour, meticulous plodder and a stickler for procedure. Maybe Heather wanted the BAA to realise just how enthusiastic and exciting a President she had been and so went for the maximum contrast in styles when choosing her successor? This sounds unlikely to me, but so many people have suggested it that I wonder if there is a grain of truth in it. Certainly, Heather had been a highly enthusiastic and energetic President and a breath of fresh air compared to some of the fuddy-duddies of years gone by! A number of councillors have suggested that she wanted Storm to nominate her back as his successor after his term as President, so that she would be the outgoing President at the BAA's

Centenary year meeting, but I have no solid proof that this is anything more than wild speculation from conspiracy theorists! Nevertheless, in the mid-1980s the BAA's 1990 Centenary and who would be the Centenary President were very high in the minds of many councillors.

However, Patrick still saw himself as the BAA's 'King-Maker' and was not amused when Storm, not Henry, was nominated by Heather to be her successor. Also, maybe there was some envy from Patrick, in as much as at that time Heather was almost as well known a TV astronomer as he was. Then of course, there was that aforementioned bizarre article featuring Heather in the bath with an inflatable Space Shuttle in a national newspaper's Sunday colour supplement…….

At the October 29th 1986 AGM Heather gave her second BAA review of the year and a splendid Presidential Address entitled 'Who discovered the Universe'. Following her address, Patrick was quickly on his feet to pay tribute to the outgoing President. Patrick said that it gave him particular pleasure that Heather's lecture had mentioned people he had met, such as Shapley, Hubble and Baade, although he admitted that even he could not claim to have met Herschel. Patrick also thanked Heather for all the work she had put in on the BAA's behalf over the previous 2 years. He pointed out that in this time she had enjoyed many honours, including a luncheon invitation from the Queen. He very much hoped that she would continue to be active in the affairs of the association for many years to come. Heather then adjourned the meeting and left the running of the BAA in the hands of Storm Dunlop. Her choice had not been contested by the council, despite a few concerns.

Fig. 29.2 Patrick, on October 18th 1986, pictured with the 10-inch Armagh Grubb refractor that he used in the 1960s. Note the Halley's Comet *tie* (Photograph by kind permission of John C McConnell)

Chapter 30
Supernova 1987A, Politics and a New Magazine

The year 1987 would prove to be a very turbulent time for the BAA, although it started calmly enough, with much to interest Patrick and the practitioners of amateur astronomy. Within the Lunar Section Gerald North was still taking spectra of the lunar surface with the Herstmonceux telescopes, where possible, during TLP alerts. However, the eventual move of the RGO to Cambridge would ultimately curtail those activities. My own aspirations to continue using the Greenwich 28-inch refractor had largely been eclipsed by my regular CCD video-taping of the lunar surface with my own equipment. Having used the large refractor once, it had proved to be a real handful to operate. My back garden equipment, with the CCD video camera, was infinitely more useful.

On January 11th 1987 Patrick was observing the Moon with his 15 inch reflector, at a magnification of 360×. While observing the crater Plato at 18.15 GMT he noted that the crater floor looked much darker than the Mare Imbrium and that there was 'a decided loss of detail over the northeast wall'. This loss of detail appeared to extend over a localised part of the floor, but 'everywhere else definition was normal.' Patrick telephoned Peter Foley and once again a TLP alert was issued. Other observers who were alerted seemed less sure that anything was 'obscured' although poor atmospheric seeing that night was a major problem. A Lunar Section meeting at the National Maritime Museum in London was scheduled for February 21st and, at that meeting, Patrick spoke about observing the Moon 'then and now' and Gerald North described how he had been using the 30 inch Coudé telescope and spectrograph at Herstmonceux to try to provide hard proof of TLP.

Interstellar Explosions!

On February 24th 1987 an event occurred which, in astronomy terms, was one of the very biggest of the twentieth century. A supernova designated SN 1987A (the first of 1987), erupted in the Large Magellanic Cloud (LMC). Although the LMC is not part

M. Mobberley, *It Came From Outer Space Wearing an RAF Blazer!: A Fan's Biography of Sir Patrick Moore*, DOI 10.1007/978-3-319-00609-3_30,

of our own galaxy it is a nearby satellite galaxy and therefore SN 1987A was about as close as you could get to having a supernova in our own Milky Way. In fact, the last supernova seen in the Milky Way itself was in 1604, so this was the closest and brightest supernova for 383 years! It was discovered by Ian Shelton and Oscar Duhalde, at Las Campanas Observatory in Chile, at magnitude 4.5, but in the coming weeks it would brighten to magnitude 2.3, a very easy object to see. As a naked eye supernova was such a staggeringly rare event Patrick decided he must go and view it first hand. So, in early April, Patrick and John Mason set off for a week in South Africa, to view the supernova for themselves. The visit had been arranged at very short notice and there was no time to arrange any trips to South African observatories. They had one decent Nikon camera with them and their sole aim was to see this once-in-a-lifetime object. On arrival in South Africa, John and Patrick travelled almost 70 miles into the veldt to get away from city lights and they took several exposures of the supernova using Kodachrome 1600 film. They only had a fixed mount with them, so even exposures from 45 to 90 seconds showed some trailing due to the earth's rotation. They estimated the supernova's brightness at magnitude 2.7–2.5 on their first night, but it appeared to have brightened to 2.4 or 2.5 by their final night.

Meanwhile, back at the BAA Council table, a different type of interstellar explosion was brewing. A major personality clash was developing between the business secretary, Mike Maunder (now a staunch ally of Patrick), and the new President, Storm Dunlop. Mike had been a very popular Council member for many years and he was well-liked by pretty much everyone around the Council table. As a supplier of his own special film developing and fixing kits (Speedibrews) he was known as a photographic chemistry genius within the BAA; almost a magician where photo powders and potions were concerned. However, despite Mike's genial nature the two men simply could not work together and, in an attempt to prove just how unreasonable Storm was, Mike had even taped his telephone conversations with the President. Mike, being a member of Croydon A.S. (who Patrick classed *en masse* as 'serpents' in the 1960s) and someone who had publicly argued with Patrick at a BAA Exhibition Meeting in the 1970s, might seem an unlikely ally in this BAA war. But, by siding with Patrick in the anti-Dunlop campaign, he went from being a serpent to a stalwart in Patrick's eyes. This is only the second case I know of where one of Patrick's serpents became re-classified as a close friend (the other case being the artist David Hardy).

In fact, Mike Maunder had been a friend of Patrick's in the 1950s before Croydon A.S. was bizarrely declared the Serpent Kingdom! It was certainly not all Dunlop versus Maunder though; there were still issues regarding Nigel Henbest being made the BAA Editor when Storm had wanted that job so badly and, in another petty squabble, others claimed Storm would not return the new computer manual from the BAA Office. There was also disagreement over what the BAA's headed notepaper should look like, and whether the President's name should be featured. Perhaps the most significant development though was Henry Hatfield's new found loathing for Storm. Henry was a tireless no-nonsense fighter for the BAA, waging many battles with banks, legal people and bureaucrats, on the association's behalf, in the 1980s and the 1990s; he once staged a sit-in protest in the RAS secretary's office until the secretary listened to his gripes about the BAA's office space! Henry hated the bureaucratic 'Twitmarshes' of the world just as much as Patrick. Once Cdr. Henry Hatfield decided to take sides it became a military naval battle that he would fight to the death and so when Henry declared that he did not think Storm was fit to be the President, well, there was no turning back. I should add that Patrick himself did not actually serve on the Council during that stormy BAA year. Since returning to the table after his mother's death he had served a year as an elected councillor, followed by 2 years as the President and then two more as the Vice-President. This was his year off. However, he was at every BAA Ordinary meeting and was in constant touch with his mates on the Council, especially with Henry Hatfield.

As with any such major personality dispute it all became very personal and bitter and very un-BAA like, but with disturbing similarities to the witch hunt declared on H. Percy Wilkins in the 1950s. The BAA had almost always been a very gentlemanly organisation, but with Patrick and his allies firmly on one side, against the President, it was going to get even messier. The BAA Council was divided and dismayed by this situation and by the fact that it had escalated to phone call recording! Some were calling the saga 'the BAA-gate scandal' (after the 1972 US Watergate scandal). On the day after the supernova erupted in the Large Magellanic Cloud, the council situation erupted too. Storm Dunlop failed to turn up at the BAA Council and Ordinary Meetings, claiming he was ill. The former President, Heather Couper, stood in for him, flanked by the two secretaries, Michael Maunder and Richard McKim. One councillor told me that by the time of the March meeting Mike Maunder's dispute with Storm had escalated to the point where the two men could barely stand the sight of each other. Another speculated that it was impractical for them to share the same platform at the Savile Row Lecture theatre without a brawl potentially breaking out and Henry Hatfield might need to bring a firearm along to keep order! In such a civilised and gentlemanly association as the BAA this sort of wild gossip was unprecedented.

Howard Miles, the BAA's monthly 'Sky Notes' speaker, agreed to stand in for Mike Maunder at the March meeting (by which time Storm had returned) and, by the late April meeting, the BAA Council had appointed Howard as the official 'Acting Business Secretary'. However, Howard had not been elected at the last BAA elections, so had no voting rights at the council table. Mike Maunder had been elected, but was unable to work with Storm. To preserve Mike's council status he was listed in the Journal as the eleventh council member. This was an illegal situation, according to the association's Bye-Laws, but Mike was unwilling to resign. Patrick, from outside the Council, was totally frustrated and said that he was utterly dismayed by this whole situation. He told his friends "It's all my fault. Oh my God! What have I done?" This was a reference to the fact that his choice of Heather Couper to succeed him had, ultimately, led to Storm Dunlop becoming the President. Behind the scenes he and Henry Hatfield were adamant that Dunlop must go and that this was not a time for building bridges.

The spring months were traditionally the time when the ballot paper for that autumn's BAA elections was compiled. This, of course, made the situation at the council table even more fraught. Mike Maunder wanted to be elected as the Business Secretary for another term, but without Storm Dunlop as the President. The acting Business Secretary

Howard Miles did not want to be the Business Secretary at all and there were others who wanted a new choice of Business Secretary entirely. As far as the Presidency was concerned, well, Storm Dunlop had his supporters and his enemies, but everyone was well aware that the posts of President and Business Secretary had never been contested before in the BAA's history, despite there being nothing in the Bye-Laws to prevent this. Every President had survived for the traditional two sessions (the maximum consecutive service allowed). If the Presidency was contested it would be a sad day for the BAA, as it would immediately be obvious that an internal feud had developed which was splitting the council. To confuse matters even further, that summer Heather Couper's partner Nigel Henbest resigned the Journal editorship, which was transferred to Jacqueline Mitton. In the preceding year a number of editions of the Journal had been late, which had annoyed Patrick considerably and must have annoyed Storm even more, as he had not been given that job which he so badly wanted. To rub salt into Storm's wounds the BAA Editor's job passed once again to someone other than him! If he had been given that post in the first place, and not the Presidency, how differently things might have turned out, because the Editor cannot realistically also be the President. [N.B. This was always considered the case in the past, but as this book was being completed, in 2012, the BAA Council decided that Hazel McGee could become the President and also continue as the Editor, from October 2013.]

Never, in the BAA's history, had there been such a period of change, or bad feeling on the council. The anti-Dunlop lobby, whose unofficial figurehead was Patrick, found the ultimate candidate to put up against Storm in the elections, namely the aforementioned lunar photographer Cdr. Henry Hatfield. Patrick put his (considerable!) weight behind Henry's election bandwagon.

Henry Hatfield's name was known to every BAA member and many were surprised that he had not been nominated for the Presidency already. As I mentioned earlier, Henry was a no-nonsense ex-Royal Navy Commander: a man of action, not words. His photographs of the Moon were exceptional for their time and his 'Hatfield Photographic Atlas', then out-of-print, was legendary. He had appeared on many of Patrick's *Sky at Night* programmes and had preceded Dunlop as the Papers Secretary from 1973 to 1980. Once his name went down on the ballot paper Storm Dunlop's days would surely be numbered. Every BAA member was familiar with Henry, partly because he had appeared on *The Sky at Night*, but to many members who just received the Journal, and never attended meetings, or observed variable stars, Storm was just someone with a very unusual name. An alternative candidate was proposed to Mike Maunder too, as Business Secretary. Many on the council dreaded another year with Dunlop and Maunder feuding, and they found a powerful alternative candidate in Rossie Atwell. Rossie was also well-known to BAA members and had been employed in the BAA Office for 11 years. Both Mike and Rossie were friends of Patrick, so he could not lose in the coming elections: he was sure that Henry Hatfield would become the next President and either Mike or Rossie would end up as the Business Secretary.

At this point I cannot resist relating a story told to me by Denis Buczynski, and others, who served on the BAA council in the 1980s. As a council member myself, from the late 1980s on, I was only too aware of how keen Patrick was to propose or second motions at the council table. Barely had the President got out the words "Who would like to propose this motion?" or "Do I have a seconder for this motion?" and Patrick's hand would shoot upward like a rocket. Like a grossly overgrown schoolboy Patrick would shout "Yes, Mr President" as he did so, usually nodding sagely as if to hammer home what a damn fine proposal it was. It was customary during the end of the council meetings, especially if business was running late, for the Royal Astronomical Society's live-in caretakers to hand out the cups of tea as the council debated its business. On a few rare occasions they were not present and some outside caterers would serve the tea and sandwiches to the BAA council instead. Towards the end of one 1980s era meeting, one caterer was passing Patrick's cup of tea over his right shoulder when, tragically, the President asked for a seconder to the latest motion. Patrick's hand went up like a Saturn V rocket heading for the Moon and caught the cup and saucer beautifully as they were passed overhead. The cup and saucer flew several feet upward, towards the ceiling, and then reversed direction, landing squarely on top of, and drenching, Rossie Atwell! Denis just could not stop laughing at this, however hard he tried. Patrick simply glared at him for the rest of the council meeting!

While all the angry council business was going on, in the spring of 1987, Patrick's *Sky at Night* programme reached its 30th Anniversary and a special programme was dedicated to celebrate the event. Patrick wrote the book *TV Astronomer – Thirty Years of the Sky at Night*, to celebrate his three decades on TV.

On May 2nd the BAA Lunar Section held a joint meeting with Torbay Astronomical Society at Torquay Central Church. As so often at these events Patrick gave a public lecture in the evening, at Torquay Boy's Grammar School in this instance. Patrick had long enjoyed connections with the Torquay/Torbay region because his lifelong friend and fellow TLP observer, J. Hedley Robinson, lived there and this had enabled Patrick to establish close links with the Grammar School and its Astronomical Society. Patrick's Public Lecture was entitled 'The Outer Solar System' and I have vivid memories of one particular aspect of that lecture in May 1987, namely the slide projector and the wretched projectionist. The slide projector used for Patrick's Torquay talk was manned by an unsuspecting helper who must have wanted to crawl under the nearest stone. The slide projector in question, which he had been lumbered with operating, had no easy way of focussing and there was no way the focus could stay fixed between slides, because all of Patrick's slides were different: some were mounted in thick plastic holders and some in cardboard frames.

However, it did have one redeeming feature I have never seen in any other slide projector, before or since. It allowed the user to totally rotate the slide in the gate, while being projected! This was an interesting and unique feature, but of no use when the slides were almost impossible to focus. As each slide came up, Patrick would bellow at the projectionist "Can we PLEASE try to focus that slide". The profusely sweating projectionist, with 200 eyes bearing down on him, would push and pull and twiddle the lens in a state of terror, and then, as he finally achieved focus, every single time, Patrick would bellow "And the NEXT slide please". By the end of Patrick's talk, the poor man was just a shaking ball of sweat! Patrick, whenever he gave major talks with many slides, carried with him a blue, totally battered suitcase, which had obviously been abroad many times and was literally on its last legs. Whenever he opened the suitcase there was virtually nothing inside it, apart from several boxes of photographic slides which some poor, hapless soul would have to load into a slide projector from hell for the great man. The reason such a big case was used was simple: if the case was smaller Patrick would certainly lose it, within minutes of putting it down! In those days the precise position of the red dot in the corner of each slide told you which way the slide should be projected. However, Patrick's logic in where he placed the spot was incomprehensible to the layman and few who had ever tried to load his slides (some of which were very thick and guaranteed to jam any slide projector) would ever repeat the favour! Remember, Patrick favoured south at the top on his lunar slides and he also favoured the classical lunar east and west system which was reversed by the IAU. Taking all these points into consideration, and bearing in mind Patrick's impatience with such matters, being Patrick's slide assistant could end up as being the job from hell!

Astronomy Now

At the same time as the BAA council was pressing the self-destruct button and the *Sky at Night* was celebrating 30 years on TV, Patrick was about to launch a completely new and ambitious venture to establish a permanent British Astronomy magazine in the marketplace. Whether the turmoil inside the BAA Council played a role in this decision I know not, but I suspect it was the biggest factor. Walk into any large UK newsagents in the mid 1980s and there were usually two astronomy magazines on sale, namely *Sky & Telescope* and *Astronomy*. Both magazines were produced in the USA and there was nothing to rival them in the UK. This was a fact that had frustrated Patrick for many years and he had tried, on at least two occasions, to launch British equivalents. Both had died a death. The first of these was a magazine called *Planetarium*, which I have never seen a copy of! The second was a cranky magazine called *Astronomy and Space* which featured some very strange pseudo-scientific articles on such topics as TLP, moonquakes and the phase of the Moon affecting Earthquakes. Both magazines died a death.

However, at the start of 1987, Patrick had gained the interest of publisher Angelo Zgorelec and the Intra Press company and had assembled an Editorial Board comprised of his best and longest duration friends and acquaintances. In fact, according to Patrick, the first serious feasibility discussion meeting about the new magazine, then without a name, was held in January 1987 in an Indian restaurant (always popular with Patrick) in West London. Patrick's team and Angelo decided to give the project a green light and they produced a one-off trial issue of a magazine entitled *Astronomy Now*. The Editorial Board, present at that first meeting, consisted of Patrick, Angelo, John Mason, Peter Cattermole, Iain Nicolson and Ron Maddison and the art editor was, of course, Paul Doherty. The first edition cover was a painting, by Paul, of the Voyager 2 fly-by of Uranus and the authors included Garry Hunt, astrophotographer H.J.P. Arnold, and the future managing director of *Orion Optics*, Barry Pemberton.

The first volume was billed as covering April to June 1987 and it was a success, although not everyone liked the rather odd title. Tagging the word 'Now' on the end seemed a bit strange at first. By the second edition, in late summer, Patrick was the Editor-in-Chief and John Mason was the Managing Editor. The magazine went monthly from October 1987. At long last Britain had an established monthly Astronomy magazine available in all newsagents. Sadly, Patrick would eventually sever all connections with the magazine and then attempt to set up rivals to it, but that lay 7 years in the future. However, *Astronomy Now* would go from strength to strength.

On the evening of September 5th, Patrick began observing the Moon at 20.25 GMT, using his number two instrument, his 12½ inch reflector; he was immediately struck by the brilliance of the small craterlet due south of the central peak of the large crater Gassendi. The craterlet was 'intensely bright and seemed to be surrounded by a luminous nimbus – much the brightest point in the area and quite anomalous.' Patrick watched the craterlet until 20.45 GMT and then alerted Peter Foley. Peter was able to confirm that the craterlet was highly luminous and surrounded by a blue halo that had a darker blue band within it. Both Patrick and Peter reported that the region had returned to normal by about 21.30 GMT. On the same night, another TLP alert, regarding the Schröters Valley/Herodotus region of the Moon, had been instigated by Marie Cook. Regarding this alert, Gerald North managed to heroically obtain a spectra of the region with the 30 inch Coudé reflector at Herstmonceux, but nothing anomalous was recorded.

The Members Vote

By October 1987 all BAA eyes were focussed on the results of the council ballot. Who would be the next President? Would Storm Dunlop survive to be the President for a second term, or would Cdr. Henry Hatfield be the first man in the BAA's 97 year history to oust an incumbent President? Also, would Mike Maunder be re-elected as Business Secretary, or would Rossie Atwell get the post? In the same week that the ballot results were counted a storm of a different kind hit southern England. One hundred and twenty mile per hour winds in the early hours of October 16th ripped off house roofs, damaged observatories and felled trees. By a strange twist of fate some of the greatest storm damage occurred in the town of Sevenoaks in Kent, where Henry Hatfield lived! Six of the famous seven oak trees that gave the town its name were felled in the gales. October 1987 would be remembered for the aftermath of The Storm in more ways than one!

As it turned out the membership voted for a complete change: they voted for Henry Hatfield as the President and Rossie Atwell as the Business Secretary. At the October 28th council meeting, AGM, and associated 'Ordinary Meeting', the incumbent President, Storm Dunlop, who knew the ballot result, failed to turn up and deliver his review of the year, or his Presidential Address. This had only happened once before in the BAA's history. Storm was, apparently, too unwell to attend. It was left to his nominator and predecessor, Heather Couper, to chair the

AGM in Storm Dunlop's absence and read the President's review of the year. By a bizarre twist of fate the incoming President Henry Hatfield was awarded the association's highest honour, the Goodacre Medal, at that meeting. So if Storm had attended he would have had to reward the man who ousted him with both a medal and the Presidency! During the Presidential Address, read out by Heather, she announced that the Societé Astronomique de France had recently granted the presentation of a special SAF Centenary award to Patrick for his popularisation of Astronomy. [I might add, that despite the Council chaos and bad blood of 1987 I never once heard Heather say one negative word about Patrick in the coming years.]

At the end of the AGM, Henry Hatfield mounted the stage and became the 50th President of the BAA. As always at BAA elections, the percentage of members voting was lethargic and out of some 3,000 members only 720 returned valid ballot papers, although this was some 200 higher than in a normal year. Henry had received 380 votes against Storm's 340; hardly a landslide, but a clear majority. A turbulent year in the BAA had ended, but Patrick was relieved that, at last, events which he felt responsible for had reached a point of closure. However, many people were saddened, and even appalled, at the ousting of Storm Dunlop, who had done so much good backroom work for the BAA and the Variable Star Section. A number of long-serving councillors and respected Section Directors resigned over Storm's treatment and the BAA Council would not be a friendly place for many months to come. Nevertheless, Patrick appeared delighted. I somehow think that if his mother had still been alive, she would not have approved of his role in the turbulent events affecting the BAA Council in 1987. Storm Dunlop never returned to the BAA Council table, but produced many excellent astronomy books in the coming years. In contrast, Mike Maunder was still a highly popular BAA Council member a quarter of a century after the tumultuous events of 1987.

Around this time, and a world apart from the war at the council table, Patrick brought out a board game (!) called, not surprisingly, 'The Patrick Moore Astronomy Game'. Up to six players could journey on a 'voyage of discovery and adventure', using a 'giant Galactic board' (70 × 48 centimetres), through the Solar System and into Outer Space. Electing their Mission Controller, they chose their own flight path, watched their fuel, avoided black holes and other hazards, visited stars and planets, and docked with other craft. The game sold for £9.95 and included a booklet of astronomical notes by Patrick, with the first 100 games sold getting a signed photograph of the man himself. The game, promoted by *Astronomy Now*, was marketed through a certain Morrison Wilpred & Co. Ltd of Shoeburyness in Essex.

As well as 1987 marking the *Sky at Night*'s 30th anniversary, October saw the 30th anniversary of the Jodrell Bank Observatory. Patrick was invited to the celebrations, which saw Professor Sir Bernard Lovell honoured by having the 250 foot diameter radio telescope named 'The Lovell Telescope'. It goes without saying that *The Sky at Night* covered the event.

Chapter 31
Business as Usual at the BBC and the BAA

During December 1987 the Lunar Section director Geoff Amery resigned from his post, citing work and family commitments, and on Patrick and Henry Hatfield's recommendation to the BAA Council, the assistant director and TLP co-ordinator Peter Foley succeeded him. Peter took over from January 2nd 1988. On that same day, the day of the BAA Christmas meeting, Patrick gave a talk on the subject of suspected naked eye variable stars.

A week later and Patrick was on prime time Saturday night television, on the ITV chat show *Aspel and Company*, as a guest of the veteran TV presenter Michael Aspel. The other guests were the writer and satirist Alan Coren and the eight times married Hungarian-American actress and socialite Zsa Zsa Gábor. Patrick was the first guest and most of the programme was about him, but in the company of the other guests he looked distinctly embarrassed and embarrassing. When Alan Coren said that he admired the Prime Minister Mrs Thatcher's legs, Patrick cut the conversation dead with the line "One things for sure, without Mrs Thatcher we'd be right in the cart". Despite being pretty negative about most women he was a big fan of Mrs Thatcher, purely for her stand on Europe and foreigners in general. Patrick played a xylophone composition of his own making on the programme, entitled 'Freefall', which he had just written. Michael Aspel asked him how many people had heard it before. "No-one" Patrick barked. "And how many will hear it afterwards?" "No-one" Patrick barked again, with, perhaps, the merest hint of a smile. Zsa Zsa seemed appalled that Patrick had not been married. "Maybe he's bright; he doesn't get married, he just lives in sin" she quipped; Patrick scowled and looked distinctly uncomfortable, as he always did if the talk got smutty. She offered to set him up with one of her girl friends, but Patrick argued that he was far too ancient. He probably didn't realise that Ms Gábor was actually 6 years older than him! Patrick looked very relieved when the programme finished and the credits rolled. Doubtless he was paid a lot for his appearance but he simply did NOT look comfortable on chat shows, especially if the topics of conversation became distinctly adult in nature.

M. Mobberley, *It Came From Outer Space Wearing an RAF Blazer!: A Fan's Biography of Sir Patrick Moore*, DOI 10.1007/978-3-319-00609-3_31,

On January 22nd 1988 the magnitude 6.5 star catalogued as ZC 3465 was due to graze the Moon's limb and maybe wink in and out between the lunar mountains, on a track that passed over the Isle of Wight and right over Patrick's village of Selsey. Unfortunately, as it turned out, the track would not quite pass over Patrick's garden, but John Mason and others successfully organised 15 groups along the roads in the region to observe the grazing occultation. Patrick used a small portable Maksutov telescope on the night and 9 of the 15 groups witnessed the Selsey occultation.

Four weeks later, on February 17th, Patrick was a guest at a Foyles Luncheon held for the launch of Alan Whicker's book 'Whicker's World Down Under'. This took place at the Dorchester Hotel in London. Most of the press photographs actually showed Patrick chatting with the naturalist Peter Scott at this event. I suppose as they were both animal lovers they must have had much to talk about.

Fig. 31.1 Patrick with the ornithologist and conservationist Sir Peter Scott and Lady Philippa Scott. The picture was taken at a Foyles Luncheon for the launch of Alan Whicker's book 'Whicker Down Under', held at the Dorchester Hotel on February 17th 1988 (By permission of the Press Association)

Back in the Lunar Section Peter Foley's first act as the new Director was to finalise arrangements for a full Lunar Section meeting at the National Maritime Museum at Greenwich on February 27th and to book Patrick to be the main speaker. John Mason and I were also confirmed as speakers. The meeting was a good one, with the Runciman Lecture theatre packed to full capacity. In his rousing talk Patrick compared the falling off in interest in the Moon, during the post Apollo era, as being identical to the situation after Beer and Madler published their detailed lunar survey in 1838. Patrick then went on to compare the lunar surface with those of the other satellites and small bodies in the solar system.

Some Personal Memories

During the 1987/1988 BAA session, and for the next 12 years, I served on the BAA council and missed only one meeting (when I was away observing Comet Hyakutake in Tenerife, in March 1996). In general, the council met on the last Wednesday of the month in the Royal Astronomical Society's council room at Burlington House, Piccadilly; at least, from October to June. The BAA Office was situated on the top floor and, for most of that time, employed two secretaries to cope with the day-to-day business. I had, since 1983, served on the Lunar Section committee and so came into contact with Patrick there on a regular basis. Getting to know someone who, since the age of 10, had been your childhood hero, was an interesting experience, but there were a number of sides to Patrick which were not quite what the average TV viewer might appreciate. What I can say, without doubt, was that Patrick was refreshingly different to just about everyone I had ever met. Patrick was a man who had no time for small talk or bureaucracy. The world to him was not made up of an infinite series of shades of grey. Everything was black or white, wrong or right, and nothing required any long discussion. Patrick always did what he said and he was a 100 % man of his word. With his physical size, booming voice and larger-than-life TV personality, few would ever cross him at the council table. However, when, they did, and when he lost a council vote, he would simply sulk, like a huge overgrown schoolboy!

Everything in Patrick's world was done in a manic fashion. There was no in-tray and nothing pending. If he agreed to do something he would do it: immediately. One example that springs to mind was when the new editor, Jacqueline Mitton, enquired as to whether anyone on the council knew a recently deceased member well enough to write an obituary for the BAA Journal. No-one did, but Patrick had met the member in question a few times and knew of his work. The council discussion on whether an obituary was really needed dragged on for many minutes. Yes, the member was well known, but nobody knew him *that* well. After some time the President decided we had discussed the matter for long enough and it was obvious we did not have enough material for an obituary. At that point Patrick waved his hand frantically. "Yes Patrick?" Patrick explained that while we had all been talking he had written an obituary for the member, based on what he knew of him. He handed several sheaves of paper to the President who forwarded them to the Editor. That was Patrick's way. Plus, everyone knew that if Patrick was asked to correct, expand, or alter anything he ever wrote, he would simply withdraw it and the subject would be history! You did things Patrick's way or did not involve him at all. That was the way he was.

Talking to Patrick on the telephone was usually a very abrupt experience. The phone would ring. I would pick it up.

"Hello".
"Martin, Patrick here".
"Oh hello Patrick, how are you".

"Ohh, wretched knee playing up. Damn shrapnel from the War of course; nothing anyone can do about it. Have you got any recent pictures of the comet?"
"Yes Patrick, I got a good one last night".
"Excellent, would you mind awfully sending me a copy, to show on *The Sky at Night*".
"No problem, I'll post one off tomorrow".
"Excellent, I'm really grateful. Must go, bye......[CLANG]"

That was about as long as you would get. Postcards were the same, generally sent to acknowledge the receipt of some photograph or other. They would arrive, obviously typed on his turn-of-the-century typewriter, with this sort of wording:

Excellent Photo.
Will use on Sundays prog.
Ever
Anon
Patrick

It was obvious when I joined the Council that Patrick rarely agreed with anything that the duo of Heather Couper and Nigel Henbest said in the post-Storm Dunlop era. The body language was a dead give-away without any other information. One altercation that always springs to mind was over the employment of a new member of staff in the BAA office. In early 1988 the office was being manned by one person: Pat Barber. She had recently been promoted to the post of Assistant Secretary, following the departure of the previous post holder, Peter Gill. However, there had always been at least two employees in the BAA Office, sometimes three. So adverts were placed for a new helper for Pat. Various candidates were interviewed and discussed by the Council. The best prospect was a young black woman of African descent called Gloria. The BAA business secretary, Rossie Atwell, read out her details to the Council. There was a huge groan from the far end of the table, which sounded like a grizzly bear in the mating season. It was Patrick. "Oh No Mr President, this won't do at all, can't we find anyone British". It was pointed out to Patrick that the new candidate *was* British. "That's not a British surname and you said she comes from Nigeria or some other godforsaken place. No, this is no good at all". Then, in a much louder voice "Does everyone around this table want to have to learn to speak Erdu? Well, I don't". At this point the arms were folded and a scarey Churchillian scowl adopted. Patrick's head was shaking in a 'not over my dead body' motion. The council sat quietly in a rather embarrassed silence. Heather Couper looked like she would have liked to say something, but thought it wiser not too. Finally, sitting next to her, Nigel Henbest looked straight at Patrick and said "You ought to be thoroughly ashamed of yourself". Patrick then opened his mouth and bellowed a huge "Ha, Ha, Ha HARRRRRRRR!" As it turned out, the council ignored Patrick's views and employed the candidate, who turned out to be very reliable.

This was not the first time that Patrick had made enemies over a racism issue. A friend of mine, who was invited to watch a *Sky at Night* being recorded in the late 1980s, told me of a furious row that had developed after the programme, between the guest (a regular on the programme) and Patrick. The subject of the row was South African apartheid, of which Patrick was totally in favour; he believed in African white supremacy and predicted that Mugabe would be the death of

Zimbabwe and that South Africa would go the same way (he was right there). After that row, the guest, who had been a regular on *The Sky at Night*, was never invited onto the programme again. In reality, Patrick's hatred of foreigners had little to do with skin colour, but everything to do with World War II. As far as Patrick was concerned hundreds of thousands of British people had died trying to keep non-British (German) invaders out and we shouldn't let foreigners take over Britain by stealth. It was an insult to the heroes who had given their lives for Britain. World War II had shown us that foreigners were not to be trusted and the more foreign someone looked, the less Patrick trusted them: it was as simple as that. If anyone objected to this view he would just shock them with out-and-out racism, as Patrick simply could not abide mindless political correctness.

Another BAA Council issue discussed in early 1988 concerned the BAA Office computer. In that era the very first IBM PC's were being brought into office life. These rarely had hard disks or colour monitors and everything was stored on a floppy disc; the processor speed was around 1 megahertz. Software was very limited too. Work had started on transferring the records of 3,000 BAA members from the old card index system onto the computer database, but there had been various technical problems. It was looking increasingly likely that the original software would need upgrading at a cost of several hundred pounds and there were other teething problems too. Midway through a highly tedious technical discussion Patrick raised his hand in the air. This debate had gone on for 20 minutes, Patrick had understood none of it, and he was not at all happy with being totally unable to contribute. The President invited Patrick to speak and the Council wondered what on Earth he could have to say on the subject of computers!

Patrick spoke (from my own notes of the meeting):

"Mr President, it is obvious that this whole computer business is just a big con. Up to last year, we had no wretched computer. Since 1890, the members' records have been on a card index and there have never been any problems, even when we had 6,000 members in the 1970s. SINCE 1890, NO PROBLEMS. Now there are loads of problems". His voice became even louder (if that were possible). "MR PRESIDENT, I propose we go upstairs to the office, grab all the bits of this infernal machine, cart them out to the Burlington House courtyard and throw the whole lot into the skips just around the corner. Then we can return to 'business as usual'. If we haven't needed a computer since 1890, we don't need one now". Once again, much to Patrick's dismay, his point of view was ignored and he sulked for the rest of the meeting!

I also recall an occasion, during the late 1980s, when the BAA Lunar Section decided to revise their small booklet entitled 'Guide for Observers of the Moon'. Getting men with day jobs, wives and families to write short articles could be like getting blood out of a stone, but Patrick invariably wrote his own contribution on the same day the project was suggested and he was usually chosen as the Guide's editor too. Woe betide anyone who took more than a few weeks to send their chapter to him, as Patrick would be phoning up to ask if they were unwell, or maybe deceased?! Obviously, having a professional quality booklet printed cost the BAA money and the cost had to be justified and quotations obtained by printers, along with a contract from the successful printer being sent to the BAA Treasurer for approval. All this sort of bureaucracy was pointless in Patrick's world. So, when Patrick announced, at the next available BAA Council meeting, that the Lunar Section would like to print a revised copy of their lunar observing guide, the Council nodded in appreciation of the idea and said that they would await further news from Patrick and the President would add it to the next agenda. One month later and the item was officially added to the Council agenda, so the President asked the enormous blazer-clad Patrick how things were progressing. "Ohhh, it's done, it's done Mr President; finished, printed, ready for distribution at the next Lunar Section meeting." The President's mouth opened and closed like a fish and eventually he enquired about the quotations and the costs; he also expressed concern that the Council had not officially approved it. "Nothing to discuss Mr President" boomed Patrick. "My local Sussex printers did an excellent job, as always. I paid by cheque. Project completed." "But we haven't approved these expenses Patrick, we can't........" Patrick interrupted "There are no expenses Mr President. I paid the bill. A trivial amount. All sales are pure profit for the BAA now. End of story." So, that was that. Job done. Patrick was a few hundred pounds out of pocket, but that was far better to him than any waffle and bureaucracy. The entire Guide had been proposed, written, edited and printed in a month without any hassle, hot air, form filling, box ticking, or pen pushing. That was how Patrick did everything. There had not even been a written contract with the Sussex printers, just a handshake from Patrick: a 'Gentlemen's agreement'.

Yet another BAA Council memory from this era is worth recounting because I simply cannot ever recall Patrick laughing so much; and it was all down to me! I was the BAA Papers Secretary for 7 years and very early on in that term I was faced with the knotty problem of an excellent paper, that had been under final revision, when the elderly author had suddenly and unexpectedly passed away. I was contacted by the member's son, who was keen for his father's paper to be completed (as per the referees' requirements). Several people around the BAA Council table, on the day I discussed this problem, had known the deceased member and spoken to him only days prior to his demise. They simply could not believe he was no longer around. The doubts surrounding his status grew and then there was pressure on me, to confirm to the councillors that I was talking about the same man! Cdr Henry Hatfield, the lunar photographer, exclaimed "Mr President, are we sure this man is dead?" The President (Colin Ronan) turned to me, eyebrows raised, but I already had my answer prepared and I replied "Mr President, the member's son informed me on Monday that he had buried his father, so I naturally assumed he was dead." Following such a solemn debate my reply brought the house down and no-one was laughing more than Patrick who appeared to be in physical pain, guffawing like a braying donkey and with tears rolling down his cheeks for many minutes.

Another Eclipse and a Visit to Chile

On March 17th/18th 1988 a total solar eclipse shadow would track from the Indian Ocean, first hitting land on the west coast of Sumatra, then crossing Sumatra, Borneo and the southern part of the Philippine Islands, passing close to the Mariana

Islands and South of the Aleutian islands, and finally ending up in the Gulf of Alaska. The maximum duration of the total phase would be 3 minutes 46 seconds. Once again the British company *Explorers Travel Club* offered an eclipse holiday to keen eclipse-chasers, but two locations were offered on this occasion. You could choose to be on Patrick's tour (he was described as the 'guest lecturer') and go to Davao Bay at Talikud Island in the Philippines, or you could go to Bangka Island in Sumatra, with Heather Couper (her first total solar eclipse). This would be Patrick's sixth total solar eclipse and, apart from his first, the 1954 eclipse in Sweden, none of the others had cost him a penny! The 1961, 1968 and 1973 trip expenses had been covered by the BBC and, for 1983 and this 1988 trip, he had been a guest lecturer. In fact, after becoming famous, Patrick never paid for any eclipse trip. When he was too old to be a guest lecturer and when, after 1999, the BBC would not fund *Sky at Night* foreign trips, he simply did not go! In fact, he also missed a few total solar eclipses in the 1970s for these same reasons. From 1961 onward he expected to travel to total solar eclipses for free, but then, as he gave so much money away, he was not a wealthy man anyway. On arrival in the Philippines Patrick was pleased to inform his colleagues that apparently the local terrorist groups had issued a joint Press announcement; this stated that they had generously decided NOT to blow up the visiting eclipse chasers, which he thought was "jolly decent of them".

As it turned out, both parties saw the eclipse. The Bangka Island group had a good view, despite thin cirrus cloud, and Patrick's group at Talikud Island saw the eclipse by the skin of their teeth! Minutes before totality a gap appeared in the clouds and the spectacle of the eclipse was seen, although the outer corona was lost in the thin cloud and hazy atmosphere. With no time to compensate for the haze Patrick simply rattled off every exposure available on his camera dial, and when the film was developed he had indeed captured one splendid shot of the corona with his wide angle lens. Phew! On that 1988 trip, while in the Philippines capital of Manila, Patrick and a few friends visited the local planetarium. According to Patrick, the lecturer had a voice like "Donald Duck lost in the fog!" Patrick also visited the Manila Observatory, where the Director, Dr Badillo and his colleague, Father Haydon, showed them around. He also went on a sightseeing trip to Taal, one of the world's active volcanoes.

Back in the UK, Patrick decided to be controversial for the May 1988 edition of the *Sky at Night*. He invited Dr Halton Arp to be the studio guest. Arp was a well known professional astronomer who had compiled a useful catalogue of unusual galaxies. However, he had been literally ex-communicated by his colleagues for daring to voice the opinion that the traditional Big-Bang explanation for the origin of the Universe might be wrong. In Arp's opinion, the extreme red shifts of quasars were not related to their great distance from us, but to their being, for example, expelled from nearby galaxies at great speed. What really worried established professionals was that Arp was not a newcomer in professional astronomy, but a highly regarded figure. It had been literally impossible for him to work as an astronomer in the USA once he voiced his opinions, but the Max Planck institute in Munich was prepared to employ him. Arp was, without doubt, one of the most controversial astronomers that Patrick ever invited as a guest on the *Sky at Night*.

At the May 25th meeting of the BAA in London, Patrick gave a short talk on how he had been observing Venus in recent weeks. The brilliant planet had been a waning crescent during the last month and, once again, Patrick had been searching for the elusive 'Ashen light', the dim glow on the un-illuminated side, dismissed by many as an optical illusion. Patrick had made a crude occulting device which jammed into his favourite eyepiece and helped hide Venus' brilliant crescent. However, he said at that meeting, that he had seen no sign of the Ashen light that apparition and wondered if this was due to the current low level of solar activity? His Venus observing note book for that year told a slightly different story though, indicating that he had 'a strong impression of the Ashen Light' on May 1st and 10th, confirmed with his crude occulting bar, and a suspicion of it on May 4th and 5th. Obviously Patrick preferred to err on the side of caution in front of such a learned audience with so many experienced observers present.

Following the long summer BAA recess, during which Patrick was photographed playing Chess alongside the actress Greta Scacchi at a local Sussex tournament, the October 1988 *Sky at Night* programme was one of those which amateur astronomers always found the most interesting: a programme about back garden observing. Patrick was again joined by the planetary artist and observer Paul Doherty who explained that he had been observing since 1957, the year that the *Sky at Night* had started. An electric fire in the studio was used to illustrate how hot air, rising from a building, can ruin planetary seeing. Paul explained that he could make a quick sketch of a planet like Mars with only a 5 minute gap between the clouds, but with 30 minutes available he could produce a very fine drawing indeed. Patrick showed a few of his own observing tricks, like a Meccano attachment which fixed his red bicycle cycle lamp onto his observing clipboard. Paul, along with BBC employee Eric Ilett, who specialised in moving models in the pre-computer graphics era, was a regular illustrator for *The Sky at Night*, but he rarely appeared in person on the programme. Patrick openly admitted on that show that he envied Paul of his artistic ability, especially when sketching Saturn.

During the summer of 1988 Patrick had received a letter telling him that he had been 'promoted' as he put it. He was in the Queen's Birthday Honours List, as a recipient of the CBE, or Commander of the British Empire. This was, in effect, an upgrade from his OBE of 1968. He received the award at Buckingham Palace on October 25th. So, the next day, at the BAA AGM Council meeting, he told the President Commander Henry Hatfield "I'm the same rank as you now!"

In November 1988 Patrick was the subject of the Sunday Observer colour supplement's 'Expert's Expert' feature. The newspaper's reporter, Heidi Kingstone, asked ten well-known astronomy figures who was their personal favourite astronomer. Most mentioned Patrick of course and Heather Couper, Henry Hatfield and Ian Robson (Lancashire Polytechnic) put Patrick at the top of their lists. Patrick gave his own favourite historical astronomer as Edmund Nevill, who wrote under the name of Edmund Neison and published the book 'The Moon' in 1875. Patrick said that he now knew Nevill's daughter. Of living astronomers, Patrick chose his good friend Professor Sir Bernard Lovell as his favourite expert astronomer.

In January 1989 Patrick spent a week in the Chilean Andes visiting the observatories of La Silla, Las Campanas and Cerro Tololo. While in the control room of the 1.5 metre Danish telescope at La Silla, Patrick was treated to a 'live' imaging session on comet Halley, then down to a feeble magnitude 24½. Patrick's attempts to make friends with the La Silla desert fox, which came to the observatory regularly for breakfast, did not get very far, but Patrick became heavily involved in anti-fox hunting activities once more in the years that followed. Patrick was accompanied by *The Sky at Night* film crew who obtained much footage for a programme televised some months later. On his return to the UK Patrick immersed himself fully in a project for the BAA Lunar Section, revising and editing, yet again, their 'Observers Guide to the Moon' booklet.

The Kolovos Flash and Cedric Allingham Returns!

Another BAA Lunar Section meeting was held at the National Maritime Museum at Greenwich on the 18th of March 1989. Patrick was a speaker at the meeting. His subject? Yes, you've guessed it: TLP! Starting his talk, Patrick said that in the 1830s, when Beer and Mädler published their maps, everyone believed that the Moon was changeless. However, in 1866 there was the Linné controversy when peoples' views changed, although Patrick stressed that he felt there was actually no real evidence that Linné had altered in any way. After the Apollo landings, once again it was felt that nothing ever happened, but Patrick said that we knew, from TLP studies, that TLP do occur, and he referred to the work being done by the BAA Lunar Section. Patrick emphasised that it was important to try to obtain spectrographic confirmation, as Kozyrev had done, in 1958. Gerald North was, of course, trying this.

Patrick then went on to describe an event that had taken place on May 23rd 1985. This particular event had achieved much prominence as it had been published in the prestigious Journal *Icarus* in 1988. The event was the photographic recording of a bright flash near to the crater Proclus C by the Greek observer George Kolovos, using a 108 mm refractor. The flash had not been seen optically but showed up as a white spot on one of seven photographic frames. The estimated area affected on the Moon (the area covered by the white blob) was 22.5 × 18 kilometres. Patrick said that not all areas were 'transient prone', but in 1970 he and Barbara Middlehurst had produced a catalogue which showed most affected areas, and Proclus exhibited a number of events. Patrick then showed a number of slides of the event in question and said that there was a theory that the flash was caused by emission of Radon gas and subsequent piezo-electric discharge. However, Patrick urged caution in the interpretation. He admitted that the flash could be a photographic effect but he felt this was unlikely. He thought that some sort of gaseous emission from below the surface was possible, although the short duration of the event was worrying. After much discussion as to the possible cause of the 'Kolovos Flash', Patrick concluded his talk. At the same meeting I had arranged a section photographic competition; Patrick, with Cdr. Henry Hatfield, did the judging at my request.

Four days later and Patrick was the main guest on the John Dunn Radio Show on BBC Radio 2. John must have surprised Patrick when he lead him into an entertaining chat about debunking UFO's, lulling Patrick into a false sense of security, and then came straight out with the direct question: "Was it you who wrote a book back in the fifties, a spoof book about Flying Saucers?" Patrick stammered "No, the only, the only, the only books I've written not under my own name were *Bureaucrats and how to annoy them* by R.T. Fishal and *The Twitmarsh Files*." But John Dunn was not beaten, he persevered: "Because they've been chasing an astronomer called Cedric Allingham….". Patrick interrupted immediately "There were several famous Flying Saucer Books, by George Adamski, who I knew, Cedric Allingham who I met once….." John interrupted "So he did exist?" "Oh Yes, I met him" said Patrick and continued…"and Truman Bethurum."

Patrick went on to explain that Truman Bethurum claimed to have had an affair with an alien female called Aura Rhanes. Bethurum's wife promptly filed for divorce, citing the alien female, and won!! "The first Interplanetary Scandal in History". With that amusing aside from Patrick, John Dunn had forgotten all about Cedric Allingham (Patrick's alter ego) and Patrick must have breathed a huge sigh of relief! Patrick also revealed in that interview that his beloved Ford Prefect car had been stolen, although it was later recovered. I have often wondered who tipped John Dunn off to interrogate Patrick about Cedric Allingham?

The Kolovos Flash was still at the forefront of Patrick's mind at the March 29th BAA Council meeting and main meeting. By this time Patrick had decided to be positive about the incident and bill it as "Proof, at last, that TLP exist". After all, very few TLP events had ever been accepted for publication in the professional scientific literature. The council meeting of March 29th was one that I remember well; Patrick was in a very vocal mood and highly excitable too. Whether this was due to the Kolovos event or something else, I know not.

At that Council meeting the President, Henry Hatfield, asked Nigel Henbest to read out his recommendations for the 'membership incentive' drive. One of Nigel's ideas (without doubt a good one) was to scrap almost all of the BAA Wednesday meetings, which were so hard for the ordinary members to attend, and concentrate mainly on Saturday meetings, which Patrick had promoted (alongside Wednesday meetings) during his Presidency. Some 13 years later this became official BAA policy for a short time, due to the falling attendances at meetings. No sooner had Nigel got his words out than Patrick was furiously waving his hand. He bellowed down the Council table: "MR PRESIDENT, I INKED IN THE LAST WEDNESDAY IN THE MONTH IN MY 1934 SCHOOLBOY DIARY & I WOULD HATE TO CHANGE THE HABIT OF A LIFETIME". So that was the end of that! Cdr Hatfield then announced that there would be a Council photograph on the steps of the RAS building at the next meeting. No Council photograph had been taken in the previous year, due to the arguments raging and the contested Presidency. Patrick's response was another bellow: "THE FIRST PHOTO SINCE THE 87 HOLOCAUST MR PRESIDENT. I PROMISE TO STAND AT THE BACK". The subject then moved onto the plans for the 1990 BAA Centenary and how we would celebrate and publicise 100 years of the Association. Once again Patrick had to dominate the proceedings. "MR PRESIDENT, THE BBC ARE TOTALLY IN FAVOUR OF A SKY AT NIGHT ABOUT THE BAA. I WILL FILM ONE. YOU HAVE MY WORD ON THAT". It never happened! The BBC changed their mind.

Although almost everyone around the BAA council table was a fan of Patrick's and had shelf-loads of his books, his domination at the table could often be a real pain. There were, in my experience, only two types of BAA Council meetings in the 1980s: those where Patrick dominated proceedings (the majority) and those where he sulked because another prominent figure had got the council's support! Fortunately, most of Patrick's views were based on common sense, but not all.

At the March BAA meeting, at 23 Savile Row, following the Council meeting, Patrick was in full, 'excitable schoolboy' mode, determined to impress upon the audience that, at last, TLP were proven, following the publication of the 'Kolovos' flash details. But, however interesting the Kolovos event had been, it was light-years away from being scientific proof. It was a single curious photograph that had been subjected to a basic analysis: no more and no less. Anyway, Patrick gave a 200 decibel microphone rupturing delivery about the Kolovos event, illustrated by some very old pictures of the region taken by Henry Brinton and others. From the audience, fellow TLP hunter Jeremy Cook asked if it could have been a head on meteor in our atmosphere. Patrick must have misheard him as he bellowed "IT DEFINITELY WASN'T A METEOR ON THE MOON AS THE FLASH MATCHES THE SURROUNDINGS WHICH, QUITE FRANKLY, IS KOLOVOS' TRUMP CARD". Heather Couper then asked whether any isophote contours had been derived from the negative. Patrick's answer was again, very swift and very loud, but far from convincing: "OH YES, KOLOVOS HAS DONE THAT ALRIGHT, I'M SURE HE HAS, ALTHOUGH I HAVEN'T SEEN THEM!!" During all this time Patrick was dashing about recording all the questioners' names as he was both the meetings recorder and the speaker! Patrick may have convinced himself about the significance of the Kolovos event, but the BAA audience was an educated one and was not swayed by the sheer volume of Patrick's argument.

On April 24th 1989 the BBC broadcast an episode of *The Reluctant Cook* (filmed the previous summer) in which chef Kevin Woodford taught Patrick how to steam cod, grill salmon, cook trout 'in a parcel' and prepare a Kedgeree meal for his Selsey friend Barney. Kevin and Patrick walked along the beach near Patrick's Selsey home and first acquired some Cod from local fisherman 'Honest Bill' whose Seafoods business had been established in 1919. Despite living a few hundred yards from the sea, cooking fish had "always been my downfall" according to Patrick. "Certain things I can cook. I think I do a reasonably good curry. Give me a chicken or a joint and I can make something of it". Patrick did not, he admitted, usually eat trout, because of the reproachful stare from the eyes of the fish. "I try to kid myself it grows on a tree. If I had to kill it, we'd all be vegetarians". Patrick apologised for the age of his oven, claiming it had once been part of a famous ship. Kevin asked which ship that might be? "Animals went into it two by two" said Patrick. Patrick also revealed that he liked his Curry "hot". His friends would confirm that "hot" often meant like molten lava. In the 1990s one of Patrick's favourite concoctions, that he could whip up on his own for visiting guests, would be his "Seafood salad";

this was despite the fact that, in general, with traditional meals, his plate would be piled high with meat and potatoes but with no greens to be seen anywhere! If forced to serve a meal for others a salad was easy to prepare, on the rare occasion when there was no other housekeeping help available.

By the summer of 1989, Patrick's monthly magazine *Astronomy Now* was fully established in newsagents and was looking like it would definitely survive, despite competing with the U.S. *Sky & Telescope* and *Astronomy* magazines. Patrick's chums were still running the magazine and John Mason had been promoted from Managing Editor to 'Deputy Editor assisting the Editor-in-Chief'. The 'Editor-in-Chief' being Patrick, of course. The magazine had a distinctly 'Patrick' appeal to it. It was British and it was quirky. A lot of the pictures were grotty and the spelling and the grammar were often rather poor. Patrick's views dominated the magazine. Nevertheless, it had survived, which was unique for a British astronomy magazine. One thing that *Astronomy Now* had definitely not mastered in the early days was reproducing night sky pictures. Frequently a photo would reproduce as a jet black box with, perhaps, one star visible! The caption would read 'The Constellation of Orion in all its glory', or similar; but NOTHING glorious could be seen at all!

On August 25th 1989 Voyager 2 passed by the final target on its long journey through the solar system: the planet Neptune. Patrick had already been at a NASA conference on optical astronomy in San Diego in the week before and he went straight from there to the Jet Propulsion Laboratory in Pasadena for the Neptune encounter. Voyager 2 showed that Neptune's atmosphere was more active than previously thought and the spacecraft discovered six new moons and also showed that the faint rings around it were very clumpy. Five hours later came the event that proved to be the highlight of the encounter, as Voyager 2 flew past the extraordinary frozen moon Triton. Patrick had a star-turn, live from Pasadena, on the August 25th BBC 1 Nine 'O Clock News.

On September 14th Patrick posed for press photographers alongside a giant inflatable asteroid outside the Cambridge Theatre in London's West End, where he performed as a bizarre Shakespeare spouting narrator in the Sci-Fi Jukebox musical 'Return to the Forbidden Planet'. The show would be revived for a UK tour 10, 12 and 13 years later, with Patrick's narration parts pre-recorded on video. Even the RAF Theatrical Association performed a version of the musical in the twenty-first century, with Patrick's parts again pre-filmed. In 2006 Brian May would be used as the narrator when this bizarre musical was resurrected, some 17 years after Patrick's first involvement with the performance.

The autumn of 1989 would see the completion of a refurbishment of the 24-inch reflector at Keele University, which as I mentioned much earlier had been part-funded by an elderly BAA Lunar Section member, Harry Thornton, some 14 years earlier. Patrick had been a great friend of Dr Ron Maddison at Keele since the 1960s and Ron had appeared on numerous *Sky at Night* episodes. So, when a Keele 'Star Party' was arranged to celebrate the telescope upgrade, Patrick was only too happy to attend, along with his friend Paul Doherty, who lived relatively nearby. The Astronomer Royal Prof. F. Graham Smith and the *Sky at Night* producer Pieter Morpurgo also attended. Much amusement was had by all when a xylophone

Fig. 31.2 Patrick poses with an inflatable asteroid outside the Cambridge Theatre in London's West End, on the 14th of September 1989. He was performing a narration role in the jukebox musical 'Return to the Forbidden Planet' (By permission of the Press Association)

suddenly appeared and Patrick agreed to give an impromptu performance. This was followed by the Astronomer Royal acting as an auctioneer when one of Paul Doherty's marvellous paintings was offered for sale! No astronomy function was ever dull when Patrick was in town.

In October 1989, Patrick had a brush with the law! While driving his ageing cream coloured P registration Triumph 2000 automatic out of a park onto the A27 Brighton-Lewes road he failed to notice another car coming and collided with it. On October 20th at a magistrates' court in Brighton, he pleaded guilty to careless driving and was fined £100. This was Patrick's first driving offence in over 30 years of owning a car. He would condemn magistrates and traffic police wholesale from that moment on!

Another event that month, which Patrick was most unhappy about, was the latest edition of *Norton's Star Atlas*, published by *Longman* as the eighteenth edition. *Norton's Star Atlas* was one of Patrick's favourite books and he still loved the edition he had owned as a schoolboy. Arthur Philip Norton (1876–1955) had been a British schoolmaster and a BAA member and in the 1970s Patrick, along with his BAA colleague Gilbert Satterthwaite, had been heavily involved in supervising the editing of the latest edition. Not surprisingly, with Patrick involved, there was not that much difference between the 1970s version and his schoolboy version and the epoch of the star charts was still based on 1950 celestial positions (in other words the precession of the Earth's axis in the intervening 20 years had been ignored). Of course, a lot had changed since then, not least better colour printing possibilities and the fact that star atlas publishers were thinking seriously about the fact that the epoch of 2000 A.D. was looming, so star charts would need redrawing. This kind of thing was not in Patrick's field of expertise at all as redrawing star charts and celestial maths was not something that he found easy to understand, or to subcontract to others.

Nevertheless, he was totally horrified when he learned that the new edition was being edited by one of those on his prime 'Serpents' list, namely Ian Ridpath, with the publisher not involving Patrick at all, in any way, shape or form! Despite this, Ridpath had roped in many other BAA members for proof reading and advice. I remember being at a BAA Council meeting in late 1989 when someone turned up with the brand new 'Epoch 2000' *Norton's Star Atlas*, in which Patrick had played no part. "Can I look at that please?" he barked. After a few seconds of flicking through he said "Looks pretty, but they've ruined it, totally ruined it. Vital data missing, paper's too thin, binding won't last one night outdoors. If it ain't broke, don't fix it. No, I tell you, THEY'VE TOTALLY RUINED IT!" With that, Patrick handed it back and his face went into an extremely sulky expression.

Wednesday October 25th 1989 was one of the most memorable BAA meetings I can remember. As well as it being the date of the AGM and the handing over of the Presidency from Henry Hatfield to Colin Ronan, there was a special guest at the meeting. Just before the start, Patrick led the science fiction writer Arthur C. Clarke into the meeting at 23 Savile Row. Arthur was normally resident in Sri Lanka but, that day, had been awarded the CBE, by the Queen, at Buckingham palace. Arthur's brother Fred and his Sri Lankan friend, Hector Ekanayake, had escorted Arthur to the palace and then on to the BAA meeting. It was a fortunate coincidence that had not escaped Patrick: Arthur would be in London on the day of the BAA AGM. He had been a BAA member since 1945 and, as we have seen, attended many BAA meetings in the 1940s and early 1950s, before he was world famous.

Patrick led Arthur through the Savile Row foyer a few minutes before the meeting started and, as the audience settled down, led him into the hall, to gasps from the audience. Patrick was a famous person, but he turned up at most BAA meetings; seeing Patrick and Arthur C. Clarke together though was a truly once-in-a-lifetime experience for most members present! Patrick's usual Churchillian scowl was replaced by a positively beaming expression. Before starting the meeting, the outgoing President Henry Hatfield introduced Arthur (as if he needed any introduction) and Arthur said that it was truly a great pleasure to be able to address the meeting, albeit briefly. He regretted he could not stay for more than a few minutes but said that he was proud to be a BAA member. He then opened the case, containing his CBE and showed it to the audience. It was the only BAA meeting where I can ever recall someone getting a spontaneous standing ovation. Following this, Patrick helped Arthur out of the hall, to his waiting limo, and then returned. Patrick and John Mason gave a talk on the Voyager probe's Neptune results, later in the meeting. A few months later, Arthur donated £100 to the BAA's Horace Dall award fund.

The Practical Joker

Patrick was always keen on playing the fool on TV, or springing surprises and practical jokes on people. He knew Esther Rantzen (her husband, Desmond Wilcox, ran the department within which *The Sky at Night* was produced in the 1970s) and so when Esther asked him to trap someone on the BBC *Hearts of Gold* programme he was only too happy to oblige. The victim was a well-liked resident of Lye in the West Midlands, a certain Tony Dyson. Mr Dyson had been tricked into meeting Patrick, so that his friends could honour him and, while a totally fictitious programme called 'Planet' was being filmed, a confused Tony was ushered onto the set, just as Patrick started the introduction. A worried and increasingly bemused Tony looked around nervously as the opening credits rolled. His expression said *What on Earth am I doing here?!* The other studio guests, Patrick's astronomy disciples since the 1950s and 1960s, Peter Cattermole and Iain Nicolson, were already in on the practical joke. In the spoof 'Planet' programme, broadcast on 4th Nov 1989, Patrick opened with the words:

"Good evening and welcome to this special edition of 'Planet'. What we are going to do this evening is discuss the spectra-photometric results from Voyager 2, now that it's by-passed Neptune and is on its way through the heliopause. I'm delighted to be joined by Dr Peter Cattermole of Sheffield University, Iain Nicolson of Hatfield Polytechnic and photometric expert Tony Dyson. Now, first of all Peter: what do you make of the recent Lyman Alpha results?"

Peter Cattermole, playing along nicely, but (I'm sure) trying not to laugh replied: "Well, they're very difficult to interpret really. I think this is something to do with the very weak plasma effect coming from the Sun. I think we're having a vortexing kind of motion of Neptune's magnetic field. It's just shifting everything over to the red!"

Patrick then turned to Tony Dyson. "Well, if I might come onto you now, with your special knowledge of photometry, what do you make of this? Do you think you'll get as far as the helium beta. I read your paper in Nature."

Tony found it hard to speak, but mumbled:

"I must be in the wrong studio."

"No, I read your paper in Nature."

"I think you've got the wrong Tony Dyson."

"No we have the right Tony Dyson, we're very conscious of your achievements and we've actually produced a special film that's coming up on the screen in one moment and it may present some views you haven't actually seen yet." A clip of Mr Dyson's achievements were then shown and the wall behind him opened up to reveal Esther Rantzen and Tony's friends from Lye, dressed in carnival gear. Esther had bagged her victim and Patrick, Peter and Iain had played along nicely. Also in November 1989, Patrick performed once more on the xylophone at the Royal Variety Performance, accompanied by the entertainer Roy Castle. It would be the first of two occasions in that month that he would work with Roy, as Patrick also appeared on *Record Breakers* on November 24th, talking about his record breaking stint as a presenter on TV.

Incidentally, despite being an amateur astronomer Patrick was often invited to give talks to university students and he earned a significant income from these duties, especially in the 1980s. Patrick was a guest Planetary Astronomy/Planetology lecturer for many years, from the late 1970s through to the early 1990s, at Queen Mary College in London (on the 'Concepts in Contemporary Astronomy' course) and often used to retire to the local Cricketers Arms pub with the students after one of his talks, where he would entertain them with amusing tales of his astronomy experiences. From the early 1980s he had been an Honorary Fellow of the college, despite his knowledge of mathematics being distinctly schoolboy standard.

In December 1989 Patrick submitted the latest revised BAA Lunar Section publication *Guide for Observers of the Moon* to his local printers and, by January, it was available to all Lunar Section members. Paul Doherty had been roped in to provide the illustrations and I had written the chapter on lunar photography. But Patrick also received sad news at the start of the year. His Selsey friend Reg Spry had died, aged 87. Reg Spry and Rossie Atwell had often been Patrick's travelling companions when he drove to Lunar Section and BAA meetings around the country. Indeed, for most of the 1980s Rossie had lived in the octogenarian Reggie's house at 60 Windsor Road in Selsey. It was quite a common sight in the 1980s to see Patrick draw up in his ageing cream Triumph 2000, prise himself out of the door, and then for Reg and Rossie to also emerge. Reg Spry had come to astronomy late in life, aged 65 in fact, but he had retired in Selsey at the same time that Patrick had arrived there from Armagh. Reg had overcome quite a few tragedies in his life, tragedies that had gained Patrick and Rossie's sympathy. He had outlived both wives and his only son had died from meningitis at the age of 39. Reg had a Professor Branestawm type character and loved assembling telescopes and observatories from the crudest of junk! Patrick was fascinated by this and encouraged Reg to write a small book on telescope and observatory construction, which he did. Patrick also persuaded Reg to serve several terms on the BAA Council and he became a well known figure at BAA meetings. Following the death of Reg Spry, Rossie Atwell (who had in many ways been a carer for Reg) moved to her final address, a flat at 56 St Peter's Crescent in Selsey, roughly half a mile north of Patrick.

Although Patrick's Triumph was well past its best, especially after it collided with another car on the Brighton – Lewes road, it was certainly not the biggest pile of junk he ever owned. His venerable Ford Prefect 'The Ark' had notched up three quarters of a million miles for him (three times the Earth-Moon distance!) and his MG Magnette had seen much service too. But undoubtedly the biggest rust bucket he ever owned, just before the Triumph, was a long, yellow, 1972 vintage, L-registration Ford Cortina estate, with a black vinyl roof and totally corroded wheel arches. That was a wreck to beat all of his wrecks and Patrick called it 'The banana' as the paint was perfectly banana coloured and the black vinyl roof looked like a bruised banana skin! The fact that the car was not perfectly straight after 300,000 miles of use made that nickname even more appropriate! He finally got rid of that heap in 1984, when he acquired the P-registration (1975 vintage) Triumph 2000. Patrick claimed that he could not fit into many cars. Well, there were many people who were taller than Patrick, but not many who had such a large girth and such ridiculously large feet! His hands were extraordinary too. By 1990 each hand, swollen by arthritis, resembled a bunch of five bananas!

At the BAA Christmas meeting of January 6th 1990 the workaholic that was Patrick, once again, acted as the meetings recorder. He had been doing this onerous job for most of the past 40 years and, although occasional replacements were found, it always kept coming back to him. To be honest, Patrick's ultimate successor as the only equally reliable and quick meetings recorder was me! I took the role on from late 1990 to late 1997, after which I became the BAA President. After 7 years I'd had enough. How Patrick did it for almost 40 years is a mystery to me! Patrick's friend Paul Murdin was the Christmas lecturer at that January 6th meeting and following Paul, the BAA comet photography expert Harold Ridley gave details of an exciting new discovery by the New Zealand amateur astronomer Rodney Austin. Harold Ridley was a BAA gentleman of Patrick's era. A man you could trust and a perfectionist when it came to meteor and comet photography. He had appeared on a number of *Sky at Night* programmes throughout the 1960s, when the programmes were broadcast live. Harold was tall and incredibly thin. When you saw him standing next to Patrick (they were a similar height) Harold looked about half Patrick's width! In fact, this was probably not far off the truth, as Patrick admitted he was 'only' 19 stone and Harold must have weighed ten stone, wringing wet! Harold had been a lifelong smoker and this ultimately would kill him off, through emphysema, some 5 years later; but in 1990 he was still in reasonable health. Harold explained to the BAA audience that the comet discovered by Austin, on December 6th, was due to reach its closest point to the Sun on April 9th and the prospects looked good for a bright, naked eye spectacle. While I was talking to Harold, after the tea break, Patrick asked him if he would like to be on *The Sky at Night* again, talking about the comet, and Harold agreed. He added, rather modestly, that Patrick could probably do just as good a job on his own, but Patrick insisted that he wanted someone with a good knowledge of photography on the programme. In 1990 it had been some 14 years since a truly great comet had been seen from the UK and that 1976 comet, West was it's name, was seen by very few due to cloud during the critical weeks. After the Kohoutek flop of 1973, it was important to be cautious, but the best prospects for Austin did look mouthwatering.

The *Sky at Night* programme with Harold and Patrick was broadcast on the 11th of March 1990. Patrick showed a picture of comet Bradfield 1987s passing near to a star cluster and with the words "The astronomer who took that is one of our leading experts on comet observation" introduced his guest. Harold explained that when he started in astronomy, as a schoolboy, he thought it would not be long before a truly great comet came along, but in fact he'd had to wait 30 years until he saw the first one, and that had been comet Arend-Roland: the very comet that ushered in *The Sky at Night*. Harold advised caution over comet Austin, but was reasonably optimistic that the comet would put on a good show. In fact, for well-equipped amateur astronomers, it was a very nice comet, peaking at fifth magnitude in the late April pre-dawn sky. However, for the general public it was another disappointment, never reaching naked eye status. Fortunately, it had not been hyped to the level of Comet Kohoutek some 17 years earlier. Five days prior to that *Sky at Night* episode

Patrick had appeared on the long running ITV series *This is Your Life* as the final guest on the programme, celebrating the life of the percussionist Evelyn Glennie. Patrick accompanied Evelyn on the xylophone as the ending credits rolled!

From April 27th to the 29th, Patrick travelled with 50 BAA members to a special meeting at Birr castle in County Offaly in Ireland; he then proceeded to Dublin. The meeting was part of the BAA's centenary year celebrations, even though the actual centenary was not until October. Patrick had visited the site of the giant and historic 72 inch telescope many times, but not accompanied by 50 other BAA members. At the University College in Dublin, Patrick gave an inspiring talk about the work of famous Irish astronomers of the past.

Two months after the Comet Austin *Sky at Night*, in May Patrick found himself live on the BBC *Wogan* chat show, except that the normal host, Terry Wogan, was on holiday and a very young Jonathon Ross was hosting the show. The main reason for Patrick being on the programme was that the Hubble Space Telescope had been launched and was sending back first light images. Unusually for Patrick he was looking exceptionally smart, dressed as he was in an immaculate white suit and black bow tie. Ross quipped that he thought it was Sean Connery coming towards him. Also on the show that night was the comedian/actor Hugh Laurie and the actress/musician/erotic artist Phoebe Legere, star of the bizarre *Toxic Avenger* films. Miss Legere was dressed in the most skin-tight white catsuit imaginable and as soon as she sat down right next to Patrick he shuffled so far along the sofa (away from her) that Hugh Laurie was barely able to cling on to his seat. However, when Miss Legere demonstrated the paintings on her skin-tight costume, Patrick's monocle was soon in place! As the programme closed, a bizarre musical spectacle could be witnessed, namely: Miss Legere on the piano, singing *Johnny be Good*, Hugh Laurie on the drums and Patrick on the xylophone!

Also in May 1990, Patrick travelled once more to the island of La Palma in the Canary islands. Patrick had made a number of *Sky at Night* recordings from the island in previous years, but it was time to revisit the site to see what new observations had been made by the 4.2 metre William Herschel Telescope. In August the *Sky at Night* covered the spectacular radar mapping of Venus by the NASA Magellan probe. Once again, Patrick was joined by the planetary geologist Peter Cattermole.

Chapter 32
A 100th Birthday and Madness in Buenos Aires

Throughout the Presidency of Henry Hatfield and into the Presidency of Patrick's great friend Colin Ronan, the BAA had been planning its centenary celebrations. The first ever meeting of the Association had been held on October 24th 1890 and, 44 years later, the 11 year-old Patrick Moore had joined the Association. Patrick had therefore been a BAA member for 56 years at the time of the centenary. The centenary meetings were organised with skill and enthusiasm by Dr John Mason, who excelled at organising these types of events. Indeed, some 9 years later, when I was the BAA President, I was indebted to John for organising an event of equal size at Truro, for the cloudy Cornwall eclipse. By the summer of 1990 two special meetings had already been held (in Ireland) and another centenary year meeting was held at Patrick's familiar haunt of Dundee, on the weekend of August 18th/19th. Patrick had chaired the meeting and also gave an evening public lecture entitled 'The Voyager Encounters' to a packed audience at Dundee University's main hall. In September Patrick had headed north to Liverpool for a BAA provincial meeting on the 15th, hosted by Liverpool Astronomical Society. This was a special meeting in the BAA's centenary year as Liverpool A.S. was actually older than the BAA by 9 years, being formed in 1881. Patrick gave a talk, to another packed audience, on the subject of the favourable opposition of Mars in 1990–1991 and encouraged those present to make observations of the planet.

For the October *Sky at Night* programme Patrick had travelled to Armagh Observatory, in Northern Ireland, where he had been the director from 1965 to 1968. By a strange coincidence, Armagh Observatory had been celebrating its 200th anniversary in 1990 while the BAA had been celebrating its 100th. Patrick had promised the BAA Council that the BBC would do a special programme on the BAA's Centenary. In fact, they did a programme on Armagh's bi-centenary instead! Also in that month, Patrick travelled to Cape Canaveral to watch the launch of the solar polar probe, Ulysses, on October 6th.

On October 16th the Royal Mail issued four astronomy themed postage stamps, and of course they wanted Patrick to promote them. The stamps, priced at 22, 26, 31 and 37 pence depicted 200 years of Armagh Observatory, Herschel's telescope,

M. Mobberley, *It Came From Outer Space Wearing an RAF Blazer!: A Fan's Biography of Sir Patrick Moore*, DOI 10.1007/978-3-319-00609-3_32,

Greenwich Observatory and Stonehenge. Patrick, dressed in a bizarre sort of Druid magician's cape and hat, at Stonehenge, played to the cameras 4 days earlier, on October 12th, and talked about the stamps and what they signified. However, when asked if he was a stamp collector, he sheepishly admitted that he was not.

Despite everything taking place in Patrick's life he was still endlessly answering letters from children, every day, on his ancient typewriter. One such letter was dispatched on October 21st 1990, to a young boy named Paul Abel, who in the twenty-first century would become a co-presenter on *The Sky at Night*. Paul shared Patrick's letter with me some 22 years later. It was typical of Patrick's style:

> Dear Paul,
> So glad you've made a start in astronomy, you'll find it the most rewarding hobby, except when Spode strikes!
> Oh dear, in a feat of supreme efficiency I managed to lock myself out of my dome which contains my beloved 15″ reflector; by the time I got in naturally the clouds came up! These things happen!!!
>
> Best wishes,
> PATRICK MOORE

The big BAA Centenary day was set for Wednesday October 24th at the Commonwealth Institute in Kensington, London. The usual venue, at 23 Savile Row, had seating capacity for 200 members, but 450 were expected for the Centenary. Colin Ronan proved to be the perfect Centenary President as he was an astronomical historian of some renown and was the ideal man to receive the BAA Centenary 'present' from the Garter King of Arms of the College of Heralds. Thanks to a generous donation from the American amateur astronomer Ben Mayer, the Association could now boast its own BAA 'Coat of Arms'! The motto, created by Mr Mayer, read: "Noumena Transcend Phenomena". This was, to Mr Mayer, supposed to mean "Vision or Ideas Transcend the Phenomena we observe". The amateur astronomer and asteroid discoverer Brian Manning, who accepted the Horace Dall medal at the Centenary Meeting, announced that asteroid 4522 had been named 'Britastra' in the association's honour.

Needless to say, Patrick was a major player in the Centenary celebrations and gave an entertaining address to the 450 assembled members. The address was entitled 'The future role of the amateur astronomer'. Patrick explained that he had been a member of the BAA for more than half of its history and, when he had joined, some of the original members were still alive. "I wasn't one of them, despite appearances" he quipped. Patrick described himself as "A dinosaur". He explained: "I don't have a computer; I don't have a photometer; I wouldn't understand a binomial theorem if you handed it to me on the end of a skewer". He continued: "I am still one of those who observes things like the Moon. Going back only a few decades we were very busy charting the lunar surface, and we were doing it visually. Then along came amateur photography." Patrick showed a lunar photograph at this point. "This photograph is by one of our leading photographers, our immediate Past President, Commander Hatfield, who produced that magnificent lunar atlas some time ago." After showing a slide of astronauts on the lunar surface, Patrick continued: "What can the amateur still do? Believe me, the answer is a very great deal. The amateur has the great advantage of being able to concentrate on time-dependent phenomena. There is still a great deal to do, particularly with regard to TLP's, or Transient Lunar Phenomena, whose reality is no longer in doubt, but which we do not yet fully understand. The Lunar Section is very busy working on that and will continue to do so in the coming years. That is only one of many aspects of lunar research that is not, in the main, covered elsewhere." Patrick then described how amateurs could contribute to solar and planetary observing as well as discover asteroids, comets and supernovae. He concluded his address by saying: "I am quite sure that in a hundred years from now, at our Bi-centenary meeting, there will be someone saying to you just what I am saying now and saying 'Now let's look forward to the BAA's third centenary!'"

Fig. 32.1 Patrick at the 1990 October BAA Centenary Council meeting with, *in the centre*, Harold Ridley (holding the envelope) and, *on the right*, Howard Miles (Photograph by the author)

My own role in BAA affairs increased dramatically after the BAA Centenary as I became the BAA Papers Secretary, the meetings recorder (Patrick's old role) and the monthly 'Sky Notes' speaker at all of the meetings. The first meeting at which I took onboard all of these roles (November 1990) was also one in which Patrick must have given his fastest talk of all time! As well as my 'Sky Notes' the 2 hour meeting featured major talks on X-Ray astronomy, meteors and comets. With only a few minutes of the meeting remaining, Colin Ronan asked Patrick if he still wished to show his slides. Patrick confirmed that he did and gave a talk on recent observations of Mars, Jupiter, Saturn (and its white spot) and Pluto in an astounding 90 seconds flat!

During 1990, Patrick had featured little in the TLP network's activities. He had reported few events on the lunar surface that he considered 'abnormal'. However, his interest in lunar matters was as keen as ever and in the January 1991 BAA Lunar Section Circular Patrick gave details of a possible historic TLP, he had stumbled across:

Fig. 32.2 Patrick, chatting with this author, at the BAA Centenary meeting exhibition in October 1990 (Photographed with the author's camera)

Recently I was reading an old book: Aurorae, by J. Rand Capron, published in 1879. In it (page 73) there is an account of what certainly seems to be a TLP. Rand Capron writes:

'A curious case, in which a patch of vapour or cloud was supposed to be detected on the Moon's surface, is reported by the Rev. J.B. Emmett in a communication to the 'Annals of Philosophy' (New Series, vol. XII, p81). It is dated 'Great Ouseburn, near Boroughbridge, July 5 1826', the observation being made with 'the greatest care with a very fine telescope.'

'On the 12th April 8h, while observing the part of the Moon called Palus Mœotis by Nevelius, with an excellent Newtonian reflector of 6 inches aperture, at a particular part of the Palus, which he minutely describes, he saw, with powers 70 and 130, 'a very conspicuous spot wholly enveloped in black, nebulous matter, which, as if carried forward by a current of air, extended itself in an easterly direction, inclining a little towards the south, rather beyond the margin of Mœotis.' April 13th 8h to 9h, the cloudy appearance was reduced both in extent and intensity, and the spot from which it seemed to issue had become more distinctly visible. On April 17th scarcely a trace of the nebulous matter remained; but so long after as June 10th 8h 'a little blackness' remained around the spot. If this observation was (as it certainly seems to be) critical and exact, there must have been a disturbance on the Moon's surface, indicating some sort of cloud – or vapour – supporting atmosphere; and probably for the purposes of Aurorae, an atmosphere of a very rarified condition would suffice.'

Patrick commented:

'Nevelius' is presumably 'Hevelius'. I am not yet sure about the identification of his Palus Mœotis, but I am working on it – and this does seem to be a clear case of a TLP seen by an experienced observer more than one hundred and fifty years ago!

In March, Patrick was, once more, headed for Keele University. His friend of the last 25 years, Dr Ron Maddison, was retiring. He was emigrating to become the Director of the Observatory at the Astronaut Memorial Planetarium in Cocoa, Florida. Patrick was more than happy to deliver the first ever 'Ron Maddison lecture' on the departure of his friend. As well as sharing a lifetime's interest in astronomy both Ron and Patrick had been recipients of the local engineering society's *Reginald Mitchell Gold Medal.* Mitchell had been the designer of the Supermarine Spitfire.

Fig. 32.3 Patrick and Ron Maddison at the first 'Ron Maddison Lecture' at Keele University in 1991, the year of Ron's retirement from Keele and his emigration to Florida (Picture by kind permission of Ron Maddison)

In April, Patrick (with John Mason and a small group of amateur astronomers and space enthusiasts) went on a trip to the Russian equivalent of Cape Canaveral: the Baikonour launch centre. It is not far from the border with Afghanistan and, for the first time, had been opened to Western tourists. To get there Patrick travelled from London to Helsinki, then on to Moscow and then, on a Russian military plane, to the airfield at Krainj. From Krainj, you can drive to Baikonur, via Leninsk. While at Baikonur, Patrick witnessed the launch of a Mir Space Station supply rocket. The visit was filmed by a BBC crew and broadcast on *The Sky at Night* in July, by which time Patrick was heading for the next total solar eclipse, in Mexico.

An Eclipse and Mayhem in Buenos Aires

The total solar eclipse of July 11th 1991 had been dubbed 'The Eclipse of the Century', although that honour had already been used twice before. The 1973 Total Solar Eclipse was given the same name, as was that of 1955! There is a tendency for very long duration eclipses to all be termed 'Eclipse of the Century'. Certainly the maximum totality duration of 6 minutes and 54 seconds was impressive and only the eclipses of 1955 and 1973 could realistically be compared with the 1991 eclipse, in Patrick's lifetime. There were two prime regions to head for to see this eclipse at its best: Baja California, Mexico or the Big Island of Hawaii. *Explorers Tours* chose Patrick and John Mason as their guest lecturers on the trip. Patrick, luckily, went on the Baja trip. I, sadly, went on the Hawaii trip and was clouded out. Despite being in Hawaii, John Mason and a few others did see the eclipse just as totality arrived. Fortunately they were on a bus which ascended the mountain road; whereas our obnoxious, fat, loud mouthed slob and bully of a tour guide made it quite clear, well in advance, that we would not be ascending any mountains, so we missed it. But I'd rather not recall that pain. However, from Hawaii, only a 4 minute totality could be seen anyway, even in clear skies. From Patrick's location, at Baja, a near maximum eclipse of almost 6½ minutes was seen.

Fig. 32.4 Patrick on July 8th 1991 in Los Angeles, at the pre-eclipse briefing, prior to the July 11th 1991 Total Solar Eclipse in Hawaii and Mexico. Photograph by the author

Both tour itineraries, comprising 750 travellers, were briefly in Los Angeles on July 4th and Patrick addressed the eclipse chasers, at several talks, with his usual entertaining, enthusiastic, but cautionary eclipse talk. On the flight from Heathrow to L.A. Patrick had been seated next to a known eccentric (barking mad to be honest) Birmingham eclipse-chaser who drove him around the twist, and he was glad to get off the plane. You have to be a little cautious when observing a total solar eclipse: it is only too easy to damage your eyesight and Patrick was keen to stress this in L.A. His friend Hedley Robinson, an experienced amateur astronomer, had damaged his own eyesight when looking through an unsafe solar filter. The July 11th 1991 eclipse would be Patrick's seventh total solar eclipse. From his site at La Paz, on the verandah balcony of a luxury hotel, he saw the eclipse, with the Sun near the zenith and a glass of wine at his side. Sheer luxury compared to the cloud that most dawn eclipse-chasers experienced in Hawaii!

Immediately after the Mexico-Hawaii eclipse, Patrick was on the move once more. The International Astronomical Union, or IAU, had a General Assembly every 3 years and Patrick and John Mason flew from their post-eclipse Mexico and Hawaii sites to get to the 1991 Assembly in Buenos Aires, Argentina. The first day of the conference was not until July 24th, but they needed time to prepare for their roles at the conference. The IAU Secretary-General had asked Patrick to be the editor of the conference newspaper, which summarised the previous day's presentations and the coming day's events. Patrick, with help from John, agreed to undertake this task. Although he and John were mere amateurs, they were well known throughout the astronomy world as people who got things done.

At a BAA meeting a few months later, by which time John was the new BAA President and so chaired the meeting, Patrick gave a hysterically funny account of that IAU meeting. As the BAA meetings recorder, I wrote his entire talk up for the Journal, but I was gutted when the BAA Officers and Patrick decided that his talk was "off the record" and could not be published! It might, apparently "offend the IAU". In 2003, when Patrick published his very patchy autobiography, he included the IAU conference farce in it, with no details spared. I asked him about this and he said he had used my write-up as the basis for the section in his autobiography! He went strangely deaf when I queried why his write-up would not offend the IAU, but mine, apparently, would! Anyway, I include below, Patrick's humorous account, as I recorded it, at the 1991 November 20th BAA meeting at 23 Savile Row in London, 4 months after the conference. As it was intended as a report for the BAA Journal I refer to Patrick as Dr Moore, but it is otherwise fairly informal.

> Dr Moore said that the IAU met every three years and in 1991 the venue was in Buenos Aires. Mr Derek McNally had asked Dr Moore and Dr Mason to edit the daily IAU newspaper which kept the delegates informed throughout the conference.
>
> Dr Moore said that just prior to the conference he had observed the total solar eclipse in Mexico from his hotel balcony with a glass of wine at hand. He tried sweeping for comets during totality and took some fine slides (shown during the meeting). After joining up with John Mason (who had observed the eclipse from Hawaii) they boarded a plane for Buenos Aires. Their plane journey was uneventful. On arrival at their hotel they played half a dozen games of pool at which they discovered they were equal in incompetence! As long as they did not mention the Falklands they were probably going to be OK (or so they thought). Their first hotel was interesting; they were on the 18th floor and the lift (which could take five people only, or four with Dr Moore!) took 24 minutes to reach the ground floor. They decided to move to the Savoy Hotel. This Hotel featured flooding toilets, and radiators which could not be turned off. The net effect of this was that the floor resembled the surface of Venus!

The intrepid duo decided to set about acquiring the facilities to produce the daily conference newspaper. They eventually found a totally incompatible computer and an equally incompatible printer at the local newspaper office. Attempts to borrow a car were fruitless and, in keeping with events, there were strangely no mains plugs in their office. A botched, potentially lethal system, of transformers and live wires stuffed into sockets was lashed together, but the printers remained stubbornly incompatible. In addition, the office assistant spoke no English and had to be communicated to by using semaphore! Dr Moore asked for a typewriter, which was duly supplied, but it did not function at all. The duo returned to their hotel to discover that 'half a million' of the local currency would not buy a round of drinks. Eventually, and out of character, Dr Moore resigned himself to using a computer word-processor to produce the daily newspaper. At this point it was discovered that the IAU press office had no fax machine or even a phone! Eventually a fax and a phone were acquired and shared between 1200 delegates!!! Unfortunately, the number of the phone was not known. The newspaper logo was bodged together from a photocopied IAU letterhead but the printer then refused to make copies unless provided with more money. Dr Moore described the local traffic as chaotic, a particularly interesting feature being the non-synchronised pedestrian crossings and traffic lights. This was probably un-important as the motorists in the city ignored traffic lights anyway! Dr Moore crashed his car in the city and Dr Mason's taxi ran over a pedestrian! The delegates eventually received identity tags, Dr Moore's tag describing him as a French delegate. He promptly produced his own tag stating categorically that he was not French (to be precise "Je ne suis pas un frog"). Events carried on in much the same vein, with the IAU computer breaking down, the Spanish helper in the office getting stuck in the lift, and the overhead projector generating large sparks. It came as no surprise to anyone when workmen arrived outside the conference venue and started digging up the road. Eventually mailboxes appeared for the delegates; Dr Moore found his mail in box 5003 (between boxes 462 and 463!) Enquiries revealed that the order of the boxes was based on the third letter of the delegate's surname. After hearing this useful information Dr Moore broke his glasses by sitting on them. The next day a room numbering error resulted in many delegates finding themselves in an Argentinian Drama Class!! The local host of the conference insisted on there being an odd number of people on the stage so he could be in the central position and pronounced that 'this was a great day for Astrology'. At this point, inexplicably, a third bed mysteriously appeared in Dr Mason's bedroom, the mailbox order became even more incomprehensible and an electrician's error crashed the computer. On the penultimate day the conference venue hotel burned to the ground! After a change of venue, the closing banquet was held and one in three delegates managed to acquire a knife and fork. The eating arrangements were unusual.... a carvery dumped in the middle of the dance floor. Finally, it was time to head back to England via *Virgin* airways. Upon boarding the plane Dr Moore remarked that he had 'never been more pleased to get into a *Virgin* in his life'. As his final slide, Dr Moore showed John Mason scaling an Argentinian lampost in an attempt to carry out a study of the local light pollution (and possibly after consuming some alcohol?) The Savile Row audience (by this time semi-paralytic after Dr Moore's talk) applauded warmly and the President remarked that he would be talking to Dr Moore about his final slide! The President then invited Mr Martin Mobberley to give his monthly round up of the night sky..........

On the first day of the IAU Conference, Patrick had arrived at the San Martin Centre in Buenos Aires and had been greeted by the former Astronomer Royal, Professor Sir Francis Graham-Smith. Professor Smith greeted Patrick with the words "We have some news about the new planet". Patrick was taken aback. He knew nothing about the claimed discovery, from Jodrell Bank, of a planet orbiting a neutron star 30,000 light-years away. Patrick put the discovery on the front page of the first day edition of the IAU Newspaper. However, Patrick immediately had doubts. He was only an amateur, but the claimed orbital period of the new planet (derived from variations in the rhythmic pulses of the neutron star) was approximately 186 days, disturbingly close to half the Earth's period around the Sun, as Patrick instantly realised; this, and the unlikelihood of a planet orbiting a neutron star, all seemed very dubious to Patrick but, nevertheless, he resolved to hold a *Sky at Night* on the subject if the discovery claim survived the IAU conference.

Maybe one reason why Patrick opted to see the July 1991 eclipse from Baja California was because he knew he would be travelling to Hawaii in late August anyway? Patrick had been to Hawaii before, most recently in 1986, but since then, the world's largest telescope, the Keck, had been under construction there. Patrick had his first view of the enormous Keck dome and the instrument inside, on Wednesday August 28th 1991. He described it as breathtaking, even though only 12 of the 10 metre diameter mirror's 36 segments were then in place. Needless to say, much Keck footage was acquired for *The Sky at Night.*

Back in the UK

The weekend of September 20th/21st saw Patrick at yet another BAA Out-of-Town meeting in the north of England. This time the venue was the University of Durham. With no-one else offering to take the meeting minutes, once again Patrick recorded them. The last time the BAA had met in Durham, Patrick had been the President. Following a talk on historical astronomy by Dr Richard Stephenson, Patrick had a few questions. He said that he was very surprised to hear that there were no records of the solar corona before 968 AD. This might indicate a prolonged solar minimum of the same type as the Maunder Minimum of 1645–1715. Patrick suggested that there could be a way to check: naked eye sunspots had been recorded in historic times, so why not check these against the times of eclipses. If it could be shown that a major sunspot was in view near the time of an eclipse, it would indicate that the sun was not undergoing a Maunder-type minimum, despite any missing descriptions of a solar corona. Dr Stephenson thanked Patrick and said this had not been investigated. He said it was a most valuable suggestion and he would look into it. Despite being, technically, an amateur astronomer, Patrick could often spot aspects of astronomy that the professionals had overlooked. But Patrick's most poignant suggestion of that weekend was yet to come. The Astronomer Royal, Professor Arnold Wolfendale, was at Durham and gave a splendid talk about the big issues and new technologies in 1991 astronomy. He mentioned the Jodrell Bank 'planet orbiting a neutron star' mystery: it was a very strange place to discover a planet. Patrick voiced his concerns after the Professor had concluded his talk. Specifically, he asked Professor Wolfendale whether it had been considered that the problem of the planet orbiting a neutron star could be related to the fact that the orbital period, of 186 days, was almost exactly half the Earth's sidereal period; could there be any misinterpretation there? The Professor said that this seemed not to be the case, because other pulsars in the same direction did not show similar effects. However, he did concede that the circular orbit of the alleged planet was puzzling. A highly elliptical orbit would be far more likely.

On October 16th Patrick was saddened to hear of the death of his dear friend J. Hedley Robinson, at the age of 86. Since the 1940s Patrick had worked with Hedley in the BAA's Mercury & Venus and Lunar Sections and Hedley had been one of Patrick's strongest supporters in the field of TLP patrolling. Patrick had been a frequent visitor to Hedley's home at Teignmouth and became a supporter of Torbay Astronomical Society and the Torquay Boys Grammar School Astronomical Society.

Wednesday October 30th saw Patrick at the 1991 BAA AGM at 23 Savile Row. At the end of this meeting two of Patrick's best friends would exchange the Presidency: Colin Ronan would retire and hand the BAA's reigns over to John Mason. As John had been the BAA's Public Relations Officer Patrick happily agreed to take on that P.R.O. role for the next 2 years, while John was the President. In fact, Patrick would hold that voluntary post for the next 9 years! Colin's outgoing Presidential address was entitled 'The Origins of the Reflecting Telescope' and included the dramatic claim that a certain Leonard Digges, in the mid sixteenth century, had invented the Reflector well before Isaac Newton. Indeed, the claim was that Digges' telescope pre-dated Galileo, making an Englishman the inventor of the telescope. Patrick was always keen to plug the work of his most loyal friends, especially the more modest ones, and Colin was one of these. He alerted the press to the BAA meeting and to Colin's dramatic historic discoveries and so various journalists from the national media attended the meeting of October 30th! The next day the front page of the *Daily Telegraph* had a banner proclaiming that the telescope was invented in England! The top headline read: 'Now it can be told: British scientists beat Galileo by 33 years', above the obviously minor headline proclaiming that President Bush wanted Israel and the Arabs to compromise over land ownership. Patrick knew how to attract press publicity: but getting the BAA onto a national newspaper's front page was a first! Around this time, in the earliest years of satellite TV, Patrick occasionally appeared on *Sky*, the satellite service launched in 1989. Towards the end of 1991 he was presented with a plaque for being the 500th guest on Derek Jameson's nightly *Sky* chat show.

In December, Patrick broadcast *The Sky at Night* about the planet, discovered from Jodrell Bank, orbiting a neutron star. Only 1 month later, Patrick's concerns about the discovery were totally vindicated. Astronomy Professor Andrew Lyne released a statement admitting that there had been errors in his calculations: the alleged planet was, as Patrick had suspected, just an artefact of the Earth's orbit around the Sun. Patrick said that this proved the absolute integrity of Professor Lyne and the Jodrell Bank team: they had admitted their error. In fact, what it did prove, was that Patrick's common sense amateur astronomer instincts often made him just as much an astronomer as the leading professionals; if only his instincts had been as good where the lunar craters and lunar 'activity' were concerned!

TLP Co-ordinator Yet Again!

As 1991 came to an end a big crisis came to a head in the BAA Lunar section. I have to admit that I was a major part of that crisis. During 1990 and 1991 a number of post-holders in the BAA Lunar Section (myself, Denis Buczynski, Rob Moseley,

Gerald North and Nick James) became increasingly at odds with the Lunar Section Director, Peter Foley. Peter was a hard working director, but seemed to us to rarely listen to the views of myself and the younger committee members, especially where the reality of recent TLP events were concerned. Over the summer months of 1991 Rob Moseley, myself and Nick James resigned our posts and Rob produced an illegal unauthorised copy of the Lunar Section publication 'The New Moon', containing criticisms of the directorship. Not surprisingly, all this came to the attention of the new BAA President, Dr John Mason. Gerald North (now a leading astronomy author) agreed to take over the editing of 'The New Moon', but as 1991 came to an end Gerald complained to the BAA Council that the director was not giving him editorial responsibility. In an attempt to salvage the rapidly collapsing Lunar Section, John Mason asked Peter Foley to attend a BAA Council meeting, so that a solution could be thrashed out. Peter Foley refused and resigned, throwing the Lunar Section into disarray, because it was Peter who produced the monthly Lunar Section Circulars that Patrick had instigated in the 1960s. Suddenly, Patrick's beloved BAA Lunar Section was virtually dead, and I was partly to blame! The director and circulars editor had resigned, three sub-section co-ordinators had resigned and Gerald North was on the verge of resigning. As the BAA Papers Secretary at the time, I was the only person who was seeing opposing sides of the situation, both from the 'rebels' standpoint and the council's standpoint.

At the BAA's 'Christmas' council meeting of January 4th 1992, the Council asked Patrick to stand in as the acting Lunar Section Director until a potential replacement for Peter Foley could be found. Patrick would need to find the material for the next Lunar Section Circular, edit it and duplicate it, as well as travelling to Tucson to give the opening address (The Discovery of Neptune) in the first "post-Neptune" Voyager conference. Rossie Atwell, a Lunar Section stalwart, agreed to gather in material for an 'emergency' Lunar Section Circular while Patrick was away. Of course, Peter Foley had also been the TLP sub-section co-ordinator, so Patrick agreed to take onboard this role too! By the time Patrick was in a position to issue the next Lunar Section Circular, January was disappearing fast, so he combined the January and February circulars into a 12 page edition, wrote a brief TLP sub-section paragraph and dispatched the copies to members. Approaching 69 years of age Patrick was, temporarily, re-burdened with all his old Lunar Section chores! In addition, at this time, Patrick was top of the bill at the first ever European *Astrofest* exhibition, organised by his *Astronomy Now* magazine in London. But, despite all this effort, Patrick still had time to observe and, on February 16th at 01.20 GMT, he even reported 'a mistiness' on the north wall of the lunar crater Langrenus, but thought it was probably not a 'genuine' TLP. At the February 26th BAA Council meeting, the Council endorsed Patrick's candidate as the new Lunar Section director. It would be Jeremy Cook, a quiet man, with no enemies: the only realistic candidate after the turmoil of late 1991. Patrick tried to persuade Peter Foley to remain as the Lunar Section's TLP co-ordinator, but without success. Peter's knowledge of the lunar surface and, especially, of suspect TLP regions, was second to none. He was also an excellent lunar photographer, but he would never return to hold a post in the lunar section again and those bizarre Ptolemaeus pictures labelled 'Top Secret', kept in his study drawer, made me doubt his sanity. Foley would die in March 2008, aged 77.

So, Patrick was, once more, the BAA TLP co-ordinator and he worked hard to get the sub-section back to an enthusiastic state. Gerald North and Jeremy and Marie Cook were still keen stalwarts of the sub-section, but others, like me, were rather disillusioned by the whole lunar scene after the events of 1991. I was fast becoming a comet photographer. Nevertheless, Patrick's enthusiasm was undiminished and, every month, a page of TLP news appeared again in the Lunar Section Circular.

Patrick and John Mason's enthusiasm for supporting out-of-town BAA meetings during John's term as President was also undiminished. March 1992 saw a BAA meeting being held at Birmingham and Patrick offering his services as the main speaker. Patrick addressed the audience with a rousing account of the life of Sir John Herschel, who was born exactly 200 years earlier, in March 1792. Patrick said that he had often been asked why the Herschel's 'other' house (Observatory House) could not have been preserved in the way 19 New King Street (Bath) was preserved. He said the answer was simple: "Wet rot, dry rot, you name it, Observatory House had it."

In 1992 Patrick became known to a whole new young audience as a disembodied head in the Channel 4 programme *Gamesmaster*. The programme was aimed at showing computer games users how to get the most out of their toys and crack the secrets to winning the games. It was hosted for most of its 6 years by Dominik Diamond who wore a red suit with a white rose. During 1993/1994 he was replaced, for some unknown reason, by a character called Dexter Fletcher, but Dominik returned in 1994. Patrick was simply a caricatured floating head in the programme, telling games console users how to crack the secrets of the most popular computer games. Patrick, of course, knew nothing about these games and he was simply reading a script (probably the only time he read any lines ever). Bizarre though it may seem Patrick was a big hit in the role, and having him as *The Gamesmaster* was an inspired decision. The programme would run for 6 years, until 1998.

Patrick's involvement with Channel 4 did not end there either. Shortly after *Gamesmaster* had started Patrick appeared on *The Big Breakfast* early morning show, during its most popular period, when Chris Evans was the main presenter on the programme. He appeared in the slot known as 'The Crunch', hosted by Chris and the two bizarre puppets Zig and Zag. Chris apologised profusely for the puppets insulting their guest and calling him an 'astrologer' to which Patrick quipped that "Astrology proves there's one born every minute". Zig and Zag then asked Patrick a series of questions preceded by, in every case: "Mr Moore, you are famous for being a legendary astronomer……" The questions included: "How loud was the Big Bang?", "Is there really a man on the Moon?" and "Have you spotted our home planet Zog?" To the last question Patrick replied that his 15-inch reflector would not show planets fainter than magnitude 14.8. The interview ended with Chris and the puppets singing "Would you like to swing on a star, carry moonbeams home in a jar, and be better off than you are, or would you rather be a pig…….." to a xylophone accompaniment from Patrick, who had brought his own portable instrument along. Bizarre!

On April 24th 1992 *The Sky at Night* celebrated 35 years of unbroken monthly coverage of astronomy on television. Every programme had featured Patrick as the presenter: an incredible achievement. Patrick had made sure that the *Sky at Night* survived by keeping it a low budget affair (no TV star salary for him, in fact, not even a contract) and by never falling out with the BBC by working for its rivals. He had realised, some 35 years ago, that rocking the boat and acting like a prima donna would see the programme off. There were those that had criticised *The Sky at Night* for not being 'exciting', but the fact that it had survived for so long proved it had a winning formula. Patrick, even into his sixties and seventies, had a superbly clear delivery and a way of getting complex concepts across to the layman. The April *Sky at Night* was an extended (but not costly) programme, summarising the developments over the last 35 years. A celebration party was held at the London Planetarium, along with *Sky at Night* guests, the producer Pieter Morpiurgo, and BBC staff. To celebrate the occasion, Madame Tussauds created a Patrick Moore waxwork: so there were two Patrick's on display! [Incidentally, some years later the well-known U.S. eclipse chaser Prof. Jay Pasachoff was in London, at an astronomical meeting held in the Planetarium, and he spotted Patrick nearby. He called out to him, but Patrick did not appear to hear him. Shortly afterwards he realised that he had been shouting at the waxwork!]

Five days later, at the BAA London meeting, on April 29th, the Editor of the UK magazine *The Astronomer*, Guy Hurst (a future BAA President), addressed the meeting on the topic of the UK Nova/Supernova patrol which he had been running for 16 years. Guy's presentation came shortly after my BAA 'Sky Notes' in which Patrick had joined me on the stage to share his recent observations of the planet Jupiter. Patrick had fallen out with *The Astronomer* (T.A) magazine many years earlier. Prior to 1975, James Muirden had been the Editor of the magazine and, as we have seen, Patrick seemed obsessed that Muirden stole facts from his books! There was no evidence for this. In fact, Guy Hurst (a more pleasant man you could never meet) had encountered a similar problem with Patrick in 1989. The astronomer Rob McNaught had supplied information to Patrick's *Astronomy Now* magazine and Guy's *The Astronomer*, regarding nova discoveries. When Patrick saw the information in *The Astronomer* he immediately deduced (wrongly) that Muirden's old publication was 'at it again'! Anyway, everyone assumed that Patrick's anti-T.A. stance would last for eternity. However, after Guy's talk at the April 29th 1992 BAA meeting he was approached by Patrick and he asked Guy if any of his observations would be "good enough" for T.A. Guy was amazed. Patrick joined T.A. and a rift going back 25 years was ended!

On the evening of May 13th, Patrick had started observing the Moon with his 15 inch reflector, but he gave up as there was just too much turbulence. Shortly afterwards he received a phone call from Gerald North who thought that, for the appalling conditions, the small Plato craterlets were oddly prominent. Patrick quickly alerted half a dozen members of the TLP network and went back out to his 15 inch reflector. Nothing was ever deduced about this report, but it was the first major alert Patrick had issued since taking charge of the TLP sub-section again.

Six days later Patrick was outside again, observing the Moon at 01.00 GMT, with his 15 inch reflector. The enthusiastic astronomical historian Allan Chapman was with Patrick on this night. Patrick observed that the south and southeast wall of Aristarchus, with the so-called 'spur', were seen to be covered with a reddish-orange glow "which was quite unmistakeable". Patrick reported (in the lunar circular) that Dr Chapman saw the effect easily, despite not being a lunar observer. Patrick continued observing until 02.05 GMT.

Although Patrick could always find a publisher for any book he chose to write and tended to avoid the use of agents, he was always happy to attend functions where a major book deal might be clinched, or where he might meet fellow authors. The literary agents A.P. Watt, with a reputation going back to 1875, were used by Patrick during the 1980s and 1990s, when the company was run by Hilary Rubinstein (retired in 1992) and Michael Horniman. A number of Science Fiction writers, such as Brian Aldiss, Ursula K. Le Guin and Harry Harrison used A.P. Watt and such people were always a source of fascination to Patrick, despite his own distinctly schoolboy level of Science Fiction. So, when A.P. Watt held a party for authors at The Orangery in Holland Park, on June 24th, he was happy to attend, especially as he knew Brian Aldiss would probably be there. Interestingly, Patrick's presence at this exact same function is briefly mentioned in Volume II of the *Edwina Currie Diaries*, although the observation is hardly flattering: 'Patrick Moore falling asleep, a lump of foam rubber in the corner'........

Chapter 33
A Nasty Accident and Shoemaker-Levy 9

On July 12th 1992 disaster struck. Patrick was, as so often on a Sunday in summer, playing cricket with the Selsey Cricket Club 1st XI. On that occasion he was fielding in the slips. A rising ball from a fast bowler hurtled towards him and caught him squarely in his left eye; this was his good eye and the one he used for observing. He was rushed to hospital in a lot of pain and with a very swollen eye. In public he was very upbeat about things: "Fielding a cricket ball with one's eye is not a good idea, but I think that eventually my sight will be all right – 90 percent at least". However, to friends he was extremely depressed, claiming "It's the end…..no more observing for me". Patrick had always used a monocle in his weaker, right eye, and after a further visit to the hospital in July he was advised that he may have to train the right eye to do the observing. The lens in the left eye had been damaged by the impact of the cricket ball. Training your weaker eye along with your brain, at the age of 69, would not be easy. Patrick had spent a lifetime training the better eye at the eyepiece: things were not looking good. Another option was trying to re-train his damaged left eye, but the lens had been distorted and glasses would not correct the problem.

Perhaps the most infuriating aspect of the cricket ball injury was that Patrick had been due to open a Space Exhibition in Minehead the next day. The exhibition had been timed to coincide with a visit to the UK by Arthur C. Clarke and had been arranged by Clarke's brother, Fred. Minehead (in Somerset) had been Clarke's childhood home. With Patrick in hospital the organisers arranged for Heather Couper to open the Exhibition. However, despite very blurry vision Patrick met Arthur a week later, in London, at a promotional function for *British Aerospace*, alongside Lord Bessborough and Sir Michael Marshall.

The serious eye injury was not Patrick's only health worry in the 1990s either. He had always had an irregular heartbeat and his weight did nothing to help him in that respect. In addition, his left leg and knee, injured (allegedly) by "shrapnel" during the war (or more probably a crash landing when he was flying the plane) was starting to let him down. On top of that arthritis was turning his hands into very insensitive claws, incapable of any delicate work. Again, his weight did him no favours where his mobility was concerned. I remember at one Hawkstone Hall BAA

M. Mobberley, *It Came From Outer Space Wearing an RAF Blazer!: A Fan's Biography of Sir Patrick Moore*, DOI 10.1007/978-3-319-00609-3_33,

Exhibition meeting, in the mid 1980s, near to the Lunar Section exhibits, Patrick bellowed (regarding his left leg): "The doctor's want me to HAVE IT OFF". There were shrieks of shock from an elderly woman in the vicinity. Patrick swung around "Oh, don't worry, they're not going to get it....I'm going to hang on to it for dear life, even if it has turned green". This was no exaggeration either, when Patrick crossed his legs and his not quite long enough trousers rode up, that leg had a distinctly gangrenous hue by the1990s! It was not pink and healthy: the tint was one of a permanent yellowy bruise.

The cricket ball in the eye incident seriously curtailed Patrick's observing activities for most of 1992 and 1993 and his very regular observing sessions came to an end after the injury. Admittedly it was also getting increasingly tricky to shoehorn his bulky frame through the little door into the oil drum dome of the 15-inch Newtonian. Eventually he would have an artificial lens surgically inserted into the damaged eye, but July 1992 marked the end of Patrick's really prolific observing period. He had been one of the BAA's top lunar and planetary observers for all of his adult life. After the lens was replaced he would still observe, but far more casually than before.

It was around this period, in the early 1990s, that Patrick's house seemed, more and more, to resemble a Travelodge hotel. After his mother had died in 1981, Patrick had soldiered on with their loyal housekeeper "Woody", or more accurately, Mrs Hester Woodward. Without her housekeeping abilities and moral support Patrick would, almost certainly, have fallen apart. She understood Patrick, his mother, and their relationship and was a huge comfort to him. Sadly, Woody also died, just over 10 years after Patrick's mother. It happened very quickly, with Woody feeling ill and passing away within a couple of days. This left Patrick all on his own, in a big house, with just hired helpers or friends and neighbours to get the daily chores done. OK, Patrick was not totally useless, but chores and hassles were not his life. He was interested in astronomy, music and cricket, not hoovering, washing and cleaning and even changing a light bulb was at the limit of his patience threshold. Fortunately, Woody had a son, Barry, and his family came to live in Selsey, and so were a great source of support. There was no shortage of people dropping in either. Patrick was famous and many visitors came just to rub shoulders with the famous man. Give one or two of them trivial chores, like wiring a plug, and it all freed up that vital astronomy writing time.

Even so, after Woody had died and when Patrick was alone with the cat in such a huge house, he did have moments when he thought about relocating to somewhere smaller, even with the hassle of moving the telescopes (assuming his vision would recover from the cricket injury). After all, he had survived the move from East Grinstead to Armagh and then Armagh to Selsey. The difference was that Farthings was a huge house and was now stuffed full of junk; a move would be a nightmare. Nevertheless, conscious of his advancing years, his dodgy heartbeat, his trouble-some knee, his recently damaged 'good eye' and his arthritis, Patrick did contact the local Estate Agents and visited a couple of cottages nearer to Chichester. Within a day or two this had made up his mind. The smaller cottages did not appeal to him, had rickety staircases and, faced with a move, it just didn't feel right. His mind was made up. He would stay in his beloved 'Farthings' at 39 West Street, for life, despite his health issues, and simply invite more friends to stay and help out and help drink his booze!

Farthings was a big house and, by the mid-1990s, it was a rare month when one of Patrick's many godsons was not lodging there. Patrick's great friend Paul Doherty was a regular visitor too. All of these people cheered Patrick up and helped get the housework done. It was a two way deal. Patrick freed himself from domestic duties and mere mortals got the chance to stay with a living legend and sometimes got a chance to appear on TV as well! After the cricket ball injury Patrick was especially grateful for some moral and physical support. One lifelong and genuine friend became a virtually permanent resident. Roger Prout, a retired Chichester lawyer and opera lover, moved in to Patrick's home in the early 1990s until Patrick started having a professional carer in the twenty-first century. Prout helped with Patrick's entertaining opera *Theseus and the Minotaur*, in which Patrick demoted the beast to more of a biscuit loving kitten! As the 1990s ticked on Patrick's house would seemingly become fuller and fuller with visitors and the door was almost always open to, well, anyone. Years later when Patrick reported the theft of a couple of valuable items to the police a close friend of his told me "Well what does he expect, his door is always open? The police must have a thousand suspects!"

The strange thing about this was that although Patrick seemed to love being sociable, and having a house full of people, his manner, with even his closest friends, was often abrupt. He simply had no interest in small talk or talking about himself, only about astronomy, or music, or cricket. He seemed to love being seen as a generous host, and loved having people on call to do tedious chores, but any form of social chit chat just bored him. Don't get me wrong, he was the most generous friend anyone could ever have, and he would try to use his influence to do anything for you, if you, or a relative of yours, was ill. But, even in a room full of his best friends, you got the impression sometimes that this British institution was a desperately lonely man, and had been ever since his mother's death. Only really copious amounts of alcohol would get him talking in the way that most people talk with their friends. He was a man who enjoyed being surrounded by his fans, but tired of their conversation almost instantly, but, having said that, he would do almost anything for them.

Swift-Tuttle, a Bottle of Whisky and Snooker

August and September 1992 were an exciting period for meteor and comet enthusiasts world-wide; all because of the comet Swift-Tuttle and its association with the Perseid meteor stream. The Perseid shower occurs in August every year, peaking in activity around August 12th/13th. Only the Geminid meteor shower in December is as good, and as reliable. One hundred and thirty years earlier, in 1862, a bright comet had been discovered by Swift and Tuttle. It proved to be a fine naked eye spectacle and its orbit was eventually linked to that of the Perseid meteor shower. In other words, bits falling off the comet, in a similar orbit, when they hit the Earth's atmosphere, would become Perseid meteors. But there was a mystery here. The orbital calculations predicted that the comet would return in about 120 years (1981 plus or minus 2 years, to be precise) and decent Perseid activity around that time led many to speculate that, in the early 1980s, comet Swift-Tuttle had returned to the Solar System and been missed. If the Earth was on the opposite side of the Sun when the comet reached perihelion this was not impossible. Remember, the relentless CCD discovery machines like LINEAR and NEAT did not exist in 1981; many comets were still discovered visually.

Patrick was convinced that Swift-Tuttle *had* been missed, but his friend John Mason, the BAA President in 1992, had never believed so. In 1991 that year's enhanced Perseid activity raised speculation that Swift-Tuttle's return was imminent. Dr Brian Marsden, of the Central Bureau for Astronomical Telegrams (CBAT) and the Minor Planet Center (MPC) issued possible new return times with two options for the comet's perihelion (closest to the Sun) time. One date was 1992 November 25th and the other was 1992 December 11th. These predictions were based on the fact that a sparsely observed comet, called Kegler, seen in 1737, might have been the same comet as Swift-Tuttle. This would make the comet's average orbital period closer to 130 than 120 years. Patrick did not believe a word of it though; he thought Brian Marsden was wrong and Swift-Tuttle would not be found before the end of 1992. However, John Mason thought it would be found; how could such a bright comet be missed, especially when astronomers were deliberately searching for it? So sure was Patrick of his view, that despite not normally being a betting man, he bet John a bottle of whisky. If Swift-Tuttle was not recovered before December 31st 1992, John would have to pay up.

At the BAA Out-of-Town meeting in Bristol, on September 26th, Patrick acted as the meetings recorder as the Lunar Section Director, Jeremy Cook, described 'What's happening on the Moon'. Little did Patrick know, but as that meeting drew to a close, on the other side of the world a Japanese comet searcher named Tsuruhiko Kiuchi had just swept up an object he hoped was Swift-Tuttle, with his 25 × 150 mm Fujinon binoculars. The upshot would be that, very soon, Patrick might have to part with a bottle of whisky! And that is exactly what happened! On October 26th, at 23 Savile Row, John Mason delivered his Presidential Address (on the Leonid Meteors and the similarly named Comet Tempel-Tuttle). I then delivered my BAA 'Sky Notes', with an update on Swift-Tuttle, and Patrick handed over the whisky! As I ended my Sky Notes Patrick stood up and addressed the audience. He admitted that he was not normally a betting man, but he had bet that Swift-Tuttle would not be discovered before the end of 1992. Patrick said that he had been totally wrong, whereas the President, John Mason, had been totally right. He apologised to John for ever doubting him. Patrick then presented John with the bottle of whisky to much applause and laughter from the audience. The next day Patrick set off for Arizona to acquire some footage of US observatories for *The Sky at Night*. Even though he was rapidly approaching 70 years of age he was permanently on the move, giving talks, doing radio and TV broadcasts, and helping out in the BAA whenever he could. In that same month of October 1992 he even found time to perform the opening ceremony at the Luton Astronomical Society's brand new Putteridgebury Observatory. The society had resurrected Horace Dall's old Dall-Kirkham Cassegrain and, of course, Patrick was the perfect man to do the honours and declare the observatory open.

As I have mentioned many times in this book Patrick was always happy to look a bit odd and play the fool on TV shows and, with his quick thinking and fast talking self-deprecating style, he was a TV game show producer's dream, at least if a barking mad eccentric guest was required. Cricket was Patrick's sporting passion, although he did play Tennis as well, was a demon at the Chess board, and no slouch at Backgammon either. However, he was game for anything, and so when Patrick was invited to appear as a contestant on the 1992 Christmas edition of the Snooker show *Big Break* (broadcast on Boxing Day) he was well up for it. The show's host Jim Davidson (dressed as Dick Whittington) introduced him as "A man who knows more about Outer Space than Jean Luc Picard of the Starship Enterprise……Mr Universe himself…..Patrick Moore." Wearing a smart black dinner jacket, with silk-edged lapels and a black bow tie, Patrick then positively bounded towards his seat and when Jim Davidson joked that the chair had been specially made for him Patrick instantly quipped back "Splendid, it has to be!" When Jim asked Patrick the big question "Is their life on Mars?" Patrick replied with his usual machine gun delivery, which had the audience laughing and Davidson swaying with data overload. Patrick also explained that, yes, he enjoyed playing Snooker, but that "for sheer, utter incompetence I am in a class of my own."

Patrick was teamed up with the female professional pool and snooker player Allison Fisher (who was dressed as Cinderella's 'Buttons'). The other guests were Ruth Madoc, teamed with Willie Thorne (dressed as the Gentle Giant from Jack and the Beanstalk) and Anthea Turner, teamed with Dennis Taylor (dressed as widow Twanky). The referee, dressed as Dick Whittington's cat, was John Virgo. Yes, it was a very bizarre Christmas scene indeed! The highlight of the programme, at least for fans of Patrick, was that despite his eye injury, he successfully mastered a trick shot set up for him by John Virgo, in which seven balls, rolling on a track created by two Snooker cues, followed each other into the side pocket. The 300 pound prize won for this feat was donated to Patrick's elected charity, the Cystic Fybrosis Trust.

BAA Highs and Lows

On January 2nd 1993 the BAA held its annual 'Christmas' meeting, which was always well attended. Patrick, who seemed to be on a Herschel 'high' in 1992/1993 gave a fascinating talk about Alexander Herschel, the grandson of William Herschel. I must admit I had never heard of Alexander Herschel prior to Patrick's talk, but Patrick could always teach, even a BAA audience, a thing or two about astronomy. There was an interesting article by Patrick in the Junior Astronomical Society (now called the Society for Popular Astronomy) magazine at that time, in which Patrick described how, when he was 11, he started doing the football pools and won £87 (I mentioned this at the start of the book). Apparently all bills in the family household were paid off and the family then went on holiday to Belgium. The remains were spent on a xylophone, but not a big telescope!

In 1992 and 1993 Patrick was still a regular attendee of all the BAA meetings and he was still a major force at the Council table. Due to the way the new BAA Bye-Laws were written, the BAA President, John Mason, had to think about his own successor very early in the term of his own 2-year Presidency; a rather strange situation. The aim of this was so that the incoming President elect could get up to speed and attend Council meetings as a Vice-President, well before he took over. However, it did not give the newly-established President much time to settle in before planning his successor. As we have already seen, the incumbent President always chose the next President, after consulting with all other living (and sane) ex-Presidents and (hopefully) after some discussion with the BAA secretaries and the Treasurer. Patrick had made it clear that he was very keen on Richard Baum being the next President.

Patrick had known Richard since the late 1940s and Richard was a visual observer of the Moon and planets, just like Patrick. He was also one of the most respected figures in the BAA as well as its 'Mercury and Venus' Section Director. Patrick was very keen that Richard's decades of BAA service should be rewarded. The only doubt regarding Richard was that he rarely travelled out of his village of Great Boughton near Chester and he had not attended a BAA London meeting for decades, let alone a council meeting. On March 31st 1993 Richard Baum travelled to London and attended the BAA Officers, Council and Ordinary meetings, giving a splendid talk at the latter. However, at the other two meetings, especially the Officers meeting, some of those present thought that making Richard the next President would be a somewhat risky strategy. The BAA Council table could get very heated at times and was not a friendly place for novices, let alone for an inexperienced President. If you were not on good terms with the majority of the council you were doomed. It would be like throwing a lamb to the slaughter, especially as Richard seemed (in a few people's eyes) to have some potentially controversial views.

The powerful Treasurer, Cdr. Henry Hatfield, was dead against Richard becoming the President (especially so soon after the 1987 council in-fighting) and, sadly, John Mason had to conclude that Patrick's lifelong friend really could not be his successor. Richard Baum could have been an excellent President, but the Officers were still nervous after the Council 'War' of 1987 and decided to play safe. Patrick was most upset, but he reluctantly accepted the decision. The last thing he wanted was to be seen as the source of yet another Presidential controversy. In the end, Council chose another Richard, Richard McKim, as the next President and an excellent President he would prove to be, despite not having the luxury of a year as Vice-President before assuming the post. Nevertheless, Richard McKim had been attending BAA Council meetings for years and, despite his youth (35 when made President), had considerable council experience.

In the BAA Journal Patrick was well-known for getting away with murder. Not surprisingly, the BAA had certain rules for its flagship publication. Papers had to be refereed by two experts and their advice had to be considered by the author. Also, authors of books were not allowed to reply to criticism by book reviewers, unless the criticism was unusually cruel. Editors and papers secretaries turned a blind-eye to this where Patrick was concerned. To most on the BAA Council Patrick was their hero, even if he could behave like a spoilt 5-year old at times. Patrick could liven up any meeting and most BAA members were only there because, years ago, they had read one of his books, or seen *The Sky at Night*. Patrick was obsessed with any form of criticism relating to his books or papers. He just *had* to reply, even to the most trivial criticism. In the December 1992 BAA Journal, in an otherwise glowing tribute to Patrick's revised *Guinness Book of Astronomy*, Dr Simon Mitton queried why only 'dead' famous astronomers were listed in the biography section. This brought a response in the form of a letter in the April 1993 Journal, from Patrick, who explained:

> There are great astronomers today, ranking with any of the past: Stephen Hawking, Martin Rees, Alec Boksenberg, Jocelyn Bell Burnell, Graham Smith… the list is endless. That is the point – the list is endless! Had I made a selection, I would have offended somebody – and justifiably so. Selecting as I have, I don't think anyone can object. So that was my reason. I hope that I will not have to include Dr Mitton himself for very many years!

On May 22nd the BAA Lunar Section met at the University of Birmingham. Patrick, once again, agreed to be the main speaker, covering one of his favourite topics: the libration regions of the Moon. The meeting went smoothly, but from an attendance point of view it was very disappointing: only 24 people turned up. Throughout the 1970s and 1980s, whenever Patrick was on the bill a good attendance was almost guaranteed, especially in a big city. Patrick could draw crowds of hundreds to public lectures and 100 was fairly typical at a BAA meeting. Maybe people were just getting bored with meetings now that e-mail, bulletin boards, computerized telescopes and CCD cameras were becoming more popular? Whether this low turnout demoralized Patrick, I know not, but in the June Lunar Section Circular he announced that he was standing down as the TLP co-ordinator and handing over the reigns to a younger man, namely Gerald North.

Only 4 days after that Birmingham meeting, Patrick was back in the President's chair for the BAA London meeting of May 26th 1993. Although it was great to see Patrick acting as the President, the reason was a sad one: John Mason's father had died and so John was unable to attend. One of the speakers that day was Jeremy Cook, the Lunar Section Director. Not surprisingly, with Patrick in control, Jeremy concentrated on TLP. By this time I was regarded as a definite TLP sceptic, despite being a former member of the TLP 'team'. Jeremy pointed out that I had been involved in, and had observed the Torricelli B 'event', some 10 years earlier. From the audience I agreed that I had been involved in a number of TLP alerts but felt that everything I had seen could be explained by spurious colour in the Earth's atmosphere, poor seeing, or a specific angle of illumination. I agreed that the Torricelli B event had been the most memorable, but even then I thought a specific illumination angle was responsible; a bit like viewing a diamond or a cat's eye at the critical angle. Jeremy felt this explanation was unlikely, but Patrick thought that, whatever the cause, it needed investigation.

On July 3rd Patrick was chairing another kind of meeting. This time he was the chairman for the first three sessions of the European Colloquium on Light Pollution, held at the University of Reading. Needless to say, amateur astronomy was well-represented.

September saw Patrick travelling north to Lancaster, for the BAA's 'Out-of-Town' meeting weekend from September 3rd to the 5th. Patrick stayed for the whole weekend and visited the Jeremiah Horrocks and Wilfred Hall Observatories. He also spent much time chatting to local celebrity and ex-Mayor of Manchester, 80 year old Dame

Kathleen Ollerenshaw, in the Bar of the Students Union at Lancaster University. Sometime later I received an e-mail from Lancaster amateur astronomer Denis Buczynski, about how Dame Kathleen, Patrick, and the astounding astronomy historian and public speaker, Allan Chapman, had turned up at Denis' Conder Brow Observatory, en route to Kathleen's Lake District cottage. Patrick drank spirits like a fish, but Allan Chapman drank tea like a whale. Denis' recollection was:

> I remember a good session at Conder Brow when Dame Kathleen turned up with Patrick and Allan Chapman on their way to Kathleen's retreat in the Lake District. Allan drank two whole teapots full of tea and Patrick finished off a bottle of Gin! It was a good laugh watching them bounce stories and experiences off each other. Each had a better tale to tell than the other, and then it started all over again. Real people with solid achievements to their name, not like the vacuous celebrities and their acolytes that abound today.

Patrick seemed to get on well with Dame Kathleen who was setting up her own observatory in the Lake District. For a woman approaching 81, deaf since the age of eight, she was in amazing control of her mental faculties and still retained an interest in Mathematics. She came across as a woman 100 % determined not to let old age affect her brain and once she had met you she never forgot your face (she reached 100 in October 2012). Dame Kathleen could also talk as rapidly and as energetically as Patrick, so when they got together, it was very noisy! Patrick had met his match, in a woman older than him! Incidentally, 13 years later, aged 93, Dame Kathleen travelled to the 2006 total solar eclipse in the Libyan Sahara; more than 10 years younger, Patrick, almost totally immobile, stayed at home.

One month after the BAA Lancaster meeting Patrick was attending the British Interplanetary Society's 'Space '93' convention at the White Rock Theatre in Hastings. The second man on the Moon, Buzz Aldrin, met up with Patrick there and, courtesy of British Telecom, both men were able to talk live, via satellite, to Arthur C. Clarke in Sri Lanka. This event marked the 60th anniversary of the BIS and photographs in the BIS archive show Patrick deep in conversation with Buzz and his wife Lois at the 60th anniversary dinner that evening.

At the October 27th BAA AGM John Mason handed over the Presidency, after the standard 2 year term, to Richard McKim. Richard would be the joint second youngest President in the BAA's 103 year history. At 35 he was the same age as Heather Couper had been when she assumed the post 9 years earlier (actually, Heather had been 4 months younger). The youngest President, Steavenson (1926–1928) was only 32 when he took office. After the problems of recent years, when proposing the next President, Richard's first major plan, on assuming office, was to revise the Bye-Laws so that the President could decide, later in his Presidency, who his successor should be. I well remember there being some opposition to Richard's plans, although Patrick appeared happy with them. There have always been very few women on the BAA Council (none in many years) and, with Patrick at the table, it was surprising that there were any! One rare (and young) female council member was Le Forbes. She had quite a few objections to Richard's plans and voiced them at the table, much to the annoyance of Patrick. "IGNORE HER, MR PRESIDENT, SHE'S JUST A WOMAN" he bellowed! Political correctness was never on Patrick's agenda! On another occasion, I remember Gina Bone, a council member and the wife of the meteor section director, Neil Bone, bringing her first child, as a baby, to the council meeting. Not surprisingly the baby became bored and would make quite a few noises from time to time. "I'M SORRY MR PRESIDENT, I CAN'T HEAR A WORD YOU'RE SAYING. THERE SEEMS TO BE A BABY, OR SOMETHING, IN HERE SOMEWHERE....YOU'LL HAVE TO SPEAK UP". Patrick scowled in Gina Bone's direction and Gina and baby would eventually leave the Council room.

Shoemaker-Levy 9

Early in 1993, in March in fact, the American astronomers Carolyn and Eugene Shoemaker, working with the amateur David Levy, had discovered an extraordinary comet with the Palomar Schmidt telescope. Named Shoemaker-Levy 9, the comet had been stretched out into a line of debris by Jupiter's gravitational pull. As 1993 progressed it became obvious that the comet was going to collide with Jupiter in July 1994. At the November 24th 1993 BAA meeting Patrick was giving a talk on 'Lunar Eclipses, past and present' but, before he started, he told the audience that when the comet collided with Jupiter he had a plan to observe the impacts: he would look at Jupiter's moons, especially if they were in the shadow of the giant planet. He thought that the flashes from the cometary impacts, on Jupiter's far side, might illuminate any eclipsed moons. It is easy to forget, but at that time no-one dreamed that a tiny comet could leave any marks on Jupiter, visible in amateur telescopes. It was a great idea of Patrick's but the bruises on Jupiter, from the impacts, would be far larger than anyone had expected.

The 1993/1994 Shoemaker-Levy 9 build-up year was exciting, but it had a much sadder side to it for Patrick. One of his most favourite young astronomers was Dr David Allen of the Anglo-Australian Observatory (who I mentioned earlier in connection with the BAA Lunar Section). He had many similarities to Patrick in that he was an only child and was so disabled in his youth, with a dislocated hip, that he was deprived of a normal school life. Throughout the 1980s and the early 1990s David had appeared on *The Sky at Night* and been a regular contributor to Patrick's *Yearbook of Astronomy*. However, tragically, he would be diagnosed with a brain tumour in August 1993, much to Patrick's dismay. I remember many BAA Council meetings during the 1993/1994 session where Patrick would give a brief update and tell us "If anyone can beat this, David can". Sadly, even he could not defeat the cancer. David Allen would pass away on July 26th 1994, 4 days after the last fragment of Shoemaker-Levy 9 would hit Jupiter and just a few days before his 48th birthday.

Although Patrick rarely smoked his pipe during the 1990s, as the dangers of tobacco were, by then, well established, he still occasionally indulged in this British passion, if only to annoy people who droned on about the health risks. He always tried to attend the annual *Pipe Smoker of the Year* event too, as he had been awarded the title at the start of 1983. Ten years later the comedian (and stuffed Emu puppet master) Rod Hull received the 1993 *Pipe Smoker of the Year* award, so at the next annual meeting on January 19th 1994 he posed for press photographers, alongside

Patrick, both smoking their pipes. Rod Hull would be declared bankrupt later that year and, 5 years later, died in a bizarre accident when he tried adjusting the TV aerial on his house roof so that he could watch a *Champions League* Manchester United match. While battling with the aerial he slipped and crashed to his death through the roof of his greenhouse.

February 24th 1994 saw Patrick and Peter Cattermole at the Science Museum in London, giving a lecture, a question and answer session, and celebrating their new joint book, entitled: 'Venus: the Geological Story'. Shortly afterwards, Patrick flew to Australia to film the major observatories at Mount Stromlo, Siding Spring and Narrabri.

Around this time Patrick entered hospital to have his 'cricket-ball damaged' cornea replaced with a plastic insert. Although his sight was never as good as it had been before the accident, it was much improved and some observing at the telescope was once again possible. On April 25th 1994, wearing half-moon tortoiseshell spectacles, a yellow tartan striped blazer with matching trousers, a purple shirt, and a black and blue bowtie, Patrick entertained the audience at the annual Royal Children's Variety Show at the London Palladium.

In May 1994 Patrick had travelled, along with the veteran BAA eclipse-chaser Mike Maunder, to view an annular solar eclipse in Texas. In an annular eclipse the sky does not go dark, because the Sun is larger than the Moon and so appears as a searing ring of light around it. Special filters must be used to view it. Patrick would never class himself as much of an astro-photographer, but on this occasion he took some really excellent pictures. He used a telephoto lens and an Inconel solar filter and his slides, shown at the May 25th BAA meeting, showed a jet-black Moon gradually drifting from second to third contact, across a yellow/orange Sun. Mike Maunder managed to secure a dramatic sequence in which, at the end, he posed in the shot, so as to 'catch' the eclipsed sun in his hand.

The predicted date for the impact of the first fragments of the Shoemaker-Levy 9 comet with Jupiter was July 16th. No-one knew exactly what to expect. Twenty-one fragments had been identified and the final fragment would hit Jupiter, 6 days after the first, on July 22nd. While the professionals expected to be able to detect the larger impacts, at infra-red wavelengths if not in the visible range, amateur astronomers were not expecting to see much at all. It was one of those situations where everyone was wiser with hindsight, after the event. How much 'damage' would a lump of dust and ice a few miles across make, impacting a ball of gas 11 times larger than the Earth? Would anything be seen? It was like dropping a small iceberg into the Pacific Ocean; would *anything* be visible from the Earth at all?

As it turned out, the impacts from the larger cometary fragments were easily visible in even the smallest amateur telescopes, and even their finder telescopes, from the evening of July 17th onwards. But Jupiter was low down in the South from the UK and a tricky object for many amateurs, especially if they had buildings or trees in the way. Patrick was not sure if his telescope would have an unobstructed view of Jupiter, but he had a great view using the old telescopes at Herstmonceux, the site of the old Royal Greenwich Observatory. His friend Paul Doherty, in Stoke on Trent, travelled to Keele University to use the 12 inch refractor there and he kept

Patrick fully informed; coincidentally Patrick received yet another honorary doctorate from that same University in the same month!

Remarkably, the whole of the comet impact week saw mainly clear skies over the UK. It was hot by day though, and Jupiter appeared mainly as a turbulent rippling blob. Nevertheless, the impact sites were easily visible. On the next *Sky at Night*, on July 31st 1994, John Mason and professional astronomers Jim Scotti and Dr Steve Miller joined Patrick to discuss the dramatic results. For the next couple of months the astronomical world was buzzing and the comet impact briefly woke even politicians up to the fact that the Solar System could be a very dangerous place. Patrick appeared in the media on a regular basis during the following weeks and Shoemaker-Levy 9 was still hot news when he appeared as the main guest on the somewhat sarcastic interview programme *Clive Anderson Talks Back*, broadcast on September 30th on Channel 4. The other guests on that episode were John Prescott (then the deputy leader of the Labour Party) and the comedian Griff Rhys-Jones.

Goodbye Astronomy Now

Throughout the summer of 1994 Patrick's creation *Astronomy Now* had developed serious financial problems which were causing him a large amount of concern. Patrick's official post in the magazine was 'Founding/Associated Editor', but the days when he ran the magazine with John Mason and a bunch of his closest chums were long gone. Patrick, at 71, had enough commitments, without running and editing a magazine. *Astronomy Now* had survived for 7 years; an amazing achievement for a minority hobby's magazine, fully covered already by American magazines like *Sky & Telescope* and *Astronomy*. In the early 1990s the magazine's circulation had been consistently higher than 15,000 copies a month and there were plenty of lucrative advertisers paying for space in the magazine. A full time managing Editor was required and this role had passed from John Mason to Martin Beech to Timothy Lyster (who moved on, briefly, to *Sky & Telescope*) and then on to Fiona Gammie and then to Steven Young, in June 1993.

In addition, the magazine had spawned the highly successful European *Astrofest* 2 day exhibition in Kensington, London, since 1992, which had been the biggest Astronomy exhibition ever seen in the UK. *Astronomy Now*'s original publisher and backer had been Angelo Zgorelec, but in April 1994 the magazine switched publishers, to a new organisation called *Hall Park Publications*, which almost immediately (if not before) developed financial problems. As 1994 progressed, and just after the bombardment of Jupiter by Shoemaker-Levy 9, contributors to *Astronomy Now* found they were not being paid for their articles or services. They started contacting Patrick, a man less interested in finance you could never meet! Particularly distressing to Patrick was that friends of his, like the artist Paul Doherty, made significant amounts of their annual income from *Astronomy Now*. These people had families and mortgages, unlike Patrick, who felt that he had not been informed of these impending developments by *Astronomy Now*'s management team. Indeed, he had always been happy with the original publisher, Angelo Zgorelec. The last thing he wanted was to be dragged down into a financial argument if *Astronomy Now's* publisher went bankrupt. Always loyal to his friends, at least, those that never criticised him, Patrick was dismayed that they were not being paid. If the magazine had stayed with the original publisher, he said this would *never* have happened. He made it abundantly clear that he did not like the current *Astronomy Now* way of doing things. Patrick's strategy had always been that *Astronomy Now* would be a happy bunch of his mates producing his kind of publication, not a magazine taking financial gambles and never listening to him, its founder. It was supposed to be a friendly astronomy magazine, not a tool used by an ailing publisher to try to rip-off writers and readers, which was how he now saw it.

By the start of August Patrick had decided that he must officially sever all ties with his 1987 creation *Astronomy Now*. He was sick with worry regarding all the financial criticism coming his way from the magazine's contributors. His last contribution to the magazine was his 'Night Sky' column, published in the September 1994 issue and entitled 'The comet and the crackpots'. He then sent out scores of photocopied letters, half with postage stamps, but when his stamps ran out, half without (!), totally dissociating himself from *Astronomy Now* and its current publishers and management.

He had acted just in time! The publishers went bankrupt weeks later and the November 1994 *Astronomy Now* never appeared. The cover of the final *Hall Park* published issue was entitled "The Closest Place to Hell" which, apparently, was not dissimilar to the magazine office in the Autumn of 1994! However, the managing editor Steven Young salvaged the magazine and transferred it, with Chris Courtiour, to the aptly named new company of *Pole Star Publications*. From December 1994 the magazine rose from the ashes and went from strength to strength and still survives, to this day. However, once Patrick had fallen out with someone, or some company, and stated that he had severed all connections, the decision was usually irreversible.

From August 1994 Patrick had nothing to do with *Astronomy Now* for more than a decade. Indeed, it would be 12 years and 4 months before he even attended *Astrofest* again, despite it being the biggest astronomy show in Europe. If he was mentioned in the magazine it was because someone else had written about him. Patrick bore these sort of grudges for life. Indeed, after severing all ties with *Astronomy Now* he was determined to start another rival publication immediately, and he did! This new magazine, *Practical Astronomy*, featured articles by many of the old *Astronomy Now* contributors and so rival issues of *Astronomy Now* and *Practical Astronomy* appeared side by side on newsagents' shelves from the end of 1994. However, Patrick was now definitely backing the wrong horse. *Practical Astronomy* would die a death a year later. But, as anyone who has crossed swords with Patrick knows, he never gives up. His dream of another astronomy magazine was still alive. However long it took he would persevere with a magazine to knock *Astronomy Now* off the shelves. He had created it and he was sure that he could destroy it too. He also thought that the manner in which *Astronomy Now* collapsed financially, but then bounced back instantly, was highly suspicious. On more than one occasion he voiced his opinion at BAA Council meetings that: "This whole business has all the fine smells of rotting fish!"

The Stargazer Bloodbath

As if all the *Astronomy Now* mayhem was not upsetting enough, there had been disturbing developments in Patrick's sleepy seaside town of Selsey, in the Autumn of 1994. One of the most popular pubs in Selsey was '*The Stargazer*', at 83 Manor Road, about half a mile north-east of Farthings. The pub was often associated, not surprisingly, with Patrick and he was proud of the association, and proud of Selsey: a small town which had always seemed far removed from the madness of the modern world. However, in mid-October 1994 Selsey's sleepy image was shattered. After a Karaoke night at the Stargazer Pub, the Beer Garden became a blood bath. At breakfast time the next morning, more than 20 police officers, many sporting semi-automatic weapons, surrounded two houses in Ruskin Close, Selsey. Four men and one woman were arrested in connection with the death of two men (scaffolder Sean Farrelly, recently released from prison on the Isle of Wight, and fisherman Jason Connors). Rumours circulated that there was a dispute over a woman at the bottom of things, along with drugs offences and a gang-land revenge attack! The local and national press interviewed Patrick about 'the bloodbath at the Patrick Moore pub'. Patrick laid the blame fairly and squarely at the door of one group of people: the police! He had, of course, been convicted of dangerous driving by the same force a few years earlier. "If the police spent more time chasing criminals and less time chasing motorists, none of this would have happened" he told the press. After the bloodbath at the 'Stargazer Pub' Patrick denied any association with it, claiming "It was named *Stargazer* long before I came here; it is nothing to do with me WHATSOEVER". Of course, prior to the bloodbath he had boasted just the opposite story.

The Eclipse that Never Happened

At the end of October 1994 Patrick embarked on a trip to South America. It was Solar Eclipse time again and, on this occasion, the path of the Moon's shadow would cross the Peru/Chile border on November 3rd. This would be Patrick's eighth eclipse trip, but one he rarely, if ever, referred to. After this eclipse he decided all such future trips would be on cruise ships. I went on the Peru/Chile eclipse trip too. It was my second eclipse trip and my first successful one. As with the 1991 Hawaii/Mexico trip, Patrick and John Mason were the tour lecturers. Patrick was assigned to the Peru eclipse group (his heart condition was not compatible with the high altitude of the Chile observation site) just north of the Peru/Chile border, while John was assigned to the Chile group, just south of the border at 4,400 metres altitude, near to Putre in the Atacama desert.

As with most of the *Explorers Tours* eclipse holidays a whistle-stop tour of the region's tourism hot spots (such as the Inca spectacle of Machhu Picchu) was arranged. Patrick briefly crossed from Peru to Chile, gave a talk at the hotel in Arica, and then went back to Peru, where everyone recalls him taking part in a bizarre dancing ritual with local girls at the famous Inca site at Machu Picchu. I never met Patrick on that trip. By some bizarre twist of fate I was in a minibus with half a dozen other people (including my father Denys, the Queen guitarist Brian May, the space artist David Hardy and a future BAA President Dr Nick Hewitt) crossing from Peru to Chile at night, just as Patrick was crossing in the other direction. Our minibuses passed in the darkness at some point, just before Brian May's Peruvian oranges were confiscated at the border post! Brian was not a friend of Patrick's in 1994 (although he had met him briefly at Imperial College in the late 1960s, when Brian was a student) but they would increasingly become the closest of friends from 1996 onwards.

Before the eclipse in Peru, disaster struck. Patrick was on the steps of his hotel waiting to board the bus, when some locals deliberately caused a distraction. People moved around the bus to see what was happening and accomplices swiped any luggage they could grab around the other side of the bus. Patrick, a man averse to bureaucracy and hassle of any kind, had his camera equipment and traveller's cheques stolen. In addition, the eclipse from the Peru site was spoiled by thin cloud. Patrick's group saw the eclipse ending, but it had not been a happy experience and, at 71 years of age, a load of hassle Patrick could have done without. Patrick would never again take part in a land-based eclipse expedition, apart from travelling to Cornwall 5 years later.

On the December 1994 *Sky at Night* Patrick showed some pictures taken by other travellers and merely stated, very quickly, "I went to Peru". As far as I know, he never mentioned the 1994 eclipse again and, unusually, even missed the January 7th 1995 BAA eclipse meeting round-up, stating that he had 'other commitments'. As on many other occasions, if something in his life went wrong, it was erased from Patrick's recollections, except when talking to his closest friends. No mention of the Peru eclipse appears in his biography either. It is erased from his life history.

Sixty Years in the BAA

At the end of November 1994 Patrick had completed 60 years of BAA membership. The BAA Council signed a large card, full of congratulations and best wishes and, at the November 30th BAA London meeting, at 23 Savile Row, the President Richard McKim presented him with the card and invited Patrick to recall some events from his 60 years of membership. Patrick confirmed that, 60 years ago, at the age of 11, he had been elected to the BAA. He had shaken the hand of the President, Harold Spencer Jones, on 1934 November 28 at the meeting of that day at Sion College, near the London embankment. Little did he imagine that he would be made President some 48 years later. Patrick recalled that there were still some original members, from 1890, around in 1934. The BAA secretaries, Sellers and Bartrum had been around for many years, and Sellers was a noted solar observer. It was Sellers who forgot to cap his finder telescope when projecting the Sun and managed to set his beard alight! Andrew Crommelin, the noted mathematician, was also around in 1934. He had worked on Halley's orbit in 1910 and used to chew his beard when faced with a knotty problem. W.H. Steavenson was also a major character in the 1930s. He was one of the few people who had been the President of both the BAA and the RAS. He knew the characteristics of every BAA telescope and was a brilliant observer, despite being blind in one eye. Patrick had been proposed for membership by Major A.E. Levin who, coincidentally, then lived in West Street, Selsey, where Patrick himself now lived. Levin was also an original BAA member and the director of the Computing Section.

Patrick recalled another couple of characters who were prominent in slightly more recent times and who I have already mentioned: Mr and Mrs Granger! Patrick said it was fair to say that they looked slightly eccentric! Mr Granger resembled a large shaggy dog and always had a cat on his shoulder! Mrs Granger was in the *Guinness Book of Records* for having the narrowest waist in the world; she looked like an hour glass and Patrick said he was always worried that she was going to snap in half! Patrick also recalled an RAS meeting in Dublin when the Grangers drove up in a motorbike and sidecar combination, with Mrs Granger driving and Mr Granger in the sidecar, complete with the cat. The effect of their dramatic arrival was positively electric. "It wouldn't have been so funny if they had been deranged, but they were not: they were perfectly sane!" Another event that stuck in Patrick's memory was the occasion when the President, the Rev. Dr Martin Davidson, had borrowed Lydia Brown's hairpin to pick the lock of the BAA library bookcase in Sion College. He had been taught how to do this by one of his parishioners, an accomplished burglar! Patrick also recounted how, as a young boy at East Grinstead, he was helped by a local B.A.A. member, W.S. Franks. Franks was another unforgettable character: "Four feet eight in height with a long white beard and a skull cap!" Patrick thought he looked exactly like a garden gnome!

Patrick then recounted how the BAA Observing Sections were slightly different from today. For example, there was a Star Colours Section and several BAA Memoirs on this subject were produced. Patrick had spent some time, as a young man, observing star colours and it had helped him learn his way around the sky. Observing had been a predominantly visual pastime in those days, with only a few members practising photography. The BAA Journal would come regularly, on the first day of the appropriate month, and often with an observing Memoir attached. Patrick said that during World War II he rarely used to get to BAA meetings. He recalled, at one such wartime meeting, that only 12 members turned up! After the War many things changed and space research, satellites, and other advances changed the face of astronomy. In conclusion, Patrick said that throughout all of his 60 years of BAA membership the BAA had remained an observational society with good links to the professionals. He said that he would not be around in another 60 years, but some members in the audience might be. In 60 years time he felt that someone in today's audience would be saying just the same, namely: "We are here, we are here to stay, we've got a very big history behind us and an equally good future ahead of us." Patrick added that on a personal level the BAA had been a part of his life since the age of 11 and he would not have missed it for anything. He thanked the audience for listening and received rapturous applause from the audience.

Six weeks later, and into a brand new year, Patrick was delighted to appear as a guest when Michael Aspel, holding the big, red, *This is Your Life* book, finally managed to capture Patrick's lifelong friend Arthur C. Clarke, on a rare trip to the UK. As well as Patrick, the *2001:A Space Odyssey* producer Stanley Kubrick paid tribute to Clarke, along with Buzz Aldrin, the composer Jean Michel Jarre and the spacecraft model maker Matt Irvine. The programme was transmitted on January 11th 1995.

Yet Another TV Astronomer!

On 22nd January 1995 the BBC launched a new six-part basic astronomy series aimed at a general adult/family audience, but with no involvement whatsoever from Patrick! The series, *Heavenly Bodies*, presented by ex *Dr Who* actor Peter Davison, was shown just after 11 p.m. on Sunday nights. As *The Sky at Night* was also transmitted on Sunday night, once a month, this meant there were two programmes on astronomy on one particular February evening! Patrick was not happy, primarily because, try as he might, *The Sky at Night* was slipping later and later in the schedules. It was now regularly on well after midnight, meaning it went out on a Monday morning and not a Sunday night. I heard him at the January 25th BAA meeting muttering: "It's being presented by a Dr Who actor; what does he know? Even worse, it's on in peak viewing time and they've shoved me so late no-one will see me". This was a slight exaggeration, as *The Sky at Night* was repeated on Saturday mornings on BBC2 and 11 p.m. could hardly be described as 'peak viewing' time. Nevertheless, Patrick was definitely miffed. He need not have worried though, the new series had a six programme run and that was it! No doubt Peter Davison wanted a normal fee for his role, and Patrick just carried on with his labour of love, a gentleman's agreement, no contract and the *Sky at Night*'s shoestring budget.

In December and January 1995 Patrick had spent much time at the telescope eyepiece, observing Mars, as well as struggling to observe the faint belt of material in Jupiter's atmosphere, following the Shoemaker-Levy 9 impacts. Because of Jupiter's low altitude, Patrick had experienced problems with a pear tree obstructing his view. Patrick described his remedy as follows: "The tree had never given me any pears and so it was not surprising that it turned into a short, stumpy thing in recent weeks".

On the evening of February 4th 1995 Patrick began observing the Moon at 6.10 p.m. with his 15 inch reflector, at a magnification of 360×. Following a general survey of the Moon he turned to the Mare Crisium area and at once thought that the whole of the Cape Agarum area appeared 'very blurred'. Seeing conditions were good and Patrick increased the magnification to ×460, which confirmed the blurred appearance. Patrick reported that he could see no trace of the usual small craterlets (once referred to as Barker's Quadrangle) and the whole area was 'very indistinct'. Patrick telephoned Gerald North, Richard Baum and (now dissociated from the BAA Lunar Section) Peter Foley. Baum was clouded out and Gerald could see nothing abnormal, but Peter Foley (according to Patrick) confirmed the appearance. Subsequently, further observations from Marie Cook and Andrew Johnson came in, but they did not notice the abnormal appearance described by Patrick.

The year 1995 saw the launch of a major new series of amateur astronomy books, launched by the publishers Springer Verlag. The UK managing director of Springer

Verlag, John Watson, an amateur astronomer himself, suggested that Patrick could be the editor of the new 'Practical Astronomy Series' and Patrick was happy to oblige. The series of books was written by various amateur astronomers, with Patrick acting as the series editor, recommending authors and content to John Watson. It would prove to be a popular series with many titles and it launched my own authoring career. The first two books in the series were entitled 'The Modern Amateur Astronomer' and 'The Observational Amateur Astronomer' and consisted of chapters by various authors, mostly on advanced techniques (CCDs and computerised telescopes) of which Patrick had little direct experience. This year also saw the creation, by the British Interplanetary Society, of 'The Patrick Moore Medal', following a generous financial donation to the Society by Patrick. The medal would be awarded for outstanding contributions to the BIS and its first recipient would be Leslie R. Shepherd (1918–2012), a former chairman of the society, a friend of Arthur C. Clarke, an expert on nuclear power and an enthusiast for making interstellar travel a reality.

500 Programmes, Goodbye Colin, Hello Hale-Bopp

In April 1995 the *Sky at Night* celebrated its 500th edition. It was roughly mid-way between the programme's 35th and 40th anniversaries, so Patrick thought it was worth a mention, although on the actual programme Patrick merely gave viewers a tour of the April night sky. Shortly after filming the 500th programme Patrick travelled to California for an exciting trip on the Kuiper Airborne Observatory, a converted C141 Cargo plane, carrying a 0.9 metre aperture telescope. The plane was being retired from service in October 1995 and Patrick wanted to film the observatory 'in action' for *The Sky at Night*.

In June 1995 Patrick was shocked to hear of the death of his lifelong friend Colin Ronan. Colin was a few days short of his 75th birthday and had been in good health, despite being badly injured in 1994 when he was knocked over by a car. Colin had been the BAA's Centenary President and an excellent editor of the association's Journal for 20 years. He had also directed the BAA's Historical Section for 12 years. Patrick and Colin had been joint presenters of an annual residential astronomy course, at Wansfell College, for over 40 years. The year 1995 seemed to be a year of bad news for Patrick. In February, the BAA comet expert Harold Ridley had died, as had Lunar Section member Bill Peters; both were well-known to Patrick. In addition, Patrick's great friend Paul Doherty would be diagnosed with cancer in 1995 too. The number of friends Patrick lost to cancer was truly staggering and it was hardly surprising that he worked so tirelessly for cancer charities.

The next month, amateur astronomers Alan Hale and Thomas Bopp discovered the monster comet C/1995 O1 Hale-Bopp. On the day of its discovery, 1995 July 23rd, the comet was a magnitude 10.5 smudge very close to the ninth magnitude globular cluster Messier 70. Alan Hale in Cloudcroft, New Mexico (a veteran comet observer) and Thomas Bopp in Stanfield, Arizona, were both observing this Deep Sky object with 0.41 and 0.44 metres reflectors respectively, when they spotted the new object.

This discovery, and that of the great comet Hyakutake, came at an interesting period in comet discovery history when professional sky patrols were at a minimum. The Shoemakers (and Levy) had been forced to wind down their patrols with the 0.45 metre Palomar Schmidt only a year or so after the discovery of Shoemaker-Levy 9. Other Near Earth Asteroid surveys such as 'NEAT' and 'LINEAR' were not fully operational although the Spacewatch patrol was operating. So the skies were relatively free from software controlled discovery machines and the likes of Bradfield, Machholz, Levy and Brewington were free to trawl in more comets in those last few years, before the machines took over. However, those famous names were not destined to discover the two Great Comets of the 1990s and neither were the well known numerous Japanese discoverers like Takamizawa, Kiuchi, Kushida and Nakamura (to name but four). As soon as Hale-Bopp was discovered it was obvious that the comet, like Kohoutek, was moving painfully slowly against the background stars; almost too slowly to believe in fact!

The period around July 20th was starting to look like a guaranteed time for astronomical excitement. In 1969, man had landed on the Moon on that day; 25 years later Shoemaker-Levy 9 fragments were bruising Jupiter, and a year after that, Hale-Bopp was discovered. [In 2009 another cometary fragment would hit Jupiter on July 19th!] But Patrick was a veteran comet observer and he knew that with almost 2 years to go, before Hale-Bopp was at its best, anything could happen. Some newspapers, including, surprisingly, the *Sunday Telegraph*, started referring to Hale-Bopp as 'The Doomsday Comet', possibly 1,000 miles across and maybe set to impact the Earth. The idea was ludicrous and Patrick was keen to debunk such nonsense!

Another subject which was classified as ludicrous and nonsense in Patrick's world was the increasing use of compulsory Metric units, rather than Imperial ones, in the UK. As the reader will already appreciate, Patrick was a highly patriotic Englishman who hated the European Union enforcing its will upon British citizens. As far as Patrick was concerned feet and inches, along with pounds and ounces, were essential to the British way of life; but in recent years Metric units had been taught in schools and shopkeepers had even been prosecuted for selling food in Imperial measures. Appalling! As a schoolboy Patrick had memorised the diameters and distances of the Sun, Moon and planets in miles and listed them all in his books. Now he was supposed to re-learn them on the whim of faceless Brussels bureaucrats. Clearly, this was simply not good enough!

So, Patrick became one of the patrons of the BWMA, the British Weights and Measures Association, which was founded by Vivian Linacre in 1995. The aim of the organisation was stated as being 'to uphold the freedom to use the Imperial System and to oppose the compulsory imposition of the Metric System in the UK.' The Patrons were listed as being Patrick, Gwyneth Dunwoody MP, Vice-Admiral Sir Louis Le Bailly and John Monson (11th Baron Monson). The BWMA supported the so-called 'Metric Martyrs', who were a group of traders prosecuted for their defiance of the Weights and Measures Act and the Price Marking Order. In the years following the creation of the society it listed an impressive set of honorary members on its website, including Ian Botham, Peter Alliss, Ranulph Fiennes, Tim Rice, Jools Holland, Max Bygraves, Christopher Martin-Jenkins, and even J.K. Rowling!

On September 5th 1995 Patrick was, once again, observing the Moon. He was joined on this occasion by Paul Doherty, who was staying at Patrick's house. Seeing conditions were good and Patrick was using ×400 on his 15 inch reflector. After checking a few of the prime TLP regions (Alphonsus, Cape Agarum, Messier, etc.) and finding them normal, he came to the crater Plato. According to Patrick's report: 'The floor of Plato appeared to be very dark, perhaps more so than usual and there were no floor craters seen, though there ought to have been. On the site of the main central crater was a very elusive white patch; Paul Doherty, who was with me, agreed.'

Four days later, Patrick attended the BAA Out-of-Town meeting at the University of Sheffield. I usually presented the monthly Sky Notes at the BAA meetings but could not attend that meeting. So, Patrick and John Mason dealt with the Sky Notes in their well-known 'Moore & Mason' double-act. In addition, Patrick gave an excellent summary of the history of the giant 72 inch reflector at Birr in southern Ireland and the plan to resurrect it in its 150th anniversary year. Just 3 weeks later and Patrick was down in the southwest of the UK, officially opening the James Lockyer Planetarium in East Devon, on September 29th.

Chapter 34
The Caldwell Catalogue, Neptune, and More Foreign Trips

During the autumn of 1995 Patrick had an idea which would prove to be a real winner and get him as much literary publicity as almost anything else he had ever undertaken. One night, or so Patrick said, he was observing the Helix nebula, NGC 7293, in Aquarius. Patrick wondered why it did not have a Messier catalogue number. The eighteenth century astronomer Charles Messier could easily have mistaken this fuzzy patch for a comet, which is why he created the Messier catalogue in the first place, so that he did not mistake galaxies and nebulae for new comets. Patrick realised that there were plenty of other 'easy' non-Messier Deep Sky objects which were under regular observation by amateurs, both in the northern and the southern hemispheres. They were covered by other historical catalogue numbers, primarily those of the NGC (New General Catalogue) but maybe a new 'additional to Messier' catalogue of the best objects was needed, now that amateurs had more access to quality telescopes. Patrick drew up a list of 109 objects (the same number as in Messier's catalogue) and listed them in descending order from north to south. From Selsey, Patrick would (in theory) just be able to see down to the 67th object, at Declination −30, but not the higher numbers than that.

But what should Patrick call this catalogue? His surname, Moore, also started with M, just like the original Messier catalogue. So, for once in his life, Patrick decided to use the first half of his double-barrelled surname of Caldwell-Moore, even if in reality you had to go back to his paternal grandfather before that name was used. He called his list the Caldwell catalogue and sent an article on objects C1 to C109 off to *Sky & Telescope* for consideration. To Patrick's amazement *Sky & Telescope* loved the idea and published his article 'Beyond Messier – The Caldwell Catalogue' in the December 1995 edition. Very soon people started using Caldwell designations and telescope manufacturers, like Celestron, even included Caldwell 'GO TO' buttons on their telescopes. A small minority of short-tempered (and envious) Deep Sky experts voiced their strong disapproval of this 'unnecessary extra catalogue', but it was only ever intended as a bit of fun. After all, Patrick was not claiming he had discovered these galaxies and nebulae, he was just drawing attention to the best Deep Sky objects that Messier had neglected or not observed. The Caldwell Catalogue was a runaway

M. Mobberley, *It Came From Outer Space Wearing an RAF Blazer!: A Fan's Biography of Sir Patrick Moore*, DOI 10.1007/978-3-319-00609-3_34,

success and the designations are still used by many amateurs to this day, although a few 'errors' in Patrick's original article have now been addressed.

In September 1995 *Practical Astronomy* magazine, Patrick's rival to *Astronomy Now*, went bust. It was less than a year since it had been formed but, despite Patrick's involvement, it sunk without a trace. The ex-editor, Adrian Ashford, advised people who to contact if they were owed money, but there was little hope for contributors who had not been paid. Nevertheless, Patrick was still determined to coax a new publisher into bringing out yet another astronomy magazine with his name on it. He was still sore about *Astronomy Now*, which was going from strength to strength without him on board. Nevertheless, despite the magazine's failure he continued to appear on TV and radio almost every week. He was the main guest on the BBC programme called *One Foot in the Past* during September, recalling, with some sadness, how the now abandoned telescopes at Herstmonceux had been used by himself and others, before the politicians messed things up.

Another Total Solar Eclipse

On October 24th 1995 it was time for the next total solar eclipse. This time the umbral track was crossing India and, after the monsoon season, the skies were almost guaranteed to be clear at the best location, namely Fatephur Sikri, just south of the Taj Mahal. However, after his distressing experience in Peru, Patrick had decided on a cruise ship holiday, like the one he had enjoyed on the good ship Monte Umbe in 1973. His days as an *Explorers Tours* lecturer were definitely over. Of course, with Patrick's fame and reputation he did not have to look far for another organisation that would hire him (and a guest) as an eclipse expert, to draw in the crowds and lecture about the eclipse. Ted Pedas, an American astronomer and Special Projects Coordinator for numerous cruise lines, who had invented astronomy themed cruises under the banner 'Science & Culture at Sea', snapped him up. He would be on the 'Voyage to Darkness' cruise aboard the Orient Line's ship *Marco Polo*.

Mind you, even without the stressful Peru trip, Patrick had never wanted to go to India. The simple fact was that Patrick was squeamish when it came to needles and could not face the thought of inoculations for Hepatitis, Typhoid, or any other disease; so, with almost all travellers to India having suffered a stomach upset (or worse) at some point, that country was one to be avoided. Remarkably, Patrick never contracted more than a cold on all of his trips abroad, but he attributed this to a diet, when in really dodgy places, of bananas, oranges and whisky! This would be Patrick's ninth total solar eclipse expedition (not counting annular eclipses). Everyone at Fatephur Sikri in India on that day, including myself, assumed that Patrick would have been clouded out. We experienced 45 seconds of totality (it seemed more like ten to be honest) and we knew that the South China Seas, near Indonesia and the Philippines, much further along the track where Patrick had gone, were quite cloudy. John Mason proposed a toast to Patrick, from the India site, just after we saw the eclipse. The Moon's shadow had now left India and was racing towards Patrick at 2,000 miles per hour, but we all thought he was unlikely to see the spectacle.

However, we were totally wrong. Onboard the good ship *Marco Polo*, accompanied by Paul Doherty's 17 year old son Chris, Patrick had a superb view. The ship's Norwegian Captain managed to position the ship dead on the eclipse track *and* in a cloud-free zone. Patrick had a much better eclipse than us because not only was it a much more leisurely event, it was a lot longer. The *Marco Polo* was close to the eclipse track centre, whereas India was near to the eclipse track start. Patrick enjoyed a full 2 minutes of totality and, on his return, was refreshed enough from his eclipse cruise to appear as the main guest on *Good Morning with Anne and Nick*, broadcast from the BBC Birmingham Pebble Mill studios on November 2nd, when he was still buzzing from the trip.

But, shortly after the eclipse cruise, Patrick was devastated to hear that his travelling companion, Chris Doherty, had been hit by a car, in a not dissimilar accident to that which had seriously injured Patrick's friend Colin Ronan in 1994. Was there no end to the bad luck that had hit Paul Doherty's family? For a time Chris' recovery was in some doubt, as he was unconscious for 5 days, but to everyone's relief he eventually did make a complete return to full health. Needless to say, on hearing the news of the accident, Patrick dropped everything to be with the Doherty family, at Chris' bedside.

Despite the collapse of *Practical Astronomy* Patrick still appeared at an associated event only days after his return from the 1995 eclipse. *Practical Astronomy* had scheduled an astronomy exhibition, similar, but much smaller to, *Astronomy Now's* amazingly popular *Astrofest* exhibition. Patrick's name had been associated with *Astrofest* for its first three formative years, namely 1992–1994. The *Practical Astronomy* event was scheduled for the *Royal Horticultural Halls* in London, but was not cancelled even after the magazine folded! Instead, Patrick found another backer, named *Top Events and Publications,* and the event proceeded as *The London Astronomy Exhibition*. Patrick was still irritated that *Astronomy Now* was doing so well, without him having any role in the magazine and, even at 72 years of age, he was determined to make another magazine survive with him as the figurehead. It just seemed so unfair to him that the magazine he had founded in 1987 was flourishing without him; but his pride prevented him from making amends with *Astronomy Now*.

In January 1996 Patrick was off on his foreign travels again, this time heading for South Africa. He had been invited to a conference held at the University of Witwatersrand, Johannesburg, on Dark Matter. While there he visited Feldhausen, Cape Town, where John Herschel had observed. Patrick also visited the observatory housing the 26½ inch refractor at Johannesburg, which had been refurbished and saved from falling into disrepair, largely due to Patrick's publicity. On an earlier visit there, in 1994, Patrick had been appalled to find that the observatory was being ruined by the holding of weddings and social functions in the dome. The telescope itself had been mechanically damaged and cigarette smoke was damaging the optics too. Patrick's advice had received much publicity and the telescope was now in far better condition thanks to his criticism. Patrick had liked South Africa, but only when it was under white supremacist rule. When apartheid finished he declared the country a very dangerous place to go "and we all know why". However, Patrick admitted that following his criticism of the dreadful state of the Johannesburg Observatory the decision to do something about it had actually been made by the Deputy Minister of Arts, Culture, Science and Technology, no less a person than Winnie Mandela!

On March 17th 1996 Patrick was at the Science Museum in London. He was an invited expert and judge at an awards ceremony for ideas by young British inventors, associated with the previous year's BBC series entitled *It'll Never Work*, with which he had been involved. Patrick would repeat this role in the coming year too, returning to the Science Museum for another awards ceremony 1 year later.

By early 1996 everyone was talking about comet Hale-Bopp. There was still a year to go, before it reached its peak, but it was brightening steadily. However, in February, March and April another comet, Hyakutake, stole the show. Everyone was hyping Hale-Bopp as 'The Comet of the Century' and Patrick was doing his best to calm things down: he had seen it all before! Hale-Bopp was not going to pass especially close to the Sun, or to the Earth. It was just an incredibly active, very large comet. But, in late January, the Japanese amateur astronomer Hyakutake swept up a comet which was in the 'Halley' category, activity-wise, and was passing very close to the Earth. Remarkably, from its discovery to its close passage was a time period of only 2 months. Unfortunately, most of the UK was clouded out at the time of close approach: the last week in March 1996. This was a great shame, as from where I had travelled to view it, namely Tenerife, it was awesome: the head was three times the size of the Moon and the tail was 60 degrees long! Once again Patrick was being quoted daily in the media and debunking claims of a 'Doomsday Comet!' However, with astronomy back in the news, it did at least allow Patrick to have a good moan about light pollution, when he was invited to talk about it on *Tomorrows World* on BBC1. Three months after Hyakutake's close fly-by Patrick was an invited celebrity at the London Planetarium, on June 20th, where a new virtual-reality presentation had been set up to show how the monuments at Stonehenge would look once English Heritage had restored the site to its prehistoric state.

On the subject of Doomsday events, the year 1996 saw the release, in June, of the blockbuster film *Independence Day*, in which aliens arrive at the Earth in giant spaceships and soon try to lay waste to large swathes of the USA. Fortunately, a number of heroes, including Jeff Goldblum and Will Smith, save the day. Someone at the BBC decided there should be a 1 hour, British, radio equivalent and, you guessed it, decided that Patrick should be the hero. With permission from the film's producer, Dean Devlin, the idea took shape and was advertised on Radio 1 with the tag line 'It's the Battle of Britain all over again, but this time our enemies aren't just unstoppable, they're inhuman!' Patrick, and Radio 1 DJ Nicky Campbell, were portrayed as being in an RAF Sentry aircraft, relaying news that a huge alien spaceship was approaching the Earth. The Prime Minister John Major and the opposition leader Tony Blair then announced the formation of an emergency coalition government, just hours before London was obliterated.

However, thanks to Patrick's knowledge of science, spacecraft and aliens, the enemy was destroyed, The Royal Family was saved and the 73 year old Patrick also managed to comprehensively duff an alien up. One of the classic comments at the end of the broadcast must surely have been written by Patrick, as it reads: "When this is all over and we've defeated these monsters, the Yanks'll take the credit for it, you wait and see….." Other celebrities involved in this ripping yarn included the pop singers Toyah Wilcox and Dannii Minogue! The 1 hour radio programme was broadcast as *Independence Day UK* on August 4th 1996 and immediately passed into obscurity, despite being quite a treat for fans of Patrick. However, August 1996 would feature a rather better remembered BBC science fiction epic, which also included a short contribution by Patrick. August 26th was declared *Star Trek* Night on BBC 2 and from 7.15 p.m. to just after midnight a whole host of *Star Trek* based programs were broadcast. At 10.35 p.m., in a snippet called 'Star Trekkers', Patrick explained why he was a fan of Mr Spock.

Neptune and Another Award

The year 1996 marked the 150th anniversary of the discovery of the eighth planet Neptune, and Patrick gave a number of talks to celebrate the anniversary, as well as writing a major paper for the Journal of the BAA. For British astronomers the history of the discovery of Neptune was rather a painful one. The seventh planet Uranus had been discovered from Bath, by William Herschel, so it was an English discovery. Neptune could also have been discovered from England, if not for an unfortunate series of events. Patrick had strong views on exactly what had happened, although of course no-one living in 1996, not even Patrick, could claim to remember what took place in 1846! To cut a long story short, two mathematicians, the young John Couch Adams in England and Urbain J.J. Le Verrier in France, set out to find an eighth planet, assuming it must be responsible for the perturbations of Uranus (in other words the gravitational pull of the undiscovered Neptune was affecting Uranus). The French mathematician Le Verrier sent his prediction to Galle in Berlin. Galle went to the Observatory Director, Encke and Encke agreed to offer time on Berlin Observatory's 9 inch refractor for the search.

On the first night of the search Galle, and his assistant D'Arrest, identified the new planet! In England, so the story goes, Adams' calculations were only taken seriously by Airy (the Astronomer Royal) after he heard of Le Verrier's similar calculations. Airy instructed an astronomer named Challis, at Cambridge, to search for the new planet, but Challis was not energetic and used a cumbersome sky checking method. If not for Airy's delay in taking Adams' mathematics seriously, and Challis' sloth, Neptune would have been discovered in England. Well, that was the accepted chain of events. For the 150th anniversary Patrick decided to delve a bit deeper. Amongst various investigations Patrick located Adam's great-great niece, a Mrs Hilda Harrison. However, Patrick did not really unearth much new information, despite deciding it was worth writing a paper on the whole episode. Essentially, Patrick claimed to have established that:

1. The mathematician Adams was in no way an observer, so did not have the confidence to search for Neptune (even though the smallest telescope would reveal it).
2. Airy was sceptical of Adam's calculations and liked order and method so much that he was loathe to disrupt the Greenwich Observatory routine.
3. Challis was painfully slow and looking for too faint an object.

However, at the BAA Out-of-London Birmingham weekend, at Aston University, held on September 21st 1996, another version of events was put forward by a Mr Dennis Rawlins, who had also spent some time investigating the Neptune controversy. Mr Rawlins' research revealed that Airy had never snubbed Adams at all. In fact, Adams' predictions involved the controversial 'Bodes Law' and Adams himself was unsure of his calculations. In addition, there was evidence that Adams, Airy, Challis and an orbit expert named Hansen, were in regular contact and that Adams had directed Challis to three possible places on the ecliptic to search. The speaker thought that all four were aware of Le Verrier homing in on Neptune and they shrouded their own research in secrecy. Concluding his talk, Mr Rawlins said that it was time to stop repeating the old myth that "Adams knew where Neptune was but Airy snubbed him."

I was sitting near to Patrick for most of that BAA meeting and, for most of the talk, Patrick was just muttering, shaking his head and going "No, no, no" in increasingly audible mutters! Patrick was immediately on his feet only seconds after Mr Rawlins talk concluded:

"One thing is for sure" Patrick bellowed, hand on hip, "Adams was a shy man, he turned down a knighthood and was overawed by Airy. Airy was pre-occupied at the time by a nasty scandal at the RGO and discussions about railway gauges. A crucial factor in the Galle & D'Arrest business was that Galle was sure the planet would show a disc. Challis had seen something that might have shown a disk, but he and Airy had been distracted by the offer of a cup of tea (!) at the critical time and the observation was never checked". A member asked Patrick what he thought of Mr Rawlins' conspiracy theory. "No comment" Patrick bellowed, firmly crossing his arms and sitting down with a sulking schoolboy grimace on his face!

One week later Patrick was at the Roman Baths in the city of Bath for the annual conference of the international Antique Telescope Society. He was delivering a talk on September 28th entitled 'The Astronomy of Birr Castle' and it was a chance to meet up with his old friend, now resident in Florida, Ron Maddison. Ron was the society's President at that time and gave a lecture entitled 'Problem Objects – Grubb Mounts'. The historian Allan Chapman was also at the conference, delivering a superb talk about 'The Brotherhood of Big Reflecting Telescopes: William Herschel to William Lassell'. A superb picture of these three men, deep in conversation at the meeting, was taken by Peter Louwman and a copy was kindly sent to me by Ron Maddison, with Peter's generous permission.

Patrick's involvement with the 150th anniversary of the discovery of Neptune did not end at the September BAA event though. Nineteen days after that meeting, on October 10th, he was in Liverpool to celebrate the discovery of Neptune's moon Triton, on October 10th 1846; and this was a British discovery. The astronomer Lassell had found the new moon with his giant 24-inch f/10 telescope, soon after Neptune's discovery. Remarkably, in a collaborative project between Royal Insurance, NMGM (National Museums & Galleries on Merseyside), astronomers from Liverpool and Oxford, Liverpool John Moores University and Liverpool Astronomical Society, Lassell's huge telescope had been recreated as a full sized working model. The telescope itself was erected at St Michael's Primary School in

Fig. 34.1 Patrick with Ron Maddison (*left*) and Allan Chapman (*right*) at the international Antique Telescope Society conference on September 28th 1996, held at the Roman Baths in Bath, England. All three men had delivered talks at the meeting, which were on: Grubb mounts (Ron), Birr Castle (Patrick) and the Brotherhood of Big Reflecting Telescopes (Allan) (Photograph by kind permission of Peter J.K. Louwman (forwarded by Ron Maddison))

Liverpool, which stood on the site of Lassell's former home in Guion Street. Patrick was taken to the school and photographed standing by the telescope, alongside his namesake: a 5 year old schoolboy, also named Patrick Moore! Just 2 days later, on October 12th, Patrick was a live guest on the BBC Saturday morning children's show 'Live and Kicking', presented by Zoe Ball and Jamie Theakston. It was Patrick's second appearance on the show (the first had been 2 years earlier on the 19th of February 1994). The other major guests on that October 12th programme, alongside Patrick, were the boy band *Boyzone*!

In 1996 BAA members Jeremy and Marie Cook decided to propose Patrick for another BAA medal: the Lydia Brown Medal for meritorious service in an honorary capacity. Patrick had accepted accolades, doctorates and an almost infinity of awards heaped onto him already, including, in 1968, the most prestigious BAA award, the Goodacre Medal and Gift. Nevertheless, the council agreed to this additional award. He had certainly done as much for the BAA as any other member in its history, so why not?

The President, Maurice Gavin, presented the award to Patrick at the October 30th BAA AGM, to thunderous applause. Patrick told the audience that the BAA had been a big part of his life for over 60 years and his membership meant a great deal to him. The last Wednesdays in each month had become 'sacred days' in his diary. Patrick said that he remembered Lydia Brown (the BAA Office Secretary for most of her life) as someone who had been part of the BAA for generations, but he felt there were far more worthy recipients for such a medal than himself. Even though his cricketing days might be coming to a close, he would always do as much for the association as he could. Patrick left the stage to further thunderous applause. He had been scheduled to give a short talk about developments at Birr Castle in Ireland at the end of that particular BAA meeting, but due to the important discovery of the UK's first supernova, by Mark Armstrong, Patrick's talk was postponed.

Only a few months earlier Patrick had accepted yet another honorary degree to add to his huge collection of them, this time from the University of Leicester. As always, in accepting the degree, Patrick was very humble and knew exactly the level at which to pitch his short address. In front of the University's academics and students and dressed in an enormous red and blue gown, plus mortar board, he made the following speech:

> Well, Vice Chancellor, ladies and gentlemen, to say that I feel honoured at being here today is an understatement. No, I've not been into space; it would take a very large rocket to launch me. I fear this will not happen, but nevertheless I feel particularly proud to be here, because after all, as we know, the University of Leicester is a leader in astronomy, and in many other things also, but clearly I am concerned mainly with astronomy. And in Leicester's own particular brand of space research and X-ray research, Leicester, I think it's fair to say, leads the astronomical world, not only in Britain, but everywhere. And, after all, astronomy today is such a changing science. I'm an astronomical dinosaur. I still look through telescopes. Most people these days do not; we've come into the modern age. I go back a long way; of course I started doing my television programs before the space age started. I suppose I've achieved a certain notoriety. Somebody came up to me the other day and said: "Yes, I've seen you on television, I'd love your autograph, you are Cyril Smith aren't you!" [There was much laughter at that joke – Cyril Smith was an absolutely enormous Liberal MP, later revealed, after his death, as a paedophile!] But astronomy is a fast changing science. I've seen a tremendous number of changes and most of you here, in fact almost all of you, will see more than I have. You'll see well into the next millennium and I will not. But Leicester has lead the way and been responsible for many fundamental advances and wherever you go, in the astronomical world, in Britain, America, Australia, Russia, anywhere, there you will find Leicester influence and Leicester graduates. It's a record to be proud of and I have no doubt whatsoever this will continue into the next millennium and that Leicester's achievements in the next millennium are going to be even greater than those in the past. And you of course will share it. You will see this, I will not. But this is a day I will never forget. I'm immensely proud to be here, immensely proud to be admitted to this great University that has done so much for science and for the world, and I thank you from the bottom of my heart. Thank you very much indeed.

Patrick then marched swiftly away from the lectern to thunderous applause. Yet another plaque from a University could be attached to the wall of his study in Selsey and yet another University could derive some publicity from granting Patrick an honorary degree.

Modern Astronomer and 40 Years on TV

In November 1996 (the month that Patrick's friend Michael Bentine succumbed to prostate cancer) Patrick's latest attempt at another high-street astronomy magazine to rival *Astronomy Now* appeared. This new magazine was entitled *Modern Astronomer*. Interestingly, this was the same title as the first book published in Springer-Verlag's

Practical Astronomy Series. This particular magazine, a rather tacky looking effort, was distributed by the aforementioned *Top Events & Publications Ltd* of Chester and listed the publisher as Roy Bird and the publishing company as *Chesterwell Management Ltd*. The editorial office was in Cambridge. Quite a geographic spread! Patrick claimed that it had the finest team of astronomers of any magazine today. The magazine would survive for a few years, with less and less involvement from Patrick and then, like *Practical Astronomy*, it sank! It would be a bitter blow. Patrick would not try again, after its failure, until 2005. The first edition of the magazine covered December 1996. Patrick's good friend John Mason, who had collaborated with him in numerous ventures before, was determined not to be involved with this publication though. He had deep concerns and disagreed with Patrick's optimism. Writing to the December 1996 BAA Journal John stressed that: 'Despite implications to the contrary, I am not involved in any editorial capacity with a new monthly periodical called *Modern Astronomer*, published by Chesterwell Management Limited'.

In January 1997 Patrick paid yet another visit to Birr Castle in Ireland, to see the restoration work that had been taking place on the historic 'Leviathan of Parsonstown': a huge alt-azimuth mounted 72 inch aperture telescope slung between two stone walls that had once been the largest telescope in the world.

Hale-Bopp excitement was at a fever pitch in early 1997 and, throughout the spring, Patrick mentioned the comet on TV, and on the radio, almost every week. It seemed almost pre-destined, with the *Sky at Night*'s 40th anniversary coming up in April 1997, that a truly great comet should appear. Arend Roland, another superb comet, had appeared in the night sky in the spring of 1957, almost as if to 'usher in' the programme. Through amateur telescopes the head of the comet was an extraordinary sight with jets and spirals of material coming off the comet's nucleus and clearly visible! In 20 years of comet observing I had never seen anything like it. Both the BBC and Patrick organised parties to celebrate the 40th Anniversary of *The Sky at Night*. It had, for many years, been the longest running programme, with the same presenter, in the world. It was, surely, always going to survive now, at least while Patrick was well enough to present it. It seemed incredible that Patrick had been a mere 34 years old when the programme had started and that he had completed four decades in the post, maintaining the same 15 and then 20-minute format for all but the landmark programmes.

The BBC celebration was held on the evening of Tuesday April 22nd 1997 in the BBC Television Centre at White City, West London. Producers, ex-producers and programme guests from many years were in attendance. I was lucky enough to get a ticket too. I had not been on the programme at that point, but Patrick had used my photographs and images on *The Sky at Night* for the past 15 years, so I was honoured to receive an invitation. I gave my spare ticket to another BAA colleague, who worked at the same Marconi site as me, Nick James. Patrick was surrounded by his best friends for most of the 40th Anniversary party and was demolishing glass after glass of gin and tonic. As everyone who has ever met Patrick will tell you, you simply can't drink him under the table. Incidentally, although he would drink almost any alcoholic wine or spirit (but not beer) his favourite tipple was the Greek wine Retsina. I managed to get a few words in with him, before Iain Nicolson, Dudley Fuller (founder of telescope maker Fullerscopes, later telescope importer Broadhurst Clarkson & Fuller) plus Paul and Chris Doherty (the latter now fully recovered from being hit by a car) and others grabbed his attention.

Once the party was in full swing the controller of BBC1 at that time, Alan Yentob, presented Patrick with a card and a present (a small Meade ETX telescope). Tragically, he described Patrick not once, but twice, as an *astrologer*, not an astronomer, which had Patrick groaning and most people highly confused. He then called the people present a room full of *astrologers* too. Now people really were starting to look insulted and there were many distinguished scientists there. That *faux pas* is still remembered to this day by all who were there. Was Yentob joking or did he really not know the difference? Certainly, no-one was laughing! In 2008 I heard an explanation from the programme producer Pieter Morpurgo. He told me he had worried that Yentob would get confused, not least because the BBC 1 controller had met the astrologer 'Mystic Meg' earlier in the day! Pieter had already stressed the difference between astronomers and astrologers prior to the party, but to no avail. Patrick was deeply worried that many astronomers would have been offended by being called astrologers and would not now attend the smaller, more private party, at his house in Selsey a few days later. For that party Patrick had a rushed set of postcards specially printed, reading 'On no account mention the word astrology', to defray any damage inflicted by Alan Yentob. In fact, when Yentob initially took over BBC 1 and carried out the usual (and very predictable) cost cutting exercise, Patrick told me that he had suggested that *The Sky at Night* might be cut from the schedules, much to Patrick and Pieter Morpurgo's horror. Fortunately, no-one else felt that way about the longest running TV programme with the same presenter in the world, and so the programme continued, no doubt helped by the fact that it had one of the smallest budgets in the corporation.

As the 40th anniversary evening wore on, many young and middle-aged people started calling it a night, but not Patrick. He was almost incoherent, soused, pickled, 'as a newt', call it what you will, but still standing and still accepting more top-ups! "I'm glad you're going to be the next President" he said to me, swaying like a huge blazer-clad wardrobe, eyes glazed and monocle in position. A woman on his other side asked how he was going to get home that night. "Oh, I'll drive, I'll drive, haven't had that much to drink, it's only 80 miles to Selsey...I'll be OK". She looked horrified. Patrick laughed. "I'm joking of course. I have a chauffeur, Ha, HA, HARRRRRRRRARARAARARR HAR HAR HARRRRRR!"

Four days later Patrick held the other, slightly smaller booze-up, at his home, on the night before the actual 40th anniversary programme. That 40th anniversary *Sky at Night*, a special 50 minute programme transmitted on Sunday April 27th, concentrated largely on the development of the telescope. Patrick wrote a friendly little book, for Springer-Verlag, to commemorate the programme. The next month's *Sky at Night*, with John Mason as the guest, was dedicated to Hale-Bopp, a definite contender now for 'Comet of the Century'. In mid-June, Thomas Bopp, co-discoverer of the comet, spent a brief stay with Patrick while on a lecture tour of the UK.

So, Patrick had survived a remarkable unbroken 40 years presenting the same TV programme on BBC 1; and they had been 40 glorious years without a co-presenter, or four, sharing the limelight! To me, this 40 year anniversary was the big one,

Fig. 34.2 Patrick at the 40th anniversary *Sky at Night* celebration on April 22nd 1997, held at the BBC TV Centre in London. Patrick's aghast expression is because the BBC 1 controller, Alan Yentob, is toasting his contribution to ASTROLOGY! *On the left*, hands clenched together, is *The Sky at Night* producer Pieter Morpurgo (Photograph by the author)

despite the other anniversaries that followed. In 1997 Patrick, at 74, was still mobile and still fully in charge of the programme. He was the boss and he called the shots and he still spoke fluently to the camera, without any prompts being needed. He was also healthy enough to travel to meetings, give talks and fly abroad if needed. The quality of his life was still high and he was fully in charge of his world. In addition, he was still working with Pieter Morpurgo at the BBC, a working relationship which had proved to be a joy. On July 5th the *Daily Express* columnist William Hickey labelled Patrick as 'National Treasure no. 40' which must surely have pleased him too. As well as passing the 40 year milestone Patrick was told that he would be receiving an academic award in a few months' time. On July 25th Patrick would be accepting yet another honorary Doctorate, this time from Portsmouth University, only a few miles from his Selsey home. However, while the first half of 1997 had gone smoothly for Patrick the rest of the year would see the loss of two good friends and the whole country in mourning for another celebrity.

On July 19th the phone in Patrick's Selsey study rang and he picked it up. It was a friend associated with the Anglo-Australian Observatory in New South Wales, ringing from the other side of the world. "I hope you are sitting down Patrick. We thought you ought to know straight away: Gene Shoemaker has been killed......" The phone slipped out of Patrick's hand and crashed onto the desk, simply because he was so shocked. Gene Shoemaker had been a friend of Patrick's for more than 30 years as he was an expert on lunar geology and a man once tipped to be the first geologist to visit the Moon.

If not for an adrenal gland disorder Shoemaker may well have flown as a geologist on one of the final Apollo missions. Instead, he had trained Apollo astronauts in geology during field trips to Arizona and had been a regular BBC studio guest, alongside Patrick and James Burke, during the Apollo Moon landing era. In later years he and his wife Carolyn, often assisted by the amateur astronomer David Levy, had discovered dozens of comets that bore the Shoemaker name, including the one that had collided with Jupiter. Gene Shoemaker had also found a number of meteor craters in the Australian Outback. It was on one of those excursions that he had been killed, in a head-on collision, between his 4×4 and a huge lorry, while driving on the Tanami Road northwest of Alice Springs, on July 18th. Patrick was totally shocked that such a great man could be wiped out, in an instant, just like that, despite having witnessed so much loss of life in the War himself. Two years later, on July 31st 1999, Gene Shoemaker's ashes were flown to the Moon by the Lunar Prospector space probe, which crash-landed in the appropriately named Shoemaker crater at the lunar South Pole. So far, he is the only person to have been 'buried' on the Moon.

Gene Shoemaker was not the only famous person to be killed in a violent car crash that summer either. August 31st 1997 was a day that has gone down in history: it was the day that Princess Diana was killed in that fateful car crash in Paris. Ten days later, on the day of the next BAA London Council meeting, I made a slight detour, with the BAA Mars Section Director Richard McKim, to Buckingham Palace. I am certainly not a great fan of royalty, but I was not going to miss the opportunity to sample the atmosphere outside the palace, which had become a sort of shrine for the nation. Richard had the same thought, as 'Buck House' was only a short walk from the RAS building at Burlington House in Piccadilly, where the BAA Office is located and where our Officers meetings and Council meetings were held. Even 10 days after Diana's death the area around the palace was totally and utterly saturated with flowers and written tributes, as far as you could see: an extraordinary sight!

It was customary, in the late 1990s, for the BAA Officers (Secretaries, President and Vice-Presidents) to have a cosy Wednesday lunch in a tiny, incredibly narrow, Italian restaurant just off Piccadilly, before the BAA Council meeting at 2 p.m. It was in Albemarle Street if I recall, not far from the Green Park Underground station. On this occasion, 10 days after Diana's death, Patrick turned up at the restaurant too and turned a few heads as he squeezed down the narrow gap to our table, set against the wall. He sat down opposite to me and I mentioned Diana's death to him. "I met her" he snapped, "at charity events". I asked what he thought of her. "A good actress, knew how to act the part. Nothing more, nothing less"; and that was the end of that conversation! On quite a few occasions when I found myself with Patrick, in a restaurant, he would pick up the menu as the waitress arrived, rapidly leaf through it, look totally confused, sigh, groan, and then say to me "What are you having Martin?" On this occasion I replied "Gammon, Egg and Chips, Patrick". As usual, Patrick said to the waitress "I'll have the same". That was his meal sorted. He could get back to thinking about astronomy! If I'd said I was having Rhinoceros droppings with a Cyanide sauce, he would have had the same. Patrick was not going to waste astronomy-time rifling through the pages of a confusing menu. The only menu he understood was the one at his local Selsey Curry House! I might add that the first time I found out that Patrick, an ultra-patriotic right wing (excluding his views on fox hunting) St George loving Brit., loved Curries, I was surprised; but maybe I shouldn't have been. Just as he could drink anyone under the table with alcohol, his Ox-like constitution was perfectly at home with hot, spicy and eye-watering food. Whichever way you looked at it Patrick's digestive system seemed to have been constructed from military grade components.

On September the 12th to the 14th 1997 the BAA held an Out-of-Town meeting at the University of East Anglia in Norwich. The University and Norwich Astronomical Society were the hosts. Patrick and John Mason were booked to give one of their entertaining double-act performances in the evening and I was asked to chair the meeting and introduce them, because the BAA President Maurice Gavin had to return to London, and in 6 weeks' time I would succeed him as President. I turned up at the lecture theatre, for the afternoon BAA meeting, and spotted that Patrick was sitting alone. He was in the foyer and well-away from the local Norwich Astronomical Society members and he was obviously lost in his own thoughts. I knew Patrick well enough by then not to be put off by his solemn Churchillian scowl. He could actually be quite a shy man in those circumstances, even if he thought nothing of speaking 'live' on TV or to audiences of hundreds of people! He might appear like he wanted to be well away from other people, but in fact he was just waiting for people to speak to him! Of course, when you are that famous and have such a serious expression, people will feel a bit nervous about starting a conversation; but he did look seriously troubled. I wondered if he was upset about his friend Paul Doherty, who I had spoken to on the telephone a few weeks earlier. It was obvious that Paul was losing his brave battle with cancer. His first operation had only temporarily eased the tumour and it was clear he only had a matter of weeks left. At the BBC's *Sky at Night* 40th anniversary celebrations Paul had looked gaunt and ill, but still had a sense of humour and some hope, but now, 5 months later, things were looking very grim indeed.

Anyway, I decided it was ludicrous that the great man was sitting there on his own, looking so depressed, so with a "Hello Patrick, are people ignoring you?" I broke him out of his silence. It turned out that Paul was very much the source of his solemn mood, but he soon brightened up after having someone he knew to talk to about it. I asked him if he had any advice for an incoming President and he was keen to help. Patrick also told me that the arthritis in his fingers now made it impossible for him to use his pocket Chess set but that a friend had sent him a folding Portland Chess Set which he could use. Then I thought I'd tell him about my decision to get a Schmidt-Cassegrain telescope and a new CCD camera. Patrick was a visual observer and not a hi-tech person, but he seemed unusually interested. Just as we were about to move into the lecture theatre for the start of the BAA meeting he turned to me and snapped "Fancy appearing on *The Sky at Night*? We've been thinking about doing a programme on CCDs and I think you might be just the man. Would you do it?" Needless to say, no amateur astronomer in the UK would be less than ecstatic at the prospect of appearing with Patrick on *The Sky at Night*, especially in those days, when Patrick was fluent and there was only one guest per programme. The world's leading professional astronomers and all sorts of celebrities had appeared with Patrick on the programme.

I was delighted. There are not many days in your life when opportunities like that turn up! Admittedly, Patrick had hinted at something similar when he saw my first lunar video tapes in 1985, 12 years earlier, but this time he really seemed to have a specific programme in mind. The rest of that day, even introducing Patrick and John for the evening lecture, was something of a blur, but I do remember one incident at the University bar, after the Moore-Mason talk. I was sitting with Patrick, his BAA friend from Selsey, Rossie Atwell, Pat Barber (BAA Office Secretary), John Mason and a few others, when a couple of young students asked if they could shake Patrick's hand and buy him a drink. Patrick was only too delighted. He was much happier now; he was surrounded by his astronomical friends and fans, had just delivered an entertaining lecture with John, signed a few books and knocked back a few shorts already! Patrick invited the students to join us. "I've been watching your programme for years" one of the students said, "almost since it first started." Patrick looked puzzled. "Oh, the *Sky at Night*'s been running for forty years" he said, "You can't possibly be old enough." "No, not the *Sky at Night, Gamesmaster*!" replied the student. Everyone laughed! As we saw earlier, Patrick, for much of the 1990s, had recorded voice-over commentaries for a computer games programme called *Gamesmaster*, on Channel 4. In the programme, a distorted characture of him appeared, telling people the tricks of specific computer games. It was this programme that the student had been watching! Of course, Patrick was not a computer user himself. I did hear a rumour once that John Mason had tried teaching Patrick the subtleties of using a computer mouse in the 1990s, but Patrick's fist had destroyed the rodent, even before Patrick's cat (who enjoyed switching Patrick's fax machine off). Anyway, the Norwich campus students meant that he was surrounded by even more fans and they bought him a drink too, so he was in his element.

Incidentally, on the subject of fax machines and feline creatures: apart from being crammed with his own astronomy books, Patrick's study contained hod loads of books on cats. He really was mad on them and, in the early years of the twenty-first century, he had two: a black and white cat called Jeannie (acquired as a kitten in 1999 by Patrick's godson Adam, who had seen an advert in the Chichester Observer) and, later, a jet black friend for Jeannie called Ptolemy, introduced to Patrick by his Selsey neighbour, Alan. Ptolemy was a dead ringer for a cat of the same name that Patrick had owned 70 years earlier, as a young boy. When Ptolemy Mk I had died, in 1933, the young Patrick had been totally devastated. Patrick would be too immobile to chase the twenty-first century cats if they escaped and was terrified that they would run into West Street and be killed by a car. So the cats were kept inside, but a small play area, fenced with chicken-wire, was made for them. This enclosure was just outside the door that led into the garden, near the boundary of Patrick's music room/TV room and his study; the play area extended over half of the bay window in Patrick's study and enclosed the house door. Patrick even had a metallic Saturn globe (installed in Xmas 2001) and a fountain in there, for them to explore! The cats could climb in and out of their own little fenced-off garden, via the bay window in Patrick's study, or via the door, but could not get out into the main garden itself. Years later, even in his immobile state, he could still keep an eye on them and let them in or out of the study window, while seated at his ancient typewriter. They could not attack birds either, while safely enclosed in their play area. The metallic sculpture of Saturn and its rings, installed in the cats' play area, was inspired by a Little Green Men ('bogies') painting by Patrick's mother Gertrude, but created by the sculptor Colin Wilkin, some 20 years after her death.

From the 1990s (when the cat in charge was Bonnie, the feline resident from 1979 to 1999) attached to the front of Patrick's porch at Farthing's you were always greeted by a note sellotaped to the inside of the glass, which read:

CAUTION: Patrick's Cat is a fiendishly clever master of escape, a veritable Papillon of the feline world. PLEASE OBSERVE THE AIRLOCK PROCEDURES

1. **Shut external door before opening the inner – and vice versa**
2. **Before opening the outer door check the little blighter is not lurking in the airlock disguised as a coat or potted plant.**

Another note declared that: **This house is run solely for the benefit of the cat.** His later moggie Jeannie used to sit on Patrick's 'in tray' or the warm fax machine, above a sign labelled **'Commanding Officer'**, inherited from Patrick's ATC Commander period, just after the war. Another wartime sign, seemingly ripped off the side of an RAF filing cabinet, was a feature of Patrick's downstairs lavatory. It was placed just above the toilet roll: **IF YOU ARE DESTROYING SECRET OR TOP SECRET MATERIAL, REMEMBER THAT AN OFFICER MUST BE PRESENT TO WITNESS THE DESTRUCTION.**

When the house was full of guests and TV crews, with doors opening and shutting all day and all night, the cats were locked inside Patrick's bedroom, with another note stuck to that door, warning: **Please Do Not Open This Door! – Don't let the cats out!**

Patrick's love of cats was well known and he was often invited to be a judge at annual cat shows. The year 1997 was no exception and he was the main celebrity at the 'Arthur's Cat Awards' event in London. Various pictures appeared in the media of Patrick clutching the West Midlands winning cat Prudence, while flanked by GMTV presenter Fiona Phillips and Wendy Turner. They, in turn, were holding cats named Tom and Jerry, the new mouse catchers at the Barbican Tube Station.

On October 29th 1997 I ascended to the dizzy heights of the BAA Presidency. The BAA AGM's had a habit of overrunning, but I was determined on that day that Patrick, who had been scheduled to speak in the second half of the meeting, would have his say. On ascending the stage and shaking hands with the outgoing President, Maurice Gavin, I wasted no time in telling everyone that this was all Patrick's fault. I had read his *Observers Book of Astronomy* in 1968 and was doomed from that moment on. It seemed incredible that 27 years earlier, in 1970 as a 12 year old boy, I had first met Patrick aged 47. Now, aged 39, here I was becoming the President and about to introduce a 74 year-old Patrick to give his talk; how time flies! As usual, Patrick fitted his delivery precisely to the time available and gave an excellent presentation on the ongoing restoration work to the Birr Castle 72 inch telescope.

Chapter 35
Death of a Friend, Selsey's Tornado and Serious Health Problems

In November 1997 Patrick's worst fears over Paul Doherty were realised. Patrick apologised to me and to the BAA Council for his absence at the November meeting, but Paul was not expected to survive for much longer. Patrick had travelled up to Stoke-on-Trent to be with Paul and his family. In fact, Paul died on the 26th, the day of the November meeting. I spoke to Patrick many times following Paul's death and Patrick never quite seemed the same person after that. It seemed to really knock the stuffing out of him, almost as much as his mother's own death. But, by mid 1998, Patrick's own health was causing him serious concern too.

November 1997 had been a bad month in other ways. While away from home it appeared that Patrick had been burgled, although precisely when a couple of highly sentimental and valuable items had disappeared was never made clear. The items in question were Patrick's father's Military Cross from the First World War and (so Patrick initially said) Neil Armstrong's Omega Speedmaster Watch: the one that Armstrong had taken to the Moon on Apollo 11. Armstrong had given the watch (or at least some sort of watch) to Patrick more than 20 years earlier. It might be thought that this was the first watch to be worn on the lunar surface, but according to Omega, due to a computer timing glitch in the Lunar Module computer, Armstrong had left his Speedmaster in the Lunar Module during the EVA (Extra Vehicular Activity). It was actually Aldrin's Speedmaster watch that was the first watch worn on the lunar surface. Bizarrely, that watch was also stolen, in 1970, en route to the Smithsonian Institute's National Air and Space Museum at Washington D.C. The loss of an astronaut's watch and Patrick's father's Military Cross was a hammer blow to Patrick, especially as it appeared likely they were pinched while he was away visiting the terminally ill Paul Doherty.

However, when it came to the police investigation Patrick was less than helpful: he simply had no time for red tape and questions. He also appeared to be confused as to whether there were two separate break-ins at his house, or just the one. Weeks later, he said it was not even Armstrong's Omega watch that had been stolen! Yes, Armstrong had given him a watch, but it was the astronaut Jim Irwin's Omega that was missing from the house [Irwin was the Apollo 15 Lunar Module pilot and had died in 1991].

M. Mobberley, *It Came From Outer Space Wearing an RAF Blazer!: A Fan's Biography of Sir Patrick Moore*, DOI 10.1007/978-3-319-00609-3_35,

According to friends of Patrick who were involved in this episode, a major problem was the sheer number of people who visited Patrick's home each month; many of these were loyal friends of Patrick, so he was loathe to make them subjects for investigation. As I mentioned earlier in the book, one friend of mine said that he told the police "There are at least one thousand suspects!" In addition, the date that the items were stolen was not clear, because Patrick had simply noticed that they had gone missing, but not precisely when. Despite Patrick's pain at the loss of such treasured items the pain of endless police questions was far greater and the matter was quickly deemed to be insoluble, with Patrick blaming the police for not apprehending the thieves instantly. What with Paul Doherty's death and the double-theft, November 1997 was one of the grimmest months Patrick had experienced since his mother's demise. I had a phone call from the great man before the year ended. When I broached the subject of the burglary he simply snapped "I wish I'd been in the house at the time Martin – I'd have laid them out – The old unarmed military combat training never leaves you……" But he also added that I should watch Angus Deayton's 'End of the Year Show' on New Year's Eve, as he was a guest on the programme.

The year 1998 also started badly for Patrick and for Selsey. High tides and gale force winds breached and all but destroyed the town's sea defences, on the weekend of January 3rd/4th, but the worst was yet to come. On Wednesday night, January 7th, Patrick was in the local Tandoori Curry House (a regular haunt of his) with, so he told me, *Modern Astronomer* publisher Roy Bird. The wind outside was gusting to storm force. Suddenly the curry house door appeared to almost be sucked out of its frame: a Tornado was passing through the town! Patrick and his guest hid under the Curry House table until the worst of it had passed. In a TV interview, Patrick (in a style reminiscent of Captain Mainwaring from *Dad's Army*) explained that he knew immediately that it was a tornado, from the drop in pressure. When the worst of the storm had subsided, Patrick went back to his house to assess the damage. He was terrified that the thatched roof might have gone. In fact he had escaped lightly compared to other Selsey residents: 1,000 buildings in the small town were damaged, with roofs and chimneys being ripped off. Patrick's small, rotating, octagonal glass observatory was damaged: the top half had lifted off and was badly smashed. The 8.5 inch Newtonian inside was damaged too, but not beyond repair. The large green metal structure, housing his 15 inch reflector, had been buckled slightly by the tornado, but it was otherwise OK. Considering the rest of the damage in the town, Patrick's property had experienced a very narrow escape. This was not the first time a tornado had passed through Selsey either. A lesser tornado had passed through the town in 1986 and another one would pass through in just under 3 years' time, on October 30th 2000. Some amusing reports in the media suggested that Patrick Moore eating a curry was the cause of the tornado! The Selsey curry house in question became quite famous after the event and changed its name, for a while, to the 'Tandoori Tornado'!

While on the subject of violent weather, I might add that one of the features of Patrick's house that the first time visitor always noticed was the distinctive amateur astronomy themed weather vane, atop the main chimney. Patrick had a special vane created for him, which depicted an amateur astronomer, peering through a tripod mounted refractor, at an eight spiked star; the star was supported on a slender metal strut from the arrow end of the vane. Surely this must have been a totally unique weather vane?

Patrick had certainly escaped lightly from the wild weather of early January 1998 and he also had something really exciting to look forward to: two cruise ship expeditions almost back-to-back; one to the Antarctic and one to the Caribbean. On both trips he would take along Chris Doherty, whose father had died only a few months earlier. The first leg of the first trip started just over a week after the Selsey Tornado for Patrick: he flew to Tierra del Fuego in Argentina to board the Orient Cruise Lines' ship *Marco Polo*. That ship, which he had travelled on for the 1995 Asia total solar eclipse, would leave Tierra del Fuego on January 18th and ultimately arrive at New Zealand on February 11th. This so-called 'Antarctic Circumnavigation' cruise was really two cruises, made up of the time spent in the Antarctic Peninsula area and the time spent in the Ross Sea area. The cruise was, in reality, only a 100 degree partial Antarctic circumnavigation. The two journeys through the different seas were separated by 9 days, while cruising from the first area to the second, because the plan was to go far north to avoid pack ice, cruise west, then go south through the Ross Sea to McMurdo Sound.

Needless to say, Patrick travelled for free: his passage was paid for by lecturing about 'The Night Sky from Antarctica', despite the fact that, at that time of year, it was permanent daylight in the frozen continent! He also entertained the passengers with various musical compositions of his own, played on his portable xylophone, given to Patrick many years ago by the former 'Goon' Michael Bentine. Patrick was not the only famous person on board though. He was, perhaps, upstaged by another famous lecturer: the explorer Sir Edmund Hillary. Patrick enjoyed the trip immensely, especially the penguins. Patrick had always liked penguins and had even performed the opening ceremony at the Penguin pool at Chester Zoo in 1984. He had also composed a xylophone solo entitled 'Penguin Parade!' On arrival back in the UK, on February 13th, he only had a few days in Selsey before he was off on his travels again.

For the next jaunt, to the total solar eclipse of February 26th 1998, he would be heading, with a small *Sky at Night* team, to the Carribean. Once again, Patrick would be travelling on a luxury cruise ship; after his 1994 Peru 'mugging' experience this was still his preferred mode of travel. This time he was on a Greek ship, the Sun Line's *Stella Solaris*, alongside many other eclipse experts including George Keene, Dr Edwin Krupp, Dr Paul Knappenberger and astronaut/astronomer Dr Ronald Parise. He was, once again, part of Ted Pedas' 'Science and Culture at Sea – Voyage to Darkness' cruises. The *Sky at Night* team accompanying Patrick included the producer for the past 17 years, Pieter Morpurgo, who had devised the ground-breaking *Crimewatch* programme, as well as numerous other quality BBC productions. The year 1998 would, sadly, be Pieter's final year of working with Patrick. Also with Patrick was, again, Chris Doherty, as well as Broadhurst Clarkson & Fuller's owner Dudley Fuller, who was notorious for various things, especially for owning an E-Type Jag., with the registration DUD 1! I went to that fantastic eclipse too, observing it from Knip Bay on the Carribean island of Curacao, with the *Explorers* company, before moving to Bonaire where I learned to Scuba dive.

Incidentally, on the beach at Knip bay, as part of our group, was the Queen guitarist Brian May and his partner, the actress Anita Dobson. Brian had been on our Chile eclipse trip in 1994 and in India in 1995. As totality approached on Knip bay, we could see a number of huge cruise ships in the distance, strung along the eclipse line. One of them may well have been the *Stella Solaris*: it was certainly not far from Curacao during the eclipse. After the big event many cruise ships that had been under the Moon's shadow docked at Curacao. On one of those ships was the *Sky & Telescope* magazine editor Leif Robinson. He had tried to get other S&T editors onto his ship but was refused for security reasons. However, Leif told me that as soon as he invited Patrick up the gangplank there were no problems. Patrick was immediately recognised and treated like royalty: he was universally famous! Six days later, on March 4th, Patrick celebrated his 75th birthday, onboard ship. He was the guest of honour and the ship's chef had even created a special 75th birthday cake with Patrick's name written on it. In addition, an artist passenger, named Robin Rector Krupp, painted a special birthday card depicting Patrick, in his huge colourful shirt, in a lecturing pose. This was Patrick's tenth total solar eclipse and the last one he witnessed. He would be clouded out for the Cornwall eclipse 18 months later.

Two weeks after that eclipse I was sitting at my desk at work and the phone rang. It was Pieter Morpurgo, the BBC's *Sky at Night* producer. He had Patrick with him and they were reliving the experience of the eclipse; "Oh and, by the way, would I still like to appear on *The Sky at Night*, talking about CCDs?!" I was obviously only too happy to say "Yes" and very relieved too. Six months after Patrick had originally raised the idea it must finally be going to happen if the producer was phoning me up. We arranged some convenient dates and agreed to aim for the early May time slot. There would be two planning meetings at the BBC White City offices and then filming at the BBC TV Centre. Patrick had been showing my pictures on *The Sky at Night* for the past 16 years, but I had never actually expected to appear on the programme. Most of the guests on *The Sky at Night* were leading professional astronomers, but I was a mere amateur; but then, so was Patrick, as he had no professional qualifications. Still, I felt that it may have been easier for Patrick to justify my appearance for three reasons. Firstly, I was the BAA President at that time. Secondly, as mentioned, Patrick had used my photographs for the past 16 years and I had never wanted any payment. Thirdly, he knew that I could cope with speaking coherently in public. Like Patrick, I could sound like I knew what I was talking about, even if I didn't! In fact, Patrick told me that he only chose people who came across clearly and coherently and could get their message across in the 20 minute programme. He said that some of the cleverest professional astronomers had a problem speaking coherently on *The Sky at Night*. There was no autocue and there never would be! However, in my case, he said he just knew I would be fine. It was fascinating simply to see what was involved in preparing a *Sky at Night* and I couldn't help wondering if Patrick really appreciated what a fantastic life he had enjoyed over the past 41 years of the programme. In a letter to the BAA Journal a few months later, the retiring producer Pieter Morpurgo said that working on *The Sky at Night* had been 'better than working for a living' for him too!

On the afternoon of April 22nd 1998 I turned up at *The Sky at Night* office in the huge BBC White City building, scarcely able to believe I was there. It seemed like a dream! This was the first planning meeting and I had fleshed out a script and brought along loads of pictures, as well as diagrams which needed 'tarting up'. Pieter Morpurgo, and his assistant Laura Vine were exceptionally friendly and efficient and I told many people afterwards that the best thing about appearing on the programme was that the BBC treated me like I was the star! Patrick was just good old Patrick but, to the crew, I was the most important person on that programme. It was a great feeling and a totally different one from my dreary job as an electronics engineer with Marconi, where I was just another dogsbody, slaving away writing software for a pittance, while arrogant managers and directors pocketed the loot from their massive salaries and, ultimately, sank the company.

Three days later I was chairing the BAA Out-of-Town meeting at the University of Birmingham in Edgbaston. It was a special meeting with a summary of the Carribean eclipse thrown in. Eclipse meetings always generated loads of attendees, namely everyone who had been to the eclipse, BAA members or not. There was a BAA Council meeting in Birmingham too. Patrick turned up at the Council meeting in considerable pain and limping badly. He had been slow on his feet, due to his 'wartime injury', on the previous Wednesday at White City, but he was now in real discomfort. "Martin, I'm going to have to get this knee seen to. It's getting too bad now. Damn thing, damn Germans. This is all down to Hitler you know, no-one else." After the Council meeting most of the councillors fancied a walk down the road to the nearby pub, which served food. Patrick, as ever, insisted he could walk it, without any help and without a stick. I strongly disagreed and we managed to get a fellow member to drive Patrick to the pub. Then I phoned a nearby taxi company, as Patrick had some local business to attend to after lunch. I insisted on their best car and best driver. "The car is for Patrick Moore, the TV Astronomer" I said. They took some convincing! In the Edgbaston pub, where the local customers' jaws had dropped on seeing Patrick walk up to the bar, I ordered gammon egg and chips (again). Yes, you've guessed it: Patrick took one glance at the menu and asked me what I was having. I said "Gammon, egg and chips Patrick". He said "Make that two". That was his meal sorted! Patrick caused much amusement during the pub lunch by vigorously shaking the tomato ketchup bottle until the top flew off coating several members with ketchup. If it had been anyone else there would have been serious words exchanged, but as it was Patrick everyone laughed, Patrick the loudest as he was *not* coated in ketchup! Having your clothes stained by ketchup hurled from Patrick Moore was an honour and a privilege that many astronomers would pay for!

My First Sky at Night Experience

Four days later (April 29th), at the next BBC meeting, I turned up at White City. Patrick was still limping, but not quite as badly. "Pain killers" he barked "Overdosed on 'em, Hah!" At that meeting we were joined by Ian Russell, the next *Sky at Night* producer. He was seeing how *The Sky at Night* was produced, before he took over in a few months' time, and he was meeting Patrick for the very first time. As we left after the meeting I had plenty of chance to talk to Patrick. He could not walk at anything other than a slow limp, so there was plenty of time to talk. Patrick was quite vocal, even in the BBC corridor. "I'm sure Ian Russell is a decent chap, but I don't see any need for a change. If it ain't broke, don't fix it. Pieter and I have had a great time running *The Sky at Night* for 17 years, but now they want to change it. It's all John Birt's fault. He says anyone over 55 must stop being a producer, because no-one over 55 can have any creative talent. You know, you could count the friends John Birt has made at the BBC on the fingers of one foot! Pieter's produced *Crimewatch* and the Dimbleby lectures and lots of other good stuff. Now he's just thrown on the scrap heap; it's disgraceful."

My big day came on Saturday May 2nd at the BBC TV Centre. The only other time I had been inside that building was a year earlier, when invited to *The Sky at Night's* 40th anniversary. I arrived at the TV Centre early in the morning, at the same time as a TV cameraman. He checked in at the reception first. "I'm either scheduled for *Noel Edmonds House Party* or *The Sky at Night*" he said to the receptionist. The receptionist made a few enquiries and then told him "You're on the *Sky at Night* apparently, Pieter Morpurgo is already here." "Oh, Thank God" he said. "No-one likes working on Noel's House Party!" After being checked by security I arrived at the main 'Studio Gate' lounge. Patrick arrived, limping badly, a few minutes afterwards and he did not look well. He seemed to have a slight cold, was slurring his words slightly and the limp was worse than ever. Fortunately he would not have to say much during the programme: I knew exactly what I had to say and to do! The schedule for the studio day on *The Sky at Night*, when it was filmed at the BBC TV Centre in the 1990s, may be of interest to readers. It went something like this:

The producer, assistant producer, video editor, floor manager, cameramen and sound man powered up the recording studio equipment, in a room well above the studio floor, at about 9 a.m. Then, at about 11 a.m., Patrick and the guest (me!) had a couple of practice run-throughs, as if the crew were actually recording the programme. *The Sky at Night* in that era actually lasted 19 minutes from the opening credits to the finishing credits. In practice, allowing for news items and the music, this left little more than 15 minutes for the guest to get the message across. Some days earlier I had asked John Mason, a regular *Sky at Night* guest, what to expect, so that I was fully prepared. Neither Patrick, nor the guest, ever had any form of autocue. They just spoke off the cuff, about their subject. However, the pictures and diagrams, the order in which they appeared and a crude subject script, had to be agreed in advance. The guest had two monitors within his eye line. One showed what the selected camera was looking at. As it often showed my face, I was terrified of looking at that monitor; it brought home the fact that I was on TV! The other monitor was much more useful. It showed the next picture in line ready for discussion. In the control room the video editor's copy of the script was marked up in pencil, with roughly where in our conversation to insert each picture. A key press number was assigned to call each picture up and then put it onto the monitor screen and then insert it into the draft recording. The run-throughs went more smoothly than I could have imagined. Yes, I was terrified, and I had no idea what I had said afterwards, but I knew my subject and had been turning the things I would say over and over in my mind for weeks; so it all went OK. Then there was a break for lunch.

Through the lunch break (just before and just after actually eating) the producer Pieter Morpurgo, the assistant producer Laura Vine, the next producer Ian Russell, the video editor, floor manager, guest (me) and Patrick had a few more informal practice runs through what we were going to say. I was amazed at just how smoothly it went once I actually started speaking, but Patrick was certainly not in the peak of health. He wondered if he had a virus, but his speech appeared to be rather slurred too, something I had not noticed at BAA Council meetings or on TV. Nevertheless, feeling less than 100 % did not stunt his appetite, although being served food by someone of an ethnic background other than pure white English stock did seem to frustrate him. The black BBC waitress, in the specially provided hospitality room, had great difficulty understanding what Patrick was asking for. Out of her earshot (just) he mumbled something along the lines of: "Well, probably lived in a mud hut not so long ago, so allowances have to be made" and then bellowed: "NO CABBAGE FOR ME, JUST MEAT AND POTATOES." It was the same choice I'd seen Patrick make dozens of times with a cooked meal: no greens, but loads of meat, a skip load of potatoes and second helpings. Of course, there was the usual qualification too: "You know, despite my size and despite my bulk, I only eat one main meal a day. I have the appetite of an anorexic sparrow. You don't get size 13 feet by over-eating, you are just born that way." No-one challenged this view, particularly as no-one thought you got large feet from over-eating! With things going so well there was plenty of time to chat, although Patrick was finding it difficult to talk as he was now ploughing through a massive bowl of BBC profiteroles, drowned in an ocean of cream. Pieter Morpurgo revealed that he had once, as a young man, been banned from the *Blue Peter* studio by producer Biddy Baxter, when he had laughed out load at a mistake during the Valerie Singleton/John Noakes/Peter Purves era!

After lunch it was back into the studio for the final recording. You had a maximum of 2 hours, with the full camera crew assembled, to bag the programme. There was no room for error, as the programme was already scheduled for broadcast on the next day, in the *Radio Times*! In practice, during the 2 hours, you had a maximum of three goes at recording a programme. Little checks by the producer on what was on the tape; the make-up lady brushing dandruff from Patrick's blazer shoulders; Patrick going to sleep while I was talking! All these distractions took away vital minutes from those 2 hours. Nevertheless, all three recordings we made were almost flawless and dead on time. I have never felt so relieved in my life. By four 'o clock we were back in the hospitality room having a celebratory cup of tea, with Patrick looking whacked and Pieter Morpurgo, Laura Vine and the incoming producer, Ian Russell, all looking pleased. I was told that the tape would be tidied up after I had gone and then transferred to the place where it was queued for Sunday night's schedule. I asked what happened if the tape was lost, but was told not to worry, they were never lost and they always had a back-up. Still, I would not be happy until I saw it on TV the next night!

Pieter Morpurgo then led Patrick, limping badly, with myself, out through a side door in the TV Centre, then out into the central hole in the main, famous, circular building, along a pathway and over to where Patrick's chauffeur-driven BBC 'people carrier' was waiting to take him home to Selsey. I then shook hands with Pieter and, scarcely believing how successful the day had been, almost floated to the White City underground station, then on to Liverpool Street and my train home. It had been a day I would never forget. My mind just boggled at the fact that Patrick had experienced the pleasure of doing this sort of thing for 41 years; but for me it would be back to the real world of electronic engineering, software and 'jobsworth' managers, with a thud, next week! I could immediately see how Patrick would miss working with Pieter Morpurgo too. He appeared to be that rare combination of a highly skilled professional who was at ease with everyone, laid back, but highly tolerant of others' failings. I could see that, however good Ian Russell proved to be, after 17 years Patrick would seriously miss Pieter. Needless to say, the programme was transmitted on BBC1 the next night and I was delighted at how it came out!

Have I Got News for You and Trafalgar Square

Only 1 week later, on Friday May the 8th 1998, Patrick was on TV again in the BBC's highly popular *Have I got News for You* satirical news quiz. He was the team colleague of Paul Merton. Ian Hislop was partnered by comedian and economist Phil Hammond. On that programme Patrick suggested that the name for the European currency should not be the 'Euro', but: "the Fiddle or the Fudge". Comedian Paul Merton thought the former was not a good idea as you might then "have a whole load of fiddle in your trousers". Angus Deayton asked Patrick about the tornado damage to his observatory, sustained while he was having a curry. Patrick said "It was a real twister", to which Angus queried "What, the curry?" Cue a Churchillian scowl from Patrick. Patrick and Paul easily answered the next 'odd one out' question: who is the odd one out between Jimmy Carter, Michael Howard, Perez De Quela and Michael Knighton (Chairman of Carlisle FC)? The answer was that one of them had not seen a UFO, namely Michael Howard. On that same programme Patrick denied that he had been elected the Monster Raving Loony Party spokesman on 'Flying Saucers'. "That would be sensible" he retorted to Angus Deayton; "No, I'm in charge of finance!"

Patrick was always happy to play the fool and playing practical jokes, or posing under a pseudonym, were almost second nature; however, in the summer of 1998 the tables were turned and Patrick had a practical joke pulled on *him*. Shortly after he had appeared on the July 6th edition of Noel Edmonds' show *Telly Addicts*, Patrick was invited to be interviewed on the subject of the asteroid impact threat, for Channel 4. Earlier in the year the astronomer Brian Marsden had stated that there was a tiny chance that the asteroid 1997 XF11 could hit the earth in October 2028, which released a tsunami of asteroid doomsday publicity. Anyway, the venue for the Channel 4 chat with Patrick was to be Trafalgar Square and the interviewer a certain Mr Dom Joly. As the interview started Patrick and Mr Joly were standing 50 yards in front of the Trafalgar Square fountains, with Patrick only too keen to play down the asteroid impact threat, while still admitting there was just a small risk of something hitting the Earth.

Dom Joly started his interview: "Patrick Moore, a lot of people talk about the problem of asteroids falling down. I'm personally absolutely terrified of being hit by an asteroid; should I be?"

Patrick: "I think not, after all we could be hit, we have been hit in the past, we could be hit again, but the Earth is a very small target in a very large area of space so the chances are slight, not nil, but, as I say, my message is DON'T PANIC!"

Dom looks up at the sky nervously: "But, I mean, just occasionally I look up and I think, ooh-err, something moving in the night sky and you just, you never know, you never know what's coming…."

Dom continues to look up into the sky very nervously.

Patrick replies: "You never know, you can't rule it out, but there's a theory, only a theory, that it did happen 65 million years ago and wiped out the dinosaurs; I'm not very happy about that but it could have done…as I say I'm not……………..."

Patrick at this point was cut short by Dom, a very large man, taller than Patrick, looking up in terror at a non-existent incoming asteroid, and absolutely bellowing: "AARGGHHHHHHHHHHHHHH." Mr Joly then dropped his microphone and started running, like a crazed madman, towards the Trafalgar Square fountains, and then he promptly hurled himself into the water!!!

Patrick's body language and the expression on his face were priceless. Initially he just continued staring straight ahead as if to say "Another complete fruitcake; why me?" Then he turns to the camera, eyebrow raised, with his sorry expression saying "There's one born every minute." Meanwhile, in the background, Dom Joly splashes around, dressed in his suit and tie, wallowing in the water and bellowing like a rutting stag!

The clip was shown on November 25th 1998 in a Channel 4 pilot program called *Comedy Lab*. The program proved very popular and the host, Dom Joly, went on to make a series called *Trigger Happy TV* where more celebrities were caught out. The clip is priceless if you ever come across it! However, in future years Patrick always denied any memory of being tricked by Channel 4 and said he could not recall the interview. Yeah, right!

Health Problems

Throughout the summer months of 1998 Patrick felt distinctly unwell. It seems to have started about the same time that I recorded that first *Sky at Night* with him, but what Patrick described as 'flu-like symptoms' turned into a disturbingly irregular heartbeat. Patrick had coped with a cardiac dysrhythmia hassle for the whole of his life, but this deterioration was now very worrying indeed and Patrick was extremely concerned that he might have a stroke. He told me that he didn't mind a heart attack that killed him outright, but "Ending up as a cabbage – I'd rather be dead." His extreme difficulty in walking was an extra burden he could do without and the lack of exercise meant he was putting on even more weight: so that meant additional problems for his heart. As well as all this, the TLP sub-section co-ordinator Gerald North was suffering serious health problems too and so in August Patrick agreed, once more, to become the BAA Lunar Section's TLP co-ordinator, despite being in no fit state to be phoned in the depths of the night and dash out to his telescopes.

One major worry that arose for me, as the BAA President, was that Patrick had agreed to be an evening lecturer at the association's September 5th/6th Carlisle weekend; but it hardly seemed likely that he would now be able to perform his celebrated double-act with John Mason. However, Out-of-London meetings, especially the furthest flung ones, had always been a favourite with Patrick and he was determined to get there somehow. Fortunately, John Mason came to the rescue and arranged for Patrick to be flown to Newcastle airport, from where John was able to get him driven to Carlisle. How many 75 year-olds with a collapsing knee joint and serious heart problems would have made the effort for no financial gain? I suspect only Patrick. Where the BAA was concerned, no effort was too much. Despite Patrick's insistence that he was OK with walking from the Carlisle hotel, where we ate our evening meal, to the lecture theatre, I insisted on the hotel laying on a taxi for Patrick. Seeing him wince with every step was just ridiculous, despite his protestations that he was perfectly fine and fighting fit! Needless to say, when in front of an audience of his fans, he performed as well as usual. However miserable Patrick felt, he could go into another gear when giving a public lecture, and he kept it up when signing books for his young fans after the performance. I gave up counting the number of Stephen's who asked him to sign their 'Patrick Moore' book. Patiently, Patrick always asked "Is that Stephen with a P-aitch or a V?"

Fig. 35.1 Patrick with John Mason at the September 5th 1998 BAA Out-of-Town meeting at Carlisle. Despite serious knee and heart problems Patrick was determined to honor his promise to entertain the audience and risked his health to attend. As always, he did the BAA talk for free (Photograph by the author)

Despite Patrick's health problems during 1998 his writing output was still considerable, as was his charity work, public speaking and generosity. With John Mason and a few other local friends he decided to devote much effort to getting a planetarium built at Chichester. This involved substantial fund-raising and much frustration with the National Lottery, who were able to provide nothing! Patrick rattled off, in a couple of days, a 62 page booklet entitled "The West Country Eclipse", about the forthcoming August 11th 1999 Total Solar Eclipse. It was plugged heavily by the BAA during my term as President and sold tens of thousands of copies, all of which was poured into the Chichester South Downs Planetarium fund. Patrick's lecture tour agent, Robert Vallier, also arranged for him to do a speaking tour of Britain during 1998 and he continued giving these talks until 2002, despite ill health. With Vallier, Patrick took part in a total of five national annual UK tours, visiting scores of lecture theatres and venues per tour. The tours (and five accompanying small books) were entitled 'Into Space!' and were highly successful. On many of these occasions Patrick would also play the xylophone.

At the end of October 1998 Patrick apologised to the BAA Council, via me, for the fact that he would not be able to attend the 1998 BAA AGM. This was a great shame, as Brian Marsden of the Minor Planet Center would be giving a leading talk on that day, as well as my own 'Presidential Address' on the subject of novae. Marsden's BAA talk attracted a lot of press attention as it was entitled '1997 XF11 – The true story' and was about what he had actually said regarding the possibility of this asteroid hitting the Earth in 2028. A few days after the AGM I had a classic postcard from Patrick. It said that the new *Sky at Night* producer, Ian Russell, was very interested in doing a programme on close-approach asteroids, with me as the guest. However, Patrick added on the postcard:

> Having heart stopped in next few days.
> Quack stops it with electrodes, then re-starts to steady beat.
> If get voltage wrong, will try contacting you via different medium.
> Anon
> Patrick

Fortunately, they did not 'get voltage wrong' and Patrick's heart was a bit steadier after the operation, "Though it will never be perfect." Unfortunately, Patrick's crisp BBC tones never seemed to return after his heart problems of 1998. His voice always had a slurred edge to it. This was partly because Patrick's mouth and brain could no longer keep up with his enthusiasm any more, but I also think that the heart problems and associated drugs were at the root of the slurring problem, together with some badly fitting dentures!

The RGO Closes but the Pro's Want Patrick Knighted

In November 1998 Patrick presented a *Sky at Night* which enabled him to air his very considerable sadness, not to say, infuriation, at the closure of the Royal Greenwich Observatory one month earlier. Patrick was joined on the programme by Dr Jasper Wall, the last Director of the RGO. As we have already seen, Patrick objected strongly to the move of the RGO from Herstmonceux to Cambridge and convened a meeting on the subject in 1986. The professional astronomical community were behind him, but of course this made no difference: when politicians want to balance the budget, any cost cutting measure is justified, as long as it doesn't affect their own wealth of course! Less than 12 years later the RGO, now at Cambridge, was axed, for the saving of mere millions of pounds a year. As Patrick pointed out some years later, the cost of the dubious Millenium Dome project could have funded the RGO well into the twenty-second century. Patrick possessed a letter, written to him in 1986 from Margaret Thatcher, stating categorically that the RGO would not be lost. He had been a supporter of Mrs Thatcher and of the Conservative Party in the 1990s, but no longer. The RGO business finally taught him that "You can NEVER trust these fly-by-night politicians." Patrick had foreseen a reduction of the RGO's importance when it moved from Herstmonceux, but even he never dreamed it would be shut down completely. That November 1998 *Sky at Night* episode was the only miserable and depressing edition of the programme in its 55 year history with Patrick at the helm. Patrick's favourite comment around this time, about the Royal Greenwich Observatory closure, was: "Created by Charles II in 1675, destroyed by Sir Humphrey Appleby in 1998." This was, of course, a reference to the scheming character played by Nigel Hawthorne in the television series' *Yes, Minister* and *Yes, Prime Minister*.

As I've already said, it was largely as a result of the RGO closure that Patrick abandoned his support for the Conservative Party. Six years earlier, in April 1992, he had been photographed applauding John Major enthusiastically, at a Tory Party rally. He was even photographed deep in conversation with Margaret Thatcher, on October 8th 1993, at the Tory Party Conference; but now things were very different indeed! Of course, by the time the RGO was finally closed the Conservatives were out of office, but it had been under them that the decision had been made. In 1999 the new UK Independence Party (UKIP) gained three seats in the European Parliament and, as getting out of Europe was a top priority for Patrick, he put his full support behind UKIP from that point on. At every election after 1999 the main gate to Farthings bore a huge UKIP banner. Patrick even produced a DVD a few years later entitled 'Britain on the Brink' where he supported the Party's anti-European stance and immigration policy. Patrick had been a solid supporter of the Conservatives, even after Margaret Thatcher had been replaced by John Major, but when the RGO closed he felt he had been lied to and it was time to switch allegiance to a party that had his stated aim: to 'Put Britain First.'

Over the years, Patrick's championing of astronomy and keeping the RGO in business had won him a lot of friends in the professional astronomical community and many top professionals thought that he was well overdue for a knighthood. Indeed, many senior professional astronomers had originally been attracted to astronomy by Patrick's *Sky at Night* programme and by his books. Many amateurs in the USA seemed to be of the opinion that Patrick was surely already 'Sir Patrick!' During 1998 the Astronomer Royal, Professor Sir Martin Rees, had launched a major campaign amongst professional astronomers to try to lobby support for Patrick to receive a knighthood. However, when the New Years Honours List was published for 1999, Patrick was not listed. Patrick had already received an OBE and a CBE for his contribution to astronomy and for his considerable charity work, but a growing body of professional astronomers, many themselves 'Knights of the Realm' felt that 40 years of presenting the *Sky at Night* and almost half a century of writing astronomy books, deserved the ultimate honour.

However, many also wondered if Patrick's complete lack of political correctness might preclude the award. Patrick was not blatantly racist in public, but he made no secret of his belief that the English were the best and that the UK should pull out of Europe. He had also been highly critical of women (especially women teachers) in recent years. Nevertheless, when Patrick failed to get a knighthood by January 1999 a new strategy was launched. Professor Sir Arnold Wolfendale, of Durham University, rallied the cause this time and sent countless letters to amateur astronomical societies (major and minor) throughout the UK during late January.

In his letter, Sir Arnold wrote:

> …it is unbelievable that after all his years of effort, Patrick Moore has not yet been knighted…
> …I am hoping that by enlisting the support of very many amateur astronomers throughout the country, we will succeed, this time…the procedure is to write a small "spontaneous" citation accompanied by as many signatories as possible, plus their names in block capitals. These should be sent to:-
> The Nominations Unit, Ashley House, 2 Monck Street, London. SW1P 2BQ.

In February 1999 I was back in the BBC TV Centre, recording a second *Sky at Night* with Patrick, this time on the risks of a Near Earth Asteroid hitting the Earth. By now, the new producer Ian Russell was fully in charge of the programme and a few cosmetic changes had been made, including a flashy *Sky at Night* studio backdrop, which was in fact a dusty piece of rag, hung from the makeshift studio partitions, but which was brought to life with multi-coloured spotlights. The format was still the same though: two planning meetings at BBC White City midweek, followed by the filming at the BBC TV Centre on the Saturday before transmission.

Patrick was finding it harder and harder to get about by now, although his heart rhythm seemed to have stabilised. He told me that he had endured an exploratory keyhole knee surgery examination and that he had decided to go for a total knee replacement operation, immediately after the August 1999 Eclipse in Cornwall. He wanted to live to experience that eclipse, cloud or no cloud: "If I drop dead on the operating table after the 1999 eclipse, that'll be fine, but not before." So, the knee operation was scheduled for a few days after the Cornwall Eclipse and after the August filming had all been completed. With luck he would be back in action a few weeks afterwards, in time for the next *Sky at Night*. When I met Patrick in the BBC TV Studio before filming the February asteroid programme I helped him get the plastic lid off his hot coffee, which Ian Russell had provided from the BBC machine in the corridor. Patrick's fingers had resembled swollen bananas for as long as I could remember, but he seemed to have severe rheumatic problems with them now and no dexterity at all. Even signing his name was a horrendously painful chore. I could not help wondering if decades hammering away at that ancient typewriter's stiff keys had been a major factor, combined with his alleged wartime spinal injury and exacerbated by the subsequent 1977 fall in the bath.

At my second planning meeting, just days before the programme recording day, Patrick had asked one of the black BBC security guards to bring his ancient typewriter up to *The Sky at Night* office. He had to type out a draft script for another programme and would not use anything other than his own typewriter, which he had hauled into the chauffeur driven car the BBC had dispatched to Selsey. Many minutes had ticked by in *The Sky at Night* office, but there was still no sign of the vintage Woodstock appearing! So, Patrick got on the phone to the front desk, to try to locate the whereabouts of his precious typewriter. The security guard found it very hard to understand what Patrick was saying over the internal phone system and Patrick could not understand the guard either. Patrick, sprawling in a chair in the BBC office, mumbled: "Niggers, Wogs and Coons, what can you do with them?" Dead on cue, another black security guard appeared at the office door, beaming and holding Patrick's beloved typewriter. "Thankyou my good man, excellent job. I really am most grateful." Patrick winked at me: "Salt of the Earth these people". I did not say a word and neither did anyone else! Patrick's views of anyone who did not look like pure English stock were set in stone after the Second World War, as with so many people of his generation.

After finalising the programme details with Ian Russell and Laura Vine we both left together, heading for the lift to take us down to the ground floor at BBC White City. Patrick was telling me that his new 'Data Book of Astronomy' was nearing completion. I had to walk very slowly as Patrick was limping badly. He was a bit concerned that he could not identify his usual BBC chauffeur in the foyer, or in the White City car park. I walked with him towards the security desk. Patrick suddenly groaned as loudly as a polar bear in the mating season. "What's wrong?" I said. "Another monkey on the desk" he gestured. I was rather worried that someone might overhear him! A very large, black security guard was manning the desk in question. I assumed that the normal staff must be on holiday, as these security guards were all new and were wearing very badly fitting uniforms and peaked caps, without any BBC logo; this implied that they all came from an external company and they were certainly not the usual BBC employees. Of course, the normal staff, indeed everyone in the building, always recognised Patrick and wished him well. He never wore a visitor's security pass, security breach or not; what was the point! A pass could be faked, but Patrick could not! Patrick limped, wincing, as he hobbled slowly up to the large security guard on the front desk, while carefully positioning his monocle.

"Can I just check that my normal chauffeur is here, I can't see him outside. Is it a different one?" Patrick asked.

"Pardon", the reply came back.

Patrick sighed "My Chauffeur, he's not here, has another car been called?"

"Who are you sir?"

Patrick's jaw dropped and he looked at me in disbelief, his monocle falling out. I think this was the first time anyone in Britain had asked him who he was for at least thirty years, and this was at the BBC, his second home!

"PATRICK MOORE" Patrick barked.

"Who?"

"PATRICK MOORE", Patrick looked at me, his face a mask of horror and disbelief.

"Can you spell that for me sir?"

Patrick slumped on the security desk, head in hands, moaning.

"Should I know who you are sir?" asked the guard very politely.

"No, not really", said Patrick. "I've only been coming here every few weeks for 42 years." Then Patrick told me not to wait for him. "I'll be OK Martin, thanks for

Fig. 35.2 Patrick and Ian Russell in the *Sky at Night* BBC office at White City, London on January 13th 1999, planning the next program with this author (Photograph M. Mobberley)

Fig. 35.3 Patrick in the BBC TV Centre recording studio on Feb 6th 1999, during the filming of a program about the asteroid impact hazard, with this author. From *left* to *right*, Mark Jones, Julie Mann, Ian Russell (then the brand new producer), Patrick and Laura Vine (Photograph by the author)

Fig. 35.4 Ian Russell and Patrick in February 1999 discussing future programme plans at lunch in the BBC TV Centre in London. Laura Vine is seen in the *background* (Photograph by the author)

your help. I'm sure I'll be out of here in a few…ermmm…days." Then, in a lower, exasperated voice: "Living in a mud hut in the Congo basin one minute, head of security the next; what do you expect?" I said nothing, but made my way to the White City tube station. I was grateful that I was fully mobile and did not have to rely on others for transportation.

On the day that we actually filmed that *Sky at Night* at the BBC TV Centre I had just received my March 1999 edition of the American *Sky & Telescope* magazine (S&T). Patrick had not received his yet, so I told him it contained a pleasant surprise. Patrick begged me to tell him more, so I did. Patrick's friend Leif Robinson, S&T's Editor, had devoted his March editorial to great astronomy popularisers and communicators, citing Patrick as the greatest. Patrick was overjoyed. Patrick was a household name in the UK, but being praised in *Sky & Telescope* still gave him a great buzz. I e-mailed Leif to tell him how much the *Sky & Telescope* praise meant to him. A few days later Patrick faxed Leif to thank him personally. Leif sent me the following E-Mail:

> Hi Martin – Pat sent me a nice fax. I had heard of his problems; glad I could brighten his day. He's one of my very favourite people. Leif

But, some 7 months later, S&T would publish an article that definitely did not meet with Patrick's approval; more on that shortly. I might add at this point that anyone who knew Patrick was well aware that he was totally averse to any kind of new technology in his study or his home. He simply did not have the patience to master any gadget, or even to wire a plug. His philosophy throughout life was: 'If it ain't broke, don't fix it'. So the reader may be surprised that Patrick owned a fax machine at all (or an ansafone). In fact, the fax machine was the one thing that the BBC had eventually insisted upon and they paid for it to be delivered and installed in Patrick's study. In practice, if a fax needed sending, Patrick would ask that day's visitor at his house to dispatch it, providing the cat could be persuaded to move off the warm lid! To speak to Patrick you dialled 01243 60 3668 and to fax him the last four digits became 7237.

I found out once more just how generous Patrick was to his friends immediately after the filming of that February 1999 *Sky at Night*. I just happened to mention, while waiting to film the programme, that I had been trying to acquire the early copies of his *Sky at Night* books. Specifically, I was missing volumes 1–3. On the Tuesday, after the Saturday filming day, the postman arrived with a parcel from Patrick. Despite his ill health he had unearthed a couple of copies of his *Sky at Night* books volumes 1 and 2 and had posted them to me, free of charge, by first class mail. He had also signed them "To Martin – Best Wishes – Patrick" and included a postcard saying "Found 1 & 2, looking for 3, Patrick." I was overjoyed as I had been looking for those missing volumes for months!

The year 1999 would be the one in which, essentially, Patrick's era of full mobility ended for good. Early on in the year he phoned me up and asked me to convey his apologies to the BAA council for the fact that he would not be standing for election to the new council later that year; he was just getting too tired and he did not think his knee, even after the operation, would allow him easy mobility. But his immobility did not stop him from writing books or moving remarkably swiftly with his weight on his good leg, when he was speaking or performing.

One 'fun' project Patrick got involved in, with the year 2000 rapidly approaching, was a joint book with the astronomy historian Allan Chapman. Entitled 'Patrick Moore's Millenium Yearbook' one might have thought that it was about the state of astronomy in the year 2000. Well, there you would be wrong. It was, in fact, a humorous yarn, in an olde worlde English font, about the state of Astronomy in the year 1000!

Also, in early 1999, Patrick appeared at, and took part in, a concert of 'Moore Music' alongside the Royal Scottish National Orchestra at the Royal Concert Hall in Glasgow. The Orchestra played ten of Patrick's marches and waltzes. A review of Patrick's musical compositions in *The Stage* stated: 'He writes marches of which Sousa would have been proud and waltzes which one of the Strausses might have laid claim to.' The concert in Glasgow, in which he played the xylophone in two of his ten compositions, was recorded and later released by *Cavendish* as a CD entitled *Moore Music*. His tour agent, Robert Vallier, was the solo pianist at the concert. Patrick stressed that he considered himself primarily a composer and not a performer.

On March 4th 1999, Patrick's 76th birthday, in front of an audience of invited friends and a few journalists (roughly 30 people in total) Patrick officially announced his first music CD: the recording of that Glasgow performance. This official launch was in the Windsor Suite on the seventh floor of London's Inter-Continental Hotel. Champagne and canapes were served in copious amounts. With many of Patrick's best friends in the room it was not long before the crowd were singing "Happy Birthday Dear Patrick", with the CD playing his rousing compositions, as he made his appearance. After shouts of "Moore, Moore" even before Patrick had played a single note, Patrick half limped, half bounded (despite his gammy knee) onto a platform with a xylophone and rattled off a few of his compositions in dramatic style, arms flying and xylophone beaters crashing down. He looked like a man having an epileptic fit such was his intensity. But, after the performance, he was happier to mingle with his friends from the music and astronomy worlds, such as eclipse chasers Mike and Wendy Maunder, telescope magnate Dudley Fuller, musician Donald Francke and many others. He was in a room full of adoring fans and friends which always transformed him into the life and soul of any party. Patrick just loved being the centre of attention.

In late March, Patrick was delivering a lecture in the far north of the UK, at the Orkney Arts Theatre in fact. Before delivering his lecture Patrick happily accepted the offer of a cup of coffee in the theatre. It did not occur to him that this was anything other than a complimentary cup. After completing his lecture, to much applause, Patrick and his 'Into Space!' lecture tour agent Robert Vallier (sometimes called Robert Vallier-Green by the media) returned to their hotel. They were unaware that the Scottish theatre staff member who had offered Patrick the coffee wanted £1.26 for it and was not prepared to take no for an answer! As Patrick tucked into his evening meal the theatre official was busy phoning every hotel in the area (at great expense relative to the cost of a cup of coffee) trying to track Patrick down. Halfway through his meal, a bemused waitress informed Patrick that he had an urgent phone call regarding £1.26 owed for a cup of coffee! Robert Vallier told the press that Patrick had been offered a drink, as he always was before giving a talk. It was naturally assumed that this was a complimentary drink. He added that to travel all that distance only to be hounded for £1.26 made them both wonder whether it was worth travelling that far north again!

Chapter 36
The 1999 Total Solar Eclipse and Ailing Health

Patrick had never been averse to a few practical jokes in his life, especially on April Fool's day. On April 1st 1999, Patrick stated, on Radio Cornwall, that it had been decided to delay the Cornwall Total Solar Eclipse by a month, to September 11th, a date that would be burned into history on that terrible day in New York 2 years later. Phone lines in the West Country were jammed by outraged planners and holiday-makers who obviously did not understand the mechanism of solar eclipses, or take note of the April 1st date. It really does make you wonder!

At the end of April 1999 Patrick, like many people, was horrified to learn of the murder of the TV newsreader and presenter Jill Dando. She was the co-presenter of the BBC *Crimewatch* programme whose producer, Pieter Morpurgo, had of course been the *Sky at Night* producer for 17 years. Indeed, Pieter Morpurgo's assistant Laura Vine worked for both programmes too and the *Sky at Night* and *Crimewatch* team even shared offices.

[By a strange coincidence an astronomy colleague of mine, Nick James, who also worked, like me, for Marconi in Chelmsford, was in the same class at school as Jill Dando in Weston-super-Mare. It really *is* a small world. In another twist of fate, 9 years later, in July and August 2008, I found myself on the Transiberian railway during an eclipse trip with Pieter Morpurgo and Peter Cattermole. This was at the same time that Jill Dando's alleged murderer was acquitted and released from prison. After hearing this, via my Dad in the UK, it was me who told Pieter Morpurgo the news, while we were both in deepest Siberia.... Sometimes I wonder if the whole of life is just some huge surreal dream!]

The date of May 15th 1999 saw Patrick's return to the Radio 4 programme *Any Questions* in which famous people sat on a panel and answered topical questions, on issues of the day, from members of a studio audience. In the late 1970s Patrick had been banned from the programme for making 'racist' remarks, but remarkably, in the ultra politically correct world of 1999, the BBC let him back on. Needless to say, Patrick lost no time putting over his strong views. On family life he proclaimed that he couldn't really comment as "Herr Hitler put paid to me getting married."

M. Mobberley, *It Came From Outer Space Wearing an RAF Blazer!: A Fan's Biography of Sir Patrick Moore*, DOI 10.1007/978-3-319-00609-3_36,

However, he made it clear that every child needed a father and a mother and he did not approve of single mothers. [Gasps from the audience!] Good God! This man's brain has not had the political correctness implant inserted! You are probably wondering what Patrick had initially said that had led to his 20 year ban from the programme? Well, when a politically correct female MP scolded Patrick that immigrants were as British as she was and as he was, Patrick had explained that if a cat has kittens in a pigsty, they don't turn into piglets!

As the big day of the Cornwall eclipse, August 11th, drew nearer, Patrick found himself appearing more and more on TV, with almost the frequency of his Apollo Moonshot days. Indeed, the Apollo days were remembered in a programme entitled 'Real Time Apollo', hosted by Jon Snow on Channel 4. Exactly 30 years after the 1969 July 20th/21st landing and moonwalk that very night was re-created as if it were happening, minute by minute, in 1999. Needless to say, Patrick was a major guest, along with other commentators like Reginald Turnill, NASA experts and even Russian cosmonauts. For a month or two, in 1999, Patrick's fame was back to its peak of the 1960s and 1970s. A local Selsey artist, Robin Lee Hall, even painted a portrait of Patrick to exhibit at the BP portrait award show. The portrait received much praise and acclaim.

With his TV appearances rocketing as August 11th approached, delaying his knee replacement operation until after the eclipse seemed increasingly like a very wise move by Patrick. Queen Rock guitarist Brian May (a keen eclipse chaser himself) appeared on the July *Sky at Night* and the viewing figures soared to 1.4 million. Patrick even began appearing on BBC trailers during the main viewing hours, starting with the words "I was four when the last total solar eclipse was seen from Britain……" A series of *Sky at Night* programmes and TV broadcasts were shown before, during, and after the big event. Patrick's friends and regular *Sky at Night* guests, Peter Cattermole and Iain Nicolson, were the main stars in the immediate pre and post eclipse *Sky at Night*'s. The pre-eclipse programme netted a staggering 2.7 million viewers; this was a *Sky at Night* record in the post-Apollo era. The standard *Sky at Night's* during the 1990s were watched by an estimated three-quarters of a million viewers.

On eclipse day, Patrick, Iain and Peter sat under umbrellas and a bank of cloud at Falmouth, near to the BBC outside broadcast studio, as it drizzled. The great Cornwall Eclipse of August 11th 1999 had been clouded out. Patrick reported that the trio uttered the words "Fiddle, fiddle, fiddle" under their breath. I suspect their actual language was a shade stronger. However, Peter and Iain did conduct some experiments, such as monitoring the drop in light levels with a light meter; this produced some worthwhile results, even under cloud.

At the same time, not too far away at Truro school, I was standing in a field with over 500 other BAA members, under the same cloud bank. Even so, to say that the day was a write-off would be wrong. The dramatic dimming of the daylight in the final seconds, as the Sun's brilliant disc was totally eclipsed by the Moon, was quite breathtaking. I had a brief feeling of what it must be like to be Patrick that week. I had interviews with Pallab Ghosh on Radio 4 and with local TV. Then, after the eclipse, I had a mad dash to the Truro offices of Radio Cornwall, where I was ushered into a small backroom and waited for a cue to speak live on Radio 4's 'World at One', with the legendary broadcaster Nick Clarke. (Sadly, Clarke would succumb to cancer in November 2006). Patrick was also on that programme, but on tape: he was still in the BBC TV studio at Falmouth. I was the only 'live' astronomer on the Radio 4 One 'O Clock news on Eclipse day! Patrick told me that he was most frustrated he could not be with the BAA party at Truro on August 11th, but the BBC wanted to do things from Falmouth "and so that was that." With his knee in such a fragile state, dashing around Cornwall was just not on. For Patrick, all that was now left was to tidy up the post-eclipse summary *Sky at Night* and then prepare for his knee operation. Patrick's description of August 11th in Cornwall summed up my feelings too: "In a curious kind of way, being under a leaden sky during a total solar eclipse is a strange, eerie experience, which I, for one, will never forget."

But for Patrick, when he got home to Selsey, there was a nasty shock in store; no, he had not been burgled, but in many ways the shock was worse than that. Patrick's study always looked like it had just been burgled anyway! As we have seen throughout this book, for all of Patrick's adult life he had been convinced that TLP, the glows and mistiness seen on the Moon by him and others, were real lunar events. Yes, you had to be careful, as spurious colour and poor seeing created by the Earth's atmosphere could produce similar effects; however, Patrick was totally convinced TLP occurred on the lunar surface. Indeed, he had spent more of his life as an amateur astronomer promoting TLP research than any other topic. He was still, in 1999, the Lunar Section's TLP co-ordinator, even though the team had very few observers by then. So, imagine Patrick's horror, on returning home to Selsey and picking up the September 1999 *Sky & Telescope* (the world's most prestigious popular astronomy magazine) to find a comprehensive debunking of his entire TLP world.

On page 118 his nightmare began: a six page article by the respected authors William Sheehan and Thomas Dobbins entitled *The TLP Myth: A Brief for the Prosecution*. It would not have been so bad if the authors had been unknown and it would have been forgettable if the article had not been in *Sky & Telescope*. But things were worse than that: it was well written and comprehensive, tearing Patrick's TLP world up a paragraph at a time. The alleged mental instability of Kozyrev (who had claimed to have obtained spectra of gas venting from Alphonsus) was detailed, as was every other flaw in the TLP argument. It was so well-written that it was hard to find fault with it, especially as, to most amateurs, TLP were a pretty controversial topic anyway. Despite Sheehan and Dobbins' comprehensive destruction of TLP, they did not criticise Patrick at all in the article. Indeed, he was never mentioned, despite him being TLP observing's greatest advocate! This was typical of most knowledgeable astronomers' views though: they were great fans of Patrick but preferred to put his embarrassing determination to stick with the TLP myth and lunar vulcanism to the backs of their minds. However, after being clouded out on August 11th and facing imminent knee surgery, this was an article Patrick could have done without. He was still the BAA Lunar Section's TLP co-ordinator, so how should he respond to such an authoritative article? He decided to do a lunar *Sky at Night*, as soon as practical, and counter the article with his own views on TLP.

The Knee Is Replaced

On September 3rd 1999 Patrick went into the Midhurst hospital in West Sussex for his knee replacement operation. He told me that the surgeon was "the best there is." He was out a week later, using sticks to hobble around. I phoned him a few days after he returned home. He was happy the operation was over, but impatient for the pain and swelling to go down so he could get back to normal. Local friends rallied around and helped him during this period. Helping a 19 stone man after a knee replacement cannot have been much fun. I spoke to Patrick quite a few times after his operation. Ian Russell had decided to proceed with my third *Sky at Night* programme entitled 'A Telescope for Christmas.' This would be the December 1999 programme and thus the last programme of the 1900s. I was very proud of having this distinction and looking forward to filming yet another *Sky at Night* with Patrick. Throughout the next 5 or 6 weeks Patrick became very frustrated at the fact that he was no more mobile now than before the operation, but he admitted that he just could not be bothered with all the physiotherapy exercises he was supposed to do to get the knee ligaments working with the new artificial joint. As far as he was concerned the knee should have worked perfectly, immediately after the operation.

On October 21st I had the first visit to the BBC offices at White City to plan the 'Telescope for Christmas' *Sky at Night*. I wondered if Patrick would be there so recently after his knee operation; he was. In fact, when I arrived I had to wait, because John Mason was with him planning the *Sky at Night* before mine, on the 1999 Leonid meteor shower. Patrick and Ian Russell were obviously killing two birds with one stone; dealing with two programmes on the same day to spare Patrick's travelling hassle. I met John in the White City foyer as his meeting ended and we both noted that it was only a week until we would meet again at the BAA AGM. When I got to the *Sky at Night* office I was pleasantly surprised: Patrick's knee was progressing well and he was very upbeat. Seven weeks after his operation he was finally reasonably mobile again. Sadly, this was not to last.

When I met Patrick in *The Sky at Night* office we both agreed on one thing: it seemed incredible that we were now living in a world, after the 1999 Cornwall eclipse. To both of us, despite the difference in our ages, the August 11th 1999 eclipse had always been an event in the far future: now it was history and we had both been clouded out! Another event that I had always felt was 'in the far future' was the predicted Leonid Storm of November 1999. If Patrick had been in better physical health he would definitely have been headed for Sinai with John Mason and many other BAA members, or maybe on yet another cruise trip. However, he was certainly not in the peak of health! On the *Sky at Night* transmitted on November 7th, John Mason, in his inimitable style, enthused about the prospects of a Leonid meteor storm in 10 days time. Patrick said he would be hoping for clear skies from Selsey. Three days later I was back in *The Sky at Night* office for the final planning phase of my December programme. On this occasion Patrick did not turn up for the meeting. Ian Russell told me that Patrick could not make it due to medical commitments. As it turned out this was not a problem: the guest was the person who did most of the work in the *Sky at Night's* during the late 1990s.

Patrick, although he was an excellent host, was quite happy to let the guests do the talking, especially when he was in bad health.

At the end of October, the firm of chartered accountants working for *Top Events and Publications*, publishers of Patrick's magazine *Modern Astronomer*, announced that *Top Events* had gone into voluntary liquidation. As the BAA President I had received quite a few letters from amateur astronomers who were owed money by the company. Patrick was very depressed. He hated to think that fellow amateurs had lost out again because of his quest to start up another astronomy magazine, after his parting with *Astronomy Now* and the failed *Practical Astronomy*. More than 5 years would elapse before Patrick would risk another astronomy magazine venture.

On the night of November 17th/18th there was, indeed, a spectacular meteor storm, especially from the Middle East. I was with John Mason and about 70 other travellers on the *Explorers* Sinai trip and we saw the display of a lifetime: almost one meteor a second at the peak of the storm. Patrick was at his home in Selsey, with *The Sky at Night* producer Ian Russell, a small camera crew and a few guests. Needless to say, it was Cornwall in August all over again. Once more, Patrick sat under an umbrella, staring at cloud and drizzle. Patrick would say later: "Observationally, 1999 was not my year."

After returning from my Sinai trip to see the Leonids it was barely 2 weeks before my own 'Buying a Telescope for Christmas' *Sky at Night*. On the big day, December 4th, I turned up very early at the BBC TV Centre, but Patrick, Ian and Clare Ashford (standing in for Laura Vine) had arrived well before me. Clare collected me from the foyer and took me to the studio gate from where Ian Russell escorted me the rest of the way. "Don't be too scared by Patrick's appearance" Ian said. "He's got a black eye." I was taken aback? Had Patrick been beaten up? Around the next corner and into Studio 2 and there was Patrick, looking dejected, complete with shiner. He brightened up briefly to greet me. "Hello Martin, have you heard?" I assumed he meant about the black eye. I started to nod, but then he said: "Yes, the damn knee's gone again, the operation hasn't worked; I'm back to square one. It's all seized up. Best surgeon in the country and he's still no good." I was horrified; all had seemingly been going so well. I was used to Patrick's 'doom and gloom' mood when things weren't going to plan, but he really was hobbling about and shuffling a step at a time. "What about the black eye?" I queried. "I slipped in the bath, because of this damn knee, cartwheeled and headbutted the soapdish." I just managed to prevent myself laughing as an image of Oliver Hardy stepping on a bar of soap on the deck of a ship came to mind. "We can cover that up with make-up" Ian quickly added. It was all strangely reminiscent of Patrick's cartwheel in the bath in 1977, although, this time, fortunately, there was no spinal damage.

For that particular *Sky at Night*, Patrick had brought along his childhood 3 inch refractor inside the BBC car, in which he had been transported from Selsey. A rather rickety looking brass-tubed thing, but something I had seen many times on television and in Patrick's books. It was fascinating to see it 'in the flesh'. Needless to say, the programme went well. The make-up lady (Anabela Dallot-Seguro) covered Patrick's black eye up so expertly that it was invisible and the programme was transmitted the next day. I was delighted that the *Radio Times* did a full article on buying a

Fig. 36.1 The author on *The Sky at Night* set during the filming of 'A Telescope for Christmas' on Dec 4th 1999 (Photographed with the author's camera)

Fig. 36.2 Patrick on the set of *The Sky at Night* on Dec 4th 1999 (Photograph by the author)

telescope, with contributions from Patrick and me; it also included free publicity for my first book, on Astronomical Equipment, written thanks to encouragement from Patrick.

After initial high-hopes in October 1999 Patrick's knee replacement failed to restore his mobility, but I have been told by his close helpers at Selsey that he had no interest whatsoever in the rigorous physiotherapy necessary to get the artificial knee in full working order. He would be hobbling about on sticks, or with a walking frame, or in a wheelchair, for the rest of his life. OK, there were days, in the following 2 years, where he could potter around his house and garden and the BBC offices, without sticks, but they were increasingly few and far between.

Two days after my third appearance on *The Sky at Night* Patrick was due to appear at the London Planetarium, in what was billed as a head-to head clash with astronomer Mark Kidger. Mark (a good friend of mine) had just published a book entitled 'The Star of Bethlehem' in which he put forward a theory that the star was a nova. Patrick disagreed (in a very amicable way) as he thought, if it was anything at all, it was two meteors. The comet expert Dr David Hughes, quoted in the *Sunday Telegraph*, had another view entirely: he thought that a planetary conjunction was the explanation. Unfortunately, due to the Planetarium charging £5,000 for hiring the venue and uncertainty amongst the sponsors, the planned event never took place outside of the newspapers. Regardless of this, Patrick and Mark agreed to differ and Patrick earmarked the subject for a future *Sky at Night*; he also decided to write his own book on the subject. According to Mark, "Patrick copied word for word whole sections of my book…if it had been anyone else I would not have been happy!" Hmmm, what was it Patrick always accused other authors of doing?

As 1999 came to a close, Patrick was involved in proof-reading the most monumental book he had ever been involved in: his *Data Book of Astronomy*. At one point he asked me if I might like to proofread parts of it, but in the end he managed the checking with an existing team of 20 friends and astronomers, amateurs and professionals, including the *Queen* guitarist Brian May. As I mentioned a few chapters ago *The Data Book of Astronomy* had grown from Patrick's *Guinness Book of Astronomy*, first published in 1979; that book had been a spin-off from Patrick's astronomy contribution to the *Guinness Book of Records*. The *Guiness Book of Astronomy* had sold over 60,000 copies in various editions; that's a lot for a single astronomy book in the 1980s and 1990s, by which time astronomy writing had become a popular 'hobby' for many amateur astronomers. However, the *Guinness Book of Astronomy* was now looking dated and simply did not contain even a fraction of the facts that the modern astronomy world had to offer. Patrick decided, along with the Institute of Physics Publishing, that it was time (while he could still type) to produce the ultimate astronomy fact book, bar none; something so comprehensive that nothing could rival it, hence the team of 20 proofreaders and data trawlers. Despite Patrick's immobility, his heart and knee treatments were now over. He was still not in good health, but he could think and type and had a good team of helpers. As the first few months of 2000 ticked away, the proofreading of the massive 530 page hardback tome edged towards completion and totally dominated Patrick's life. However, as well as the monthly *Sky at Night* programmes there was another pleasant distraction for Patrick during the spring of 2000.

He was asked to present an Outstanding Achievement Award to the British born astronaut Dr Michael Foale at the annual *Pride of Britain Awards* ceremony, held in London's Hilton Hotel. On April 11th, he duly presented Michael with his trophy.

At this time Patrick was still sore about the September 1999 TLP de-bunking by *Sky & Telescope* and had decided to counter the comments from others in the astronomy world that it had marked the death knell of TLP studies. For the May 2000 *Sky at Night* he was joined by Douglas Arnold, a regular *Sky at Night* contributor from the programme's early days and an expert on lunar photography. The subject for the programme would be 'The Moon in Focus.' But how could Patrick counter such a powerful attack by *Sky & Telescope*? In a recent article in the December 1999 BAA Journal, Richard McKim, the Mars Section Director, had drawn attention to an article published by the veteran French astronomer, Audouin Dollfus (1924–2010). The Frenchman had discovered Saturn's moon Janus in December 1966, when the rings of Saturn were edge-on to the Earth. [To recap, Patrick had also claimed he recorded Janus through the Armagh 10 inch refractor, but had not realised it for what it was…although many found his claim rather hard to accept.] Also, Dollfus had, more controversially, reported a tenuous atmosphere around Mercury (of a few millibars) in the 1960s. In 1999, in an article entitled *Lueurs sporadiques sur la lune*, Dollfus had described video polarimetric measurements he had made with the 1 metre Meudon Cassegrain telescope, while aimed at the lunar crater Langrenus. He claimed that the images obtained showed changes in the crater caused by moondust levitated from the lunar surface by outgassing from the lunar interior. Many French amateurs regarded Dollfus as their nearest equivalent to Patrick Moore, even though, outside the astronomy world, he was nowhere near as famous. Richard McKim drew BAA members' attention to these very observations in that December 1999 Journal. The only problem was that they had been made 7 years earlier, in December 1992 and January 1993: they were old news, reported in a recent paper.

However, this was the sort of report Patrick had been hoping for. On the May 2000 *Sky at Night* he triumphantly announced: "TLP, Transient Lunar Phenomena: tiny disturbances in the Moon's crust, reported by amateurs over the years and rather greeted with scepticism by the professionals. Well, now, for the first time, we do have proof. Audouin Dollfus, using the Meudon telescope, has obtained pictures of a disturbance inside the walled plain Langrenus and here are his pictures: see the dust there at the top being disturbed by gas coming out from below the crust. So, the amateurs were right after all and although the Moon may be quiet it is not completely inert." This interpretation of Dollfus' isolated 7 year old images was somewhat over the top, to say the least, but Patrick had kept the TLP myth alive, just a bit longer. However, it did little to boost observations to the BAA Lunar Section TLP network, which were at an all time low, despite the fact that Patrick was still the co-ordinator.

On July 19th 2000 Patrick accepted a royal invitation to be part of a human celebration in the Queen Mother's 100th birthday pageant. Alongside many other celebrities in Horse Guards Parade that day he symbolised the changes that had occurred in astronomy during the last 100 years, positioned next to a bizarre brown coloured mock alt-azimuth telescope. However, due to his mobility problems Patrick was provided with an armchair, so he could sit and observe the pageant in comfort.

With Patrick's health and mobility remaining highly fragile, in mid-2000 a plan was hatched to minimise his journeys to and from London to record *The Sky at Night*. The July-August and September-October programmes were made into two part programmes, so the monthly format continued, but each topic was covered in two programmes instead of one. Cambridge undergraduate Chris Lintott was the guest for the first double-programme and Iain Nicolson did the second. Incidentally, it was me who first suggested Chris Lintott as a guest on *The Sky at Night* to producer Ian Russell (in an e-mail of September 7th 1999). He was a regular guest from 2000 onwards and eventually became the co-presenter. I had suggested to Ian that Chris, and Chris' friend David Reid, might be filmed testing telescopes for my Christmas 1999 programme. In the end the pre-programme testing idea was shelved, but as one of Patrick's protégés, from his contacts with J. Hedley Robinson and Torquay Boys Grammar School, Chris Lintott was surely destined to eventually become a regular presenter on the programme anyway.

Patrick had first met Chris when he was just an 11 year old schoolboy and I vividly recall BAA meetings in the mid 1990s where Patrick was accompanied by the teenagers Lintott and Reid for the whole day. According to Chris himself he first saw Patrick at the school giving a lecture about the outer planets and it was when Patrick explained that we don't yet know everything about Uranus, Neptune and Pluto that the young Lintott realised there was still work to be done. He decided to be an astronomer and became one of Patrick's protégés, along with his friend David. I suppose it was history repeating itself really, as go back to the 1950s and 1960s and you would have found Patrick accompanied by the young Iain Nicolson and Peter Cattermole. From 1965 to 1968 it would have been his young apprentices Terry Moseley and Pat Corvan at Armagh and throughout the late 1970s, 1980s and into the 1990s Patrick was often accompanied by a young John Mason.

As I have already mentioned in this book Patrick would "rather be found dead in a ditch" than do any form of paid product advertising or endorsement. However, there were some exceptions where he loaned his distinctive voice to a worthy cause and gave his fee to charity. In September he appeared in a local radio commercial for Brighton based Mind's Eye's latest radio campaign for electric fuel. Patrick had been approached by Mind's Eye and Chichester radio production company Mango. He donated his entire fee towards keeping the RGO telescopes at Herstmonceux preserved. Of course, as Dudley Fuller was a personal friend of Patrick and Fuller's engineers had modified Patrick's 15-inch 'scope he always had a good word to say about Broadhurst Clarkson & Fuller and he happily endorsed that brand name. In the twenty-first century you occasionally saw Patrick's name linked to other telescope brands too, but as he was by then totally immobile and did no observing, such endorsements were totally meaningless. Patrick was, in his eighties, too old and tired to complain about his name being exploited to infinity, based on one or two sentences he had uttered when drunk!

In the early hours of October 30th 2000 Patrick was in the shower at home after a long day at the typewriter. With his knee problems and history of falling over in the bath, this was a dangerous place to be. Suddenly, Patrick heard a whooshing sound which he had heard before, almost 3 years earlier. It rose to almost deafening proportions. Patrick could not believe his ears: another Tornado was battering Selsey. Once again, caravans on the West Sands, not far from Patrick, had taken a pounding, with some 200 sustaining damage. Interviewed the next day by the local press, Patrick said he was "Utterly stunned" that a tornado had struck the area twice. "I can't explain it" he said. "It's said that lightning never strikes twice, but clearly the same is not true of tornadoes and Selsey. It makes me quite nervous to go outside. Thankfully my observatories escaped the battering they got last time."

Worse was yet to come though. Patrick was still composing music and a few months earlier he had completed (for free) a rousing march for the Band of the Royal Parachute Regiment, entitled 'Out of the Sky'; he had many more musical excursions planned. On one evening in mid-November he was sitting at his piano at Selsey, composing a wedding march for a friend. Satisfied that it sounded OK, he closed the piano lid and went to bed. The next morning he woke up and found he had lost the use of his right hand: there was barely any feeling or movement there. For many years Patrick's hands had been swollen and arthritic, but this was something much, much worse. Tests at the hospital showed that his upper spine had crumbled in the neck area. It was impossible to tell whether the condition was temporary or permanent. This was the same area that Patrick said he had injured in a crash landing during the war and had aggravated again when he fell in the bath in 1977. On that latter occasion he had endured spinal surgery, but he was now much older and an operation was too risky and unlikely to help. Patrick sank into a mood of despair. First the heart problems, then the failure of the knee replacement, now semi-paralysis: was life still worth living? He had just started a major book on the history of spaceflight and astronomy combined. That would surely have to be shelved. Without his right hand he could not type and could not observe. Even opening up his observatories was impossible, so observing the Moon and planets, his lifelong pleasure, was surely finished? Admittedly, he had been here before, when the cricket ball hit his eye, but this time *everything* was going wrong: legs, hands, heart; the lot. Patrick seriously considered whether he should even carry on with *The Sky at Night*. He had certainly never considered giving that up before.

Chapter 37
A Knighthood, a BAFTA and a Fellow of the Royal Society

They say every cloud has a silver lining and there was some light at the end of the tunnel for Patrick. A week after his partial paralysis, with Patrick still in the depths of despair, on Tuesday Nov 21st 2000 he received a letter from 10 Downing Street. Unfortunately, with his right hand virtually dead, he could not open it. Fortunately, a local taxi driver who had come to collect Patrick was able to open the letter. Patrick could scarcely believe the contents. He was to be knighted in the New Years Honours List. From January he would be 'Sir Patrick'. Arnold Wolfendale's campaign had paid off. Of course, Patrick could not tell anybody about it yet, but it was a huge boost in a month of utter despair. I received a phone call from Patrick a few days later, on the 26th in fact. He mentioned nothing of the knighthood, just the despair about having to give up observing and maybe having to stop being the BAA's Public Relations Officer. He had taken up that voluntary P.R.O. role 9 years earlier; initially just for the 2 years when John Mason had been the BAA President, but he was still happily doing it aged 77. However, when Patrick mentioned considering giving up *The Sky at Night* I told him in no uncertain terms that while there was a breath left in his body he should never give that up: otherwise the rest of the astronomy world would have nothing to live for. He seemed to like that!

On December 15th 2000 the UK's greatest amateur astronomer, George Alcock, died in a hospital in Peterborough. During his 88 years he had discovered five comets and five novae, using nothing but binoculars. He had also discovered an outburst of a recurrent nova and made countless observations of meteors, comets and planets. I phoned Patrick the next day to tell him the news. Despite the fact that Patrick knew he was soon to be 'Sir Patrick' he was as depressed as I had ever heard him. He still had virtually no use of his right hand and, at one point, he said he envied Alcock being dead: "It's better than being like this." I had no idea how to reply to that, but had learned, over the years, that Patrick could lose all hope when a situation was totally out of his control, especially after his mother's death. I knew he would always present a more optimistic face to the press though, especially after he had downed a few Gin & Tonics.

M. Mobberley, *It Came From Outer Space Wearing an RAF Blazer!: A Fan's Biography of Sir Patrick Moore*, DOI 10.1007/978-3-319-00609-3_37,

News of Patrick's impending knighthood was released by the media on December 30th and he then became the subject of a flurry of TV interviews over the next few days. Knighthoods for the 'Goon' Spike Milligan and Olympic rower Steve Redgrave were announced on the same day. In various TV interviews Patrick said that he felt "very, very pleased indeed" but he also felt that there were so many people who deserved a knighthood more than him. "After all my only contribution has been mapping the Moon in the early days and of course presenting *The Sky at Night* for almost 44 years." With all this attention and praise on the national TV news he seemed to cheer up, not least because of all the congratulations. I phoned him on the 30th and there was obviously a party going on at Selsey! There were certainly plenty of friends in his study and Patrick said the phone had never stopped ringing with messages of congratulations. Over the coming weeks he received over 1,000 congratulatory letters! The press publicity also mentioned Patrick's current paralysis and many of the letters sympathised with his condition and wished him well. If not for this barrage of good feeling, as 2000 turned to 2001, I do wonder if Patrick would have lost the will to live. On the first *Sky at Night* of the new millennium, on Jan 8th 2001, Patrick thanked everyone for their good wishes after his knighthood had been announced.

However grim Patrick felt though, he rarely cancelled appointments. He just could not face disappointing his fans, especially his youngest fans. So, despite his paralysis, he kept his promise to be at the 2001 Young Tie Designer of the Year Awards on February 8th. The winner was 17 year old Jo Rea, whose design for Vivienne Westwood won the top prize. But, it was the runner-up who Patrick was really there for. The 14 year old Louise Chambers had designed a spectacular and colourful space themed tie, especially for Patrick, complete with loads of planets. The press photographs showed Patrick posing with the two girls and the truly gigantic tie based on Louise Chambers' eye-catching design.

Although Patrick did slowly regain a limited use of his right hand in early 2001 and was well enough to travel to Buckingham Palace to be knighted by Prince Charles at the March 2nd investiture, accompanied by godsons Chris and Adam, his mobility and stamina were still very low and he rarely attended any BAA meetings in that year. He told friends that in the brief period after Prince Charles had 'done the business with the sword' they had chatted about the much missed 'Goon' Michael Bentine, a mutual friend. In that same month, after a lifetime of championing lunar observing and TLP hunting, Patrick finally stepped down as the Lunar Section's TLP co-ordinator; he simply did not have the energy to continue even in that role. That must have been a heart-rending decision as he knew he could never return to that position, or play any further part in lunar observing.

Nevertheless, his disabilities were not going to stop him campaigning for causes dear to his heart and so, only 3 weeks after he was knighted, Sir Patrick was back in London on March 28th, this time on the doorstep of Tony Blair, at No. 10 Downing Street. The chance to ban fox hunting in the current Parliament was ebbing away, as an election was imminent. Despite the House of Commons voting in favour of banning fox hunting, the House of Lords, on the previous day (March 27th) had voted to allow it under self-regulation. So, flanked by other anti-hunt campaigners, Patrick delivered five huge boxes of petitions, containing 84,885 signatures, to Tony Blair.

Interviewed outside the doorway of No. 10, supported by two walking sticks, Patrick stated: "I am presenting this petition because I know that to chase intelligent animals to exhaustion and then see them torn to pieces by dogs, is morally wrong."

The trips to London and the accolades were not over for Patrick either. On May 13th 2001 he was sitting in the audience, amongst distinguished invited guests from the world of showbusiness, at the BAFTA British Academy Television awards, when Angus Deayton invited the second man to set foot on the Moon, Buzz Aldrin, to take the stage. Patrick, despite looking surprised, was surely in little doubt that this had to be about him. After all, the BAFTA organisers had insisted that he should be there and he must have suspected why. Aldrin began his tribute to Patrick with the words:

"I'm pleased to say that this special award from the Academy tonight is being presented to my good friend, the presenter of *The Sky at Night* on BBC 1, Sir Patrick Moore. *The Sky at Night* began somewhat on the whim of a BBC Executive called Paul Johnstone, who wanted to make a programme or two, then to be called 'Star Map'. Forty-four years later Sir Patrick is still at it and *still* has no contract with the BBC. [Gasps from the audience]. He says it is just a gentlemen's agreement."

Aldrin then went on to show clips from *The Sky at Night* over the years and clips of people impersonating Patrick, namely, Ronnie Barker, Mike Yarwood and Spike Milligan. Some of Patrick's more eccentric interviews with people speaking Venusian (from his BBC 'One Pair of Eyes' program) and the classic 50th *Sky at Night* (trying to observe with George Hole's 24 inch reflector through cloud) were also shown. After the clips ended, to thunderous applause and a standing ovation, Patrick gingerly ascended the steps to the stage (despite his knee) and shook Aldrin's hand. Patrick was, quite obviously, somewhat overwhelmed by the tribute and the presentation of this lifetime achievement award. In accepting the award Patrick said:

"I must say I feel really overwhelmed. There are so many people here who have done so much more than I have; after all I've merely done some commenting. I did help, I suppose, in mapping the Moon, but I have a sort of feeling that Buzz knows a bit more about the Moon than I do. I have not myself been in space; it would take a very massive rocket to launch me! I feel indeed very honoured that Buzz is here to present this award; it's great of him and, after all, if Buzz's trip in Apollo 11 had gone wrong the Space Age would not be where it is now. He was the first and therefore all honour to him; his name will live for all time; mine certainly won't! All I can say is: I don't think for one moment I deserve this award but I am more than grateful. All I can say therefore is thankyou very much indeed, it's been one of the great days of my life. Thankyou." To further rapturous applause Patrick left the stage. It was, without a doubt, the most popular award of the night.

Many of Patrick's awards and medals, such as his RAF regalia, OBE, CBE and the Knighthood, were kept in a glass case on the wall, alongside his bureau and ancient typewriter. However, it was always very difficult to get Patrick to explain what they all were, especially the alleged RAF stuff. "I got this one from a Christmas Cracker and that one from a Cornflake packet" he would quip, under interrogation about the medals' military origin!

But the accolades did not end with the BAFTA. On the very next day, May 14th, an announcement was made that Patrick had been elected as an honorary fellow of the Royal Society, the country's oldest and most prestigious scientific body. The

Vice President of the Royal Society, Professor John Enderby, said that Patrick's election was a timely and appropriate recognition of his great contribution to science in this country. The fellowship was officially awarded in recognition of Patrick's *44 years of presenting the Sky at Night and incredible history of astronomy publications, inspiring generations of young and old alike with his enthusiasm for the subject*. This was, quite possibly, the greatest honour that could ever have been bestowed upon Patrick, but he was not going to rest on his laurels or even take a rest. A General Election was looming and he wanted people to vote for UKIP, so on May 25th he travelled to the Methodist Central Hall in Central London, where UKIP were holding their pre-election rally.

Although Patrick's right hand did regain some movement in 2001 he was still frustrated by his inability to use it properly; this was especially frustrating when he was asked to sign books, something he had always enjoyed immensely. Arthritis had been bad enough, but this was much worse. But despite his mobility and general health problems Patrick still managed to get out and about during 2001 if there were friends ready to chauffeur him or escort him on long journeys. Of course he had to visit the BBC TV Centre once a month or so, but even this would not continue for much longer. Patrick had hoped that the BBC might fund him to travel to the southern hemisphere for a trip to see the June 21st Total Solar Eclipse, but it was not to be. He moaned bitterly, but no-one believed he was fit enough to make the trip anyway.

Around this period, on July 1st to be precise, Patrick appeared in a short TV programme entitled *Alter Ego* where the impressionist Jon Culshaw interviewed famous TV personalities while he was impersonating them, in other words it was Patrick Moore interviewing Patrick Moore!! Culshaw also appeared in a xylophone duet (Patrick with the real Patrick) playing Prodigy's Firestarter! A video clip of that has to be seen by all Patrick Moore fans, although personally I think that Culshaw's impersonation, although very good, is midway between being Patrick and Patrick's good friend Cdr. Henry Hatfield! Talking of xylophones, some month's later an amusing animation, entitled 'Patrick Moore plays the xylophone', was created by Rob Manuel & Jonti Picking and became very popular amongst Patrick's fans and even Patrick himself. Showing Patrick gliding through the cosmos with his xylophone, it was the sort of harmless fun that Patrick enjoyed.

Patrick was always happy to perform the opening ceremony at new observatories or astronomical facilities and on August 19th he performed the grand opening ceremony at the UK supernova discoverer Tom Boles' observatory, at Coddenham near Ipswich. Tom had already discovered 11 supernovae, mainly from his old site near Northampton, so was already established as a leading UK amateur. Needless to say, with Patrick performing the opening ceremony there was no shortage of guests and a *Look East* film crew televised the event for local TV. Tom would discover more than 140 additional supernovae from the observatory during the remaining years of Patrick's life.

That same month Patrick signed up to the Democracy Movement, who were a group of patriotic celebrities determined to stop Britain dumping the pound sterling and adopting the Euro. Some of the characters were the same crowd who had, 6 years earlier, formed the BWMA (the British Weights and Measures Association,

Fig. 37.1 Patrick at the opening of Tom Boles Observatory at Coddenham, Suffolk, on the 19th of August 2001. Despite serious mobility problems Patrick made it to Suffolk and declared the observatory open (Photograph by the author)

founded by Vivian Linacre). At a gala dinner, held at an Indian restaurant in central London, 200 guests paid £50 a head to kickstart the campaign and Patrick was right behind them, despite his disabilities.

Patrick's old friend Arthur C. Clarke had predicted that an alien monolith would be unearthed on the Moon in his Space Odyssey film script, based in the year 2001; such an event would surely have changed the world. As everyone knows though, the horrific events of September 11th 2001 really did change the world, in a different way; but then in Patrick's view if you let foreigners into your country something like that may well happen at some stage. However, his biggest fear was how George W. Bush was going to react and Patrick's comment at the time was that if Ronald Reagan had been in charge 911 would never have happened at all. Reagan was one of the many famous people who Patrick claimed to have met, although he never fully explained how…..

The Star of Bethlehem, Room 101 and a Big Party

On December 9th 2001 Patrick finally broadcast a *Sky at Night* about the Star of Bethlehem, a subject he had been due to go 'head to head' with Mark Kidger about at the London Planetarium 2 years earlier, before the sponsors pulled out. However, there was an extra incentive this time. Patrick had jumped on the bandwagon and written a book on the subject too, even if much of the book had been pinched from Mark and comet expert David Hughes! Patrick, Mark and David debated the subject and the BBC held a website vote. The options for the Star of Bethlehem were that it was a nova in Aquila (Mark's theory) a conjunction of Jupiter and Saturn (David's theory) or two bright meteors (Patrick's theory) or that none of these were right. The result of the vote was that Mark's idea won with 36.8 % of the vote; David came second with 29.4 %; 24.8 % of participants thought another explanation was likely; and Patrick came a very poor fourth with 9 %!

As we have seen numerous times in this biography, Patrick appeared on many strange TV programmes over the decades, but perhaps none stranger than the Channel 4 Comedy Lab episode of 'The Pooters' in which he was a guest on the episode broadcast on February 1st 2002. At the start of the programme Patrick was featured looking through a large refractor at the night sky; then, seeing an alien craft, he exclaimed that he must inform the International Astronomical Union. However, he then glanced at his watch and realised that he would soon be missing an episode of 'The Pooters' and so his message to the IAU would have to wait! During the episode Patrick arrives in a Rolls Royce at a stately home and becomes part of a distinctly Victorian looking dysfunctional reading circle. One of his roles in this reading circle is to explain why *Simply Red* pop star Mick Hucknell would burn up if he tried re-entering the Earth's atmosphere. However, when the reading circle decided they wanted to know more about the constellation of Perseus they chose a member of their group impersonating George Michael to explain this aspect of the night sky. Utterly bizarre! Patrick looked suitably bemused throughout the entire episode!

During the late 1990s Patrick had come up with the idea of writing a comic play entitled *Galileo: The True Story*. Not surprisingly he completed the project rapidly and it became a four scene production incorporating 14 songs with all of the script, overture and songs composed entirely by Patrick himself. Part of the plot involved Galileo marrying the Pope! The first version of this play was performed at Torquay Boys Grammar School, the school where Patrick had long enjoyed connections via his BAA friend J. Hedley Robinson. However, in March 2002, with help from former Torquay pupil Chris Lintott and the Cambridge University Astronomical Society, it was given a full production in the Playroom, Corpus Christi College, Cambridge and later, in June 2002, at Downing College in Cambridge. In August the show would also be staged a few miles north of Selsey, in Chichester and on September 1st it was even performed by the Cambridge cast in Patrick's own garden! Remarkably, Patrick always performed the on-stage narration whenever his play was performed. The enthusiasm and drive of Patrick, even at the age of 79 and in poor health, was utterly amazing.

The year 2002 would mark Patrick's 45th anniversary of presenting *The Sky at Night*. At previous anniversaries the BBC had held big parties at the TV Centre and a world tour of major observatories had even been funded to celebrate the 25th and 40th anniversaries. Patrick felt that, in all likelihood, the 45th birthday would be the last milestone. To still be presenting the programme in April 2007 for the 50th anniversary, when he would be 84, seemed unlikely, given his current state of health. Patrick was therefore appalled when he was told that the corporation had no plans for a 45th anniversary world observatory tour (not that his health would have stood it) or even for a party! Patrick was not to be outdone. He usually held a smaller party of his own at Selsey for the major *Sky at Night* landmarks, but this time his own party would be a whopper! Like his mother, Patrick loved holding regular parties and filling his house with fans, friends and everyone he knew in Selsey, such as the entire Cricket Club, the local drama group, the postman, the milkman, the butcher and anyone local with a high military rank.

Patrick was very angry (a rare emotion for him) about the BBC being too mean to even afford the standard *Sky at Night* party after his amazing 45 year run on TV; a run that, to all intents and purposes, had cost the BBC nothing compared to 'fly-by-night' TV shows hosted by vastly overpaid media celebrities, who invariably ended up disgraced or with a drug habit. He also knew that his health would soon prevent him travelling up to London. Maybe it was time to vent his spleen live on air?

As a sort of 'final politically incorrect fling' just months before Patrick abandoned the BBC TV Centre in London for the very last time, he appeared, on March 21st 2002, on BBC Question Time. He was alongside four guests with a political view or two, namely: Tim Collins, Evan Harris, Dawn Primarolo and Mary Ann Sieghart. I guess the BBC 'powers' thought that, despite Patrick's forthright and non politically correct views, he was a knight and a BAFTA winner and a Fellow of the Royal Society, so what harm could such a distinguished man of 79, in chronically bad health do? They would soon find out! After making clear his views that America's post-911 problems with Afghanistan were not Britain's, especially after the USA's late entry into World War II ("I was in the War, you were NOT" he told the female panellists) and that we must pull out of Europe "NOW", he moved onto the cruelty of fox hunting. As the programme drew to a close, David Dimbleby asked Patrick if he had anything to say, to fill the final few seconds of the programme. "Yes", replied Patrick. "BAN ALL WOMEN TEACHERS." That was the last time that Patrick appeared on Question Time or, to my knowledge, broadcast live from the BBC TV Centre!!

Four days later, on March 25th, in a show recorded weeks earlier, Patrick appeared on the humorous BBC programme *Room 101*, hosted by Paul Merton. This was a show in which guests listed the things they most hated, so that those items could be banished from the Earth or 'dispatched to Room 101', provided Paul agreed with the reasons cited. Patrick listed six items he wanted to dispatch: the Archbishop of Canterbury, George Carey (for refusing to condemn fox hunting despite being a patron of the RSPCA); junk inserts in newspapers; Maybugs; Loud pop music in restaurants; untearable cellophane food wrappers; and, finally, female radio announcers!

At the end of March, with the BBC still unmovable on the subject of major 45th anniversary *Sky at Night* funding, or even a party at the BBC TV Centre, Patrick sent invitations out to almost 300 people to attend his own celebration on Sunday April 28th, at his home, *Farthings*. Meanwhile, he was so hurt by the BBC's attitude that he complained bitterly in the press. In various news reports, repeated by the BBC, Patrick made the following comments:

> For our 40th anniversary we did a world trip. This time there is no money. Instead we are going to look back over the programme since 1957. Five years ago we squirreled some money away and saved it for a big, big programme. But the BBC found that loophole and closed it. It is a shame, but I expected it, because TV now is all about soap operas, cooking, police dramas and ethnic minorities. The *Sky at Night* must be one of the cheapest programmes the BBC makes. It gets in nobody's way and has a faithful following.

Patrick also revealed that *ITN*, the *Discovery Channel* and *Sky* had all approached him but he had his 'Gentlemans agreement' with the BBC and he had never broken a 'Gentlemans agreement'. He said he was involved with the BBC but had never been an employee.

The fifth of April 2002 marked a proud moment in Patrick's life: the official opening of the South Downs Planetarium at Chichester, a few miles up the road from Selsey, by the Astronomer Royal Professor Sir Martin Rees. Undoubtedly, Patrick's fund-raising, his presence and his mere backing of the project were major factors in the Planetarium succeeding, but as Patrick emphasised, the real workers who had made this possible were John Mason, John Green, Peter Fray and Roger Prout. Patrick stressed that the building's naming as 'The Sir Patrick Moore Building' was certainly not his idea. Nevertheless, Patrick's ability to fund raise was legendary and for once he had agreed to do a TV advertisement, for *Abbey National* ISAs. The advert was only 1 minute long and Patrick freely admitted they had offered him £20,000. He told them to write the cheque direct to the South Down's Planetarium. Patrick also lambasted his local Tory MP, Andrew Tyrie, for not initially supporting his Planetarium plans. 'Wouldn't it have been nice to have had just a flicker of interest from Mr Tyrie, our MP?' he said in the local newspaper.

There was to be a sad footnote to the opening of the Planetarium though. Patrick's lady friend of over 40 years, Miss Rossie Atwell, insisted on being at the opening ceremony. She was 80 years of age and had become very frail, following a mystery illness which struck her in late 1997. She was in a wheelchair at the ceremony and caught a chill that day. Sadly, she died a week later. She had lived in Selsey for many years, literally ever since Patrick's mother died and with Patrick's full encouragement. She had also been a regular visitor to Farthings and had often accompanied Patrick to various BAA meetings.

On the morning of April 28th 2002 I arrived at Patrick's 45th *Sky at Night* anniversary party at around 11 a.m., 1 hour before it started. But already there were literally hordes of people in his driveway, in his hallway and in almost every room of his house. A giant marquee had been erected on the west side of the house, joining on to the side doorway. The whole of Patrick's west lawn, right up to his 'Summerhouse for the book overspill' was filled with the marquee. There were hordes of catering staff in there too. Patrick was dressed in a sort of vivid purple coloured blazer and was constantly shaking everyone's hands while holding on to his two walking sticks. I commented on the blazer's colour. "Bought it in the Far East" he said "but in the shop lighting it looked like Oxford Blue – HAH!"

His famous study was full of people, as was his living room. People like Lembit Öpik MP (still with his TV weather presenter girlfriend Siân Lloyd – his 'Cheeky girl' era was in the future!) were over-spilling onto the large lawns as well and sitting by the cylindrical 'dome' housing his 15 inch reflector. For April it was a gloriously warm, sunny and dry day, but I was astounded that anyone would want so many people in their house! Former guests of *The Sky at Night* from years gone by were there, people from the BAA and loads of friends from Selsey, the cricket club and the local amateur dramatics group. Food and drink were in plentiful supply and the party went on well into the evening, when the BBC actually screened the special 50 minute 45th anniversary programme. Patrick certainly knew how to lay on a good bash!

One month later, Patrick's guest on the May *Sky at Night* was Professor Fred Watson of the Anglo-Australian Observatory. This would be Patrick's last *Sky at Night* with producer Ian Russell and the last one to be filmed at the BBC TV Centre studio. In many ways that programme marked the end of an era for Patrick; the end of 45 years travelling to the BBC in London. After that, apart from a few programmes 'on location', the recording studio essentially became Patrick's house. The BBC TV Centre era had been a great one, with the low budget *Sky at Night* hidden away in the large 'Features and Events' department of the BBC where it could share the resources of bigger budget programmes like *Crimewatch* and next to which the astronomy programme's tiny expenditure would not raise any eyebrows. Patrick even used to joke that *The Sky at Night* used to steal the *Crimewatch* team's tea and biscuits from under their noses! Nevertheless, as the BBC's longest running programme *The Sky at Night* still had a lot going for it and would survive even with the programmes being recorded at Patrick's home and the monthly budget cut to £20,000.

In 2002, for the first time, Patrick started thinking seriously about writing an autobiography. He had, not surprisingly, been asked to write one many times. John Watson of the publishers Springer-Verlag had tried to coax him to write one. "Don't think I haven't tried – he just won't have it!" John told me. In a sense, Patrick had already written a limited one. His book *TV Astronomer: Thirty Years of the Sky at Night*, published in 1987, was described by Patrick as "In no way meant to be an autobiography"; but it covered aspects of his TV career from 1957 to 1987 fairly well. However, Patrick did not like anyone delving into his embellished war years, his 'only child' upbringing, his 'mother's boy' image and the enigmatic fiancé, who he had no pictures of and could not even supply (up to 2003) a first name for; what publisher would publish an autobiography if huge chunks of his life were missing? Well, *Contender Books* would and the price was right. So, Patrick started on the project. Maybe he feared someone else would attempt it if he did not; one of his 'Serpents' perhaps?

In the course of delving into his past and what had started him off in astronomy, Patrick embarked on another project too. The eccentric garden gnome-like character, W.S. Franks, who had inspired the teenage Patrick, was a subject he had thought of speaking about at a BAA meeting, so he decided he would go ahead and offer the subject as a talk. Needless to say, the BAA Meetings Secretary Nick Hewitt jumped at the chance of a talk by Patrick. On May 29th, to the delight of all BAA members and to a packed house, Patrick delivered a brief but fascinating account of the man who had inspired the teenage Patrick Moore. Sadly, Patrick told the Savile Row audience that he felt this might be the last talk he ever gave to a BAA meeting. Patrick wrote the talk up and it also appeared, some months later, in the association's Journal.

Nine days earlier, on May 20th, Patrick had been at Lancaster University, opening the Dame Kathleen Ollerenshaw Observatory. Dame Kathleen, who would be 90 that October, was in good form at the ceremony. In praising the Physics Department for their initiative in building the observatory, Patrick speculated that a future

Astronomer Royal might, one day, have a first look at the night sky from the facility. Following the opening ceremony Patrick gave a lecture entitled 'The Sky at Night' to a packed Faraday Lecture Theatre, in which he described highlights of man's exploration of the Solar System. Professor Keith Wigmore presented Patrick with a Laser Pointer and he and Dame Kathleen were guests of honour at a dinner in the University's Cartmel College.

At this point I cannot resist mentioning that at the start of May, at Patrick's request, I had sent him a list of ideas which I thought would be good for *Sky at Night* programmes, under the new producer. One of these was to have Stephen Hawking on the programme. It seemed ludicrous that Patrick had never had Hawking as a guest. I received a garbled message on my ansafone a few days later. "Patrick here. Re. your idea, Hawking, well, we've got him, we've got him!" They had indeed. Patrick had agreed it was crazy that Hawking had never appeared on the programme and the first programme (in June) under the new producer Jane Fletcher was an interview with Hawking at Cambridge. I was delighted I had such an influence, or maybe it was just a coincidence? But if it was, why had Patrick phoned me?

For the July *Sky at Night* Patrick introduced the programme from inside the dome of the new Chichester South Downs Planetarium, with John Mason explaining how it all worked. Some 35 years earlier, as I described nearer the start of this book, Patrick had tried doing this in the Armagh Planetarium dome, but had forgotten just how faint projected stars looked to insensitive 1960s TV cameras. Now, with an image intensifier, it was easily possible.

Just as the May *Sky at Night* had marked the end of an era for Patrick, the programme aired in August marked the dawn of a new one. That programme, with asteroid hazard expert Duncan Steel, was filmed entirely in Patrick's study at Selsey. This new arrangement meant that he no longer had to travel up to London. The new producer, Jane Fletcher, worked from the BBC offices at Birmingham and with a basic camera crew it was possible (just) to shoot *The Sky at Night* in Patrick's study, even if things were rather cramped. Patrick preferred the 'single guest in the studio' format that he had been used to for all of his TV life, but the new producer favoured the 'multiple guests crammed into the study' approach. Nevertheless, Patrick certainly preferred not having to travel and to be able to broadcast from his home. His appearance became somewhat stranger than usual during this period. Viewers now became used to his hair sticking up, Patrick being unshaven, dandruff on his shoulders and nasal hair sticking out of each nostril in abundance! Viewers were also used to him sounding a little drunk too, but in addition he now had the most extraordinary pair of trousers ever seen, which had such a humungous waist band that his trouser belt was now not far below his tie knot! It was often hard to tell if he was really wearing a waistcoat, but no, his trousers almost reached up to his neck! If they'd been much higher he surely could have stuck his head through his trouser fly, presumably thereby clinching a comedy BAFTA too?

What with Patrick's knighthood, BAFTA award, medals galore, seven honorary doctorates and his fellowship of the Royal Society, it might have been thought that there were no more accolades to come. However, in August 2002 Patrick received one more. Maybe the BBC had, at last, realised how stingy it had been with money for *The Sky at Night*'s 45th anniversary and had decided to make amends? On Tuesday August 25th 2002 the BBC *Tomorrows World* programme presented its awards and Patrick was in the audience. A veteran of broadcasting, with a history going back into the 1950s (like Patrick) was chosen to present the programme's Science Communication award. That veteran was, of course, the BBC's Raymond Baxter (1922–2006) after whom the award was named. Baxter explained that *Tomorrow's World* had been running for 37 years (with a variety of presenters) but, compared to the winner of the award, this was relatively short.

As soon as Baxter mentioned that the winner was in the *Guinness Book of World Records* as the longest running presenter of a programme in the world, the cameras trained on Patrick, who looked genuinely gob-smacked. A series of clips of Patrick over the years were then shown; they were a very similar set of clips to those shown when he received his BAFTA award. Patrick was then helped up the steps to receive the new award from Raymond Baxter. Patrick said that he was taken aback. He had no idea he was receiving the award and felt "very, very honoured." However, he thought that he would be remembered for one thing: "The man who swallowed a large bluebottle on *The Sky at Night*". That was not the only surprise for Patrick though, because fresh off a plane from L.A., rock guitarist Brian May was waiting in the wings to congratulate him. The programme was transmitted exactly 1 month later, on September 25th. Eight days earlier Patrick had delivered one of his final public performances, at Salisbury's City Hall, as part of that year's 'Into Space!' lectures that he had presented around the UK.

It was around this time that Brian May had persuaded Patrick to appear as a performer in a bizarre TV advert for a luxury 3 CD boxed set entitled 'The best of the best Air Guitar albums in the world ever'. Patrick's mobility was highly compromised at that time but he still did a hilarious job. The ad would appear on TV screens in November, prompting many viewers to do a double take or choke on whatever they were eating.

At the start of October 2002 Patrick used part of his new film studio, namely his garden, to interview the famous ex-monk and telescope builder John Dobson, who was coming to the end of a lecture tour in the UK and heading to Ireland a couple of days later. Dobson was 87 at the time, 8 years older than Patrick, but considerably slimmer and far more spritely. The intention was to film a whole *Sky at Night* about John's invention of the Dobsonian telescope, a simplified version of the Newtonian. However, the programme was never made. I once mentioned this to Patrick, mystified that his interview with Dobson had not translated into a *Sky at Night* programme. Patrick scowled at me and simply said "Ohhh yersss, John Dobson. Nutty as a fruit cake." I decided not to press him any further! In the twelfth *Sky at Night* book the October programme was a write-up of the Dobsonian story, despite the fact that the actual programme, screened in October 2002, was about amateur astrophotography, with the main guest being Patrick's friend John Fletcher. Almost a year later the amateur astronomer Pete Seiden wrote up an account of John Dobson and included pictures of Patrick with John, Richard Fleet and Nicky Fleet in Patrick's garden. Patrick was not amused that *Astronomy Now* had 'resurrected' part of that deleted experience. Dobson would finally appear on the *Sky at Night* almost 5 years later, on the 2007 July 1st/2nd programme, only 2 months short of his 92nd birthday. However, he would not be in Patrick's garden at that time, but with his 'Sidewalk Astronomers', showing passers-by the Moon through Dobsonian telescopes, in Monrovia, near Los Angeles, while being interviewed by Chris Lintott.

With so many awards being showered upon him (even Chichester council gave him an award for services to the town) Patrick obviously decided he could now say what he liked and get away with it. At an *Age Concern* meeting on October 22nd, at Winchester, Patrick criticised the amount of Government and National Lottery money given to immigrants, compared with that spent on the elderly. As the National Lottery had turned down a request for money from the South Downs Planetarium he had a big grudge against their money distribution policy. "We are being swamped by people who come here because we are a soft touch; and, yes, I know people will say 'racist, racist, racist' but we are being swamped by parasites. Call me a racist, but I would send them all back to where they came from. Everything we do for them takes away from what we can do for ourselves." Needless to say the *National Assembly Against Racism* and a spokesman for *Age Concern* were far from happy about such a well known public figure daring to speak his mind and voicing the frank views of many of his generation!

Of course, where other famous personalities were concerned such comments might have ended their careers and they would have rapidly backtracked, but not Patrick. He liked winding such people up even more. On Patrick's standard letter notepaper he included his name (Sir Patrick Moore CBE FRS) and his address, the latitude and longitude of his house Farthings and his telephone and fax numbers. He decided to wind people up with a modified Heraldic Crest on the notepaper too. He told me that the Coat of Arms was genuine and apparently arranged by a large group of his friends after his knighthood, headed by yet another godson, Ian Makins. Not surprisingly the Crest was highly astronomical and highly British, comprising a chevron, three crescents (for the Moon, Venus and Mercury) plus three dragons (as targets for St George to slay) and a black cat at the top, peering through a small refractor. For the motto Patrick had a popular one, namely *Signum Pacis Amor*, which is Latin for 'Love is the token of peace'. However, with a touch of the pen to the 'P' of Pacis, it became 'Racis' on his notepaper, which I guess Patrick thought would wind up his politically correct and racism obsessed opponents into a state of total apoplexy!

The December 2002 *Sky at Night* was the 600th edition of the programme and for that anniversary Patrick welcomed the Astronomer Royal, Professor Sir Martin Rees, to his home. Professor Rees had opened the South Downs Planetarium at Chichester only 8 months earlier. One month later, in January 2003, Patrick was chauffered to a meeting to celebrate the 50th anniversary of the Society for Popular Astronomy in London. As we have seen, as a young man Patrick was heavily involved in the formation of the SPA, then called the JAS (Junior Astronomical Society). In that same month, on January 15th, Patrick welcomed the British born astronaut Piers Sellers and his entire Space Shuttle Atlantis crew to the National Space Centre at Leicester. The happiness of that day would soon turn to horror though, as just 2 weeks later, on February 1st, the Space Shuttle Columbia would burn up and disintegrate on re-entry. After the loss of another seven astronauts the Shuttle program was halted for 2½ years.

For the February 3rd 2003 *Sky at Night*, a programme about astronomical art, the illustrator of Patrick's book, *Suns, Myth's and Men*, the book which led to *The Sky at Night* being created, finally appeared on the programme! It had taken 46 years, but *The Sky at Night*'s original illustrator and the man who played a key part in that book 50 years ago, David Hardy, finally made it onto the programme. Three weeks later Patrick was an invited celebrity author at the British Book Awards, held at *Le Meridien*, Grosvenor House in Park Lane. Press photographs showed him with another Moore, namely Michael Moore, who had won the Book of the Year Award for his work *Stupid White Men*.

Chapter 38
A Biography, a Co-Presenter, a Venus Transit and a Deadly Goose Egg

In 2002 Patrick had finally completed the biography he said he would never write, for Contender Books. One week before Patrick's 80th birthday (March 4th 2003) the book was published. Called '80 Not Out', in celebration of his age and his love of cricket, the book was soon snapped up by fans and amateur astronomers alike. However, while it was an entertaining read it only really covered Patrick's period as an author, TV star and a globetrotter in any detail. Hopefully, the book you are now reading fills in the gaps, especially regarding his life as an amateur astronomer in the BAA, his obsession with TLP, and those elusive war years. Needless to say, after almost 46 years of Patrick appearing on TV, and with him being a much-loved national institution, the biography received much publicity in the press. Patrick chose to launch the book at The Athenaeum Club in London's Pall Mall, on the day after his 80th birthday. Clearly, both the biography and reaching 80 were a cause for celebration. The Athenaeum Club suited Patrick well as it had a bias towards admitting members with a scientific, literary or artistic reputation. Accompanying Patrick at the book launch was Michelle Brown, the Director of Publishing for Contender Books.

The first interviewer to visit Patrick after '80 Not Out' was published was Penny Wark from *The Times* newspaper. Judging by the account of her visit to *Farthings* (published on March 31st 2003 in *The Times*) she had latched onto the huge omissions in the biography at the first reading. Yes, you've guessed it, virtually nothing about those wartime years, the RAF, or the fiancée. Also, because of her determination to get at the truth, and just get some sort of answers, she immediately encountered Patrick's stonewalling on those subjects. Personally, I have never read such a realistic description of Patrick's squirming, when faced with such questions, in any press interview and it must have made Patrick most unhappy to read about it in a national newspaper. He could not, of course, go into any detailed explanation of why he had skipped the War Years when writing the book, simply because he had embellished the stories of that era so much that he would have had to admit that he had lied about when he joined and what he did; ditto regarding the fiancée.

The easiest way out would maybe have been to say "Look, Miss Wark, I was a total mother's boy OK. My mother and housekeepers looked after me my whole life.

M. Mobberley, *It Came From Outer Space Wearing an RAF Blazer!: A Fan's Biography of Sir Patrick Moore*, DOI 10.1007/978-3-319-00609-3_38,

I have never done any domestic chores and my cooking, washing, cleaning and gardening have virtually all been done by others. I am not interested in being nagged about how useless I am outside astronomy and music and the only girl I fancied after the War (Percy Wilkins' daughter) wouldn't touch me with a barge pole. Plus, I've told a few fibs about the War Years and my fiancée, so I can't elaborate. Please avoid these subjects if you want to be kind to a sick, partially paralysed old man, who has done so much for amateur astronomy and raised millions for charity." Possibly such a strategy was a bit too risky when talking to a determined journalist, but then so was just stonewalling. Unfortunately for Patrick, Miss Wark kept pressing on with her questions about the War. To try to keep her at bay, Patrick (who, as always, had plenty of visitors in the house, especially when a journalist was coming) made sure the dining room door was open and then moved back into his study to try to break up or tame Miss Wark's questioning. But she had sussed him out, and her write up of him in *The Times*, on March 31st, was spot on. Patrick must have absolutely squirmed when he saw the write up headlined (reminiscent of Basil Fawlty) "Please don't mention the war" in the paper. Plus, I dare say he would have been destroyed when he actually read the article and saw an accurate character assassination. However, Patrick, like an elephant with a grudge, never forgot or forgave and he would time his revenge on Miss Wark for December 2004, when he wrote the foreword to the paperback edition of his biography.

Patrick, as we have seen, rarely had anything to do with *Astronomy Now* after 1994. However, he agreed to an interview with one of his best friends, Iain Nicolson, for the magazine's March 2003 edition, timed to coincide with his biography's release. One line in the interview was especially interesting: 'I thought they [the lunar craters] were volcanic, but I was wrong. We now know they are due to impacts.' Never before had I heard or read Patrick admitting so comprehensively that his lifelong belief, that the lunar craters were volcanic, was wrong! He had, at last, accepted defeat on one of the scientific pillars of his life.

With Patrick becoming increasingly immobile he rarely attended functions of any kind after 2001, and he rarely travelled outside his Selsey home to record *The Sky at Night*. However there were a few exceptions. On Friday April 25th 2003 he was happy to present the 12 year old East Sussex Chess genius David Howell with the Best Young Chess Player award on that year's *Britain's Brilliant Prodigies* programme at Carlton Television's LWT studios. Patrick, as we have seen, was a Sussex County Chess player in his youth. The programme, hosted by Gabby Logan and the Duchess of York, was transmitted 10 days later on ITV1 and other awards were presented by Robin Gibb, Bob Geldof and Evelyn Glennie. Another rare excursion was also looming for Patrick: a trip to the northernmost regions of mainland Britain. In the early hours of May 31st 2003 an annular solar eclipse occurred in Scotland. This would not be an event like the August 1999 Cornwall Eclipse of course, as the sky would never become dark. However, a bright ring of light around the Sun should have been seen, through suitable solar filters, just after the Sun rose from Scotland. The optimum place to see the event would be Iceland. However, Patrick was in no fit state to travel abroad and he wanted to see the event from British soil. The *Sky at Night* covered the Scottish event and Patrick, with a few friends, including Iain Nicolson and Brian May, almost saw the annularity phase. As the 'ring of fire' Sun rose behind the clouds they just managed to see half an arc before the annular phase ended.

August 2003 was the month in which Mars came closer to the Earth than at any time in the last 60,000 years. Actually, this was purely a technicality. To all intents and purposes Mars comes as close every 15 years or so. Even so, Patrick was determined to see the Red Planet through his big 15-inch reflector and friends and colleagues made sure that this was possible, although getting his huge immobile torso through the low observatory door was a major exercise in logistics! To coincide with this event Patrick wrote a novel entitled *Voyage to Mars*, but it was no Sci-Fi classic, just 1950s style fiction that any schoolboy could have dreamed up. The BBC arranged a live astronomy programme called *All Night Star Party* on the evening of Saturday August 23rd, to celebrate the proximity of Mars and to point a few telescopes 'live' at objects in the night sky. Patrick was the guest of honour on the programme, hosted at Jodrell Bank. Around this time Patrick also agreed to take part in a new Radio 4 version of *The Hitchhiker's Guide to the Galaxy*, entitled *The Tertiary to Quintessential Phases*. His speaking parts, acting as himself in the program, were recorded in late 2003 and broadcast the following year, with more episodes following in 2005.

From October 2003 one of Patrick's young protégés, Chris Lintott (yes, that pupil from Torquay Boys Grammar School, now a Cambridge graduate) joined *The Sky at Night* team as the programme's astronomy researcher. Patrick had always liked the young Chris and, for the next few years, he would become firmly established as the co-presenter on the programme, as well as being one of Patrick's almost endless 'honorary godson's. "I would have loved to have had a son, but Herr Hitler put a stop to that". Ermmm, no he didn't Patrick, because your fiancée never existed!

December 2003 was an exciting time for Mars mission observers and Patrick was very busy just keeping up with events. The NASA Mars Rovers Spirit and Opportunity landed successfully and returned spectacular pictures. However, sadly, the British Beagle probe disappeared without a trace. It should have been a great Christmas Present for Professor Colin Pillinger and his UK team, but after presumably landing at around 02.45 GMT on Christmas Day, no contact could be established with the probe. On Christmas morning BBC Radio Solent's Richard Cartridge phoned Patrick up at his home to seek his views. The conversation started with a rather abrupt Patrick, as Richard wished him a very happy Christmas, but a somewhat slurred Patrick snapped back "Well, it's not a very happy one so far is it? I mean this doesn't look good news I'm afraid...." Patrick went on to say that later that night Jodrell Bank might try to pick up the signal from the Beagle lander and there was a chance the following night, but it wasn't looking good. However, the live radio interview then descended into farce when someone knocked on Patrick's front door. With his mobility compromised, and two walking sticks being the norm, Patrick did not want to limp from his study to the front door and his porch 'cat airlock'. But the knocking continued and Patrick, live on air, shouted "Oh Damn. Who the hell? COME IN! Ohhhhh...Hold on ten seconds. COME IN! Will you....... ten seconds right......." Richard Cartright then descended into laughter as Patrick hobbled to his front door bellowing "COME IN!" with his voice getting steadily fainter. Then, sometime later, with Richard padding the gap, Patrick reappeared. "I'm so sorry, I'm back…hammering on the door……and I'm alone in the house……alone in the house and they were…………at the door…." Trying to regain his composure Richard recovered and Patrick said that he thought there might be primitive life on Mars, but not Little Green Men, and he reiterated that the Beagle news was looking very grim.

The January 2004 *Sky at Night* was, arguably, one of the least astronomical *Sky at Night's* of all time. 'Astronomy & Music', broadcast on the night of January 4th/5th, had guests who were singers, musicians and historians, but there was precious little about astronomy. Nevertheless, it was a programme Patrick had always wanted to do and maybe he felt time was running out for him? Being cynical, I wonder if it was merely timed to coincide with another of Patrick's musical CD's being released?

From March 2004, with Patrick obviously struggling to present a whole programme (his speech was becoming increasingly slurred) the trend was to have as many guests on *The Sky at Night* as possible, to minimise the workload for him. Also from March, Chris Lintott, the programme's astronomy researcher, became the permanent co-presenter. Patrick was becoming virtually incoherent when he spoke even moderately fast during 2004. This was partly due to the medication he was taking (and the odd Gin & Tonic may have been a factor too) but also due to the fact that his chronic arthritis was now even affecting his jaw. On *The Sky at Night* Patrick had always crammed as much news as possible into the 16 or 17 minutes of speaking time. This was now Chris' role, or that of the numerous guests. But despite his increasing disabilities Patrick still had a very active mind and many pursuits and charity interests. His huge garden and home were often used to host charity events, ranging from raising money for medical good causes, such as *Cancer Research UK*, the local St Wilfrid's Hospice and the *Alzheimer's Society*, to promoting animal charities such as *Cats Protection*. At the end of March he was even the judge at London's Savoy Hotel for the *Rescue Cat of the Year Awards'* 'Cat with the best personality!' Patrick's garden was also becoming increasingly used to host informal meetings and summer picnics for space related organisations, big or small, such as the British Interplanetary Society or the Society for the History of Astronomy. Patrick was frustrated by his mobility problems which were preventing him from attending astronomy meetings, but if the members came to his garden he could still meet up with all his best astronomy mates and adoring fans.

Patrick was not totally housebound yet though. There was a rare *Sky at Night* outing for him in May, due to an outside broadcast from a salt mine at Boulby, near Whitby, in Yorkshire, where astronomers were trying to detect Dark Matter from space. Being down a mine allows the intervening solid rock to shield the detectors from the less penetrative radiation that is of no interest to the researchers. It was good to see that Patrick could still get out and about, even if it was only a day trip down a very deep mine.

The Transit of Venus

June the 8th 2004 marked a major event in Patrick's life and one in which I was proud to play a part. A transit of Venus, where that planet passes in front of the Sun's face, as seen from the Earth, had not occurred since 1882. No-one living had seen one. The 2004 Venus transit was an event which had been in a select group of 'things to look forward to in your lifetime' for many astronomers. For the younger Patrick, the big four had always been: the return of Halley's Comet (1985/1986); the Cornwall Total Solar Eclipse (1999 Aug 11th); the 1999 Leonid meteor storm (Nov 16/17) and, finally, the 2004 Transit of Venus. Well, Patrick had seen comet Halley well (despite breaking his shoulder) but had sat under clouds for the Cornwall eclipse and had been totally clouded out (and rained out) for the 1999 Leonids. It was about time Patrick's skies cleared for that fourth and final event.

On June 7th a dozen of the UK's top astro-imagers, who had accepted an invitation from Patrick and the BBC *Sky at Night* team to assemble at Selsey, arrived in Patrick's village, home and garden, to celebrate the event that had not been witnessed in 122 years: that long awaited Transit of Venus. Astro-imagers who attended included Damian Peach, Gordon Rogers, John Fletcher, Dave Tyler, Jamie Cooper, Ninian Boyle and myself. A few professional astronomers, like the Saturn expert John Zarnecki, also turned up, as did the Queen guitarist Brian May. In addition to these astronomers various Selsey neighbours came and went during the day, as well as an entourage of BBC cameramen, *Sky at Night* staff and satellite dish technicians. There were plenty of other visitors too, including one chap (Douglas Richardson) who Patrick had taught to speak French when he was a schoolteacher in the pre *Sky at Night* years. On the afternoon of the 7th a few of us (who were lucky enough to stay overnight at Farthings) arrived and started setting out our equipment. It was obvious that there was only one optimum spot for the Venus ingress at 5:20 UT, namely along the hedge to the southwest of Patrick's big green dome and his 5 inch refractor's observatory. There would be competition for this spot come the morning: I was sure of it.

A battery of German equatorial mountings, long and short refractors, Maksutov's, Maksutov-Newtonians, webcams, laptops and even mainframe PC's soon appeared on Patrick's lawn that afternoon. His 5 inch Cooke refractor's run-off roof was opened up and used to project a huge solar image, in anticipation of the big event. Chris Lintott was a busy man, ensuring that everyone was happy, fed and watered and finding somewhere to store their equipment. Chris even opened up the observatory of the big 15 inch reflector so that Damian Peach and I could really have a good inspection of this famous telescope. Patrick was semi-mobile on that day, but he always needed his two sticks with him when pottering around the garden and he vastly preferred to be sitting down whenever possible.

At 7 p.m. we all assembled at Selsey's New Magna Tandoori Restaurant, a few hundred yards from Patrick's house, for a pre-transit curry. Every one was in high spirits: how often does a rare astronomical spectacle occur in Britain when the weather forecast is for wall-to-wall sunshine from dawn until late afternoon? In the case of a Venus Transit 'rare' seems a totally inadequate word: no-one had seen one of these for 122 years, from anywhere on Earth. Everyone I spoke to about this day agreed on one thing: there was a feeling about this event, namely that we were in the right place at the right time. Farthings, at 39 West Street, Selsey, had the Gods smiling upon it on that occasion: Spode had gone on holiday!

Returning to Patrick's after the curry, at 11 p.m. Damian and Dave Tyler polar aligned Dave's 90 mm f/19 refractor and 80 mm Vixen Flourite apochromat. Damian located the famous double-double star Epsilon Lyrae as well as the magnificent coloured double star Albireo through the Vixen. Rarely, if ever, had I seen such exquisite textbook diffraction patterns around any stars; it was a crystal clear night, seeing was good and top class optics were available.

I managed almost 5 hours of fitful sleep before getting up just before 5 a.m. to guard my spot for the Venus ingress; but, early as that was, my spot had been pinched by non-other than Brian May, with his H-Alpha Solarscope! Fortunately, there was enough room for me to squeeze in between Brian, Jamie Cooper and Ninian Boyle. I must say, I had NEVER seen so many H-alpha solar telescopes in one place, not even at a solar eclipse!

Well, to cut to the chase, it was clear from ingress to egress and we all had plenty of time to look through everyone's telescopes. Jamie Cooper's 18 centimetres Maksutov-Newtonian gave the sharpest view. There was wall-to-wall sunshine for 6 hours and, where we were, a few 100 yards from the coast, the temperature was a very pleasant 25 °C. Further inland temperatures reached 30 °C in the afternoon. BBC 1 and BBC News 24 were running updates on the transit throughout the day, co-ordinated mainly by Adam Hart-Davis. Every now and then the producer, Jane Fletcher, would shout for quiet as Patrick, Chris and Brian went live on the air. For the last few updates we were asked to stand behind Patrick with loads of telescopes to show the scene at Selsey. When Patrick bellowed to the News 24 anchorman Philip Hayton: "Well, I was just going to explain that if you hadn't interrupted me" a big cheer went up from the crowd! Throughout the day Patrick projected the Sun's disc onto white card with his famous, childhood, brass-tubed, 3 inch refractor, which had cost him £7 and 10 shillings back in the 1930s. The refractor never left his side and the piece of white card was always close at hand, as well as a glass of tomato juice. I seemed to be the only person there who realised what a great picture could be snapped of Patrick, alongside his childhood telescope and with the black dot of Venus crossing the Sun being projected onto his white card. I still regard that shot of mine as the most memorable of the day.

John Fletcher, an amateur astronomer who had spent much time with Patrick over the last few months, even cooking meals for him, found that he was clouded out for the ingress, so he drove all the way down from the Cotswolds, to Selsey, just so that he could witness the middle and end of the transit. John had been given a telescope by Patrick in 2004: the late Paul Doherty's old 12″ reflector.

Of course, there was much speculation beforehand about whether we would see the 'Black Drop' linking Venus to the Sun's limb at second and third contact (when the planet has just entered or is just about to leave the solar disc). Those projecting the image saw it, and many with small filtered apertures, observing visually, saw it; but when those with the highest quality instruments and webcams examined their

Fig. 38.1 Chris Lintott, Brian May (Queen guitarist) and Patrick, in Patrick's garden at Selsey for the June 8th 2004 Transit of Venus (Patrick was broadcasting live on TV while this photograph was snapped by the author)

Fig. 38.2 Patrick using his childhood 3 inch refractor to project the image of the Venus Transit on June 8th 2004 (Photograph by the author)

video frames it appeared to be nothing more than atmospheric seeing related. In the rarest sharp moments, when the atmosphere was still, there was no Black Drop at all; just a sharp black Venusian disc, with a clear thin channel of Sun right to the solar edge, which was fascinating! A historical enigma had been solved. Mind you, it has to be said, that at times the 6 hours between ingress and egress seemed to drag, especially with virtually no sunspots being visible on the solar disc; but there were plenty of things to discuss with so many fellow amateurs being there. As Patrick said "It can hardly be called spectacular, but it is exceptionally rare: no-one alive has seen one of these before today."

As the last dent in the Sun disappeared, at around 12:24 BST, Patrick recorded his final words to the camera and a big cheer went up from the crowd. This had been a historic day, at a historic place, with great friends from the astronomy world and Patrick had been on top form. It was a day that none of us would ever forget and a day that the heavens truly smiled on Patrick and on 39 West Street, Selsey and on all who were there!

The Deadly (Almost) Goose Egg

There was no doubt that June 8th had been a great success for Patrick but disaster would strike only 2 days later. On June 10th Patrick was taken seriously ill at his home. He thought he was having a stroke: he could not breathe and his heart was beating irregularly. "I thought I was dying" he told me. Patrick was rushed to hospital and the illness was diagnosed as food poisoning, not a stroke at all. Patrick had eaten a goose egg the day before and salmonella from this was almost certainly the cause. Patrick spent 3 weeks in hospital and, for the first time in over 47 years, was unable to appear at all on *The Sky at Night*, which went out on the Sunday Night/Monday morning of July 4th/5th at 12.15 a.m. As the opening credits of *The Sky at Night* faded, Chris Lintott simply said that Patrick "could not be with us this evening" and left it at that. It was a unique opening. Since April 1957 EVERY *Sky at Night* had been introduced by Patrick, including the one in 1977, just after he had damaged his spine in a fall. At the time the programme was filmed no-one really knew if Patrick would survive. This unique Patrick-less programme was not repeated on the following Saturday. Chris was joined on the programme by one of *The Sky at Night*'s most regular guests over the years, Dr John Mason (another one of Patrick's protégés, although 25 years older than Chris) and by Dr John Zarnecki, a Saturn/Cassini mission expert. The programme was about the preliminary results as Cassini flew past Saturn. Patrick spoke to the press from his bed the next day, saying that he had "Very, very, nearly died…it was awful." During the 3 weeks that Patrick had been in hospital the amateur astronomer John Fletcher, an old friend and a retired postman from Tuffley in Gloucestershire, had lived in Farthings and looked after Patrick's two cats.

Patrick's friends and neighbours had always done their best for the great man in the early years of the twenty-first century, forever dropping in to help with routine chores. Patrick had always shunned any full-time professional help up to July 2004, but medics and friends alike stressed that it was not sensible for a man of his age, immobile and with heart, spine, rheumatic and now food poisoning problems, to carry on without help. He reluctantly agreed, and an official carer was assigned to Patrick on his return from hospital, one that would live at Farthings and look after him. The improvement in Patrick's health after this point was notable. He'd been too proud to ask for help before the goose egg poisoning, but having full-time professional help perked him up considerably. However, in the years to come the cost of this care would push him to the brink of bankruptcy.

A few weeks after he returned, John Fletcher was back at Selsey, helping to sort out Patrick's slide and photo collection. The task proved so immense (more than 7,000 images!) that eventually another Selsey man, Trevor Little, took the project onboard with help from various other people. Patrick's near brush with death had made him think seriously about trying to archive his slides in case other astronomers ever wanted to use them. Despite his immobility, John told me that Patrick was updating his massive *Data Book of Astronomy* as soon as he was half-recovered from the salmonella scare. He would sit at his ancient typewriter for typically 5½ hours every single day hammering with his swollen banana fingers as fast as his spinal and arthritis problems would allow him to, on those ancient and heavy keys. When his phone rang, as it frequently did, Patrick would shout "Damn, blast and hell" and then pick up the phone with the cheery words "Hello there, Patrick here!"

It was around this time that Patrick had finally conceded to having a stair-lift installed in Farthings, as it took him so long to haul his bulk up the stairs to bed, even with helpers. Initially, Patrick had been totally opposed to this, citing it as "the beginning of the end" and "they'll put me in a home next". He was less than happy with the cost too, as any stair-lift capable of carrying his vast bulk up the ancient stairs would have to be a serious piece of engineering. He told me that John Fletcher had raised the issue of what would happen if there was a power cut when he was half way up or down. "I could be marooned in mid-air for hours, maybe days……." Patrick moaned! But, as he was getting less mobile every year, he eventually had to concede defeat and approve its installation.

October 4th 2004 saw a first for *The Sky at Night*. The previous evening's BBC1 programme was repeated on the digital channel BBC4, with an extra 10 minutes to allow Patrick and Chris Lintott to have a brief look around the current night sky. Patrick was delighted to have extended air time only months after he had nearly died. Extended BBC 4 repeats would become the norm from then on. In the same month Patrick agreed to be part of the cast in a new UK Science Fiction/Comedy venture called *Starhyke*. Obviously Patrick's mobility meant that his amateur dramatics roles, such as the Demon King, were well behind him, but he was simply asked to perform the role of 'The Oracle' in episode three of this series (entitled 'Kill Jill'); a not dissimilar performance to his role in the *Gamesmaster* series. Unfortunately, after the six pilot episodes of *Starhyke* were shot and the post-production was completed and after a further 4 years (!) the media reported that both *Sky* and *Virgin* had turned down the series. All six episodes were eventually released as a DVD in 2009.

Seventy Years in the BAA

On November 28th 2004 Patrick completed 70 years as a BAA member. Remarkably, he was not quite the longest serving BAA member at that time. The 89 year old Dr Henry King, the BAA President from 1958 to 1960 and a sworn enemy of Patrick's (as we have seen, the feeling was mutual and had been for 50 years) was a member for 21 months longer (but King would eventually die in 2005). However, Patrick's contribution to the BAA over those 70 years had been truly staggering. He had been an incredibly active visual observer in the association for all but the last 7 years and he had also served as the President, the Lunar Section Director and the Mercury & Venus Section Director. He had also given hundreds of talks at BAA meetings and written most of the meeting reports throughout the 1950s, 1960s, 1970s and 1980s, without usually even signing his name as the recorder at the end! In addition, he had published over 200 papers, letters and articles in the BAA Journal over those 70 years. In almost every book that he had written the BAA received free publicity and he must have been directly responsible for hundreds (if not thousands) of BAA members joining over the years. However, there can be no doubt that Patrick's whole lifestyle and TV career owed everything to the BAA. He was a product of the BAA: a twentieth century observer who had been inspired by the Victorian doctors, military men, reverends and greybeards who dominated the association when he had joined in 1934. His BAA mentors of Francks, Barker, H. Percy Wilkins, Major Levin and others, such as Guy Porter and Will Hay, had inspired the childhood Patrick and made him the great astronomer and entertainer he had become.

As we have seen, the BAA paid tributes to Patrick at the 50th and 60th anniversaries of his membership and the 70th would be no different. I kept reminding the President in 2004, Tom Boles, of the impending anniversary, just in case no-one else reminded him. However, the November 2004 BAA meeting was a joint meeting with the Royal Geological Society, not a uniquely 'BAA' event. The Christmas meeting would be a different affair though. After a gap of 2 years the Association was back at its standard meeting place of 23 Savile Row. That lecture theatre had been the BAA's meeting place since 1970, although it was now renamed 'The English Heritage Lecture Theatre' instead of 'The Civil Service Scientific Societies Lecture Theatre'. It seemed especially fitting that Patrick was back there. He had not attended a BAA meeting at 23 Savile Row since May 29th 2002.

The BAA's 2004 Christmas meeting took place on December 18th 2004 to a packed house. As the meeting started, Patrick, accompanied by his friend the Gloucester amateur John Fletcher, as well as two other helpers, arrived and managed to walk, with two sticks, to an especially reserved large armchair. As soon as he appeared in the theatre there were cheers and applause. The BAA had really missed Patrick not being around at its meetings. The President Tom Boles had wondered what to buy Patrick as a mark of his 70 years as a member. Chris Lintott had suggested a Moon globe, but only 'cheap and nasty' ones were available. Then it came to light that *Broadhurst Clarkson & Fuller* of Farringdon Road, who had modified Patrick's 15 inch reflector at his request in 1972, were agents for some superb gold Orreries of the Solar System. These were manufactured by the *Devon Clock Company*. Normally these Orreries would sell for £580, but BC&F only wanted £475 as it was for Patrick. It was the perfect gift and the Orrery featured a plaque inscribed 'Sir Patrick Moore CBE. 70 Years membership of the British Astronomical Association. 28th November 2004.'

At 14.50 GMT in that December 18th meeting Tom Boles presented Patrick with the Orrery and congratulated him on his 70 years of membership. Tom started by saying that 70 years ago a fresh-faced 11 year old had joined the BAA. Patrick, standing by his reserved armchair, near to the front of the theatre, told the audience how he had known lots of entertaining BAA characters over the years. We had all heard the stories before, but no-one minded at all. Patrick specifically mentioned two former Presidents, Dr W.H. Steavenson and the Reverend Martin Davidson (the latter an East End Vicar who the reader might recall had learned how to pick locks from a criminal parishioner) as well as Miss Cicely M. Botley, who had an encyclopaedic memory. Patrick thanked the President and the BAA and said that the association had always been at the forefront of astronomy and always would be. The audience applauded enthusiastically and Patrick thanked them again at the very end of the meeting, when the vice president, Guy Hurst, thanked Patrick once more.

Fig. 38.3 Tom Boles with Patrick on December 18th 2004 presenting the great man with an orrery to celebrate his 70 years of BAA membership. This was at the BAA Christmas meeting inside the familiar venue of 23 Savile Row in London (Photograph by the author)

The next day I e-mailed pictures of the celebration to John Fletcher who was staying with Patrick at Selsey. John said that Patrick was delighted by the day and he (John) was just cooking a Sunday Roast while they watched England beating South Africa in the Test match on the *Sky Sports* channel. John had placed my pictures on Patrick's home computer, a PC that his helpers and their children operated for him.

All things considered November and December 2004 had been successful months for Patrick. As well as being honoured after 70 years of BAA membership, a bill to ban fox hunting had finally been passed by the Labour Government on November 18th. In the years to come hunt supporters would find ways to bend this law, much to Patrick's frustration, but the passing of the bill had been a breakthrough.

Another Clash with BBC Officialdom

Despite Patrick's increasing immobility and his reliance on two walking sticks he still made every effort to appear on TV programmes for the BBC whenever he could, even if he found himself increasingly at odds with the corporation on a number of issues. On February 25th 2005 he appeared on the long running children's programme *Blue Peter* to explain how, 75 years ago, the ninth planet Pluto had been discovered by Clyde Tombaugh. Patrick had been an occasional guest on *Blue Peter* for 32 years, although most of his appearances occurred during the 1970s and 1980s. The BBC records list this 2005 *Blue Peter* appearance as his tenth (since December 1973) although I am sure there must have been far more. To commemorate Patrick's contribution to astronomy the *Blue Peter* presenter Simon Thomas awarded him a prestigious gold *Blue Peter* badge on that February 2005 show. It was also announced that the programme now had minor planet 16197 officially named as 'Bluepeter' by the International Astronomical Union.

The success of Patrick's somewhat patchy autobiography *80 not out* led to a paperback version being released in March 2005. This cheaper copy, published by Sutton Publishing, was essentially the same as *80 not out*, except that Patrick had included a mention of his 70 years of BAA membership, inserted more illustrations, and a few corrections had been incorporated. For example, the incorrectly labelled picture of him with Eric Morecambe was correctly titled 'Cliff Michelmore' and the misspelling of his 'Caldwell' double-barrel surname was corrected (it had been Cladwell in the original!) He also found time to write a foreword (written in December 2004) in which he mainly had a dig at *The Times* journalist Penny Wark, who had insisted on continually questioning him about his war years when the first version of the biography was published. Patrick said, in that foreword to the paperback edition: 'She came down to see me, didn't get the answers she may well have wanted after her oft-repeated questions and departed looking furious (at least I thought she looked furious, it may of course have been her natural expression).' Miss Wark responded in her column in a separate article about Patrick more than 2 years later, on May 9th 2007, when she said: 'He added a tetchy (and untrue) reference to me in the foreword to his autobiography. I didn't think that was very nice.'

The original hardback biography *80 Not Out* had sold well and so Patrick and a friend who was an independent filmmaker (Charles Wallace of the Moving Image company) decided to make a DVD biography of Patrick's life. Patrick told me that the BBC had "messed us about" over the biography. According to Patrick they first liked the plan and wanted to fund it, then they wanted to fund 50 % of it and then they wanted to own all the rights to it without paying for anything. The last straw, again according to Patrick, was that the BBC wouldn't pay for extracting material from the TV archives for the biography. "So", he told me proudly some time later (on July 16th 2005) "WE will make the DVD ourselves." Part of the problem was concerned with Patrick wanting clips from when he had appeared with *Morecambe & Wise* to be included in the DVD of his life. Patrick felt that, after 48 years with the BBC, in which he had only missed one episode and had literally done it all as a labour of love, they might at least pay the *Morecambe & Wise* royalties as a sign of gratitude. But this was not offered and when Patrick was contacted by a female producer about doing a BBC biography of him Patrick told her: "You have as much chance of getting my co-operation as having Halley's Comet turn up next year in West Sussex." In fact, Patrick very unfairly blamed *The Sky at Night* producer Jane Fletcher for not being able to use BBC archive material. He even complained to her superiors, who told him he was wrong, she was right and archival royalties had to be paid for, but the BBC weren't prepared to pay for them, despite Patrick's service to the BBC. It seemed incredible to Patrick that numerous so-called entertainers were literally paid millions every year (each!) by the corporation "for spouting filth" whereas he had contributed 48 years-worth of science for virtually nothing; yet the BBC would not even fork out a tiny amount for the *Morecambe & Wise* royalties, despite the fact that all the clips he wanted to use had him included as a performer!

Anyway, light-years away from all this BBC 'Twitmarsh' bureaucracy: inspired by the June 2004 success of the *Sky at Night* Venus Transit Party, held in the *Farthings* garden, Patrick and Jane had decided to hold another Selsey party 9 months later. This time it would be a Star Party, with invited amateurs using the telescopes in Patrick's back garden. Patrick had been unable to use his telescopes properly for many years and, even then, he had been helped by friends to get to the eyepiece when conditions were right. As a result, by 2004, the telescopes were in a very sorry state. Fortunately, his local Selsey and Ham astronomers Pete Lawrence, Mark Parrish and Ian Sharp did an excellent job of restoring the 15 inch reflector (including removing the mirror and getting it re-aluminised) so that the big 'scope was available for the Star Party. I was invited to that Star Party and appeared on the programme, talking about my observations of Saturn and Comet Machholz (very close to the celestial North Pole) with Patrick's telescopes. Remarkably, despite very cloudy weather in the UK in February and early March 2005, the Star Party, on the evenings of March 11/12 and 12/13 worked well. The first night was partially clear and the second night totally clear! It almost always seemed to be clear whenever I visited Patrick at Selsey! I stayed in a nearby bed & breakfast with Tom Boles, Damian Peach and Dave Tyler. Patrick had an appalling cold and cough and we all tried to dissuade him from hobbling outside. Mercifully, he agreed! Patrick was, by then, under 100 % supervision, with a carer living at Farthings full-time as well as other helpers always available.

To be brutally honest Patrick was a very sorry sight when not in front of the TV cameras, even in 2005. He could not get out of a chair unaided and could hardly grip even a piece of paper in his hands. But, despite all of this, his brain was still very active and he was still typing (painfully slowly) all day (and all night sometimes). Patrick's lack of mobility did not stop him attending a few functions though, as friends and helpers were always on call. These functions were numerous and diverse. For example, at the end of February 2005 he was pictured with Sussex Junior Chess champions after their win against Ireland in the 'Sir Patrick Moore Sussex Challenge Cup' event. Patrick had been a county Chess player in his youth and was the Vice President of the Sussex Junior Chess team.

On the second night of the March 2005 Star Party I had experienced enough by 2 a.m. and decided to walk back to my B&B and a nice warm bed. However, Patrick stayed awake (dozing in his chair occasionally) and was still drinking champagne at 6 a.m. in the morning with the BBC *Sky at Night* Producer Jane Fletcher, Damian Peach, Dave Tyler and others. After they left him at 6 a.m. his carer finally managed to put him to bed! Patrick's ability to down alcohol was quite legendary, although he said he had only been drunk twice in his life. God only knows how much alcohol that involved?! At the star party he revealed he had perfected a very good "banana wine", although he described the taste of a friend's "mint wine" as "utterly revolting" and "unspeakably horrid!"

Fig. 38.4 The author with Patrick, discussing his life and listening to many of his entertaining stories, in his study at Selsey on March 11th 2005 (Photographed by Damian Peach with the author's camera)

Fig. 38.5 Patrick, Jane Fletcher (producer) and another member of the *Sky at Night* TV crew in Patrick's music room on March 12th 2005, during filming of a *Sky at Night* 'Star Party' (Photograph by the author)

Chapter 39
A New Magazine, a Pacemaker and BANG!

During the March 2005 *Sky at Night* Star Party Patrick revealed that, more than 5 years after the demise of his rather sad *Modern Astronomer* publication, he was once more launching a British magazine rival to *Astronomy Now* (the magazine he had founded in 1987 and then fallen out with 7 years later). However, this time he was publishing his new magazine with some substantial backing. It was to be a BBC magazine called *BBC Sky at Night*, published in Bristol by *Genesis Publications*. The first magazine would be the June 2005 edition and it appeared on bookshelves in late May. Patrick had assembled a formidable team for this new magazine, including Adrian Ashford (who had been involved in his previous ventures, as well as with *Astronomy Now* and *Sky & Telescope*) and Paul Parsons (former *Astronomy Now* and *Focus* editor). A number of other friends, including *The Sky at Night* co-presenter Chris Lintott, were also drafted in. I was asked to write a software review for the second issue of the magazine and a further review for the next issue. The magazine would feature a give-away CD/DVD with each copy, occasionally containing archival *Sky at Night* footage clips.

On August 6th 2005, BBC4, the new digital channel, ran an all night long *Sky at Night* and space evening celebrating British amateur astronomy. Patrick was filmed in his garden chairing the whole event, which included a celebration of 48 years of the programme, as well as showing Patrick's favourite 1950s Flying Saucer film. I appeared, at Patrick's insistence, in a separate programme entitled 'Star Men'. Despite Patrick's dispute with the BBC, about using *Morecambe & Wise* footage in his planned home-made DVD, they did include the clips in the *Sky at Night* compilation and, presumably, as it was BBC produced there was no problem regarding the royalties? Indeed, Patrick happily introduced the clips on the programme as if there was no dispute at all! Nevertheless, the *Moving Image Company*, namely Patrick's friend Charles Wallace, released their version of Patrick's life, with no help from the BBC, in the autumn of 2005. Entitled 'The Astronomical Patrick Moore – the authorised version – First Draft' and later subtitled 'Revised & Extended' it was pure unadulterated Patrick with, of course, his best friends paying tribute to him.

M. Mobberley, *It Came From Outer Space Wearing an RAF Blazer!: A Fan's Biography of Sir Patrick Moore*, DOI 10.1007/978-3-319-00609-3_39,

Around this time *The Sun* newspaper announced that astronomer Mark Garlick had discovered the constellation of Victor Meldrew in space! They consulted Patrick who said he was not surprised, as he had seen stellar patterns resembling the abominable snowman, a petrol pump and an aardvark!

Dark Side of the Moore

I visited Patrick again for three other *Sky at Night* sessions in 2005: two were in his garden and concerned with meteor observing and viewing Mars. On both occasions Patrick was determined to get out onto his back lawn and be seen observing the night sky and on both occasions I grabbed one of his upper arms and another amateur, Bruce Kingsley, grabbed the other, so we supported him as he shuffled with his walking frame from his study to his garden and back again. He was breathless after each walk in the dark and it was brave of him to even attempt the short trek in his condition: dodgy knee, morbidly obese, partially paralysed and with a dodgy heart too. After the August 2005 meteor observing and filming session Patrick was totally exhausted and very stressed indeed; to make matters worse, his carer had gone for a quick lie down at around 3.00 a.m. too. Everyone decided to go to bed, even though Patrick was keen to stay up and drink the dawn away. The BBC producer Jane Fletcher kindly tried to get his new stair-lift to work to help get him upstairs to bed. Patrick was a proud man though. He did not want Jane's help and when she pressed the button on the stair-lift, at 4.20 a.m., Patrick secretly held the STOP button down to make it look as if she had broken the system! Jane then, rightly, decided to wake Patrick's carer to help get him up to bed. An exasperated Patrick, frustrated by his total dependence on others, could be heard bellowing throughout the house: "Just go, go, GO you dreadful woman. JUST LEAVE ME ALONE AND GO DAMN YOU." Jane, looking somewhat bemused, eventually left for her Selsey B&B (Greenacres) at just before 5.00 a.m., but they both made up and were friends again the next day! It was just like watching a patient mother coping with a gigantic toddler having a tantrum! It was also the only time that I had ever seen Patrick get really angry.

The increasingly old, cantankerous, paralysed, drunk and disabled Patrick could be a really spiteful person at times to Jane Fletcher, openly telling people "She is THICK" in her presence. It was a huge tribute to Jane's patience that she did not resign from the job of *The Sky at Night* producer in her first year. Indeed, amateur astronomers owe her a huge debt! In his eighties Patrick's hatred of women who could thwart his plans could be scary at times. His disabilities could make him a very bitter man. He had been a powerhouse of energy in his youth, but now his spine, heart, arthritis and knee problems meant that he was virtually immobile, even though his brain was still buzzing with energy and ideas. After a few too many Gin and Tonics and a bad day, mixed with his medication, he could have an acid tongue, but mainly directed at the influx of foreigners into the U.K. Patrick was having to spend eye-watering amounts of money on his carers' fees now, yet asylum seekers and trainee terrorists merely had to break into the UK and money was showered upon them and on their enormous numbers of illegitimate offspring. At least, that was how he saw things. Patrick had always worked hard, paid taxes, been unbelievably generous and raised millions for charity. Yet he was a second class citizen, entitled to less help than the potential terrorists invited into our shores with welcome arms and showered with benefits! Had all those who died in the two World Wars trying to keep foreign invaders out died in vain? It appeared so, from his perspective. Around this time The Sun newspaper announced that astronomer Mark Garlick had discovered the constellation of Victor Meldrew in space! They consulted Patrick who said he was not surprised, as he had seen stellar patterns resembling the abominable snowman, a petrol pump and an aardvark!

Even Patrick's closest friends would wince at some of his racist comments when he was in a foul mood, but it was his way of venting his frustration at his ailing body and the state of the country. If he had repeated some of them in public, or on air, he would have been arrested! "Do you know what to do if you see a Pakistani drowning?" he would bark, well after midnight. "No, Patrick" his guests would groan. "GOOD!!" he would squeal with glee. And he wouldn't stop there. "Do you know what you would *throw* a Pakistani if he was drowning?" he'd continue. "No, Patrick, really, you've said enough……". "HIS WIFE AND KIDS……HARRRRR, HARRRRR, HARRRRRRRR" he would bellow. Everyone would look at the floor in an embarrassing silence, not sure what to say, except quickly change the subject to astronomy or cricket. It was hard to reconcile the unbelievably generous and kind Patrick, who had raised so much money for charity, with the one that sometimes emerged after booze, medicine and his growing list of disabilities altered the balance of his mind.

One also has to remember that Patrick was brought up in a wealthy, white, middle-class environment by Victorian era parents who had spent many years living in Kenya. After the War his views of Germans and then all foreigners were pretty extreme and fixed for life. Hell, even one of his favourite childhood adventure books, *Biggles Flies Again*, has openly racist lines! Rewiring the neural net of his porridge-like octogenarian brain to have more politically correct views was just not possible: his grey matter was now hard-wired. In addition, no-one really thought that he wanted to see any person drowning. He hated unkindness and cruelty. The simple fact was that Patrick despised political correctness so much that he would say the most shocking things, just as a nuclear option counter attack to the 'Thought police' that he also hated so much. On a similar theme, namely refusing to fit in with current political correctness or accepted thinking, Patrick refused to bow to the evidence for Global Warming, declaring: "I simply don't believe a word of it; these same people were saying the next Ice Age was coming a few years ago! The exact same people!"

In the twenty-first century there seemed, in Britain and elsewhere, to be an increasingly popular tribe of people best described as 'The Offenderati'. In other words, people whose whole life consisted of scouring the media and the Internet for things to be hair-trigger offended by, while pretending to be perfect people themselves. In fact, such people are rarely what they seem; they simply say the politically correct thing while hiding their true (often deeply envious) thoughts and using their allegedly caring views simply to generate self-publicity. Patrick was just the opposite, because he spoke his mind and said what he thought, as well as exaggerating his views to offend those who were obsessed with political correctness. To him and many other people this was far better than being a hypocrite. To some Patrick might have seemed an out-and-out racist and sexist, but to anyone who knew him well, especially before he became confined to a wheelchair and sentenced to a regime of drugs and painkillers, he was the kindest and most generous person they could ever meet, especially where children and animals were concerned. The fact that Patrick wrote the books 'Bureaucrats: How to Annoy Them' and 'The Twitmarsh files' shows how much he hated the ranks of the bureaucrats and politically correct pen pushers, as well as professional scientists who were "trying to brainwash us all!"

Patrick also had various legal gripes with all sorts of people (including the BBC on copyright matters) as his health failed and so his solicitor Peter Rouse's Blue Bentley was often parked in the *Farthing*s driveway. In addition, his finances were in a hopeless mess and he literally had no idea how much money he had coming in, going out, or owed, let alone what the tax man wanted from him! With so many carers and helpers over the years and Patrick's amazing generosity, money was flowing out of Farthings in bucket-loads, but little was coming in. Of course, with anyone else in his situation, the house would have been seized to pay for Patrick's care and he would have been carted into an old people's home; but there was no way Patrick would let that happen. "I will leave this house in a wooden box" he often said. However, even with lots of friends and his status in the community something had to give financially. These worries, and his crumbling body, were making him a very bitter man, totally unlike the younger Patrick of the twentieth century.

Around this time Patrick intimated to me and to a few others he could trust, that his dear friend the rock star Brian May had offered to buy Patrick's estate, to raise cash and to keep *Farthings* financially afloat, if that was what it took. Patrick said that equity release on the house was looking like the only way he could financially survive now. Brian was happy to lease the house back to Patrick at the 'One Peppercorn if demanded' rate; in other words, for free! He was also happy to pay for Patrick's carers. What a splendid chap that Brian May is! With Patrick being semi-paralysed, immobile and approaching 20 stones in weight, the necessary carers' fees were draining thousands of pounds every month from his dwindling resources. "I would have been staring into the abyss of the Black Hole of Poverty, if not for Brian, it's as simple as that!"

Patrick's finances were in utter turmoil, but with the equity release offer and a potential solution to funding the carers, he still felt compelled to be generous, to write and to front *The Sky at Night* every month, for peanuts. After all, how much longer could he possibly survive in this condition? Maybe the equity release would not be needed if he keeled over in the next year or two? But then he desperately wanted to reach the 50th anniversary of *The Sky at Night*. He was still keen on charitable work as well and in 2005 he even produced a fund-raising little booklet called 'Asteroid', with 87 year old Arthur C. Clarke, to raise money for the victims of the 2004 Boxing Day tsunami in Sri Lanka. As an asteroid impact in the ocean would result in an even bigger tsunami it was an appropriate title. Patrick asked me for a picture of asteroid 2602 Moore for the booklet, but then he forgot to use it, despite mentioning me in the relevant chapter! Far from wanting to see foreigners drown (as he often quipped) he was actually raising money for flood victims in a far-flung corner of the world, even though his own cash reserves were fast running out. The joint Moore/Clarke venture raised £13,000.

While Brian May's genuine concern for Patrick was most welcome and totally honourable during this time, there were other much more dubious characters eyeing up Patrick's situation in the twenty-first century. Patrick was a British Institution and even in his fragile state there were those who could see ways of making huge amounts of media-generated money and fame out of the great man in his hour of need, with Patrick benefiting little from these ventures. Middlemen could see ways of maybe tricking him into endorsing products, astronomical or otherwise, selling his unwanted spare possessions (signed or not), or pretending to be acting as an agent between Patrick and a lucrative media event. The less said about these slimy characters the better, but fortunately Patrick still had an active brain and so instinctively he usually knew when characters were genuine, like Brian May, or dodgy.

83 and the Pacemaker Goes In

As 2005 came to an end Patrick was appalled to hear that the London Planetarium was closing. It had been deemed that astronomy was not a big enough money-spinner and so the space would become a sort of museum to A-list celebrities! Needless to say, in the coming months Patrick was very vocal in the press about the closure. He had been involved in numerous clashes with the London Planetarium and even the Science Museum during his life, but he argued, quite rightly, that if he and his Sussex friends had managed to set up and run a successful planetarium in Chichester, in the last few years, the subject must surely be more popular than ever.

As Patrick started 2006 his physical appearance on TV was getting worse than ever and he was looking very old indeed. His speech was now so slurred that it was becoming incomprehensible and the guests on *The Sky at Night* often looked worried and puzzled by his appearance. To me at least, the overall impression was that I was watching Private Godfrey, from *Dad's Army*, after a night on the booze. I attended his 83rd birthday party on March 4th and, unusually for Patrick, he was finding it difficult to put a brave face on things. At midnight he made it clear that he was planning on retiring and strongly suggested we should retire ourselves, or go to our Bed & Breakfast accommodation. The birthday had been a happy affair, with numerous cakes brought along, a bagpipe accompaniment to the cakes being cut and the cakes invariably being iced with either a telescope theme or a cat theme! Even Patrick's 86 year old cousin Eileen Tanner turned up, although I did not see one word pass between them all evening! However, by midnight Patrick really had endured enough. In the morning he felt really rough and a doctor from the hospital came to fit a monitor to record his irregular heartbeat. The next day Patrick was told, in no uncertain terms, to "Have a pacemaker fitted NOW or you are a dead man!" On March 7th, 3 days after his 83rd birthday, the pacemaker was fitted at St Richard's Hospital in Chichester. Quite why Patrick had not agreed to this before, after a lifetime of heartbeat irregularity, is a mystery, but he was a squeamish man and as with his knee replacement, he seemed to wait until his body was falling apart before he would trust medics to poke around inside him. However, after the pacemaker was fitted Patrick was a new man. The effect on his health was dramatic. He had been taking pills to steady his heartbeat all his life and now a pacemaker had turned him into a different person. Even he admitted that he had not felt this good for many years.

Despite his improved health Patrick still had to accept that he was never going to use his main telescopes again. He was fully aware that he may not have long to live and he wanted to donate his telescopes to people who would use them. He had offered his 12.5″ alt-azimuth Newtonian to Peter Cattermole who, according to Patrick, would soon take it abroad, to his home. However, Peter did not take it off his hands, which was probably just as well as it would play a starring role in the 50th *Sky at Night* anniversary programme! The 8.5-inch equatorially mounted Wildey/With-Browning was another instrument that Patrick could never see being used again, even on TV. It was badly rusted and dilapidated and needed giving away to someone who would properly restore it and its dome. Amateur astronomer Bruce Kingsley, from Kent, told Patrick that he would restore it and Patrick was happy to give it to a good home. In June 2006 Bruce removed the telescope, followed by the famous rotating octagonal dome in October.

A Kinder Side to Patrick

I will digress briefly at this point to a very personal matter and a true indication of just how considerate Patrick could be to those who he regarded as ‘true friends’ rather than ‘hangers-on’ who just craved rubbing shoulders with famous people. On June 14th 2006 my mother Barbara was diagnosed with cancer. Only a handful of UK astronomers knew about this and I had not told Patrick. Within days, Patrick had sent me a kind and lengthy letter wishing my Mum all the best (he had spoken to her many times on the phone when trying to get hold of me). Patrick was clearly far from being a well man himself, despite his new pacemaker. He was virtually immobile, semi-paralysed due to his spinal problems and barely a month went by without him suffering some sort of fall from his Zimmer frame. During these periods John Fletcher often travelled down to stay with Patrick for a week or two until he felt better. Physically, Patrick was a wreck, but despite this he offered to do anything he could to help me and my parents once he heard that my mother was terminally ill.

As my mother’s health declined and her death inched nearer Patrick was on the phone each week, offering me support. He assured me that: “Take my word for it, it will only be a temporary parting; you WILL meet again and she WILL always be there for you. I am NOT a religious man Martin but, my mother is still there for me and we WILL meet again. I do know this, but I can’t say any more.” Whether Patrick was just being very thoughtful, or whether he had experienced some kind of spiritual encounter, I know not. His religious views were non-existent, but he often qualified this with “We are here for a purpose, I am sure of that.” In addition, after his own mother had died in January 1981 he became very interested in spiritualism and, as I mentioned much earlier, this was in no small way due to his friend Michael Bentine. When my Mum died, a matter of hours after the crystal clear moonrise UK lunar eclipse of September 7th/8th, Patrick’s letter of condolence was one of the first to arrive. As I mentioned at the start of the book, both our mothers were born on June 27th: Patrick’s in 1886, mine, 43 years later, in 1929. By coincidence Patrick’s good friend Douglas (H.J.P.) Arnold died, after a short illness, on my mother’s final birthday. Patrick was very saddened to hear of Douglas' demise. He had been one of the most enduring guests on *The Sky at Night* and had appeared on the programme since the 1960s.

On September 14th 2006 Patrick's house welcomed a special visitor, Mrs Venetia Phair (nee Burney). This was not the first time that Patrick had met the woman who was responsible for the naming of Pluto, over 76 years earlier, when she was 11 years of age! In 1930 Miss Burney had been an Oxford schoolgirl at the time that Clyde Tombaugh had discovered the ninth planet (as it was regarded then). Her grandfather had told her about the new planet and as she knew that Pluto was the Roman god of the underworld and that mythological name started with Percival Lowell's initials (after which Tombaugh's observatory was named) she suggested the new planet should be named Pluto. Venetia's Great Uncle, Henry Madan, contacted a former Astronomer Royal, Herbert H. Turner, with the suggestion. Turner telexed Lowell Observatory and, on May 1st 1930, the Lowell Observatory astronomer Vesto Slipher announced that the schoolgirl Venetia's suggestion of 'Pluto' had been accepted. Just over 76 years later the 88 year old Venetia had a good day out at Patrick's home in the company of a few friends, who included, amongst others, Patrick's helper John Fletcher. She would die in 2009, aged 90.

BANG!

Revitalised with energy after his pacemaker insertion, Patrick resumed work on a new project he had recently started. Impressed by the sales of Stephen Hawking's 'A Brief History of Time' but well aware that most people only survived the first few chapters, Patrick had decided to write his own equivalent. However, cosmology was definitely NOT Patrick's strongpoint, so he roped in Brian May and Chris Lintott as co-authors. According to Brian May's website (brianmay.com) the idea of co-authoring such a book was first broached to him a full 3 years earlier, during the May 31st 2003 Scottish annular eclipse trip, when, during an outdoor snack, Patrick told Brian he must co-author a book with him about 'The Complete History of the Universe'. Despite Brian's scepticism Patrick apparently insisted that Brian would co-author such a book and also should complete the Ph.D. that his Queen career had put on hold! Anyway, the book, entitled 'BANG! – The Complete History of the Universe' went on sale in November 2006 and quickly made it into the top 100 best-selling books in the UK, ultimately selling 180,000 copies. This generated much needed income for Patrick, who was sliding further into debt. Patrick held a party at his house on November 9th, to celebrate the publication of *BANG!* and close friends and contributors to the book were invited. On arrival they were presented with complimentary copies of the book and a free T-Shirt. While the success of *BANG!* provided vital funds for Patrick it was a lavish, pricey, full colour production and in the 21st century most book publishing profits, even for successful books, were way below what Patrick had enjoyed in the late 1960s and 1970s. Yes, the income was welcome, but it would not solve his financial problems. I should add that I provided Patrick with my picture of the three authors together in the book's preface, but never received a single penny for my contribution from Canopus Publishing! Three days before the *BANG!* party, on November 6th, Patrick had presented the British born astronaut Piers Sellers with an Outstanding Achievement Award at the annual *Pride of Britain* awards ceremony in London. Six years earlier Patrick had presented Piers' NASA colleague, Michael Foale, with the same award.

At the end of 2006 Patrick was boosted by yet another 'lunar activity' paper (they seemed to crop up every 5 years or so) that seemed to indicate that the Moon was not totally dead. So, the TLP myth had a another dying gasp left in it, maybe? At the start of the December 3rd/4th late night *Sky at Night* Patrick announced (for the nth time) that now there was proof that the Moon was still active and so he gave TLP yet another plug. However, the evidence was somewhat flimsy. A paper had been published in the prestigious magazine *Nature* by Peter Schultz of Brown University and by two of his colleagues. They reported that the Moon might still be venting gas from its deep interior, near to a 2.8 kilometre wide volcanic depression known as the crater Ina (18.6° N, 5.3° East) which had razor sharp edges and only two associated craterlets bigger than 30 metres in diameter (evidence that Ina was young). Apparently spectra showed that Ina contained numerous patches of freshly exposed material. Not surprisingly Patrick was 100 % convinced that this flimsy evidence vindicated all of his TLP research. Others were certainly not so enthusiastic! On December 15th, after filming for the 650th episode of the *Sky at Night* was completed, Patrick, Jane Fletcher and a few guests celebrated the historic 650 episode milestone with a small party in Patrick's house. 2007 was looming too, and that would be a very special anniversary year indeed. By the end of 2006, in the four years that Jane Fletcher had been in charge of *The Sky at Night*, she had slowly won Patrick's affections over. Admittedly, she had never got him to agree to any background music to the programme, which he would have loathed, but Patrick now called her "Jane, Dear", whereas in 2002 he had simply referred to her as "That WRETCHED woman!" Major Progress!

Chapter 40
50 Years of the Sky at Night

As 2006 turned to 2007, fans of Patrick started to realise that, with luck, he really was going to make it to the 50th anniversary of the programme. With his heart, knee, arthritis and spine problems, not to mention the infamous salmonella goose egg incident, even his most optimistic supporters had started to doubt whether Patrick could survive until April 2007, but as the year opened things were looking good. The producer Jane Fletcher had been heard to say that "Sheer cussedness will get him there."

The TV media latched on to this remarkable anniversary quickly, but it would appear that the BBC 1 programme controllers did not. The first *Sky at Night* for 2007, broadcast on the night of Jan 7th/8th had featured a very special guest, the astronaut Piers Sellers. Patrick was proud to have a 'British' astronaut on the programme, especially one that was a big fan of the great Sir Patrick Moore. In addition, it was the 650th episode of the programme. So what did the BBC 1 schedulers do to mark this occasion? They broadcast the programme at 1.55 a.m! O.K., in truth it was repeated at 8 p.m. the next evening, on a Freeview/satellite channel (BBC 4) and at lunchtime on BBC 2 a week later, but Patrick was appalled and angry. The programme was in its 50th anniversary year, the guest was a British astronaut and so they put it out at 1.55 a.m., after a grotty old American film!

Patrick was quickly onto BBC News 24, live on the phone, even before the broadcast went out. He was not a happy man and made it clear to the rather shaken female newsreader that it was the programme scheduler's fault and, of course, "The scheduler is a WOMAN." The next month Patrick was given a much earlier timeslot and a 1 hour long programme. However, he was definitely NOT going to forget the incident! *The Sky at Night* had always been a labour of love for Patrick and, as I've said so many times, he had never signed a contract with the BBC. It had always been a "gentleman's agreement" and Patrick had (unlike most people as famous as him) no interest in demanding millions. Perhaps because he was feeling like he had been treated shabbily, in early 2007, at a *Sky at Night* filming at his house, he told some leading amateurs "You know what I get paid...have a guess....have a guess....... HAH! it's £250 per programme. Hard to believe isn't it?!" Mind you, many of the amateurs there thought he actually did it for free.

M. Mobberley, *It Came From Outer Space Wearing an RAF Blazer!: A Fan's Biography of Sir Patrick Moore*, DOI 10.1007/978-3-319-00609-3_40,

In an attempt, perhaps, to pacify Patrick he was interviewed by the *Radio Times* (formerly the BBC's flagship TV listings publication) for the February 3rd to 9th issue by the magazine's columnist Benji Wilson. It was, of course, pure Patrick and he managed to get a dig in about the BBC having lost the programmes where he was commentating live on the Moon landings. He did say that the answer to life, the Universe and everything was 43 though, not 42!

Ever since Patrick's split with *Astronomy Now*, in late 1994, he had stubbornly refused to have anything to do with the magazine or its massive annual astronomy and telescope exhibition *Astrofest*. He had never been there after the first 3 years (1992–1994) despite it being the biggest astronomy gathering in the country and despite him being the biggest name in astronomy. However, this year was different. It was a special year for both *Astronomy Now* and for *The Sky at Night*. The programme was 50 years old in April and the magazine was 20 years old and both had originated from Patrick. In addition, the *Astrofest* organisers knew that there was no bigger draw than Patrick; with his new book 'Bang!' co-authored by Brian May and Chris Lintott selling well, having all three there would be a major coup. Of course, there was going to be a problem getting Patrick there in his physical state. He had conceded in early 2007 that pretending to be able to walk, even with a Zimmer frame, was not a good idea. He was now wheelchair-bound. But even in a wheelchair, getting him up to London would be a major undertaking.

Nevertheless, as the organisers, Brian May, Chris Lintott and the 'Bang' publishers (Canopus publishing) were keen to give it a go and because his many fans wanted to see him there, Patrick agreed to travel to the 2007 *Astrofest*. *Astronomy Now* started plugging the appearance of Patrick, Brian and Chris as soon as there was even a hint that they might be there. The December 2006 issue of the magazine alerted potential ticket purchasers that all three *would* be there. In private, Patrick simply stated to me that "I am only going to support Chris and Brian and for no other reason." But even after the three were advertised to appear there were disagreements with the arrangements, made via an agent who *Astronomy Now* were forced to deal with. On January 11th, Chris Lintott's website announced the following:

> A short announcement to say that, contrary to advertising, Brian, Patrick and I will not be appearing at *Astrofest* this year. We hope to arrange some public appearances shortly, however.

This seemed to prompt a very swift compromise from the organizers as, a few days later, Chris Lintott posted an update:

> Update: We have managed to reach an agreement with the organisers. There will be a short presentation, a chance to ask questions and a book signing.

As it turned out the event was a great success. On February 10th Patrick was wheeled up to the stage by the UK supernova discoverer Mark Armstrong, in an impressive looking cogged, tracked and tilting motorised wheelchair and Patrick, Brian May and Chris Lintott plugged *Bang!* and answered countless questions from the audience. Brian paid an amusing and touching tribute to Patrick and some video clips of the funniest moments from *The Sky at Night* were shown, including, of course, the famous 50th *Sky at Night* in George Hole's garden when they were clouded out! At the end Patrick received a rousing standing ovation from the 600 astronomers packed into the hall. It was, most people thought, probably Patrick's last appearance at such an event and it was in the *Sky at Night's* 50th anniversary year. It was an emotional farewell and a few tears were shed as the presentation ended and Patrick said goodbye. However, like Frank Sinatra, it was not Patrick's last *Astrofest* bow. He was back a year later, on February 9th 2008, with another question and answer session and again in February 2009, 2010, 2011 and 2012!

To anyone who knew Patrick well though, the reason he finally agreed to turn up at *Astrofest* had little to do with the fee, or the simultaneous 20th and 50th anniversaries of the magazine and the TV programme, or even the book *Bang!* As far as Patrick's pride was concerned, he *was* 'Mr Astronomy UK' and he should be the biggest name in UK Astronomy and the face on the biggest circulation UK Astronomy magazine too. This latter title had not been his since 1994, but now, 2 years after its launch and with sales of 25,000 issues monthly, he had his face back on the biggest circulation UK Astronomy magazine, *BBC Sky at Night*. He could turn up at *Astrofest*, in front of 600 fans, with Brian May and Chris Lintott alongside him, with a smug look on his face, as if to say: "Even after thirteen years and in a wheelchair, I still don't give up.........your days are numbered......I'm the Daddy.......my new magazine outsells yours."

In the same month, to celebrate the impending fiftieth anniversary of *The Sky at Night*, the Royal Mail issued six stamps containing images of deep sky objects which Patrick had included in his Caldwell catalogue. The objects were the Saturn Nebula, the Eskimo Nebula, the Cat's Eye Nebula, the Helix Nebula, the Flaming Star Nebula and the Spindle Galaxy. These corresponded to Patrick's Caldwell numbers C55, C39, C6, C63, C31 and C53 and the C numbers actually appeared on the stamps. A disaster was narrowly averted when, just before they were issued, various amateur astronomers contacted the Royal Mail to tell them that they had mis-labelled objects on the Flaming Star Nebula stamp. Although the stamps did not actually mention Patrick or his TV program, just having his Caldwell numbers on British postage stamps was quite a source of pride for him.

With the 50th anniversary date of *The Sky at Night* approaching and a date now set for a BBC funded bash in Patrick's garden (April 24th, exactly 50 years after the first broadcast) the members of the South Downs Astronomical Society decided to organise a special tribute for invited guests inside the South Downs Planetarium at Chichester. The date of March 17th 2007 seemed most appropriate, as it was well clear of the April celebrations and was fairly close (13 days later) to Patrick's 84th birthday. It was also, highly appropriately, St Patrick's Day! Patrick had actually held a lunar eclipse star party in his garden on the evening of March 3rd, in the hours before his birthday. Many friends, including Brian May, turned up, armed with telescopes and the total phase ended just 1 minute before Patrick officially became 84.

Patrick was unaware of the impending South Downs Planetarium celebration (although highly suspicious) but, of course, he had to be informed that his presence was required inside the 'Sir Patrick Moore building' on St Patrick's day. It was a great day out, full of Patrick's genuine friends who had been carefully selected to give him the most pleasure. Many of the people there went back a long way with

Patrick. John Mason, one of the organisers and trustees of the planetarium, was a major player in organising the day. Friends from as far back as the 1950s, like former BAA Saturn Section director Gilbert Satterthwaite, were there. Other BAA stalwarts, like Mike Maunder and his wife Wendy, turned up and Patrick's old mate from his Armagh days, Terry Moseley, made it from Northern Ireland. Entertaining tributes were paid inside the Planetarium dome, by John, Terry, Allan Chapman and Scotland's Astronomer Royal and part-time magician, Professor John Brown. At the end of the formal proceedings Patrick's original 1987 Madame Tussaud's waxwork, recovered by the new Planetarium, was unveiled to much amusement from the audience. Best of all, Patrick was not expected to do much, except just lap up the tributes!

Fig. 40.1 Patrick at a joint surprise celebration of his 84th birthday and 50 years of *The Sky at Night* at the South Downs Planetarium, Chichester, on March 17th 2007. Allan Chapman and Rachel Chapman toast his 50 years on TV. Patrick's 1980s waxwork and a female astronaut mannequin also attended the meeting! (Photograph by the author)

The 50th Anniversary

The BBC screened a total of three TV shows in 2007 to celebrate *The Sky at Night* programme's 50th anniversary. The first, appropriately screened on April 1st, featured impressionist Jon Culshaw impersonating Patrick, as he would have looked on the first transmission date of April 24th 1957. The real 2007 version Patrick was able to converse with his younger self via a special crystal ball time-travel device. For me, as someone very familiar with Patrick's life, the most memorable part of that programme was the mention of the Mare Orientale. Patrick knew, in 2007, that the imminent BAA paper by his former mates Richard Baum and Ewen Whitaker, on the subject of who really discovered the Mare Orientale, would contain no mention of him. Yet Patrick, in past years, had declared that his mentor Percy Wilkins had discovered it, then that he had been working with Wilkins who discovered it, then that he, Patrick, had discovered it (and he threw in the discovery of the limb crater Einstein for good measure). As we have seen on several occasions throughout this book, in the late 1940s and 1950s Patrick and Wilkins had been the most ardent and vocal enthusiasts about the Mare Orientale and the libration regions, but neither had actually *discovered* the Mare Orientale. Wilkins had, possibly, independently *re-discovered* it, but that was as much as one could say.

Patrick managed to get in a pointed dig at this stage of the programme. When his 1957 persona played by Jon Culshaw mentioned he had been observing the Mare Orientale, the real 2007 Patrick offered a word of warning to him, namely that "The Americans would re-discover that." "Re-discover it?" Culshaw repeated with a puzzled look on his 1957 Patrick clone countenance. Patrick knew that with one million people watching that *Sky at Night*, he had a lot more clout to get his viewpoint over than did the authors of a paper that would be read by only 3,000 people. The 'Americans' jibe was a dig at his old friend and adversary Ewen Whitaker, who now lived in the USA but had known Wilkins and Patrick as a young man. Also in that programme, an artificially aged Brian May and Chris Lintott, on Mars in 2057, appeared alongside a disembodied head of Patrick who had been "Uploaded to the BBC computer in 2015". It was an entertaining spoof and the first of the three programmes which marked the anniversary. The two other programmes were a tribute to the 50th anniversary consisting of a (rather hidden in the schedules) 'We love the Sky at Night' programme, where big name celebrities said what they thought of Patrick and then the film of the 50th Anniversary party in the huge marquee on Patrick's lawn.

The Party

The 50th anniversary *Sky at Night* Party, held on Tuesday April 24th 2007, was a big event. Strictly speaking it was a 'BBC guests only' event although, as always, there were more people there than officially invited. In addition, Patrick extended the event to the Wednesday where he, not the BBC, would be in charge. By BBC budget standards this was a major undertaking. For the 40th anniversary the BBC had funded a party at the BBC TV Centre and Patrick had funded one at his house the following weekend. For the 45th anniversary Patrick had funded a massive party at his house, with a gigantic marquee on the smaller west lawn. As we saw earlier, he had strongly criticised the BBC for not supporting that event. For the 50th anniversary the BBC arranged for the erection of a massive marquee on the much larger east lawn at Farthings. This marquee was so huge it went right up to the old site of the rotating octagonal observatory and an extension to the tent on the north side included Patrick's entire 12.5-inch telescope and run-off shed, as part of the studio backdrop. This, along with a replica of the 1950s and 1960s backdrop planisphere and the somewhat hilly floor (laid over Patrick's somewhat hilly lawn), created a dramatic and surreal atmosphere.

On the morning of the party, as usual, Patrick was roused at about midday. He often stayed awake till the early hours, but slept till noon. I was told by his closest helpers that Patrick always slept naked in his later years and so getting him out of bed in his eighties was not for the squeamish. Now around 20 stones, Patrick was not a pretty sight and he had to be washed, his bed sores greased and cleaned and then hauled from his slumbers using a mechanical crane-like winch mounted above the bed. For no obvious reason Patrick always referred to the winch as 'George!' On that Tuesday morning, midway through the cleaning and greasing process, a phone call came through from a woman at BBC Wales, wanting Patrick's comments about his 50 years on TV. A friend of Patrick's, Peter Cattermole, who was in his bedroom at the time, told me: "She must have had no idea she was talking to a stark naked 20 stone Patrick with all his bits hanging out!" I heard a similar tale from the astro-photographer Dave Tyler, who some months later had arrived with the world renowned planetary imager Damian Peach, at Patrick's home. The front door was nearly always open and they went in and shouted for Patrick. To quote Dave precisely:

> My funniest moment at Patrick's was when Damian and me went in and there was no one about; so we called upstairs and he shouted back to come on up. We went up the stairs and stood in his bedroom doorway and he leaned around his wing-back chair, which had its back to us. He was watching telly STARK NAKED! We said we'd go down and wait for him. "OK" he said. His helper was getting him dressed but had gone somewhere.

At the 50th Anniversary party many video projections of the good and the not-so-good and the famous, paying tribute to Patrick, were played early on in the evening. These included Arthur C. Clarke, Brian Blessed, Jonathon Ross, Richard Branson, Jon Culshaw, MP Lembit Opik (or Optic Lemsip if you prefer) who was there in person with his 'Cheeky Girl' (!) companion Gabriella Irimia, percussionist Evelyn Glennie, MP Charles Kennedy, Discworld author Terry Pratchett, John Craven, Adam Hart-Davis, Myleene Klass, Alex James (bass guitarist from the band *Blur*) and others. A near infinity of people presented awards, verbal tributes and gifts to Patrick during the day with, most of the time, the astronaut Piers Sellers acting as the anchor man. Various people posed for pictures of themselves alongside celebrities such as the *Rolling Stone* Bill Wyman and Myleene Klass. What on Earth was Patrick's connection with Bill Wyman you might ask? Well, a track of Patrick playing the Xylophone had been included on Wyman's 2005 DVD entitled *Digital Daydreams*. No, honestly, I'm not making this up!

All five living *Sky at Night* producers were at the party too, namely Jane Fletcher, Ian Russell, Pieter Morpurgo, Pat Wood and Pat Owtram. Jerry Stone presented the Arthur C. Clarke award to *The Sky at Night* and the space artist David Hardy presented an 'eclipse clock' to Patrick. Terry Pratchett asked if he could name his observatory after Patrick, who, of course, agreed. In addition, the British Astronomical Association's youngest member, Philip Jennings, provided Patrick with a card containing messages from well-wishers, on behalf of the BAA. A big map of the Moon, published by Patrick, was sitting in Patrick's hallway for people to sign, but unusually the BBC had hired security guards to keep people from venturing further into Patrick's house. On this subject Patrick commented that the

BBC had been "watching too many American films." Asteroid expert Professor Iwan Williams then announced that, in honour of the *Sky at Night*, the International Astronomical Union had named asteroid 57424 (the same digits as 1957 April 24th) as Caelumnoctu, which was Latin for 'Sky at Night'. The asteroid had been discovered on September 16th 2001 by the LINEAR telescope in Soccorro, New Mexico.

Fig. 40.2 Patrick and the 4 previous *Sky at Night* producers at the 50th anniversary BBC party at Selsey on April 24th 2007. Patrick seated (sort of!). Standing, from *left to right*: Pat Owtram, Pat Wood, Pieter Morpurgo and Ian Russell (Photograph by the author)

The BBC 1 controller Peter Fincham also said a few words, introducing himself as "the man who paid for this event" which got a big cheer. He said that he had recently attended a celebration of the BBC comedy *Last of the Summer Wine* which would soon be celebrating 35 years on TV and that seemed a remarkable enough period of time. Fortunately for Patrick and all of the guests who had been at the 40th *Sky at Night* Party at the BBC TV Centre in 1997, Mr Fincham did not repeat the dreadful gaffe of his predecessor Alan Yentob, of continually saying "astrology" when he meant "astronomy!"

The 50th anniversary party continued (without the BBC there) the next day with more friends coming and going, including Brian May the Queen guitarist. Unfortunately, following a BBC row about how the real Queen was misrepresented in a BBC documentary, Fincham would resign as the BBC controller less than 6 months later. He had lasted 18 months in his top BBC job, while Patrick had survived more than 50 years in his.

The *Sky at Night*, during Jane Fletcher's reign as its producer, was edited at BBC Birmingham and, with that Midlands connection, Patrick ended up with yet another

Fig. 40.3 Jerry Stone presents the Arthur C. Clarke award to Patrick and Jane Fletcher at the 50th anniversary BBC party at Selsey on April 24th 2007 (Photograph by the author)

Fig. 40.4 Two of Patrick's most favourite people: Peter Cattermole (*left*) and Pieter Morpurgo (*right*). Both attended the 50th anniversary *Sky at Night* party in April 2007 (Photographed here by the author 1 year later in Moscow, on a Jules Verne eclipse trip)

plaque to put in his study to mark 50 years on TV. The Royal Television Society (RTV) Midland Centre awarded Patrick their 2007 Baird Medal to 'recognise a truly outstanding and unsurpassed contribution to television within a Midland based programme.'

So, remarkably, Patrick had survived the half century *Sky at Night* celebrations and awards despite heart, knee, arthritis and spinal problems. But was he going to mellow and rest on his laurels after that milestone was passed……ABSOLUTELY NOT! He made the headlines a few weeks later, in mid-May, because of his comments about women and how they were wrecking TV, specifically 'jokey women newsreaders' and the fact that the BBC was now 'run by women'. He said that he was still not over the fact that the 650th episode with astronaut Piers Sellers had been put out at 1.55 a.m. in January. Patrick complained, in an interview for the *Radio Times*, that the whole of TV these days was just soap operas, cooking and quizzes and he thought men and women needed different channels entirely. When speaking on the subject of *Eastenders* he described it as 'Diarrhoea……..which may be true to life, but not what you want to see on TV.' As a final comment he said he would rather be dead in a ditch than be a celebrity in the reality show 'Big Brother.' Even his co-presenter Chris Lintott felt obliged to disassociate himself from Patrick's latest attack on women, in his blog at http://chrislintott.net. A BBC spokesman commented: 'He is, as always, forthright in his personal views and that's what we all love about him.' Once again, his favourite newspaper journalist (NOT) Penny Wark, of *The Times*, responded with a piece on Patrick in the May 9th edition of that famous newspaper, entitled: 'Sir Patrick Moore: a dark cantankerous star.'

In June 2007 the British Astronomical Association Journal, containing the article about the real history of the Mare Orientale, was published. This, of course, was the limb region on the Moon which Patrick's mentor Wilkins claimed to have discovered, then Patrick claimed to have discovered, but in fact it had been discovered much earlier. By all accounts Patrick was not amused, but as the authors Baum and Whitaker had not called him a liar, he could possibly just have shrugged the article off. As I was one of the referees of that paper I half expected a phone call from Patrick, and to be labelled as a 'serpent', but it never happened. However, Patrick decided not to ignore it and he wrote a very brief and carefully worded letter to the BAA Journal, which would be published 4 months later in the October edition. His letter read as follows:

> My first drawing of the Mare Orientale was made in 1948, and I reported it to the BAA Lunar Section, because it was not on any of the maps in my possession and I thought (naively) that it was new. I did believe it to be a limb-sea of the Mare Smythii type, and I discussed it with Wilkins – suggesting the name 'Mare Orientale' or the Eastern Sea. It seems therefore that others had seen the edge of the sea before I did in 1948. I didn't have Franz' map in 1948, and I can't read German anyway.

Anyone having read this book will know that this letter was really a total load of utter hogwash. Patrick must have known the true story of the Mare Orientale's discovery for almost seven decades, not least because his mentor Wilkins wrote a paper on his claimed discovery of it in the December 1937 BAA Journal and it was declared by the BAA Lunar Section Director MacDonald, 4 months later in April 1938, as being 'described once before by Franz!' Patrick had seen the feature himself, after reading

Wilkins' paper, as early as 1940, not 1948! Indeed, Wilkins had laid claim to it as his own discovery, for two decades before Patrick decided he would claim it for himself, following Wilkins' death! Neither man could have been ignorant of the earlier observation of the Mare Orientale by Franz. So, to those of us in the know, Patrick's letter implying the Baum/Whitaker paper had suddenly revealed new evidence to him is extraordinary! A man with Patrick's fascination with the Moon must have known that neither Wilkins, nor himself, could conceivably have any claim to the Mare Orientale's discovery. It would probably have been better if Patrick had just ignored the 2007 Baum/Whitaker paper and not written a response at all.

While Patrick's letter to the BAA Journal was presumably on its way to the editor, on the July 1st/2nd edition of *The Sky at Night* the program finally featured the 91 years old John Dobson, footage of whom, in Patrick's garden some 5 years earlier, had never been aired. But this time Chris Lintott had been to see the legendary inventor of the 'Dobsonian' telescope in Monrovia, California. Patrick appeared to have a new monocle on that program, featuring some kind of permanent adhesive to keep the thing from falling out. When I next saw Patrick he said that the TV people had been frustrated with the monocle constantly dropping from his eye socket and so "various types of sticky stuff that stuck to flesh" had been tried out on the interface between eyebrow and monocle rim and "now we've got the problem licked!"

The Sunday/Monday night of August 12th/13th saw yet another 'open night' at Selsey, because it was the maximum of the Perseid meteor shower and at a New Moon too, so skies would be nice and dark. A host of neighbours, friends, astronomers, fame-seekers and celebrities descended on Patrick's back lawn and Patrick was wheeled outside too. The celebrities included guitarist Brian May, impressionist Jon Culshaw and Dr Allan Chapman. One highlight during the evening was Allan Chapman's attempts at getting onto a low sunbed in the dark. Apparently it was his first experience of a sunbed and after several failed attempts and curses and then expressing his need for lessons in 'climbing into a bed', Patrick bellowed into the night "Ask a German!"

For the October 7th *Sky at Night* Patrick was transported to Jodrell Bank to mark that telescope's own 50th anniversary and meet up with his old friend Sir Bernard Lovell. With Patrick now 84, and Sir Bernard 94, it must have been the oldest TV interview ever conducted! Patrick had been visiting the Jodrell Bank facility even before the main dish had been constructed and before his *Sky at Night* programme had started, when the main role of that facility had simply been recording meteors from their effect on the ionosphere.

On the night of October 23rd/24th 2007 a truly amazing celestial event occurred: the comet 17P/Holmes suffered a colossal outburst in brightness, unprecedented in comet observing history. Well, it was unprecedented except for the same comet doing the same thing in the autumn of 1892 and early 1893! It brightened almost a million-fold. Despite now being wheelchair-bound and with his comet observing days surely over (so he had thought) Patrick was wheeled out of his front door by the *Sky at Night* regular, Selsey resident and amateur astronomer Pete Lawrence, where he could view the naked eye comet for himself. It was still possible for Patrick to make astronomical observations, despite the appalling state of his body! You just never knew what was coming next in astronomy.

Chapter 41
Still Alive Despite All the Odds!

A Chat with a Time Lord and Another Orientale Apology

Many people would have bet serious money on Patrick never making it to that 50th *Sky at Night* anniversary year of 2007, but as things turned out he sailed straight through it. "Surely", people were saying, "This man has nuclear hardened, military grade DNA; every possible thing is wrong with his body but he just keeps on going!" He was still in demand too, even occasionally by advertisers. As I have mentioned before, Patrick held the view that product endorsement for money was totally immoral. If he thought something was good he would freely say so, but surely if you pocketed the tens of thousands of pounds on offer this was not only rampant greed, but accepting a bribe as well? Despite fellow celebrities 'filling their boots' in this way Patrick held firm, but in old age he did agree to do a few adverts, where he was just an actor, or doing a voiceover and where his fee was donated to charity.

One example of this was the January 2008 voiceover he recorded for Post Office Savings. In the advert, heard on many commercial radio stations, Patrick played a Post Office customer approaching a cashier called Ken, played by the *Early Doors* actor John Henshaw. As Patrick approached the desk Henshaw was heard to say "Ohhhhh, it's The Master of the Universe!" Patrick then says "Ken, the stars seem to be telling me that your Instant Saver account is one of the best on the High Street?" Henshaw replies "Spot on clever clogs; it has a gross interest rate of 5.75 %". Definitely the most amusing part though was the ending, where a digitally speeded up Patrick then rattled out the small print terms and conditions, at twice the speed even he spoke in his prime. One month later, on February 9th 2008, and for the second year running, Patrick once more appeared at *Astrofest*, in a wheelchair, although there was no mention of savings interest rates there, just astronomy.

In 2008 there seemed to be even more *Sky at Night* specials on the Freeview/Satellite channel BBC 4, which already carried the extended version of the monthly programme. Admittedly many of these 'specials' were actually just off-cuts from the hours and hours of surplus programme material filmed by Jane Fletcher for the normal programmes. Nevertheless, whichever way you looked at it, Patrick seemed

M. Mobberley, *It Came From Outer Space Wearing an RAF Blazer!: A Fan's Biography of Sir Patrick Moore*, DOI 10.1007/978-3-319-00609-3_41,

to be increasing his air time as his body got more and more frail. Like the Russian presidents of the cold war era he seemed to just be dusted off and propped up for the cameras once a month, but spent most of his life being hauled out of bed, then washed and fed, by various carers.

As I have already mentioned, the Queen rock guitarist Brian May had offered to alleviate Patrick's financial problems by paying carers fees and raising equity from purchasing Patrick's two-thirds of an acre estate of Farthings, while still allowing him to live there until his death. Patrick was living far longer than anyone had expected for a man in his semi-paralysed state, with a lifelong heart condition and the cost of looking after him was relentless and frightening. So, equity release on the entire property was inevitable. Patrick officially sold 39 West Street to Brian on February 13th 2008 for £480,000. Its market value would have been far greater if the house (and especially the thatched roof) had been in good condition, but it had become rather dilapidated in recent years and of course the actual sale price was academic. Patrick simply wanted to live there until he died and present *The Sky at Night* from his home too. With the rent charged by Brian being "One peppercorn when demanded" this was now feasible. It was a huge relief to Patrick. At last his financial problems had gone away.

Patrick's body was in a truly appalling state by February 2008, but nevertheless his brain was as active as ever and so when the impressionist Jon Culshaw phoned him up live in front of a studio audience that month, pretending to be Dr Who in the style of Tom Baker, Patrick played along from the very first second that he picked up the phone!

This is what the audience heard:

Sound of Patrick's phone ringing…………

"Hello this is Patrick Moore's ansafone. Sorry I can't take your call at the moment. If you leave your name and your number I'll ring you back as soon as I can. This is Patrick Moore."

[Beepabeepabeep. Rapid beeping tone, then Patrick picks the phone up himself]

PM: "Hello"

JC: "Um Hello"

PM: "Hello"

JC (impersonating Dr Who Tom Baker): "Oh hello, is that Patrick, Sir Patrick Moore?"

PM: "Yes, who's there?"

JC: "Oh, Sir Patrick, My Dear Fellow, I AM…. THE DOC-TOR….."

PM: "Yerrss." [Laughs from the studio audience]

JC: "Oh, hello, can I just say before I make my request, you certainly are one of the greatest Earthmen. I've admired your pioneering work for many years."

PM: [Laughs something inaudible] "What can I do for you?"

JC: "Well, my dear fellow I just thought that you're the best person to ask for a little advice. You see, what it is, the Time Lords have just sent me on a mission. I've got to go and thwart Sutekh."

PM: "I beg your pardon?"

JC: "Sutekh, Last of the Osirans. He's trying to break free from his ancient bonds deep in the eye of Horus, so that he can break free and destroy the world. Anyway, to stop him I need to find a good place to land the Tardis on Mars and, er, Patrick I wonder if you can think of anywhere suitable."

PM: [Instantly] "I would say probably somewhere in the Syrtis Major I would think, that's a good area." [Laughs from studio audience]

JC: "But it can be affected by the most terrible sandstorms."

PM: "Ah no, but all Mars is like that. Syrtis Major is a good one, near the equator, due south of Hellas somewhere."

JC: "Yeah, you know in one of my former selves, when I had to defeat the Ice Warriors I landed near the Valles Marineris."
PM: "Ah, not, not, not, not a good area, I recommend the Syrtis Major, you want somewhere north of the Hellas basin." [More laughs from the studio audience]
JC [Laughing]: "Yes, that's probably right actually. Thankyou. What if I was to land north of Olympus Mons?"
PM: "Well, you could do, getting a bit down near Tempe. No, I would certainly go for Syrtis Major, due north of Hellas, I wish you the best."
JC: "Patrick that information is invaluable to me……"
PM: "I'm delighted."
JC: "….and can I say thank you for all your pioneering work mapping the Moon. That helped us defeat the Cybermen."
PM: "Yes, I'm glad about that."
JC: "Well thank you very much indeed Sir Patrick for that information on Mars, that's invaluable, with that information we can certainly thwart Sutekh."
PM: "Delighted to hear it."
JC: "Have you heard of the planet Gallifrey?"
PM: "Of course I have; 84 light years away, close in the sky to Epsilon Eridani."
JC: "Oh, absolutely, in the constellation of Casterbus."
PM: "No, no, no it's in Eridanus. No your maps are wrong." [Laughs from audience]
JC: "Even the Time Lords can be wrong you know."
PM: "Not orften, but they can be."
JC: "Goodbye Sir Patrick. Take care my friend."

On the night of February 20th/21st the same Jon Culshaw (being himself this time), Terry Pratchett and Allan Chapman were just three of the guests who had assembled on Patrick's lawn to watch a decidedly cloudy total lunar eclipse for *The Sky at Night*. In that same week Patrick was appalled to hear that an 88 year old former WWII RAF Wellington pilot, Jack Tagg, had been told that he would have to go blind in one eye before he could get NHS treatment! Patrick sent a 'substantial' donation to Mr Tagg, saying that he was 'furious' at the lack of NHS treatment. Speaking in the *Mail on Sunday* Patrick said: 'Here we have a man who fought for us against Hitler, who's going blind, whose sight can be saved by injections and the people won't pay for it. I want to launch a nationwide appeal to save it.'

When *The Sky at Night* covering the February total lunar eclipse was shown, Patrick, speaking to Allan Chapman, seemed to feel he had to admit, on TV, that he had not discovered the Mare Orientale or the crater Einstein. With Allan holding Patrick's lunar observing notes from the 1940s, Patrick said:

"I'll put one thing right I may say: I've been quite innocently claiming a discovery I didn't make. In 1948 I was observing the Moon and on the edge of the Moon I saw something that hadn't been seen before. It looked like the edge of a small Mare or Sea and I called it the Mare Orientale – there's my drawing of it, and I thought then that it hadn't been seen before…wasn't on any of my maps…must tell you I found out it *had* been seen before…it's in a German Atlas of 1909 of which I was totally ignorant, so I claimed a discovery I didn't make …and there's another crater on the edge there….then called Caramuel, and now called Einstein – This is Einstein right on the edge of the Moon and I thought this was new…..and it may have been…I don't know, not that it matters…….."

Hmmmm. I think it would have been far better if he hadn't dug himself into an even deeper hole on this issue!

Arthur's Gamma Ray Exit, Metrication and Big Brother!

On the night of March 18th/19th 2008 Patrick heard the sad news that one of his longest surviving friends, the legendary Arthur C. Clarke, had died, aged 90, in Sri Lanka. He was phoned by BBC News 24 live that night and said:

"He was ahead of his time in so many ways. Quite apart from artificial satellites there were other things too. A great science fiction writer, a very good scientist, a great prophet and a very dear friend, I'm very, very sad that he's gone."

As if to signify Clarke's passing (Patrick said Arthur must have arranged it) a host of high energy Gamma Ray Bursts were detected at the edge of the observable universe within hours of his death, including the highest energy event ever detected! On a completely different topic, if anyone ever brought up the subject of the allegations that Clarke was a paedophile (claimed by the Sunday Mirror in 1998) Patrick would simply retort something like: "Hah! Someone tried that on me too. If you're famous and single they'll come up with that one at some point." As it had turned out the Sri Lankan police found that there was no substance at all in the claims made against Clarke, clearing the way for him to receive his knighthood. The day after Arthur C. Clarke passed away Patrick was driven to the new Intech Planetarium at Winchester, where the bestselling author Terry Pratchett and Patrick were the chief guests at the opening of the new facility.

While it was goodbye to an old friend it was welcome back to another a few weeks later, as Patrick's 5-inch f/12 Cooke refractor, which had been away for 5 months being cleaned and restored by Steve Collingwood of Broadhurst Clarkson and Fuller, finally returned in a gleaming state to his Selsey garden. It had become a bit of a rust heap in recent years and although Patrick was now unable to use his telescopes the refractor was often used by visitors to his garden during televised Star Parties. Collingwood performed the same service on Patrick's 3 inch childhood refractor 2 years later. It was all good publicity for BC&F and its proprietor, Patrick's old mate Dudley Fuller.

In May the *Sky at Night* held a rather spooky programme entitled 'What we don't know' to celebrate program number 666. A week later he appeared on TV inspecting an astronomical garden (designed by Barry Mayled) at the Chelsea flower show, along with Brian May. The garden featured rocket-shaped conifers and galaxies of colourful bedding plants. Patrick commented that "This beautiful garden shows how even a comparatively small area can be transformed into an amateur astronomer's lair."

At the end of the month, with Patrick as the guest of honour, Bruce Kingsley officially re-opened the renovated dome and 8.5 inch Newtonian which Patrick had previously owned for almost 50 years. This ceremony took place at Bruce's observatory at White Waltham near Maidenhead. The telescope, with its With/Wildey mirror and Browning mount, had never looked in such excellent condition and Patrick was very pleased that he had given it to Bruce.

In August 2008 I visited Patrick as he had asked me to talk about the recent Total Solar Eclipse, which I had observed from Siberia, on *The Sky at Night*. I had an amusing chat with him about his own 1968 trip to see a Total Solar Eclipse there.

Patrick proudly showed me his new music room coffee table: a 4 foot diameter glass table top supported on a perfect bronze replica, almost 2 feet high, of the Apollo Lunar Module. It was quite spectacular. I nosily asked him what it had cost and he said "It was free, made for me by a fan, the designer Mark Stoddart!" He was as happy as I had seen him in a long time. Apparently, Buzz Aldrin owned an identical Stoddart creation. Patrick had also arranged for his publishing friend Robin Rees to help reprint some of his old 1950s Sci-Fi novels like *Quest of the Spaceways*, *Crater of Fear*, *Peril on Mars* and *The Frozen Planet*, to raise money for a good cause. He was proudly showing the books to anyone who turned up at his house and Brian May drew attention to them on his own web site.

Patrick also told me that he was delighted he had recently managed to persuade the International Astronomical Union to name three asteroids (also called minor planets) after friends of his; this had only been confirmed 2 months before I visited him in August. The names were announced on consecutive Minor Planet Circulars 63172 and 63173. He was particularly pleased that the minor planet 6137 was named 'Johnfletcher' after his friend, the retired postman John Fletcher, who had been a frequent visitor and helper at Patrick's house, ever since he had suffered his debilitating spinal problems. The other two names were 'Lintott' (after Chris Lintott) allocated to minor planet 4937 and 'Brianmay' (after Brian May) allocated to minor planet 52665. To have three of your suggested names all accepted in that June 18 batch of Minor Planet Circulars was quite an achievement. In September, Patrick's house was, once more, full of astronomers and a few TV celebrities, including Jon Culshaw, when yet another *Sky at Night* Star Party was held at Farthings.

During 2008 Patrick had been back on his hobby horse about scrapping metric units and getting back to the old imperial measurements. Incensed that it was illegal for British shops to sell items in old pounds and ounces he and his collaborator Roger Prout wrote a letter to the *Daily Telegraph* which was published on October 26th:

> **Rendering metric enforcers harmless**
> My collaborator Roger Prout and I have come up with a solution to enforced metrication, which remains the law (report, October 19). Traders would have their archaic weights engraved with a metric measure. If you went into a shop to buy 4lb of potatoes, a large notice above the counter would inform you that 4lb is 1.816 kilograms. You can't order 4lb – naughty, naughty – so you would order 1.816 kg, and the vendor would use his 1.816 weight. That this is exactly the same as 4lb is mere coincidence. The prosecuting officials would be helpless. We would not have made them look fools (nature has done that already) but we would have rendered them harmless.
>
> Sir Patrick Moore
> Selsey, West Sussex

The first 2 weeks of December 2008 saw one of the most ludicrous periods of newspaper, TV and website journalism regarding Patrick (or anyone) that I can ever remember. As we saw earlier, Patrick hated modern television. He hated soap operas, female newsreaders, smut, the lack of any moral standards, programs about ethnic minorities and, above all, reality TV programs. So you can imagine the reaction of his fans when it was widely announced that the 85 year old, pacemaker powered, wheel-chair bound Patrick would be appearing on Channel Four's 'Celebrity Big Brother', along with a 2 foot 8 inch tall dwarf, a rapper, a cocaine addict and a female performer with the group *The Satanic Sluts*! One of the websites described Patrick as 'very spry for his age.' Hmmm, 20 stones, paralysed and having to be winched out of bed every day? Errrmm, no, I don't think so somehow! Anyone who knew anything about Patrick knew that there was no way he would appear on such a program, for *any* financial reward! I knew exactly what Patrick would say about this: "I would Rather be Dead in a Ditch!" Well, I have no doubt the experience would have killed him anyway. I was just about to phone Patrick on December 15th, as the media websites were so insistent that he was appearing on the programme, when I decided to check the website of Patrick's friend, Brian May. Well, he hadn't believed it either and so he had already phoned Patrick up. Brian said, on his website, on that same day:

> I have been asked by so many people today about Patrick Moore (apparently) going on some TV show called Celebrity Big Brother. I was shocked, but of course was not inclined to believe anything that the Newspapers print, from bitter experience! Having just talked to the inimitable Dr. Moore, I am pleased to be able to relate to you all his comment ... he says simply: "I Would Rather be Dead in a Ditch!"

Phew! My sanity was restored and Patrick's comment was exactly as I had expected! But surely there had never been such rubbish and downright lies talked by the media at any point in Patrick's 50 years in the spotlight? At the start of 2009 Patrick, once more, agreed to appear at the annual 'Astrofest' in London, on February 7th, despite only just having recovered from a severe cold. His popularity at this event in the previous 2 years must surely have pleased him greatly. This time he would be alongside his regular 1970s, 1980s and 1990s guest on the *Sky at Night*, Dr Garry Hunt, as well as the nonagenarian Reginald Turnill, who had been the BBC Space correspondent from Patrick's prime Apollo years. The main topic was that 2009 marked 40 years since Apollo 11, but a question and answer session followed with many interesting questions from the floor, covering everything from the future of lunar exploration to the reason for methane being found on Mars.

On March 11th 2009 Patrick appeared live on the BBC program *Daily Politics*. He was invited onto the programme (part of which featured pensioners from the Bournemouth area) because he was known to be angry that the ban on fox hunting was not being enforced properly. Hunt supporters were still roaming the countryside on horseback, hunting dogs in attendance, but they were claiming foxes were not being killed, and no-one had been prosecuted. Also on that programme were the Labour MP Liam Byrne, who had voted for the fox hunting ban, and Conservative Party Work and Pensions spokesperson Theresa May, who had voted against it. Patrick's main ally in Conservative ranks, on this subject, had been Anne Widdecombe. Patrick had, on numerous occasions, criticized the previous Archbishop of Canterbury, George Carey, for being a patron of the RSPCA while refusing to condemn fox hunting. This time Patrick directed his wrath at the present Archbishop of Canterbury, Rowan Williams, who Patrick said had also refused to condemn fox hunting. Patrick suggested: "Rather than running the Church of England I think that be-whiskered old coot would be better off running a whelk stall!" He directed more straight language at Theresa May, interrupting her in full flow. "May I ask you one question? There is a law there that was intended to stop cruelty to animals; it's not done that. Are you really happy with that, because if so, frankly, you should be ashamed of yourself?" Patrick added: "Whoever the next government is they must put this act in order, to stop these unpleasant people wriggling through, even if they cover themselves in slime." When Ms May tried to offer an answer, Patrick's reply was blunt: "Dear lady, this is an adult discussion, please don't be childish." Rather predictably the 86 year old Patrick also got the final words in: "We'll win in the end, we must do. Decency must prevail in the end, despite all these unpleasant people."

On March 31st 2009, *The Sky at Night* producer Jane Fletcher suddenly resigned, stating in an e-mail to all of the programme's contributors that: 'After 20 years of being at the BBC, I am leaving to spend more time with my family. Mark Scott is the new producer for the programme.' Jane had been the sixth *Sky at Night* producer and had been in charge for the last 7 years, in fact the whole of the era where Patrick no longer traveled to the BBC TV centre in London, but instead filmed each programme from his home at Selsey, under direction from BBC Birmingham. Few female producers would surely have put up with the cantankerous misogynist that Patrick occasionally became and *Sky at Night* viewers will forever be indebted to Jane for cheerily persevering with her job, despite Patrick's verbal abuse! Having said all that, I never met any woman who knew Patrick well who actually disliked him, whatever he said! Deep down they knew he was a kind and generous man, with Victorian standards of fair play and decency, who wouldn't hurt a fly; in fact many women actually agreed with his views on falling TV standards due to endless cookery programmes, soap operas and 'jokey' women newsreaders. Many who had met Jane were very sad to hear that she was leaving and Patrick was no exception. However, by the August programme the new man Mark Scott was no longer listed as the series producer. Paul Vanezis acquired that role, alongside executive producer William Lyons, with Jane Fletcher reappearing as the permanent series consultant. Patrick admitted that he was very relieved Jane was back in the team; not as the series producer, but at least she was involved again.

Around this time, after having had a variety of live-in carers, some good, some not so good, some very dodgy, over the last few years, Patrick acquired a more permanent carer named Dawn. She seemed more focused and attentive than any of the previous incumbents and made sure that Patrick had some help with phone calls and answering letters too. Patrick told his closest friends that, after talking to Brian May, it had been agreed that his new carer could live in *Farthings* after Patrick's death and this was now mentioned in his will. While this was, to some, an unexpected development, Patrick, now totally dependent on full time care, thought it was the best plan for him. Mind you, as far as Selsey's residents and Patrick were concerned, whatever help he needed would always be available; carers in residence or not. In extreme cases good friends like John Fletcher (the former Gloucestershire postman) would travel hundreds of miles just to look after Patrick's cats, if he was in hospital.

One good example of how the West Sussex locals regarded Patrick was his own Selsey postman, Adrian. Most Royal Mail postmen or postwomen shove stuff through the letterbox and that is pretty much the end of the story. Patrick's postman had a key to enter Farthings and was welcome to make himself a cup of tea in the kitchen; that was usually just before he went out into the garden to read the rain gauge and max/min thermometer, so that Patrick could send the data to the Met Office! Back inside he would open all of Patrick's mail and mark it in importance, sort it into piles and, if one of the carers was acting as a secretary that day, he would give her the piles so she could help Patrick deal with the correspondence when he woke up. Remarkable!

The book Patrick had co-authored with Chris Lintott and Brian May, namely *BANG! – The Complete History of the Universe*, had enjoyed steady sales in 2008 and 2009 and in September it was re-released as a paperback edition. Despite Patrick's highly immobile condition he took part in a book-signing ceremony at the re-launch, organised by Matthew Clark of Torbay Book Shop. This was literally a local event for Chris Lintott, a former student of Torquay Boy's Grammar School, through which Patrick had been acquainted in the 1960s, 1970s and 1980s via his friend J. Hedley Robinson. In fact, all three authors were familiar with the Paignton/Torbay region and, as we saw at the start of the book, Patrick was at RAF Paignton during the war and was very familiar with the Palace Theatre where the book signing took place. I should perhaps qualify the term 'book signing' here. Following Patrick's spinal problems and his increasingly debilitating arthritis he was rarely able to even hold a knife, fork or spoon in his eighties. This was, quite obviously, a source of much frustration, as it affected his ability to type and to sign books. So, he had a rubber stamp made, which was based on his signature from the days when he still had some manual dexterity. Anyone having a book signed by Patrick from his early eighties onward was likely to have a rubber stamped signature rather than the genuine article.

75 Years with the BAA

Regarding books and on a personal note, in October 2009 my own book regarding Patrick's Caldwell Objects was published by Springer. In the front of the book I gave a short tribute to Patrick and his life in astronomy. I posted him one of the first copies to be printed and, typical of Patrick, he phoned me the instant it arrived at Selsey, telling me how much he had enjoyed seeing it. He told me that he had just had a minor cataract operation but was otherwise chugging along and looking forward to the BAA event in a few weeks time, at which he was the guest of honour. This event, brilliantly organized by Patrick's friend John Mason, was to be a BAA celebration of Patrick's 75 years of membership. It turned out to be a splendid and memorable day. The theme of the event was modern developments in amateur astronomy, but with Patrick's 75 year anniversary (as we saw earlier he joined aged 11) in everyone's mind he was constantly referred to during the presentations.

In total 11 speakers, including myself, gave talks throughout the day. The venue was highly memorable: the famous Faraday Theatre at the Royal Institution in London's Albemarle Street, used for all the Royal Institution Christmas Lectures. The first talk, about lunar craters by Peter Cattermole, was conducted jointly with Patrick, who made amusing contributions from his wheelchair. I was the second speaker on the agenda. At lunch I sat next to Patrick during the speakers' meal, held at the plush nearby Brown's Hotel in Mayfair. His arthritis seemed worse than ever and he could only drink his soup when it was placed in a large mug, which he could just about grip with both his swollen, arthritic hands. Nevertheless, he was upbeat and very skeptical about the alleged discovery of water on the Moon. There was certainly nothing wrong with his brain that day, even if the rest of him was falling apart! A trio of carer, helper and chauffeur had driven him up from Selsey, to London, at 6 a.m. Remarkably, for a man in such poor health, aged 86, he listened to all the talks without falling asleep and obviously enjoyed the day.

At 5.30 p.m., at the end of a very long day, the new BAA President David Boyd concluded the meeting, paying tribute to Patrick's contributions over 75 years. Virtually all of the 400 BAA members present at that meeting had been inspired into astronomy because of Patrick's books and TV programmes. As the President concluded his talk Patrick's hand waved frantically. He wanted the last word: of course! Patrick was wheeled to the centre of the floor in the dramatically sculpted and historic Faraday Lecture Theatre and, obviously quite tired now, gave these closing remarks:

"If I may, can I take your time for just a minute or so. After all, I've seen so much, all through the BAA that I've belonged to and, when I joined, the world was a very different place: no television, no electronics, almost no radio. I remember trying to tune my radio and listen in to the test scores in Australia, amid all the interference. [It was] a different kind of country altogether, a different country. We were governed by ourselves, our House of Commons governed us; totally honest politicians, not the present bunch of crooks!" [Huge laughter and applause at that jibe towards M.P.'s fiddling their expense claims in 2009] "Skies were darker, that's how our astronomy evolved. And I'd like to say first of all I don't think all of you realise how much influence the BAA has had, not only here but all over the world. It's been, I think, more inspirational than any other society, even the RAS, and there's nothing like it. In my own case I've been immensely grateful to the BAA for all it's done for me and I'm only one of many. And there's past astronomers, like Steavenson, people like that, and they'll go on.

I first spoke officially to the BAA at the age of eleven, when I went to the front to shake hands with the President, the Astronomer Royal Harold Spencer Jones, when I became a member. This may well be the last occasion I will speak to you, I don't know. Certainly, I've had immense help and no-one realises, I think, how influential the BAA has been. So, I would like, as I say, to thank you all; and again it's been a great honour to be here today, and particularly to my lifelong friend John Mason. So, thank you again, it's been a great day and one that I certainly won't forget in the years I've got left. And, the BAA has a glorious past and it will go on to have a glorious future; it's existed now for 120 years and in 120 years time it will still be going strong. It's done an immense amount of good, all the time. So, thank you all, thank you very much and clear skies to you all. Thank you."

There was a thunderous applause after that from the 400 BAA members present and a standing ovation that lasted for almost a minute. Patrick was quite moved and after the ovation he added: "I'm not usually stuck for words but I am now and once again, thank you. You've done so much for me….I've tried to give a little back…. I've tried to give a little back, I don't know, I've done what I can, anyway thank you and clear skies…."

To more sustained applause the meeting then officially ended.

Patrick was certainly delighted at how smoothly the BAA's celebration of his 75 years of membership had gone, but would, perhaps, have been less impressed if he had spotted a cartoon in the 2010 *Viz* Annual (sub-titled 'The Council Gritter') published that same month. In the Viz cartoon a fat, monocle-wearing character, with hair sticking up (unmistakeable for anyone else) had trained his astronomical telescope on the hairy Elastoplast covered arse of a sailor, who was undressing in front of the window of a brand new building, next to his garden. Patrick exclaims: "Looks like a Full Moon tonight. But Good Gwacious, this is incredible – the Moon appears to have gone all hairy and there's an Elastoplast on it!" Meanwhile, a high-ranking Naval Officer, standing next to a sign saying NEWLY BUILT SEAMANS MISSION, bellows over Patrick's garden wall: "So! Spying on my sailors getting undressed now are you?!" Somehow, at that time, I didn't think anyone would be sending Patrick the Viz Annual for his 2009 Christmas present.

As 2009 came to a close, in the final days of December, Patrick was back in St Richard's hospital in Chichester with a serious chest infection. Only days before he started feeling unwell he had spent most of a bitterly cold December evening filming a *Sky at Night* Star Party in his back garden. Everyone else was wrapped up well but, as usual, Patrick just had a few token garments wrapped around his gigantic torso. Patrick was famously immune to the cold but it would appear that he had pushed his luck too far that night and several hours in subzero temperatures had taken its toll on his 86 year old lungs.

However, despite his carers and friends insisting that he must go into hospital, Patrick initially refused point blank. He knew that he was so weak that if he went in he may never come out again; and, if he was going to die, he wanted to die in his beloved Farthings and nowhere else. Worse than dying though, he might miss another *Sky at Night* (as with the salmonella incident in 2004) and he could see the BBC mandarins immediately replacing him with the younger presenters; suddenly 52 years of presenting would count for nothing! When Patrick refused point blank to do something he became the classic stone wall immovable object, but he was getting dangerously frail on this occasion. Fortunately his GP convinced him that hospital was his only option and so Patrick spent the Christmas and New Year period incarcerated and with intravenous antibiotic drip-feeds battling to cure his lung infection. But he was determined to get back home for the next *Sky at Night* filming day. Eventually, a few days into the New Year, Patrick was released from hospital and the ambulance men helped get him into the stair-lift and then into his own bedroom. Bliss! However, it was a false dawn. After the drug regime and immobility Patrick's already enormous legs had ballooned with fluid retention. He was now larger and heavier than ever and his carers found it virtually impossible to get him from his bed to his wheelchair and then into the stair-lift. He was bedridden and now over 20 stones (280 pound) in weight and could not get to his beloved study where he had worked on his books for the past 42 years. Fortunately they finally got him downstairs in time for that next BBC filming day. Phew! Amateur astronomer Bruce Kingsley, who Patrick had donated his old eight and a half inch reflector and merry-go-round observatory to, a few years earlier, spent much time visiting the great man and helping him out in this period.

Remarkably, only weeks after being in intensive care, Patrick was transported by helpers to London again, to the annual *Astrofest* meeting organised by *Astronomy Now*. At that event, on Saturday February 6th 2010, he was once again featured with Brian May, answering questions from amateur astronomers and promoting their joint *Bang!* book, written with their co-author Chris Lintott. The main feature, entitled 'An audience with Brian May and Patrick Moore', was chaired by Patrick's friend (for half a century) Iain Nicolson and there were plenty of Patrick's favorite one-liners on the agenda when the audience asked a few questions. A few examples will help set the general tone:

Question: "Is there common ground between your type of music and Brian's music" Patrick, scowling and bellowing: "NO!" [Hilarious laughter from the audience].

On the subject of space exploration Patrick said: "Mr Obama has just cancelled all NASA's Moon plans: One small step for a President – One giant leap backward for mankind". [More applause].

On whether astronomy should be taught in schools, Patrick's long held beliefs were unchanged: "Some teachers can make any subject boring, and many of them do!" [More hilarity from the audience].

Jane Fletcher to the Rescue!

Less than 11 months after resigning as the producer of *The Sky at Night*, Jane Fletcher returned as a one-off producer for the February 2010 programme about Mars. Paul Vanezis was still the 'series producer' and William Lyons the 'executive producer'. Almost as soon as Jane had resigned her role in the programme it was realized by Patrick and the others that, despite knowing little about astronomy when she had taken on the job and despite some skepticism from Patrick about her abilities, the loss of her expertise had been a major blow to the programme. She also knew how to deal with Patrick and had quickly learned how to film people well when they were observing the night sky in the freezing cold, at well after midnight; quite a tricky thing to get right. So, almost from the month when Jane had resigned, she had been recalled in a consultative role and now she was the producer for the February programme; but by the year's end she would be back in full control once more.

There had been quite a few subtle changes to *The Sky at Night* format around this period. Chris Lintott had not always been able to appear on the programme, mainly due to his 'Galaxy Zoo' work commitments and trips abroad, and so local Selsey amateur astronomy imaging expert Pete Lawrence had taken on an increasingly larger role in explaining the night sky, as seen from Patrick's garden. In addition, Paul Abel, a keen BAA planetary observer from Leicester, as well as being an astronomy researcher specializing in relativity and Hawking radiation, joined Pete outdoors to give the visual observer's perspective on things. Paul's excitable delivery, 'Old School' appearance and love of visual observing had caught Patrick's eye, which is why he would become a regular on *The Sky at Night*. The programme's researcher, Chris North, also appeared more frequently on the programme, talking to Patrick about recent news when his friend Chris Lintott was not around.

Away from the monthly filming schedule Patrick was as busy as ever in 2010, despite being reliant on others to transport him. His 87th birthday on March 4th passed relatively smoothly, with the usual mountain of cards arriving, along with a bottle of Jameson's Irish Whiskey from Brian May. A small St Patrick's Day (March 17th) party at *Farthings* had to be cancelled when Patrick was laid low with a stomach bug but, to compensate, a much bigger St George's Day (April 23rd) party was laid on. Patrick's carer Dawn was handed his huge address book and asked to invite dozens of people. If they didn't respond, Patrick would then phone a second time. Patrick's St George's day parties were fairly regular events with him sometimes dressing up with a 'red cross on a white background' smock and waving a long cardboard sword to slay the dragon! As St. George is the patron saint of England, with his cross forming the national flag, it was hardly any surprise that the 'English to the core' Patrick liked his St George's day parties.

By the start of April 2010 Patrick had appeared on the first *Dr Who* episode featuring the new Doctor, Matt Smith. Patrick was one of six faces on a laptop screen of important space authorities, to be contacted in an interplanetary emergency. He was ranked as equal in importance alongside facilities such as NASA, Jodrell Bank and the Tokyo Space Centre! In the clip the actress Annette Crosbie says she likes Patrick Moore, when his face appears on the monitor. The new Doctor replies "I'll give you his number, but watch him, he's a devil!" Later, Patrick asks "Who is your lady friend?" and the Doctor scolds him with "Patrick behave!" Bizarre!

In May, Patrick and his carers decided that packing him off to bed every night on his stair-lift, and getting him back down again each morning, was just creating too much hassle. It was decided that the music room adjoining the study, where he used to play the piano or the xylophone, or simply watch the TV, would have to double as a bedroom too. This would make life a lot easier and also mean that the 20 stone behemoth that was Patrick Moore would not have to strain the Stannah Stair-lift to the limit of its design specification any longer. Even though Patrick's ancient staircase had been reinforced to bear his weight and that of the chair, it was obviously working on its limit when hauling his bulk upstairs and down. Also, the Farthings ghost 'Claude' had been quite noisy recently and tended to roam downstairs (and throw glassware about) so maybe if Patrick was on the ground floor it would be a bit more restrained……!! Obviously Patrick was absolutely dead against this permanent move downstairs, at least initially. Over the phone he told me "They'd have one foot in the door then Martin – it would be the beginning of the end – No Way!" Now, quite who 'they' were, I did not ask, but eventually he agreed to the plan. Carting Patrick downstairs and back upstairs each day was just becoming far too time consuming.

Also in May, Patrick wrote a rare article for *Astronomy Now* magazine about his book writing over almost 60 years. In response the magazine launched an appeal to identify all of Patrick's titles. Little did they know that I had compiled the ultimate list many years earlier! Patrick's article would appear in the July issue of the magazine.

In June Patrick was asked by the BBC *Focus* magazine to do a road test of some popular children's 'space food' samples to see what he thought of them. The resulting video was uploaded to *Youtube*. Patrick was asked to sample the savoury, ice-cream and chocolate delicacies of Space Raiders ("Tasteless Onion"), Space Dust ("Taste's like sweet dust"), Mars Bars and Milky Way ("Everybody knows these"), Flying Saucers ("Sherbert India Rubber!") and Astronaut's Ice Cream ("Ice Cream").

Patrick and members of *The Sky at Night* team were asked to give a presentation at the British Film Institute (BFI) on London's Southbank on June 30th 2010, commencing at 18:20 BST. It was a special live presentation to celebrate the programme's record-breaking 53 year run on TV with the same presenter. However, the last week in June saw very hot and humid conditions across the southern UK. Throughout his entire life Patrick had been immune to cold and even bitterly cold weather, but hot and humid weather always made him feel unwell. As he got older this type of hot, humid weather made him dangerously overheat. On June 29th, after a very rough night with no sleep, Patrick was taken into hospital to be re-hydrated and to try and make him more comfortable. As a result the BFI event had to go ahead without the star of the show. The other *Sky at Night* members turned up and Chris Lintott hosted the event with Pete, Paul and Jane Fletcher; but, without Patrick, it was simply not the show the audience had expected. It was, of course, the one and only Patrick that they had come to see! Fortunately the July *Sky at Night*, about the Atlantis Space Shuttle (featuring Cdr. Kenneth Ham and astronaut Piers Sellers in Patrick's study) had been filmed many weeks earlier.

In August and September Patrick was a star guest at two major non-astronomical events. The first of these was held on August 20th when Patrick was driven up to London for a Nintendo celebration, to mark the 25th anniversary of the *Mario* computer game character. As we saw earlier in the book, in the 1990s Patrick was the *Gamesmaster*. His voice and graphically modified face were used to depict an all-knowing being who knew all the tricks you needed to get the best from video games of that era. An advert was filmed celebrating the 25th anniversary of *Mario* and it was first aired during ITV's *Coronation Street* on September 13th. Celebrities Ant and Dec, Jonathon Ross, Jamie Redknapp and Jamie Cullum were all in the broadcast, but Patrick stole the show. Appearing as himself, not as the *Gamesmaster*, he just said, in an all-knowing way: "Jump on top of the level, run across, and jump into the Warp Zone……. Try that."

On September 23rd Patrick (and, so he claimed, his famous Ford Prefect car!) were taken to Goodwood for a 70th anniversary of the Battle of Britain tribute. The race course, in its former guise, was RAF Westhampnett, during World War II.

Patrick had not been happy with the way things had been running on *The Sky at Night* by the summer of 2010 and one programme in particular, about Star Birth, he had felt had been too dull. He made his feelings clearly known to *The Sky at Night* team and it was decided that the best way to please Patrick was to bring back Jane Fletcher as the Series Producer for the October programme about 'Light Echoes'. Jane agreed and Patrick was relieved, as were many of the regular guests. Jane would now be back at Selsey, once a month, expertly directing the camera crews as she had last done 19 months previously. She would also be keeping Patrick happy, as she had done so well ever since the programmes had started being filmed from his house in 2002. Well, apart from the odd rant from Patrick that is! There were still a few gripes, especially as some involved with the show had suggested there was now too much amateur astronomy on the programme, but with Jane back in charge Patrick was a much happier man.

On the night of October 16th/17th I was delighted, once again, to be invited to play a small part in the programme (scheduled for November) about Comet Hartley. Patrick wanted to be outside under the stars for the part of the programme where he talked to me about Great Comets of the past, but so that he didn't catch a chill (as he had at the end of 2009) his wheelchair was now equipped with a heated blanket that was draped over him, making him look even more humungous than normal! For the first time ever, on that visit to Selsey, I noticed Patrick's vintage Woodstock typewriter had been moved off his writing desk, to be replaced with a huge flat screen PC monitor and a keyboard incorporating enormous keys. These were big enough that even Patrick's horrendously gnarled and swollen banana-like fingers could use them.

On the book front, the whole of the closing months of 2010 had seen a team of Patrick's trusted friends helping him and the Canopus publishing Director Robin Rees. They were all roped in to check and revise Patrick's massive Data Book of Astronomy for Cambridge University Press, ready for a new edition being published.

Chapter 42
Life, the Universe and Everything……

Bad Vibrations

Despite Jane Fletcher's very welcome return in her role as *The Sky at Night* Series Producer the Farthings film set could be a rather stormy place in late 2010. The problem was that it had seemed pretty clear when Chris Lintott had first joined the programme that he was Patrick's intended successor. Indeed, Patrick had said so himself to me and others, in private. However, he had said the same about John Mason for a full 10 years before Chris came along! In truth, Patrick wanted to be the main presenter as long as he still drew breath and he knew that the programme's many fans also wanted him in that role. Many thought that when Patrick died, the programme would die with him, but it did seem quite likely, in the early years of the twenty-first century, that Chris would be a worthy successor. In the fickle world of television he fitted the stereotype requirement of being young, tall and very confident in front of a camera. He also had a full grasp of the subject.

However, as the years had gone by, with Chris Lintott as the co-presenter, more and more people were becoming involved with every year that passed, both as guests and reporters. As Chris was a busy professional astronomer these days and local Selsey man Pete Lawrence was a highly accomplished amateur astronomer, like Patrick, if the programme was allowed to drift more in the amateur direction then, clearly, Chris Lintott's role as Patrick's automatic successor could be under threat. I heard quite a few disturbing stories from Patrick, while sitting in his study, about heated arguments at *Farthings* during the closing months of 2010. All were quite obviously rooted in this professional/amateur power struggle. On a couple of occasions, either in reaction to Patrick's increasingly racist jokes, or other matters, Chris Lintott had, according to Patrick, simply walked out of the house. It became increasingly obvious to all the viewers that the presenters were being split into two groups, with amateur astronomy being well represented by Pete Lawrence and Paul Abel, whereas professional astronomy was being covered by Chris Lintott, Chris North, a guest or two and Patrick as the anchor; but the two groups were, whenever possible, being kept well apart.

M. Mobberley, *It Came From Outer Space Wearing an RAF Blazer!: A Fan's Biography of Sir Patrick Moore*, DOI 10.1007/978-3-319-00609-3_42,

On several occasions, in late 2010 and early 2011, I spoke to Patrick at Selsey and then on the phone and was rather saddened when he told me: "The Sky at Night is not a happy place anymore Martin……. I think I've created a monster, but Jane is keeping things under control……I don't think the viewers will have any idea of what goes on here……It's like a War Zone sometimes……I feel like I'm John Simpson……HAH!"

From my own perspective, as someone who had watched *The Sky at Night* for more than 40 years and appeared as a guest on ten occasions, I have to say that the Pete Lawrence/Paul Abel double act brought an entertaining breath of fresh air to the programme which, with Patrick so frail, had become rather depressing at times. In comparison, the sections of the programme devoted to professional astronomy were, to me at least, invariably boring by 2010. Plenty of glossy TV programmes on BBC1, BBC2, BBC4 and the satellite channels were covering professional astronomy to death by that time, but the only programme regularly encouraging people to actually go outside and look at the heavens was *The Sky at Night*, during Pete and Paul's section.

As things happened though another BBC issue was to totally eclipse this pro/am problem, not least because Chris Lintott had decided to spend a year working from the Adler Planetarium in Chicago and so would not be able to get to Selsey for many *Sky at Night* filming days.

The other BBC issue was a series of three programmes scheduled for the first week in January 2011, entitled *Stargazing Live*. Hosting these programmes was the rising star of science broadcasting on the BBC, Professor Brian Cox. The idea was that live observations from local astronomical societies would be shown on the nightly programmes and that they were timed to coincide with astronomical events hyped as 'spectacular' by the BBC. However, to most amateur astronomers, including Patrick, the timing was crazy. January is usually the coldest and cloudiest month imaginable in the UK and while the sunrise solar eclipse on January 4 was certainly unusual, it was a daytime event. As for the 'spectacular' Quadrantid meteor shower, well I don't know any amateur astronomers who regard it as spectacular and most UK observers only ever see a few Quadrantids in their entire lives, such is the likelihood of cloud in January.

Also, while amateur astronomers may be eccentric, they are not mad enough to want to spend the entire night outside in January. Look at the huge star party events held in the USA and the UK over the last couple of decades; they are mainly held in the autumn months, not in mid-winter and the organisers of these events are highly experienced. However, it appeared that the BBC was not interested in such hard facts! One source told me that the BBC timing was solely down to a competitor's Darts coverage and nothing to do with astronomy at all! Worse still was the fact that Patrick and the entire *Sky at Night* team were not invited to appear on the programme. It was quite obviously a deliberate snub and Patrick was quite obviously very upset by it. When a man has written hundreds of astronomy books and presented a programme for almost 54 years for virtually no money, solely for the BBC, and when he is regarded as a national institution, well, he rightly expects better treatment.

The entire *Sky at Night* team were shocked by the fact that the *Stargazing Live* programme did not involve Patrick at all and the subject so dominated conversations at *Farthings* that Jane Fletcher banned it from being mentioned on filming days! Yes, Brian Cox was a big star on TV these days and in some respects was the real TV successor to Patrick, but he was a professional particle physicist with no amateur astronomy experience! Indeed, no-one on *Stargazing Live* had any serious reputation whatsoever as an experienced observer in amateur astronomy, which was truly astounding! Rather predictably skies were almost totally cloudy in the *Stargazing Live* week and most of the show was simply recorded footage. There was certainly a brief burst of interest in amateur astronomy in the UK, following the programmes, but it quickly faded away. Nevertheless, Patrick was pretty upset by such shabby treatment. Surely, as a minimum, an interview could have been filmed with him in his study? Apparently not! It was obvious that someone very high up in the BBC echelons of power did not like Sir Patrick Moore. Whether this had something to do with Patrick's politically incorrect views or just ageism discrimination (which the BBC had been found guilty of concerning its *Countryfile* presenters that year) was unclear, but as Patrick was still fronting *The Sky at Night* it seemed like a good way to irritate that programme's three-quarters of a million regular viewers. As someone at the BBC told me, the huge *Stargazing Live* budget was big enough to keep *The Sky at Night* running for 7 years!

Despite this snub Patrick soldiered on and once more was transported, with helpers in attendance, to *Astrofest* 2011, held in Kensington on February 5th. Brian May was also there, promoting the *Starmus* Astronomy and Music Festival being hosted in Tenerife in June. The other Brian, Prof. Cox, was at *Astrofest* too and so Patrick was, at last, able to meet the man who many thought would be his true successor as Mr Astronomy and Space UK. However, amongst the UK's real amateur astronomers, the opinion would always be that no-one, absolutely NO-ONE, could ever replace Patrick! Around this time the Royal Astronomical Society (RAS) created a Patrick Moore Medal for secondary level teachers who had made an outstanding contribution to the teaching of astronomy or geophysics. Nominations for the first candidate had to be received by September and the first recipient would be Dr Becky Parker, Head of Physics at Simon Langton Grammar School.

88th and 700th Birthdays!

In cricketing parlance Patrick reached '88 not out' on 2011 March 4th. Readers may be interested to know that Brian Cox reached less than half that age, 43, the day before, on March 3rd. It was a remarkable achievement for Patrick, a man who had endured a lifelong heart condition, a crumbling spine, was morbidly obese, rarely ate any vegetables until his carers changed his ways and washed all his daily pills down with copious amounts of Gin & Tonic! However, the big party at *Farthings* was held on the next day, Saturday March 5th, and it was a double party, because the 700th episode of *The Sky at Night* would be transmitted on the following night.

Coincidentally Brian Cox's new series, *Wonders of the Universe*, would also go out on the same night. Whether this was deliberate planning by someone at the BBC, no-one knew. Guests on the 700th programme, filmed in the weeks beforehand, were Brian Cox, impressionist Jon Culshaw, The Astronomer Royal Professor Sir Martin Rees and Brian May, who Sir Martin Rees quipped "looks exactly like Sir Isaac Newton!"

One hundred and fifty invited guests turned up at Patrick's double celebration bash including Brian May, Terry Pratchett, Tim Rice, Garry Hunt and Jon Culshaw. Bizarrely there was a ban on photography in the garden marquee at the event. No-one could ever remember such a thing ever happening before at *Farthings* and Patrick was as mystified as anyone! The most popular theory was that Brian May had stopped dying his trademark 'big hair' black and was letting it go grey and that he wanted it kept a secret. However, that seemed like a pretty unlikely and feeble excuse, because plenty of people had photographed him a month earlier at *Astrofest*. Still, astronomers are resourceful people and security guards are invariably gorillas, whose knuckles scrape along the pavement, while their tiny brains possess an IQ in single figures. So, although many cameras were confiscated, until the owners reclaimed them on leaving the party, the guards were apparently too thick to realise that most mobile phones have built-in cameras; so many pictures of the event did emerge anyway…how hilarious!

A few weeks later Patrick was well enough to travel to London with *The Sky at Night* team to visit a group of city-based amateurs called 'The Baker Street Astronomers'. On April 7th 2011 he played host to the visiting Australian planetary imager Anthony Wesley, along with the planetary imaging expert Damian Peach, who had moved to live in Selsey during the past year. Despite the socialising it was quite clear to all those present that Patrick was feeling pretty grim in the first weeks of April and his carers were often having to ensure that he did not exhaust himself. Much of the time was spent with Patrick asleep, either in his chair, or in his bed. Eventually he felt so grim that it was time to re-visit the local hospital. Patrick had contracted an infection sometime in late March or early April and, in his fragile medical state, had been totally unable to shake it off. His carers sent letters out to all those who had been invited to Patrick's St George's Day Party on April 23rd, advising them that due to Patrick's health the party was cancelled until further notice. After a week of treatment Patrick was back at *Farthings* and able to stop worrying about his cats and the filming for the next *Sky at Night*, but he was still in a fragile state. However, his health did not stop him from adding a voice over to a poem about the stars (Glittering Sky) by a Northumberland musician and composer Carl Cape.

Unfortunately, just 2 months later, Patrick was back in St Richard's hospital in Chichester, shortly after filming an outside broadcast with John Mason and Paul Abel at the South Downs Planetarium. This time it looked even more serious as he was diagnosed with pneumonia. Patrick had been due to attend the 'Weed and Wildflower Festival' in Bignor, a few miles from Selsey, where he was supposed to be reading from his new book of children's poems (!) entitled 'Within the Glade: A Collection of Poems Written to Amuse Children (of all Ages).' His publisher read the poems in his absence. This was a new departure for Patrick and, admittedly, the book only contained 23 poems. They were written in the nonsense style of Edward Lear. According to Patrick he wrote his first poem, about a leaping frog, while in his garden many years earlier with two children. They loved the poem and so he then wrote 22 more over the following years. After showing his poems to the publisher Sandra Saer, who he had known at the BBC, she encouraged him to have them properly published, but due to his illness he just had no alternative but to miss the book launch at the Wildflower Festival.

Fortunately, after a week-long stay in hospital at the start of July, Patrick was, once again, released back to his home; but it was obvious that he was worryingly frail now. On the following month's *Sky at Night* Patrick's voice was weak and strained. Even when he had been seriously unwell on previous occasions he still had a powerful, if somewhat slurred voice. He just did not sound like the same Patrick at all now; but despite all the odds he ploughed on and was able to film yet another *Sky at Night* with the team, including Chris Lintott, who had just completed his stay at the Adler Planetarium in Chicago. He was also happy to open his garden up yet again for the latest *Farthings* functions, namely the British Interplanetary Society picnic on August 6th and the Chichester Cats summer party on August 21st, which raised over £400 for the *Cats Protection* national charity for moggy welfare.

Nevertheless, there had been a definite change in Patrick's well-being. He had always been largely immune to cold weather and positively hated hot weather; but now, even in August (admittedly a cool August in 2011) he felt cold. Part of the problem was *Farthing's* ageing oil fired central heating system. It had been installed on Patrick's request after he had purchased the huge thatched house in 1968, but it had always gurgled alarmingly (some said 'terrifyingly') and it now did little to heat up the house. Patrick asked amateur astronomer and central heating expert Bruce Kingsley, who the reader will recall acquired Patrick's 8.5-inch reflector and merry-go-round dome, to sort the system out. Bruce and his team did such a good job that a number of visitors to *Farthings* in the following weeks declared that the inside of the house was now as hot as the solar photosphere.

But just as Patrick was feeling stronger and a lot warmer, disaster struck. His beloved cat Jeannie, acquired by his godson Adam in 1999, became ill with a failing kidney in October 2011; it had to be put out of its misery by the vet. When Patrick's cat Bonnie had died, after inhabiting Farthings from 1979 to 1999, he had sworn he would never have another cat because their deaths were just too painful for him to bear. His godson Adam had bought Jeannie and changed Patrick's mind. However, Patrick had always believed that Jeannie and the younger feline inhabitant of Farthings in the twenty-first century, Ptolemy, would outlive him. They were even made provision for in his will. Once again, Patrick had outlived one of his feline companions. Boy and man he had seen seven of them die or be put down. Only 'Ptolemy the second' was still alive now.

Ironically, Jeannie had died just as Patrick was putting the finishing touches to his short literary effort entitled 'Miaow! Cats really are nicer than people!' and so he devoted the final pages of the 64 page booklet to his departed feline companion in a section simply called 'Dear Jeannie.' Patrick's arthritis, combined with his crumbling upper spine, meant that he was now so disabled that even typing on the giant keyboard in his study was virtually impossible. Everything now had to be dictated or co-authored, with the co-author doing all of the donkey work. Gripping anything at all in his claw like hands was a major effort and life was virtually unbearable at times. Even grappling with his regular Gin & Tonic, delivered in a mug at 12:30 p.m. every lunchtime, was becoming close to impossible and he could barely grip the rubber stamp he used to sign his name in books with. By the end of 2011 it was becoming hard to tell who was more disabled, Patrick or Stephen Hawking; yet both these remarkable men seemed able to stay alive by sheer willpower alone.

Still, despite Patrick's condition during that autumn of 2011 he was amused and delighted when he received a request for a racehorse to be named after him. The Berkshire trainer Harry Dunlop purchased the grey colt yearling (foaled March 18th 2010) in October 2011, at Deauville, for 16,000 Euros. Sired by a stallion named 'Astronomer Royal' and with the mare aptly named 'America Nova' what else could the offspring possibly be called? Just 8 months later the equine 'Sir Patrick Moore' would win the Compeigne prize in northern France and earn 26,000 Euros for its owners.

89th and 55th Milestones Looming!

In Patrick's final years the prize that most spurred him on, in a physical state where many would have mentally thrown in the towel, was passing another milestone. April 2012 would, remarkably, see the 55th Anniversary of *The Sky at Night* and with that deadline in sight there was something to live for; likewise his 89th birthday. As the year 2011 came to a close the BBC *Stargazing Live* team finally agreed that Patrick could have a minor role in the second three night extravaganza, staged for January 16/17/18 in the New Year. It was not much of a role, for a man who had fronted the same astronomy programme for 55 years on the BBC and who had written hundreds of books, but it was *something* and far better than the obvious snub of the previous year. Patrick would have a role not dissimilar to his old 'Gamesmaster' persona in the discussion programme that followed immediately after *Stargazing Live*. He was filmed weeks before posing a set of space themed 'trick questions' which Brian Cox and the bizarre collection of non-astronomer 'celebrities' would grapple with 'Live' in the studio. For example, one question Patrick was filmed asking was "Has Buzz Lightyear ever been into space?" This, of course, was a reference to the childrens' toy of the same name, an example of which had indeed been taken up on a Space Shuttle Mission.

However, despite successfully filming the 'trick questions' for *Stargazing Live* well in advance of the programme, Patrick's health was worryingly fragile by December 2011. Senior people in the BBC were querying his appearance and his dreadfully slurred, breathless, immobile and weak delivery. His lips were a frighteningly dark blue colour, indicating serious cardiovascular problems and he was right on the limit of being able to perform at all for the January edition of *The Sky at*

Night, recorded on December 14th. One member of the team, desperately concerned for his wellbeing, told me: “He was in a terrible state for that show. To be honest, I don’t think they should have used him. He was weak and kept falling asleep. He was so bad that he couldn’t perform various other recording duties later on. In that situation, by all means get him to open the show, but then give him a rest. They film him at such unflattering angles these days too.” But Patrick was determined to carry on, only too aware of how quickly he could be booted off the show that was his, if the BBC decided he was just too old. Appearing on the monthly programmes was all that he had to live for.

So, given Patrick’s health problems, I was amazed when, just after Christmas, I received a phone call from one of Patrick’s many carers asking if I could attend a New Year’s Eve party at *Farthings*, in a few days’ time. Well, this invitation was far too short notice for me and, living 160 miles from Selsey, it would have been a real hassle to travel to. This view seems to have been shared by many others as well. I received an e-mail from Damian Peach the next day saying that he had dropped in to Patrick’s New Year’s Eve party but there were no astronomers there at all, just ‘loads of old biddies!’ All rather sad really……

Celebrities, Twitterers and Fame-Seekers

Well, the *Stargazing Live* set of three shows were broadcast as planned in mid-January 2012 and Patrick’s pre-recorded trick questions were shown in each of the discussion programmes, which pleased him greatly. However, as in the previous year, there was precious little live stargazing on the programme. It seems incredible, that in a country renowned for its amateur astronomers, like the supernova discoverer Tom Boles, asteroid tracker Peter Birtwhistle and variable star observer extraordinaire Gary Poyner, that the programme’s researchers were too lazy to have some of the country’s leading amateurs involved, even in minor roles, in the show; but there was a glimmer of hope. The aforementioned brilliant planetary imager Damian Peach had won the RGO astrophotography prize some weeks earlier and, remarkably, researchers from the show had actually contacted him about appearing on the programme. This looked far more encouraging.

However, as every amateur astronomer who has ever had dealings with television people knows, nothing is straightforward and, unless you are a flavour of the month celebrity, you will simply be messed about and receive, at most, a pittance (or nothing) for your appearance! First they told Damian he might be interviewed by Brian Cox, then he might be interviewed by a ‘halfwit numpty’, then it would be an interview in his back garden, then at Jodrell Bank. Even as the first *Stargazing Live* programme was being broadcast Damian still had no idea what was happening and we are talking here about an amateur astronomer internationally renowned and revered, by amateurs and professionals alike, for his planetary images. What a shambles! Finally, with 1 day to go, the BBC told Damian he would be on the final show and ‘would he please make his way to Jodrell Bank?’ So, Damian drove from Selsey to

Jodrell Bank, but he then found out that he wasn't on the main programme at all! His only appearance would be standing up for an hour as part of the background studio audience and then, in answer to a comment by Brian Cox, introducing himself as the winner of the RGO planetary competition.

When Damian explained that he had lugged all his gear to Barbados to get better planetary imaging conditions this just resulted in a tone of ridicule from one of the celebrity jokers present. As Patrick had confidently predicted, the programmes had nothing whatsoever to do with the fine tradition of amateur astronomy that had existed in the UK for decades and it was more about vacuous 'celebrities' self-promoting themselves. The one true amateur astronomer on the show, Damian, was messed about and then laughed at, after driving hundreds of miles to be there and after having to take a day off work too. Appalling! In my experience *The Sky at Night* team always treated people infinitely better than that, despite the show's microscopic budget.

On the subject of self-promotion, the appearance of *Stargazing Live* had produced a very unpleasant aspect to the astronomy scene, when combined with the explosion in social media such as *Facebook* and, especially, *Twitter*. Suddenly, people on the fringes of amateur astronomy, who craved TV fame, but who had little hands-on observing experience, were able to self-promote themselves to infinity and impress the gullible. For them *Stargazing Live* was seen as the astronomical equivalent of *Britain's Got Talent*, or rather No Talent Whatsoever! Disturbingly, many of these Twitterers' most gullible victims appeared to be media science researchers and editors, too lazy to check out someone's credentials, so they were just going by the number of apparent *Twitter* followers they seemed to possess. Pathetic! The techniques followed by a few of these 'TV star wannabees' was very similar and often proceeded as follows:

1. Find a hot astronomical topic of the day, or hour, or minute, or some important topical observations by real (but modest) observers.
2. Make or fake observations of the same objects and Tweet, with lots of hype, your observations (not mentioning the real observers at all) to lots of your 'Followers' and magazine editors.
3. As soon as someone takes the bait and mentions you on an Astronomy website, hype the link to the National Press and National TV and hope it goes semi-viral. Then alter your *Twitter* and *Facebook* profiles to read 'consultant to BBC, Discovery Channel, Sky TV etc etc....'

   The lengths some people will go to for a bit of TV fame is truly stomach churning.

The techniques used by some of the lowest forms of life were even dodgier. Patrick often held functions in his garden for various causes, ranging from cancer charities to cat charities to meetings of astronomical societies. So, if you were a fame seeker and found out one of these events was due to take place, simply sign up to the society, turn up in Patrick's garden, say "Hi Patrick, yes, I've been raising money for this charity for years... Oh, by the way, I'm an amateur astronomer too, didn't you know?" You then Tweet an image to the world of yourself and Patrick (making sure no-one else is in the frame) saying: "This is me, currently staying with my good friend Patrick." Well, you're destined for stardom surely? Hmmm, possibly, but after 55 years of TV fame Patrick was nobody's fool. Well aware of the fact that many people had seen him as a 'Kingmaker', since 1957, he could see these people coming a mile off, whereas many young TV researchers were still wet behind the ears.

Also, to the naïve it is easy to assume that if an alleged astronomer has tens of thousands of *Twitter* followers and writes articles for a big astronomy website they must be the real deal; but what if the *Twitter* followers were stolen from a big local astronomical society by changing the password and putting the odd link to pornography in the tweets? What if all the fake astronomer's articles were plagiarised from genuine astronomers' writings, with occasional words altered to disguise the fact. Also, what if even more *Twitter* followers were acquired by purchasing virtual followers online illegally? It sounds incredible, but all this and more was taking place in 2011 and 2012, simply because of the *Stargazing Live* effect and the sudden power of *Twitter* to create the illusion of apparent astronomical expertise, where none existed. However, while not having any social media presence himself, Patrick had good advice from *The Sky at Night* team, as did his carers, who often answered his telephone. Next to the phone was posted a list of 'nutters, fame-seekers, conmen and plagiarists' so that the carers and Patrick would know who was a publicity seeking nutter and who was not. When the editor of a large circulation U.S. astronomy magazine phoned Patrick up, with serious doubts about one self-obsessed fame-seeker, who claimed to be a great friend of Patrick's and who could, allegedly, clinch an exclusive interview with him, Patrick left him in no doubt: this was no friend of his, just someone who had been trying to gate-crash his way onto *The Sky at Night* for years.

There were some much more sinister goings-on than just fame-seeking though and I've mentioned some of the issues already in this biography. Patrick was a very trusting man where his friends were concerned and Farthings was an open house to them all. However, when you are famous you tend to attract a few percent of slime into your house, along with the 97 % of genuine friends. Patrick was now totally dependent on others for everything. By late 2011 he could not even type, so anyone who he could dictate a book too, even children, were welcomed into Farthings. Children tend to bring their friends along and sometimes their friends see opportunities to pocket something. However, Patrick would never have a bad word said about any child in his house, whatever the evidence. There were plenty of dodgy adults wandering in and out of Farthings too though. For some, the constant availability of free booze (remember Patrick's famous 'Colonel Iremonger's house rules') was like a magnet to them.

While the majority of temporary fill-in carers were, indeed, very caring, some came and went rapidly and there were times when it was difficult for Patrick to guess whether they had ulterior motives, or not. One temporary carer even seduced a married friend of Patrick's at Farthings, while simultaneously trying to swindle Patrick into the bargain! Needless to say, both ended up being blacklisted and barred from Patrick's house. Also, dodgy middlemen, offering to act as an agent for Patrick, had sometimes come and gone, at the same time as items that were not for sale had vanished from his home. One individual, claiming to be a staunch 'Put Britain First' man and a member of UKIP (which would get anyone through Patrick's door) vanished at the same time as a valuable Orrery had disappeared. A photograph of Patrick with Neil Armstrong mysteriously vanished too.

Then there were the complete fruitcakes, like the local guy claiming to be a former member of a famous disbanded 1960s pop group; the only problem with that was he had taken the name of a band member who had died in 1972! The Twittering fame seekers created by the *Stargazing Live* effect seemed positively tame in comparison to the outright criminals trying to pinch booze and memorabilia from Patrick's house, while flirting with the carers and while Patrick was too frail to stamp any authority down. Fortunately Patrick's main and trusted carer, Dawn, was very much on the ball (some of the dodgy carers had even tried to topple her authority) and so major thieving and swindling activities were prevented. Without Dawn and Julia (the housekeeper) running Farthings at this time I shudder to think what else might have occurred in Patrick's house during 2009–2012.

Still, one positive upshot of the Brian Cox effect was that there were a lot of people talking about astronomy in the coming weeks and, in addition, Patrick was feeling a bit more alive than he had been in December. He was asked in a number of interviews, on the radio, online, and in newspapers, what he actually thought of Prof. Cox? Patrick's answers were polite, but perhaps less than glowing: "These young astro-media people come and go. Can't say I've seen his show. Seems a nice enough chap, but he's NOT an astronomer…. He's a particle physicist and he's apparently a wannabee rock star too."

By late January 2012 Patrick had not been out of Selsey for months. Once again, various people were raising doubts as to whether a man who was so obviously frail should still be on the programme at all, but Jane Fletcher e-mailed the presenters with her official line, just in case any potential trouble makers in the media started spreading rumours. She told the team (on January 24th): 'As far as the BBC are concerned Patrick is presenting the programme for the foreseeable future, there are no other plans.' Even so, over recent months Patrick had not left his home, his health was simply too fragile. However, by the end of January he was feeling strong enough for a major trip to Leicester, to open the 'Sir Patrick Moore Planetarium' at Leicester's Space Centre, on the 26th of the month. It was a good day out with scores of Patrick fans and school kids delighted to see the legendary TV astronomer. Of course, any trip like this had to be arranged like a military exercise, with Patrick's wheelchair being locked into a people carrier and a professional chauffeur and a team of carers travelling with him. Patrick's only physical chore on that day was to grab the cord that opened the curtains, covering the revamped Planetarium's plaque, in his claw-like hand. Nevertheless, he greatly enjoyed the trip. It had also proved that he could consider another trip to *Astrofest* in a few weeks' time.

As Patrick had, quite obviously, survived the Leicester day out, maybe he could survive a trip out to celebrate the 55th anniversary of *The Sky at Night* too? It was looking possible and so, on February 3rd, *The Sky at Night* producer Jane Fletcher e-mailed a number of former guests on the programme, including myself, with the following message, headed 'SURPRISE 55th party for Sir Patrick':

> Dear Friends,
> I am trying to organise a SURPRISE S@N party for Patrick, to help celebrate the 55th anniversary of the programme which is this April.
> I hope to gather as many people from the programme as I can for Friday the 27th April, 5pm. Venue is Seals in Selsey.
> I will be in touch nearer the day with details, but hope you can be there.
> I have no doubt missed some people off this list, and it's by no means exclusive, so please let me know if this should go to someone else and also if you can come, so I can keep a note of numbers.
>
> Best,
> Jane
>
> PS ITS A SURPRISE!!!!

This was not the only celebration Jane had in mind either, as she was also working on getting approval from the BBC for an official celebration in London, at the brand new 'Media Café' in the equally pristine 'New Broadcasting House' building, built onto the famous original Broadcasting House, the home of BBC Radio.

On February 11th Patrick was off on his travels yet again, fulfilling his promise to be at *Astrofest*. This time he was simply rubber-stamp signing 'The Yearbook of Astronomy 2012' along with his co-editor John Mason and accompanied, once more, by Brian May. Patrick's good friend John Fletcher, as he had done on so many other occasions, helped wheel Patrick from the people carrier to the book signing area and although Patrick was dozing off at times during the day, he survived.

Nine days later, on February 20th, and at my suggestion, *The Sky at Night* team invited the remarkable Gary Poyner onto the programme to talk about the 250,000 variable star magnitude estimates he had made and the 500 variable stars he monitored from light-polluted Birmingham. Gary e-mailed after the filming to tell me that Patrick had been very frail, but he had greatly enjoyed his experience at Selsey. The programme was aired the next month.

Of course, March 4th 2012 saw yet another birthday for Patrick. Against all the odds he was now 89 years old and the 55th anniversary of *The Sky at Night* was just 1 month away. Quite how a man weighing more than 19 stones, with a lifelong cardiac dysrhythmia problem, virtually paralysed, crippled with arthritis, who washed his pills down with Gin & Tonics each day, had got to within 1 year of 90 amazed even Patrick. In various media interviews he said that the last 10 years had been 10 years too many and that all the arrangements had been made for his demise. His body was left to medical science and he couldn't give a damn what bits were thrown away. A candle would be lit in his house after his death and, at a predetermined time, people would gather to watch him blow the candle out from the afterlife, while a tape recorded message would play his voice saying "I'll blow that candle out if it kills me!"

One month later and the first of two 55th anniversary *Sky at Night* programmes was broadcast, on the night of April 1st/2nd. It was entitled 'Warp Factor 55' and was billed as taking the viewers 'at the speed of light to the ends of the observable Universe, stopping, en-route to look at stars, nebulae and galaxies.'

The 55th Anniversary Parties

A few days after the transmission of the first of those anniversary programmes it became obvious that Jane Fletcher had succeeded in getting the BBC to agree to an official 55th anniversary party; this would be funded by the corporation and the event would be held at the Media Café, inside New Broadcasting House. My own invitation arrived on April 5th (just 11 days before the event) and read:

> Mark Thompson, Director-General, would like to invite you to a reception to celebrate The Sky at Night which has been presented by Sir Patrick Moore for 55 years. Monday 16 April, 6pm – 8pm. The Media Café, New Broadcasting House. London. W1A 1AA.

The invitation card featured an amusing cartoon on the front, by Andy Watt of Folio Art, showing Patrick, with a rather strange looking telescope, perched on a ludicrously flimsy tripod! Surely, not a telescope that Patrick would even want to be seen dead with? It was obvious though, that unlike for the 50th anniversary event in Patrick's garden, this was going to be a much smaller celebration. I was told by a friend connected to the programme that the aim was to invite Patrick's 55 favourite astronomers and *Sky at Night* guests to the 55th anniversary London party, but with even 3 days to go the BBC were still e-mailing potential guests, which was clearly ludicrously short notice!

Of course, with Patrick's fragile health it was always going to be risky arranging such an event, but the celebration would go ahead, even if he could not turn up. However, as things turned out Patrick's carers and driver got him there yet again, once more seated in his huge wheelchair, locked into the suitably equipped people carrier. Various programme guests and friends from over the years turned up, including the usual faces of Brian May, Terry Pratchett, Jon Culshaw, Tim Rice, Colin Pillinger, Allan Chapman, Pieter Morpurgo, Garry Hunt and around 40 others. However, Patrick's co-presenter Chris Lintott was not there; he was, apparently, in Chicago. The BBC D.G. gave a nice speech in which he was very appreciative of Patrick's contribution and he seemed genuinely thrilled the corporation should have the ownership of such a record-breaking programme. Even the rival broadcasters' ITV were there, and they interviewed an amazingly smart looking Patrick, dressed in a brand new white shirt, brand new black suit and a brand new bow tie! Patrick's comments to the ITV crew were: "I'm absolutely staggered. I never thought when I began doing television shows that I'd be on for another year, let alone 55 years. I didn't know if I was going to be good enough, or if the subject matter would hold up."

On Friday April 27th, just 11 days later, following on from the day of filming at *Farthings* for the second 55th anniversary 'Moore Marathon' programme, the Selsey party itself took place. The venue, as planned, was The Seal Pub & Restaurant at 6 Hillfield Road, just a quarter of a mile from Patrick's house. Patrick had no idea that this second party had been arranged, even though so many people surrounding him were fully informed. One of his closest friends, the amateur astronomer John Fletcher, had arrived as early as Wednesday to help out with the busy period. The whole of Thursday had seen much activity at *Farthings* with film crews setting up posters,

Fig. 42.1 Patrick with Ian Sharp (*centre*) and Damian Peach (*left*) at the 55th Anniversary Sky at Night party inside the brand new Media Café at New Broadcasting House, London, on April 16th 2012. Just visible, standing in the *background*, is Patrick's friend John Fletcher (By kind permission of Ian Sharp)

cameras and filming areas. Then there was the hectic filming day itself. Around 5 p.m. all of the filming activities were completed and a bottle of champagne was brought in to toast Patrick's remarkable 55 years of presenting *The Sky at Night*. At this stage Patrick was still totally unaware of the looming second party and so he was told that the Seal Pub (known simply as 'Seals') could not deliver the usual round of food and drinks that they always dispatched to *Farthings* on filming days. So, Jane told Patrick that they had decided to go to Seals to have the food there. Whether Patrick suspected something fishy was going on at this stage I know not, but even at 89 there was little wrong with his brain, especially the part that knew when something was being concealed. Anyway, the BBC crew and presenters then all walked to Seals, with John Fletcher pushing Patrick's heavy wheelchair up the West Street incline and along Hillfield Road. Patrick was absolutely delighted when he saw that the pub was filled with scores of his closest friends, including many well-known astronomers.

Of course, many of the attendees were the same as those who had gathered at New Broadcasting House on the 16th, but no-one was too bothered about that. The guests who attended the Selsey party included the co-presenter Chris Lintott and also John Mason; neither had made it to the London event. Patrick was especially delighted that two of his most favourite people were both able to make it to The Seal, namely Peter Cattermole and Iain Nicolson. The impressionist Jon Culshaw made the second party too, as, of course, did the producer and organiser Jane Fletcher, along with the presenters Pete Lawrence and Paul Abel and also Damian Peach, who lived in Selsey. In fact, more than 80 people crammed into the pub that evening, just to toast Patrick and *The Sky at Night*. Patrick wore a huge multi-coloured T-shirt decorated with a solar system theme and no-one seemed to mind that his trousers seemed to be virtually falling off while he was seated in his wheelchair; fortunately the T-shirt was enormous enough to prevent the police being called to arrest a flasher! One feature of the party, which many guests would probably rather forget, were the attempts by various astronomers (professional and amateur) and by various BBC people to sing along using the pub Karaoke machine, as the night wore on! There were certainly a lot of sore heads in the morning, but amazingly the frail 89 year old Patrick survived the experience.

It was certainly a good job that the crew had filmed the 'Moore Marathon' anniversary show (Patrick's favourite 55 astronomical objects) *before* the Selsey party! Quite what type of programme might have emerged if they had filmed it afterwards does not bear thinking about, especially as the Friday evening bash had started at 5 p.m.! On the Saturday, when the BBC were still packing all of their kit away, more than 50 of Patrick's friends, who had stayed at local hotels and B&Bs overnight, dropped into Farthings during the day, to wish him well. He told many people in the coming weeks that the BBC and Selsey 55th anniversary celebrations were the best days of his life and the best surprises of his life, especially the second event, which he knew nothing about until he was wheeled into the pub.

Now, you would have thought, 5 years on, that the *Radio Times* magazine would have learnt from their mistake in 2007, when they interviewed Patrick just before the 50th anniversary *Sky at Night* programme and they reported his verbal attack on women, claiming they were wrecking television. Well, they did not appear to have learnt anything! Patrick could be a cunning old codger in his 80s and, when he had the attention of the media and the praise of the BBC hierarchy for reaching yet another ludicrous TV milestone, he knew that this was the time for a sexist or racist rant that he could get away with. This time women were not the target of his venting spleen, but the entire German nation. In the *Radio Times* for 5th–11th May, published on May 1st, despite the extremely politically correct world we now live in, Patrick made it clear that he still hated all Germans, well after the second World War. 'The only good Kraut is a dead Kraut' he stated, as well as: 'The Germans tried to conquer us. The French betrayed us. The Belgians did very little and the Italians made us our ice cream.' Quite possibly, anyone else saying these words in 2012 would have been arrested, but at 89, and being a public institution, Patrick could get away with it. Surely he could not possibly live much longer, so why not have a rant, now? You certainly can't have one when you're dead! At least, that must have been his reasoning.

Another feature of the interview was that Patrick's already highly embellished wartime recollections were getting even more elaborate. As well as mentioning, yet again, the death of his fictional wartime fiancée, to justify his hatred of the Germans, he added another twist to his 'Foggy Dewhirst' mode yarns about the war, slipping in a story about climbing over the dead bodies of his pilot and co-pilot to land a Lancaster bomber (a plane Patrick admitted to me he had never flown in, only Wellingtons) and, yet again, promoting himself to Flight Lieutenant, a rank he never achieved! Then he threw in a tantalising rumour of a hidden career in military intelligence, which he could never talk about, because he'd promised his Group Captain! All total and utter hogwash, along with so many other tall tales Patrick had invented (and probably had even come to believe himself, within his 89 year old grey matter) when the media spotlight was upon him; but the press lapped his rants up. This time the backlash against his extreme German views was far less severe than for his 2007 rant about women, but once again the BBC mandarins must have been groaning at his antics. However, they should, perhaps, have considered themselves lucky compared to the utter maelstrom of chaos that the BBC would descend into 6 months later……but let's not go down that particular avenue!

Well, the second of the 55th Anniversary *Sky at Night* programmes was broadcast on the Sunday night of May 6th/7th 2012 and featured viewers' attempts to observe 55 objects visible in the April night sky. This was a rare occasion where Chris Lintott and Pete Lawrence sat next to each other, on Patrick's left hand side. On the other side of Patrick, the monocle side, sat Paul Abel and the impressionist Jon Culshaw. It was an entertaining show, largely due to the presence of Culshaw and Patrick's admission that "He knows me better than I know myself!" So, yet again, another incredible TV milestone had been passed by Patrick and *The Sky at Night*. Despite his horrendous health issues he seemed now to almost possess a 'Davros' form of invincibility. I might add that years ago, when I was an Electronics Engineering student at Brunel University, I rented lodgings in a house owned by an elderly landlady in Hillingdon. She had various ailments, took lots of tablets, and was in her 80s. Her favourite expression used to be: "A creaking gate hangs longest!" That certainly seemed to be the case with Patrick in 2012.

A Second Transit of Venus

As I mentioned much earlier in this book, when Patrick had been a young man there were three huge astronomical events which he hoped he would live to witness from the UK; these events were the Cornwall Total Solar Eclipse of August 11th 1999, the Leonid Meteor Storm of November 17th/18th 1999 and the Transit of Venus on June 8th 2004. As we have already seen, the only one of these three events that had provided clear skies for Patrick was the Transit of Venus in 2004, for which the Sun had shone on Selsey for the entire morning and a very memorable day was had by all in Patrick's garden.

Transits of Venus are very rare events and the barren gaps in which there are none are way longer than the average human lifespan, being 121.5 and 105.5 years in length. Prior to the 2004 event the previous transit had occurred in December 1882! However, Venusian Transits also come in pairs, so when one transit eventually occurs, there will be another just 8 years later. Thus, the June 8th 2004 event would be followed by a second, on June 6th 2012. The 2004 event had been close to optimum as seen from the UK, because the whole of the 6 hour transit had been visible with the Sun well above the horizon and at a healthy altitude from the middle to the end of the spectacle. The weather on that memorable day was, as we have already seen, absolutely perfect. The circumstances were completely different in 2012 though. The perfect place to see the entire event, at a high altitude, would be the Pacific Ocean: in other words, the opposite side of the world to West Sussex!

From Selsey the event would almost be over even at sunrise, just before 5 a.m. British Summer Time, with the black disc of Venus having cleared the disk completely less than 1 h later. Even in the final stages the Sun would only be 7° above the local horizon; and if that was not bad enough the weather forecast, as the event approached, looked very grim indeed. In 2004 the West Sussex June weather had been clear and warm, with a pleasant 25–30°C experienced across the sunny southern UK on Transit Day. Eight years later and the temperature at sunrise in Selsey was just 10°C. Nevertheless, it was decided that an attempt to view the event would be made. A special Transit Day cake, with a blue background icing (the sky) and a yellow Sun, plus a black Venus silhouette, had even been made for Patrick! Black icing was used to spell out the words 'Transit of Venus 6th June 2012.' Of course, the 89 year old Patrick was extremely frail now and so wheeling him onto Selsey's eastern beach, pre-dawn, was rather risky, despite his apparent immortality and invincibility. Two days earlier the Duke of Edinburgh, a week prior to his 91st birthday, had been rushed to hospital with a bladder infection, after spending 4 hours aboard a boat on the chilly Thames, as part of the Queen's 60th Anniversary celebrations. The Duke (Prince Philip) was in far better overall health than his astronomical counterpart and so wheeling Patrick out to a cold Selsey beachfront was a bit worrying. [As an aside, I might add that another 91 year old, the Science Fiction author Ray Bradbury, a contemporary of Arthur C. Clarke, actually died *during* the 2012 Transit of Venus.]

Anyway, Patrick was up for it, so, accompanied by numerous fans and *Sky at Night* friends (except Chris Lintott and Pete Lawrence who were in Svalbard) and after some pre-event pies, mash and booze, a motley crew braved the cloudy pre-dawn Selsey sky in the hope of a miracle. Those assembled near the beach were the astrophotographers Damian Peach, Jamie Cooper and Ninian Boyle (who had all been in Patrick's garden for the 2004 event) as well as *The Sky at Night* presenter Paul Abel, impressionist Jon Culshaw, Derby astronomer Chris Newsome, amateur astronomer Robin Flegg (who provided the cake) and the 1980s Dr Who assistant Nicola Bryant (who had played the occasionally skimpily dressed Perpugilliam 'Peri' Brown in that TV series).

Of course, with Patrick's and Selsey's reputation for delivering the clear sky goods, even in the direst weather situations, the motley crew were hopeful that something might be seen, but the density of cloud at sunrise was truly depressing. Patrick's childhood 3 in. refractor was carted along to the lifeboat ramp near the beach and Patrick was trussed up in numerous layers of clothing, with a massive black upper garment and what looked like a double bed sized quilted sleeping bag covering his chest, lower body and legs! Well, the Sun rose but nothing was seen for a good 40 minutes. However, the Selsey magic soon started to work and, barely 10 minutes from the end of the Transit, there was a miraculous thinning in the dense cloud. The cloud thickness was just sufficient to dim the dangerous solar light levels by enough to see the black dot of Venus safely crossing the disc. Truly, a miracle had, yet again, occurred at Selsey! Damian Peach, normally a fairly reserved chap, was heard to exclaim: "There it is, THERE IT IS !!…I CAN SEE VENUS, I DON'T BELIEVE IT, I CAN SEE VENUS!!!"

A quarter of an hour later, with the Sun once more covered in cloud, and with the event now over, but with everyone happy, it was back to Farthings for Champagne and cake for breakfast. Who would have possibly believed it!? Not only had the wheelchair bound Patrick survived long enough to celebrate the 55th *Sky at Night* anniversary, he had even lived long enough to see a second Transit of Venus from Selsey. Absolutely remarkable!

Continuing Health Problems

However, just as things had developed a week after the 2004 Venus Transit (when Patrick had gone down with food poisoning) a week after the 2012 Venus Transit Patrick started feeling unwell again. This time it was severe pain in his left knee, the one that had been replaced in 1999. Patrick absolutely hated being carted off to hospital because he knew he would not get his own way there, he would have no idea what might be happening at *Farthings*, and he also knew that the NHS always seemed to want to keep him in there for far too long. He also had an understandable terror of the local authorities suddenly deciding 'enough is enough' and carting him off to an old folks home where he would be unable to present *The Sky at Night* (maybe there would be a coup in his absence by the other presenters?) and where he would end his days amongst snoring, farting and dribbling cabbages whose daily highlight was dunking biscuits in their tea. While Patrick was certainly physically disabled his brain was still sharp and there was no way he was going to be carted off to that sort of prison. He preferred to be safe in his castle with the doors bolted against local authority bureaucrats and his own fans and carers manning the drawbridge to prevent any kidnap attempts. If he was going to die he wanted it to happen in his own home. Also, when he was in residence at *Farthings*, friends and fans were always dropping by to chat with him, have a Gin and Tonic with him and worship him, whereas he simply had to do what he was told to do when lying in a hospital bed. Ever since his childhood days obeying orders had never been a part of Patrick's way of life. So, if he felt unwell, he would avoid phoning the NHS if at all possible.

Nevertheless, even for a man used to the constant gnawing of arthritis the knee pain was becoming intolerable. His carer Dawn phoned the local surgery and, very shortly, a total of four (!) doctors turned up at *Farthings*, one after the other. Strange how the NHS works en masse when you are famous! Bizarrely, the first three doctors seemed to have no idea what the problem might be, with the second declaring: "He's dying, prepare for the worst!" Clearly, that doctor had no idea whatsoever of Patrick's sheer cussedness to stay alive. However, the fourth doctor advised that the hospital had to carry out tests on the knee and, on June 15th, Patrick was, yet again, reluctantly carted off to St Richards Hospital in Chichester, where they diagnosed and treated this latest medical crisis. This was all intensely irritating to Patrick as Sunday June 17th (Father's day) had already been designated as another charity open day in his garden, with guests paying to visit his famous house and see the telescopes and with all the proceeds going to the local St Wilfrid's Hospice, the

Alzheimer's Society and the Selsey Manhood Wildlife and Heritage Group. Damian Peach was asked to show the many visitors around Patrick's telescopes and explain how they worked, while projecting an image of the Sun with Patrick's 5-inch Cooke refractor. It was a successful day, but the advertised star of the show could not be there in person.

One week later an ambulance brought Patrick back to his house, his carer Dawn, and Ptolemy the cat, much to his relief. He could now think about filming the next *Sky at Night* and the best time for the BBC camera crew to, once more, descend upon Selsey. With Patrick's increasing health problems even recording the opening and closing sections of *The Sky at Night* programme required a very flexible schedule, with filming way in advance of the broadcast date. Patrick had reasonable weeks, bad weeks, very bad weeks and also weeks when he was in hospital and the BBC bosses were concerned at his physical appearance and the slurring of his words. Due to the extreme regime of blood thinning drugs that Patrick was prescribed (to prevent blood clots, heart attacks and a stroke) his arms were becoming increasingly covered with brown and purple bruises, resulting from the slightest knocks, which took months to heal. On screen his arms and wrists were often bandaged to conceal these ugly bruises and throughout 2012 the monocle was never glued on anywhere near to his actual eyeball! Choosing the right time to film Patrick's contribution was becoming fraught and time consuming. It was all becoming a delicate juggling game and the other presenters were filling more of the time, with Paul Abel often travelling to BBC Birmingham to add voice-overs, which in happier times Patrick would have coped with in one take, and with no script.

Fortunately the wall-to-wall BBC coverage of the London 2012 Olympics meant that there would be a 6 week gap between the July and August programmes and much of the backyard observatory part of the planned mid-August programme had been filmed as early as May. As things turned out the August programme would be the last one to be produced by BBC Birmingham; from September the production would shift to BBC Bristol. Changes of this type always made Patrick very suspicious, but at least with Jane Fletcher still involved he felt someone in the BBC hierarchy was on his side, to prevent anything drastic happening, like him being ousted.

Frail, but Seemingly Indestructible!

One month after his last hospital incarceration Patrick was in reasonable shape for the next event in his garden, the annual 'Patrick's Picnic' British Interplanetary Society day at Farthings, held on July 21st. Fortunately, it was a warm summer's day (a rare event in the cool and very wet British summer of 2012) and the picnic featured some games, a quiz and a few guest speakers. A few entertaining comments were made by Patrick while holding the microphone in his claw-like hand and there were speakers too, most notably the former BBC spacecraft model builder Mat Irvine, who brought many of his superb models of spacecraft along for BIS members to study. On previous occasions Patrick's home had been an open house to all astronomers and spaceflight enthusiasts and the guests had been invited to 'help themselves' to alcohol, but not this time. With Patrick's health on a permanent knife-edge and with solid evidence that items in Patrick's house went missing with such an open-door policy, Farthings was off bounds and no visitors stayed there overnight. A marquee with a portaloo was erected in the garden so that the usual visitors to the indoor toilets could not enter the house and, inevitably, wander around. This had been a major concern at other recent 2012 events hosted at 39 West Street, Selsey.

On August 6th, 2 weeks after the BIS 'Patrick's Picnic' event, yet another of Patrick's long term friends died; this time it was one of his most favourite people, Professor Sir Bernard Lovell. Sir Bernard was just a few weeks short of his 99th birthday and easily the oldest of Patrick's long term acquaintances. Patrick freely admitted that if not for Sir Bernard's Jodrell Bank telescope and Sputnik, *The Sky at Night* might never have survived more than 6 months on TV. The BBC News 24 summaries showed a clip of Patrick interviewing Sir Bernard on *The Sky at Night* from October 2007, the 50th anniversary of Jodrell Bank and Sputnik. Sir Bernard was shown telling Patrick that "In 1957 we thought we understood the Universe but radio astronomy showed that there was much more that we did not understand". Only hours before Sir Bernard's passing the NASA Mars Curiosity Rover had successfully landed on the Red Planet. Somehow, despite the news summaries being wall-to-wall coverage of the London Olympics, both these astronomical events were covered in detail. Patrick had lived to see another spectacular NASA achievement, but also another death of a friend he had known for decades; he seemed to be outliving all of his friends…

I know I am repeating myself here, but quite how Patrick was outliving all of his friends into the twenty-first century was a total mystery that caused much amazement and bafflement in this period. Patrick had never been health conscious. He had smoked a pipe for much of his adult life and his lifelong cardiac dysrhythmia had meant that he was on tablets to keep his heartbeat regular until the pacemaker was installed in 2006. During the 1980s and 1990s, whenever I had a meal with him, he never, ever, had greens or vegetables (describing them as rabbit food) and he despised brown bread with a passion, admitting this was a childhood hatred that he could not alter. Admittedly, under the control of his carers, his diet had changed somewhat, but he was still grossly overweight, totally immobile and, let's not forget, aged 89 in 2012! In addition, he was drinking Gin & Tonic and Whisky like it was going out of fashion, which was another reason why he hated his trips to hospital so much; he wasn't allowed booze in a hospital bed. Yet, despite this most unhealthy of lifestyles he was still refusing to keel over! But, even for Sir Patrick, time was running out.

One story from this time may be of interest and amusement to the reader. On August 20th 2012 the brilliant astrophotographers Damian Peach (resident in Selsey) and Dave Tyler (on holiday in Selsey) were invited to spend an evening with Patrick at his house Farthings. Damian sent me an e-mail about the evening which was pretty typical of anyone's evening with Patrick in his later years, or his earlier years come to that! I have included Damian's e-mail below, without alteration:

Hi Martin,
Dave and myself went around PMs last night to spend the evening with him. We ended up getting completely wasted on Gin. When I got home I stumbled through the door and collapsed on the bed. Next thing I remember I woke up at 4am with the headache from hell – like my skull was going to explode! Went down and passed out on the sofa for a bit. Finally crawled into work at 10am which is the earliest I could physically manage! I still feel decidedly under the weather. Of course Patrick just kept going with no obvious effects! The man is not human!
This is how it went last night:
PM: What would you like to drink?
Dave/Me: G&T would be nice.
We then drunk the first one, upon which PM piped up:
"Your glass appears to be empty!"...he then calls for the carer.
She appears and Patrick says "Could you bring the bottle of Gin in and the tonic water?"
From there on in it was rapidly downhill. I still don't feel right even now! For 89 he just puts everyone else to shame. Even now he can drink anyone under the table!
I've got a talk this evening at the Solent Astronomical Society 40th anniversary meeting. Needless to say I'd rather just go home and crawl into bed....
Oh the pain of it all!....
Damian

Just 5 days after that drinking session and after a very welcome visit to Farthings by one of Patrick's best friends, John Fletcher, it was time for Patrick to, once more, outlive someone famous and talk about them on the BBC News. This time it was the departure of the first man to set foot upon the Moon. Neil Armstrong, aged 82, had suffered from blocked coronary arteries and had undergone a quadruple heart bypass operation just 1 week earlier, but complications had arisen and he had passed away. Of course, 43 years earlier the landing of Apollo 11 had made Patrick one of the most famous names in Britain as he, James Burke and Cliff Michelmore had covered the Moon landings on BBC TV, while watched by a record-breaking TV audience. The news broke in the UK during the early evening of Saturday August 25th and Patrick's carers were able to wheel the frail, immobile and huge (but seemingly indestructible) 89 year old to the phone, to speak live on BBC News 24 about the death of this most famous astronaut, some 7 years his junior. Patrick, voice weak and rather slurred, said the following, at close to midnight: "He inspired the whole science of space travel and, you know, if you had to choose somebody to be the first man on the Moon, and to set an example, I think Neil Armstrong would be the man. Somebody described him as a reluctant hero and I agree with that entirely, a reluctant hero, because he was world famous and remained world famous and he will be much missed by everybody. He had every possible qualification for this type of thing and he was a very worthy first man on the Moon."

Just 1 week later, on September 1st, the BBC were phoning Patrick up again about Neil Armstrong, but on a slightly different theme. A BBC viewer, named Philip Longden, had audio tape recorded some of the BBC 1 Moon Landing coverage on July 21st 1969, when he was just 12 years old. The actual BBC video recordings had, tragically, been erased many years earlier, but Philip had recorded some 18 min of coverage of the historic Apollo 11 commentary, featuring Patrick and James Burke. The BBC telephoned Patrick and played him part of the recording over the phone and he was delighted; "Good for him" he said of the viewer who, 43 years earlier, had taped the historic event.

The Final Weeks......

As I have already mentioned, Patrick's deteriorating physical condition meant that typing, however slowly, was now totally impossible and had been for the whole of 2012. His claw like hands could not even punch the huge letter keys of the special computer keyboard that friends had supplied for him a couple of years earlier. So, understandably, Patrick's book production had slumped in recent years. Of course, to most men approaching 90, this would not be a worry, but Patrick's cussed determination to churn out another *Sky at Night* episode each month and another book every few months, had become his entire purpose in life, whether simply to please his fans or to infuriate his enemies.

So, his solution to this arthritis typing crisis was twofold, namely dictation and collaboration. He could dictate to any adult or child who was a regular visitor to his study and collaborate with the other *Sky at Night* presenters, just as he collaborated with John Mason on the annual Yearbook (with John doing virtually all of the editing work). With this strategy in place 2012 had seen a resurgence of books with Patrick's name on the front cover. As well as his humorous and very soppy *Miaow!: Cats really are nicer than people!* booklet he worked with Chris North on *The Sky at Night: Answers to Questions from Across the Universe* and somehow stuck his name on Pete Lawrence's book *The New Astronomy Guide: Star Gazing in the Digital Age*, despite Patrick knowing nothing about digital imaging! In addition, after the success of the *BANG!* book, which had been a collaboration between himself, Brian May and Chris Lintott, the trio combined for yet another production entitled *The Cosmic Tourist: The 100 Most Awe-inspiring Destinations in the Universe*. Again, Patrick's dictated contribution was more inspirational than anything else, but it still got his name back on yet another book.

To celebrate *The Cosmic Tourist* book launch, Patrick asked his carer Dawn to arrange a party and to phone around all those on his 'Selsey Bash' invitation list. The party was scheduled for the evening of Wednesday October 17th 2012 at Farthings, commencing at 7 p.m., and it took place, despite Patrick's increasingly frail state. Crate loads of *Malbec* red wine were unloaded into Farthings to keep the guests happy. Most of the people at the party seemed to be newspaper reporters, as many interviews with the authors appeared in the next week's columns. At the time of the party Brian May had been in the middle of an intense week at Westminster, shuffling between meetings between politicians to try to avert a planned badger cull to reduce the spread of T.B. in cattle. As animal lovers Brian and Patrick were adamant that vaccination was the only sensible approach and, remarkably, the former Queen guitarist did manage to put the planned cull on hold, much to Patrick's delight. So, half of the press reports emerging from Selsey seemed to be about the *Cosmic Tourist* book and half seemed to be about badgers!

During this period Patrick's increasingly frail appearance on *The Sky at Night* programme had prompted a number of concerned viewers to query his health online, via astronomy forums and Twitter. Patrick was clearly losing weight during 2012, but not in a healthy sense. His face looked gaunt and his eyes were starting to stare out of their sockets, as if it was taking an enormous effort to speak just a few words, which it was. His hospital visits were becoming far more frequent in this period too.

On the Moore Winter Marathon programme, broadcast on October 7th/8th, all of the co-presenters had assembled alongside Patrick to discuss various objects in the winter night sky. Such a gathering was extremely rare as Pete Lawrence and Paul Abel almost always worked as a double act outdoors, with Patrick, Chris Lintott and Chris North indoors with a guest. The problem was that Patrick was becoming increasingly forgetful and mentally sluggish by the closing months of 2012 and, when in a group, with a guest, he forgot what to say, or just repeated the same comments. When the camera was simply pointed at him, as for the opening and closing remarks, numerous retakes could make Patrick just about look OK, even if the voice was horribly slurred. However, amongst a group of people, rapidly exchanging ideas and astro banter, his ageing brain did not stand a chance. This was very obvious on the October show, with Patrick saying virtually nothing, looking like he was on the verge of falling asleep and only managing the odd few words. It was clear that filming Patrick amongst other presenters and guests, if he was seriously weak, might not work anymore. In fact, on the next month's programme, for the first time ever (excluding the 2004 food poisoning episode) Patrick was not involved with the guest Carlos Frenk at all. Chris Lintott and Chris North questioned Carlos about Dark Matter, while sitting in Patrick's study, without Patrick being present.

Patrick was not happy with his diminishing role in the programme throughout 2012, but clearly there was no other option. I heard about a number of mini-tantrums where Patrick had moaned "I'm hardly in the programme anymore. Fifty-five years of effort for no money is obviously not good enough......thrown on the scrap heap. No point me being in the programme at all these days. I'm not wanted. Rampant ageism is alive and well at the BBC!" His co-presenters were quick to reassure Patrick that he was wanted, but clearly his frailty was becoming an increasing talking point amongst viewers. Nevertheless, to the newspaper reporters still interviewing him about his *Cosmic Tourist* book and his impending 90th birthday, he was upbeat: "Am I planning to retire at 90? Heavens no! I'm planning a big birthday party and I've got many more years of presenting left to do!"

Despite his increasing frailty, exhaustion, befuddlement and short term memory problems, Patrick had been pleased to learn that *The Sky at Night* programme was, once more, going to receive an award. It was a Special Award from the British Interplanetary Society/Arthur C. Clarke Awards committee and was presented to Pete Lawrence and Paul Abel on behalf of *The Sky at Night* team. The presentation took place on the terrace of the House of Lords on October 26th and, bizarrely, the Master of Ceremonies was the impressionist Jon Culshaw, who had appeared on the programme a number of times in the past year. Of course, with Jon around you had a number of other celebrities present too, if only vocally. So, those gathered to witness the award of 'The Arthurs', were treated to impressions of Brian Cox, Patrick himself, and the former *Dr Who* from the 1970s, Tom Baker. As the reader may recall, *The Sky at Night* had won another 'Arthur', 5 years earlier, at its 50th anniversary, when few would have bet money that Patrick would still be on television in 2012. Jon Culshaw tweeted after the presentation: "And thanks to The British Interplanetary Society (aye) for their award to #bbcskyatnight. We'll take it to Sir Patrick."

Patrick's health continued to be very poor indeed during November 2012 and it was becoming common for him to be rushed to hospital for a week and then released just in time for him to say a few words to a *Sky at Night* camera, or to do a magazine interview, before he became too tired, or just too befuddled. Any other person approaching 90 in Patrick's state would have been permanently hospitalized, but in hospital Patrick could not have endless friends visit him, could not do media interviews, could not have a Gin & Tonic, and could not have his own way, as he did in his Selsey castle. Also, his surviving cat Ptolemy would not be able to visit him either.

The root cause of all Patrick's problems in 2012 was his increasing Congestive Heart Failure. Although his pacemaker was telling the heart to beat regularly the cardiac muscles were now far too weak to cope with pumping blood around his massive body, with its deteriorating arteries. Most people were amazed that a man this old and this heavy, with a lifetime of heart and arthritis issues and such a huge food and alcohol intake, could possibly still be alive at all. Patrick's Congestive Heart Failure was causing severe life-threatening water retention issues to be a permanent part of his life and his doctors instructed him that if he insisted on going home he must spend several hours each day with his legs elevated, to reduce the water retention problems. This was in addition to a host of other measures and even more drugs. Increasingly his legs resembled Zeppelins, which in turn was causing an issue with his knee joints and the spread of fluid and infections to other organs. This, in turn, was making him tired and befuddled, even without the constant pain from the arthritis. However, he was determined to be released from hospital as soon as possible and simply would not spend most of the day with his legs elevated at home. Patrick had always been an impatient workaholic and so sitting with his legs up on a bed, even if he had a mug of Gin & Tonic to hand, was simply not on. So, the hospital trips to reduce his water retention became more and more regular and Patrick became weaker and more incomprehensible. The few minutes he was seen for on *The Sky at Night* each month were the highpoint and the focus of his life, which was increasingly dominated by him being washed, fed and dressed by his carers, or by hospital staff.

Almost all of *The Sky at Night* programmes in recent years had been filmed 2 or 3 weeks before the show was broadcast and this was the case with the edition transmitted at midnight on BBC1 on Sunday night/Monday morning, December 2nd/3rd 2012. It was recorded on the afternoon and evening of November 19th/20th. Unlike for the previous programme, Patrick was back in the study with guests Dave Rothery and Katie Joy, talking about Mercury and looking slightly stronger than in the last two shows. When the subject of possible water at the poles of Mercury was mentioned Patrick said that he was a total skeptic, and quipped that if someone brought him a cup full of this water he would believe it! Chris Lintott also presented a piece commemorating that it was now 40 years since the last Apollo went to the Moon. In a poignant exchange between Patrick and Chris, when the November Total Solar Eclipse seen from Australia was mentioned, Chris said to Patrick "Don't you wish we'd been there?" "I wish we had" replied Patrick. "Next Time" said Chris....... *The Sky at Night* producer Jane Fletcher had the foresight, while Patrick was still coherent, to film an extra set of opening and closing sequences in that

November session, for the January show, as it was obvious things were looking very grim. Five days later, on November 24th, Patrick's friend Damian Peach left Selsey for good and moved house, as he was sick of the daily commute to work.

Sadly, time was running out for a 'Next Time' of anything for Patrick, let alone a trip to a Total Solar Eclipse. As mentioned earlier, the December Mercury & the Moon programme had been filmed on the afternoon and evening of November 19th/20th and the next 2 weeks saw Patrick back in hospital at Chichester, with more fluid retention problems and associated infections. Patrick's heart was now so weak that even speaking coherently for a few minutes was sapping all of his energy. On November 30th, well aware of Patrick's worsening state and very conscious of the media circus and even criminal activity that could result at Farthings if Patrick died, Brian May went to visit him. He told his followers that Patrick was 'A little poorly, but still managing to tell a joke or two. He says "Best Wishes to everyone"'.

On December 3rd, the day after the Mercury & the Moon edition of *The Sky at Night* was transmitted on BBC1, the camera crews were back at Farthings, filming for the next programme: the first planned for 2013. Patrick's back lawn was filled with amateur telescopes that might have been received as Christmas presents by the time the next show was due to be aired. The previous weeks and months in Britain had been horribly wet and cloudy and the giant planet Jupiter was now at opposition, closest to the Earth and at the highest altitude it had been in the sky for 10 years. After severe flooding in various parts of the U.K. the weather had suddenly turned very cold indeed and Pete Lawrence e-mailed that it had been a long, cold, but clear night filming all the telescopes on Sir Patrick's back lawn. However, Patrick was back in hospital again, his condition almost critical, so he was not at Farthings when the telescope testing was being filmed in his garden for the January 2013 show; his opening and closing lines for that show were already recorded.

This time the doctors told Patrick that there was nothing more they could do to help him now and as he was absolutely committed to ending his days in his beloved Farthings house, where his mother had died on January 7th 1981, he was sent by ambulance, back to his beloved home and Ptolemy the cat, on December 6th.

Over the next 2 days Patrick's condition slowly deteriorated and, at the crack of dawn on December 8th, Brian May headed back down to Selsey. He told his followers that Patrick was 'fighting back but still poorly' but it was obvious to all of Patrick's friends that this looked like the end; he was barely able to stay conscious. Patrick's housekeeper Julia Knight started phoning friends who Patrick had wished to be there in his final hours, like Peter Cattermole, Iain Nicolson, Jane Fletcher and several of his closest godsons. Those that could make it quickly arrived to keep vigil alongside Brian May, a friend and legal man, David McCahearty, Patrick's carers and Ptolemy the cat. Pete Lawrence e-mailed me and a few others at 6.50 p.m. from Selsey, on that grim Saturday evening, to say that this really did look like the end of the legend that was Sir Patrick Moore.

Throughout Saturday night and Sunday evening a few reports arrived, via Brian May, from inside Farthings. Around 11 p.m. on December 8th the messages said: 'It looks like hours rather than days, but of course you can never be sure. Just thought you should be prepared. David McCahearty is here. Peter Cattermole just arrived and

Julia is in command, doing a great job. Patrick is very peaceful, but mostly, seemingly, not conscious. We've prepared a press statement and are working on a Memorial website. Donations to Cats Protection.' This was followed by another message in the early hours of Sunday December 9th: 'Effectively he is gone from our world. And – thank Heavens – peacefully. We all gave him what we could when we could. It's time to let him go, gratefully.'

Remarkably, in this social media obsessed world, no-one sneaked the information to the press and so those gathered at Farthings just waited until Patrick finally stopped breathing and his pacemaker could no longer encourage his heart to work. Patrick died at Sunday lunchtime on December 9th 2012, at 12.25 p.m. The BBC News in the hours preceding Patrick's demise had been dominated by the apparent suicide of a nurse at the King Edward VII hospital in London, following a hoax call from an Australian radio station's presenters, about the Duchess of Cambridge's morning sickness. Other news items had been the ongoing row over same sex marriages in Churches, Egyptian President Morsi's extra powers and Nelson Mandela spending the night in hospital. The imminent Manchester City/Manchester United derby match was also making the headlines. However, seconds before the lunchtime weather forecast on BBC News 24, at 12.57 p.m., the BBC newsreader Clive Myrie announced the sad news. The red BREAKING NEWS banner scrolling across the bottom of the screen blared out SIR PATRICK MOORE DIES. It had taken just 32 min from Patrick's final breath for the news to get from Patrick's house to be the main BBC headline. After the weather forecast the BBC newsreader Annita McVeigh ran with the story as the first item, and it dominated the News channels for the next 24 h.

I must admit that even though I knew the sad news was imminent the sight of that red 'BREAKING NEWS SIR PATRICK MOORE DIES' headline scrolling across the screen was deeply upsetting. Suddenly it hit me that the most influential figure in my entire life, apart from my own parents, had departed. A man whose book, 'The Observer's Book of Astronomy', I had picked up as a 10 year old boy in 1968 and had been influenced by ever since, was gone. It was hard to comprehend, even though I had known it was coming. Somehow, seeing it announced on the BBC News, made it horribly, terribly real.

Brian May tweeted 'I took your kind messages down to him yesterday afternoon folks. Not sure how much he took in, but he felt the love. THANKS. Bri.' This was followed by: 'Patrick was very peaceful in his last hours. With his beloved cat Ptolemy purring on his bed. And his pals all around. A great innings.'

Of course, tributes then started to pour in. Within hours 1,110 comments had been received by the BBC website and they closed it to further traffic. An infinity of tributes also piled into the web page set up by Brian May. Hundreds of Tweets flew around Twitter and within hours #PatrickMoore and #RIPSirPatrickMoore were trending. It goes without saying that the next two nights in the UK were two of the clearest nights anyone could ever recall seeing, with Jupiter and Orion dominating the evening sky. Colourful stratospheric clouds were seen over northern Europe that night and a brief period of impressive auroral activity was seen from parts of Northern Ireland, Northern England and Scotland. All purely coincidental of course....

Also, just the same as happened after the death of Sir Arthur C. Clarke, some 9 h after Patrick's passing a Gamma Ray Burst, GRB 121209A, was detected, close to the ecliptic, on the borders of Aquarius and Capricornus. Again, as I say, purely a coincidence; but I just thought I'd mention these things.

However, few of the astronomers I knew were observing on the crystal clear nights of December 9th or 10th, as they were all in a state of total shock. We had begun to believe that Patrick was immortal and now he was gone. The last 5 months of 2012 had certainly been grim ones for astronomy and space fans, especially in the UK. Patrick's dear friend Prof. Sir Bernard Lovell had died on August 6th and Neil Armstrong had passed away on August 25th. Another BAA friend of Patrick's, Richard (Dick) Chambers, some 8 years his junior, had died on November 21st.

The BAA President Bill Leatherbarrow, who had known Patrick for almost half a century, summed the Association's feelings up when he wrote: 'Many of us felt that this day could never come, and that the normal laws of nature would somehow be suspended in this case. Sadly and inevitably, that was not to be and we shall all miss a presence that has enriched British astronomy and the lives of most of us, for as long as we can remember.'

The BBC quickly announced that there would be a tribute programme for Patrick 2 days after his death, at 10.35 p.m. on BBC1 and the personal tributes just kept on coming in. Brian May announced that Patrick's internment would be a very private affair, but that some sort of memorial party would be held on what would have been his 90th birthday, on March 4th 2013. The next morning Chris Lintott and Brian May (via a video link in Brian's case) appeared on the BBC Breakfast programme, talking about Patrick. Brian installed security guards at Farthings as soon as Patrick passed away and just in time too! Soon after Patrick's death was announced at least one opportunistic vulture tried to gain entry to the house, but was turned away. One piece of lowlife even advertised one of Patrick's ancient typewriters for sale on eBay, for £1,000, within hours of his passing, until the item was withdrawn after complaints.

Six days after Patrick's death, on Saturday December 15th, all of the BAA members at the association's Christmas Meeting stood in silent memory for Dick Chambers and for Patrick. It was 78 years and 17 days since Patrick's name had first been announced at a BAA meeting, as a brand new elected member on November 28th 1934. I received a phone call from *The Sky at Night* co-presenter Paul Abel 2 weeks after Patrick's death. Paul had been a speaker at that BAA Christmas meeting (talking about Black Holes) and his final slide had shown Patrick, dressed as a Druid magician, at Stonehenge in 1990. The presenters and a small group of Patrick's closest friends had attended his cremation at Chichester, 10 days after his passing. It was, not surprisingly, both a sad and a moving experience and the Selsey Branch of the Royal British Legion had saluted as the hearse headed from Patrick's home to Chichester. Paul said that they all returned to Patrick's (now Brian May's) beloved house Farthings after the cremation, but as he told me: "Without Patrick there, it was spooky. It just didn't feel right. I headed back to Leicester as soon as I could." He also confirmed that Patrick's cat Ptolemy had gone to a caring new owner. Did Patrick blow out the candle, as promised, at the appointed time, from the afterlife?

Well, ermmm, no! No-one could remember where the audio tape Patrick had stored for this event (saying "I'll blow this candle out if it kills me!") was kept! So, a candle was lit and after a while the mourners blew it out....... Even the Farthings ghost Claude did not lend his services. All a bit of an anti-climax really.......

Patrick's final appearance on *The Sky at Night* would be in the programme transmitted at midnight on January 6th/7th 2013, using the extra footage filmed in November 2012. Patrick had, just, extended his TV career into yet another year! It was also rather fitting that the only relevant letter published in the *Radio Times* for that week was from David Hardy, whose splendid illustrations for *Suns, Myths and Men*, almost 60 years earlier, had played such a crucial role in the first *Sky at Night* producer already knowing who Patrick was, before he even saw him perform in front of a TV camera. I find it just a bit spooky to think back to the November 1994 Total Solar Eclipse when David, Brian May and myself, relatively unknown to one another, occupied the same small mini-bus while we crossed the Peru/Chile border at night. Somewhere along that road Patrick was heading in the other direction and we passed in the darkness..........

So, that was the story of the life of Sir Patrick Moore. Born March 4th 1923 and died December 9th 2012. He died, like his father 65 years earlier, just before the Winter Solstice. Patrick's lifespan was 89 years, 9 months and 5 days. He lived 5 years longer than it takes the seventh planet Uranus to orbit the Sun, which is a pretty good innings. He also, undoubtedly, encouraged more human beings to look up into the night sky than any other man in history: a pretty impressive achievement I'd say.

Rest in Peace Sir Patrick. We will never see your like again!

Chapter 43
Epilogue: Patrick's Legacy

How will history regard Patrick Moore? Surely, more than 55 years of presenting a regular monthly program on astronomy will never be beaten? However, astounding though that achievement is, his everlasting legacy will be his books. His output as an author was truly colossal, especially in the genre of popular science writing. Ironically, the causes he pursued most vigorously, namely drawing the Moon, believing its craters were volcanic and his belief in Transient Lunar Phenomena, were a total waste of time. His drawing work with Wilkins was totally superseded well before the Apollo lunar landings and the lunar craters are caused by impacts. As for TLP, well, in most astronomers' view they were almost certainly just figments of the Earth's atmosphere or the product of vivid imaginations, a romantic legacy from the days when changes on the Moon and Martian canals were considered quite reasonable. But when Patrick turned to broadcasting and book writing he became a legend. He was also, surely, one of the most generous men to ever walk the surface of the Earth. He could have been wealthy but he gave all his talks for free or for charity, offered complete strangers a look through his telescopes (and as much food and alcohol as they could swallow) and never did advertising for commercial gain. His cars and telescopes were all old and second hand, like his typewriter!

The big break of clinching *The Sky at Night* in 1957 and the subsequent launch of Sputnik and then the race to the Moon in the 1960s, made his name a household one. That break, as we have seen, was the result of a number of fortunate factors, not least Patrick's ability to speak clearly, concisely and without errors, while on live television. He was, without doubt, the right man, in the right place, at the right time.

There were other critical issues too. During the 1950s Patrick was one of the youngest BAA members, amongst a sea of elderly greybeards. He looked and sounded the part as far as television presenting was concerned and his enthusiasm shone through. Single-handedly he (with help from the Apollo Moon programmes) lowered the age of the average BAA member substantially and those members stayed as his fans for decades. By the time Patrick died the average age of the typical BAA member was back in the fifties: the children of Patrick's religion who had

M. Mobberley, *It Came From Outer Space Wearing an RAF Blazer!: A Fan's Biography of Sir Patrick Moore*, DOI 10.1007/978-3-319-00609-3_43,

just aged, but stayed loyal to his hobby and personality. Also, when Patrick started broadcasting each month he had more than 20 years of meticulous backyard observing experience to back him up, something which seems to have no bearing at all amongst twenty-first century TV astronomers (so-called)!

In addition, it is often who you know, not what you know, that gets you into a good position. In Patrick's case he had already been partly inspired by Arthur C. Clarke's success at writing science fiction. When you see someone of your own generation making it as an author, it helps you to make the big leap from employment (as a teacher) to self-employment. Being reported as a good lecturer speaking about the Moon clinched him that critical *Guide to the Moon* book contract too. Patrick also knew Desmond Leslie and Lord Dowding (the UFO fanatics) which secured his first TV appearance as a UFO critic. Perhaps the most crucial slice of luck though was Paul Johnstone, the first producer of *The Sky at Night*, stumbling across Patrick's book *Sun's Myths and Men*, just before meeting Patrick as a UFO sceptic. Another critical connection was that Major Levin, who proposed the young Patrick for BAA membership in 1934, was a leading light in the BAA Computing Section, as was Dr J.G. Porter, the BBC's 'voice of astronomy'. Living within a few 100 yards of an astronomical observatory with a kindly astronomer (Franks) and owner (Hanbury) in the 1930s was another massive slice of luck.

All these factors played a part in clinching Patrick's appearance on that experimental astronomy programme back in 1957. The naked eye comet Arend-Roland appearing at the same time and the Soviets launching Sputnik less than 6 months after that critical first programme kept the show going too. Suddenly the public were very interested in space, space travel and the Soviet nuclear threat.

Critically, at that point Patrick did not get greedy, he just cashed in on his fame, his lovable eccentricity and his writing stamina, but without any form of financial contract with the BBC. He realised exactly how lucky he had been to get those big breaks in 1957 and, unlike so many self-obsessed cringe-worthy 'TV personalities', it did not go to his head: he kept his feet firmly on the ground, even if his brain was amongst the stars. Without the distractions of a wife, a family, or a boring day job and with a mother and housekeepers to look after him, he became what he was: a living, breathing, astronomy writing, broadcasting and observing machine. Nice work if you can get it!

Patrick could justifiably be called the greatest astronomy populariser of all time on the basis of his literary output alone, over so many years, even without him spending more than 55 years on British TV. But of course, one thing led to another. When you are a household name publishers come knocking on your door. The only person who could really rival Patrick as the all-time *science* populariser would be Isaac Asimov (1920–1992). Asimov's general science knowledge was far more extensive and his science fiction was infinitely better, but in terms of astronomy, well, Patrick really had no rivals. How many books did Patrick actually write? For some obscure reason he is often credited in the media as being the author of a mere 60 books. How utterly ridiculous! He had passed that milestone by 1964! Even Patrick did not know how many books he should be attributed with having written. His study at Selsey was filled with his books: bookshelf after bookshelf. A summer house on the west side of his house was totally filled with the overspill, as well as spare typewriters!

In assessing Patrick's total literary output, one has to be very careful. His name is associated with over 700 different titles in one way or another, either as the author, the editor, the series editor or simply by writing a foreword. The *Yearbook of Astronomy* ran for half a century with Patrick Moore as the main or deputy editor. He was the principal editor of that book for the majority of that period, apart from during the early years, when he was the deputy to J.G. Porter and apart from when he worked alongside John Mason, who carried out the donkey work as Patrick became increasingly frail. Thirteen books documented *The Sky at Night* programmes, with a further three (*The Unfolding Universe, TV Astronomer* and *Eyes on the Universe*) being associated with the series reaching 25, 30 and 40 year milestones.

Then there were his 'Guide to' books covering the Moon, the Planets, Mars and Comets. Much of the text from these was transferred to other books in revisions over the years. Thus, *Guide to the Moon* became *Survey of the Moon* then back to *Guide to the Moon* and finally *Patrick Moore on the Moon.* As a series editor Patrick inspired other, more technically minded amateurs, to write books about advanced back-garden astronomy which he could not, thus retaining a share of the book and partly shedding his self-confessed 'dinosaur' image. Patrick started this 'Editor in Chief' process with Lutterworth Press in the 1960s and continued it into the 1990s and twenty-first century with Springer-Verlag's Practical Astronomy Series. If one includes the annual Yearbooks and books where Patrick had a significant editorial input, but exclude books where he was just 'series editor', then Patrick can realistically be called an author of well over 300 popular science books. A quick look at the books in my study confirms this. Okay, many of these were revisions of earlier books, but a lot were complete revisions. On top of this remarkable tally we have his works of schoolboy fiction, amongst which I feel his alter-ego Cedric Allingham's book belongs! Then there was *Bureaucrats: How to Annoy Them* and *The Twitmarsh Files*, under the pseudonym of R.T. Fishall. How does one categorise them?

Essentially, when Patrick was at home in his study and it was cloudy, his fingers were hammering away at that Woodstock keyboard; and they were hammering away for more than 50 years, almost continuously! In his final years he was dictating books or co-authors were doing the bulk of the work. The astute reader will easily deduce that allotting over 300 books to Patrick in his main half-century of typing means that he was churning out about six books per year, *on average*. Surely, you might think, this is impossible?! In fact, when you look at the publication dates, in some years he had far more than this. For example, 12 new books with him as the main author appeared in 1977 alone! I have attempted to list his books, by year, in the appendix, where I repeat some of these amazing statistics. It makes staggering viewing. Put a famous man with no wife, children, DIY chores, or boss in front of a typewriter, equip him with a boundless enthusiasm for astronomy; then instil in him the view that no-one *ever* tells him what he can and can't write, because there is always another publisher who will take his work. Then let him go for over half a century: the result is Patrick Moore! He wallowed in his greatest love of astronomy, while friends, mother and housekeepers did the tedious chores. Result: over 300 books to your name. In many ways Patrick was like a precocious excitable schoolboy, who just never grew up once he realised that he could carry on with his childhood ways throughout his entire life. Despite his heart problems, he had boundless reserves of energy for writing, speaking and observing the night sky.

Even though early footage of Patrick on the BBC is very limited and, tragically, this applies to the Apollo Moon landing tapes too, his books leave a legacy which will survive for centuries. This guarantees that Patrick will never be forgotten. One thing is for sure: we shall not see his like again, unless someone clones him! Now there's an idea....

My final advice to the reader is this. When I read my first Patrick Moore book, *The Observer's Book of Astronomy*, in 1968, I noticed the address of the British Astronomical Association in the back. The next year, aged 11, I signed up. If not for that mention in Patrick's book I would not have joined the BAA or, ultimately, written this book. Patrick was always keen for people to join the Association that changed his life, so I too will now include the BAA contact details below. Without the BAA there would have been no Patrick Moore and no programme called The Sky at Night. It was the BAA that made Patrick what he was, not the BBC!

The British Astronomical Association
Burlington House
Piccadilly
London
W1J 0DU
http://www.britastro.org/

Appendix

Patrick's Variable Stars

This is a list of Patrick's 48 favourite (and not so favourite) variable stars, in order of how frequently he observed them from 1967 to 1994. Of course, he started observing as a child, but these are the official BAA database records covering observations submitted from when he was aged 44–71. His BAA Variable Star Section I.D. code was MR. The numbers in brackets denote the number of magnitude estimates that Patrick made. Many of these stars are observable through binoculars or small aperture telescopes, or in the case of novae, like HR Delphini, were binocular objects at their peak. In total he made 5,612 variable star estimates in this time period: not bad for a mainly lunar and planetary observer! As well as these BAA database records, Patrick also sent many observations to the AAVSO (American Association of Variable Star Observers) between 1968 and 1990. The official AAVSO records show that he paid his $5 membership and $1 application fee on May 20, 1968 and submitted observations to them backdated to 1954 (when he was 31) and until 1990 (when he was 67). His AAVSO code was MOP. In total Patrick sent 2,183 variable star magnitude estimates to the AAVSO. Anyway, here are the stars he observed for the BAA with the number of observations in parentheses:

HR Delphini (700); SS Cygni (467); R Coronae Borealis (439); SU Tauri (369); U Cygni (304); W Cygni (285); T Corona Borealis (267); U Orionis (265); RU Pegasi (252); R Leonis (217); R Cygni (206); TZ Persei (198); Gamma Cas (170); T Tauri (156); RX And (155); Chi Cygni (153); Alpha Orionis (148); R Arietis (109); R Serpentis (78); AY Lyrae (77); S Delphini (76); W Cassiopeia (68); Eta Geminorum (46); V1500 Cygni (45); S Coronae Borealis (42); Rho Cassiopeia (41); P Cygni (39); T Cephei (38); Omicron Ceti (35); SS Aurigae (21); U Geminorum (18); R Andromedae (17); S Bootis (12); U Delphini (13); EU Delphini (12); X Aurigae (11); Epsilon Aurigae (9); Mu Cephei (9); Beta Pegasi (9); Rho Persei (7); S Virginis (6); X Persei (5); AF Camelopardali (4); X Leonis (4); R Lyrae (4); CH Ursae Majoris (3); S Cephei (2); R Scuti (1)

M. Mobberley, *It Came From Outer Space Wearing an RAF Blazer!: A Fan's Biography of Sir Patrick Moore*, DOI 10.1007/978-3-319-00609-3,

Patrick's Voluntary BAA Roles

Year (Oct/Oct)	Observing section posts and other BAA posts	BAA council status	Miscellaneous comments
1934/46	Ordinary member/observer		From childhood to RAF navigator
1946/47	Lunar assistant to Wilkins		Rejoined RAF as ATC commander
1947/48	Lunar assistant to Wilkins		Schoolteacher. Father died Dec 1947
1948/49	Lunar Section secretary		Schoolteacher
1949/50	Lunar Section secretary		Schoolteacher
1950/51	Lunar Section secretary		Schoolteacher
1951/52	Lunar Section secretary	Elected	Schoolteacher
1952/53	Lunar Section secretary	Elected	Schoolteacher/Guide to the Moon
1953/54	Lunar Section secretary	Elected	Became a full time author
1954/55	Lunar Section secretary	Absent	Started churning out books!
1955/56	M&V Director/Lunar secr.	Ex officio	BIS *Spaceflight* Ed. 1956–1959
1956/57	Mercury & Venus Director	Ex officio	Started *The Sky at Night*
1957/58	Mercury & Venus Director	Ex officio	& JAS President from 1956–1958
1958/59	Mercury & Venus Director	Ex officio	
1959/60	Mercury & Venus Director	Ex officio	
1960/61	Mercury & Venus Director	Ex officio	
1961/62	Mercury & Venus Director	Ex officio	
1962/63	Mercury & Venus Director	Ex officio	
1963/64		Absent	
1964/65	Lunar Section Director/TLP	Ex officio	Created Lunar TLP sub-section
Patrick moved from East Grinstead to Armagh Planetarium in 1965 while Lunar Sec. Director			
1965/66	Lunar Section Director/TLP	Ex officio	
1966/67	Lunar Section Director/TLP	Ex officio	
1967/68	Lunar Section Director/TLP	Ex officio	OBE & Walter Goodacre Award
Patrick moved from Armagh to Selsey in 1968 in his final months as Lunar Section Director			
1968/69		Vice-President	Apollo Moon Landing era
1969/70	Variable Star Binoc. co-ord.	Vice-President	Apollo Moon Landing era
1970/71	Variable Star Binoc. co-ord.	Vice-President	Apollo Moon Landing era
1971/72	Lunar Section Director/TLP	Ex officio	Apollo Moon Landing era
1972/73	Lunar Section Director/TLP	Ex officio	Apollo Moon Landing era
1973/74	Lunar Section Director	Ex officio	
1974/75	Lunar Section Director	Ex officio	
1975/76	Lunar Section Director	Ex officio	Resigned as Lunar Dir. in Feb 1976

(continued)

(continued)

Year (Oct/Oct)	Observing section posts and other BAA posts	BAA council status	Miscellaneous comments
1976/77		Vice-President	Serious spinal injury May 1977
1977/78	Lunar Sec. TLP co-ord.	Absent	Absence due to Mother's illness
1978/79		Absent	Absence due to Mother's illness
1979/80		Absent	Absence due to Mother's illness
1980/81		Absent	Gertrude Moore died Jan 7th 1981
1981/82		Elected	
1982/83		PRESIDENT	
1983/84		PRESIDENT	
1984/85		Vice-President	
1985/86		Vice-President	
1986/87	COUNCIL TURMOIL!	Absent	Founded *Astronomy Now*
1987/88		Elected	Awarded CBE
1988/89		Vice-President	
1989/90		Vice-President	
1990/91		Absent	
1991/92	Public Relations (+TLP)	Elected	July 1992 Serious eye injury
1992/93	Public Relations Officer	Vice-President	
1993/94	Public Relations Officer	Vice-President	Severed all links to *Astronomy Now*
1994/95	Public Relations Officer	PRO role	*Practical Astronomy* rose and fell!
1995/96	Public Relations Officer	Vice-President	Rec'd BAA Lydia Brown Medal
1996/97	Public Relations Officer	PRO role	*Modern Astronomer* rose (fell in 99)
1997/98	Public Relations Officer	Elected	Last solar eclipse jaunt, Feb 1998
1998/99	Public Relations (+ TLP)	Elected	Cornwall eclipse/Knee replaced
1999/00	Public Relations (+ TLP)	PRO role	Serious mobility problems start
2000/01	Made a Life Hon. Vice-Pres.	Vice-President	Knighted, but partially paralysed
2001/02		Vice-President	
2002/03		Vice-President	Autobiog. '*80 not out*' published
2003/04		Vice-President	Venus transit & food poisoning
2004/05		Vice-President	BBC *Sky at Night* magazine appears
2005/06		Vice-President	
2006/07		Vice-President	April 2007: 50 year of *The Sky at Night*

(continued)

(continued)

Year (Oct/Oct)	Observing section posts and other BAA posts	BAA council status	Miscellaneous comments
2007/08		Vice-President	
2008/09		Vice-President	
2009/10		Vice-President	75 years of BAA membership!
2010/11		Vice-President	
2011/12		Vice-President	Witnessed 2nd Venus transit!
2012/13	From 2012 Oct to 2012 Dec	Vice-President	V.P. for 6 weeks up to Dec 9th 2012

A table depicting Patrick's numerous voluntary roles within the British Astronomical Association (BAA). He never, *ever*, claimed any expenses whatsoever in any of these roles and they were all unpaid. The BAA year runs from October to October and the table generally reflects this as well as it can do, given such a rigid row and column format. In reality Patrick's BAA roles were far more complex than depicted, so I have listed his most prominent roles for each year. He was an almost permanent BAA Lunar Section and Planetary Section committee member of one sort or another from the late 1940s to the late 1990s. Arguably his most onerous role is not obvious at all in the above table. This was his ceaseless and remorseless writing up of the BAA minutes after each Ordinary, Special or Annual General meeting. He started as the meetings recorder from the late 1940s and, more often than not, recorded all the meetings he attended for the Association, right through to the period when his mother became ill in the late 1970s. He never requested that his name be put at the base of the report either, as he considered this irrelevant! From the 1980s, after Patrick's term as BAA President, various other reluctant volunteers did this chore. I was the meetings recorder for about 7 years in the 1990s and that was enough for me! The term 'Ex officio' in the table means that by virtue of another significant post (such as a Section Directorship) Patrick was a Council member with a vote and so did not have to be elected that year. The years when he was absent from the Council table were, with the exception of the years when his mother was ill, enforced by the BAA bye-laws. These rules changed over the decades, but they limited the number of consecutive years that someone could serve on the Council if they were not a Secretary, the Treasurer, or a Section Director. In addition, when Patrick served as the Lunar Section Director he often performed multiple duties, as he was also invariably the TLP sub-section co-ordinator, the Lunar Section meetings recorder, a regular speaker and even the Lunar Section Circulars editor! On a couple of occasions during the 1990s he opted to just be the Public Relations Officer (PRO) but not stand for an elected Council post, as he knew he could not attend every meeting and so did not want to stop other younger councillors from being elected. As 'Sir Patrick' he was given a permanent 'Honorary Vice-President' post so he did not have any chores and did not have to enter his name into the annual Council ballot either.

Total Solar Eclipse Trips Experienced by Patrick

Year/Date	PM age	Total solar eclipse location	Duration	Comments
1954 June 30	31	Stromstad, Sweden	2 m 32 s	BAA trip
1961 Feb 15	37	Mt. Jastrebac, Yugoslavia	2 m 27 s	BBC trip
1968 Sep 22	45	Yurgamysh, Siberia	0 m 39 s	BBC trip
1973 Jun 30	50	Monte Umbe, off Mauritania	5 m 56 s	Transolar/BAA/BBC
1983 Jun 11	60	Surabaya, East Java	5 m 6 s	Explorers
1988 Mar 18	65	Talikud Island, Philippines	3 m 21 s	Explorers
1991 Jul 11	68	La Paz, Baja California Sur	6 m 35 s	Explorers
1994 Nov 3	71	Peru/Chile border region	2 m 55 s	Explorers
1995 Oct 24	72	South China Seas	2 m 10s	Marco Polo Cruise
1998 Feb 26	74	Caribbean nr Curacao	3 m 39 s	BBC/Solaris Cruise
1999 Aug 11	76	Falmouth, Cornwall	2 m 2 s	BBC UK broadcast

Patrick's 11 total solar eclipse trips and durations (deduced from NASA data) spent under the umbral shadow, whether clear or cloudy. Almost all of these eclipses were seen in clear skies, or through haze with very thin cloud. In fact, only Patrick's final 1999 Cornwall eclipse, when he was 76, was 100 % cloudy at totality, although his 1994 eclipse suffered from cloud and the theft of Patrick's possessions meant he never referred to that particular South American eclipse trip, ever! Patrick's autobiography mistakenly lists a 1986 eclipse in the index under 'Solar Eclipses', but this is incorrect, as that was a total lunar eclipse, witnessed after observing Comet Halley. The only eclipse trip Patrick ever paid for was his first, in 1954, when he was 31! The large gap between 1973 and 1983 was mainly caused by Patrick's reluctance to travel abroad when his mother became frail. Adding all the times together gives Patrick a cumulative time under the Moon's shadow of 37 minutes 22 seconds.

Patrick's 300+ Books and Other Media Offerings

Compiling any list of Patrick's books is a daunting task. As I have already mentioned, apart from the sheer number of them it is often tricky to draw the line between a slight revision, a major revision and a complete rewrite. Patrick's 60 year writing career means that his 1953 book *Guide to the Moon* bears no resemblance to his 2001 book *Patrick Moore on the Moon*, yet the former gradually evolved into the latter. I have used my judgement (I own many of these books!) to draw a dividing line. All of Patrick's "Guide to" books of the 1950s had various revisions and rewrites. The same is true for his *Atlas of the Universe* and all the "Atlas of" books. Patrick's *Guinness Book of Astronomy Facts and Feats*, as another example, evolved almost every year and then, after sales of 60,000 copies, transformed into the infinitely larger *Data Book of Astronomy*. I have not included (except with his own autobiography) simple reprints of the same book or trivial revisions; but I have classed substantial revisions and total rewrites as new books, even if the title is the same. Of course, much of the basic material in Patrick's books is very similar. There are eight planets plus Pluto in our solar system. Then there are comets. There are the stars too, and the Universe as a whole. Guides to these basic categories encompass more than 80 % of Patrick's output. So there is, inevitably, a lot of overlap in many of his books. Nevertheless, his total output is truly staggering consisting, as I say again, of more than 300 books. Roughly 15 % of Patrick's total output were co-authored books, where Patrick wrote the lions share but another author contributed and, sometimes, wrote the tricky 'hard science' bits. In many cases where he collaborated with a good friend (such as Paul Doherty doing the artwork) the book would be listed as being co-authored despite Patrick writing the bulk of the words.

As I mentioned earlier, for 40 years Patrick was the sole editor of the annual *Yearbook of Astronomy* and he was the deputy editor for 2 years in the 1960s (with Guy Porter); he was also the joint editor in the twenty-first century, with John Mason. I have included all of the Yearbooks in the list as the editing task would have been significant, even if there were so many contributors in each volume. Then there is Patrick's fiction output; a total of 30 books, plus books written under pseudonyms to consider. The fiction was schoolboy stuff that stands no comparison with the best, or even the second best S.F. writers, but it still resulted in another book on the shelves, so I have included them in the tally. Patrick was the series editor of many books as well, mainly for Lutterworth in the 1960s and Springer in the 1990s; I have not included those Springer books in the count as he was simply a name on the books and contributed very little, except a nod of approval to the title. Neither have I included books where Patrick simply wrote a foreword or an introduction, of which there are dozens more! Eliminating fiction books, the Yearbooks and books where Patrick was an editor still leaves him with well over 200 titles. Including everything credits him with that figure of 300, which is simply the result of a man at a turn-of-the-century typewriter, hammering out an average of six books a year for over half a century. Extraordinary!

I will repeat myself here, but as someone who has used Patrick's typewriter and felt how stiff those keys are, I am not surprised his swollen banana sized fingers were crippled with arthritis from his 1970s onwards! Consider this too.....there was no word processing on Patrick's typewriter. You cannot backspace and edit a line. You have to get the sentence right *first time*. Modern authors have it so much easier, but Patrick shunned the electronic route, much preferring a machine he knew and trusted! Patrick wrote as he spoke. Occasionally he would make a typo as he hit the wrong key, but the grammar was pretty much perfect as he typed it. When you use a traditional typewriter this is the only way to proceed. As I stated earlier, compared to Patrick's peak years the word processor authors of today don't appreciate how easy it is now, and Patrick could still churn out a more readable book ten times faster! In addition, while modern authors can trawl endless facts using Google, and 'cut-and-paste' stuff from *Wikipedia*, for the Patrick Moore of the twentieth century it all came out of his brain. Absolutely incredible!

Patrick's books in chronological order:

1950

TRANSLATION:
The Planet Mars by Gerard De Vaucouleurs; translated by Patrick.

1951

No books

1952

FICTION:
The Master of the Moon: An enthralling science-fiction story.

1953

FACTUAL:
Guide to the Moon

1954

FICTION:
Frozen planet
The island of fear: A story for boys
Out into space (co-authored with A.L. Helm)
SPOOF:
Flying Saucer from Mars by Cedric Allingham (alias Patrick Moore)
FACTUAL:
The boy's book of space
Suns, Myths and Men

1955

FICTION:
Destination Luna by Patrick Moore and Patricia Cullen (artwork)
Mission to Mars (The first book in the Maurice Gray series)
Quest of the Spaceways
FACTUAL:
Guide to the planets
Earth satellite: The new satellite projects explained
The Moon: A complete description of the surface of the Moon, by Wilkins & Moore

1956

FICTION:
Wheel in space
The Domes of Mars (Maurice Gray book 2)
World of Mists
FACTUAL:
The planet Venus
Guide to Mars

1957

FICTION:
The Voices of Mars (Maurice Gray book 3)
FACTUAL:
Isaac Newton
The true book about earthquakes and volcanoes
Making and Using A Telescope by H. Percy Wilkins & Patrick Moore
Space Research and Exploration – Associate editor Patrick Moore.

1958

FICTION:
Mission to Mars
Peril on Mars (Maurice Gray book 4)
FACTUAL:
The true book about the Earth
Science and Fiction
The Solar System
Space exploration (National Book League. Reader's guides, 3rd series)

1959

FICTION:
Raiders of Mars (Maurice Gray book 5)
FACTUAL:
The true book about man
Rockets and earth satellites

1960

FICTION:
Captives of the Moon
FACTUAL:
Astronautics. Illustrated by David Hardy.
Guide to the Stars

1961

FICTION:
Wanderer in space
FACTUAL:
Astronomy
The Amateur Astronomer (Amateur Astronomer's Library Volume 1)
Conquest of the air: The Wright Brothers
Navigation by Henry Brinton and Patrick Moore
Yearbook of Astronomy 1962 (Associate Editor to J.G. Porter)

1962

FICTION:
Crater of Fear

FACTUAL:
The Observer's Book of Astronomy. Illustrated by L. F. Ball.
Life in the Universe by Francis Jackson and Patrick Moore
The true book about worlds around us by Patrick Moore and A.L. Helm
The astronomer's telescope
The Stars (Young enthusiast library – Young Scientist Series)
Telescopes and Observatories (Finding Out Books)
Exploring Time by Henry Brinton and Patrick Moore. Illustrated by Cyril Deakins
Yearbook of Astronomy 1963 (Associate Editor to J.G. Porter)
The Planets (revised 'Guide to the Planets')

1963

FICTION:
Invader from Space
FACTUAL:
Practical Amateur Astronomy (Amateur Astronomer's Library Volume 4 – Editor)
Survey of the Moon (similar to Guide to the Moon)
Space in the sixties
Yearbook of Astronomy 1964 (Associate Editor to J.G. Porter)

1964

FICTION:
Caverns of the Moon
Adventure in Space by Patrick Moore and Angus MacVicar
FACTUAL:
The Sky at Night Volume 1
Exploring Weather by Patrick Moore and Henry Brinton
The Picture History of Astronomy
Astronomy
Boy's Book of Astronomy
Exploring the moon
The true book about Roman Britain
Your Book of Astronomy
Yearbook of Astronomy 1965 (Editor)

1965

FICTION:
Pearl on Mars
FACTUAL:
Life on Mars by Francis L. Jackson and Patrick Moore
Against hunting
Exploring other planets
Yearbook of Astronomy 1966 (Editor)

1966

FACTUAL:
Exploring the World

Moore Life on Mars
Naked-eye astronomy (Amateur Astronomer's Library Volume 5)
Exploring the planetarium
The new look of the universe)
Stars and space (Junior reference books)
Yearbook of Astronomy 1967 (Editor)

1967

FACTUAL:
Amateur Astronomer's Glossary (Amateur Astronomer's Library Volume 6)
Basic Astronomy
The craters of the moon: An observational approach
Exploring rocks
Exploring Maps by Henry Brinton and Patrick Moore. Illustrated by Cyril Deakins
Exploring earth history by Patrick Moore and Henry Brinton
Armagh Observatory 1790–1967: A History.
Yearbook of Astronomy 1968 (Editor)

1968

FACTUAL:
The Sky at Night Volume 2
Exploring the stars
Exploring the Galaxies
Space – the Story of Man's Greatest Feat of Exploration
The Sun and its influence. (Revised, not written, by Patrick)
The Sun (Norton Amateur Astronomer's Library)
Yearbook of Astronomy 1969 (Editor)

1969

FICTION:
Planet of Fire
FACTUAL:
Moon Flight Atlas
Astronomy and Space Research: An annotated list
Development of Astronomical Thought
Yearbook of Astronomy 1970 (Editor)

1970

FACTUAL:
The Sky at Night Volume 3
The Atlas of the Universe
Seeing Stars (BBC Series)
Gunpowder Treason (When & Why) by Henry Brinton and Patrick Moore
Yearbook of Astronomy 1971 (Editor)

1971

FACTUAL:
Mars: The Red World
Yearbook of Astronomy 1972 (Editor)
Guide to the Planets (complete rewrite of 1955 book)

1972

SPOOF:
How Britain Won the Space Race by Desmond Leslie and Patrick Moore
FACTUAL:
The Sky at Night Volume 4
Astronomy of Birr Castle
Astronomy and Space: Volume 1
Stories of Science and Invention
How to Recognise the Stars (Mini Books) by Patrick Moore and Lawrence Clarke
Challenge of the Stars (with David A. Hardy)
Black Holes in Space (with Iain Nicolson)
Yearbook of Astronomy 1973 (Editor)
HUMOROUS:
Can You Speak Venusian? A Guide to the Independent Thinkers

1973

FACTUAL:
Astronomy and Space: Volume 2
Legends of the Stars
Astronomical Telescopes and Observatories for Amateurs
Mars by Patrick Moore and Charles Cross
The Comets: Visitors from Space
Colour Star Atlas
Man, the Astronomer
Astronomy for Ordinary Level
Guide to Comets
Yearbook of Astronomy 1974 (Editor)

1974

FACTUAL:
Watchers of the Stars: The Scientific Revolution
The Young Astronomer and His Telescope
Astronomy Quiz Book
Yearbook of Astronomy 1975 (Editor)
Concise Atlas of the Universe (revised and more manageable version of 1970 monster)

HUMOROUS/ART:
Mrs. Moore in Space by Patrick Moore and Gertrude Moore (his mother)

1975

FACTUAL:
Let's Look at the Sky: Planets
Yearbook of Astronomy 1976 (Editor)

1976

FACTUAL:
The Sky at Night Volume 5
A. to Z. of Astronomy
Next 50 Years in Space
The next 10,000 Years: A vision of man's future in the Universe (with Adrian Berry)
Legends of the planets
The New Guide to the Stars
Guide to the Moon (total post Apollo rewrite of the 1953 book)
Yearbook of Astronomy 1977 (Editor)

1977

FICTION:
Spy in Space (Scott Saunders novel 1)
Planet of Fear (Scott Saunders novel 2)
FACTUAL:
Comets
Guide to Comets (substantial revision of 1973 book)
Guide to Mars (total rewrite of 1956 book)
Astronomy of Southern Africa by Patrick Moore and Pete Collins
Modern Astronomy
Atlas of Mercury by Charles A Cross and Patrick Moore
Planets (Wonder Why)
The Earth (Wonder Why)
The Stars Above
Yearbook of Astronomy 1978 (Editor)

1978

FICTION:
The Moon Raiders (Scott Saunders novel 3)
Killer Comet (Scott Saunders novel 4)
FACTUAL:
The Sky at Night Volume 6
Story of Astronomy
Beginner's Book of Astronomy
New Concise Atlas of the Universe (revision of 1974 edition)
Stars (Wonder Why)
Yearbook of Astronomy 1979 (Editor)

1979

FICTION:
The Terror Star (Scott Saunders novel 5)
FACTUAL:
Fun to Know About the Mysteries of Space by Patrick Moore and Lawrence Clarke
The Guinness Book of Astronomy, Facts and Feats
Yearbook of Astronomy 1980 (Editor)
Basic Astronomy (very poorly produced revision of 1967 book)

1980

FICTION:
Secret of the Black Hole. (Scott Saunders novel 6)
FACTUAL:
The Sky at Night Volume 7
The Pocket Guide to Astronomy
Out of the Darkness the Planet Pluto by Patrick Moore and Clyde Tombaugh
Yearbook of Astronomy 1981 (Editor)

1981

FACTUAL:
Moon
The Moon (Mitchell Beazley Atlas)
William Herschel: Astronomer and Musician of 19 New King Street, Bath
Everyman's Scientific Facts and Feats by Magnus Pyke and Sir Patrick Moore
Yearbook of Astronomy 1982 (Editor)
HUMOROUS:
Bureaucrats: How to annoy them by R.T. Fishall (Pseudonym)

1982

FACTUAL:
The Unfolding Universe
Pocket Guide to the Stars and Planets
Atlas of Saturn by Garry E. Hunt and Patrick Moore
Saturn by Patrick Moore, Ian Nicholson, and Garry Hunt
The Planet Venus by Garry Hunt and Patrick Moore
Yearbook of Astronomy 1983 (Editor)

1983

FACTUAL:
Travellers in Space and Time
The Atlas of the Solar System
History of Astronomy
What's New in Space
Countdown! or, How nigh is the end?
The Space Shuttle Action Book (Pop-up book)
Yearbook of Astronomy 1984 (Editor)

1984

FACTUAL:
Patrick Moore's Armchair Astronomy
The Return of Halley's Comet by Patrick Moore and John Mason
Yearbook of Astronomy 1985 (Editor)

1985

FACTUAL:
The Sky at Night Volume 8
The Story of the Earth by Peter Cattermole and Patrick Moore
Halley's Comet Pop-Up Book by Patrick Moore and Heather Couper
The Universe by Patrick Moore and Iain Nicholson
Yearbook of Astronomy 1986 (Editor)
HUMOROUS:
The Twitmarsh files by R.T. Fishall (pseudonym)

1986

FACTUAL:
Patrick Moore's A-Z of Astronomy (similar to 1976 book 'A-Z of Astronomy')
Men of the Stars
Explorers of Space
Yearbook of Astronomy 1987 (Editor)

1987

FACTUAL:
TV Astronomer: 30 Years of the "Sky at Night"
The International Encyclopedia of Astronomy (Edited by Patrick Moore)
Yearbook of Astronomy 1988 (Editor)

1988

FACTUAL:
The Planet Neptune
Space Travel for the Under Tens
Stars and Planets
Yearbook of Astronomy 1989 (Editor)

1989

FACTUAL:
The Sky at Night Volume 9
The Nature of the Universe by Patrick Moore and Iain Nicolson (Editors)
Astronomy: Selected Topics (Extending Science) by Bernard Abrams and Patrick Moore
Astronomer's Stars
Atlas of Uranus by Garry E. Hunt and Patrick Moore
Astronomy for the Under Tens
Yearbook of Astronomy 1990 (Editor)

1990

FACTUAL:
Man's Future in Space
Exploring the Earth and the Moon
Astronomy for GCSE
The Universe for the Under Tens
Yearbook of Astronomy 1991 (Editor)

1991

FACTUAL:
Guide to the Night Sky
Philip's Stargazer: The Complete Astronomy Map and Guide Pack: N. Hemisphere
Patrick Moore's Passion for Astronomy
Yearbook of Astronomy 1992 (Editor)

1992

FACTUAL:
The Sky at Night Volume 10
Space Travel for the Beginner
The Race into Space
Patrick Moore's Story of the Earth for the Under 10s
Yearbook of Astronomy 1993 (Editor)

1993

FACTUAL:
New Guide to the Planets (fully revised version of 1971 rewrite of 1955 book)
Fireside Astronomy: An Anecdotal Tour Through the History and Lore of Astronomy
Yearbook of Astronomy 1994 (Editor)

1994

FACTUAL:
Atlas of Neptune by Garry E. Hunt and Patrick Moore
The Starry Sky
The Great Astronomical Revolution: 1543–1687 and the Space Age Epilogue
Stars of the Southern Skies
Mission to the Planets: The Illustrated Story of Man's Exploration of the Solar System
Yearbook of Astronomy 1995 (Editor)
VIDEO:
Patrick Moore's Guide to Astronomy (VHS Tape)

1995

FACTUAL:
Teach Yourself Astronomy
Let's Look at the Sky: the Stars
The Stars (Starry Sky) Patrick Moore and Paul Doherty
Sun and Moon (Starry Sky) by Patrick Moore and Paul Doherty

The Planets (Starry Sky) by Patrick Moore and Paul Doherty
Comets and Shooting Stars (Starry Sky) by Patrick Moore and Paul Doherty
The Observational Amateur Astronomer (Editor)
The Modern Amateur Astronomer (Editor)
(plus Series Editor of Springer Practical Astronomy Series)
Yearbook of Astronomy 1996 (Editor)
Patrick Moore's piano collection: An edition of music composed by Patrick Moore

VIDEO/AUDIO:

Patrick Moore's Astronomy (Audio Cassette)
The Apollo Story (VHS Tape)

1996

FACTUAL:

Yearbook of Astronomy 1997 (Editor)

AUDIO:

Planet of Fear (Audio Cassette)

1997

FACTUAL:

Patrick Moore's Beginner's Guide To Astronomy
Into Space! – A journey through the solar system and beyond)
Eyes on the Universe
The Sun in Eclipse (Practical Astronomy) by Michael Maunder and Sir Patrick Moore
Atlas of Venus by Peter Cattermole and Patrick Moore
The Observer's Year: 366 Nights in the Universe
Brilliant Stars
Yearbook of Astronomy 1998 (Editor)

VIDEO/AUDIO:

Patrick Moore's Guide To Astronomy. (VHS Tape)

1998

FACTUAL:

Atlas of the Solar System by Smith, Hunt, Nicholson and Patrick Moore,
West Country Eclipse: 11 August 1999 (proceeds donated to Chichester Planetarium)
Into Space 2! – The Great Universe
Yearbook of Astronomy 1999 (Editor)

1999

FACTUAL:

Patrick Moore's Guide to the 1999 Total Eclipse
The Caldwell Catalogue by Patrick Moore, Paul Doherty, and H.J.P. Arnold
The Starry Sky Collection (Starry Sky) by Patrick Moore and Paul Doherty
Patrick Moore on Mars
Astronomy before the Telescope by Patrick Moore and Christopher Walker
Transit: When Planets Cross the Sun by Michael Maunder and Sir Patrick Moore

Patrick Moore's Millennium Yearbook: The View from AD 1001 (with Allan Chapman)
The Photographic Atlas of the Stars by H.J.P. Arnold, Paul Doherty, and Patrick Moore
The Wandering Astronomer
Into Space 3! Breaking the Barriers
Yearbook of Astronomy 2000 (Editor)
VIDEO:
An Astronauts View Of Earth by Patrick Moore and William Shatner (VHS Tape)

2000

FACTUAL:
The Data Book of Astronomy
Into Space 4! First Contact
Stargazing: Astronomy Without a Telescope
Exploring the Night Sky with Binoculars
Yearbook of Astronomy 2001 (Editor)
Philips Atlas of the Universe (complete rewrite of 1970s Mitchell Beazley Atlas)

2001

FACTUAL:
The Star of Bethlehem
Voyages of Discovery
Astronomy for GCSE (revised) by Patrick Moore & Chris Lintott
Patrick Moore on the Moon
Yearbook of Astronomy 2002 (Editor with John Mason)
Children's charity book with a small Patrick contribution:
2001 A Space Oddity, 100 Short Stories on the theme of.....
AUDIO:
Moore Music – Patrick Moore on the Xylophone with the Royal Scottish National Orchestra at the Royal Concert Hall Glasgow in 1999. Conductor Derek Carden and Pianist Robert Vallier. (CD)

2002

FACTUAL:
The Sky at Night Volume 11
Astronomy with a Budget Telescope by Patrick Moore and John Watson
Astronomy Encyclopedia
Yearbook of Astronomy 2003 (Editor with John Mason)

2003

FICTION:
Voyage to Mars
FACTUAL:
Handbook of Astronomy
Into Space 5! Mars, the next frontier

Patrick Moore: 80 Not Out – The Autobiography
Yearbook of Astronomy 2004 (Editor with John Mason)
VIDEO:
Moore On The Moon (DVD)

2004

FACTUAL:
Guide to Stars and Planets
Stars of Destiny: A Scientific Look at Astrology
Futures: 50 Years in Space: The Challenge of the Stars with David A. Hardy
Yearbook of Astronomy 2005 (Editor with John Mason)

2005

FACTUAL:
The Sky at Night Volume 12
Guide to the Night Sky: A Guided Tour of the Stars and Constellations
Venus
Stars & Planets by Patrick Moore and Carole Stott
The Amateur Astronomer
Yearbook of Astronomy 2006 (Editor with John Mason)
Patrick Moore: The Autobiography (paperback edition of 80 not out – minor revisions)
Booklet to raise money for the Sri-Lankan Tsunami victims:
Asteroid by Sir Patrick Moore and Arthur C. Clarke
VIDEO
'The Astronomical Patrick Moore – the authorised version' (DVD)

2006

FACTUAL:
50 Years in Space: What We Thought Then What We Know Now…with David A. Hardy
Moore on Mercury: The Planet and the Missions
Yearbook of Astronomy 2007 (Editor with John Mason)
VIDEO
Outer Space with Sir Patrick Moore (DVD)

2007

FACTUAL:
Bang! The Complete History of the Universe by Brian May, Patrick and Chris Lintott
Our Universe: Facts, Figures and Fun
Sir Patrick Moore – Journey to the Stars
Space: The First 50 Years – A Celebration with Sir Patrick Moore and H.J.P. Arnold
Yearbook of Astronomy 2008 (Editor with John Mason)
VIDEO
Journey to the stars (DVD)

2008

FACTUAL:
Saturn: The Complete Guide by Garry E. Hunt and Patrick Moore
Yearbook of Astronomy 2009 (Editor with John Mason)
Can you play cricket on Mars? – and other scientific questions answered

2009

FACTUAL:
Yearbook of Astronomy 2010 (Editor with John Mason)
VIDEO
Apollo 11: a night to remember (Sky at Night DVD)

2010

FACTUAL:
First Book of Stars
Yearbook of Astronomy 2011 (Editor with John Mason)
The Sky at Night Volume 13

2011

FACTUAL:
Yearbook of Astronomy 2012 (Editor with John Mason)
HUMOROUS:
Within the Glade: A Collection of Poems Written to Amuse Children (of all ages)

2012

FACTUAL:
The Sky at Night: Answers to Questions from Across the Universe, with Chris North.
The New Astronomy Guide: Star Gazing in the Digital Age, with Pete Lawrence.
The Cosmic Tourist, by Brian May, Patrick Moore and Chris Lintott.
Yearbook of Astronomy 2013 (Editor with John Mason)
HUMOROUS:
Miaow! – Cats really are nicer than people!

Index

M. Mobberley, *It Came From Outer Space Wearing an RAF Blazer!: A Fan's Biography of Sir Patrick Moore*, DOI 10.1007/978-3-319-00609-3,

D

K

L

M

Printed by Printforce, the Netherlands